Springer Proceedings in Mathematics & Statistics

Volume 415

This book series features volumes composed of selected contributions from workshops and conferences in all areas of current research in mathematics and statistics, including data science, operations research and optimization. In addition to an overall evaluation of the interest, scientific quality, and timeliness of each proposal at the hands of the publisher, individual contributions are all refereed to the high quality standards of leading journals in the field. Thus, this series provides the research community with well-edited, authoritative reports on developments in the most exciting areas of mathematical and statistical research today.

B. Rushi Kumar · S. Ponnusamy · Debasis Giri ·
Bhavani Thuraisingham · Christopher W. Clifton ·
Barbara Carminati
Editors

# Mathematics and Computing

## ICMC 2022, Vellore, India, January 6–8

Springer

*Editors*
B. Rushi Kumar
Department of Mathematics
School of Advanced Sciences
Vellore Institute of Technology
Vellore, Tamil Nadu, India

Debasis Giri
Department of Information Technology
Maulana Abul Kalam Azad University
of Technology
Haringhata, West Bengal, India

Christopher W. Clifton
Department of Computer Science
Purdue University
West Lafayette, IN, USA

S. Ponnusamy
Department of Mathematics
Indian Institute of Technology Madras
Chennai, Tamil Nadu, India

Bhavani Thuraisingham
Department of Computer Science
The University of Texas at Dallas
Richardson, TX, USA

Barbara Carminati
Department of Theoretical and Applied
Sciences
University of Insubria
Varese, Italy

ISSN 2194-1009 ISSN 2194-1017 (electronic)
Springer Proceedings in Mathematics & Statistics
ISBN 978-981-19-9309-1 ISBN 978-981-19-9307-7 (eBook)
https://doi.org/10.1007/978-981-19-9307-7

Mathematics Subject Classification: 97R20, 94C15, 90B50, 93C10, 94D05, 05C90, 12D05, 92E20



This Springer imprint is published by the registered company Springer Nature Singapore Pte Ltd.
The registered company address is: 152 Beach Road, #21-01/04 Gateway East, Singapore 189721, Singapore

# Contents

introduced initially for Texture Analysis (TA), and since then it is utilized auspiciously in various applications. In LBP, neighbors are thresholded to 0 or 1, by comparing their values with the center pixel, which is further transformed to LBP code by supplying binomial weights. The basic LBP descriptor has attracted the attention of different research groups. As a result, various LBP variants were launched after the LBP proposal. Some of these LBP variants are explored from [2–6]. All these LBP variants achieve stupendous outcomes with respect to the application they were introduced. The grayscale-based descriptors have earned huge respect in the form of discriminativity achieved by them in different applications. But research has significantly progressed from grayscale-based descriptors to color-based descriptors. The color-based descriptors achieved notable attention in the last few years. By using different color space formats, more complementary information is extracted as compared to grayscale counterparts. In Sect. 2, some of the color-based LBP variants are discussed which achieve excellent outcomes.

In [7], the authors develop a new LBP variant for Face Recognition (FR) called MB-ZZLBP. In MB-ZZLBP, a novel method was introduced for feature extraction. Precisely, there is the usage of mean filter first in 9 different regions of the $6 \times 6$ image patch to suppress the effect of image noise. After the mean generation, the zigzag-oriented pixels are compared. After comparison, the code of MB-ZZLBP is built. Forming code in all places creates the MB-ZZLBP image, which forms the size of 256. MB-ZZLBP is tested on 2 challenges (i.e. light and expression variations) and it achieves good results in these conditions.

After carefully analyzing MB-ZZLBP, it is found that MB-ZZLBP discriminativity is limited by not including color features for classification. To remedy this challenge, the proposed work introduces the advanced version of MB-ZZLBP, the so-called Color MB-ZZLBP (CMB-ZZLBP) for FR. In CMB-ZZLBP, the MB-ZZLBP features are extracted from each channel of RGB color format. Further, all three channel features are integrated to build CMB-ZZLBP size. FLDA [8] is used to achieve compressed feature representation, and matching is performed from SVM [9] and NN [10]. Experiments clearly justify the effectiveness of CMB-ZZLBP against MB-ZZLBP on GTFD [11]. CMB-ZZLBP proves its dominance against numerous literature techniques also. CMB-ZZLBP secures the best ACC of 96.66% on a training size of 9.

*Road map*: Sect. 2 gives related works, MB-ZZLBP and CMB-ZZLBP are presented in Sect. 3, results are portrayed in Sect. 4 and conclusion with future prospects are pasted in Sect. 5.

## 2 Related Works

This part covers up some LBP variants developed by using different color space formats. Shu et al. [12] introduced Multiple Channels LBP (MCLBP) for color TA. In MCLBP, the single channel and multichannel details are fused by using RGB color format. The resultant feature reflects dependency and correlation among distinct

# Color Multiscale Block-ZigZag LBP (CMB-ZZLBP): An Efficient and Discriminant Face Descriptor

**Shekhar Karanwal**

**Abstract** Literature reports numerous local descriptors based on extracting rich information from color space formats. The color scale format provides more robustness as compared to grayscale counterparts. This work introduces such descriptors called Color Multiscale Block-ZigZag LBP (CMB-ZZLBP) for Face Recognition (FR). CMB-ZZLBP is the advanced method of MB-ZZLBP. In MB-ZZLBP, first mean patch is generated (from 9 regions of the $6 \times 6$ patch) and then zigzag pixels are compared to develop MB-ZZLBP code. MB-ZZLBP forms the histogram representation of 256, by computing MB-ZZLBP code in each position. The major issue with MB-ZZLBP is that it restricts its robustness due to grayscale feature extraction. By introducing CMB-ZZLBP, this issue is resolved effectively. In CMB-ZZLBP, the MB-ZZLBP feature extraction is done from each component of the RGB color space format. Further, all three channel features are integrated to build the CMB-ZZLBP feature size. FLDA is used to achieve compressed feature representation, and classification is performed from SVM and NN. Experiments justify the effectiveness of CMB-ZZLBP against MB-ZZLBP on Georgia Technology Face Dataset (GTFD). CMB-ZZLBP proves its dominance against various literature techniques also. CMB-ZZLBP secures the best ACC of 96.66% on a training size of 9.

**Keywords** Local feature · Advanced local feature · Feature compression · Classification · Dataset · Gray samples · Color samples

## 1 Introduction

The emergence of local descriptors has given new directions for feature extraction pertaining to computer vision and pattern recognition. In the last 2 decades, numerous local descriptors were introduced for distinct applications. Among all these, LBP [1] is regarded as the most prolific and discriminant descriptor. LBP was

---

S. Karanwal (✉)
Department of CSE, Graphic Era University (Deemed), Dehradun, UK 248002, India
e-mail: shekhar.karanwal@gmail.com

 
B. Rushi Kumar et al. (eds.), *Mathematics and Computing*, Springer Proceedings in Mathematics & Statistics 415, https://doi.org/10.1007/978-981-19-9307-7_1

# Computer Science

channels. Furthermore, MCLBP is expanded to MCLBP + M, in which local color differences are decomposed into signs and magnitude color differences. Results on 5 texture datasets prove the prominence of developed descriptors. Agarwal et al. [13] proposed the MCLTCoP for Image Retrieval (IR). In MCLTCoP, each channel neighborhood (of RGB format) is utilized with the co-occurrence of the V channel (of HSV format), to create the feature size. Specifically, the difference values derived from neighborhoods and centers are used for making MCLTCoP size. Experiments conducted on the Corel-1 k dataset confirm the potency of MCLTCoP. Tiecheng et al. [14] discovered the Color Context Binary Pattern (CCBP) for TA and Scene Classification (SC). In CCBP, the neighbor and scale context are progressively employed for correcting the encoded bits. Initially, the intra-channel neighbor details are encoded in 3 states (valued) of scale space and then majority voting is utilized for state correction over different scales. Further distances existing between color features are computed, and uncertain bits are corrected across neighboring bit propagation. Ultimately, all histograms are joined for developing the size. For all this evaluation, the RGB color format is utilized. Experiments on 3 datasets confirm the efficacy of CCBP.

Karanwal et al. [15] invented the DCD for FR. In DCD, the color features of HELBP, LBP and LPQ are derived by using RGB color space format, and further all color features are joined to form DCD feature size. The color form of HELBP, LBP and LPQ is called CHELBP, CLBP and CLPQ. DCD attains superb results on the GTFD face dataset, by defeating the accuracy of numerous techniques. Vipparthi et al. [16] invented the CDLQP for IR. By using the RGB format, the edge features (directional) are acquired among neighbors and reference pixels in 4 directions from each color channel. CDLQP secure extraordinary results on MIT and Corel-5000 datasets. Karanwal et al. [17] presented 2 novel descriptors for FR, so-called CZZBP and CMBZZBP. In the former descriptor, 3 different zigzag-oriented designs are created for 3 channels of RGB color format. Then depending on the designed structure, zigzag pixels are compared to build the size of the respective color channel. Ultimately, all 3 features (of the channel) are joined to build CZZBP size. For CMBZZBP, the median window is used for making size. Both descriptors achieve great results on GTFD and Faces94 datasets.

Jebali et al. [18] introduced Local Binary Quaternion Rotation Pattern (LBQRP) for TA. In LBQRP, the color texture is represented by using the Quaternion concept. The distance within 2 color can be defined as the rotation angle among 2 quaternion units utilizing the geodesic distance. The local histograms generated after LBQRP coding are used as the features. The color format utilized is RGB. Experiments on 3 datasets reflect the potential of LBQRP. Sotoodeh et al. [19] proposed 2 descriptors for IR CRMCLBP and PDM. In CRMCLBP, the RMCLBP concept is applied on 3 channels of the RGB format. Extracted features from 3 channels are joined to form the CRMCLBP size. In PDM, the ideal set of features is selected from the k-mean clustering algorithm. Experiments confirm the ability of the proposed descriptors. Agarwal et al. [20] invented MCLTP for different applications. In MCLTP, cross-channel information (of the color texture) is captured by integrating H-V, S-V and V-V components (acquired from HSV image presentation). The texture details derived in such a way contain the color details, and local texture details would also include in such presentations. Experiments done on 5 datasets shows the capability of MCLTP.

Umer et al. [21] developed the biometric recognition structure by using all phases of FR. During pre-processing, facial landmarks are detected which is followed by SIFT for feature extraction. The task of classification is assigned to SVMs. Experiments confirm the potency of the proposed method on 5 challenging datasets. Umer et al. [22] developed the FR system using all phases of FR. In the pre-processing step, facial landmarks are detected from which facial regions are extracted. Then SIFT is utilized for feature extraction which undergoes distinct learning methods to produce different feature representations. Eventually, the matching is done by SVMs. Results illustrate the developed method's efficacy.

## 3 Description of Descriptors

This section provides the explanation of grayscale MB-ZZLBP and the proposed one Color MB-ZZLBP.

### *3.1 Multiscale Block-ZigZag LBP (MB-ZZLBP)*

In [7], Karanwal et al. introduced the MB-ZZLBP descriptor for FR. MB-ZZLBP is the grayscale-based descriptor. In MB-ZZLBP, the first filtration of the mean is used in 9 regions of the $6 \times 6$ image patch. Each region reflects the dimension of $2 \times 2$. After mean generation, a $3 \times 3$ patch evolves. By using mean in regions, the unwanted image noise is reduced. Then pixels placed as per zigzag orientation are compared. Precisely, the immediate neighbor (as per zigzag orientation, of current/previous pixel) is differentiated from the current/previous pixel. This procedure is conducted in eight locations of the $3 \times 3$ patch. As a consequence, eight difference values are obtained in eight positions of the $3 \times 3$ patch. Those difference values that are higher or equal to 0 are supplied with label 1 else label 0 is supplied. Imposing the binomial weights to a generated binary pattern forms the MB-ZZLBP code after adding values, for one pixel location. Developing MB-ZZLBP code in all positions forms the MB-ZZLBP map image, whose size (histogram) is 256. Equations 1 and 2 show the MB-ZZLBP code procedure for a single location. In Eq. 1, the mean is generated from 9 regions. $L_{i,j}$ is equipped with region values and $U_{i,j}$ contain mean values. Equation 2 computes the MB-ZZLBP code. S, N, $U_{N,s}$ and $U_{N,s+1}$ signify the total size, radius, current pixel and neighbor pixel (as per zigzag orientation). Figure 1 shows the MB-ZZLBP illustration.

$$U_{i,j} = \text{mean}\left(L_{i,j}\right) \tag{1}$$

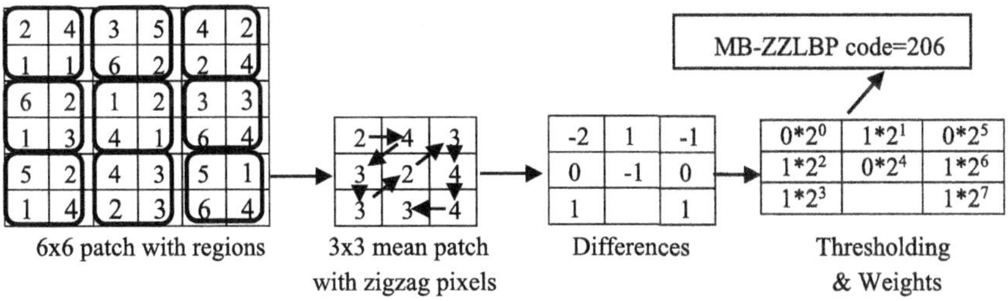


**Fig. 1** MB-ZZLBP illustration

$$\mathrm{MB-ZZLBP}_{S,N}(x_c) = \sum_{s=0}^{S-1} i(U_{N,s} - U_{N,s+1})2^s,\ i(c) = \begin{pmatrix} 1 & c \geq 0 \\ 0 & c < 0 \end{pmatrix} \quad (2)$$

## 3.2 Color MB-ZZLBP (CMB-ZZLBP)

The authors of [7] have introduced the novel local (face) descriptor MB-ZZLBP. MB-ZZLBP was tested on two challenges, i.e. light and expression variations. MB-ZZLBP achieved good results under these conditions by using EYB and Faces94 datasets. The major part which is missing in [7] is the color form of MB-ZZLBP, which adds significant discriminativity if used. This work takes this challenge and proposes the novel descriptor CMB-ZZLBP for FR. Literature reveals that the color form generates more discriminativity as compared to the gray form, therefore color-based descriptor is proposed. The detailed explanation of CMB-ZZLBP is as follows.

In CMB-ZZLBP, the MB-ZZLBP concept is deployed on three channels of RGB color format. Feature sizes generated from all three channels are integrated to form the feature size of the CMB-ZZLBP descriptor. Each channel forms the size of 256, therefore CMB-ZZLBP size is 768. Figure 2 shows the illustration of CMB-ZZLBP through the transformed images with their histograms. To make a more efficient feature descriptor, this feature is projected in lower dimensions by using FLDA. Details pertaining to FLDA size are elaborated in the experiments section. Then the performance of the compact feature descriptor is evaluated by three matching algorithms. Two are RBF and POLY, which are SVM-based classification techniques. The other one is the Exhaustive Search Concept (ESC), which is an NN-based classification technique. In this concept, cosine distance is utilized for measuring similarity. Figure 3 displays a schematic diagram/framework of the proposed work. The phases of feature compression and classification are also conducted on MB-ZZLBP.

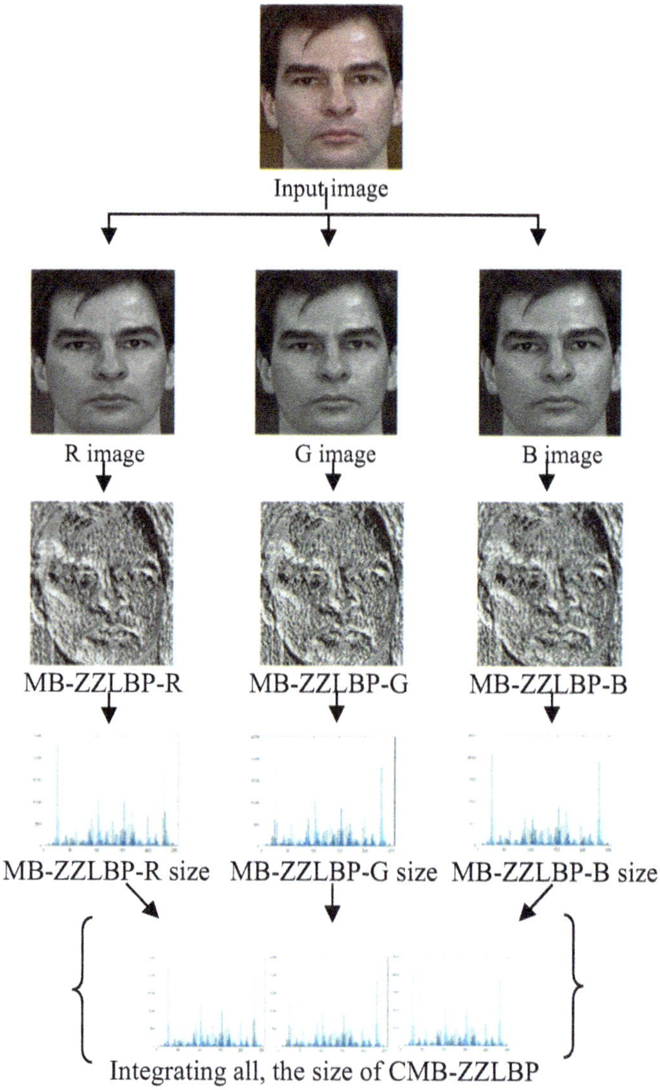


**Fig. 2** CMB-ZZLBP illustration

## 4 Experiments

This section initiates by introducing the description of the dataset used for accuracy analysis. Then feature size essentials are discussed pertaining to evaluated descriptors. Next accuracy is analyzed on four subsets (of the GTFD [11] dataset) and finally, the proposed descriptor is compared with the numerous techniques from the literature.

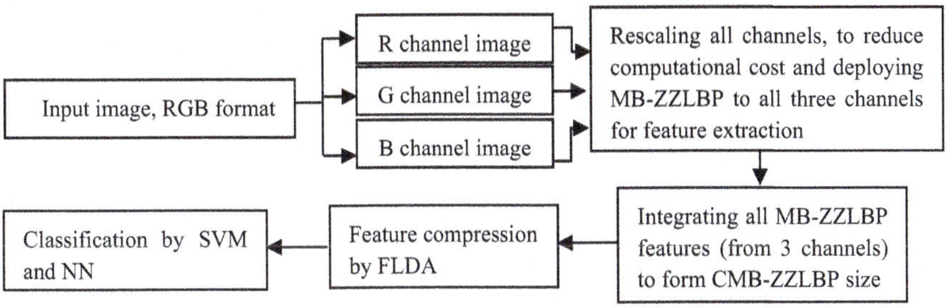


**Fig. 3** Block diagram of proposed work

**Fig. 4** Some samples of GTFD dataset

## *4.1 Dataset Information*

The dataset taken for accuracy analysis is GTFD. GTFD dataset is equipped with 750 color images with respect to 50 subjects. Each subject contributes 15 samples taken in 4 different conditions. These are scale, pose, expression and light variations. Resolution of samples in GTFD dataset is not consistent throughout. Some samples of GTFD dataset are furnished in Fig. 4. Figure 4 delivers some of the facial images of the GTFD face dataset. Precisely, the 3 subject images are displayed in Fig. 4.

## *4.2 Feature Size Essentials with Respect to Evaluated Descriptors*

For MB-ZZLBP evaluation, the color samples are transformed to grayscale and then they are downsampled to $51 \times 47$. Size generated by MB-ZZLBP is 256. For CMB-ZZLBP evaluation, 3 channels are extracted (separate) from the RGB image, and then each channel is downsampled to $51 \times 47$. The rescaling motive is to lower the cost of computation. This is the face pre-processing step applied before feature extraction is performed. Figure 5 shows the face pre-processing step applied before deploying feature extraction. From each rescaled channel, MB-ZZLBP is deployed

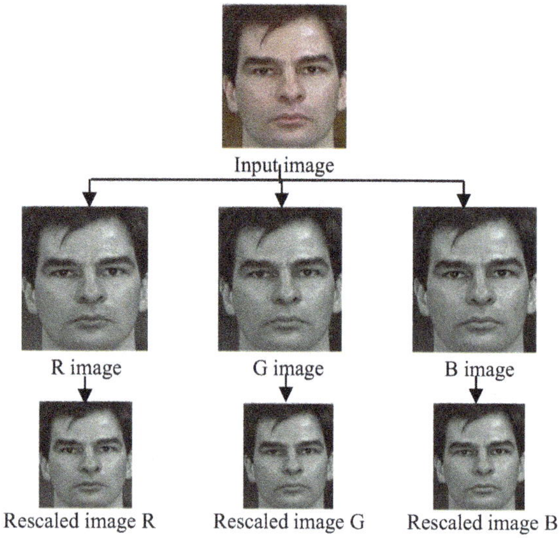


**Fig. 5** Pre-processing steps before feature extraction

for feature extraction. Ultimately, all three channel features are merged to generate CMB-ZZLBP feature size. Therefore, CMB-ZZLBP builds the feature size of 768. On both MB-ZZLBP and CMB-ZZLBP, the sub-space technique FLDA is imposed for the feature compression. After PCA compression, the size produced is 190 and after LDA compression, the size generated is 12. LDA size is utilized for accuracy analysis. MATLAB R2018a is used for testing.

## *4.3 Accuracy Analysis on Different Subsets*

The accuracy is analyzed by the formula depicted in Eq. 3. In Eq. 3 the elements ACC, $T_{stse}$ and FMS imply the evaluated accuracy, test size/set and false matched samples. Another element $T_{rgse}$ implies the training size, used for the specifications of training details. On each subset (created from $T_{rgse}/T_{stse}$), the ACC is analyzed after recording FMS on $T_{stse}$. FMS are those which are falsely classified. Suppose FMS generate 5 samples on $T_{stse} = 400$ (means $T_{rgse} = 7$ and $T_{stse} = 8$, per subject), then analyzed ACC = 395/400 = 98.75%. Similarly, ACC is analyzed on every subset:

$$\left[\mathrm{ACC} = \frac{\mathrm{T_{stse}} - \mathrm{FMS}}{\mathrm{T_{stse}}} * 100\right] \quad (3)$$

For this work, descriptors are evaluated on $T_{rgse} = 6:9$ and $T_{stse} = 9:6$. On each subset, the finest ACC is analyzed after executing the classifier 15 times. All ACC obtained are placed in Table 1. Table 1 confirms the capacity of color descriptors

**Table 1** ACC analysis through table on GTFD

| | $T_{rgse}$ essentials | | | |
|---|---|---|---|---|
| | $T_{rgse} = 6$ | $T_{rgse} = 7$ | $T_{rgse} = 8$ | $T_{rgse} = 9$ |
| All approaches | FMS/ACC | | | |
| MB-ZZLBP + FLDA + SVM (RBF) | 42/90.66 | 30/92.50 | 25/92.85 | 17/94.33 |
| MB-ZZLBP + FLDA + SVM (POLY) | 63/86.00 | 52/87.00 | 40/88.57 | 31/89.66 |
| MB-ZZLBP + FLDA + NN (ESC Cosine) | 52/88.44 | 36/88.50 | 38/89.14 | 28/90.66 |
| **CMB-ZZLBP + FLDA + SVM (RBF)** | **26/94.22** | **20/95.00** | **16/95.42** | **10/96.66** |
| **CMB-ZZLBP + FLDA + SVM (POLY)** | **36/92.00** | **28/93.00** | **23/93.42** | **16/94.66** |
| **CMB-ZZLBP + FLDA + NN (ESC Cosine)** | **42/90.66** | **34/91.50** | **27/92.28** | **23/92.33** |

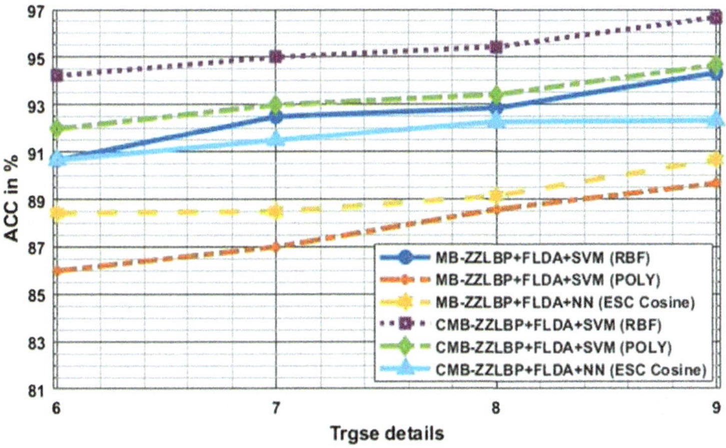


**Fig. 6** ACC analysis through graph on GTFD

against the grayscale descriptors. CMB-ZZLBP reflects its potential more than MB-ZZLBP on all subsets. The best ACC of CMB-ZZLBP is extracted from the RBF classification. The ACC analysis through the graph is shown in Fig. 6.

## *4.4 Accuracy Comparison with Literature Techniques*

As SVM (RBF) extracts the finest ACC from CMB-ZZLBP, RBF classification results are used for comparison against 10 other techniques from the literature. These 10 techniques pertain to Local, Global/DR, Sparse Representation (SR) Classification, Regression Classification and Dictionary-Based. The ACC attained from these 12 are as follows.

**Table 2** ACC comparison on GTFD

| | | $T_{rgse}$ essentials | | | |
|---|---|---|---|---|---|
| | | $T_{rgse} = 6$ | $T_{rgse} = 7$ | $T_{rgse} = 8$ | $T_{rgse} = 9$ |
| Techniques | Technique type | ACC in % | | | |
| DCD [15] | Local | 88.66 | 90.25 | 92.57 | 93.33 |
| CLPQ [15] | Local | 80.66 | 83.50 | 84.57 | 87.33 |
| FDLPP [23] | Dimension Reduction | NE | 83.08 | 86.19 | 86.33 |
| FLPP [23] | Dimension Reduction | NE | 72.50 | 73.52 | 76.22 |
| GBSBP [24] | Local | 87.55 | 89.75 | 91.71 | 92.33 |
| GBSBP + LPQ [24] | Local | 90.66 | 91.50 | 93.14 | 94.33 |
| ICS_DLSR [25] | SR Classification | 78.45 | 81.47 | NE | NE |
| RGLRR [25] | Regression Classification | 79.47 | 82.62 | NE | NE |
| FKESRC [26] | SR Classification | 70.29 | NE | NE | NE |
| KED [26] | Dictionary-Based | 65.07 | NE | NE | NE |
| **CMB-ZZLBP** | **Local** | **94.22** | **95.00** | **95.42** | **96.66** |

*NE-Not Evaluated

DCD [15], CLPQ [15], GBSBP [24] and GBSBP + LPQ [24] obtain the ACC of [88.66 90.25 92.57 93.33%], [80.66 83.50 84.57 87.33%], [87.55 89.75 91.71 92.33%] and [90.66 91.50 93.14 94.33%] when $T_{rgse} = 6 : 9$. FDLPP [23] and FLPP [23] secure ACC of [83.08 86.19 86.33%] and [72.50 73.52 76.22%] on $T_{rgse} = 7 : 9$. ICS_DLSR [25] and RGLRR [25] attain ACC of [78.45 81.47%] and [79.47 82.62%] when $T_{rgse} = 6 : 7$. FKESRC [26] and KED [26] procure the ACC of 70.29 and 65.07% when $T_{rgse} = 6$. Table 2 presents all the ACC. It is judged from Table 2 that CMB-ZZLBP is the best among all all subsets.

## 5 Conclusion and Future Prospect

This work proposed a novel descriptor CMB-ZZLBP for FR. CMB-ZZLBP is the advancement of MB-ZZLBP. In MB-ZZLBP, the first mean patch is generated (from 9 regions of the $6 \times 6$ patch) and then zigzag pixels are compared to develop MB-ZZLBP code. MB-ZZLBP forms the histogram representation of 256, by computing MB-ZZLBP code in each position. The major issue with MB-ZZLBP is that it restricts its robustness due to grayscale feature extraction. By introducing CMB-ZZLBP, this issue is resolved effectively. In CMB-ZZLBP, the MB-ZZLBP feature extraction is done from each channel of RGB color format. Further, all three channel features are integrated to build the CMB-ZZLBP feature size. FLDA is used to achieve compressed feature representation, and matching is conducted from SVM and NN. Experiments clearly justify the efficacy of CMB-ZZLBP against MB-ZZLBP on the

GTFD face dataset. CMB-ZZLBP proves its dominance against numerous literature techniques also.

The proposed work can be extended to futuristic research by incorporating some points which remain uncovered in the proposed work. First, some other hybrid color format can be utilized for improving the accuracy. Second, the extraction of regional features would immensely improve the accuracy and third integration of global and regional features in 1 framework. All these points build a proposal for future research.

## References


1. Karanwal, S.: Robust LBP for face recognition in different challenges. Multi. Tool Appl. **81**, 29405–29421 (2022)
2. Saidi, I.A., Rziza, M., Debayle, J.: A novel texture descriptor: circular parts LBP. Img. Ana. Ster. **40**(2), 105–114 (2021)
3. Kar, C., Banerjee, S.: Tropical cyclones classification from satellite images using BLBP and histogram analysis. In: SCTA, pp. 399–407 (2021)
4. Kartheek, M.N., Prasad, M.V.N.K., Bhukya, R.: RMP: a handcrafted feature descriptor for FER. J. Ambient. Intell. Humanized Comput. (2021)
5. Song, T., Xin, L., Gao, C., Zhang, T., Huang, Y.: QELBP with adaptive structural pyramid pooling for color image representation. Pattern Recognit. (2021)
6. Bhattacharjee, D., Roy, H.: PLGF: a novel local image descriptor. IEEE Trans. Pattern Ana. Mac. Intell. **43**(2), 595–607 (2021)
7. Karanwal, S., Diwakar, M.: MB-ZZLBP: multiscale block ZigZag LBP for face recognition. In: MARC, pp. 613–622 (2021)
8. Qin, X., Wang, S., Chen, B., Zhang, K.: R-FLDA with GCLF. In: CAC (2020)
9. Junior, P.R.M., Boult, T.E., Wainer, J., Rocha, A.: Open-set SVM. IEEE Trans. Syst. Man Cyb.: Syst. 1–14 (2021)
10. Rastin, N., Jahromi, M.Z., Taheri, M.: A GWD k-NN for multi-label problems. Pattern Recognit. **114** (2021)
11. http://www.anefian.com/research/face_reco.htm.
12. Shu, X., Song, Z., Shi, J., Huang, S., Wu, X.J.: Multiple channels LBP for color texture representation and classification. Sig. Proc. Img. Com. **98** (2021)
13. Agarwal, M., Maheshwari, R.P.: MLTCP for CBIR. Ira. J. Sci. Tech. **44**, 495–504 (2020)
14. Tiecheng, S., Jie, F., Shuang, L., Tianqi, Z.: Color CBP using progressive bit correction for image classification. Chi. J. Electron. **30**(3), 471–481 (2021)
15. Karanwal, S.: DCD by the fusion of 3 novel color descriptors. Optik **244** (2021)
16. Vipparthi, S.K., Nagar, S.K.: CD-LQP for CBIR. Hum. Cent. Comp. Inf. Sci. **4**(6) (2014)
17. Karanwal, S., Diwakar, M.: Two novel color local descriptors for face recognition. Optik 1–15 (2021)
18. Jebali, H., Richard, N., Naouai, M.: LBQRP for CTR. In: ICPR, pp. 3698–3705 (2021)
19. Sotoodeh, M., Moosavi, M.R., Boostani, R.: A novel adaptive LBP-based descriptor for color image retrieval. Expert. Syst. Appl. **127**, 342–352 (2019)
20. Agarwal, M., Singhal, A., Lall, B.: Multi-channel LTP for CBIR. Pattern Anal. Appl. **22**, 1585–1596 (2019)
21. Umer, S., Dhara, B.C., Chanda, B.: Biometric recognition system for challenging faces. In: NCVPRIPG (2015)
22. Umer, S., Dhara, B.C., Chanda, B.: Face recognition using fusion of FLT. Measurement **146**, 43–54 (2019)
23. Ran, R., Feng, J., Zhang, S., Fang, B.: A GMF DR framework and extension for manifold learning. IEEE Trans. Cyb. 1–12 (2020)

24. Karanwal, S.: Graph based structure binary pattern for face analysis. Optik **241** (2021)
25. Wang, S., Ge, H., Yang, J., Tong, Y.: Relaxed group low rank regression model for multi-class classification. Multimed. Tools Appl. **80**, 9459–9477 (2021)
26. Fan, Z., Wei, C.: Fast kernel SRC for undersampling problem in face recognition. Multimed. Tools Appl. **79**, 7319–7337 (2020)

# Effect of Noise in the Quantum Network Implementation of Cop and Robber Game

**Anjali Dhiman and S. Balakrishnan**

**Abstract** The quantum network designed for the implementation of quantum cop and robber game in the presence of a noisy environment is investigated. In particular, the amplitude damping noise model and phase damping noise model are studied thoroughly by calculating the fidelity of the quantum states. From the analysis of fidelity graphs, we have observed that there exist suitable entangling operators which can suppress the noises in the quantum network.

**Keywords** Game theory · Quantum games · Quantum networks · Noise models

## 1 Introduction

Since the emergence of modern physics, quantum mechanics has remained the centerpiece of interest among the research community. Quantum entanglement is one of the most fascinating quantum phenomena which has revolutionized the field of modern computation and has gifted modern science with a variety of paradigms of potential fields like quantum computation [1, 2], quantum cryptography [3, 4] and many more. Especially, the field of quantum information and transmission is gaining immense popularity and progress due to the revolutionary results that are unveiled. With the increasing introduction of quantum physics in almost every field, even the game theory has not remained untouched. In 1962, John von Neumann and Oskar Morgenstern were the first to bring the notion of game theory, which is the study of decision-making when two or more parties are fighting for the same interests [5]. When game theory is studied from the quantum perspective, it increases the strategic space of the players due to the principle of superposition of states, which allowed them to achieve optimal results. However, the quantum game theory gained prominence in 1999, when Eisert, Wilkens and Lewenstein introduced the EWL scheme to quantize the simultaneous game, namely the Prisoner's dilemma [6].

A. Dhiman · S. Balakrishnan (✉)
Department of Physics, School of Advanced Sciences, Vellore Institute of Technology, Vellore 632104, India
e-mail: physicsbalki@gmail.com

 
B. Rushi Kumar et al. (eds.), *Mathematics and Computing*, Springer Proceedings in Mathematics & Statistics 415, https://doi.org/10.1007/978-981-19-9307-7_2

Though the game theory finds its application in various fields such as economics, biology and social science [7], the aspect of data transmission between distant parties has always remained a concern in quantum communication. Therefore, quantum networks have become fundamentally important to develop a system for secure quantum communication. There are schemes like the peer-to-peer scheme and the client–server scheme which are available for investigating the quantum games on quantum networks [8]. Although both the schemes are originally designed for the implementation of simultaneous games, which are more popular among game theorists; recent work [9] has emphasized on a sequential game, namely the cop and robber game which was first defined by Nowakowski and Winkler [10].

Practically, when the quantum network is exposed to the environment, it becomes inevitable to prevent the quantum data from outside disturbance. The direct interaction of the quantum network with the surroundings causes decoherence in the data transmission, which is a serious concern in quantum communication [2]. There are several noise models which are categorized based on their distinctive properties to study decoherence. For instance, collective noise is caused by the specific kind of symmetry arising between the environment and qubits, coupling together without any distinction [11], on the other hand, Pauli's noise channels are the set of noise processes such as depolarizing channel, bit-flip channel and phase bit-flip channel [12]. However, the noise models like amplitude damping (AD) and phase damping (PD) have been widely studied and have special importance as they have the potential to cause entanglement sudden death (ESD) [13–16].

In this work, we intend to study the quantum network designed for the implementation of the quantum version of the cop and robber game, when it is exposed to the external environment. Thus, we have incorporated two noise forms, namely amplitude damping and phase damping in the quantum circuit [17]. In this work, we have considered the case where noise is acting on only one channel of the circuit, while another channel remains unaffected. In order to get a comprehensive view of the damping caused by the noise, we have calculated the fidelity of the quantum states. Interesting inferences are obtained from the analysis of the fidelity graphs. Implications of the results are discussed in the conclusion.

## 2 Quantum Version of Cop and Robber Game

The cop and robber game is a popular sequential game. In the classical version of this game, the player has certain alternatives which he/she chooses to make consistent decisions. The other player has the choice to play rationally or not [18]. The game has also been studied in [19], where the players play their moves on the graph, and the position of players is represented by the vertices on the graph. The quantum version of the graphically studied cop and robber game has been achieved using graph-preserving quantum operations in [20]. However, we have considered the quantum version of the cop and robber game studied in [9], where one of the players is the quantum player and another is the classical player. The quantum player is free to

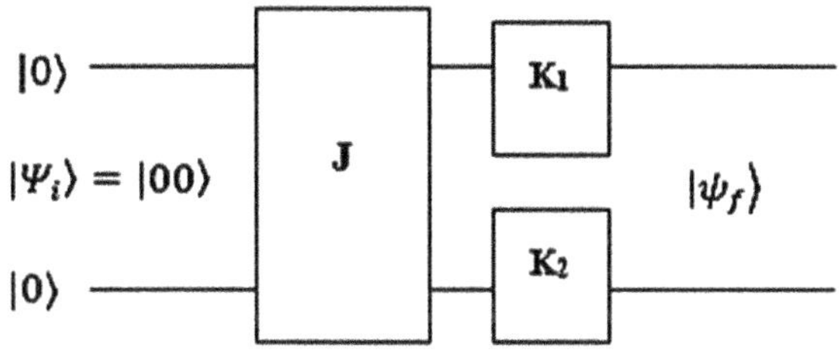


**Fig. 1** The quantum player playing first by applying non-local operator 'J' and the classical player applying local operators '$K_1$' and '$K_2$'

apply the quantum entangling operators, while the classical player can use only local operators. One of the players begins with the initial state |0 and |0, and thereafter applies their respective strategy one by one to reach their desired final state. The different cases are considered in the game such as robber playing as the quantum player and allowed to apply a non-local operator on the initial qubits, and hence entangled state is obtained. On the entangled state, the classical player, cop applies the local operators. Similarly, in another case, the cop is considered as the quantum player, hence, he/she applies non-local operator, whereas the robber is considered as a classical player and applies local operator on the given state. The general form of the quantum version of the game is represented in Fig. 1 using a block diagram. The quantum player who applies two-qubit non-local operator J is given as [21]

$$\mathrm{J} = \begin{pmatrix} x_1 & 0 & 0 & -iy_1 \\ 0 & x_2 & -iy_2 & 0 \\ 0 & -iy_2 & x_2 & 0 \\ -iy_1 & 0 & 0 & x_1 \end{pmatrix} \tag{1}$$

where

$$x_1 = e^{\frac{-iC_3}{2}} cos\frac{C_1 - C_2}{2}; \qquad x_2 = e^{\frac{iC_3}{2}} cos\frac{C_1 + C_2}{2}$$
$$y_1 = e^{\frac{-iC_3}{2}} sin\frac{C_1 - C_2}{2}; \qquad y_2 = e^{\frac{iC_3}{2}} sin\frac{C_1 + C_2}{2}$$

where $C_1, C_2$ and $C_3$ are the entangling parameters such that $\frac{\pi}{2} \geq \mathrm{C}_1 \geq \mathrm{C}_2 \geq C_3 \geq 0$. The classical player is bound to use single-qubit operators whose general form is given as [2]

$$\mathrm{K}_1 = \begin{pmatrix} a_1 & -b_1 \\ b_1^* & a_1^* \end{pmatrix}, \quad \mathrm{K}_2 = \begin{pmatrix} a_2 & -b_2 \\ b_2^* & a_2^* \end{pmatrix} \tag{2}$$

where

$$a_1 = cos\left(\frac{\gamma_1}{2}\right)e^{\frac{-i}{2}(\delta_1+\beta_1)}; \qquad b_1 = sin\left(\frac{\gamma_1}{2}\right)e^{\frac{i}{2}(\delta_1-\beta_1)}$$
$$a_2 = cos\left(\frac{\gamma_2}{2}\right)e^{\frac{-i}{2}(\delta_2+\beta_2)}; \qquad b_2 = sin\left(\frac{\gamma_2}{2}\right)e^{\frac{i}{2}(\delta_2-\beta_2)}$$

where $0 \le \gamma_i \le \pi$ and $-\pi \le \delta_i, \beta_i \le \pi$ for $i = 1, 2$.

Consider the case in which the quantum player first applies entangling operator J on the initial state $|\psi_i\rangle = |00\rangle$ and afterwards the classical player applies the local operator $(K_1 \otimes K_2)$ to the given state and reach the favorable final state $|\psi_f\rangle$. If the quantum player is the cop, then the desired final state for him to reach is either $|00\rangle$ or $|11\rangle$, whereas if a robber is the quantum player, then he desires to achieve the final state $|01\rangle$ or $|10\rangle$. The various winning strategies for the quantum player are obtained and analyzed thoroughly in [9]. The final state which the player wants to reach is dependent on the quantum player. The game has been analyzed from the perspective of a quantum player. It is observed that the quantum player is able to reach his favorable final state only when a certain set of local operators is adapted by the classical player. There exists no universal entangling operator that can take a quantum player to his desired state irrespective of the local strategies of the classical player [9]. Further, the quantum version of the cop and robber game is implemented on the quantum circuit using various unitary operators, and the final quantum state $|\psi_f\rangle$ is achieved using the peer-to-peer scheme, as shown in Fig. 2.

The first block represents the operations applied by the quantum player, starting from the initial qubits A and B, while the second block is representing the classical player applying the local operations on the qubits A and B. Player 1 sends the state $|\psi\rangle$ to player 2 which is given as

$$|\psi\rangle = \frac{1}{\sqrt{2}}\left[i\cos\left(\frac{C_1 - C_2}{2}\right)|00\rangle_{AB} - \sin\left(\frac{C_1 - C_2}{2}\right)|11\rangle_{AB}\right] \otimes |0f\rangle_{A1B1} \quad (3)$$

Here, $f$ is the measurement on qubit B1 which can be 0 or 1. It can be noticed from Fig. 2 that the quantum state $|\psi\rangle$ would interact with the environment during the

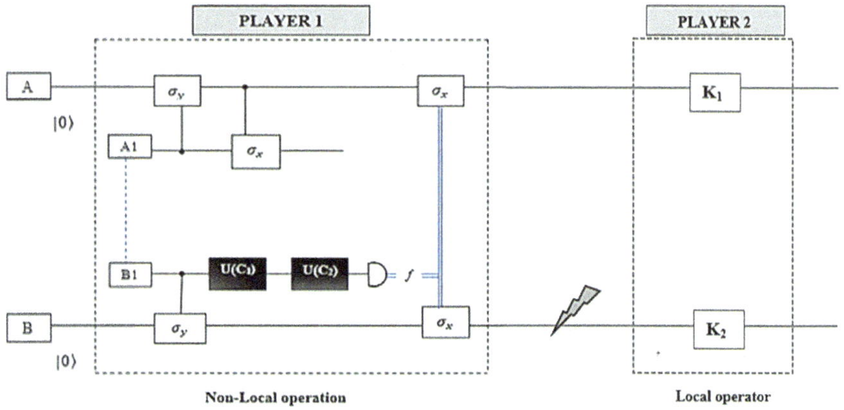


**Fig. 2** The quantum circuit affected by noise when player 1 transfers the quantum state $|\psi\rangle$ to player 2. $\sigma_x$, $\sigma_y$ and $\sigma_z$ are the Pauli operators and $U(C_1)$, $U(C_2)$, $K_1$ and $K_2$ are the single-qubit operators. The lightning sign on channel B indicates the noise in the quantum network

transmission from player 1 to player 2. Details of arriving at Eq. (3) can be found in [9].

## 3 Noise Models

The quantum circuit implementation of the cop and robber game is considered to be ideal. In reality, any practical quantum transfer always admits the interaction of the qubit with the environment. These interactions cause noise in the quantum channels leading to the loss of information. To understand the effect of the external environment on the dynamics of the open quantum system, the noise is distinguished according to its unique properties. In this work, we intend to focus on the two noise models, namely the amplitude damping model and the phase damping model.

### *3.1 Amplitude Damping (AD) Noise Model*

In the process of amplitude damping, the loss of quantum information is caused by the dissipation of energy when the system interacts with the environment which acts as a vacuum bath. The dynamics of amplitude damping is characterized by a general unitary operator known as the Kraus operator [2], given as

$$E_0 = \begin{bmatrix} 1 & 0 \\ 0 & \sqrt{1-\eta} \end{bmatrix}; \quad E_1 = \begin{bmatrix} 0 & \sqrt{\eta} \\ 0 & 0 \end{bmatrix} \tag{4}$$

Here, $\eta$ is the decoherence rate ranging from 0 to 1 ($0 \leq \eta \leq 1$). Note that Kraus operators mentioned above are for the single-qubit noise channel.

### *3.2 Phase Damping (PD) Noise Model*

The process of loss of quantum information exclusively occurring through the quantum mechanical process and without the loss of energy is called phase damping. The Kraus operators that describe the phase damping are given as [2]

$$E_0 = \sqrt{1-\eta}\begin{bmatrix} 1 & 0 \\ 0 & 1 \end{bmatrix}; \quad E_1 = \sqrt{\eta}\begin{bmatrix} 1 & 0 \\ 0 & 0 \end{bmatrix}; \quad E_2 = \sqrt{\eta}\begin{bmatrix} 0 & 0 \\ 0 & 1 \end{bmatrix} \tag{5}$$

Here $\eta$ is the decoherence rate such that $0 \leq \eta \leq 1$.

The effect of noise can be investigated by determining the fidelity between the initial state and the final state in the presence of noise. The steps to calculate fidelity are as follows [17]:

1. Determine the density matrix $\rho$ for the initial quantum state $|\psi\rangle$

$$\rho = |\psi\rangle\langle\psi| \tag{6}$$

   where $|\psi\rangle$ is the initial pure state.
2. Apply the Kraus operator of the particular noise model considered on the state such as

$$\rho_k = \sum_i E_i \rho E_i^\dagger \tag{7}$$

   where $E_i$ are the Kraus operators of the selected noise model.
3. Calculate fidelity by comparing the initial state and final state after the noise is incorporated using

$$F = \langle\psi| \rho_k |\psi\rangle \tag{8}$$

   This is the expression used to calculate the fidelity of the network.

## 4 Calculation and Analysis of Fidelity

During the transmission of quantum state $|\psi\rangle$ from the quantum player to the classical player, noise can be created due to the interaction between the state and the environment. Figure 2 depicts this scenario, where the lightning symbol is used to indicate the noise. As a result of noise, the state $|\psi\rangle$ given by Eq. (3) is affected. It can be noticed in the quantum circuit that the noise is affecting only channel B, whereas channel A remains unaffected. We consider the amplitude and phase damping models to understand the effect of noise in the quantum circuit. We can do so, by calculating the fidelity of the quantum states.

### *4.1 Effect of Amplitude Damping (AD)*

Firstly, we consider the case of amplitude damping introduced in the quantum circuit. The fidelity between the initial state $|\psi\rangle$, transferred by player 1 to player 2, and the final state, $|\psi\rangle$ getting affected by the noise channel before reaching player 2, is calculated using the steps mentioned in the previous section. The initial quantum state $|\psi\rangle$ is expressed in matrix form as

$$|\psi\rangle = \begin{bmatrix} icos\left(\frac{C_1-C_2}{2}\right) \\ 0 \\ 0 \\ -sin\left(\frac{C_1-C_2}{2}\right) \end{bmatrix}_{AB} \tag{9}$$

Following the steps to calculate fidelity as discussed in Sect. 2,

1. The density matrix $\rho$ for the initial state $|\psi\rangle$ is obtained as

$$\rho = \begin{bmatrix} \cos^2\theta & 0\ 0 & -i\sin\theta\cos\theta \\ 0 & 0\ 0 & 0 \\ 0 & 0\ 0 & 0 \\ i\sin\theta\cos\theta & 0\ 0 & \sin^2\theta \end{bmatrix} \tag{10}$$

where $\theta = \left(\frac{C_1-C_2}{2}\right)$. Note that $C_1$ and $C_2$ are parameters of entangling operator J such that $\frac{\pi}{2} \geq C_1 \geq C_2 \geq 0$. To minimize the mathematical complexity, we have assumed $C_3 = 0$.

2. Apply Kraus operators on the above state such that $\rho_k^{AD} = \sum_i E_i\rho(E_i)^\dagger$. Note that $E_i$ are the single-qubit Kraus operators which are assumed to act on the second qubit. Therefore, we have

$$\rho_k^{AD} = \begin{bmatrix} \cos^2\theta & 0\ 0 & -(\sqrt{1-\eta})i\sin\theta\cos\theta \\ 0 & 0\ 0 & 0 \\ 0 & 0\ 0 & 0 \\ (\sqrt{1-\eta})i\sin\theta\cos\theta & 0\ 0 & (1-\eta)\sin^2\theta \end{bmatrix} \tag{11}$$

3. The expression obtained for fidelity, by taking the inner product between the initial state $|\psi\rangle$ and final state $\rho_k^{AD}|\psi\rangle$, is given by

$$F_{AD} = \left[\cos^2\theta + (1-\eta)\sin^2\theta\right]^2 \tag{12}$$

This is the expression for fidelity due to amplitude damping in the quantum network. Note that if the decoherence rate $\eta$ becomes zero, then the fidelity reaches the maximum. This suggests that the transmission of the quantum state is achieved with maximum accuracy in the absence of noise. It is important to mention that the same expression of fidelity is obtained when the amplitude damping acts on channel A. Using the expression for fidelity $F_{AD}$, given by Eq. (12), a graph is plotted as shown in Fig. 3. Note that we consider the term $\theta = \left(\frac{C_1-C_2}{2}\right)$ as an effective entangling parameter. From the graph, it can be observed that the fidelity of the quantum states decreases with the increase in the rate of decoherence. Thus, it is evident that the higher the decoherence rate ($\eta$), the lower is the fidelity. When entangling parameters are equal, $C_1 = C_2$, $\theta$ becomes 0, thus the fidelity of the network remains maximum irrespective of the decoherence rate.

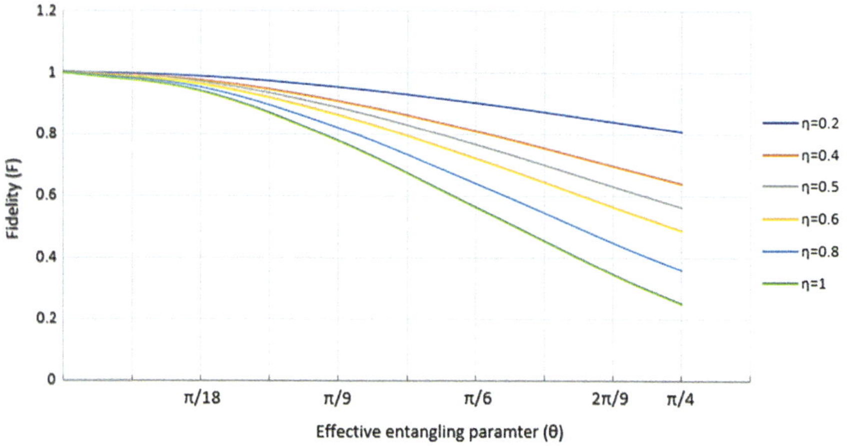


**Fig. 3** Effect of amplitude damping on the fidelity of quantum circuit, with respect to entangling parameter term $\theta$ and decoherence rate $\eta$

This implies that the set of entangling operators with parameters $C_1 = C_2$ supresses the noise in the quantum network, even if the damping reaches the maximum value. Further, the effective entangling parameter $\theta$ becomes $\frac{\pi}{4}$, when the entangling parameters are $C_1 = \frac{\pi}{2}$ and $C_2 = 0$ corresponding to the CNOT operator, $J\left(\frac{\pi}{2}, 0, 0\right)$, the effect of damping on the circuit is maximum. It is clear from the graph that the fidelity decreases with effective entangling parameters other than $\theta = 0$.

## 4.2 *Effect of Phase Damping (PD)*

In this case, we study the effect of phase damping on the quantum network by calculating the fidelity of the quantum state by adopting the following steps:

1. The initial state $|\psi\rangle$, given by Eq. (9), sent by the quantum player to the classical player is the same, therefore, the density matrix $\rho$ is the same.
2. Apply Kraus operators of PD noise, such that $\rho_k^{PD} = \sum_i E_i \rho (E_i)^\dagger$. Hence,

$$\rho_k^{PD} = \begin{bmatrix} \cos^2\theta & 0\ 0 & -(1-\eta) i \sin\theta\cos\theta \\ 0 & 0\ 0 & 0 \\ 0 & 0\ 0 & 0 \\ (1-\eta) i \sin\theta\cos\theta & 0\ 0 & \sin^2\theta \end{bmatrix} \quad (13)$$

3. The expression obtained for fidelity, by taking the inner product between the initial state $|\psi\rangle$ and final state $\rho_k^{PD}|\psi\rangle$, is given by

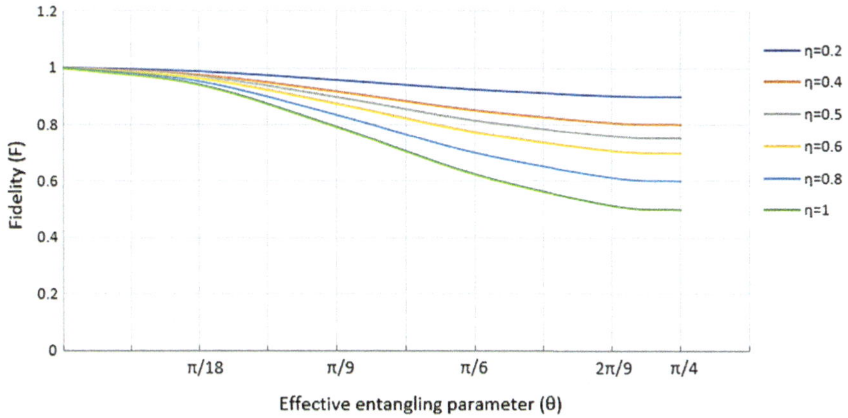


**Fig. 4** Effect of phase damping on the fidelity of quantum state, with respect to effective entangling parameter θ and decoherence rate η

$$F_{PD} = \left[\cos^4\theta + \sin^4\theta + 2(1-\eta)\sin^2\theta\cos^2\theta\right] \tag{14}$$

Note that in the absence of damping ($\eta = 0$), the fidelity of the quantum state reaches the maximum. The same expression of fidelity is obtained when phase damping acts on channel A. The graph shown in Fig. 4 indicates the variation of fidelity with respect to effective entangling parameters for the different decoherence rates. The graph exhibits a similar pattern as in the case of amplitude damping. The fidelity under the effect of phase damping decreases with the decoherence rate and entangling parameters. However, it can be noted that the fidelity reduces more under amplitude damping as compared to phase damping. The lowest value of fidelity in amplitude damping is 0.2499, while in phase damping, the lowest level is 0.4998 for the choice of $\theta = \frac{\pi}{4}$, which corresponds to CNOT gate, $J\left(\frac{\pi}{2}, 0, 0\right)$.

## 5 Conclusion

In this work, we have investigated the effect of noise on the quantum network which has been designed to implement a quantum version of the cop and robber game. The two important noise models, amplitude damping noise model and phase damping noise models, are incorporated into the quantum circuit, and their effects are analyzed by calculating fidelity. Further, graphs are plotted to realize the variation in the fidelity of quantum states with respect to entangling parameters $C_1$, $C_2$, and decoherence rate $\eta$. It is observed that the fidelity of the quantum states decreases as the decoherence rate increases, thus affecting the accuracy of quantum data transmission.

Entangling operators $J(C_1, C_1, 0)$ corresponding to the case $\theta = 0$ prevent the quantum state from the influence of decoherence. This suggests that this set of entangling operators can suppress the amplitude damping and phase damping and thus provides secure data transmission. Moreover, at $\theta = \frac{\pi}{4}$, which corresponds to the CNOT operator, $J\left(\frac{\pi}{2}, 0, 0\right)$, the fidelity reduces to its minimum value. This effect on fidelity is observed in the cases of amplitude damping and phase damping for the CNOT operator.

It is interesting to note that the effect of amplitude damping and phase damping on the quantum circuit is similar in all aspects. In both cases, the fidelity reaches its minimum value at $\theta = \frac{\pi}{4}$. However, amplitude damping causes more distortion in the transmission as compared to phase damping. Therefore, amplitude damping results in more loss of information as compared to phase damping.

It is evident from this work that the appropriate quantum strategy can completely suppress the noise effect on the transmitted quantum state. However, this result is quite different from the observation that the classical strategy performs better than the quantum strategy when the noise in the quantum circuit is more than 50% [22]. Moreover, the payoffs of the players get affected when the decoherence rate increases. In our work, it is worth investigating the effects of noise on the payoffs of the players in the cop and robber game. In the present work, we have considered damping in only one channel. We can extend our analysis of damping in both communication channels. Nonetheless, this work presents significant observations on the significance of quantum strategy to reduce the effects of amplitude damping and phase damping on the transmitted quantum state. To summarize, the appropriate application of entangling operators can suppress the noise in the quantum network implementation of the cop and robber game.

## References


1. Preskill, J.: Lecture Notes for Physics 229: Quantum information and computation. California Institution of Technology, Pasadena (1998)
2. Nielsen, M.A., Chuang, I.L.: Quantum Computation and Quantum Information. Cambridge University Press, New Delhi (2008)
3. Bennett, C.H., Brassard, G.: Quantum cryptography: public key distribution and coin tossing. In: Proceedings of the IEEE International Conference on Computers, Systems, and Signal Processing, pp. 175–179. India (1984)
4. Ekert, A.K.: Quantum cryptography based on Bell's theorem. Phys. Rev. Lett. **67**, 661 (1991)
5. Neumann, J., Morgenstern, O.: Theory of Games and Economic Behavior. Wiley, New York (1967)
6. Eisert, J., Wilkens, M., Lewenstein, M.: Quantum games and quantum strategies. Phys. Rev. Lett. **83**, 3077 (1999)
7. Guo, H., Zhang, J., Koehler, G.J.: A survey of quantum games. Decis. Support Syst. **46**, 318 (2008)
8. Liu, B., Dai, H., Zhang, M.: Playing distributed two-party quantum games on quantum networks. Quantum Inf. Process. **16**, 290 (2017)
9. Dhiman, A., Uttam, T., Balakrishnan, S.: Implementation of sequential game on quantum circuits. Quant. Inf. Process. **19**, 109 (2020)

10. Nowakowski, R.J., Winkler, P.: Vertex-to-vertex pursuit in graph. Discr. Math. **43** (1983)
11. Bourennane, M., Eibl, M., Gaertner, S., Kurtsiefer, C., Cabello, A., Weinfurter, H.: Decoherence-free quantum information processing with four-photon entangled states. Phys. Rev. Lett. **92**, 107901 (2004)
12. Chiuri, A., Rosati, V., Vallone, G., Padua, S., Imai, H., Giacomini, S., Macchiavello, C., Mataloni, P.: Experimental realization of optimal noise estimation for a general Pauli channel. Phys. Rev. Lett. **107**, 253602 (2011)
13. Banerjee, S., Ghosh, R.: Dynamics of decoherence without dissipation in a squeezed thermal bath. J. Phys. A: Math. Theor. **40**, 13735 (2007)
14. Omkar, S., Srikanth, R., Banerjee, S.: Dissipative and non-dissipative single-qubit channels: dynamics and geometry. Quant. Inf. Process. **12**, 3725 (2013)
15. Huang, J.H., Zhu, S.Y.: Necessary and sufficient conditions for the entanglement sudden death under amplitude damping and phase damping. Phys. Rev. A **76**, 062322 (2007)
16. Yu, T., Eberly, J.H.: Qubit disentanglement and decoherence via dephasing. Phys. Rev. B **98**, 165322 (2003)
17. Sharma, V., Thapliyal, K., Pathak, A., Banerjee, S.: A comparative study of protocols for secure quantum communication under noisy environment: single-qubit-based protocols versus entangled-state-based protocols. Quantum Inf. Process. **15**, 11 (2020)
18. Quilliot, A.: A short note about pursuit games played on a graph with a given genus. J. Comb. Theory **38**(1), 89–92 (1985)
19. Bonito, A., Nowakowski, R.J.: The Game of Cops and Robbers on Graphs. American Mathematical Society, Providence, Rhode Island (2011)
20. Glos, A., Miszczak, J.A.: The role of quantum correlations in Cop and Robber game. Quantum Stud.: Math. Found. **6,** 1 (2017)
21. Rezakhani, A.T.: Characterization of two-qubit perfect entanglers. Phys. Rev. A **70**, 052313 (2004)
22. Kairon, P., Thapliyal, K., Srikanth, R., Pathak, A.: Noisy three-player dilemma game: robustness of the quantum advantage. Quantum Inf. Process **19**, 327 (2020)

# Study of Decoherence in Quantum Cournot Duopoly Game Using Modified EWL Scheme

A. V. S. Kameshwari and S. Balakrishnan

**Abstract** The newly proposed modified Eisert-Wilkens-Lewenstein scheme can be widely used to understand any quantum game through quantum operators. This scheme provides a vast range of two-qubit entangling operators which is otherwise not possible using the traditional Eisert-Wilkens-Lewenstein scheme (EWL) and the Marinatto-Weber scheme (MW). In our work, the proposed modified EWL scheme is further explored in noisy market games. Decoherence commonly known as noise is an unavoidable interaction of the system with the surroundings. We analyze the effects of decoherence in the Cournot duopoly game when amplitude damping is present in either of the communication channels. We find an interesting result that decoherence affects channel 2 but does not affect channel 1. Furthermore, we discover that the effect of decoherence can be partially mitigated by selecting an appropriate entangling operator.

**Keywords** Game theory · Quantization scheme · Cournot duopoly game · Decoherence

## 1 Introduction

Basically, a game is any competitive activity that involves two or more rational players or multiple agents who compete to maximize their respective payoffs according to a fixed set of rules [1]. The players are also called the decision makers who interact with one another. Game theory provides a mathematical model to understand the competitive situation between the decision makers [1]. The publication, "Theory of Games and Economic Behaviour" by Oskar Morgenstern and John von Neumann, in the year 1944 led to the foundation of the present-day understanding of game theory [2]. Game theory became more popular in the year 1994 when three famous game theorists won the Nobel Prize in Economic Sciences [1]. A game theoretic model

A. V. S. Kameshwari · S. Balakrishnan (✉)
Department of Physics, School of Advanced Sciences, Vellore Institute of Technology, Tamil Nadu, Vellore 632014, India
e-mail: physicsbalki@gmail.com



B. Rushi Kumar et al. (eds.), *Mathematics and Computing*, Springer Proceedings in Mathematics & Statistics 415, https://doi.org/10.1007/978-981-19-9307-7_3

has a wide range of applications in the field of economics, politics, social sciences and biology [3].

The motivation to implement the ideas of classical game theory in the quantum domain came with the advancement of quantum mechanics and its application in different fields. David Meyer, in the year 1999, was the first to introduce the concept of quantum games in his work [4]. In his work, he exclusively showed how powerful are the players who adopt quantum strategies over the players with classical strategies. Meyer's work motivated a significant amount of research in quantum game theory; refer to [5–8]. In principle, any classical game can be extended to a quantum game with the help of quantization schemes. The Eisert, Wilkens and Lewenstein (EWL) [9], Marinatto and Weber (MW) [10], and modified EWL schemes [11–13] are the notable quantization schemes. Both EWL and MW schemes faced criticism about their inability to show the quantumness in quantum games [6]. Mostly, quantum games are widely studied using EWL and MW schemes which exploit only controlled unitary operators.

In this paper, we use a modified EWL scheme to study the competitive market games also known as duopoly games in economics [14]. We analyze the Cournot duopoly game in the presence of decoherence in the communication channels with amplitude damping. The Cournot duopoly is a simultaneous game introduced by Augustin Cournot where both firms move at a time with no information of the opponent's move [15]. Decoherence is the interaction of any physical system with the environment which causes loss of information [16–19]. In reality, no system is closed, i.e., devoid of interaction with the environment and hence decoherence is inevitable. In quantum games, decoherence is well known to lower the player's average payoff; refer to [13, 16–19]. Recent works on duopoly games in the presence of decoherence show the effect of decoherence in sequential market games [20–22]. Taking this as a reference, we explicitly show how decoherence with amplitude damping channel affects the simultaneous market game.

## 2 Quantum Cournot Duopoly Game with Decoherence

Cournot duopoly is a competition between two manufacturing firms; let them be called Firm 1 and Firm 2 which provide certain goods that are strategic substitutes for the other [15]. The quantity of goods produced by Firm 1 and Firm 2 is given as $q_1$ and $q_2$, respectively. In this work, we introduce decoherence in the quantum form of the Cournot duopoly game. Decoherence or noise is the inevitable process which occurs when any physical system is in contact with the surroundings [16–22]. To observe how decoherence affects the Cournot duopoly game, amplitude damping is taken which is described by the following Kraus operator [13, 16]:

$$m_0 = \begin{pmatrix} 1 & 0 \\ 0 & \sqrt{1-p} \end{pmatrix} \tag{1}$$

where $p$ is the amount of noise present in the channel in the range, $0 \leq p \leq 1$. The maximum amount of noise in the channel is represented as $p = 1$. The most general form of the entangling operator which is non-local and produces entanglement is given as [11–13]

$$J(c_1, c_2, c_3) = \begin{pmatrix} e^{\frac{-ic_3}{2}}c^- & 0 & 0 & -ie^{\frac{-ic_3}{2}}s^- \\ 0 & e^{\frac{ic_3}{2}}c^+ & -ie^{\frac{ic_3}{2}}s^+ & 0 \\ 0 & -ie^{\frac{ic_3}{2}}s^+ & e^{\frac{ic_3}{2}}c^+ & 0 \\ -ie^{\frac{-ic_3}{2}}s^- & 0 & 0 & e^{\frac{-ic_3}{2}}c^- \end{pmatrix} \tag{2}$$

where $c^{\pm} = \cos\left(\frac{c_1 \pm c_2}{2}\right)$, $s^{\pm} = \sin\left(\frac{c_1 \pm c_2}{2}\right)$ and $c_1, c_2, c_3$ are the geometrical points of the Weyl chamber satisfying the condition $c_1 \geq c_2 \geq c_3 \geq 0$ [23, 24]. Further, we have considered $c_3 = 0$ for the mathematical simplicity as it reduces the three-dimensional Weyl chamber to two dimensions. This further reduces the entangling operator $J$ of the form:

$$J = \begin{pmatrix} c^- & 0 & 0 & -is^- \\ 0 & c^+ & -is^+ & 0 \\ 0 & -is^+ & c^+ & 0 \\ -is^- & 0 & 0 & c^- \end{pmatrix} \tag{3}$$

The game initially is in the state $|00\rangle$ that is the state is further entangled using Eq. (3). The player strategies are the local operator of the form as given below [25]

$$U_A = \begin{pmatrix} \cos\frac{\theta_1}{2} & -\sin\frac{\theta_1}{2} \\ \sin\frac{\theta_1}{2} & \cos\frac{\theta_1}{2} \end{pmatrix} \quad \text{and} \quad U_B = \begin{pmatrix} \cos\frac{\theta_2}{2} & -\sin\frac{\theta_2}{2} \\ \sin\frac{\theta_2}{2} & \cos\frac{\theta_2}{2} \end{pmatrix} \tag{4}$$

where $0 \leq \theta_j \leq \pi$, $j = 1, 2$. Equivalently, the strategies $U_A$ and $U_B$ adapted by Firm 1 and 2 are parameterized by $\theta_1$ and $\theta_2$, respectively. Using the Kraus operator of the amplitude damping channel as mentioned earlier, the Cournot duopoly game is analyzed when noise is present in either of the channels.

## 2.1 Decoherence in Channel 1

In this section, the quantum Cournot duopoly game is analyzed when channel 1 is noisy. The final state of this game for the amplitude damping channel is represented as [13]

$$\left|\psi_f\right\rangle = J^{\dagger}(U_A \otimes U_B)(m_0 \otimes m_0)J\left|00\right\rangle \tag{5}$$

The tensor product of the strategies of the firms can be written as

$$U_A \otimes U_B = \begin{pmatrix} X & -Y & -W & Z \\ Y & X & -Z & -W \\ W & -Z & X & -Y \\ Z & W & Y & X \end{pmatrix} \tag{6}$$

where $X = \cos\left(\frac{\theta_1}{2}\right)\cos\left(\frac{\theta_2}{2}\right)$, $Y = \cos\left(\frac{\theta_1}{2}\right)\sin\left(\frac{\theta_2}{2}\right)$, $W = \sin\left(\frac{\theta_1}{2}\right)\cos\left(\frac{\theta_2}{2}\right)$ and $Z = \sin\left(\frac{\theta_1}{2}\right)\sin\left(\frac{\theta_2}{2}\right)$ are the entries of a two-qubit local operator.

We substitute Eqs. (1), (6) and (2) in Eq. (5) to attain final state $|\psi_f$ of the form:

$$|\psi_f\rangle = \begin{pmatrix} X + p\left(iZc^-s^- - Xs^-s^-\right) \\ Y\cos c_1 + iW\sin c_1 + p\left(Ys^+s^- - iWc^+s^-\right) \\ iY\sin c_1 + W\cos c_1 + p\left(Ws^+s^- - iYc^+s^-\right) \\ Z + p\left(iXs^-c^- - Zs^-s^-\right) \end{pmatrix} \tag{7}$$

Measurement of the final state is performed using the measurement operators $M_1$ and $M_2$ of the form [26]:

$$M_j(x_1, x_2) = \begin{cases} (x_1|0\rangle\langle 0| + x_2|1\rangle\langle 1|) \otimes (|0\rangle\langle 0| + |1\rangle\langle 1|) \text{for } j = 1 \\ (x_2|0\rangle\langle 0| + x_1|1\rangle\langle 1|) \otimes (|0\rangle\langle 0| + |1\rangle\langle 1|) \text{for } j = 2 \end{cases} \tag{8}$$

where $x_i \in [0, \infty)$ is a set of continuous strategies adopted by both firms. On performing the measurement, the quantities of the firms are obtained using the formula given below [26]

$$q_1 = tr(M_1\rho),\ q_2 = tr(M_2\rho) \tag{9}$$

where $\rho = |\psi_f\rangle\langle\psi_f|$ represents the density matrix of the final state. On substituting Eq. (8) in Eq. (9), the quantities of the firms become

$$\begin{aligned} q_1 = &\left(\left(pXs^-s^- - X\right)^2 + \left(pZs^-c^-\right)^2 + \left(pYs^+s^- + Y\cos c_1\right)^2 \right. \\ &\left. + \left(pWc^+s^- - W\sin c_1\right)^2\right)x_1 \\ &+ \left(\left(pYc^+s^- - Y\sin c_1\right)^2 + \left(pZs^-s^- - Z\right)^2 + \left(pXs^-c^-\right)^2\right)x_2 \\ q_2 = &\left(\left(pXs^-s^- - X\right)^2 + \left(pZs^-c^-\right)^2 + \left(pYs^+s^- + Y\cos c_1\right)^2 \right. \\ &\left. + \left(pWc^+s^- - W\sin c_1\right)^2\right)x_2 \\ &+ \left(\left(pWs^+s^- + W\cos c_1\right)^2 + \left(pYc^+s^- - Y\sin c_1\right)^2 + \left(pZs^-s^- - Z\right)^2 + \left(pXc^-s^-\right)^2\right)x_1 \end{aligned} \tag{10}$$

In the above equation, on substituting $c_1 = c_2$, i.e., for the choice of entangling operator $J(c_1, c_2 = c_1, c_3 = 0)$, the quantities of the firms become [26]

$$\begin{aligned} q_1 &= \left(X^2 + Y^2 cos^2 c_1 + W^2 sin^2 c_1\right)x_1 + \left(W^2 cos^2 c_1 + Y^2 sin^2 c_1 + Z^2\right)x_2 \\ q_2 &= \left(X^2 + Y^2 cos^2 c_1 + W^2 sin^2 c_1\right)x_2 + \left(W^2 cos^2 c_1 + Y^2 sin^2 c_1 + Z^2\right)x_1 \end{aligned} \quad (11)$$

The quantities of the firms obtained in Eq. (11) satisfy the condition $q_1 + q_2 = x_1 + x_2$ similar to that of the noiseless game. The general expression for the firms' profit is given as [26–28]

$$u_1 = q_1[k - (q_1 + q_2)] \text{ and } u_2 = q_2[k - (q_1 + q_2)] \quad (12)$$

where $k = a - c$, $a$ is the market's net capacity and $c$ represents cost [26–28]. The Nash equilibrium of the Cournot duopoly game is obtained by calculating the best response of both firms. The best response of the firms is calculated by differentiating the firm's profit function and then equating it to zero [25]. The Nash equilibrium so obtained is given below [26]

$$x_1^* = x_2^* = \frac{k\left(X^2 + Y^2 \cos^2 c_1 + W^2 \sin^2 c_1\right)}{2\left(X^2 + Y^2 \cos^2 c_1 + W^2 \sin^2 c_1\right) + 1} \quad (13)$$

Profit of the firms at the Nash equilibrium is [26]

$$u_1^* = u_2^* = \frac{k^2\left(X^2 + Y^2 \cos^2 c_1 + W^2 \sin^2 c_1\right)}{\left[2\left(X^2 + Y^2 \cos^2 c_1 + W^2 \sin^2 c_1\right) + 1\right]^2} \quad (14)$$

From the above equation, it is observed that for the choice of entangling operator $J\left(0 \leq c_1 \leq \frac{\pi}{2}, c_2 = c_1, c_3 = 0\right)$, the profit from the Cournot duopoly game with decoherence is identical to the noiseless case; refer to [26]. Further, from Eq. (14) it can be seen that the profit of the firm is independent of the level of noise in channel 1 as it only depends upon the strategies and the entangling operator. Also, the quantities $q_1$ and $q_2$ are independent of decoherence ($p$) for the mentioned entangling operator. This observation is due to the presence of noise in the communication channel before the application of the strategies adopted by the firms. The effect of noise in channel 1 for the quantum Cournot duopoly game can be killed with an appropriate choice of entangling operator.

## 2.2 *Decoherence in Channel 2*

In this section, the quantum Cournot duopoly game is analyzed when channel 2 is noisy. The final state of the game for the amplitude damping channel becomes [13]

$$\left|\psi_f\right\rangle = J^\dagger(m_0 \otimes m_0)(U_A \otimes U_B)J\left|00\right\rangle \quad (15)$$

We substitute Eqs. (1), (6) and (2) in Eq. (15), to attain the final state of the form:

$$|\psi_f\rangle = \begin{pmatrix} X + p\left(iZc^-s^- - Xs^-s^-\right) \\ Y\sqrt{1-p}\cos c_1 + iW\sqrt{1-p}\sin c_1 \\ iY\sqrt{1-p}\sin c_1 + W\sqrt{1-p}\cos c_1 \\ Z + p\left(iXs^-c^- - Zc^-c^-\right) \end{pmatrix} \tag{16}$$

The measurement operator given by Eq. (8) is applied to this final state, and the following quantities $q_1$ and $q_2$ of Firms 1 and 2 are obtained by using Eq. (9):

$$\begin{aligned} q_1 &= \left((X - pXs^-s^-)^2 + (pZc^-s^-)^2 + Y^2(1-p)cos^2c_1 + W^2(1-p)sin^2c_1\right)x_1 \\ &\quad + \left(Y^2(1-p)sin^2c_1 + W^2(1-p)cos^2c_1 + (Z - pZc^-c^-)^2 + (pXc^-s^-)^2\right)x_2 \\ q_2 &= \left((X - pXs^-s^-)^2 + (pZc^-s^-)^2 + Y^2(1-p)cos^2c_1 + W^2(1-p)sin^2c_1\right)x_2 \\ &\quad + \left(Y^2(1-p)sin^2c_1 + W^2(1-p)cos^2c_1 + (Z - pZc^-c^-)^2 + (pXc^-s^-)^2\right)x_1 \end{aligned} \tag{17}$$

For the choice of entangling operator $J(c_1, c_2 = c_1, c_3 = 0)$, the quantities of the firms become

$$\begin{aligned} q_1 &= \left(X^2 + Y^2(1-p)cos^2c_1 + W^2(1-p)sin^2c_1\right)x_1 \\ &\quad + \left(W^2(1-p)cos^2c_1 + Y^2(1-p)sin^2c_1 + Z^2(1-p)^2\right)x_2 \\ q_2 &= \left(X^2 + Y^2(1-p)cos^2c_1 + W^2(1-p)sin^2c_1\right)x_2 \\ &\quad + \left(W^2(1-p)cos^2c_1 + Y^2(1-p)sin^2c_1 + Z^2(1-p)^2\right)x_1 \end{aligned} \tag{18}$$

Using the expressions of the quantities of the firms from Eq. (18), we attain Nash equilibrium as

$$x_1^* = x_2^* = \frac{ka}{b(2a+b)} \tag{19}$$

The firm's profit at Nash equilibrium is

$$u_1^* = u_2^* = \frac{k^2ab}{(2a+b)^2} \tag{20}$$

where $a = X^2 + (1-p)Y^2\cos^2 c_1 + (1-p)W^2\sin^2 c_1$, $b = X^2 + (1-p)Y^2 + (1-p)W^2 + (1-p)^2Z^2$ and $0 \le p \le 1$.

From the above equation, it can be identified that the profit of firms depends upon three parameters, namely strategies adopted by the firms, entangling operator $J\left(0 \le c_1 \le \frac{\pi}{2}, c_2 = c_1, c_3 = 0\right)$ and decoherence $(p)$. Further, when noise is present after the application of the strategies of the players, no choice of player strategies and entanglement can eliminate the effects of decoherence. Therefore, noise in channel 2 affects the profit function of both firms independent of the choice of entangling operator $J\left(0 \le c_1 \le \frac{\pi}{2}, c_2 = c_1, c_3 = 0\right)$.

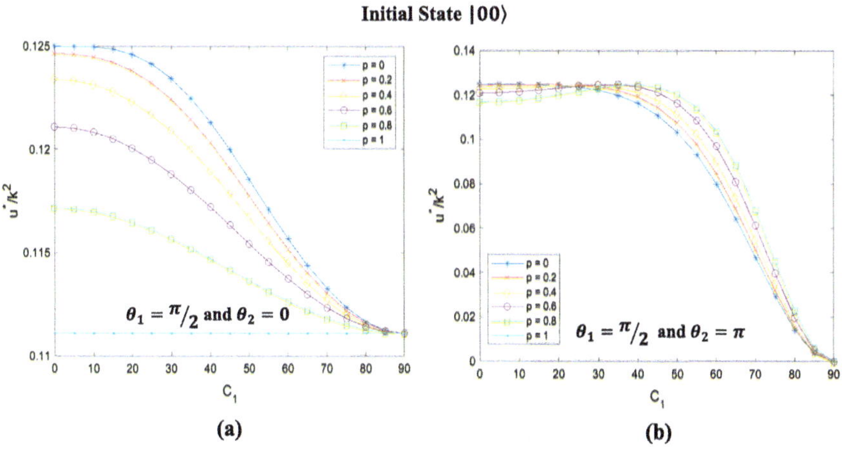


**Fig. 1** Effect of noise on the profit of the firms for a particular strategic combination of the firms

The impact of noise in channel 2 parameterized by $p$ on the firm's profit is analyzed from the graphs. In Fig. 1a, the strategic choice of Firms 1 and 2 is taken as mixed strategy and identity, whereas in Fig. 1b the strategic choice is taken as mixed strategy and flip. The graphs show that decoherence has an effect on the firm's profit. Observe from Fig. 1a that as decoherence increases from $p = 0$ to $p = 1$, the profit of the firms decreases. The decrease in the profit of the firms due to decoherence or noise can be seen evidently for lower levels of entanglement. A similar observation can also be found in Fig. 1b where an increase in decoherence decreases the profit of the firms for a particular choice of entangling operator. From this observation, we can state that the entangling operator parameterized by $c_1$ either increases or decreases the profit of the firms for a given value of $p$. Such an observation has arrived at for the noisy prisoner's dilemma with an amplitude damping channel using a modified EWL scheme [13]. Further from the analysis carried out in Sects. 2.1 and 2.2, analytically it can be said that the presence of noise in channel 2 affects the profit of the firms more than that of channel 1. A similar observation can also be found using a modified EWL scheme in the prisoner's dilemma game [13] and the EWL scheme [16].

## 3 Conclusion

It is well understood that a certain amount of decoherence in the communication channels reduces the outcome of the players. We found that due to decoherence, the quantum version of the Cournot duopoly game lowers the original profit of the manufacturing firms. Such an observation can also be found in the works of Zhu and Kuang [20, 21] and Khan et al. [22], for the Stackelberg duopoly game which is a well-known sequential game. We found that when decoherence in the form of amplitude damping is present in the channels; its effect on the profit of the firms can be modulated with an appropriate choice of entangling operator. According to Sect. 2.1, when noise is present in channel 1, its effect on the firm's profit can be eliminated by using the entangling operator $J\left(0 \leq c_1 \leq \frac{\pi}{2}, c_2 = c_1, c_3 = 0\right)$. However, this is not the case for amplitude damping in channel 2. Furthermore, when damping is present in channel 2 for a given value of decoherence (p), the profit of the firms is found to either increase or decrease for a given entangling operator, depending on the strategies used by both firms. Our analysis shows that the effect of decoherence in channel 2 is greater than that in channel 1. This behavior is caused by the presence of decoherence in the channel after the players' respective strategies have been applied, whereas in the case of channel 1, decoherence is present before the application of player strategies. Such an observation is consistent with the available literature [13, 16]. In this work, we have discussed only amplitude damping in either channel 1 or channel 2. Further, amplitude damping in both channel 1 and channel 2 can also be analyzed for the same quantum Cournot duopoly game. The entire analysis to calculate the noisy duopoly games can also be approached using the density matrix formalism.

## References

1. Osborne, M.J.: An Introduction to Game theory. Oxford University Press, Oxford (2004)
2. Von Neumann, J., Morgenstern, O.: Theory of Games and Economic Behavior. Princeton University Press (1947)
3. Colman, A.M.: Game Theory and Its Applications in the Social and Biological Sciences. Butterworth-Heinemam, Oxford (1995)
4. Meyer, D.A.: Quantum strategies. Phys. Rev. Lett. **82**, 1052–1055 (1999)
5. Guo, H., Zhang, J., Koehler, G.J.: A survey of quantum games. Decis. Support Syst. **46**(1), 318 (2008)
6. Khan, F.S., Solmeyer, N., Balu, R., et al.: Quantum games: a review of the history, current state, and interpretation. Quantum Inf. Process. **17**, 309 (2018)
7. Flitney, A.P., Abbott, D.: An introduction to quantum game theory. Fluct. Noise Lett. **2**, R175 (2002)
8. Piotrowski, E.W., Sladkowski, J.: An invitation to quantum game theory. Int. J. Theor. Phys. **42**, 1089 (2003)
9. Eisert, J., Wilkens, M., Lewenstein, M.: Quantum games and quantum strategies. Phys. Rev. Lett. **83**, 3077–3080 (1999)
10. Marinatto, L., Weber, T.: A quantum approach to static games of complete information. Phys. Lett. A **272**, 291–303 (2000)
11. Vijayakrishnan, V., Balakrishnan, S.: Role of two-qubit entangling operators in the modified Eisert-Wilkens-Lewenstein approach of quantization. Quantum Inf. Process. **18**, 112 (2019)
12. Vijayakrishnan, V., Balakrishnan, S.: Significance of entangling operators in the purview of modified EWL scheme. Quantum Inf. Process. **19**, 315 (2020)
13. Kameshwari, A.V.S., Balakrishnan, S.: Study of decoherence and memory in modified Eisert-Wilkens-Lewenstein scheme. Quantum Inf. Process. **20**, 282 (2021)
14. Gibbons, R.: Game Theory for Applied Economists. Princeton University Press (1992)
15. Cournot, A.: Researches into the Mathematical Principles of the Theory of Wealth. In: Bacon, N., Macmillan, R. (eds.) New York (1897)

16. Chen, L.K., Ang, H., Kiang, D., Kwek, L.C., Lo, C.F.: Quantum prisoner dilemma under decoherence. Phys. Letts. A **316**, 317–323 (2003)
17. Flitney, A.P., Hollenberg, L.C.L.: Multiplayer quantum minority game with decoherence. Quantum Inf. Comput. **7**, 111 (2007)
18. Nawaz, A.: The generalized quantization schemes for games and its application to quantum information, Ph.D. thesis, Quaid-I-Azam University, Islamabad, Pakistan (2007). arXiv:quant-ph/1012.1933
19. Khan, S., Ramzan, M., Khan, M.K.: Quantum Parrondo's game under decoherence. Int. J Theor. Phys **49**, 31 (2010)
20. Zhu, X., Kuang, Le-Man.: The influence of entanglement and decoherence on the quantum Stackelberg duopoly game. J. Phys. A: Math. Theor. **40**, 7729 (2007)
21. Zhu, X., Kuang, Le-Man.: Quantum Stackelberg duopoly game in depolarizing channel. Commun. Theor. Phys. **49**, 111 (2008)
22. Khan, S., Ramzan, M., Khan, M.K.: Quantum Stackelberg duopoly in the presence of correlated noise. J. Phys. A: Math. Theor. **43**, 375301 (2010)
23. Zhang, J., Vala, J., Whaley, K.B., Sastry, S.: Geometric theory of nonlocal two-qubit operations. Phys. Rev. A **67**, 042313 (2003)
24. Rezakhani, A.T.: Characterization of two-qubit perfect entanglers. Phys. Rev. A **70**, 052313 (2004)
25. Nielsen, M.A., Chuang, I.L.: Quantum Computation and Quantum Information. Cambridge University Press, Cambridge (2000)
26. Kameshwari, A.V.S., Balakrishnan, S.: Cournot and Stackelberg duopoly game in the purview of modified EWL scheme. Quantum Inf. Process **20**, 337 (2021)
27. Frackiewicz, P.: Remarks on quantum duopoly schemes. Quantum Inf. Process. **15**, 121–136 (2016)
28. Shi, L., Xu, F., Chen, Y.: Quantum Cournot duopoly game with isoelastic demand function. Phys. A **566**, 125614 (2021)

# A New Aggregation Operator for Single-Valued Triangular Neutrosophic Number in Decision-Making

**G. Tamilarasi and S. Paulraj**

**Abstract** In the last few years, many researchers have established Multi-Attribute Decision-Making (MADM) in fuzzy, intuitionistic fuzzy, and neutrosophic sets. This research paper develops a Single-Valued Triangular Neutrosophic Generalized Ordered Weighted Harmonic Averaging (SVTrNGOWHA) operator to aggregate Single-Valued Triangular Neutrosophic (SVTrN) number information and all properties are discussed in detail. Further, the MADM method depends on SVTrN-GOWHA operator and score function for SVTrN numbers in ranking the alternatives. Lastly, the proposed approach for numerical example is tested and it gives the most effective of the proposed approach.

**Keywords** SVTrNGOWHA operator · Multi-attribute decision-making

## 1 Introduction

MADM problem is an important role in decision-making situations. In 1988, Yager [31] introduced the concept of ordered weighting averaging operators which assigns weight to the greatest input value and solving multi-criteria decision-making (MCDM) problems. Yager [33] proposed a Generalized Ordered Weighted Averaging operator (GOWA) that is combined the OWA operator with generalized mean operator. Xu and Da [29] established the ordered weighted geometric (OWG) operators, combing with OWA operators and GM operators. Yager [32] proposed the power average operator. Chen et al. [3] developed an ordered weighted harmonic averaging operator and applied to the method of combination forecasting. Many researchers proposed various aggregation operators for MADM problems based on uncertain environment. In 1965, Zadeh [34] discovered fuzzy set, which deals with

G. Tamilarasi (✉)
Research Scholar, Department of Mathematics, Anna University, Chennai, TN, India
e-mail: tamiltara5@gmail.com

S. Paulraj
Professor, Department of Mathematics, Anna University, Chennai, TN, India
e-mail: profspaulraj@gmail.com



B. Rushi Kumar et al. (eds.), *Mathematics and Computing*, Springer Proceedings in Mathematics & Statistics 415, https://doi.org/10.1007/978-981-19-9307-7_4

uncertainty situations. Wang and Fan [28] developed ordered weighted averaging operator in fuzzy environment and applied to decision-making problems. Wei and Yi [16] proposed harmonic aggregation operators and applied to MAGDM software selection problems with triangular fuzzy linguistic variables. In 1986, Atanassov [2] established an intuitionistic fuzzy set that is described by fuzzy set. Wang and Zhong [15] developed weighted aggregation operators under intuitionistic fuzzy situations and solving MCDM problems.

In 1998, Smarandache [25] introduced neutrosophic sets which deals with membership, non-membership, and indeterminacy membership functions. Deli and Subas [17] extended Single-Valued Trapezoidal Neutrosophic Weighted Aggregation Operator (SVTNWAO) and solving for MCDM problem. Jun Ye [21, 22] developed Trapezoidal Neutrosophic Number Weighted Arithmetic (TNNWAA) and Geometric Averaging (TNNWGA) operators to deal with MADM problems. Jun Ye [11] proposed similarity measures depending on interval neutrosophic sets applied to MCDM problems. Jun Ye [12] established TNNWAA and TNNWGA operators described in trapezoidal neutrosophic numbers applied to MADM problems. Zhao et al. [35] developed generalized weighted aggregation operator for solving MADM problems depending on Interval-Valued Neutrosophic Sets (IVNSs). Liu and Tang [14] developed generalization power aggregation operators with IVNSs to handle decision-making problems. Xu and Wei [30] established a minimum deviation method for neutrosophic MADM problems. Harish and Nancy [4] investigated MCDM problems handle with hybrid weighted aggregation operators in neutrosophic environment. Surapati and Mallick [26] extended trapezoidal neutrosophic weighted averaging operator and Hamming distance to deal with VIKOR (VIekriterijumsko KOmpromisno Rangiranje) strategy to decision-making problems. Sahin et al. [23] developed a new solution for solving MADM problems.

Bharatraj and Anand [20] developed MCDM problem for power harmonic aggregation operator under SVTN number and interval-valued neutrosophic numbers. Jana et al. [7] established Hamacher operation laws in SVTN arithmetic and geometric operator for solving MADM problems. Jana et al. [6] developed Interval Trapezoidal Neutrosophic Number Weighted Arithmetic Averaging (ITNNWAA) operator and Geometric Averaging (ITNNWGA) operator for solving MADM problems. Garai et al. [27] extended the possibility mean ranking technique for neutrosophic numbers and applied to MADM. Paulraj and Tamilarasi [24] developed some new harmonic averaging operators for SVTN environment and apply with MADM problems. Aliya et al. [1] developed Triangular Neutrosophic Cubic Fuzzy Weighted Arithmetic Averaging (TNCFWAA) and Geometric Averaging (TNCFWGA) operator for solving MADM problems. Jana and Pal [8] developed Dombi operations and power averaging operators for solving MCDM problems under a neutrosophic environment. Jana et al. [10] introduced arithmetic and geometric averaging operators using Dombi operations on SVTN numbers for solving MCDM problems. Jana and Pal [9] developed dynamic intuitionistic fuzzy aggregation operators to apply to gray relational analysis approach for solving multiple attribute problems.

Upon investigating the literature, no research work has achieved harmonic averaging operators with triangular neutrosophic numbers for MADM. To merge this gap, harmonic averaging operator that deals with SVTrN numbers is presented. In order to analyze the harmonic averaging operators in SVTrN numbers, to simplify comparison and application for MADM problems and this research paper attempts do to the following:

1. To define some SVTrN aggregation operators, that is, the SVTrN Weighted Harmonic Averaging (SVTrNWHA), SVTrN Ordered Weighted Harmonic Averaging (SVTrNOWHA), SVTrN Generalized Ordered Weighted Harmonic Averaging (SVTrNGOWHA) operators.
2. To propose an easy and straightforward technique for solving MADM problems where the ratings of the performance are expressed in SVTrN numbers and also investigate some of their properties.
3. The main goal of this developed approach chosen the best one and the SVTrN-GOWHA operator considers the position of input argument for any stage which does not focus on the degree of input argument.

Connecting harmonic operators and the MADM problems when the attribute values are SVTrN numbers, this paper developed SVTrN generalized ordered weighted harmonic averaging (SVTrNGOWHA) operator. The organization of this paper contents is as follows: Sect. 2, the preliminary concepts of SVTrN numbers are presented. Section 3, SVTrN weighted harmonic averaging operator is derived and SVTrN-GOWHA operator is proposed. Section 4, a new MADM approach is proposed and applied to MADM problem. Section 5, the conclusion is presented.

## 2 Preliminaries

This section reviews basic definitions about the concept of SVTrN numbers.

**Definition 1** ([25]) Let Z be a non-empty set. Then a neutrosophic set $N$ of Z is defined as $N = \{< z, T_N(z), I_N(z), F_N(z) > |z \in Z\}$, $T_N : N \to [0, 1]$, $I_N : N \to [0, 1]$, $F_N : N \to [0, 1]$ for satisfy the condition $0 \leq T_N(z) + I_N(z) + F_N(z) \leq 3$ for every $z \in N$. The function $T_N$, $I_N$, and $F_N$ are said to be the degree of truth, indeterminacy, and falsity-membership functions of $N$, respectively.

**Definition 2** ([18]) Let $n_l, n_m, n_u \in R$ such that $n_l \leq n_m \leq n_u$. A SVTrN number $\tilde{n} =< (n_l, n_m, n_u); \alpha_{\tilde{n}}, \beta_{\tilde{n}}, \gamma_{\tilde{n}} >$ is a special neutrosophic set on the real number set $R$, whose truth $T_{\tilde{n}}(z)$, indeterminacy $I_{\tilde{n}}(z)$ , and falsity $F_{\tilde{n}}(z)$ membership functions are defined as follows:

$$T_{\tilde{n}}(z) = \begin{cases} \frac{z-n_l}{n_m-n_l}\alpha_{\tilde{n}}, & \text{for } n_l \leq z \leq n_m \\ \alpha_{\tilde{n}}, & \text{for } z = n_m \\ \frac{n_u-z}{n_u-n_m}\alpha_{\tilde{n}}, & \text{for } n_m \leq z \leq n_u \\ 0, & \text{otherwise.} \end{cases} \tag{1}$$

$$I_{\tilde{n}}(z)=\begin{cases}\frac{n_m-z+(z-n_l)\beta_{\tilde{n}}}{n_m-n_l}, & \text{for } n_l\leq z\leq n_m\\ \beta_{\tilde{n}}, & \text{for } z=n_m\\ \frac{z-n_m+(n_u-z)\beta_{\tilde{n}}}{n_u-n_m}, & \text{for } n_m\leq z\leq n_u\\ 0, & \text{otherwise.}\end{cases} \tag{2}$$

$$F_{\tilde{n}}(z)=\begin{cases}\frac{n_m-z+(z-n_l)\gamma_{\tilde{n}}}{n_m-n_l}, & \text{for } n_l\leq z\leq n_m\\ \gamma_{\tilde{n}}, & \text{for } z=n_m\\ \frac{z-n_m+(n_u-z)\gamma_{\tilde{n}}}{n_u-n_m}, & \text{for } n_m\leq z\leq n_u\\ 0, & \text{otherwise.}\end{cases} \tag{3}$$

**Definition 3** ([18]) Let $\tilde{m} =< (m_l, m_m, m_u); \alpha_{\tilde{m}}, \beta_{\tilde{m}}, \gamma_{\tilde{m}} >$ and $\tilde{n} =< (n_l, n_m, n_u); \alpha_{\tilde{n}}, \beta_{\tilde{n}}, \gamma_{\tilde{n}} >$ be two SVTrN numbers and $r \neq 0$, then

1. $\tilde{m}+\tilde{n} =< (m_l+n_l, m_m+n_m, m_u+n_u); \alpha_{\tilde{m}}\wedge\alpha_{\tilde{n}}, \beta_{\tilde{m}}\vee\beta_{\tilde{n}}, \gamma_{\tilde{m}}\vee\gamma_{\tilde{n}} >$
2. $\tilde{m}-\tilde{n} =< (m_l-n_u, m_m-n_m, m_u-n_l); \alpha_{\tilde{m}}\wedge\alpha_{\tilde{n}}, \beta_{\tilde{m}}\vee\beta_{\tilde{n}}, \gamma_{\tilde{m}}\vee\gamma_{\tilde{n}} >$
3. $\tilde{m}\tilde{n} = \begin{cases} < (m_ln_l, m_mn_m, m_un_u); \alpha_{\tilde{m}}\wedge\alpha_{\tilde{n}}, \beta_{\tilde{m}}\vee\beta_{\tilde{n}}, \gamma_{\tilde{m}}\vee\gamma_{\tilde{n}} >, (m_u>0, n_u>0)\\ < (m_ln_u, m_mn_m, m_un_l); \alpha_{\tilde{m}}\wedge\alpha_{\tilde{n}}, \beta_{\tilde{m}}\vee\beta_{\tilde{n}}, \gamma_{\tilde{m}}\vee\gamma_{\tilde{n}} >, (m_u<0, n_u>0)\\ < (m_un_u, m_mn_m, m_ln_l); \alpha_{\tilde{m}}\wedge\alpha_{\tilde{n}}, \beta_{\tilde{m}}\vee\beta_{\tilde{n}}, \gamma_{\tilde{m}}\vee\gamma_{\tilde{n}} >, (m_u<0, n_u<0)\end{cases}$
4. $\frac{\tilde{m}}{\tilde{n}} = \begin{cases} < (\frac{m_l}{n_u}, \frac{m_m}{n_m}, \frac{m_u}{n_l}); \alpha_{\tilde{m}}\wedge\alpha_{\tilde{n}}, \beta_{\tilde{m}}\vee\beta_{\tilde{n}}, \gamma_{\tilde{m}}\vee\gamma_{\tilde{n}} >, (m_u>0, n_u>0)\\ < (\frac{m_u}{n_u}, \frac{m_m}{n_m}, \frac{m_l}{n_l}); \alpha_{\tilde{m}}\wedge\alpha_{\tilde{n}}, \beta_{\tilde{m}}\vee\beta_{\tilde{n}}, \gamma_{\tilde{m}}\vee\gamma_{\tilde{n}} >, (m_u<0, n_u>0)\\ < (\frac{m_u}{n_l}, \frac{m_m}{n_m}, \frac{m_l}{n_u}); \alpha_{\tilde{m}}\wedge\alpha_{\tilde{n}}, \beta_{\tilde{m}}\vee\beta_{\tilde{n}}, \gamma_{\tilde{m}}\vee\gamma_{\tilde{n}} >, (m_u<0, n_u<0)\end{cases}$
5. $r\tilde{m} = \begin{cases} < (rm_l, rm_m, rm_u); \alpha_{\tilde{m}}, \beta_{\tilde{m}}, \gamma_{\tilde{m}} >, (r>0)\\ < (rm_u, rm_m, rm_l); \alpha_{\tilde{m}}, \beta_{\tilde{m}}, \gamma_{\tilde{m}} >, (r<0)\end{cases}$
6. $\tilde{m}^{-1} =< (\frac{1}{m_u}, \frac{1}{m_m}, \frac{1}{m_l}); \alpha_{\tilde{m}}, \beta_{\tilde{m}}, \gamma_{\tilde{m}} > (\tilde{m}\neq\tilde{0})$

**Definition 4** ([13]) Let $\tilde{n} =< (n_l, n_m, n_u); \alpha_{\tilde{n}}, \beta_{\tilde{n}}, \gamma_{\tilde{n}} >$ be a SVTrN number. Then the score function of $\tilde{n}$ represented as follows:

$$S(\tilde{n}) = \frac{1}{8}(n_l+n_m+n_u)(2+\alpha_{\tilde{n}}-\beta_{\tilde{n}}-\gamma_{\tilde{n}}) \tag{4}$$

where $n_l, n_m, n_u \in R$ and $0 \leq \alpha_{\tilde{n}}+\beta_{\tilde{n}}+\gamma_{\tilde{n}} \leq 3$.

## 3 SVTrN Generalized Ordered Weighted Harmonic Averaging Operator

This section is based on harmonic averaging operators to establish SVTrNWHA, SVTrNOWHA, and SVTrNGOWHA operators.

**Definition 5** Let $\tilde{n}_k =< (n_{kl}, n_{km}, n_{ku}); \alpha_{\tilde{n}k}, \beta_{\tilde{n}k}, \gamma_{\tilde{n}k} >, k = 1, 2, ..., n$ be a set of SVTrN numbers. Then, the SVTrNWHA operator is defined as

$$SVTrNWHA(\tilde{n}_1, \tilde{n}_2, ..., \tilde{n}_k) = \frac{1}{\left(\sum_{k=1}^{n} \frac{w_k}{\tilde{n}_k}\right)} \quad (5)$$

where $w = (w_1, w_2, ..., w_k)^T$ is the weight of $\tilde{n}_k$ such that $w_k > 0$ and $\sum_{k=1}^{n} w_k = 1$.

**Definition 6** Let $\tilde{n}_k =< (n_{kl}, n_{km}, n_{ku}); \alpha_{\tilde{n}k}, \beta_{\tilde{n}k}, \gamma_{\tilde{n}k} >$ be a set of SVTrN numbers. Then, the SVTrNOWHA operator is defined as

$$SVTrNOWHA(\tilde{n}_1, \tilde{n}_2, ..., \tilde{n}_k) = \frac{1}{\left(\sum_{k=1}^{n} \frac{w_k}{\tilde{m}_k}\right)} \quad (6)$$

where $w = (w_1, w_2, ..., w_k)^T$ is the weight of $\tilde{n}_k$ such that $w_k \in [0, 1]$ and $\sum_{k=1}^{n} w_k = 1$, where $\tilde{m}_k$ is the largest $k$th element in the collection of $\tilde{n}_k$, $k = (1, 2, ..., n)$.

**Definition 7** Let $\tilde{n}_k =< (n_{kl}, n_{km}, n_{ku}); \alpha_{\tilde{n}k}, \beta_{\tilde{n}k}, \gamma_{\tilde{n}k} >$ be a set of SVTrN numbers. Then, an SVTrNGOWHA operator is defined as

$$SVTrNGOWHA(\tilde{n}_1, \tilde{n}_2, ..., \tilde{n}_k) = \frac{1}{\left(\sum_{k=1}^{n} \frac{w_k}{\tilde{m}_k^r}\right)^{\frac{1}{r}}} \quad (7)$$

where $w = (w_1, w_2, ..., w_k)^T$ is the weight of $\tilde{n}_k$ such that $w_k > 0$ and $\sum_{k=1}^{n} w_k = 1$. and $\tilde{m}_k$ is the largest $k$th element in the collection of $\tilde{n}_k$.
$\tilde{m}_k =< (m_{kl}, m_{km}, m_{ku}); \alpha_{\tilde{m}k}, \beta_{\tilde{m}k}, \gamma_{\tilde{m}k} >$ is reordering of the collection of $\tilde{n}_k$, where $r \neq 0, r \in R$ is a parameter.

**Special Cases of SVTrNGOWHA Operator**

(i) If $r = 1$, then SVTrNGOWHA operator reduces to SVTrNOWHA operator.
$SVTrNOWHA(\tilde{n}_1, \tilde{n}_2, ..., \tilde{n}_k) = \frac{1}{\left(\sum_{k=1}^{n} \frac{w_k}{\tilde{m}_k}\right)}$

(ii) If $r = 2$, then SVTrNGOWHA operator reduces to GOWQHA operator with SVTrN number.
$SVTrNGOWQHA(\tilde{n}_1, \tilde{n}_2, ..., \tilde{n}_k) = \frac{1}{\left(\sum_{k=1}^{n} \frac{w_k}{\tilde{m}_k^2}\right)^{\frac{1}{2}}}$

(iii) If $r = -1$, then SVTrNGOWHA operator reduces to OWA operator with SVTrN number.
$SVTrNOWA(\tilde{n}_1, \tilde{n}_2, ..., \tilde{n}_k) = \left(\sum_{k=1}^{n} \frac{w_k}{\tilde{m}_k}\right)$

**Theorem 1** *Let* $\tilde{n}_k =< (n_{kl}, n_{km}, n_{ku}); \alpha_{\tilde{n}k}, \beta_{\tilde{n}k}, \gamma_{\tilde{n}k} >, (k = 1, 2, ..., n)$ *be a set of SVTrN number and* $w = (w_1, w_2, ..., w_k)^T$ *be a weighted vector of* $\tilde{n}_k$, $w_k \in [0, 1]$, $\sum_{k=1}^{n} w_k = 1$ *and the parameter* $r \in R$, *then the aggregation value by utilizing the operator is defined as*

$$SVTrNGOWHA\,(\tilde{n}_1, \tilde{n}_2, ..., \tilde{n}_k) = \frac{1}{\left(\sum_{k=1}^{n} \frac{w_k}{\tilde{m}_k^r}\right)^{\frac{1}{r}}}$$

$$= \left\langle \left( \frac{1}{\left(\sum_{k=1}^{n} \frac{w_k}{(\tilde{m}_{kl})^r}\right)^{\frac{1}{r}}}, \frac{1}{\left(\sum_{k=1}^{n} \frac{w_k}{(\tilde{m}_{km})^r}\right)^{\frac{1}{r}}}, \frac{1}{\left(\sum_{k=1}^{n} \frac{w_k}{(\tilde{m}_{ku})^r}\right)^{\frac{1}{r}}} \right); \min_k(\alpha_{\tilde{m}k}), \max_k(\beta_{\tilde{m}k}), \max_k(\gamma_{\tilde{m}k}) \right\rangle$$

**Properties**

(i) **Monotonicity:** Let $\tilde{n}_k =< (n_{kl}, n_{km}, n_{ku}); \alpha_{\tilde{n}k}, \beta_{\tilde{n}k}, \gamma_{\tilde{n}k} >$ and $\tilde{n}_k' =< (n_{kl}', n_{km}', n_{ku}'); \alpha_{\tilde{n}k}', \beta_{\tilde{n}k}', \gamma_{\tilde{n}k}' >, (k = 1, 2, ..., n)$ be set of SVTrN numbers. If $\tilde{m}_k \leq \tilde{m}_k'$ for k=1,2, ..., n. Then SVTrNGOWHA$(\tilde{n}_1, \tilde{n}_2, ..., \tilde{n}_k)$ $\leq$SVTrNGOWHA $(\tilde{n}_1', \tilde{n}_2', ..., \tilde{n}_k')$

(ii) **Idempotency:** Let $\tilde{n}_k =< (n_{kl}, n_{km}, n_{ku}); \alpha_{\tilde{n}k}, \beta_{\tilde{n}k}, \gamma_{\tilde{n}k} >, (k = 1, 2, ..., n)$ be a set of SVTrN number. If all $\tilde{n}_j$ are equal, $\tilde{n}_j = \tilde{n}, (k = 1, 2, ...n)$, then SVTrNGOWHA$(\tilde{n}_1, \tilde{n}_2, ..., \tilde{n}_k)$ =SVTrNGOWHA$(\tilde{n}, \tilde{n}, ..., \tilde{n}) = \tilde{n}$.

(iii) **Commutativity:** If $(\tilde{n}_1', \tilde{n}_2', ..., \tilde{n}_k')$ is any permutation of $(\tilde{n}_1, \tilde{n}_2, ..., \tilde{n}_k)$, then SVTrNGOWHA$(\tilde{n}_1, \tilde{n}_2, ..., \tilde{n}_k)$ =SVTrNGOWHA$(\tilde{n}_1', \tilde{n}_2', ..., \tilde{n}_k')$.

(iv) **Boundedness:** Let $\tilde{n}_k =< (n_{kl}, n_{km}, n_{ku}); \alpha_{\tilde{n}k}, \beta_{\tilde{n}k}, \gamma_{\tilde{n}k} >, (k = 1, 2, ..., n)$ be a set of SVTrN numbers and Let $\tilde{n}_k^+ = \langle(\min_k(m_{kl}), \min_k(m_{km}), \min_k(m_{ku}));$ $\min_k (\alpha_{\tilde{m}k}), \max_k (\beta_{\tilde{m}k}), \max_k (\gamma_{\tilde{m}k})\rangle$ $\tilde{n}_k^- = \langle(\max_k(m_{kl}), \max_k(m_{km}), \max_k(m_{ku}));$ $\min_k (\alpha_{\tilde{m}k}), \max_k (\beta_{\tilde{m}k}), \max_k (\gamma_{\tilde{m}k})\rangle$. Then $\tilde{n}^- \leq$ SVTrNGOWHA$(\tilde{n}_1, \tilde{n}_2, ..., \tilde{n}_k) \leq \tilde{n}^+$.

## 4 MADM Problems with Neutrosophic Numbers

Considering the MADM problem, assume the set of alternatives $A_i$ and attributes $C_k$. The rating of an alternative $A_i$ with an attribute $C_k$ then the neutrosophic decision-makers can be describe as $(\tilde{S}_{ik})_{m\times n}, i = 1, 2, ..., m$ *and* $k = 1, 2, ..., n$. The weights of the attribute are $w = (w_1, w_2, ..., w_k)^T$, satisfying $0 \leq w_k \leq 1 (k = 1, 2, ..., n)$ and $\sum_{k=1}^{n} w_k = 1$.

The following algorithm is to obtain the solution of multi-attribute decision-making problem with SVTrN number information by using SVTrNGOWHA operators with score function.

Let $\tilde{S} = (\tilde{S}_{ik})_{m\times n}$ be the decision matrix provided by decision expert $D$ which can be expressed as

$$D = \left(\tilde{S}_{ik}\right)_{(m\times n)} = \begin{matrix} & C_1 & C_2 & \dots & C_n \\ A_1 \\ A_2 \\ \vdots \\ A_m \end{matrix} \begin{pmatrix} \tilde{S}_{11} & \tilde{S}_{12} & \dots & \tilde{S}_{1n} \\ \tilde{S}_{21} & \tilde{S}_{22} & \dots & \tilde{S}_{2n} \\ \vdots & \vdots & \ddots & \vdots \\ \tilde{S}_{m1} & \tilde{S}_{m2} & \dots & \tilde{S}_{mn} \end{pmatrix}$$

where $\tilde{S}_{ik} =< (S_{ikl}, S_{ikm}, S_{iku}); \alpha_{ik}, \beta_{ik}, \gamma_{ik} >, i = 1, 2, ..., m, k = 1, 2, ..., n.$

**Step 1: Computation of the normalized given matrix**
If all the ratings are either profit or cost, then there is no need of normalization. Otherwise, the normalized decision matrix is constructed.

$$(\tilde{R}_{ik}) = \begin{cases} \dfrac{\tilde{S}_{ik}}{\sum\limits_{k=1}^{n} \tilde{S}_{ik}}, & \text{if the rating is profit} \\ \dfrac{\frac{1}{\tilde{S}_{ik}}}{\sum\limits_{k=1}^{n} (\frac{1}{\tilde{S}_{ik}})}, & \text{if the rating is cost} \end{cases}$$

Therefore, every SVTrN decision matrix $D$ is converted into a normalized decision matrix $N$ represented by

$$N = (\tilde{R}_{ik})_{m\times n} = \begin{matrix} & C_1 & C_2 & \dots & C_n \\ A_1 \\ A_2 \\ \vdots \\ A_m \end{matrix} \begin{pmatrix} \tilde{R}_{11} & \tilde{R}_{12} & \dots & \tilde{R}_{1n} \\ \tilde{R}_{21} & \tilde{R}_{22} & \dots & \tilde{R}_{2n} \\ \vdots & \vdots & \ddots & \vdots \\ \tilde{R}_{m1} & \tilde{R}_{m2} & \dots & \tilde{R}_{mn} \end{pmatrix}$$

where $\tilde{R}_{ik} =< (R_{ikl}, R_{ikm}, R_{iku}); \alpha_{ik}, \beta_{ik}, \gamma_{ik} >, i = 1, 2, ..., m, k = 1, 2, ..., n.$

**Step 2: Construct expert ratings for each alternative**
To aggregate expert ratings and utilize GOWHA operator for neutrosophic environment, aggregated value can be represented by $R_i$
$r_i = SVTrNGOWHA(\tilde{R}_{i1}, \tilde{R}_{i2}, ..., \tilde{R}_{in}), (i = 1, 2, ..., m)$
**Step 3: Ranking of the alternatives**
To find the rank of the alternatives based on Definition 4.
**Step 4: End**

## 4.1 Numerical Example

In this section, we discussed the effectiveness of the proposed approach applied to system analyst problem. The proposed SVTrNGOWHA operator to deal with MADM problem is adapted from [18]. Let us consider a software company desire to hire a system analyst and, after the basic screen process, three candidates to take arbitrary computer science background students from 3 sources are State University ($A_1$), Deemed University ($A_2$), and Central University ($A_3$). The experts assess the three candidates with respect to five attributes, which are emotional steadiness ($C_1$), oral communication skill ($C_2$), personality ($C_3$), past experience ($C_4$), and self-confidence ($C_5$). The weighted vector of the five attributes is $w = (0.15, 0.25, 0.20, 0.25, 0.15)^T$ and according to the decision matrix shown in Table 1.

**Step 1: Computing the normalized given matrix**
Calculate the normalized matrix $N = (\tilde{R}_{ij})_{(m\times n)}$ of the matrix shown in Table 2.
**Step 2: Construct aggregate ratings for each alternative**
To aggregate expert ratings for each $A_i$ w.r.to each $C_k$.
If the decision-maker, alternatives $k = 1, 2, 3, i = 1, 2, 3, 4, 5$ and parameter $\lambda = 1$.
$\tilde{r}_1 = SVTrNGOWHA(C_1, C_2, C_3, C_4, C_5)$
$\tilde{r}_1 =< (\frac{1}{\frac{0.15}{0.152}+\frac{0.25}{0.136}+\frac{0.20}{0.124}+\frac{0.25}{0.108}+\frac{0.15}{0.103}}, \frac{1}{\frac{0.15}{0.216}+\frac{0.25}{0.216}+\frac{0.20}{0.21}+\frac{0.25}{0.185}+\frac{0.15}{0.172}},$
$\frac{1}{\frac{0.15}{0.372}+\frac{0.25}{0.323}+\frac{0.20}{0.32}+\frac{0.25}{0.32}+\frac{0.15}{0.272}}); 0.3, 0.8, 0.3 >$
$\Longrightarrow < (\frac{1}{8.2157}, \frac{1}{5.0231}, \frac{1}{3.1389}); 0.3, 0.8, 0.3 >$
$\tilde{r}_1 =< (0.1217, 0.1991, 0.3186); 0.3, 0.8, 0.3 >$
In the same way, find $\tilde{R}_2, \tilde{R}_3$
$\tilde{r}_2 =< (0.1535, 0.1982, 0.2577); 0.4, 0.5, 0.6 >$
$\tilde{r}_3 =< (0.1321, 0.1975, 0.2951); 0.5, 0.2, 0.8 >$
**Step 3: Ranking of the alternatives**
Finally, to calculate the ranking result of alternatives $A_i$.
$S_1$= Score value of alternative $A_1 = S(\tilde{R}_1) = 0.0959$,
$S_2$= Score value of alternative $A_2 = S(\tilde{R}_2) = 0.0990$,
$S_3$= Score value of alternative $A_3 = S(\tilde{R}_3) = 0.1171$
Since $S_3 > S_2 > S_1$, the third alternative source $A_3$ is best.

**Analyzing Different Variation of $\lambda$ on Results of Alternatives**
Furthermore, to analyze different variation $\lambda$ that deals with harmonic aggregation operator based on SVTrN number provided by desirable alternative. A complete variation of the ranking value of each alternative with respect to $\lambda$ is shown in Fig. 1. From this figure, to observe the value of $\lambda$ that decreases when the score value of each alternative increases and also increases the value of parameter $\lambda$ when the score value of alternative decreases but the rank order of these alternative remains the same. Hence the best alternative is $A_3$.

**Effectiveness of the Developed Method**
Advantages of the proposed approach are pointed out after comparing the results of the existing works with the proposed work. First, our proposed work is compared with Jun Ye [22] and it is considered with the same problem under neutrosophic environment to obtain the similar alternative. The existing work considers arithmetic and geometric aggregation operators to solve MADM problems under neutrosophic environment. Similarly, consider our proposed work compared with existing work Deli and Subas [18] to deal with value and ambiguity de-neutrosophication for solving MADM problems obtained similar ranking result. The comparative results for

**Table 1** The aggregated matrix given by the expert

| | $C_1$ | $C_2$ | $C_3$ | $C_4$ | $C_5$ |
|---|---|---|---|---|---|
| $A_1$ | $< (4.6, 5.5, 8.6); 0.4, 0.7, 0.2 >$ | $< (5.8, 6.9, 8.5); 0.6, 0.2, 0.3 >$ | $< (5.3, 6.7, 9.9); 0.3, 0.5, 0.2 >$ | $< (4.4, 5.9, 7.2); 0.7, 0.2, 0.3 >$ | $< (6.5, 6.9, 8.5); 0.6, 0.8, 0.1 >$ |
| $A_2$ | $< (6.2, 7.6, 8.2); 0.4, 0.1, 0.3 >$ | $< (7.1, 7.7, 8.3); 0.5, 0.2, 0.4 >$ | $< (6.2, 8.9, 9.1); 0.6, 0.3, 0.5 >$ | $< (6.3, 7.5, 8.9); 0.7, 0.4, 0.6 >$ | $< (7.5, 7.9, 8.5); 0.8, 0.5, 0.4 >$ |
| $A_3$ | $< (5.5, 6.2, 7.3); 0.8, 0.1, 0.2 >$ | $< (4.7, 6.9, 8.5); 0.7, 0.2, 0.6 >$ | $< (7.1, 8.5, 8.9); 0.5, 0.2, 0.7 >$ | $< (6.6, 8.8, 10); 0.6, 0.2, 0.2 >$ | $< (5.3, 7.3, 8.7); 0.7, 0.2, 0.8 >$ |

**Table 2** The normalized matrix of the expert

| | $C_1$ | $C_2$ | $C_3$ | $C_4$ | $C_5$ |
|---|---|---|---|---|---|
| $A_1$ | $< (0.108, 0.172, 0.323); 0.3, 0.8, 0.3 >$ | $< (0.136, 0.216, 0.32); 0.3, 0.8, 0.3 >$ | $< (0.152, 0.216, 0.32); 0.3, 0.8, 0.3 >$ | $< (0.124, 0.21, 0.372); 0.3, 0.8, 0.3 >$ | $< (0.103, 0.185, 0.272); 0.3, 0.8, 0.3 >$ |
| $A_2$ | $< (0.144, 0.192, 0.246); 0.4, 0.5, 0.6 >$ | $< (0.165, 0.194, 0.25); 0.4, 0.5, 0.6 >$ | $< (0.174, 0.199, 0.255); 0.4, 0.5, 0.6 >$ | $< (0.144, 0.225, 0.273); 0.4, 0.5, 0.6 >$ | $< (0.147, 0.186, 0.267); 0.4, 0.5, 0.6 >$ |
| $A_3$ | $< (0.127, 0.165, 0.25); 0.5, 0.2, 0.8 >$ | $< (0.108, 0.183, 0.291); 0.5, 0.2, 0.8 >$ | $< (0.122, 0.19, 0.298); 0.5, 0.2, 0.8 >$ | $< (0.164, 0.226, 0.305); 0.5, 0.2, 0.8 >$ | $< (0.152, 0.233, 0.343); 0.5, 0.2, 0.8 >$ |

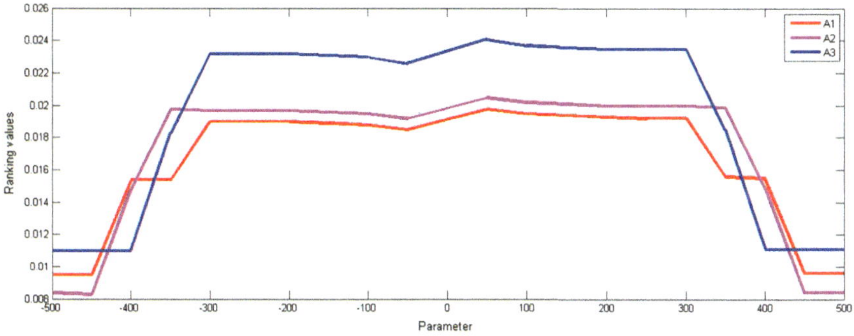


**Fig. 1** Different variation values of alternatives w.r.to parameter $\lambda$ in SVTrNGOWHA operator

**Table 3** Decision-making results of different aggregation operators

| Method | Operator | $A_1$ | $A_2$ | $A_3$ | Best alternative |
|---|---|---|---|---|---|
| Jun Ye [22] | SVTrNWAA | 0.0984 | 0.1 | 0.1185 | $A_3$ |
| | SVTrNWGA | 0.0981 | 0.0999 | 0.116 | $A_3$ |
| Deli and Subas [18] | Value and Ambiguity | 0.059 | 0.127 | 0.189 | $A_3$ |
| Proposed method | SVTrNWHA | 0.098 | 0.0997 | 0.1292 | $A_3$ |
| | SVTrNGOWHA | 0.0959 | 0.099 | 0.117 | $A_3$ |

different aggregation operators are shown in Table 3. Furthermore, the proposed operator SVTrNGOWHA involves a different parameter $\lambda$, which makes it flexible in the process of information and is more adequate to model practical MADM problems.

## 5 Conclusion

The objective of this paper the MADM problem with attribute values in SVTrN numbers form has been investigated. Some SVTrN number operations and the corresponding operation laws have been established based on harmonic averaging operations. Then a method based on operators (SVTrNWHA, SVTrNOWHA, and SVTrNGOWHA) has been constructed to effectively deal with the MADM problem under neutrosophic environment. The procedure has been clearly explained with the help of an illustration. In future research, we aim to extend the proposed operator and applied to several examples such as information material, project selection, and many other areas of decision-making problems.

## References

1. Fahm, A., Amin, F., Ullah, H.: Multiple attribute group decision making based on weighted aggregation operators of triangular neutrosophic cubic fuzzy numbers. Granul. Comput. **6**(2) (2021). https://doi.org/10.1007/s41066-019-00205-2
2. Atanassov, K.T.: Intuitionistic fuzzy sets. Fuzzy Sets Syst. **20**(1), 87–96 (1986)
3. Chen, H., Liu, C., Sheng, Z.: Induced ordered weighted harmonic averaging (IOWHA) operator and its application to combination forecasting method. Chinese J. Manag. Sci. **12**(5), 35–40 (2004)
4. Garg, H., Nancy: Some hybrid weighted aggregation operators under neutrosophic set environment and their applications to multicriteria decision-making. Appl. Intell. (2018). https://doi.org/10.1007/s10489-018-1244-9
5. Jana, C., Pal, M.: A robust single-valued neutrosophic soft aggregation operators in multi-criteria decision making. Symmetry **11**, 110 (2019). https://doi.org/10.3390/sym11010110
6. Jana, C., Pal, M., Karaaslan, F., Wang, J.Q.: Trapezoidal neutrosophic aggregation operators and their application to the multi-attribute decision-making process. Scientia Iranica E **27**(3), 1655–1673 (2020). https://doi.org/10.24200/sci.2018.51136.2024
7. Jana, C., Muhiuddin, G., Pal, M.: Multiple-attribute decision making problems based on SVTNH methods. J. Ambient Intell. Humaniz. Comput. Springer 3717–3733 (2020). https://doi.org/10.1007/s12652-019-01568-9
8. Jana, C., Pal, M.: Multi-criteria decision making process based on some single-valued neutrosophic Dombi power aggregation operators. Soft Comput. **25**, 5055–5072 (2021). https://doi.org/10.1007/s00500-020-05509-z
9. Jana, C., Pal, M.: A dynamical hybrid method to design decision making process based on GRA approach for multiple attributes problem. Eng. Appl. Artif. Intell. **100**, 104203 (2021). https://doi.org/10.1016/j.engappai.2021.104203
10. Jana, C., Muhiuddin, G., Pal, M.: Multi-criteria decision making approach based on SVTrN Dombi aggregation functions. Artificial Intelligence Review. Springer (2021). https://doi.org/10.1007/s10462-020-09936-0
11. Ye, J.: Similarity measures between interval neutrosophic sets and their applications in multicriteria decision-making. J. Intell. Fuzzy Syst. **26**, 165–172 (2014). https://doi.org/10.3233/IFS-120724, IOS Press
12. Jun Ye, Trapezoidal neutrosophic set and its application to multiple attribute decision-making. Neural Comput. & Appl. **26**, 1157–1166 (2015). https://doi.org/10.1007/s00521-014-1787-6
13. Hezam, I.M., Nayeem, M.K., Foul, A., Alrasheedi, A.F.: COVID-19 vaccine: a neutrosophic MCDM approach for determining the priority groups. Results Phys. **20**, 103654 (2020). https://doi.org/10.1016/j.rinp.2020.103654
14. Liu, P., Tang, G.: Some power generalized aggregation operators based on the interval neutrosophic sets and their application to decision making. J. Intell. Fuzzy Syst. **30**, 2517–2528 (2016). https://doi.org/10.3233/IFS-151782, IOS Press
15. Wang, J., Zhong, Z.: Aggregation operators on intuitionistic trapezoidal fuzzy number and its application to multi-criteria decision making problems. J. Syst. Eng. Electron. **20**(2), 321–326 (2009)
16. Wei, G., Yi, W.: Fuzzy linguistic hybrid harmonic mean operator and its application to software selection. J. Softw. **4**, No. 9 (2009)
17. Deli, I., Subas, Y.: Single valued neutrosophic numbers and their applications to multicriteria decision making problem. Neutrosophic. Sets Syst. (2014)
18. Deli, I., Subas, Y.: A ranking method of single valued neutrosophic numbers and its applications to multi-attribute decision making problems. Int. J. Mach. Learn. Cyber. (2015). https://doi.org/10.1007/s13042-016-0505-3
19. Deli, I.: Operators on single valued trapezoidal neutrosophic numbers and SVTN-Group decision making. Neutrosophic Sets Syst. **22** (2018)

20. Bharatraj, J., Clement Joe Anand, M.: Power harmonic weighted aggregation operator on single valued trapezoidal neutrosophic numbers and interval-valued neutrosophic sets. Fuzzy Multi-criteria Dec. Mak. Neutrosophic Sets Stud. Fuzziness Soft Comput. **369** (2019). https://doi.org/10.1007/978-3-030-00045-5-3
21. Ye, J.: Trapezoidal Neutrosophic set and its application to multiple attribute decision making. Neural Comput. Appl. **26**, 1157–1166 (2015). https://doi.org/10.1007/s00521-0140-1787-6
22. Ye, J.: Some weighted aggregation operators of trapezoidal neutrosophic numbers and their multiple attribute decision making method. Informatica (2016)
23. Sahin, M., Kargin, A., Smarandache, F.: Generalized single valued triangular neutrosophic numbers and aggregation operators for application to multi-attribute group decision making. New Trends in Neutrosophic Theory and Applications, vol. II (2018)
24. Paulraj, S., Tamilarasi, G.: Generalized ordered weighted harmonic averaging operator with trapezoidal neutrosophic numbers for solving MADM problems. J. Ambient Intell. Hum. Comput. (2021)
25. Smarandache, F.: Unifying field in logics. Neutrosophy: Neutrosophic Probability Set and Logic. American Research Press, Rehoboth (1998)
26. Pramanik, S., Mallick, R.: VIKOR based MAGDM strategy with trapezoidal neutrosophic numbers. Neutrosophic Sets Syst. **22** (2018)
27. Garai, T., Garg, H., Roy, T.K.: A ranking method based on possibility mean for multi-attribute decision making with single valued neutrosophic numbers. J. Ambient Intell. Hum. Comput. **11**, 5245–5258 (2020). https://doi.org/10.1007/s12652-020-01853-y
28. Wang, X., Fan, Z.: Fuzzy ordered weighted averaging (FOWA) operator and its application. Fuzzy Syst. Math. **17**(4), 67–72 (2003)
29. Xu, Z., Da, Q.: The ordered weighted geometric averaging operators. Int. J. Intell. Syst. **17**(7), 709–716 (2002)
30. Xu, D.-S., Wei, C.: Minimum deviation method for single-valued neutrosophic multiple attribute decision making with preference information on alternatives. J. Intell. Comput. **9**, No. 2 (2018). https://doi.org/10.6025/jic/2018/9/1/54-75
31. Yager, R.R.: On ordered weighted averaging aggregation operators in multicriteria decision making. IEEE. Trans. Syst. Man Cybern. **18**, 183–190 (1988)
32. Yager, R.R.: The power average operator. IEEE Trans. Syst. Man Cybern. Part A: Syst. Hum. **31**(6), 724–731 (2001)
33. Yager, R.R.: Generalized OWA aggregation operators. Fuzzy Optim. Dec. Mak. **3**, 93–107 (2004)
34. Zadeh, L.A.: Fuzzy Sets Inf. Control **8**(3), 338–353 (1965)
35. Aiwu, Z., Du, J., Hongjun, G.: Interval valued neutrosophic sets and multi-attribute decision-making based on generalized weighted aggregation operator. J. Intell. Fuzzy Syst. **29**, 2697–2706 (2015). https://doi.org/10.3233/IFS-151973, IOS Press

# Redundancy of Codes with Graph Constraints

**Ghurumuruhan Ganesan**

**Abstract** In this paper, we study linear code redundancy in the presence of graph constraints. First, we describe linear parity check codes with graphical constraints and employ the probabilistic method to achieve the Gilbert-Varshamov redundancy bound. Next, we define a fractional version of graph capacity and obtain bounds for arbitrary graphs, again using the probabilistic method.

**Keywords** Linear codes · Bipartite graphs · Fractional graph capacity

## 1 Introduction

Codes based on graphs arise often in both theory and applications and it is important to understand redundancies of such codes. Typically, redundancy bounds like Gilbert-Varshamov, Hamming and Singleton are obtained under the Hamming distance measure with no restrictions on the codes themselves. In many applications, the code itself might have additional graph constraints.

In this paper we are interested in linear and non-linear codes with graph constraints. Graph based linear codes like low density parity check (LDPC) codes [15] are used extensively in communication systems with the constraint that the left and right vertex degrees in the bipartite graph representation are small compared to the total number of vertices. Similarly, expander codes [14] are also popular because of their inherent *expansion* property and algorithms for encoding and decoding in linear time is presented for such codes in [14]. Recently, [7] has described a localized decoding procedure for expander codes capable of correcting a fraction of errors using a (relatively) few number of symbols from corrupted codeword. In Sect. 2,

An abridged version of this paper was presented in 23rd-Thailand-Japan Conference on Discrete and Computational Geometry, Graphs and Games (TJCDCGGG, 2021) and is available in the book of abstracts at https://www.math.science.cmu.ac.th/tjcdcggg/.

G. Ganesan (✉)
IISER, Bhopal 462066, India
e-mail: gganesan82@gmail.com



B. Rushi Kumar et al. (eds.), *Mathematics and Computing*, Springer Proceedings in Mathematics & Statistics 415, https://doi.org/10.1007/978-981-19-9307-7_5

we obtain linear codes with graph constraints that also attain the Gilbert-Varshamov bound.

The capacity of a graph [13] measures the capability communication without errors for channels modelled by a confusion graph. Many bounds for the graph capacity are known: Lovász [11] used projection techniques to determine the capacity of the cycle $C_5$ on 5 vertices and [6] who studied analogues of Shannon capacity and its connections with the ultimate chromatic number. Marton [12] obtained expressions for graph capacities for a sequence of graphs based on typical sequences and more recently, [1] investigated the problem of approximating graph capacity by finite graph products.

In this paper, we study and establish bounds for a fractional version of the graph capacity in terms of its structural parameters: We obtain an upper bound for the fractional capacity in terms of the full graph capacity and a lower bound in terms of the average and maximum vertex degrees.

The paper is organized as follows: In Sect. 2 we study linear parity check codes with graphical constraints and use random bipartite graphs to determine achievability of the Gilbert-Varshamov bound. Next in Sect. 3, we define and obtain bounds for the fractional capacity of a graph.

## 2 Linear Parity Check Codes with Graph Constraints

We begin with some general definitions. Let $\mathcal{Y}$ be any finite set. An $n$-length word is an element of $\mathcal{Y}^n$ and an $n$-length code $\mathcal{C}$ is a subset of $\mathcal{Y}^n$. If $\mathcal{Y}$ is a finite field, then we have the concept of linear codes: we say that $\mathcal{C}$ is *linear* if for any $\mathbf{c}, \mathbf{d} \in \mathcal{C}$ and $x, y \in \mathcal{Y}$, the word $x \cdot \mathbf{c} + y \cdot \mathbf{d} \in \mathcal{C}$.

We define the Hamming distance between $\mathbf{x} = (x_1, \ldots, x_n)$ and $\mathbf{y} = (y_1, \ldots, y_n)$ in $\mathcal{Y}^n$, as

$$d(\mathbf{x}, \mathbf{y}) = \sum_{i=1}^{n} 1\!\!1(x_i \neq y_i), \tag{2.1}$$

where $1\!\!1(.)$ is the indicator function. In this section, all distances are Hamming and the minimum distance between any two words in $\mathcal{C}$ is denoted as $d_H(\mathcal{C})$. The relative distance, rate and redundancy of $\mathcal{C}$ are respectively defined as

$$\delta_H(\mathcal{C}) := \frac{d_H(\mathcal{C}) - 1}{n}, \; R(\mathcal{C}) := \frac{\log(\#\mathcal{C})}{n \log(\#\mathcal{Y})} \text{ and } \xi(\mathcal{C}) := 1 - R(\mathcal{C}), \tag{2.2}$$

where $\#\mathcal{C}$ is the size of $\mathcal{C}$ and logarithms are always to the base 2.

In this section we set $\mathcal{Y} = \{0, 1\}$ and begin with a description of the random graph construction of linear parity check codes. Let $U = \{u_i\}_{1 \le i \le n}$ and $V = \{v_j\}_{1 \le j \le m}$ be the left and right vertex sets of the complete bipartite graph $K_{n,m}$ and let $\{Z_f\}_{f \in K_{n,m}}$ be independent and identically distributed (i.i.d.) Bernoulli random variables with indices from the edge set of $K_{n,m}$ and satisfying

$$\mathbb{P}(X_f = 1) = p = 1 - \mathbb{P}(X_f = 0)$$

for $0 < p < \frac{1}{2}$ a constant not depending on $n$. Let $G$ be the random subgraph consisting of the edges $f$ satisfying $X_f = 1$ and define $G$ on the space $(\Omega, \mathcal{F}, \mathbb{P})$.

For $1 \leq j \leq m$ let $\mathcal{V}_j \subset X$ be the neighbour set of $v_j \in Y$ and let $\mathcal{C}$ be the set of all words $\mathbf{x} = (x_1, \ldots, x_n) \in \mathcal{C}$ satisfying

$$\bigoplus_{i \in \mathcal{V}_j} x_i = 0 \text{ for all } 1 \leq j \leq m. \tag{2.3}$$

Setting $m = n\epsilon$, we see that the code $\mathcal{C}$ is linear with rate at least $1 - \frac{m}{n} = 1 - \epsilon$ [14].

We now introduce expansion and graph constraints on $\mathcal{C}$. For $0 < \gamma < 1$ we say that $\mathcal{C}$ has a *diversity index* of at least $\gamma$ if

$$\#\left(\mathcal{R}_x \setminus \mathcal{R}_y\right) \geq \gamma \# \mathcal{R}_x \text{ for any } b_x, b_y \in Y. \tag{2.4}$$

Thus any two parity nodes have at least a fraction $\gamma$ of different neighbours and this could be interpreted as an expansion property with respect to the right vertices of $K_{n,m}$. We remark that in the usual construction via expander graphs, the condition for expansion is with regards to the (left) codeword index nodes of the bipartite graphs (see [14]).

A $n$-length *constraint* $E$ is simply an event in the collection $\mathcal{F}$. For example, the event $F_n$ that $v_{i-1}$ and $v_{i+1}$ are both adjacent to $u_i$ for each $i$, is a constraint. The random graph $G$ is said to satisfy the constraint $E$ if $G \in E$. Given a sequence of constraints $\mathcal{E}$ and real numbers $0 < \delta, \gamma < 1$, we ask if there is a linear code having relative distance $\delta$ and diversity index $\gamma$ that also satisfies the constraints. If so, what would be the redundancy of such a code?

If there were no constraints or diversity, then the Gilbert-Varshamov bound (Theorem 4.2.1, [5]) implies that codes with redundancy at most $H(\delta) + o(1)$ are available, where $o(1) \longrightarrow 0$ as $n \to \infty$ and

$$H(x) := -x \cdot \log x - (1 - x) \cdot \log(1 - x) \tag{2.5}$$

is the binary entropy function. Does imposing diversity and constraints increase the redundancy of a linear code? The following result says that if the constraints are not too strict, then we can still get linear parity check codes satisfying the Gilbert-Varshamov redundancy bound and with a given diversity index.

**Theorem 1** *Let $0 < \delta < \frac{1}{2}$ and $0 < \gamma < 1$ be any two constants and let $\{E_n\}$ be a sequence of constraints with probability $p_n = \mathbb{P}(\mathcal{E}_n)$ satisfying*

$$\frac{\log\left(\frac{1}{p_n}\right)}{n} \longrightarrow 0 \tag{2.6}$$

*as $n \to \infty$.*

*For all $n$ large, there is a $n$-length linear parity check code $W_n$ that satisfies the constraint $E_n$ and has relative distance at least $\delta$, redundancy $H(\delta) + o(1)$ and diversity index $\geq \gamma$.*

The condition (2.6) is satisfied, for example, if $\mathbb{P}(E_n) \geq e^{-f(n)}$ for some sublinear function $f$, i.e. if $\frac{f(n)}{n} \longrightarrow 0$ as $n \to \infty$. For all $n$ large, the code $W_n$ then attains the Gilbert-Varshamov bound and satisfies the constraint $E_n$. For example, the probability of the event $F_n$ described prior to the statement is $\geq p^{2\sqrt{n}}$ and so (2.6) holds in this case.

*Proof of Theorem* 1: We obtain our proof in three steps. First we choose the edge probability $p$ to be an appropriate constant so that the diversity condition is ensured and in the second step, we ensure that the resulting code $\mathcal{C}$ obtained from (2.3) has a minimum distance of at least $\delta n + 1$ with large probability, with probability $\to 1$ as $n \to \infty$. We also set $\epsilon > H(\delta)$ so that $\#\mathcal{C} \geq 2^{n(1-\epsilon)}$. Finally, in the third step, we add the constraints into $\mathcal{C}$.

*Step* 1 (*Ensuring diversity*): Let $\mathcal{C}$ be the linear code as obtained in (2.3). To ensure that $\mathcal{C}$ satisfies the diversity property, we argue as follows. Let $b_x$ and $b_y$ be any two right vertex nodes. We have that a left vertex $a_i$ is present in $\mathcal{R}_x$ with probability $p$ and is present in $\mathcal{R}_x \cap \mathcal{R}_y$ with probability $p^2$. Therefore by standard deviation estimates (Corollary $A.1.14$, pp. 312, [2]), for $0 < \theta < \frac{1}{4}$ we get that

$$\mathbb{P}\left(|\#\mathcal{R}_x - np| \geq np\theta\right) \leq \exp\left(-\frac{\theta^2}{4}np\right) \tag{2.7}$$

and that

$$\mathbb{P}\left(\left|\#(\mathcal{R}_x \cap \mathcal{R}_y) - np^2\right| \geq np^2\theta\right) \leq \exp\left(-\frac{\theta^2}{4}np^2\right). \tag{2.8}$$

Letting

$$R_{tot} := \bigcap_x \{|\#\mathcal{R}_x - np| \geq np\theta\}$$

we then get that

$$\mathbb{P}(R_{tot}) \geq 1 - 2m^2 e^{-\frac{\theta^2}{4}np^2}. \tag{2.9}$$

Similarly, using $np^2 < np$, we get from (2.7) and (2.8) that the event

$$F_{x,y} := \left\{\#\left(\mathcal{R}_x \setminus \mathcal{R}_y\right) \geq np(1-\theta) - np^2(1+\theta)\right\}$$

occurs with probability at least $1 - 2e^{-\frac{\theta^2}{4}np^2}$ and so letting $F_{tot} := \bigcap_{x,y} F_{x,y}$, we then get that

$$\mathbb{P}(F_{tot}) \geq 1 - 2m^2 e^{-\frac{\theta^2}{4}np^2}. \tag{2.10}$$

From (2.9), (2.10) and the union bound, we therefore we get that the event $E_{div} := R_{tot} \cap F_{tot}$ occurs with probability

$$\mathbb{P}(E_{div}) \geq 1 - 4m^2 e^{-\frac{\theta^2}{4}np^2}. \tag{2.11}$$

If $E_{div}$ occurs, then for any right nodes $b_x, b_y$ we have

$$\begin{aligned} \frac{\#\left(\mathcal{R}_x \setminus \mathcal{R}_y\right)}{\#\mathcal{R}_x} &\geq \frac{np(1-\theta) - np^2(1+\theta)}{np(1+\theta)} \\ &= \frac{1-\theta}{1+\theta} - p \end{aligned} \tag{2.12}$$

which is at least $\gamma$ provided $\theta$, $p$ are sufficiently small constants. We henceforth fix such a $p$.

<u>*Step 2 (Estimating the minimum distance)*</u>: For the left vertex set $\mathcal{J} = \{u_1, \ldots, u_b\}$, we upper bound the probability the word $c(\mathcal{J}) = (c_1, \ldots, c_n)$ defined by $c_i = 1$ if $u_i \in \mathcal{J}$ and $c_i = 0$ else, is present in the code $\mathcal{C}$. We split our analysis into two possibilities based on the cardinality $\#\mathcal{J} \leq t$ or not, where integer $t \geq 1$ will be determined later.

*Case I* ($\#\mathcal{J} = b \leq t$): The probability that any vertex in $\mathcal{J}$ is adjacent to the right vertex $v_1$ is $p$. Therefore with probability $p(1-p)^{b-1}$ the right vertex $v_1$ is adjacent to $u_i$ and no other vertex in $\mathcal{J}$. Thus with probability $(1 - p(1-p)^{b-1})^m$, there is no "unique" right vertex adjacent only to $u_i$ and no other vertex in $\mathcal{J}$. If $E_{uni}(\mathcal{J})$ is the event that each vertex in $\mathcal{J}$ contains a unique right neighbour, then by the union bound

$$\mathbb{P}\left(E_{uni}^c(\mathcal{J})\right) \leq g(1 - p(1-p)^{b-1})^m \leq g e^{-mp(1-p)^{b-1}}. \tag{2.13}$$

If $E_{uni}(\mathcal{J})$ occurs, then the word $c(\mathcal{J}) \notin \mathcal{C}$ because the parity constraints will not be satisfied. Therefore if the event

$$E_{tot} := \bigcap_{\mathcal{S}} E_{uni}(\mathcal{J}) \tag{2.14}$$

occurs, where the intersection is over all subsets $\mathcal{J}$ of size $g \leq t$, then any word in $\mathcal{C}$ has weight $\geq t+1$ where weight is defined to be the number of indices with 1 as the entry. But $\mathcal{C}$ is linear and so $d_H(\mathcal{C}) \geq t+1$. From (2.13) we have

$$\mathbb{P}\left(E_{tot}^c\right) \leq \sum_{b=1}^{t} b\binom{n}{b} e^{-mp(1-p)^{b-1}} \leq t^2 \binom{n}{t} e^{-mp(1-p)^{t-1}}$$

provided $t < \frac{n}{2}$. Using $\binom{n}{t} \leq \left(\frac{ne}{t}\right)^t$ we further get that

$$\mathbb{P}(E_{tot}^c) \leq e^{-\theta_0} \tag{2.15}$$

where

$$\theta_0 := mp(1-p)^{t-1} - t\log\left(\frac{ne}{t}\right) - 2\log t \geq \frac{m}{2}p(1-p)^{t-1} \geq 4C\cdot n, \tag{2.16}$$

for some constant $C > 0$, since $m = \epsilon n$ and $p > 0$ is a constant.

*Case II* ($t+1 \leq \#\mathcal{J} = b \leq \delta n$): For vertex $v_j \in Y$, we recall that $\mathcal{V}_j$ is its (left) neighbour set in $G$ and define

$$H_j(\mathcal{J}) := \{\#\left(\mathcal{V}_j \cap \mathcal{J}\right) \text{ is odd}\}.$$

If $H_j(\mathcal{J})$ occurs, then $c(\mathcal{J})$ would not satisfy the parity constraints at $v_j$ and so the occurrence of the event $\bigcup_{1\leq j\leq m} H_j(\mathcal{J})$ implies that $c(\mathcal{J}) \notin \mathcal{C}$. Set

$$E_{comb} := \bigcap_{\mathcal{J}}\left(\bigcup_{1\leq j\leq m} H_j(\mathcal{J})\right) \tag{2.17}$$

where the intersection is over all $\mathcal{J}$ satisfying $t+1 \leq \#\mathcal{J} \leq \delta n$. Under $E_{comb}$, no word in $\mathcal{C}$ has weight between $t+1$ and $\delta n$ and so together with case $I$, we get that if $E_{tot} \cap E_{comb}$ occurs, then $d_H(\mathcal{C}) \geq \delta n + 1$.

It remains to estimate $\mathbb{P}(E_{comb})$. We know that $\#\mathcal{V}_j$ has a Binomial $(n, p)$ distribution and so $\#(\mathcal{V}_j \cap \mathcal{J})$ is Binomial $(b, p)$ distribution. Consequently,

$$\mathbb{P}(H_j^c(\mathcal{J})) = \sum_{\substack{0\leq k\leq b \\ k \text{ even}}} \binom{b}{k} p^k \cdot (1-p)^{b-k} = \frac{1}{2}\left(1 + (1-2p)^b\right). \tag{2.18}$$

and so for any $0 < \eta < \frac{1}{2}$ we can choose $t \leq g$ large enough so that $\mathbb{P}\left(H_j^c(\mathcal{J})\right) \leq \frac{1}{2^{1-\eta}}$. In turn this gives

$$\mathbb{P}\left(\bigcap_{l=1}^{m} H_l^c(\mathcal{J})\right) \leq \frac{1}{2^{m(1-\eta)}} = \frac{1}{2^{(1-\eta)\epsilon n}}.$$

The number of choices for $\mathcal{J}$ is $\binom{n}{g}$ and so by the union bound

$$\mathbb{P}\left(E_{comb}^c\right) \leq \left(\sum_{b=t+1}^{\delta n}\binom{n}{g}\right)\cdot\frac{1}{2^{(1-\eta)\epsilon n}} \leq \frac{1}{2^{\beta n}}, \tag{2.19}$$

where $\beta := (1-\eta)\epsilon - H(\delta)$ and the final estimate in (2.19) is due to the Hamming ball size bounds in Proposition 3.3.1 [5]. Recalling that we have chosen $\epsilon > H(\delta)$ strictly, we now set $\eta > 0$ smaller if necessary so that $\beta > 0$ strictly. Setting $E_{dist} := E_{tot} \cap E_{comb}$ we get from (2.19), (2.15) and (2.16) that

$$\mathbb{P}(E_{dist}) \geq 1 - e^{-4Cn} - \frac{1}{2^{\beta n}} \geq 1 - e^{-3Cn} \tag{2.20}$$

for all $n$ large.

Combining (2.11) and (2.20) and using the fact that $m = \epsilon n$, we get from a union bound that

$$\mathbb{P}(E_{div} \cap E_{dist}) \geq 1 - 4m^2 e^{-\frac{\theta^2}{4}np^2} - e^{-3Cn} \geq 1 - e^{-2Dn} \tag{2.21}$$

for some constant $D > 0$.

<u>*Step 3 (Incorporating constraints)*</u>: By the relation (2.6), we see that $\mathbb{P}(E_n) \geq e^{-Dn}$ where $D > 0$ is as in (2.21). Therefore from (2.21) we have

$$\mathbb{P}\left(E_n \cap E_{div} \cap E_{dist}\right) \geq e^{-Dn} - e^{-2Dn} > 0$$

and so the probabilistic method establishes the existence of our desired an $n$-length linear code satisfying the diversity, graphical and redundancy constraints. □

## 3 Fractional Graph Capacity

For a connected graph $H$ with vertex set $\{1, 2, \ldots, n\}$, let $\mathcal{N}_H[u]$ be the closed neighbourhood of $u$ consisting of all neighbours of $u$, including $u$. A stable set in $V$ is a set $\mathcal{I}$ such that no edge of $H$ has both endvertices in $\mathcal{I}$ and we the maximum size of a stable set in $H$ is denoted as $\alpha(H)$.

For integers $1 \leq k \leq r$, we define $H(r, k)$ with vertex set $\{1, 2, \ldots, n\}^r$ as follows. Two vertices $\mathbf{u} = (u_1, \ldots, u_r)$ and $\mathbf{v} = (v_1, \ldots, v_r)$ are said to be adjacent in $H(r, k)$ if $u_i \in \mathcal{N}_G[v_i]$ for each $1 \leq i \leq r$ and

$$\sum_{i=1}^{r} 1\!\!1(u_i \neq v_i) \leq k,$$

where $1\!\!1(.)$ is the indicator function. For $k = n$, the above definition is simply the strong graph product.

**Definition 1** For $0 < \zeta \leq 1$ the $\zeta$-fractional capacity of $H$ is defined as

$$\Theta_\zeta(H) := \sup_{r \geq \frac{1}{\zeta}} \left(\alpha(H(r, \zeta r))\right)^{\frac{1}{r}}. \tag{3.1}$$

For $\zeta = 1$, this reduces to the graph capacity as defined in [13] and we refer to $\Theta(H)$ and $\Theta_\zeta(H)$ as the *full* and *fractional* graph capacities, respectively.

Letting $H(.)$ be the entropy function as in (2.5), we have the following:

**Theorem 2** *Let $d_{av}$ and $\Delta$ be the average and maximum vertex degree of a connected graph $H$. For any $0 < \zeta \leq 1$ the fractional graph capacity satisfies*

$$n \cdot \max(W(\zeta, d_{av}), W(\zeta, \Delta), W(\zeta, n-1)) \leq \Theta_\zeta(H) \leq n \cdot \left(\frac{\Theta(H)}{n}\right)^\zeta \tag{3.2}$$

*where*

$$W(a, y) := \begin{cases} \left(2^{H(a)} \cdot y^a\right)^{-1} & \text{if } 0 < a < \frac{y}{y+1} \\ (y+1)^{-1} & \text{if } \frac{y}{y+1} \leq a \leq 1. \end{cases} \tag{3.3}$$

We have the following remarks:

**Remark 1** For example if $H = C_5$ then $d_{av} = \Delta = 2$ and from [11] we have that $\Theta(H) = \sqrt{5}$. For $\zeta = \frac{1}{2}$, we get the following estimates for the "half" capacity of $C_5$ :

$$\frac{5}{2\sqrt{2}} \leq \Theta_{\frac{1}{2}}(C_5) \leq \frac{5}{\sqrt[4]{5}}.$$

**Remark 2** In general, we see from (3.2) that as $\gamma \to 0$, the fractional graph capacity $\Theta_\gamma(G) \to n$, the maximum possible value. On the other end, setting $\gamma = 1$ in the lower bound (3.3), we get that the full graph capacity

$$\Theta(G) \geq \frac{n}{d_{av}+1},$$

the Turán's bound [16].

*Proof of Theorem* 2: The lower bound is obtained using a combination of probabilistic method and Gilbert-Varshamov argument [8] and for the upper bound we use a recursive relation analogous to the Singleton bound [8].

We begin with the lower bound. Let $\mathbf{w} = (w_1, \ldots, w_r)$ have independent entries that are uniform in $\{1, 2, \ldots, n\}$ so that the expected degree of each $w_j$ is $d_{av}$. If $\mathcal{B}_\zeta(\mathbf{w})$ is the set of vertices adjacent to $\mathbf{w}$ in $H(r, \zeta r)$, then

$$\mathbb{E}\#\mathcal{B}_\zeta(\mathbf{w}) = \sum_{l=0}^{\zeta r} \binom{r}{l} d_{av}^l. \tag{3.4}$$

For $0 < \zeta < 1 - \frac{1}{d_{av}+1}$, the Hamming ball estimates in Proposition 3.3.1 [5] gives

$$\mathbb{E}\#\mathcal{B}_\zeta(\mathbf{w}) \leq n^{yr} \tag{3.5}$$

where $y = \frac{H(\zeta)+\zeta \log d_{av}}{\log n}$ satisfies $0 < y < 1$. Setting

$$\mathcal{Z}(\eta) := \{\mathbf{v} : \#\mathcal{B}_{\zeta}(\mathbf{v}) \leq n^{r(y+\eta)}\} \tag{3.6}$$

for $\eta > 0$ to be determined later, we use the Markov inequality to obtain

$$\#\mathcal{Z}(\eta) \geq n^r \left(1 - \frac{1}{n^{r\eta}}\right). \tag{3.7}$$

Let $\mathcal{D} := \{\mathbf{w}_1, \ldots, \mathbf{w}_M\} \subseteq (\eta)$ be a stable set of maximum size in $H(r, \zeta r)$. By the maximality, we must have that the union $\bigcup_{i=1}^{M} \mathcal{B}_{\zeta}(\mathbf{w}_i) = \mathcal{Z}(\eta)$ and so from (3.7) and (3.6), we have

$$n^r \left(1 - \frac{1}{n^{r\eta}}\right) \leq \#\mathcal{Z}(\eta) \leq \sum_{i=1}^{M} \#\mathcal{B}_{\zeta}(\mathbf{w}_i) \leq M \cdot n^{r(\theta+\eta)}. \tag{3.8}$$

Consequently $M \geq n^{(1-\theta-\eta)r} \left(1 - \frac{1}{n^{r\eta}}\right)$ and choosing $\eta = \frac{1}{\sqrt{r}}$, taking $r$th roots and allowing $r \to \infty$, we get that

$$\Theta_{\zeta}(H) \geq W(\zeta, d_{av}) \tag{3.9}$$

for $0 < \zeta < 1 - \frac{1}{d_{av}+1}$.

For $1 - \frac{1}{d_{av}+1} \leq \zeta \leq 1$, we have

$$\mathbb{E}\#\mathcal{B}_{\zeta}(\mathbf{w}) \leq \sum_{k=0}^{r} \binom{r}{k} d_{av}^k = (d_{av} + 1)^r. \tag{3.10}$$

Again the Gilbert-Varshamov argument implies that (3.9) is true for $1 - \frac{1}{d_{av}+1} \leq \gamma \leq 1$. Similarly, the deterministic estimate

$$\#\mathcal{B}_{\zeta}(\mathbf{w}) \leq \sum_{k=0}^{\zeta r} \binom{r}{k} \Delta^r \tag{3.11}$$

gives $\Theta_{\zeta}(H) \geq W(\gamma, \Delta)$ and using $\Delta \leq n - 1$ we get that $\Theta_{\gamma}(G) \geq W(\gamma, n - 1)$. This proves the lower bound in (3.2).

For the upper bound, we use recursion. Letting $s(r, z)$ be the maximum size of a stable set in $H(r, z)$, we first obtain a recursion for $s(r, z)$ in terms of $r$. Let $\mathcal{D}$ be a maximum stable set in $H(r, z)$ and let $\mathcal{D}(w) \subset \mathcal{D}$ be those vertices containing $w$ as the last entry (i.e. $r$th component. There are $n$ choices for $w$ and so the pigeonhole principle implies that $\#\mathcal{D}(w_0) \geq \frac{s(r,z)}{n}$ for some $w_0$. Removing $w_0$ from each vertex in $\mathcal{D}(w_0)$ gives us a new vertex set $\mathcal{C}(w_0) \subset H(r - 1, z)$. By definition $\mathcal{C}(w_0)$ is stable in $H(r-1,z)$ as well and so $s(r,z) \leq n \cdot s(r-1,d)$. Subsequent iterations gives us

$$s(r,z) \leq n^{r-z} \cdot s(z,z). \tag{3.12}$$

Setting $z = \zeta r$, taking $r$th roots and using

$$\sup_{r \geq \frac{1}{\zeta}} (s(\zeta r, \zeta r))^{\frac{1}{r}} = \sup_{j \geq 1} (s(j,j))^{\zeta} = (\Theta(H))^{\zeta},$$

we then obtain the upper bound in (3.2). □

**Acknowledgements** I thank Professors V. Guruswami, C. R. Subramanian and the referees for crucial comments that led to an improvement of the paper. I also thank IMSc and IISER Bhopal for my fellowships.

## References

1. Alon, N., Lubetzky, E.: The Shannon capacity of a graph and the independence numbers of its powers. IEEE Trans. Inf. Theory **52**, 2172–2176 (2006)
2. Alon, N., Spencer, J.: The Probabilistic Method. Wiley Interscience (2008)
3. Dougherty, S.T.: Algebraic Coding Theory Over Finite Commutative Rings. Springer Briefs in Mathematics (2017)
4. Greferath, M., Schmidt, S.E.: Linear codes and rings of matrices. In: Proceedings of AAECC 13 Hawaii, Springer LNCS 1719, pp. 160–169 (1999)
5. Guruswami, V., Rudra, A., Sudan, M.: Essential Coding Theory (2019). https://cse.buffalo.edu/faculty/atri/courses/coding-theory/book/web-coding-book.pdf
6. Hell, P., Roberts, F.S.: Analogues of the Shannon capacity of a graph. North-Holland Math. Stud. **60**, 155–168 (1982)
7. Hemenway, B., Ostrovsky, R., Wootters, M.: Local correctability of expander codes. Inf. Comput. **243**, 178–190 (2015)
8. Huffman, W.C., Pless, V.: Fundamentals of Error Correcting Codes. Cambridge University Press (2003)
9. Irwansyah, D.S.: Structure of linear codes over the ring $B_k$. J. Appl. Math. Comput. **58**, 755–775 (2018)
10. Liu, Z., Wang, J.: Linear complementary dual codes over rings. Des. Codes Cryptogr. **87**, 3077–3086 (2019)
11. Lovász, L.: On the Shannon capacity of a graph. IEEE Trans. Inf. Theory **25**, 1–7 (1979)
12. Marton, K.: On the Shannon capacity of probabilistic graphs. J. Comb. Theory Ser. B **57**, 183–195 (1993)
13. Shannon, C.E.: The zero-error capacity of a noisy channel. IRE Trans. Inf. Theory **22**, 8–19 (1956)
14. Sipser, M., Spielman, D.: Expander codes. IEEE Trans. Inf. Theory **42**, 1710–1722 (1996)
15. Urbanke, R., Richardson, T.: Modern Coding Theory. Cambridge University Press (2008)
16. West, D.B.: Introduction to Graph Theory. Pearson (2001)

# Tree Parity Machine-Based Symmetric Encryption: A Hybrid Approach

**Ishak Meraouche, Sabyasachi Dutta, Haowen Tan, and Kouichi Sakurai**

**Abstract** In a symmetric key encryption the sender and the receiver must possess the same pre-distributed key in order to encrypt or decrypt the exchanged messages. Exchanging symmetric keys is a challenging issue in cryptography. In this paper, we put forward a symmetric key encryption technique that does not require any common pre-shared "knowledge" between the parties. More specifically, we use a type of neural network called Tree Parity Machines (TPMs) which, when synchronized, enable two parties to reach a common state. The common state can be used to establish a common secret key. Our method makes use of the Tree Parity Machines to reach a common state between the parties communicating and encrypt the communications with an ElGamal-type encryption methodology. The advantage of our implementation is that the initial key exchange method is fast, lightweight and believed to become a post-quantum candidate. We have analyzed the randomness of the produced ciphertexts from our system using NIST randomness tests and the results are included in the paper. We also demonstrate security against chosen plaintext attacks.

**Keywords** Encryption · Symmetric key · Key exchange · Neural networks

I. Meraouche (✉) · K. Sakurai
Kyushu University, Fukuoka, Japan
e-mail: ishak.meraouche@gmail.com

K. Sakurai
e-mail: sakurai@inf.kyushu-u.ac.jp

S. Dutta
University of Calgary, Calgary, Canada

H. Tan
Cyber Security Center, Kyushu University, Fukuoka, Japan

 
B. Rushi Kumar et al. (eds.), *Mathematics and Computing*, Springer Proceedings in Mathematics & Statistics 415, https://doi.org/10.1007/978-981-19-9307-7_6

## 1 Introduction

Cryptographically secure key exchange protocols or public key encryption schemes such as RSA [13], Diffie-Hellman (DH) key exchange [3] or ElGamal [4] base their security on problems which are "hard" to solve by probabilistic polynomial time adversaries. However, the Diffie-Hellman key exchange technique, which constitutes the main building block of ElGamal, is a key exchange protocol. It allows two parties to mutually compute the same secret key in presence of adversaries who are eavesdropping on the channel. ElGamal [4] extended the idea of the Diffie-Hellman key exchange protocol to get an asymmetric encryption protocol.

With the advent of a novel technique proposed by Abadi and Andersen [1] which allows two neural networks to mutually learn to encrypt communication in the presence of an eavesdropping neural network, researchers are exploring alternative ways of securing communications with the hope of creating lightweight post-quantum-safe primitives. Abadi and Andersen [1] borrow ideas from generative adversarial networks and put forward a methodology which they termed as adversarial neural cryptography. The work of [1] has seen a decent amount of follow-up works such as [6, 11, 17] where the methodology has been adapted to achieve steganographic techniques and also extensive security analysis in [18] and security improvements in [2, 10]. The works along this direction are purely based on the philosophy where two or more neural networks compete against each other to learn and achieve a goal, e.g. learning to encrypt a communication. The security provided does not depend on any well-defined "hard" problem which is improbable to solve by any probabilistic polynomial time adversary.

**Motivation of the work.** The existing methodology does not immediately provide any sort of provable security, e.g. *semantic security* which is widely believed to be the minimum security requirement nowadays. The motivation behind our work is to implement a secure symmetric key encryption scheme using Tree Parity Machines and standard hardness assumptions like decisional Diffie-Hellman (DDH). Using Tree Parity Machines removes the need of having a pre-shared state/information in order to generate a secret key as in [1]. On the other hand, using an ElGamal-type encryption method over a DDH group ensures some reasonable provable security against chosen plaintext attacks.

**Our Contribution.** We aim to take initial steps into hybridizing existing cryptography techniques with recent neural networks-based cryptography techniques. Concretely, we aim to realize a symmetric key encryption scheme based on the hardness of discrete logarithm problem and using the Tree Parity Machine proposed by Kanter, Kinzel and Kanter [8]. Tree Parity Machines are neural networks composed of three layers: `Input layer`, `Hidden layer` and `Output layer`. Kanter et al. [8] show that two Tree Parity Machines can be synchronized and obtain the same state that can be later used to generate a common secret key.

We choose the Tree Parity Machines to establish key(s) in order to gain more speed and flexibility in the key generation process. Additionally, the Tree Parity

Machines exchange does not rely on a problem which is hard to solve by probabilistic polynomial time adversaries and therefore can be seen as a potential quantum-safe candidate. To the best of our knowledge, such a hybrid approach toward constructing a symmetric key encryption has not been considered in the literature before.

## *1.1 Related Works*

**Key Exchange using Tree Parity Machines** Kanter et al. [8] proposed a model of two neural networks that can synchronize and learn to exchange a secret key on a public channel just like the Diffie-Hellman protocol [3]. The two parties are neural networks with the same structure. They have the following structure as shown in Fig. 1.

We can see that the neural networks are composed of an input layer $I$ containing $K \cdot N$ neurons to read the input, a hidden layer $W$ containing $K$ neurons to mix and transform the input and finally a single neuron output layer.

To mutually obtain the same weights vector $W$ which will be used to generate the secret key, the two neural networks need to publicly exchange random vectors. Alice will initiate the communication by generating a random input vector and send it publicly to Bob. Alice and Bob will pass the same vector through their neural network in order to obtain the output. Alice and Bob will now publicly compare their outputs. If they have different outputs, they use Hebbian Learning [7] in order to update their parameters and repeat the process (i.e. continue training). If they have equal outputs, they can stop the training.

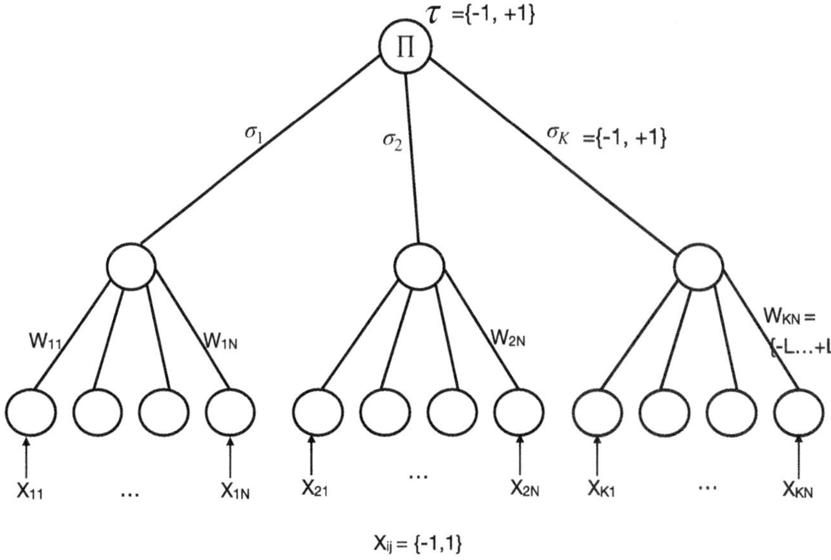


**Fig. 1** Neural Network structure of a Tree Parity Machine

Hebbian learning [7] (proposed in a different context) is an unsupervised learning technique which allows two neural networks to synchronize and obtain the same training weights remotely. The formula for updating the parameters is detailed in Eq. 1

$$W_{i,j} = Lf(W_{i,j} + \tau_A \cdot \Theta(\sigma_i, \tau_1) \cdot \Theta(\tau_A, \tau_B)) \quad (1)$$

where $W_{i,j}$ is the current weight value we are updating, $Lf$ is a function that limits the value it receives as parameter in $\{-L, \ldots, +L\}$, $\tau_A$ and $\tau_B$ are Alice's and Bob's output, respectively, $\sigma_i$ is the output of the $i$th hidden neuron and $\theta$ is a function that returns 0 if its two parameters are not equal and 1 otherwise.

However, this exchange has been proven weak against attacks as shown in [9]. The authors have shown that three different attacks can be done on the Tree Parity Machine. These attacks either bruteforce all possible structures for the TPMsor eavesdrop the communication to mimic Bob's behavior and synchronize passively. The authors in [9] also show probabilistic-based attacks.

The authors in [15] show that using a value of $K = 8$ and $N = 16$ exponentially increases the possible weights and makes it difficult for an attacker to realize probabilistic or brute force attacks. They also show that using these values, a small change in the parameters increases the time required to perform a passive attack polynomially. Therefore, we will be using these parameters for the structure of our neural networks to provide maximum security for the exchange. On the other hand, [5] has a concrete instantiation of a symmetric key protocol via. DES which we believe falls short of achieving semantic security.

**Secure Communication using Adversarial Neural Cryptography** [1] This GANs-based neural network model proposed in late 2016 by Abadi and Andersen [1] shows that two neural networks can learn to encrypt a communication in the presence of an eavesdropper by training in a GANs setup. We aim to compare our work with this work in Sect. 5

In a GANs setup, two neural networks compete against each other in order to generate data that is statistically similar to the original training Data. We usually find the Generator that tries to generate data as similar as possible to the original data and the Discriminator has to tell generated data and original data apart.

In the model proposed by Abadi and Andersen [1], the setup is also based on GANs but is slightly different. The Generator Alice will be in charge of generating ciphertexts that are easy to decrypt for Bob and difficult to decrypt by the eavesdropping adversary Eve.

The two parties need to have a pre-shared secret key to encrypt data with.

As for the training, at each iteration, Alice will generate a ciphertext using a random key and send it to Bob. Bob will try to decrypt the ciphertext using the same random key (We assume that Bob can get the secret key safely). A third party, the eavesdropper Eve, will intercept each ciphertext and try to decrypt it without the key.

Encryption and Decryption are done by passing the key and the plaintext and ciphertext, respectively, through the party's neural network and processing it.

The neural network structure is composed of a fully connected layer that reads the plaintext $P$ and the secret key $Key$ and will then be followed by multiple convolutions and activation to get the ciphertext $C$ as an output. The decryption follows the same procedure with the exception that the receiving party uses $C$ instead of $P$ as input along with $Key$. On the other hand, the eavesdropper only has $C$ as input to her neural network.

In our experiments with an average computer, establishing a connection takes between 15 and 30 min (in order to train the neural networks for the first time), and after that, encryption and decryption processes are mostly instant. This means that this model can be used in production environments without any worries about performance even on resource-limited devices such as IoT devices. The authors in [2, 10] proposed some changes to the neural network structure to get ciphertexts that are close to uniformly random. To cope with the problem of long training time, Researchers in [12] proposed training and saving the neural network parameters in advance and using them directly when a communication is needed.

## 2 A Hybrid Method for Symmetric Encryption

Existing neural networks-based symmetric key encryption schemes arising from TPMs [5, 8] or from adversarial neural networks [1, 12] do not provide any provable security. The security of the aforementioned proposals was not based on any hard problem like discrete log or factorization problem. In fact, [9, 18] analyzed the shortcomings of the proposals which imply vulnerability against multiple attacks including chosen plaintext attacks.

In this section, we present a hybrid approach to develop a symmetric key encryption which uses a TPMs to generate common randomness between two parties and a setup for ElGamal-type encryption. The main advantage of our model is that there is no need to share a common (secret) state in advance like the work in [1]. The synchronization of the Tree Parity Machines of the two parties will allow them to reach a common secret state that can be used to generate secret keys.

For the sake of completeness, we summarize the Diffie-Hellman key exchange protocol below (between Alice and Bob):

1. $q$ is a prime, $G$ is a cyclic group of order $q$ in which DDH is hard and $g$ is a generator of $G$. The information $(G, g, q)$ is made public.
2. Alice uniformly selects $r_A \in \mathbb{Z}_q - \{0\}$; computes $K_A = g^{r_A} \mod q$; sends $K_A$ to Bob.
3. Bob uniformly selects $r_B \in \mathbb{Z}_q - \{0\}$; computes $K_B = g^{r_B} \mod q$; sends $K_B$ to Alice.
4. Alice computes $K_B^{r_A}$ and Bob computes $K_A^{r_B}$ locally.
5. Both output $g^{r_A \cdot r_B}$ as their common secret key.

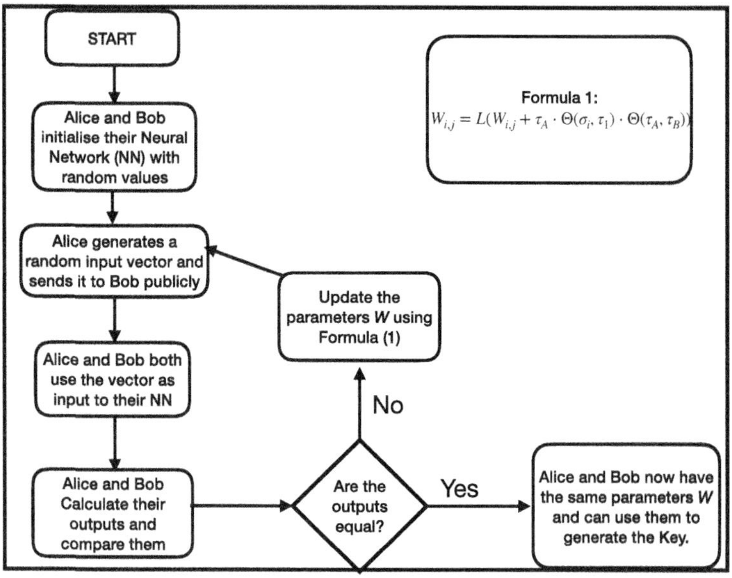


**Fig. 2** Synchronization process between Alice and Bob

In our proposal, we let the parties Alice and Bob use TPMs to generate the common randomness which serves as the exponent of $g$ to establish the secret key. The details are given below.

1. $(G, g, q)$ are public values such that $G$ is a DDH group.
2. Alice and Bob synchronize their TPMs as shown before in Sect. 1.1 and Fig. 2 to obtain the same vector of parameters $W$.
3. Alice and Bob generate a common secret Key from the weight-vector $W$ (Details in Sect. 2.1).

Figure 2 gives a pictorial depiction of the steps during the synchronization process of TPMs.

## 2.1 Key Generation

The generation of Key from the weight-vector $W$ and the public generator $g$ are as follows. When the neural networks are synchronized, both the parties have the same parameters $W$ which is an array of integers $W = [a_0, a_1, \cdots, a_{K \cdot N}]$ with $a$ taking values between $-L$ and $+L$.

All the elements of $W$ are converted to binary and concatenated together transforming negative numbers into positive numbers. The result of the concatenation $\mathsf{CR}$ is then used to calculate the secret key in binary.

The common secret key is calculated as $\mathsf{Key} = \widetilde{\mathsf{CR}}$, a conversion of $\mathsf{CR}$ into an element of $\mathbb{Z}_q$. In the following, we describe the encryption scheme.

**Method 1**. In the first method, once the key is obtained by Alice and Bob, encryption and decryption algorithms are similar to the ElGamal.

1. Alice (or Bob) randomly chooses $r \leftarrow_{\$} \mathbb{Z}_q$.
2. Alice (or Bob) computes
   - $g^r$ and outputs $(C_1, C_2) \longleftarrow (g^r, (g^r)^{\mathsf{Key}} \cdot M)$ as ciphertext.
3. Bob (or Alice) receives $(C_1, C_2)$.
4. Bob (or Alice) computes $M \leftarrow \frac{C_2}{C_1^{\mathsf{Key}}}$.

*Security against chosen plaintext attacks.* It is well known that over a DDH group the ElGamal encryption provides IND-CPA (Indistinguishability under Chosen Plaintext Attack) security. The IND-CPA security of the above-mentioned scheme can be proved using the same proof strategy and is omitted from this draft.

**Method 2** The second method generates a mutual random vector between Alice and Bob and uses it to transform the secret key every time a communication is needed.

The encryption and decryption work as follows for the second method:

1. Alice (or Bob) chooses a random vector $r$ as long as the input of their Neural Network and sends it to the other party.
2. Alice and Bob both pass the vector r through their neural network and calculate the output without the activation (in order to get a vector of integers).
3. Now they both have the same output vector $R$.
4. In order to encrypt a message, Alice (or Bob) does the following:
   (a) Choose the next unused element $e$ with index $i$ inside the vector $R$.
   (b) Calculate $Key' = Key * e$, i.e. multiply the key by the element $e$.
   (c) To encrypt a message $M$ as $C$, the sender calculates $C = M * Key' \mod q$.
   (d) The sender sends the pair $(C, i)$ where $i$ is the index of the element $e$ used to transform the key.
5. Bob (or Alice) receives $(C, i)$.
6. Bob (or Alice) gets the $ith$ element $e$ from the vector $R$.
7. Bob (or Alice) calculates $Key' = Key * e$.
8. Bob (or Alice) can now decrypt $C$ using $Key'$ with the following formula: $M = C * Key'^{-1} \mod q$.

*Security against chosen plaintext attacks.* Similarly to method one, randomness is introduced to the key at every exchange. The only inconvenience of this method is that the two parties will need a new random vector $R$ every time they have used all the elements of the current vector $R$. However, this method provides slightly faster encryption as it is only a multiplication whereas in the first method calculating $(g^r)^{\mathsf{Key}}$ is required and is more expensive to calculate. This method is more appropriate for IoT devices such as sensors that do not need to communicate too often with a server. For example, sensors that send average weather during the day every 24 h.

## 3 Results, Training Time and Encryption Time

We conducted hundreds of simulations training Alice and Bob to perform the synchronization of TPMsand get a common secret key to use it for encryption as defined in the previous section. Table 1 summarizes our results over 3 sets of 200 training simulations. Each simulation contains an adversary Eve with the same structure as Bob. Eve will try to get the secret key by mimicking Bob's behavior to synchronize with Alice and Bob without their knowledge. Technically speaking, Eve updates her parameters when her output is the same as Bob's and Alice's.

As shown in Table 1, we performed 900 simulations each divided into groups of 300. We can see that the average number of exchanges required to synchronize is around 230 exchanges. The synchronization takes on average around 300 milliseconds with a minimum of 73 ms and a maximum of 610 ms synchronization time recorded. During all of our 900 simulations, there has been no simulation where Eve has been able to secretly synchronize with Alice and Bob and end up with the same key as them. We have used $K = 8, N = 16, L = 8$ as parameters for the Tree Parity Machine. This means that the Tree Parity Machine has $K \cdot N = 128$ input neurons and $K = 8$ hidden neurons that can take a value between $-L$ and $+L$ (i.e. Between $-8$ and $+8$).

We have chosen these values as they have been found to be optimal in the experiments conducted in [15]. The authors state that among all the different structures they used, the structure that uses the values $K = 8, N = 16, L = 8$ is the most secure and out of the 1 million simulations they have performed, only 1 successful synchronization by Eve has been recorded. This is the reason why have used this technique as it is the safest according to the work done in [15].

**Table 1** Table summarizing the time and number of exchanges required to synchronize

| Sim. | # Sim. | Avg. # exchange | Min. Sync time (ms) | Max. Sync time (ms) | Avg. Sync time (ms) |
|---|---|---|---|---|---|
| 1 | 300 | 222 | 124 | 610 | 266 |
| 2 | 300 | 229 | 110 | 606 | 278 |
| 3 | 300 | 225 | 73 | 580 | 273 |

# 4 Security Analysis

The security of the encryption algorithm is dependent on the randomness of the key generated using the Tree Parity Machine. If the generated key is close to uniformly random, then the security of the protocol can be reduced to the security of the key exchange.

## 4.1 NIST Statistical Test Results

To test the randomness of the keys generated and therefore the security of our proposed method, we have conducted the NIST statistical test on a series of ciphertexts generated by our implementation of the proposed model using the two methods.

We have generated approximately two sets of 500 ciphertexts each with a unique key. The first set being generated with the first method and the second set with the second method. All the ciphertexts had a size of 2048 bits.

Both the two methods got approximately the same results which are detailed in Table 2.

In order to compare our results, we also conducted the NIST randomness test on ElGamal with the same amount of sample ciphertexts as our first encryption method is similar to that of ElGamal.

**Table 2** NIST randomness test results on our model and ElGamal

| Test name | Result with Method 1 | Result with Method 2 | Result with original ElGamal scheme |
|---|---|---|---|
| Frequency | SUCCESS | SUCCESS | SUCCESS |
| BlockFrequency | SUCCESS | SUCCESS | SUCCESS |
| CumulativeSums | SUCCESS | SUCCESS | SUCCESS |
| Runs | SUCCESS | SUCCESS | SUCCESS |
| LongestRun | SUCCESS | SUCCESS | SUCCESS |
| Rank | SUCCESS | SUCCESS | SUCCESS |
| FFT | SUCCESS | SUCCESS | SUCCESS |
| Non Overlapping Template | FAILED | FAILED | FAILED |
| OverlappingTemplate | SUCCESS | SUCCESS | SUCCESS |
| Universal | FAILED | FAILED | FAILED |
| ApproximateEntropy | FAILED | FAILED | FAILED |
| Serial | SUCCESS | SUCCESS | SUCCESS |
| LinearComplexity | FAILED | FAILED | FAILED |

We can see that the method has passed most of the tests performed by the NIST statistical test. According to [16], the samples are considered random and therefore secure if they pass at least 7 NIST statistical tests which is our case.

As we can see, the NIST randomness test shows that our ciphertexts are uniformly random which implies that our method can be used at a production level securely.

### 4.2 Security Against Chosen Plaintext Attacks

We have already mentioned the IND-CPA security of the encryption scheme described in this paper. Due to lack of space, we do not give the details. However, in order to prove that the scheme is secure against chosen plaintext attack, we must analyze the randomness/unpredictability of the key $\widetilde{\mathsf{CR}}$ in $\mathbb{Z}_q$ generated by the $\mathsf{TPMs}$.

### 4.3 Other Attacks

Additionally to the NIST Statistical test, we have noticed that the key exchange using the Tree Parity Machines has been proven to be vulnerable against multiple attacks as shown in [9]. The authors in [9] show three different attacks that can be applied in order to allow a third party Eve to simulate the exchange between Alice and Bob and end up with the same weights array $W$. However, as shown in [14, 15], this can be avoided by increasing the size of the neural networks (i.e. number of input neurons and neurons in the hidden layer).

## 5 Comparison with Existing Works

We compare our proposed implementation with the model of Abadi and Andersen [1] in terms of multiple factors as shown in Table 3.

We can see that our model outperforms the model proposed by Abadi and Andersen [1] in terms of synchronization time. This is mainly due to the large CNN (Convolutional Neural Network) structure used by Abadi and Andersen versus a relatively smaller unique hidden layer neural network structure used in our model. Our model also has the advantage of not relying on any initial common state or pre-shared secret such as a secret key. As for the key length, ElGamal encryption needs large keys; therefore, the key generated with our model is quite large versus a small 32 or 64 bits key in the model by Abadi and Andersen [1]. The encryption time is roughly the same for both the techniques as it is a simple neural network feed-forward operation in the model used by Abadi and Andersen [1] and a simple mathematical multiplication in our proposed technique. However, the encryption technique learned by the neural networks in [1] is blackboxed and cannot be known. The messages in

**Table 3** Comparison of our work with the work by Abadi and Andersen [1]. *The encryption time in our model does not require multiple iterations as opposed to [1]

| Model | Min and Max Sync. time recorded* | Pre-shared info. | Key length | Enc. type | Message needs to be as long as the key? | Ciphertext size |
|---|---|---|---|---|---|---|
| Our model | $73 - 10^3$ ms | No | Varies depending on the values of K, L and N. In our model, the key length is 500 bits | ElGamal -based affine cipher | No | Large |
| Abadi and Andersen [1] | 15–30 mins | Yes | Chosen by the user. The authors used 32bits. However, the longer the length, the longer training will be | Blackboxed | Yes | Small (As long as the message) |

our model do not need to be as long as the key in contrary to the model in [1] but this comes for the price of larger ciphertexts in our model versus ciphertexts as long as the message in the model in [1]. Lastly, it has not been verified that the Tree Parity Machines can be synchronized with multiple parties simultaneously in contrary to the model in [1] where it is possible as shown by the researchers in [12].

Additionally, the authors in [18] have shown that the original model by Abadi and Andersen [1] has only passed the `BlockFrequency` and the `NonOverlapping` Template and failed the rest. However, the improved versions in [2, 10] achieve better results.

## 6 Conclusion and Future Work

We have proposed an encryption that makes use of the Tree Parity Machines [8] in order to reach a common state between two parties and encrypt the communication using an ElGamal-like encryption. The technique allows fast and lightweight synchronization and encryption of messages between two parties. The key generation is purely based on the common state established between the two Tree Parity Machines in synchronization and does not require any initial common knowledge. Once the key is generated, it can be used to encrypt the messages as explained in Sect. 2.1.

As a future work, we aim to do the following:

- Enable synchronization between more than two parties to exchange the same key.
- Find the optimal values for the parameters $K$, $L$, $N$ in order to build a Tree Parity Machine that is resistant to all the attacks shown in [9] by experimenting with more values as done in [15] but using larger numbers.
- Investigate the reasons behind failing the NonOverlappingTemplate, Universal, Approximate Entropy and the LinearComplexity NIST statistical tests.
- Study the possibility of using multiple hidden layers and whether it provides more security or not.
- Study the resistance against quantum attacks in the key exchange phase.

**Acknowledgements** Ishak Meraouche is financially supported by the Ministry of Education, Culture, Sports, Science and Technology (MEXT) for his studies in Japan. S. Dutta is grateful to NICT, Japan, for financially supporting his research at Kyushu University during 2018–2019. This project was originally conceived during that period and a preliminary draft was discussed with the first author. Haowen Tan is supported by the Collaboration Hubs for International Program (CHIRP) of Strategic International Collaborative Research Program (SICORP) through the Japan Science and Technology Agency (JST).

## References

1. Abadi, M., Andersen, D.G.: Learning to protect communications with adversarial neural cryptography (2016). arxiv:1610.06918
2. Coutinho, M., de Oliveira Albuquerqueand Fábio Borges, R., García-Villalba, L.J., Kim, T.: Learning perfectly secure cryptography to protect communications with adversarial neural cryptography. Sensors **18**(5), 1306 (2018)
3. Diffie, W., Hellman, M.: New directions in cryptography. IEEE Trans. Inf. Theor. **22**(6), 644–654 (2006). https://doi.org/10.1109/TIT.1976.1055638
4. Elgamal, T.: A public key cryptosystem and a signature scheme based on discrete logarithms. IEEE Trans. Inf. Theory **31**(4), 469–472 (1985)
5. Godhavari, T., Alamelu, N., Soundararajan, R.: Cryptography using neural network. In: 2005 Annual IEEE India Conference-Indicon, pp. 258–261. IEEE (2005)
6. Hayes, J., Danezis, G.: Generating steganographic images via adversarial training. Adv. Neural Inf. Process. Syst. **30**, 1954–1963 (2017)
7. Hebb, D.: The organization of behavior. EmphNew york (1949)
8. Kanter, I., Kinzel, W., Kanter, E.: Secure exchange of information by synchronization of neural networks. EPL (Europhysics Letters) **57** (2002)
9. Klimov, A., Mityagin, A., Shamir, A.: Analysis of neural cryptography. In: Zheng, Y. (ed.) Advances in Cryptology—ASIACRYPT 2002, Proceedings of the 8th International Conference on the Theory and Application of Cryptology and Information Security, Queenstown, New Zealand, 1–5 Dec. 2002. Lecture Notes in Computer Science, vol. 2501, pp. 288–298. Springer (2002). https://doi.org/10.1007/3-540-36178-2_18
10. Li, Z., Yang, X., Shen, K., Zhu, R., Jiang, J.: Information encryption communication system based on the adversarial networks foundation. Neurocomputing **415**, 347–357 (2020). https://www.sciencedirect.com/science/article/pii/S0925231220313175

11. Meraouche, I., Dutta, S., Sakurai, K.: 3-party adversarial cryptography. In: International Conference on Emerging Internetworking, Data & Web Technologies, pp. 247–258. Springer (2020)
12. Meraouche, I., Dutta, S., Sakurai, K.: 3-party adversarial cryptography. In: Barolli, L., Okada, Y., Amato, F. (eds.) Advances in Internet, Data and Web Technologies, pp. 247–258. Springer International Publishing, Cham (2020)
13. Rivest, R.L., Shamir, A., Adleman, L.: A method for obtaining digital signatures and public-key cryptosystems. Commun. ACM **21**(2), 120–126 (1978). https://doi.org/10.1145/359340.359342
14. Ruttor, A., Kinzel, W., Kanter, I.: Dynamics of neural cryptography. Phys. Rev. E **75**(5), 056104 (2007)
15. Salguero, E., Fuertes, W., Lascano, J.: On the development of an optimal structure of tree parity machine for the establishment of a cryptographic key. Secur. Commun. Netw. **2019**, 1–10 (2019)
16. Sýs, M., Riha, Z., Matyas, V., Marton, K., Suciu, A.: On the interpretation of results from the nist statistical test suite, vol. 18, pp. 18–32, Jan. 2015
17. Yedroudj, M., Comby, F., Chaumont, M.: Steganography using a 3-player game. J. Vis. Commun. Image Represent. **72**, 102910 (2020)
18. Zhou, L., Chen, J., Zhang, Y., Su, C., James, M.A.: Security analysis and new models on the intelligent symmetric key encryption. Comput. Secur. **80**, 14–24 (2019)

# Metadata Analysis of Web Images for Source Authentication in Online Social Media

**Mohd Shaliyar and Khurram Mustafa**

**Abstract** Camera mobility-based embedded devices now facilitate easy creation, manipulation, and sharing of digital content on social media. While doing so, the shareable misinformation or disinformation content may harm society. Such contents may be variously forwarded without verified integrity and authenticity of the source on Online Social Networks (OSN). Thus, it is inevitable to trace the epicenter and the kind of information being spread on social networks. In this paper, we have investigated the types of metadata linked with digital images and analyzed the different attributes that are susceptible to squandering the integrity of source authentication with the easy availability of online tools and mobile-based apps. Finally, we accentuate protecting the metadata through the watermarking technique to reveal a piece of important information in the digital forensic investigation.

**Keywords** Digital forensic · Digital image · DWT · EXiF · IPTC · Metadata · XMP · Watermarking

## 1 Introduction

With the dramatic advancement in the field of OSN has come of age with Facebook, YouTube, Twitter, WhatsApp, Tinder, Instagram, etc. OSN has become a cost-effective way to share digital content. It is an easy-to-access platform to obtain information about the globe, weather, people, politics, finance, etc. As per [1], OSM will have roughly 3.78 billion users in 2021, reaching 4.41 billion by 2025. Because sharing digital content on social media is so easily accessible, users are often ignorant about the source's integrity and authenticity. Such lapses and practices may harm society. To prevent such activities, we need to look at digital content metadata.

---

M. Shaliyar (✉) · K. Mustafa
Department of Computer Science, Jamia Millia Islamia, New Delhi 110025, India
e-mail: mohdshaliyar@yahoo.com

K. Mustafa
e-mail: kmustafa@jmi.ac.in



B. Rushi Kumar et al. (eds.), *Mathematics and Computing*, Springer Proceedings in Mathematics & Statistics 415, https://doi.org/10.1007/978-981-19-9307-7_7

Metadata describes data about data. It describes the entity via its attribute values. A file stored on a computer system may have associated metadata (name, creation date) and system-specific metadata (accessed date). Both types of metadata are susceptible to modifications. So, every metadata must be preserved to protect the data's integrity and the source's authenticity. Moreover, unaltered metadata content serves as a backbone of digital forensic research. However, image analysis for forensic purposes comprises two phases: image authentication and source identification [2]. The first determines whether the image has gone through any process of modification [3]. The latter determines the source. Even so, it is based on the device's features and image-making techniques [4].

However, the Exif format used by each device maker varies. This information includes the image source, technical data, and GPS location. Metadata techniques are among the simplest ones. However, they depend on the manufacturer's rules to determine what metadata should be included in an image. Because metadata is so simple, it can be changed with existing tools. However, proof of no alteration in an image's metadata can be very beneficial in forensic investigations.

Moreover, images are the most shareable digital resource on OSN [5]. Pixel values compose images visually, but image metadata contains a vast quantity of information. Image metadata describes everything from the camera's maker and model to the image's GPS location. Metadata attributes reveal vital information about an image, such as copyright, GPS position, and date/time. However, this, too, has severe results.

By publishing a group of images on social media, an interloper can determine the time and location of a photographer by their GPS coordinates. In 2017, four apache choppers were attacked by insurgents in Iraq by having metadata coordinates of web-published images by an unaware soldier [6]. Besides this, numerous collections of GPS users' locations were recorded through metadata of online published videos that enable larcenists to further their aims [7].

## *1.1 State of the Art*

Metadata is a collection of data about an entity. With the ease of access to digital cameras and the Internet, users may instantly share images with geotagging. However, geotagging attaches GPS locations to photographs, jeopardizing privacy.

As per [8], the authors outlined the image creation life cycle, emphasizing the value of metadata. In recent years, much research has been conducted to protect image metadata. In [6], the authors try to secure metadata through symmetric key cryptography, but their methodology proved inadequate. They could not securely exchange keys and were susceptible to a man-in-the-middle attack.

The authors employed online tools to validate image authenticity using Exif metadata [9]. The automated insertion of the APP0 marker, absence of the IFD1 marker, and shortening of the Huffman code implied morphing. The authors of [10] carried out a deep analysis of 4000 images and 10 mobile phone manufacturers. They test images from Flickr and other mobile devices. Their study found 10 types of errors in Exif metadata. They conclude that many manufacturers do not strictly follow Exif specifications. However, they also tested five image metadata tools for robustness.

The authors [11] used metadata in provenance analysis to protect the authenticity and copyright of creative work such as paintings or memes. Provenance analysis gives a timeline and validity of uploaded, re-uploaded, or modified content. In [12], the authors have created an Android application to check the legitimacy of digitally embedded images using Error Level Analysis (ELA) and metadata features. They also provided a scale to assess a modification percentage. It defined significant modification as less than 40%, perceptible modification as 40–60%, suspicious modification as 60–80%, and no modification as 80–99%.

In [13], the authors developed a way to detect an individual's location using geotags in uploaded images or videos. They downloaded an image from social media, extracted their Exif metadata, created a table of GPS, and compared it with a GPS mapper. Finally, they used the EXIF tool to locate the geographic position. In [14], the authors utilized a digital certificate as an Exif metadata item to assure an image's authenticity and integrity.

Metadata can contain copyright information in addition to manufacturer and GPS location data. Facebook was prosecuted by a German photographer for deleting metadata and violating the German Copyright Act [2]. ***So metadata is a double-edged sword. It must be available publicly for authenticity and copyright, but it must also be unalterable to prevent misuse and abuse.***

According to the research mentioned above, the morphed image can be detected by the additional metadata attributes or measuring scales. However, "how to authenticate similar images with different metadata" is challenging.

The paper is divided into 6 sections. Section 1 consists of an introduction. Section 2 describes the types of image metadata. We have shown experimental validation in Sect. 3. The proposed solution is in Sect. 4. Results and discussion are carried out in Sect. 5, and in Sect. 6, we conclude the paper.

## 2 Image Metadata

As is seen from the above-mentioned work on image metadata, we can conclude that it is highly significant to secure metadata. Our work is to authenticate the source of information or misinformation in online social media. To accomplish this, we first discovered image-related metadata. Second, we identified metadata attributes that are susceptible to alteration. There exist three types of metadata: EXIF, IPTC, and XMP. An introductory detail on each is given as follows.

## 2.1 EXIF Metadata

EXIF (Exchangeable Image File Format) metadata falls in the class of Technical metadata. Technical metadata may contain a collection of technical details of digital objects such as object maker, copyright information, brand, model, date of image shot, sensitivity, focal distance, and GPS coordinate. In 1992, JPEG created the first format named JFIF, which allowed the exchange of JPEG bit streams between applications. Later, in 1998, JPEG introduced a new standard called EXIF. It enables camera makers or photographers to directly insert camera and image data in JPEG and TIFF files. This metadata may include detailed sources of authentication information, such as the time and location of the image click and the device utilized.

Moreover, the JPEG file format is divided into several markers. Marker FDD8 defines a Start-Of-Image (SOI), whereas FDD9 defines an End-Of-Image (EOI). In between these markers, data is divided into different segments. As per JPEG flexibility, one can add more markers and segments as metadata. The authors [14] conduct a deep study on the errors of Exif metadata in mobile devices.

## 2.2 IPTC Metadata

IPTC (International Press Telecommunications Council) comes in a descriptive metadata class. Descriptive metadata contains details about digital devices such as the author's name, email address, copyright, license, address, and contact number. Initially, IPTC was used as a standard for information interchange among news organizations. In 1994, Adobe Photoshop allowed users to update metadata for digital images via file information. Moreover, many photo and publishing organizations adopted the IPTC standard. Besides these, IPTC metadata provides rich information during a forensic investigation, utilizing attributes such as the author's name, address, copyright, and caption.

## 2.3 XMP Metadata

The administrative class of metadata includes XMP (Extensible Metadata Platform). Administrative metadata identifies administrative information of a file, such as when the file was created, determines access rights, title, author name, and intellectual property rights, and preserves metadata details. In 2001, Adobe developed XML-based XMP. Moreover, in 2005, Adobe developed an "IPTC core schema for XMP" by incorporating the old IPTC headers into the new XMP framework. The specific XMP fields are title, author, date of creation, and subject. It also contains metadata of embedded images.

# 3 Experimental Validation

In this section, we experiment to ensure the integrity of data and the authenticity of the source of information. We also highlight the image metadata attributes that can be modified using free Android apps or computer software. First, we identify attributes that have potential in forensic investigation. Along with this, we shall analyze their authenticity by various computer-based tools such as Metadata++, EXIF tool, File date changer, and Android-based app EXIF editor for GPS alteration. In Fig. 1, we have shown an original image downloaded from Flickr with its genuine metadata values, as shown in Figs. 2a, 3a, 4a, and 5a.

## *3.1 Date and Time*

This attribute has tremendous significance in forensic investigation. This signifies when the factor of investigation. These attributes give information about the date and time of access, creation, and modification of an image. Figure 2a and b show the metadata with its original and modified date and time values, whereas for a file system, the original access, create, and modify data is shown in Fig. 3a, and the modified version of access, create, and modify data is shown in Fig. 3b. As a result of this experiment, we concluded that the metadata of date and time is vulnerable to maintaining an image's integrity.

**Fig. 1** Original image

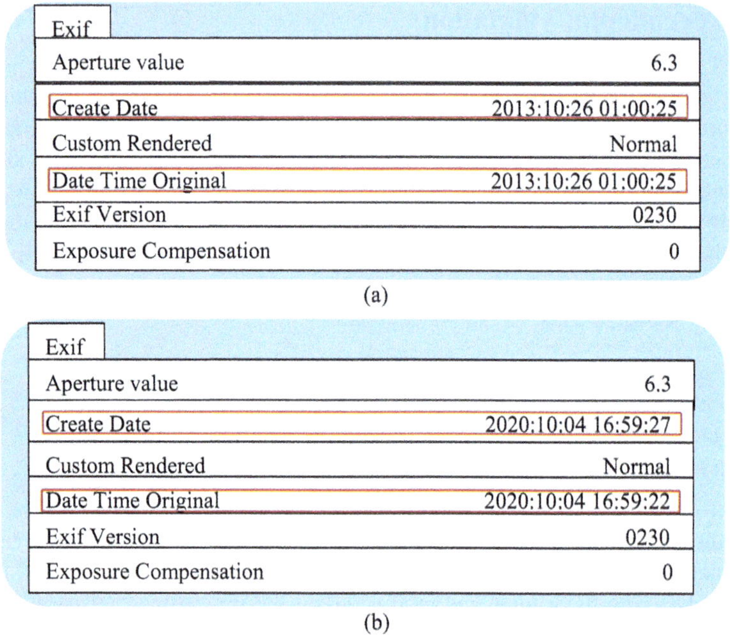


**Fig. 2** **a** Original metadata with original date and time value, **b** Modified metadata with altered date and time values

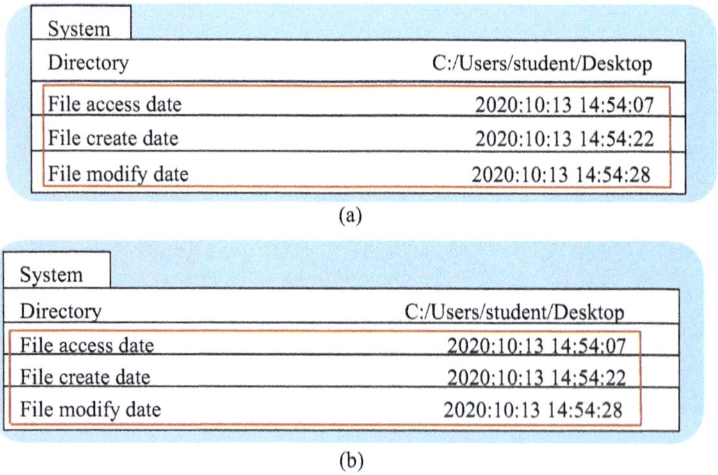


**Fig. 3** **a** Original values of access create and modify the date, **b** Modified values of access, create, and modify the date

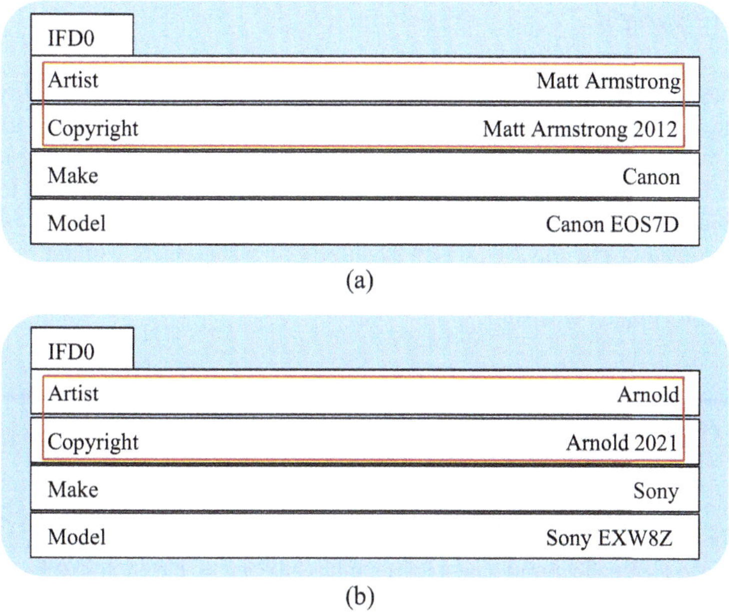


**Fig. 4** **a** Original values of name and copyright, **b** Modified values of name and copyright

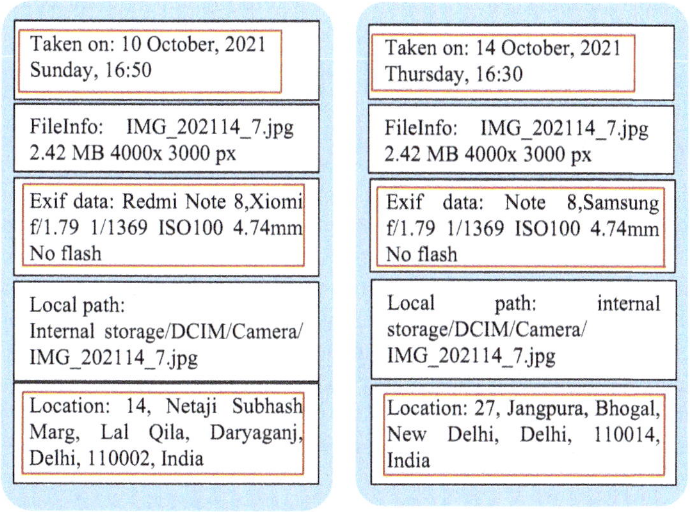


**Fig. 5** **a** Original GPS information of the image, **b** Modified GPS information of the image

### *3.2 Copyright*

This attribute gives details about the ownership or the creator of the entity. This signifies the ***who*** factor of investigation. This may reveal important information about the creator in a forensic investigation until and unless it is unalterable. As shown in Fig. 4a, the original copyright information has been modified to other details in Fig. 4b. This type of change may lead to severe consequences, such as publishing an image on OSN, which negatively impacts society with the copyright information of others. So, the copyright information should be preserved to have the authenticity of the ownership.

### *3.3 GPS Coordinate*

It is momentous to know "where the crime took place" in a forensic study. This contributes to ***where*** factors. When an image is clicked, some GPS coordinates get attached to the image through geotagging if it is enabled by the photographer. Before the alteration, the genuine GPS information of the original image of Red Fort (New Delhi, India) and modified values of metadata attributes are shown in Fig. 5a and b, respectively. We have also shown the same image with modified GPS details and other modified metadata values. To know the true information about the epicenter of the initial message, it is crucial to know the correct GPS location of the source.

### *3.4 Contact Details*

Contact details also play a significant role in forensic examination. The perpetrator can be traced out effortlessly as long as their accurate contact details are accessible. To trace out the malicious user on OSN, the contact details should be embedded in an image of the user who is going to share it initially; through this, it becomes an easy and effective way to trace out the malicious users on OSN.

In the above experimental work, it has been seen that all potential attributes that may lead to source authentication in OSN are susceptible to modification. Therefore, the authors proposed a solution to the aforementioned issues in light of *watermarking*.

## 4 Proposed Solution

The digital watermark is a technique for source identification, authentication, copyright protection, disclosure, e-medical services, and e-voting, among other things. The information for authentication or copyright in the form of text, audio, images, or videos is embedded in the host/carrier signal in digital watermarking. Watermarking was typically done in three stages: (a) generation, (b) insertion, and (c) extraction [15]. During all watermarking operations, however, the insertion and extraction of the watermark should not jeopardize the quality of the original host.

The watermarking technique is divided into spatial and transform domains based on the domain. In the spatial domain, the watermark bits are directly embedded into the coefficients of host signals. It is, however, simple to implement and has a lower level of complexity. Further, it has a high embedding capacity and imperceptibility. In the transform domain, the carrier signal is transformed into the frequency domain before the watermark is embedded. Due to this, the watermark bits are uniformly spread across the host signal. As a result, the watermarked becomes more resilient against various attacks while also being more complex than the spatial domain. As per [16], embedding capacity is relatively high in the spatial domain, but processing time, imperceptibility, and resilience are all relatively low. While the embedding capacity in the transform domain is limited, it is highly imperceptible and robust.

Further, the transform domain has a high level of computational complexity and processing time. In this review article [17], the authors described the research on digital forensic analysis of multimedia data transmitted via online social networks. The study is divided into three classes: source identification via forensic analysis, the credibility of uploaded multimedia, and identification of sharing platforms. They have also discussed various challenges and issues in spreading misinformation in OSM.

According to its robustness, watermarking is characterized as fragile, semi-fragile, and robust. Furthermore, it is also classified as visible or invisible depending on its visibility. The watermarking process is categorized as video, picture, text, and audio watermarking based on a carrier signal.

In this section, we demonstrate the feasibility of a solution to the above-stated issues by using the Discrete Wavelet Transform (**DWT**) watermarking technique to preserve the metadata of an image through a Matlab environment. For experimental purposes, the image has been downloaded from [18]. The proposed solution is divided into two phases: insertion and extraction of a watermark. Among the different metadata attributes, we used the contact/phone number information of a user as a watermarked in the cover image of Lena. Through this, we can resolve the issue of source authentication. Whenever a user is trying to upload an image on OSN, the model will check whether the image which is going to be uploaded has a contact number as a watermark on it. If the watermark is already present, then the model will not overwrite the current sender's contact number. If it is not found, then the model will insert the contact number as the watermark of the current sender. In this way, we can trace out the user who shared the information on OSN for the first time. Hence, the source will be authenticated. However, instead of a contact number, many other metadata attributes may exist that can be utilized as a watermark in the source authentication.

(a) (b)

**Fig. 6** **a** Original image, **b** watermark image

## *4.1 Watermark Insertion*

To embed a watermark, we extract a green channel of the original cover image of Lena Fig. 6a. We split the green channel into four sub-bands, LL1, HL1, LH1, and HH1, by using DWT with Haar wavelet. As the LL1 sub-band contains low-frequency components and carries the maximum energy in the image, we further split it into the LL2 sub-band. Hence, the LL2 sub-band is selected for the watermark embedding process. To insert a contact number as a watermark in the image, first, we convert each decimal digit of the contact number into an 8-bit binary number. As a result, we have a $10 \times 8$ matrix (DB) of binary numbers. In the matrix, we assign the value of $-1$ to every corresponding 0. Using Eq. 1, we embed a watermark into the new sub-band NLL2, where i and j represent corresponding pixel values. K is an embedding strength ranging from 0.005 to 0.5, as shown in Table 1:

$$\mathrm{NLL2}\,(\mathrm{i,j}) = \mathrm{LL2}(\mathrm{i,j}) * (1 + \mathrm{K} * \mathrm{DB}) \tag{1}$$

After embedding, the NLL2 is inserted in LL1 through inverse Discrete Wavelet Transform (IDWT). Further, LL1 is injected into the green channel of the cover image with IDWT. Hence, as shown in Fig. 6b, we have a watermarked image.

## *4.2 Watermark Extraction*

In the watermark extraction process, we extract the LL1 sub-band from the green channel of the watermarked image through DWT. From LL1, we derive the ELL2 sub-band, which contains the watermark. In Eq. 2, NLL2 is an embedded watermark, ELL2 is an extracted watermark, K is an embedding strength, and EBit is a $10 \times 8$ matrix consisting of 1 or $-1$ corresponding to each pixel value denoted by i and j. Further, we change the values of EBit from $-1$ to 0 to have a binary matrix. Finally, EBit is converted into their corresponding decimal numbers:

**Table 1** Embedding strength K, PSNR, and BER values

| K | PSNR | BER | SSIM |
|---|---|---|---|
| 0.005 | 76.0153 | 80 | 1 |
| 0.006 | 76.0153 | 80 | 1 |
| 0.007 | 76.0153 | 75 | 1 |
| 0.008 | 76.0153 | 0 | 1 |
| 0.009 | 76.0153 | 0 | 1 |
| 0.01 | 76.0153 | 0 | 1 |
| 0.02 | 66.4729 | 80 | 0.9999 |
| 0.03 | 63.9741 | 51 | 0.9999 |
| 0.04 | 61.7808 | 11 | 0.9999 |
| 0.05 | 59.4736 | 47 | 0.9999 |
| 0.06 | 57.86758 | 57 | 0.9999 |
| 0.07 | 56.7314 | 19 | 0.9999 |
| 0.08 | 55.4658 | 40 | 0.9999 |
| 0.09 | 54.3733 | 53 | 0.9999 |
| 0.1 | 53.5479 | 37 | 0.9999 |
| 0.2 | 47.5167 | 39 | 0.9998 |
| 0.3 | 43.9877 | 42 | 0.9997 |
| 0.4 | 41.4838 | 46 | 0.9994 |
| 0.5 | 39.5521 | 39 | 0.9992 |

$$\text{EBit}(i, j) = (\text{NLL2}(i, j) - \text{ELL2}(i, j))/(k * \text{ELL2}(i, j)) \quad (2)$$

In the process of watermarking, we used the LL2 sub-band of the Lena image to insert the contact number of a user as a watermark. From Table 1, it can be observed that with the increased value of embedding strength K, the imperceptibility (PSNR) starts to decrease. For the optimal solution, we choose an embedding strength of 0.01 for three similar values of PSNR, SSIM, and for minimum values of BER.

## 5 Result and Discussion

In the conducted experiment, it has been seen that the attributes that were selected for forensic investigation are found to be susceptible. The alteration of metadata is easily feasible, and hence source and copyright authenticity are at risk. Moreover, the integrity of data and the authenticity of the source are very much dependent on these attributes. Further, an attacker may advance their aim due to the free availability of tools to alter metadata attributes. So, it is a strong necessity to preserve the values of the attributes so that they can be used to maintain the integrity and trace out the source.

**Fig. 7** **a** Graph of Embedding strength (K) versus PSNR, **b** Graph of Embedding strength (K) versus BER

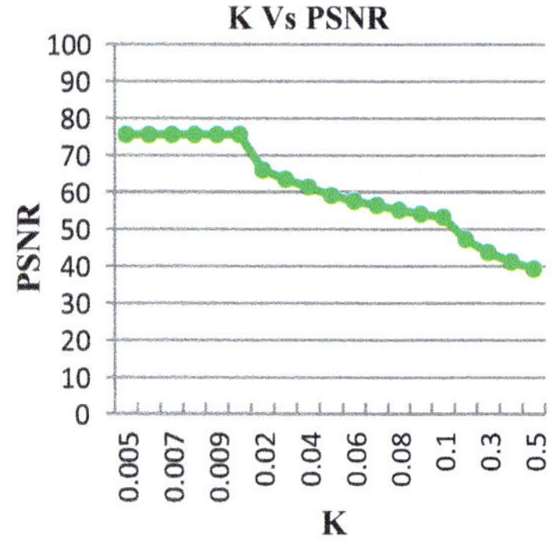


(a)

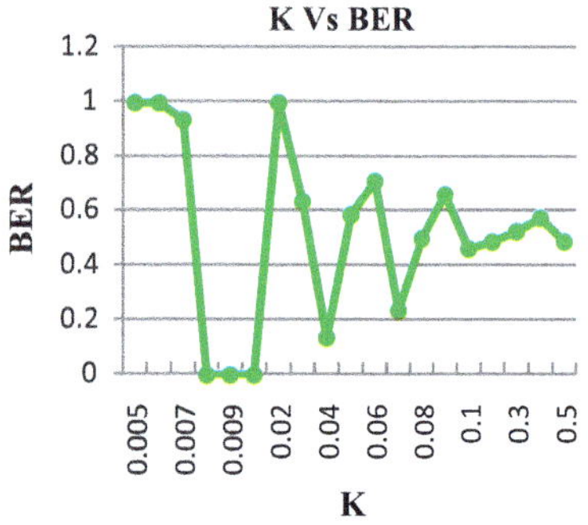


(b)

We proposed a solution based on the watermarking technique. This reveals that the source can be authenticated if the metadata values can be embedded as a watermark in an image uploaded on OSN. With the values of embedding strength = 0.01, PSNR = 76.0153, structural similarity index (SSIM) = 1, and BER = 0, we have an optimal solution as shown in Fig. 7a and b.

## *5.1 Limitation*

In this study, the proposed solution for source authentication through watermarking has certain limitations. The findings we have obtained are free from any kind of attack on the watermark image but limited to JPEG compression, cropping, rotating, filtering, etc.

## 6 Conclusion

In this paper, we have explored various aspects of image metadata and the pertinent attributes of metadata specific to forensic investigation. The attributes, including date and time, and copyright, to name a few, were studied for source authentication and data integrity. It is found that these are vulnerable and can be easily altered with freely available tools. These vulnerabilities can cause severe social, political, or financial threats to genuine users. Further, we have proposed a simple but effective watermarking solution in a Matlab environment for source authentication in online social media by embedding a contact number as one of the metadata attributes. With the embedding factor of 0.01, the results were encouraging in terms of PSNR, SSIM, and BER as 76.0153, 1, and 0, respectively. In the future, the study can be extended with various watermarking methods, attacks, and optimization techniques by using multiple watermarks, such as GPS coordinates and Aadhaar, to authenticate the source of information in online social media.

## References


1. Number of social media users 2025 | Statista. https://www.statista.com/statistics/278414/number-of-worldwide-social-network-users/. Accessed 16 Nov 2021
2. Gloe, T., Kirchner, M., Winkler, A., Böhme, R.: Can we trust digital image forensics? Proc. ACM Int. Multimed. Conf. Exhib. 78–86 (2007). https://doi.org/10.1145/1291233.1291252
3. Chamlawi, R., Khan, A.: Digital image authentication and recovery: Employing integer transform based information embedding and extraction. Inf. Sci. (Ny) **180**(24), 4909–4928 (2010). https://doi.org/10.1016/j.ins.2010.08.039
4. Thing, V.L.L., Ng, K.Y., Chang, E.C.: Live memory forensics of mobile phones. Digit. Investig. **7**, S74–S82 (2010). https://doi.org/10.1016/j.diin.2010.05.010
5. Kim, What Do People Love to Share on Social Media? https://kimgarst.com/what-do-people love-to-share-on-social-media/ Accessed 1 Oct 2020
6. Bhangale, R.: Securing Image Metadata using Advanced Encryption Standard (Doctoral dissertation, Dublin, National College of Ireland) (2020). http://norma.ncirl.ie/id/eprint/4149
7. Salama, U., Varadharajan, V., Hitchens, M.: Metadata based forensic analysis of digital information in the web. Annu. Symp. Inf. Assur. Secur. Knowl. Manag. 9–15 (2012)
8. Girona, C.J., Delgado, J.: Silvia Llorente Distributed Multimedia Applications Group ( DMAG )—Departament Arquitectura Computadors ( DAC ) Universitat Politècnica de Catalunya (UPC) BarcelonaTECH, pp. 1–3
9. Gangwar, D.P., Pathania, A.: Authentication of digital image using Exif metadata and decoding properties. Int. J. Sci. Res. Comput. Sci. Eng. Inf. Technol. January 2019, 335–341 (2018). https://doi.org/10.32628/cseit183815
10. Sandoval Orozco, A.L., Arenas González, D.M., García Villalba, L.J., Hernández-Castro, J.: Analysis of errors in Exif metadata on mobile devices. Multimed. Tools Appl. **74**(13), 4735–4763 (2015). https://doi.org/10.1007/s11042-013-1837-6
11. Bharati, A., et al.: Beyond pixels: Image provenance analysis leveraging metadata. In: Proceedings—2019 IEEE Winter Conference on Applications of Computer Vision, WACV 2019, pp. 1692–1702 (2019). https://doi.org/10.1109/WACV.2019.00185

12. Shichkina, Y.A., Tishchenko, V.T., Fatkieva, R.R.: Synthesis of the method of operative image analysis based on metadata and methods of searching for embedded images. In: 2020 9th Mediterranean Conference on Embedded Computing, MECO 2020, pp. 8–11 (2020). https://doi.org/10.1109/MECO49872.2020.9134145
13. Kumar, P.R., Srikanth, C., Sailaja, K.L.: Location Identification of the Individual based on Image Metadata. Procedia Comput. Sci. **85**(Cms), 451–454 (2016). https://doi.org/10.1016/j.procs.2016.05.191
14. Harran, M., Farrelly, W., Curran, K.: A method for verifying integrity & authenticating digital media. Appl. Comput. Inform. **14**(2), 145–158 (2018). https://doi.org/10.1016/j.aci.2017.05.006
15. Agarwal, N., Singh, A.K., Singh, P.K.: Survey of robust and imperceptible watermarking. Multimed. Tools Appl. **78**(7), 8603–8633 (2019). https://doi.org/10.1007/s11042-018-7128-5
16. Singh, P.: A survey of digital watermarking techniques, applications and attacks. Certif. Int. J. Eng. Innov. Technol. (IJEIT) **9001**(9) (2008)
17. Pasquini, C., Amerini, I., Boato, G.: Media forensics on social media platforms: a survey. Eurasip J. Inf. Secur. **2021**(1), 1–19 (2021). https://doi.org/10.1186/S13635-021-00117-2/TABLES/7
18. Image Databases. https://imageprocessingplace.com/root_files_V3/image_databases.htm. Accessed 16 Nov 2021

# A Computational Diffie–Hellman-Based Insider Secure Signcryption with Non-interactive Non-repudiation

**Ngarenon Togde and Augustin P. Sarr**

**Abstract** An important advantage of signcryption schemes compared to one pass key exchange protocols is non-interactive non-repudiation (NINR). This attribute offers to the receiver of a signcrypted ciphertext the ability to generate a non-repudiation evidence, that can be verified by a third party without executing a costly multi-round protocol. We propose a computational Diffie–Hellman based insider secure signcryption scheme with non-interactive non-repudiation. Namely, we show that under the computational Diffie–Hellman assumption and the random oracle model, our scheme is *tightly* insider secure, provided the underlying encryption scheme is semantically secure. Compared to a large majority of the previously proposed signcryption schemes with NINR, our construction is more efficient and it does not use any specificity of the underlying group, such as pairings. The communication overhead of our construction, compared to Chevallier Mâmes' signature scheme is one group element.

**Keywords** Signcryption · Non-interactive non-repudiation · Insider security · Computational Diffie–Hellman · Random oracle model

## 1 Introduction

A signcryption scheme provides simultaneously the functionalities of encryption and signature schemes [24]. A natural use of a signcryption scheme is to build an asynchronous secure channel i.e., a confidential and authenticated asynchronous channel. Given the similar uses of signcryption and (one pass) Key Exchange Protocols (KEP), to build confidential and authenticated channels, it appears, from a real world perspective, that the right security definition for signcryption schemes is insider security [3]. Informally, insider security ensures (i) *confidentiality* even if the

N. Togde · A. P. Sarr (✉)
Laboratoire ACCA, UFR SAT, Université Gaston Berger, Saint-Louis, Senegal
e-mail: augustin-pathe.sarr@ugb.edu.sn

N. Togde
e-mail: ngarenon.togde@ugb.edu.sn



B. Rushi Kumar et al. (eds.), *Mathematics and Computing*, Springer Proceedings in Mathematics & Statistics 415, https://doi.org/10.1007/978-981-19-9307-7_8

sender's static private key is revealed to the attacker, and (ii) *unforgeability* even if the receiver's static private key is disclosed.

A signcryption scheme is said to provide *non-repudiation*, if the receiver of a signcrypted ciphertext has the ability to generate a non-repudiation evidence, that can be verified by a third party (a judge, for instance); as a result, a message sender cannot deny having signcrypted the message. The non-repudiation attribute is said to be *non-interactive*, if a non-repudiation evidence can be *generated and verified without executing a multi-round protocol*. An important advantage of signcryption schemes, compared to one pass KEP, which often outperforms signcryption schemes, is non-interactive non-repudiation (NINR).

A signcryption scheme with the aim to provide NINR was proposed for the first time by Bao and Deng [5]; unfortunately their design fails in achieving confidentiality [19]. Malone–Lee [19] proposes an efficient design with NINR he analyzes in the Random Oracle (RO) model. The scheme achieves confidentiality under the computational Diffie–Hellman (cDH) assumption, and unforgeability under the gap Diffie–Hellman Assumption. Unfortunately, the security model he uses is closer to the outsider than to the insider model. Indeed, the scheme fails in providing insider confidentiality. In [8], Bjørstad and Dent (BD) propose a design based on Chevallier Mâmes' (CM) signature scheme they show to tightly achieve insider unforgeability under the cDH assumption and *outsider* confidentiality under the gap DH assumption. Unfortunately, as for the ML scheme, the BD scheme does not achieve insider confidentiality.

In subsequent works [2, 13, 14, 20, 23], several insider secure schemes with NINR have been proposed. The designs offer a superior security, compared to the ML or BD schemes. However, they are less efficient and often assume some specificities of the underlying groups, such as the existence of a bilinear pairing. In [2], Arriaga et al. propose a generic insider secure signcryption scheme, with randomness reuse, in the standard model. They exhibit an insider secure instantiation of their design, under the Decisional Bilinear and the $q$-Strong Diffie–Hellman (DBDH and $q$-sDH) assumptions. Unfortunately, the unforgeability is achieved in the registered key model [20], wherein an attacker is required to register the *keys pairs* it uses in its attack. Matsuda et al. [20] propose a generic composition of signature and tag-based encryption schemes, which yields to different shades of security depending on the security attributes of the base schemes. They exhibit two constructions with NINR that fully achieve insider confidentiality (under the cDH and the gap DH assumptions respectively) and unforgeability (under the co-cDH assumption). Chiba et al. [13] propose a generic construction of signcryption schemes, and exhibit two insider secure constructions with NINR under the DBDH and the $q$-sDH assumptions. In [14], Fan et al. propose a signcryption scheme with non-interactive non-repudiation (SCNINR), based on Boneh et al.'s signature scheme [10], they show to be insider secure under the DBDH assumption, without resorting the RO model. Sarr et al. [23] propose, over the group of signed quadratic residues, a SCNINR, based on a signature scheme of their own design, they show to be insider secure under the RSA assumption and the RO model.

The basic design principle in the SCNINR schemes from [8, 14, 19, 23], is (i) a Diffie–Hellman (DH) secret derivation, using ephemeral keys from the sender and the receiver's static public key, followed by (ii) an encryption using some part of the derived secret, and (iii) a signature generation, using the sender's private key, on the plain text and some part of the derived DH secret. One may notice also that these schemes assume rather specific groups or have loose security reductions. As tightly secure cDH-based signature schemes exist [12, 15, 17], we investigate whether such schemes can be leveraged as building blocks for tightly (multi-user) insider secure cDH based SCNINR schemes. As we aim at an efficient design, we use the random oracle (RO) model. We propose a new SCNINR, termed $\mathcal{SC}_{\mathsf{edl}}$, based on a variant of Chevallier–Mâmes' signature scheme [12], tailored to (i) be combined with Cash et al.'s twin Diffie–Hellman key exchange [11], (ii) and to allow a use of the same randomness in the DH key exchange and in the signature generation.

And, using the trapdoor test technique [11], we show that $\mathcal{SC}_{\mathsf{edl}}$ is tightly insider secure under the cDH assumption and the RO model, provided the underlying symmetric encryption scheme is semantically secure. Even better, we show the insider confidentiality attribute in the *secret key ignorant* multi-user model, i.e., when the sender public key is chosen by the adversary and the challenger does not know the corresponding private key. Compared to the ML and BD schemes, which do not require any specificity of the underlying group and do not achieve insider security, $\mathcal{SC}_{\mathsf{edl}}$ offers a stronger security, even if it is less efficient. And, compared to the schemes from [2, 13, 14, 20, 23], $\mathcal{SC}_{\mathsf{edl}}$ offers a tight security reduction, a better efficiency, and a comparable or a superior security.

This paper is organized as follows. In Sect. 2, we present some preliminaries on the syntax of SCNINR schemes and the insider security definitions for SCNINR. In Sect. 3, we propose the $\mathcal{SC}_{\mathsf{edl}}$ scheme. We propose our security results in Sect. 4, and compare our design with previous constructions in Sect. 5.

## 2 Preliminaries

*Notations.* $\mathcal{G} = \langle G \rangle$ is a cyclic group of prime order $p$, $\mathcal{G}^*$ denotes the set $\mathcal{G} \setminus \{1\}$. We denote by $\mathsf{Exp}(\mathcal{G}, t)$ the computational effort required to perform $t$ exponentiations with $|p|$-bits exponents in $\mathcal{G}$; $\mathsf{Exp}(\mathcal{G})$ denotes $\mathsf{Exp}(\mathcal{G}, 1)$. For an integer $n$, $[n]$ denotes the set $\{0, \ldots, n\}$. If $S$ is a set, $a \leftarrow_{\mathrm{R}} S$ means that $a$ is chosen uniformly at random from $S$; we write $a, b, c, \ldots \leftarrow_{\mathrm{R}} S$ as a shorthand for $a \leftarrow_{\mathrm{R}} S$; $b \leftarrow_{\mathrm{R}} S$, etc. We denote by $\mathsf{sz}(S)$ the number of bits required to represent $a \in S$. For a probabilistic algorithm $\mathcal{A}$ with parameters $u_1, \ldots, u_n$ and output $V \in \mathbf{V}$, we write $V \leftarrow_{\mathrm{R}} \mathcal{A}(u_1, \ldots, u_n)$. We denote by $\{\mathcal{A}(u_1, \ldots, u_n)\}$ the set $\{v \in \mathbf{V} : \Pr(V = v) \neq 0\}$. If $x_1, x_2, \ldots, x_k$ are objects belonging to different structures (group, bit-string, etc.) $(x_1, x_2, \ldots, x_k)$ denotes a representation as a bit-string of the tuple such that each element can be unequivocally parsed.

*The* cDH *Assumption.* We assume the existence of an algorithm $\mathsf{Setup}_{\mathsf{grp}}(\cdot)$, which on input a security parameter $k$ outputs a system parameter $\Pi_k$ which fully identifies a group $\mathcal{G} = \langle G \rangle$ together with its order. For $X \in \mathcal{G}$, we denote the smallest non-negative integer $x$ such that $G^x = X$ by $\log_G X$. For, $X, Y \in \mathcal{G}$, we denote $G^{(\log_G X)(\log_G Y)}$ by $\mathsf{cDH}(X, Y)$; if $B \in \mathcal{G}$, we denote $(\mathsf{cDH}(X, B), \mathsf{cDH}(Y, B))$ by $\mathsf{2DH}(X, Y, B)$. The $\mathsf{cDH}$ assumption is said to hold in $\mathcal{G}$ if for all efficient algorithms $\mathcal{A}$, $\mathsf{Adv}_{\mathcal{A}}^{\mathsf{cDH}}(\mathcal{G}) = \Pr[X, Y \leftarrow_{\mathrm{R}} \mathcal{G}; Z \leftarrow_{\mathrm{R}} \mathcal{A}(G, X, Y) : Z = \mathsf{cDH}(X, Y)]$ is negligible in $k$.

*A Symmetric Encryption* scheme $\mathcal{E} = (\mathsf{E}, \mathsf{D}, \mathbf{K}, \mathbf{M}, \mathbf{C})$ is a pair of efficient algorithms $(\mathsf{E}, \mathsf{D})$ together with a triple of sets $(\mathbf{K}, \mathbf{M}, \mathbf{C})$, which depend on the security parameter $k$, such that for all $\tau \in \mathbf{K}$ and all $m \in \mathbf{M}$, it holds that $\mathsf{E}(\tau, m) \in \mathbf{C}$ and $m = \mathsf{D}(\tau, \mathsf{E}(\tau, m))$. Let $\mathcal{A} = (\mathcal{A}_1, \mathcal{A}_2)$ be an adversary against $\mathcal{E}$ and let $\Pr(O_{i,i=0,1}) = \Pr\left[\begin{array}{l}(m_0, m_1, st) \leftarrow_{\mathrm{R}} \mathcal{A}_1(k); \tau \leftarrow_{\mathrm{R}} \mathbf{K}; c \leftarrow_{\mathrm{R}} \mathsf{E}(\tau, m_i); \\ \hat{b} \leftarrow_{\mathrm{R}} \mathcal{A}_2(k, c, st)\end{array} : \hat{b} = 1\right]$; then $\mathsf{Adv}_{\mathcal{A},\mathcal{E}}^{\mathsf{ss}}(k)$ denotes the quantity $\mathsf{Adv}_{\mathcal{A},\mathcal{E}}^{\mathsf{ss}}(k) = |\Pr(O_0) - \Pr(O_1)|$, where $m_0$, $m_1 \in \mathbf{M}$ are distinct equal length messages. The scheme $\mathcal{E}$ is said to be $(t, \varepsilon(k))$-*semantically secure* if for all adversaries $\mathcal{A}$ running in time $t$, $\mathsf{Adv}_{\mathcal{A},\mathcal{E}}^{\mathsf{ss}}(k) \leqslant \varepsilon(k)$.

## 2.1 Insider Security for SCNINR

We recall the syntax of a SCNINR scheme and the insider security definitions in the Flexible Signcryption / Flexible Unsigncryption Oracle (FSO/FUO) model [4], also termed dynamic Multi-user model [2].

**Definition 1** A *signcryption scheme* is a quintuple of algorithms $\mathcal{SC} = (\mathsf{Setup}, \mathsf{Gen}_S, \mathsf{Gen}_R, \mathsf{Sc}, \mathsf{Usc})$ where

(a) $\mathsf{Setup}$ takes a security parameter $k$ as input, and outputs a public domain parameter $dp$.
(b) $\mathsf{Gen}_S$ is the sender key pair generation algorithm. It takes as input $dp$ (an implicit parameter) and outputs a key pair $(sk_S, pk_S)$, wherein $sk_S$ is the signcrypting key.
(c) $\mathsf{Gen}_R$ is the receiver key pair generation algorithm; it takes $dp$ as input and outputs a key pair $(sk_R, pk_R)$.
(d) $\mathsf{Sc}$ takes as inputs $dp$, a sender private key $sk_S$, a receiver public key $pk_R$, and a message $m$, and outputs a signcryptext $C$. We write $C \leftarrow_{\mathrm{R}} \mathsf{Sc}(sk_S, pk_R, m)$.
(e) $\mathsf{Usc}$ is a deterministic algorithm. It takes as inputs $dp$, a receiver secret key $sk_R$, a sender public key $pk_S$, and a signcryptext $C$, and outputs either a valid message $m \in \mathbf{M}$ or an error symbol $\perp \notin \mathbf{M}$.

And, for all $dp \in \{\mathsf{Setup}(k)\}$, all $m \in \mathbf{M}$, all $(sk_S, pk_S) \in \{\mathsf{Gen}_S(dp)\}$, and all $(sk_R, pk_R) \in \{\mathsf{Gen}_R(dp)\}$, $m = \mathsf{Usc}(sk_R, pk_S, \mathsf{Sc}(sk_S, pk_R, m))$. The scheme is said to provide NINR if there are two algorithms $\mathsf{N}$ and $\mathsf{PV}$, termed *non-repudiation evidence generation* and *pubic verification algorithms* such that:

- N takes as inputs a receiver secret key $sk_R$, a sender public key $pk_S$, and a signcrypted ciphertext $C$, and outputs a *non-repudiation evidence* $nr$ or a failure symbol $\perp$; we write $nr \leftarrow \mathsf{N}(sk_R, pk_S, C)$.
- PV takes as inputs a signcryptext $C$, a message $m$, a non-repudiation evidence $nr$, a sender public key $pk_S$, and a receiver public key $pk_R$, and outputs $d \in \{0, 1\}$; we write $d \leftarrow \mathsf{PV}(C, m, nr, pk_S, pk_R)$.
- For all $dp \in \{\mathsf{Setup}(k)\}$, all $C \in \{0, 1\}^*$, all $(sk_S, pk_S) \in \{\mathsf{Gen}_S(dp)\}$, and all $(sk_R, pk_R) \in \{\mathsf{Gen}_R(dp)\}$, if $\perp \neq m \leftarrow \mathsf{Usc}(sk_R, pk_S, C)$ and $nr \leftarrow \mathsf{N}(sk_R, pk_S, C)$ then $1 = d \leftarrow \mathsf{PV}(C, m, nr, pk_S, pk_R)$.

---

**Game 1** SKI–MU Insider Confidentiality in the FSO/FUO–IND–CCA2 sense

We consider the experiments $E_0$ and $E_1$, described hereunder, wherein $\mathcal{A} = (\mathcal{A}_1, \mathcal{A}_2)$ is a two–stage adversary against a SCNINR scheme $\mathcal{SC}$;

(1) The challenger generates $dp \leftarrow_R \mathsf{Setup}(k)$ and $(sk_R, pk_R) \leftarrow_R \mathsf{Gen}_R(dp)$;
(2) $\mathcal{A}_1$ is provided with $dp$ and $pk_R$, and is given access to:
(a) an unsigncryption oracle $\mathcal{O}_{\mathsf{Usc}}(\cdot, \cdot)$, which takes as inputs a public key $pk$ and a signcrypted ciphertext $C$, and outputs $m \leftarrow \mathsf{Usc}(sk_R, pk, C)$, and (b) a non–repudiation evidence generation oracle $\mathcal{O}_{\mathsf{N}}(\cdot, \cdot)$ which takes as inputs a public key $pk$ and a signcrypted ciphertext $C$ and outputs $nr \leftarrow \mathsf{N}(sk_R, pk, C)$.
(3) $\mathcal{A}_1$ outputs $(m_0, m_1, pk_S, st) \leftarrow_R \mathcal{A}_1^{\mathcal{O}_{\mathsf{Usc}}(\cdot,\cdot), \mathcal{O}_{\mathsf{N}}(\cdot,\cdot)}(pk_R)$ where $m_0, m_1 \in \mathbf{M}$ are distinct equal length messages, $st$ is a state, and $pk_S$ is the attacked sender public key ($sk_S$ is unknown to the challenger).
(4) In the experiment $E_{b,b=0,1}$, the challenger computes $C^* \leftarrow_R \mathsf{Sc}(sk_S, pk_R, m_b)$.
(5) $\mathcal{A}_2$ outputs $b' \leftarrow_R \mathcal{A}_2^{\mathcal{O}_{\mathsf{Usc}}(\cdot,\cdot), \mathcal{O}_{\mathsf{N}}(\cdot,\cdot)}(C^*, st)$ ($\mathcal{O}_{\mathsf{Usc}}(\cdot, \cdot)$ and $\mathcal{O}_{\mathsf{N}}(\cdot, \cdot)$ are as in step 2).
(6) For $E_{b,b=0,1}$, $\mathsf{out}_b$ denotes the event: (i) $\mathcal{A}_2$ never issued $\mathcal{O}_{\mathsf{Usc}}(pk_S, C^*)$ or $\mathcal{O}_{\mathsf{N}}(pk_S, C^*)$, and (ii) $b' = 1$.

And, $\mathsf{Adv}^{\mathsf{cca2}}_{\mathcal{A},\mathcal{SC}}(k) = \mid \Pr(\mathsf{out}_0) - \Pr(\mathsf{out}_1) \mid$ denotes $\mathcal{A}$'s CCA2 insider security advantage.

---

**Definition 2** (*Secret Key Ignorant Multi-user Insider Confidentiality*) A SCNINR $\mathcal{SC}$ is said to be $(t, q_{\mathsf{Usc}}, q_{\mathsf{N}}, \varepsilon)$-secure in the Secret Key Ignorant Multi-user (SKI–MU) insider confidentiality in the FSO/FUO IND–CCA2 sense, if for all adversaries $\mathcal{A}$ playing Game 1, running in time $t$, and issuing respectively $q_{\mathsf{Usc}}$ and $q_{\mathsf{N}}$ queries to the unsigncryption and non-repudiation evidence generation oracles, $\mathsf{Adv}^{\mathsf{cca2}}_{\mathcal{A},\mathcal{SC}}(k) \leqslant \varepsilon$.

**Definition 3** (*Multi-user Strong Insider Unforgeability*) A SCNINR is said to be $(t, q_{\mathsf{Sc}}, \varepsilon)$ *Multi-user Insider Unforgeable in the FSO/FUO–sUF–CMA sense* if for all attackers $\mathcal{A}$ playing Game 2, running in time $t$, and issuing $q_{\mathsf{Sc}}$ queries to the signcryption oracle, $\mathsf{Adv}^{\mathsf{suf}}_{\mathcal{A},\mathcal{SC}}(k) \leqslant \varepsilon$.

Confidentiality and unforgeability are natural security goals for signcryption schemes. The soundness and unforgeability of non-repudiation evidence attributes are specific to SCNINR schemes.

**Game 2** MU Insider Unforgeability in the FSO/FUO–sUF–CMA sense

$\mathcal{A}$ is a forger, $dp \leftarrow_{\mathrm{R}} \mathsf{Setup}(k)$ still denotes the public domain parameter.

(1) The challenger computes $(sk_S, pk_S) \leftarrow_{\mathrm{R}} \mathsf{Gen}_S(dp)$.
(2) $\mathcal{A}$ runs with inputs $(dp, pk_S)$ and is given a FSO $\mathcal{O}_{\mathsf{Sc}}(\cdot,\cdot)$, which takes as inputs a valid public receiver key $pk$ and a message $m$ and outputs $C \leftarrow_{\mathrm{R}} \mathsf{Sc}(sk_S, pk, m)$.
(3) $\mathcal{A}$ outputs $((sk_R, pk_R), C^*) \leftarrow_{\mathrm{R}} \mathcal{A}^{\mathcal{O}_{\mathsf{Sc}}(\cdot,\cdot)}(dp, pk_S)$. It succeeds if:
(i) $\perp \neq m \leftarrow \mathsf{Usc}(sk_R, pk_S, C^*)$, and
(ii) it never received $C^*$ from $\mathcal{O}_{\mathsf{Sc}}(\cdot,\cdot)$ on a query on $(pk_R, m)$.

$\mathsf{Adv}^{\mathsf{suf}}_{\mathcal{A},\mathcal{SC}}(k) = \Pr(\mathsf{Succ}^{\mathsf{suf}}_{\mathcal{A}})$ denotes the probability that $\mathcal{A}$ wins the game.

**Game 3** Soundness of non–repudiation

(1) The challenger computes $dp \leftarrow_{\mathrm{R}} \mathsf{Setup}(k)$.
(2) $\mathcal{A}$ runs with input $dp$ and outputs $(C^*, pk_S, sk_R, pk_R, m', nr) \leftarrow_{\mathrm{R}} \mathcal{A}(dp)$.
(3) $\mathcal{A}$ wins the game if:
(i) $\perp \neq m \leftarrow \mathsf{Usc}(sk_R, pk_S, C^*)$, and
(ii) $m \neq m'$ and $1 = d \leftarrow \mathsf{PV}(C^*, m', nr, pk_S, pk_R)$.

$\mathsf{Adv}^{\mathsf{snr}}_{\mathcal{A},\mathcal{SC}}(k)$ denotes the probability that $\mathcal{A}$ wins the game.

**Definition 4** (*Soundness of non-repudiation*) A SCNINR is said to achieve $(t, \varepsilon)$-*computational soundness of non-repudiation* if for all attackers $\mathcal{A}$ playing Game 3 and running in time $t$, $\mathsf{Adv}^{\mathsf{snr}}_{\mathcal{A},\mathcal{SC}}(k) \leqslant \varepsilon$.

**Game 4** Unforgeability of non–repudiation evidence

$\mathcal{A}$ is an attacker against $\mathcal{SC}$, $dp \leftarrow_{\mathrm{R}} \mathsf{Setup}(k)$ is the domain parameter.

(1) The challenger computes $(sk_S, pk_S) \leftarrow_{\mathrm{R}} \mathsf{Gen}_S(dp)$; $(sk_R, pk_R) \leftarrow_{\mathrm{R}} \mathsf{Gen}_R(dp)$;
(2) $\mathcal{A}$ runs with inputs $(dp, pk_S, pk_R)$, and outputs $(C^*, m^*, nr^*) \leftarrow_{\mathrm{R}} \mathcal{A}^{\mathcal{O}_{\mathsf{Sc}}(\cdot,\cdot), \mathcal{O}_{\mathsf{Usc}}(\cdot,\cdot), \mathcal{O}_{\mathsf{N}}(\cdot,\cdot)}(dp, pk_S, pk_R)$.
(3) $\mathcal{A}$ wins if:
(i) $C^*$ was generated through the $\mathcal{O}_{\mathsf{Sc}}(\cdot,\cdot)$ oracle on inputs $(pk_R, m)$ for some $m$,
(ii) $1 = d \leftarrow \mathsf{PV}(C^*, m^*, nr^*, pk_S, pk_R)$, and
(iii) $nr^*$ was not generated by the oracle $\mathcal{O}_{\mathsf{N}}(\cdot,\cdot)$ on a query on $(pk_S, C^*)$.

$\mathsf{Adv}^{\mathsf{unr}}_{\mathcal{A},\mathcal{SC}}(k)$ denotes the probability that $\mathcal{A}$ wins the game.

**Definition 5** (*Unforgeability of non-repudiation evidence*) A SCNINR is said to achieve $(t, q_{\mathsf{Sc}}, q_{\mathsf{Usc}}, q_{\mathsf{N}}, \varepsilon)$ *unforgeability of non-repudiation evidence* if for all adversaries $\mathcal{A}$ playing Game 4, running in time $t$, and issuing respectively $q_{\mathsf{Sc}}$, $q_{\mathsf{Usc}}$, and $q_{\mathsf{N}}$ queries to the signcryption, unsigncryption, and non-repudiation evidence generation oracles, $\mathsf{Adv}^{\mathsf{unr}}_{\mathcal{A},\mathcal{SC}}(k) \leqslant \varepsilon$.

## 3 The New Construction

We consider the following variant of Chevallier–Mâmes' (CM) signature scheme [12]; $\mathsf{H}_1 : \{0,1\}^* \rightarrow \mathcal{G}$, $\mathsf{H}_2 : \{0,1\}^* \rightarrow \mathbf{K}$, and $\mathsf{H}_3 : \{0,1\}^* \rightarrow [p-1]$ are hash functions, $\mathsf{aux}$ denotes some auxiliary information.

**A Variant of Chevallier–Mâmes' signature scheme**

$\mathsf{Setup}_{\mathsf{Sign}}(k)$: the setup outputs a description of the group $\mathcal{G}$, a generator $G$ of $\mathcal{G}$, its prime order $p$, together with descriptions of the hash functions $\mathsf{H}_{i,i=1,2,3}$.
$\mathsf{Gen}(dp)$: $sk \leftarrow_{\mathrm{R}} [p-1]$; $pk \leftarrow G^{sk}$; **return** $(sk, pk)$;
$\mathsf{Sign}(sk, m)$: $x_1, x_2 \leftarrow_{\mathrm{R}} [p-1]$; $X_1 \leftarrow G^{x_1}$; $X_2 \leftarrow G^{x_2}$; $R \leftarrow \mathsf{H}_1(X_1, X_2)$; $V \leftarrow R^{x_1}$;
$W \leftarrow R^{sk}$; $h \leftarrow \mathsf{H}_3(m, X_1, X_2, G, R, V, W, pk, \mathsf{aux})$; $\sigma \leftarrow x_1 + h \cdot sk$; **return** $(X_2, W, \sigma, h)$;
$\mathsf{Vrfy}(pk, (X_2, W, \sigma, h), m)$: $X_1 \leftarrow G^{\sigma} pk^{-h}$; $R \leftarrow \mathsf{H}_1(X_1, X_2)$; $V \leftarrow R^{\sigma} W^{-h}$;
**if** $h = \mathsf{H}_3(m, X_1, X_2, G, R, V, W, pk, \mathsf{aux})$ **then return** 1; **else return** 0;

As for CM, in the RO model, the signature generation can be efficiently simulated, and the scheme can be shown to be unforgeable under $\mathsf{cDH}$ assumption. An interesting property of this scheme is that when it comes to extend it to a SCNINR, in a simulation of a signcrypted ciphertext generation, we can generate $X_1, X_2 \leftarrow_{\mathrm{R}} \mathcal{G}$ such that for all $(B, Z_1, Z_2) \in \mathcal{G}^3$, using the trapdoor test technique [11], we can efficiently decide whether $\mathsf{2DH}(X_1, X_2, B) = (Z_1, Z_2)$ or not. Then, if $(B_1, B_2) \in \mathcal{G}^2$ is a receiver public key, and a twin Diffie–Hellman key exchange [11] is performed using $(X_1, X_2)$ and $(B_1, B_2)$, we can use a trapdoor test at both the sender and the receiver. Then, as for the signature scheme's unforgeability, we can show the signcryption scheme to be tightly insider secure.

**The $\mathcal{SC}_{\mathsf{edl}}$ Scheme**

$\mathsf{Setup}(k)$: the algorithm defines a group $\mathcal{G} = \langle G \rangle$ of prime order $p$, together with an encryption scheme $\mathcal{E} = (\mathsf{E}, \mathsf{D}, \mathbf{K}, \mathbf{M}, \mathbf{C})$ and the hash functions $\mathsf{H}_1 : \{0,1\}^* \rightarrow \mathcal{G}$, $\mathsf{H}_2 : \{0,1\}^* \rightarrow \mathbf{K}$, and $\mathsf{H}_3 : \{0,1\}^* \rightarrow [p-1]$. The domain parameter is $dp = (\mathcal{G}, \mathcal{E}, \mathsf{H}_1, \mathsf{H}_2, \mathsf{H}_3)$. We assume $p \geqslant |\mathbf{K}|$.
$\mathsf{Gen}_S(dp)$: $a \leftarrow_{\mathrm{R}} [p-1]$; $(sk_S, pk_S) \leftarrow (a, G^a)$; **return** $(sk_S, pk_S)$;
$\mathsf{Gen}_R(dp)$: $b_1, b_2 \leftarrow_{\mathrm{R}} [p-1]$; $(sk_R, pk_R) \leftarrow \left((b_1, b_2), (G^{b_1}, G^{b_2})\right)$; **return** $(sk_R, pk_R)$;
$\mathsf{Sc}(sk_S, pk_R, m)$: Parse $pk_R$ as $(B_1, B_2)$; $x_1, x_2 \leftarrow_{\mathrm{R}} [p-1]$; $X_1 \leftarrow G^{x_1}$; $X_2 \leftarrow G^{x_2}$;
$R \leftarrow \mathsf{H}_1(X_1, X_2)$; $V \leftarrow R^{x_1}$; $W \leftarrow R^{sk_S}$;
$Z_1 \leftarrow B_1^{x_1}$; $Z_2 \leftarrow B_2^{x_1}$; $Z_3 \leftarrow B_1^{x_2}$; $Z_4 \leftarrow B_2^{x_2}$;
$\tau_1 \leftarrow \mathsf{H}_2(X_1, X_2, Z_1, Z_2, Z_3, Z_4, pk_S, pk_R)$; $\tau_2 \leftarrow \mathsf{H}_2(X_2, X_1, Z_3, Z_4, Z_1, Z_2, pk_S, pk_R)$;
$c \leftarrow \mathsf{E}(\tau_2, m)$; $h \leftarrow \mathsf{H}_3(m, \tau_1, c, X_1, X_2, G, R, V, W, pk_S, pk_R)$;
$\sigma \leftarrow x_1 + h \cdot sk_S \mod p$; **return** $(X_2, W, \sigma, h, c)$;
$\mathsf{Usc}(sk_R, pk_S, C)$: Parse $sk_R$ as $(b_1, b_2) \in [p-1]^2$;
Parse $C$ as $(X_2, W, \sigma, h, c) \in \mathcal{G}^2 \times [p-1]^2 \times \mathbf{C}$.
$X_1 \leftarrow G^{\sigma} pk_S^{-h}$; $Z_1 \leftarrow X_1^{b_1}$; $Z_2 \leftarrow X_1^{b_2}$; $Z_3 \leftarrow X_2^{b_1}$; $Z_4 \leftarrow X_2^{b_2}$;
$\tau_1 \leftarrow \mathsf{H}_2(X_1, X_2, Z_1, Z_2, Z_3, Z_4, pk_S, pk_R)$; $\tau_2 \leftarrow \mathsf{H}_2(X_2, X_1, Z_3, Z_4, Z_1, Z_2, pk_S, pk_R)$;
$m \leftarrow \mathsf{D}(\tau_2, c)$; $R \leftarrow \mathsf{H}_1(X_1, X_2)$; $V \leftarrow R^{\sigma} W^{-h}$;
**if** $h = \mathsf{H}_3(m, \tau_1, c, X_1, X_2, G, R, V, W, pk_S, pk_R)$ **then return** $m$; **else return** $\perp$;
$\mathsf{N}(sk_R, pk_S, C)$: Parse $sk_R$ as $(b_1, b_2)$; Parse $C$ as $(X_2, W, \sigma, h, c)$.

$X_1 \leftarrow G^{\sigma} pk_S^{-h}$; $Z_1 \leftarrow X_1^{b_1}$; $Z_2 \leftarrow X_1^{b_2}$; $Z_3 \leftarrow X_2^{b_1}$; $Z_4 \leftarrow X_2^{b_2}$;
$\tau_1 \leftarrow \mathsf{H}_2(X_1, X_2, Z_1, Z_2, Z_3, Z_4, pk_S, pk_R)$; $\tau_2 \leftarrow \mathsf{H}_2(X_2, X_1, Z_3, Z_4, Z_1, Z_2, pk_S, pk_R)$;
$m \leftarrow \mathsf{D}(\tau_2, c)$; $R \leftarrow \mathsf{H}_1(X_1, X_2)$; $V \leftarrow R^{\sigma} W^{-h}$;
**if** $h = \mathsf{H}_3(m, \tau_1, c, X_1, X_2, G, R, V, W, pk_S, pk_R)$ **then return** $(\tau_1, \tau_2)$; **else return** $\perp$;
$\mathsf{PV}(C, m, nr, pk_S, pk_R)$: Parse $C$ as $(X_2, W, \sigma, h, c)$ and $nr$ as $(\tau_1, \tau_2)$;
$m' \leftarrow \mathsf{D}(\tau_2, c)$;
**if** $m' \neq m$ **then return** 0;
$X_1 \leftarrow G^{\sigma} pk_S^{-h}$; $R \leftarrow \mathsf{H}_1(X_1, X_2)$; $V \leftarrow R^{\sigma} W^{-h}$;
**if** $h = \mathsf{H}_3(m, \tau_1, c, X_1, X_2, G, R, V, W, pk_S, pk_R)$ **then return** 1; **else return** 0;

For the consistency of $\mathcal{SC}_{\mathsf{edl}}$, one can observe that, as $\sigma = x_1 + h \cdot sk_S$, $G^{\sigma} pk_S^{-h}$ yields $X_1$; similarly $R^{\sigma} W^{-h}$ yields $V$. Then, if $C \leftarrow_{\mathrm{R}} \mathsf{Sc}(sk_S, pk_R, m)$ the same $Z_i$'s are computed in the signcryption and unsigncryption algorithms. And, the same values of $\tau_1$ and $\tau_2$ are derived both in $\mathsf{Sc}(sk_S, pk_R, m)$ and $\mathsf{Usc}(sk_R, pk_S, C)$. The remaining part in the definition of $\mathsf{Sc}$ (resp. $\mathsf{Usc}$) is essentially a proof (resp. verification) of equality of discrete logarithms ($\mathsf{edl}$) modified to include $m$, $\tau_1$ and $c$. Doing so, for all $dp \in \{\mathsf{Setup}(k)\}$, all $m \in \mathcal{M}$, all $(sk_S, pk_S) \in \{\mathsf{Gen}_S(dp)\}$, and all $(sk_R, pk_R) \in \{\mathsf{Gen}_R(dp)\}$, $m = \mathsf{Usc}(sk_R, pk_S, \mathsf{Sc}(sk_S, pk_R, m))$. Moreover, if $nr \leftarrow \mathsf{N}(sk_R, pk_S, \mathsf{Sc}(sk_S, pk_R, m))$ then $1 = d \leftarrow \mathsf{PV}(C, m, nr, pk_S, pk_R)$.

## 4 Security of the $\mathcal{SC}_{\mathsf{edl}}$ Scheme

We have the following results; detailed proofs are given in [21].

**Theorem 1** *We assume the RO model. If $q_X$, with $X \in \{\mathsf{H}_2, \mathsf{Usc}, \mathsf{N}\}$, is an upper bound on the number of times $\mathcal{A}$ issues the $\mathcal{O}_X$ oracle in Game 1, the* cDH *problem is $(t(k), \varepsilon_{\mathsf{cDH}}(k))$-hard in $\mathcal{G}$, and the encryption scheme $\mathcal{E}$ is $(t(k), \varepsilon_{\mathsf{ss}}(k))$-semantically secure, then $\mathcal{SC}_{\mathsf{edl}}$ is $(t(k), q_{\mathsf{Usc}}, q_{\mathsf{N}}, \varepsilon(k))$-secure in the SKI–MU insider confidentiality in the FSO/FUO–IND–CCA2 sense, where*

$$\varepsilon(k) \leqslant \varepsilon_{\mathsf{cDH}}(k) + \varepsilon_{\mathsf{ss}}(k) + 4(q_{\mathsf{H}_2} + 2q_{\mathsf{Usc}} + 2q_{\mathsf{N}} + 1)/p + 2q_{\mathsf{H}_3}/|\mathbf{K}|. \quad (1)$$

**Theorem 2** *Let $q_X$, where $X \in \{\mathsf{H}_1, \mathsf{H}_2, \mathsf{H}_3, \mathsf{Sc}\}$, be an upper bound on the number of times $\mathcal{A}$ issues the $\mathcal{O}_X$ oracle in Game 2. Under the RO model, if the* cDH *problem is $(t(k), \varepsilon_{\mathsf{cDH}}(k))$-hard in $\mathcal{G}$, then $\mathcal{SC}_{\mathsf{edl}}$ is $(t(k), q_{\mathsf{Sc}}(k), \varepsilon(k))$-MU insider unforgeable in the FSO/FUO–sUF–CMA sense, where $\varepsilon \leqslant \varepsilon_{\mathsf{cDH}} + ((q_{\mathsf{Sc}} + q_{\mathsf{H}_3})^2 + q_{\mathsf{Sc}}^2)/2p + (q_{\mathsf{H}_3} + 2q_{\mathsf{H}_2} + 1)/p$.*

**Theorem 3** *Under the RO model, the $\mathcal{SC}_{\mathsf{edl}}$ scheme achieves $(t, \varepsilon)$-computational soundness of non-repudiation, where $\varepsilon \leqslant q_{\mathsf{H}_3}^2/2p$ wherein $q_{\mathsf{H}_3}$, is an upper bound on the number of times $\mathcal{A}$ issues queries to the $\mathcal{O}_{\mathsf{H}_3}$ oracle.*

**Theorem 4** *Under the RO model, if the* cDH *problem is $(t(k), \varepsilon_{\mathsf{cDH}}(k))$ hard, then $\mathcal{SC}_{\mathsf{edl}}$ achieves $(t, q_{\mathsf{Sc}}, q_{\mathsf{Usc}}, q_{\mathsf{N}}, \varepsilon)$ unforgeability of non-repudiation evidence wherein $\varepsilon \leqslant \varepsilon_{\mathsf{cDH}} + 1/|\mathbf{K}| + 3/(2p)$.*

## 4.1 On the Concrete Choice of the Set of Domain Parameters

A concrete instance of a cryptographic problem is said to have $k$-bits of security if any adversary $\mathcal{A}$ running in time $t$ and trying to solve the problem succeeds with probability $\varepsilon \leqslant t/2^k$. A cryptographic scheme is said to have $k$-bits of security with respect to some security attribute, if any attacker playing the security game that defines the attribute and running in time $t$, succeeds with probability $\varepsilon \leqslant t/2^k$.

In $\mathcal{SC}_{\mathsf{edl}}$, if the underlying group $\mathcal{G}$ and the encryption scheme $\mathcal{E}$ are chosen such that the $\mathsf{cDH}$ problem in $\mathcal{G}$ has $(k+1)$-bits of security and $\mathcal{E}$ has $(k+3)$-bits of security then, from (1), it follows that $\mathcal{SC}_{\mathsf{edl}}$ is $(t, q_{\mathsf{Sc}}, q_{\mathsf{Usc}}, q_{\mathsf{N}}, \varepsilon)$-secure in the SKI–MU insider confidentiality in the FSO/FUO–IND–CCA2 sense, where $\varepsilon \leqslant t/2^{k+1} + t/2^{k+3} + 4(q_{\mathsf{H}_2} + 2q_{\mathsf{Usc}} + 2q_{\mathsf{N}} + 1)/p + 2t/|\mathbf{K}|$. As an $\mathcal{O}(\sqrt{p})$ algorithm is known for the discrete logarithm problem, $\alpha\sqrt{p} \geqslant 2^{k+1}$ for some "moderate" constant $\alpha$. As $q_{\mathsf{H}_2} + 2q_{\mathsf{Usc}} + 2q_{\mathsf{N}} + 1 \leqslant 2t$ and $|\mathbf{K}| \geqslant 2^{k+3}$, we obtain $\varepsilon \leqslant t/2^k$. Hence, $\mathcal{SC}_{\mathsf{edl}}$ has $k$-bits of security in the SKI–MU insider confidentiality in the FSO/FUO–IND–CCA2 sense. A similar analysis shows that under the same assumptions, $\mathcal{SC}_{\mathsf{edl}}$ has $k$-bits of security with regard to (i) (ii) the MU insider strong unforgeability in the FSO/FUO–sUF–CMA sense, (iii) the soundness of non-repudiation, and (vi) the unforgeability of non-repudiation evidence.

# 5 Comparison with Other Schemes

*The design of* $\mathcal{SC}_{\mathsf{edl}}$ integrates the randomness reuse idea suggested in [2, 20]. A $\mathcal{SC}_{\mathsf{edl}}$ sender (resp. receiver) key pair generation requires one (resp. two) exponentiations. An execution of the $\mathsf{Sc}$ algorithm requires $\mathsf{Exp}(\mathcal{G}, 8)$. Four of the 8 exponentiations can be performed *offline*, before the receiver public key and the plain text are provided. If the receiver public key is provided before the plain text (this may occur in email systems where the recipient is often typed before email text) *all* the 8 exponentiations can be performed before the plain text is provided The $\mathsf{Usc}$ and $\mathsf{N}$ algorithms require $\mathsf{Exp}(\mathcal{G}, 4)$ (two pairs of exponentiations with the same exponent) and two multi-exponentiations. The public verification algorithm requires two multi-exponentiations. If the encryption scheme $\mathcal{E}$ is such that a clear text and a corresponding ciphertext have the same length, the communication overhead of $\mathcal{SC}_{\mathsf{edl}}$, compared to the CM signature scheme is one group element. Notice that we neglected the group membership tests, as they have a negligible cost in $\mathbb{Z}_q^*$ and elliptic curve groups.

In [19], Malone–Lee (ML) proposes a very efficient design with NINR. Unfortunately, the design, which is analyzed in the RO model under de $\mathsf{cDH}$ assumption, does not achieve insider security. Also the reduction uses the Forking Lemma [6, 22]. Assuming $q_{\mathsf{H}} = 2^{32}$, for a security target of 128-bits, the underlying group $\mathcal{G}'$ must be chosen to offer 160-bits of security. In the case $\mathcal{G}'$ is a (sub)group of the rational points of an elliptic curve $\mathcal{G}' = E(\mathbb{F}_{q'})$, $q'$ has to be chosen such that

$|q'| \approx 320$. An execution of the ML $\mathsf{Sc}$ or $\mathsf{Usc}$ algorithm requires two exponentiations. As a modular multiplication (performed with the Karatsuba–Ofman algorithm) in $\mathbb{F}_{q'}$ has complexity $\approx |q'|^{1.585}$. Given the tightness of our reduction, in ECC, we need $|q| = 256$ to have 128 bits of security. As $\mathsf{Mult}(\mathbb{F}_{q'}) \approx 1.42 \cdot \mathsf{Mult}(\mathbb{F}_q)$, assuming that a group operation in $\mathcal{G}'$ requires $14 \cdot \mathsf{Mult}(\mathbb{F}_{q'})$ (see[1] [16, p. 96]), $\mathsf{Exp}(\mathcal{G}') \approx 6720 \cdot \mathsf{Mult}(\mathbb{F}_{q'}) \approx 9570 \cdot \mathsf{Mult}(\mathbb{F}_q) \approx 1.78 \cdot \mathsf{Exp}(\mathcal{G})$. The ML design is about (a) 2.25 times faster for signcryption, and (b) 1.25 times faster for unsigncryption than ours.

Bjørstad and Dent's (BD) design [8] tightly achieves, in the RO model, insider unforgeability under the $\mathsf{cDH}$ assumption and *outsider* confidentiality under the gap DH assumption. The scheme does not achieve insider confidentiality. The $\mathsf{Sc}$ algorithm requires $\mathsf{Exp}(\mathcal{G}, 3)$ operations, the $\mathsf{Usc}$ algorithm requires two multi-exponentiations. The BD construction is about 2.5 times faster than $\mathcal{SC}_{\mathsf{edl}}$ for signcrypted ciphertext generation and about 3 times faster for unsigncryption.

Some of the designs we consider hereunder assume the existence of groups $\mathbb{G}_1, \mathbb{G}_2, \mathbb{G}_T$ together with a bilinear pairing $e : \mathbb{G}_1 \times \mathbb{G}_2 \to \mathbb{G}_T$. Recall that for a choice of the groups $\mathcal{G}$, $\mathbb{G}_1$, $\mathbb{G}_2$, and $\mathbb{G}_T$ (where $\mathcal{G}$ is a classical ECC group), with a target of 128-bits of security, the cost of a pairing evaluation is about $\approx \mathsf{Exp}(\mathcal{G}, 8)$, $\mathsf{Exp}(\mathbb{G}_1) \approx \mathsf{Exp}(\mathcal{G}, 3)$, and $\mathsf{Exp}(\mathbb{G}_2) \approx \mathsf{Exp}(\mathcal{G}, 6)$ [7, p. 126].

Arriaga et al.'s generic construction with NINR [2] is insider secure in the standard model. They propose an instantiation of their design which assumes the Decisional Bilinear and the $q$-Strong DH assumptions. Unfortunately, the unforgeability is achieved in the registered key model [20], wherein an attacker needs to register to the challenger the keys pairs it uses in its attack. The design assumes the existence of groups $\mathbb{G}, \mathbb{G}_1, \mathbb{G}_2, \mathbb{G}_T$ such that (i) $\mathbb{G}_1, \mathbb{G}_2, \mathbb{G}_T$ are of order $q$, (ii) there is a bilinear pairing $e : \mathbb{G}_1 \times \mathbb{G}_2 \to \mathbb{G}_T$ and (iii) a one to one and efficiently invertible mapping from $\mathbb{G}$ to $\mathbb{Z}_q$.

An evaluation of the $\mathsf{Sc}$ algorithm requires $\mathsf{Exp}(\mathbb{G}, 2) + \mathsf{Exp}(\mathbb{G}_1)$ and one multi-exponentiation in $\mathbb{G}$. The $\mathsf{Usc}$ algorithm requires two multi-exponentiations, one in $\mathbb{G}$ and one in $\mathbb{G}_2$, and a pairing evaluation. For a target of 128 bits of security, we expect $\mathcal{SC}_{\mathsf{edl}}$ to be 1.5 times faster for signcryption and 2.8 times faster for unsigncryption.

Matsuda et al. [20]'s two generic constructions with NINR are insider secure in the FSO/FUO model. The security reduction is provided in the RO model. The most efficient among the instanciations that achieve insider security in the FSO/FUO model uses as base schemes, the DHIES encryption scheme [1] and the BLS signature scheme [9]. The construction assumes the existence of groups $\mathbb{G}_1, \mathbb{G}_2, \mathbb{G}_T$ together with a bilinear pairing $e : \mathbb{G}_1 \times \mathbb{G}_2 \to \mathbb{G}_T$. A $\mathsf{Sc}$ operation requires $\mathsf{Exp}(\mathbb{G}_1, 3)$, an $\mathsf{Usc}$ operation requires $\mathsf{Exp}(\mathbb{G}_2)$ and two pairing evaluations. Compared to $\mathcal{SC}_{\mathsf{edl}}$, for a target of 128 bits of security (given that $\mathsf{Exp}(\mathbb{G}_1) \approx \mathsf{Exp}(\mathcal{G}, 3)$, $\mathsf{Exp}(\mathbb{G}_2) \approx$

[1] If $|\mathcal{G}| = 2^\lambda$, the cost of $\mathsf{Exp}(\mathcal{G})$ using the classical square-and-multiply algorithm is $\approx 1.5 \cdot \lambda$ operations in $\mathcal{G}$. And if $\mathcal{G}$ is such that the multiplication of two of its elements requires 14 multiplications in $\mathbb{F}_q$ then the computational cost of an exponentiation is $14 \cdot 1.5 \cdot \lambda$ multiplications in $\mathbb{F}_q$.

$\mathsf{Exp}(\mathcal{G}, 6)$ and the cost of a pairing evaluation $\approx \mathsf{Exp}(\mathcal{G}, 8)$) we expect our design to be 1.12 times faster for signcryption, and about 3.6 times faster for unsigncryption.

For a comparison with Chiba et al.'s generic construction with NINR [13], we consider the most efficient among the instantiations they propose. It achieves insider security in the FSO/FUO model, under the Decisional Bilinear and the $q$-strong DH assumptions. Although the insider security is shown in the standard model, the unforgeability is achieved in the registered key model. Besides, the scheme assumes the existence of a pairing $e : \mathbb{G}_1 \times \mathbb{G}_2 \rightarrow \mathbb{G}_T$, with $\mathbb{G}_1 = \mathbb{G}_2$. The $\mathsf{Sc}$ algorithm requires $\mathsf{Exp}(\mathbb{G}_1, 3)$ together with a multi-exponentiation. The $\mathsf{Usc}$ operation requires one exponentiation, one multi-exponentiation, and one pairing evaluation. We expect $\mathcal{SC}_{\mathsf{edl}}$ to be about 1.5 times faster for signcryption, and about 2.3 times faster for unsigncryption.

Fan et al.'s design [14] assumes the existence of a bilinear map $e : \mathbb{G} \times \mathbb{G} \rightarrow \mathbb{G}_T$, where $\mathbb{G}$ and $\mathbb{G}_T$ are multiplicative cyclic groups. The $\mathsf{Sc}$ algorithm requires one pairing, $\mathsf{Exp}(\mathbb{G}, 4) + \mathsf{Exp}(\mathbb{G}_T)$, and $(n + 1)/2$ group operations in $\mathbb{G}$, where $n$ is the bit-length output of some collision resistant hash function $\mathsf{H} : \mathbb{G} \rightarrow \{0, 1\}^n$ used in the design. The unsigncryption algorithm requires 3 pairings, $\mathsf{Exp}(\mathbb{G}, 2)$, and $(n/2 + 1)$ group operations in $\mathbb{G}$. A signcrypted ciphertext is an element of $\mathbb{G}_T \times \mathbb{G}^3$. For a choice of the groups $\mathcal{G}$, $\mathbb{G}$, and $\mathbb{G}_T$, with target 128-bits of security, we expect our design to be about (a) (b) 2.5 times faster for signcryption, and (c) 7.5 times faster for unsigncryption than Fan et al.'s construction, in addition to having shorter signcrypted ciphertexts.

In the scheme from [23], defined over the (RSA based) group of signed quadratic residues $\mathbb{J}_N^+$, the $\mathsf{Sc}$ algorithm requires $\mathsf{Exp}(\mathbb{J}_N^+, 6)$ and the $\mathsf{Usc}$ algorithm requires $\mathsf{Exp}(\mathbb{Z}_N, 3)$ (we ignore the exponentiation with the RSA public exponent, which is often small and sparse). Unfortunately, the security reduction uses the Forking Lemma, which implies a $1/q_{\mathsf{H}}$ security degradation, where $q_{\mathsf{H}}$ is the number of digest queries the attacker issues. For $q_{\mathsf{H}} = 2^{32}$, if the target security is 128-bits, the RSA modulus needs to have a bitlength $|N| \approx 7864$ [18][2]. Then, considering a square-and-multiply based exponentiation, $\mathsf{Exp}(\mathbb{J}_N^+) \approx 11796 \cdot \mathsf{Mult}(\mathbb{Z}_N)$, where $\mathsf{Mult}(\mathbb{Z}_N)$ denotes the cost of a multiplication in $\mathbb{Z}_N$. In contrast $\mathcal{SC}_{\mathsf{edl}}$ can be instantiated over an elliptic curve (sub)group $\mathcal{G} = E(\mathbb{F}_q)$ such that $|q| \approx 256$ and $\mathcal{G}$ has 128-bits of security. Assuming that a group operation in $\mathcal{G}$ requires $14 \cdot \mathsf{Mult}(\mathbb{F}_q)$ [16, p. 96], $\mathsf{Exp}(\mathcal{G}) \approx 5376 \cdot \mathsf{Mult}(\mathbb{F}_q)$. As $\mathsf{Mult}(\mathbb{Z}_N) > 30 \cdot \mathsf{Mult}(\mathbb{F}_q)$, for a 128-bits security target, we expect $\mathcal{SC}_{\mathsf{edl}}$ over $\mathcal{G}$ to be at least 13 times faster (for key generation, signcryption, unsigncryption, etc.) than the design from [23].

Compared to the ML and BD schemes, which do not require any specificity of the underlying group and do not achieve insider security, $\mathcal{SC}_{\mathsf{edl}}$ offers a stronger security, even if it is less efficient. And, compared to the schemes from [2, 13, 14, 20, 23], $\mathcal{SC}_{\mathsf{edl}}$ offers a tight security reduction, a better efficiency and a comparable or a superior security. We summarize in Table 1 some elements of comparisons. The column Assumptions indicates the computational assumptions used in the security reductions; FL and IS stand respectively for Forking Lemma and

[2] see also www.keylength.com

**Table 1** Comparison of the proposed signcryption schemes with some SCNINR schemes from the litterature

| Scheme | Assumptions | FL | IS | Computations | Overhead |
|---|---|---|---|---|---|
| ML [19] | RO, cDH | y | n | [2, 0, 0] [2, 0, 0] | $2 \cdot \mathsf{sz}(\mathbb{Z}_p)$ |
| BD [8] | RO, cDH | n | p | [2, 0, 0] [0, 2, 0] | $\mathsf{sz}(\mathcal{G}) + \mathsf{sz}(\mathbb{Z}_p)$ |
| ABF [2] | DBDH, $q$-sDH | n | y | [3, 1, 0] [0, 2, 1] | $\mathsf{sz}(\mathbb{G}) + \mathsf{sz}(\mathbb{G}_1)$ |
| MMS [20] | RO, GDH, co–cDH | n | y | [3, 0, 0] [1, 0, 2] | $\mathsf{sz}(\mathbb{G}_1) + \mathsf{sz}(\mathbb{G}_2)$ |
| CMSM [13] | DBDH, $q$-sDH | n | y | [3, 1, 0] [1, 1, 2] | $\mathsf{sz}(\mathbb{Z}_p) + 4 \cdot \mathsf{sz}(\mathbb{G}_1)$ |
| FZT [14] | DBDH, DL | n | y | [5, 0, 1] [2, 0, 3] | $\mathsf{sz}(\mathbb{Z}_p) + 2 \cdot \mathsf{sz}(\mathbb{G}_1)$ |
| SSN [23] | RO, RSA | y | y | [6, 0, 0] [3, 0, 0] | $\mathsf{sz}(\mathbb{Z}_p) + 2 \cdot \mathsf{sz}(\mathbb{Z}_N)$ |
| Ours: $\mathcal{SC}_{\mathsf{edl}}$ | RO, cDH | n | y | [8, 0, 0] [4, 2, 0] | $2 \cdot \mathsf{sz}(\mathbb{Z}_p) + 2 \cdot \mathsf{sz}(\mathcal{G})$ |

Insider Security (in the FSO/FUO model). The letters 'y' and 'n' stand for "yes" and "no", respectively; 'p' stands for "partial" (BD achieves insider unforgeability, but *outsider* confidentiality). In the column Computations $[a, b, c][a', b', c']$ means that a Sc (resp. Usc) operation requires $a$ (resp. $a'$) exponentiations, $b$ (resp. $b'$) multi-exponentiations, and $c$ (resp. $c'$) pairing evaluations. We recall that the number of exponentiations has to be considered in conjunction with the underlying mathematical structure. For instance, as previously said, if a scheme requires a bilinear pairing $e : \mathbb{G}_1 \times \mathbb{G}_2 \rightarrow \mathbb{G}_T$, for a target of 128 bits of security, it holds $\mathsf{Exp}(\mathbb{G}_1) \approx \mathsf{Exp}(\mathcal{G}, 3)$ and $\mathsf{Exp}(\mathbb{G}_2) \approx \mathsf{Exp}(\mathcal{G}, 6)$. The column Overhead indicates the signcrypted ciphertext overhead compared to the *cleartext*.

# References


1. Abdalla, M., Bellare, M., Rogaway, P.: The oracle Diffie-Hellman assumptions and an analysis of DHIES. In: Naccache, D. (ed.) CT-RSA 2001. LNCS, vol. 2020, pp. 143–158. Springer, Heidelberg (2001)
2. Arriaga, A., Barbosa, M., Farshim, P.: On the joint security of signature and encryption schemes under randomness reuse: efficiency and security amplification. In: Bao, F., Samarati, P., Zhou, J. (eds.), Applied Cryptography and Network Security. ACNS 2012. LNCS, vol 7341. Springer, Berlin, Heidelberg (2012)
3. Badertscher, C., Banfi, F., Maurer, U.: A constructive perspective on signcryption security. In: Catalano, D., De Prisco, R. (eds.), Security and Cryptography for Networks. SCN 2018. LNCS, vol. 11035. Springer, Cham (2018)
4. Baek, J., Steinfeld, R., Zheng, Y.: Formal proofs for the security of signcryption. J. Cryptol. **20**(2), 203–235 (2007)
5. Bao, F., Deng, R.H.: A signcryption scheme with signature directly verifiable by public key. In: Imai, H., Zheng, Y. (eds.), Public Key Cryptography. PKC 1998. LNCS, vol. 1431. Springer, Berlin, Heidelberg (1998)
6. Bellare, M., Neven, G.: Multi–signatures in the plain public–key model and a general forking lemma. In: Proceedings of the 13th ACM Conference on Computer and Communications Security, pp. 390–399. ACM (2006)

7. Benhamouda, F., Couteau, G., Pointcheval, D., Wee, H.: Implicit zero-knowledge arguments and applications to the malicious setting. In: Gennaro, R., Robshaw, M. (eds.), Advances in Cryptology—CRYPTO 2015. CRYPTO 2015. LNCS, vol. 9216. Springer (2015)
8. Bjørstad, T.E., Dent, A.W.: Building better signcryption schemes with Tag-KEMs. In: Yung, M., Dodis, Y., Kiayias, A., Malkin, T. (eds.), Public Key Cryptography—PKC 2006. PKC 2006. LNCS, vol. 3958. Springer, Berlin, Heidelberg (2006)
9. Boneh, D., Lynn, B., Shacham, H.: Short signatures from the Weil pairing. J. Cryptol. **17**(4), 297–319 (2004)
10. Boneh, D., Shen, E., Waters, B.: Strongly unforgeable signatures based on computational Diffie–Hellman. In: Yung, M., Dodis, Y., Kiayias, A., Malkin, T. (eds.), Public Key Cryptography—PKC 2006. PKC 2006. LNCS, vol. 3958. Springer, Berlin, Heidelberg (2006)
11. Cash, D., Kiltz, E., Shoup, V.: The twin Diffie-Hellman problem and applications. J. Cryptol. **22**(4), 470–504 (2009)
12. Chevallier–Mames, B.: An efficient CDH–Based signature scheme with a tight security reduction. In: Shoup, V. (eds.), Advances in Cryptology—CRYPTO 2005. CRYPTO 2005. LNCS, vol. 3621. Springer, Berlin, Heidelberg (2005)
13. Chiba, D., Matsuda, T., Schuldt, J.C.N., Matsuura, K.: Efficient generic constructions of signcryption with insider security in the multi-user setting. In: Lopez, J., Tsudik, G. (eds.), Applied Cryptography and Network Security. ACNS 2011. LNCS, vol. 6715. Springer, Berlin, Heidelberg (2011)
14. Fan, J., Zheng, Y., Tang, X.: Signcryption with non–interactive non–repudiation without random oracles. In: Transactions on Computational Science X, pp. 202–230. Springer, Berlin, Heidelberg (2010)
15. Goh, E.J., Jarecki, S.: A signature scheme as secure as the Diffie–Hellman problem. In: Biham, E. (eds.), Advances in Cryptology—EUROCRYPT' 03. EUROCRYPT 2003. LNCS, vol. 2656. Springer, Berlin, Heidelberg (2003)
16. Hankerson, D., Menezes, A.J., Vanstone, S.: Guide to Elliptic Curve Cryptography. Springer (2004)
17. Katz, J., Wang, N.: Efficiency improvements for signature schemes with tight security reductions. In: Proceedings of the 10th ACM Conference on Computer and Communications Security, pp. 155–164. ACM (2003)
18. Lenstra, A.K.: Key lengths. Handbook of Information Security, vol. 2, pp. 617–635. Wiley (2005)
19. Malone–Lee, J.: Signcryption with non–interactive non–repudiation. Designs, Codes and Cryptography, vol. 37, no. 1, pp. 81–109. Springer (2005)
20. Matsuda, T., Matsuura, K., Schuldt, J.C.N.: Efficient constructions of signcryption schemes and signcryption composability. In: Roy, B., Sendrier, N. (eds.), Progress in Cryptology—INDOCRYPT 2009. INDOCRYPT 2009. LNCS, vol. 5922. Springer, Berlin, Heidelberg (2009)
21. Ngarenon, T., Sarr, A.P.: A Computational Diffie–Hellman based Insider Secure Signcryption with Non Interactive Non Repudiation (full version) (2022). https://hal.archives-ouvertes.fr/hal-03628351/
22. Pointcheval, D., Stern, J.: Security proofs for signature schemes. In: Maurer, U. (eds.), Advances in Cryptology—EUROCRYPT'96. EUROCRYPT 1996. LNCS, vol. 1070. Springer, Berlin, Heidelberg (1996)
23. Sarr A.P., Seye P.B., Ngarenon T.: A Practical and Insider Secure Signcryption with Non-interactive Non-repudiation. In: Carlet C., Guilley S., Nitaj A., Souidi E. (eds.), Codes, Cryptology and Information Security. C2SI 2019. LNCS, vol. 11445. Springer, Cham (2019)
24. Zheng, Y.: Digital signcryption or how to achieve cost(signature & encryption) $\ll$ cost(signature) + cost(encryption). In: Kaliski, B.S. (eds.), Advances in Cryptology—CRYPTO '97. CRYPTO 1997. LNCS, vol. 1294. Springer, Berlin, Heidelberg (1997)

# *k*NN-SVM with Deep Features for COVID-19 Pneumonia Detection from Chest X-ray

**Aman Bahuguna, Deepak Yadav, Apurbalal Senapati, and Baidya Nath Saha**

**Abstract** Most attention has been paid to chest Computed Tomography (CT) in this burgeoning crisis because many cases of COVID-19 demonstrate respiratory illness clinically resembling viral pneumonia which persists in prominent visual signatures on high-resolution CT befitting of viruses that damage lungs. However, CT is very expensive, time-consuming, and inaccessible in remote hospitals. As an important complement, this research proposes a novel *k*NN-regularized Support Vector Machine (*k*NN-SVM) algorithm for identifying COVID-induced pneumonia from inexpensive and simple frontal chest X-ray (CXR). To compute the deep features, we used transfer learning on the standard VGG16 model. Then the autoencoder algorithm is used for dimensionality reduction. Finally, a novel *k*NN-regularized Support Vector Machine algorithm is developed and implemented which can successfully classify the three classes: Normal, Pneumonia, and COVID-19 on a benchmark chest X-ray dataset. *k*NN-SVM combines the properties of two well-known formalisms: *k*-Nearest Neighbors (*k*NN) and Support Vector Machines (SVMs). Our approach extends the total-margin SVM, which considers the distance of all points from the margin; each point is weighted based on its *k* nearest neighbors. The intuition is that examples that are mostly surrounded by similar neighbors, i.e., of their own class, are given more priority to minimize the influence of drastic outliers and improve generalization and robustness. Thus, our approach combines the local sensitivity of *k*NN with the global stability of the total-margin SVM. Extensive experimental results demonstrate that the proposed *k*NN-SVM can detect COVID-19-induced pneumonia from chest X-ray with greater or comparable accuracy relative to human radiologists.

A. Bahuguna · D. Yadav
Chandigarh University, Chandigarh 140413, PB, India

A. Senapati
Central Institute of Technology, Kokrajhar 783370, AS, India
e-mail: a.senapati@cit.ac.in

B. Nath Saha (✉)
Concordia University of Edmonton, Edmonton, AB T5B 4E4, Canada
e-mail: baidya.saha@concordia.ab.ca

 
B. Rushi Kumar et al. (eds.), *Mathematics and Computing*, Springer Proceedings in Mathematics & Statistics 415, https://doi.org/10.1007/978-981-19-9307-7_9

**Keywords** *k*NN · SVM · COVID-19 · Pneumonia · Chest X-ray · Deep learning · VGG16

## 1 Introduction

The impact of the recent outbreak of novel coronavirus disease, COVID-19, on the global healthcare systems is unprecedentedly enormous [22]. Many COVID-19 infections have included respiratory illness manifesting such as fever and cough, developing pulmonary symptoms like chest discomfort and shortness of breath, and clinically resembling viral pneumonia which preserves the hallmark characteristics of COVID-19 infection as bilateral, peripheral ground-glass, and consolidative pulmonary opacities on CT [6]. High-resolution CT, which combines many X-ray images from multiple angles into a single picture, can image in-depth visual signatures. Unfortunately, a CT scan to manage the pandemic is not practical because it is expensive, time-consuming, labor-intensive, inaccessible in remote hospitals, scanning equipment needs prolonged deep sterilization after each potentially infected patient is scanned, and there is always the risk of transmission of the virus to healthcare workers [3]. To better address these issues, CXR, though less sensitive to detect the lung pathology caused by the coronavirus, has taken center stage as a front line diagnostic test because X-ray machines are widely available, scans are relatively low cost, are ubiquitous in both emergency and rural hospital settings, can be installed on a mobile platform, relatively easy to disinfect, and are one of the most affordable ways to respond the outbreak [25].

Deep learning has been extensively employed in medical imaging over the past decade, and it has surpassed the performance of medical professionals in many cases [28]. Finding the presence of pneumonia in the chest X-ray can be interpreted as a classification problem. Several Convolutional Neural Networks (CNNs)-based deep learning models show great performance on various image classification tasks, and VGG16 is one of them. However, deep learning models rely on large-scale datasets to train and evaluate classifiers. In this context, transfer learning is preferred due to the limited availability of COVID-19 chest X-ray samples. In this study, we used the VGG16 model from Keras, which was pretrained on a large-scale ImageNet dataset. Transfer learning avoids the training of the deep models from scratch and also the lack of training data, and it takes advantage of the extraction of knowledge achieved through visual recognition from large-scale ImageNet.

To improve the performance of the VGG16 model, this study performs four sequential steps: (a) first we extract the deep or bottleneck features from the VGG16 model's second last dense layer, (b) then reduce the dimension through the Autoencoder algorithm, and (c) cluster the reduced features through K-Means algorithm and aggregate the cluster information in the reduced feature sets, and finally (d) classify through *k*NN-SVM. Experimental results demonstrate that clustering information improves the performance of classification results.

SVMs [15] have been widely used in many applications due to their robustness, especially for small training sets, adaptability to various classification and regression problems through the incorporation of appropriate kernel functions, as well as the ability to obtain a global optimal solution via quadratic programming. However, it is well-known that minimizing solely the empirical training error can result in poor generalization due to overfitting [38]. Regularization methods such as 1-norm [42] or manifold regularization [5] and loss functions such as least-square SVM [32] have been proposed as a solution to this problem. However, a persistent problem endemic to these approaches is their inability to perform under noisy conditions. This issue is especially exacerbated when these methods are confronted with extreme outliers. As the approaches are margin-based, they tend to weigh the outliers very strongly, which leads to overfitting and a loss of robustness; this ultimately affects generalization.

To mitigate the effects of outliers and inspired by previous attempts to weigh individual examples differently during training, we propose to use a weighting scheme based on *k*NN [10, 16]. We propose to assign weights to training examples proportional to the number of neighbors of the same class. The intuition is that, locally, a data point will be surrounded by similar neighbors (i.e., of the same class), and consequently, extreme outliers will be weighted less. This prevents such outliers from exerting too much influence on the final classifier and improves robustness. However, depending on the choice of $k$ and the density of the data, *k*NN itself can be very locally sensitive. This motivates the adoption of total-margin SVMs [40] as the formalism underlying our approach. The total-margin SVM extends classical SVMs by adding extra surplus terms in the objective and constraints which measure the deviation of *all* data points from the classification hyperplane [20, 24]. Thus, while the slack variables measure the deviation of miscategorized points, the surplus variables measure the deviation of the correctly categorized points. Training a classifier that maximizes the *total margin* requires minimizing error (measured by slack variables) and maximizing "right classification" (measured by surplus variables). By weighting the data points in the total-margin SVM with *k*NN-based weights, our method combines the kNN's local sensitivity with the total-margin SVM's global stability. We refer to this algorithm as *k*NN-weighted SVM. Finally, *k*NN-SVM has the benefit of allowing the use of a wide range of different distance metrics including those learned via metric learning approaches such as LMNN [37] and MDML [18].

On a synthetic example, Fig. 1 compares the behavior of our proposed *k*NN-weighted SVM to that of classical SVMs. The dataset (Fig. 1, left) consists of uniformly randomly generated linearly separable data. The dataset also contains six outliers, three for each class, which are circled. Figure 1 (center) shows the weights of the training examples for a standard soft-margin SVM. More importantly, for soft-margin SVM, the badly misclassified outliers have the same weights relative to the correctly classified within-margin examples. Figure 1 (right) shows the weights of the training examples for the *k*NN-weighted SVM. Because *k*NN-weighted SVM assigns weights proportional to similar neighbors, the outliers' influence on the classifier is greatly reduced. The reduced overfitting enables the maximization of the total margin. As shown in Fig. 1, the *k*NN-weighted SVM has a larger margin than the classical SVM.

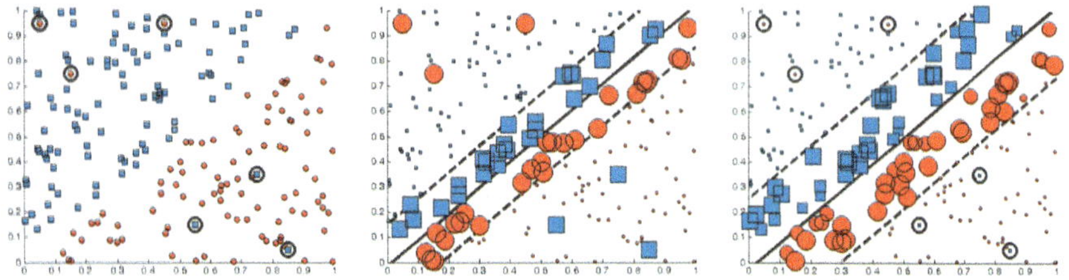

**Fig. 1** A simple synthetic example to compare the behavior of the proposed *k*NN-SVM (right) with classical soft-margin SVM (center). The dataset (left) is almost linearly separable except for the six circled outliers. These outliers receive high weights under the classical soft-margin SVM. In contrast, *k*NN-SVM minimizes their effect as they have few neighbors of the same class as themselves. As a result, the effects of overfitting are reduced, and the *k*NN-SVM can **achieve a greater margin**. We show theoretically and empirically that this corresponds to better generalization

There has previously been research on combining the power of *k*NNs and SVMs [8, 14, 19, 30]. These strategies all seek to "localize" by picking a few training examples that are close to a test example and designing a SVM for these chosen training examples [14]. At a high level, these methods can be thought of as learning one SVM for each partition of the data space. The class of these approaches in general, and the SVM-KNN approach [41] in particular, is closely related to our approach where it finds the neighbors to a query and then trains a local SVM. On the contrary, we train the SVM on all the examples from the training set by incorporating the *k*NN distance function directly into our optimization problem. Our method "localizes" SVMs by employing a data-driven regularization approach. Instead of focusing on a test point, we weigh the input training space based on their feature locations. Instead of using multiple localized SVM models [41], we use a single SVM model to capture both "global" and "local" information.

## 2 Deep Learning for CXR-Based Covid Detection

Panwar et al. [26] developed a deep model called "nConvNet" which employs transfer learning on a pretrained VGG16 network for fast detection of COVID-19 patients from chest X-rays. Similarly, Das et al. [12] utilized CNNs and Xception model to build deep transfer learning-based COVID-19 detection model from chest X-rays. Bassi et al. [4] proposed a novel twice transfer learning method called "output neuron keeping" which performs better than both twice and simple transfer learning and simple transfer learning model. Twice transfer learning method is a two-stage simple transfer learning method where in the first stage it trains the pretrained VGG16 model on a large CXR dataset first and then trains it with a smaller COVID-19 CXR dataset. Brunese et al. [7] developed a two-stage transfer learning approach by pretrained VGG16 model. In the first stage, they detect whether a chest X-ray is of a healthy patient or of a patient with generic pulmonary disease. Then, if the X-ray image is of a patient with generic pulmonary disease, they pass this image to another model which detects whether this pulmonary disease is COVID-19 or not. Salman et al. [29] exploited the pretrained InceptionV3 model for feature extraction using a transfer learning approach for covid pneumonia detection. Jaiswal et al. [17] created a new model called "CovidPen" that detects COVID-19 infection from chest X-ray images using transfer learning on a Pruned EfficientNet model. Pham [27] et al. reported that three pretrained CNN Models named AlexNet, GoogLeNet, and Squeezenet demonstrate higher accuracy for COVID-19 classification from chest X-ray and they take less time than other pretrained models as well. Al-Waisy et al. [2] devised a new deep learning system called Covid-CheXNet by combining two different deep learning models, ResNet34 and HRNet, and exploited the CLAHE method and Butterworth bandpass filter to enhance the poor image quality and reduce the noise level, respectively. Waheed et al. [35] proposed a special type of network called CovidGAN which uses an Auxiliary Classifier Generative Adversarial Network (ACGAN) to generate synthetic Chest X-ray images. This research showed that the training dataset augmented with CovidGAN offers better classification results with deep neural networks. Ahmed et al. [1] proposed ReCoNet, an end-to-end CNN architecture that used two loss functions—Multi-tasking learn loss function and a joint weighted cross-entropy loss function for improving its performance. LV et al. [21] proposed cascade-SEMEnet which employed SEME-ResNet50 for detecting the type of lung infection and a DenseNet169 for the subdivision of viral pneumonia, used to diagnose lung disease. This research also utilized Contrast Limited Adaptive Histogram Equalization (CLAHE) to improve the contrast of chest X-rays and U-Net to remove the non-pathological features on the chest X-rays.

## 3 Proposed Methodology for COVID-Induced Pneumonia Detection from Chest X-ray

To improve the performance of transfer learning of the pretrained VGG16 model, this study explores a new avenue that follows the four sequential steps which are demonstrated in four different subsections.

### *3.1 Deep Feature Extraction Using VGG16*

We used VGG16 [31] for extracting deep features from the chest X-ray images. VGG16 is a 16-layer convolutional neural network. We used a pretrained version of the VGG16 network trained on the ImageNet dataset [13], which is a dataset of over 14 million images: the pretrained network can classify images into 1,000 different classes, for instance, cat, dog, and other objects. VGG16 has a very simple architecture that consists of convolution layers of 3 × 3 filter with a stride 1 and always uses the same padding and max pool layer of 2 × 2 filter with stride 2. This arrangement of convolution and max pool is repeated throughout the architecture. In the end, it has two Fully connected layers followed by a softmax for the multiclass classification [31]. The 16 in VGG16 refers to the number of layers that have weights. This network is fairly large and has around 138 million (approximately) parameters. Using pretrained VGG16 as a fixed feature extractor, we utilize the transfer learning approach. We load the VGG16 network with weights pretrained on ImageNet, only keeping the convolutional base and truncating the fully connected layer head. Then, we construct our new fully connected layer head, which consists of Global Average Pooling layer, Dense layer with 64 neurons and ReLU activation, Dropout with 0.2 rate, and a last Dense layer (output layer) with the softmax activation and 3 neurons for classification and append it on top of the VGG16 convolutional base. We then freeze the convolutional base of VGG16 such that only the fully connected layer head is going to be trained. After the model has been trained, we pass the entire training and test data through the model and collect the deep features from the last Dense layer before the output layer.

### *3.2 Dimensionality Reduction Using Autoencoders*

To reduce the dimensionality of the extracted deep features, we exploited Autoencoder [36]. Autoencoder is a self-supervised deep learning model that is used to reduce the dimensionality of input data. Firstly, an encoder which is a compression unit that compresses the input data. And secondly, a decoder which decompresses the compressed input by reconstructing it. Each Dense layer, except for the last one, is followed by a BatchNormalization layer and a LeakyReLU activation layer with the value of alpha = 0.3. The last Dense layer uses a linear activation function. We train the autoencoder with Adam as an optimizer.

### *3.3 Incorporating Clustering Information into Classification Tasks*

To improve the performance of the classification tasks, we cluster the reduced features computed above using the $K$-means algorithm [15] as illustrated in Fig. 2. The goal of K-means clustering is to divide $n$ observations into $k$ clusters, with each observation belonging to the cluster with the closest mean (cluster centers or cluster centroid). K-means clustering reduces within-cluster variances (squared Euclidean distances). Optimal values of $K$ are determined in this study using the Silhouette Method in combination with the Elbow Method.

(a) Clustering before Classification.

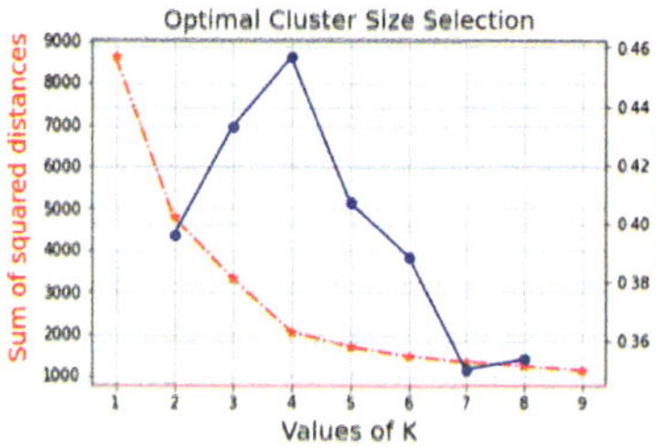


(b) Optimal Values of K for K-Means Algorithm.

**Fig. 2** Incorporating clustering for improving classification tasks

## *3.4 kNN-Regularized Support Vector Machine (kNN-SVM)*

Let the training data consists of $l$ pairs $(\boldsymbol{x}_i, y_i), i = 1, ..., l$, with $x_i \in \mathbb{R}^p$ and $y_i \in \{-1, 1\}$. Typically in SVMs, the input space is mapped into a high dimensional feature space using the mapping function $\phi(\boldsymbol{x})$ to increase the linear separability. The following Quadratic Programming (QP) problem [15] is used to find the optimal separating hyperplane of the $k$NN-SVM:

$$\min_{\boldsymbol{w},\xi} \frac{1}{2}||\boldsymbol{w}||_2^2 + C\sum_{i=1}^{l} D_i\xi_i \\ \text{subject to} \quad y_i(\boldsymbol{w}^t z_i + b) \geq D_i(1 - \xi_i), \forall i, \xi_i \geq 0 \tag{1}$$

where $z_i = \phi(\boldsymbol{x}_i)$, $\boldsymbol{w}$ is a weight vector, and $C$ is the margin parameter that determines the tradeoff between the margin maximization$(2/||\boldsymbol{w}||)$ and error minimization. $\xi_i(i = 1, ..., l)$ are non-negative variables called slacks which measure the distance of the example $(\boldsymbol{x}_i, y_i)$ from the optimal separating hyperplane. $D_i = \frac{|N_i|}{k}$, and $N_i \in \{N(i) : C(N(i)) = C(i)\}$, where $C(i)$ and $N(i)$ are the class and the neighbors of $i$, respectively. Given that we are using a $k$NN formulation, $N(i) = k, \ \forall i$. $D_i$ explains the significance of the slack and surplus variables. The key intuition in this framework is that instances surrounded by instances in the training data from its own class get more importance than the instances surrounded by the members of the opposite class. For traditional SVM, $D_i = 1, \forall i$. We now present some key features of this framework as a function of the number of neighbors ($k$).

- It is necessary that $k \geq 1$.
- When $k == l$, where $l$ is the number of training instances, every $D_i$ becomes the fraction of the number of examples in its class. Then this formulation reduces to the total-margin SVM as proposed by Yoon et al. [40].
- If $k$ is sufficiently large, the formulation is robust to outliers and noise and works well.

- If $k$ is quite small, say $k \longrightarrow 1$, the formulation can become potentially infeasible. This is due to the fact that outliers can have their $D$ values to be 0 (as no neighbors of their class may be within $k$). To avoid such cases, we perform a pre-processing step and identify the number of neighbors that will mitigate the effects of the outliers.
- While it is potentially possible to add a small non-zero term, say $\epsilon$, to each $D_i$, it does not work well. First when $\epsilon \longrightarrow 0$, the problem can become infeasible. But if $\epsilon$ is higher, the final solution can become very sensitive to the choice of this parameter. A more principled way of defining $D_i$ is to employ Laplace correction, a standard method for probability estimation to ensure that the probabilities do not go to 0 or 1. Hence, $D_i = \frac{|N_i|+1}{k+n}$ where $n$ is the number of classes. This will ensure that outliers get a weight of 1/2 while the rest of the points will end up with a reasonable weight.

As long as the value of $k$ is reasonable, our method is quite robust. Typically, we introduce Lagrange multipliers, $\alpha_i$, and obtain the dual:

$$\begin{aligned} &\max_{\alpha_i} \sum_{i=1}^{l} \alpha_i D_i - \frac{1}{2} \sum_{i=1}^{l} \sum_{j=1}^{l} \alpha_i \alpha_j y_i y_j K(\boldsymbol{x}_i, \boldsymbol{x}_j) \\ &\text{subject to} \quad 0 \leq \alpha_i \leq C, \forall i, \quad \sum_{i=1}^{l} \alpha_i y_i = 0 \end{aligned} \tag{2}$$

where $K(\boldsymbol{x}_i, \boldsymbol{x}_j) = \phi(\boldsymbol{x}_i)^t \phi(\boldsymbol{x}_j)$ is any kernel function. Solving the above dual problem, we obtain the decision function needed to predict the classification of a new data point $\boldsymbol{x}'$: $f(\boldsymbol{x}') = sign(\sum_{i=1}^{l} \alpha_i^* y_i K(\boldsymbol{x}_i, \boldsymbol{x}') + b^*)$, where '*' denotes the optimal solution. The value of $b^*$ can be obtained from the constraint of Eq. 1 using Karush-Kuhn-Tucker (KKT) conditions: $b^* = \frac{1}{N_s} \sum_{s \in S} (y_s - \sum_{m \in S} \alpha_m y_m K(\boldsymbol{x}_m, \boldsymbol{x}_s))$, where $S$ is the set of indices of support vectors and $N_s$ is the number of support vectors.

**Multiclass *k*NN-Weighted SVM** We now extend the QP formulation of *k*NN-weighted SVM for two-class classification into general multiclass settings [11]. Assume the given training data consists of $l$ pairs $(\boldsymbol{x}_i, y_i)$, $i = 1, ..., l$ with $\boldsymbol{x}_i \in \mathbb{R}^p$, where each example is assigned a label $y_i$ from a fixed finite set $\mathcal{Y} \in \{1, ..., m\}$. A feature function $\psi(\boldsymbol{x}, \boldsymbol{y})$ can be defined in such a way that it explicitly includes the $y$-labels and allows for a separate weight vector $\boldsymbol{w}_k$ for each class $k$ [33]. We define $\psi(\boldsymbol{x}, y)$ for our experiment as

$$\psi(\boldsymbol{x}, y) = \frac{1}{\sqrt{(2\pi)^k |\boldsymbol{\Sigma}|}} \exp\left( -\frac{1}{2} (\boldsymbol{x} - \boldsymbol{\mu}_y)' \boldsymbol{\Sigma}_y^{-1} (\boldsymbol{x} - \boldsymbol{\mu}_y) \right). \tag{3}$$

We now consider the following Quadratic Programming (QP)-based optimization problem [34] to define the multiclass *k*NN-SVM.

$$
\begin{aligned}
&\min_{w,\zeta,\eta} \frac{1}{2}||\boldsymbol{w}||_2^2 + C\sum_{i=1}^{l} D_i\zeta_i \\
&\text{subject to} \quad \boldsymbol{w}^t\delta\psi_i(\boldsymbol{y}) \geq D_i(1-\zeta_i), \\
&\forall i, \forall y \in \mathcal{Y}\setminus y_i, \zeta_i \geq 0, \eta_i \geq 0.
\end{aligned}
\tag{4}
$$

$\delta\psi_i(\boldsymbol{y})$ is defined as $\delta\psi_i(\boldsymbol{y}) \equiv \psi(\boldsymbol{x}_i, y_i) - \psi(\boldsymbol{x}_i, \boldsymbol{y})$. Variables $\zeta_i$ and $D_i$ are defined above. Once a complete weight vector is learned, a new test example $\boldsymbol{x}'$ is classified as $f(\boldsymbol{x}') = \arg\max_y \boldsymbol{w}^t\psi(\boldsymbol{x}', y)$ [39]. The subsequent dual formulation is

$$
\begin{aligned}
&\max_{\alpha} \sum_{i,y\neq y_i} \alpha_{iy} D_i - \frac{1}{2}\sum_{\substack{i,y\neq y_i \\ j,\bar{y}\neq y_j}} \alpha_{iy}\alpha_{j\bar{y}}\delta\psi_i^t(\boldsymbol{y})\delta\psi_j(\bar{\boldsymbol{y}}) \\
&\text{subject to } \alpha_{iy} \leq C, \forall i, \forall y \in \mathcal{Y}\setminus y_i.
\end{aligned}
\tag{5}
$$

# 4 Experimental Results and Discussions

## *4.1 Dataset Description*

The dataset used in this experiment consists of chest X-rays of normal, COVID-19, and pneumonia patients. The dataset includes 69 confirmed COVID-19, 79 confirmed pneumonia, and 158 normal images, i.e., overall 306 images. Again, the pneumonia images consist of 79 bacterial pneumonia and 79 viral pneumonia cases. Images have been accumulated in this dataset from diverse sources. The normal and pneumonia X-ray images are collected from the Kaggle Chest X-ray dataset [23]. This dataset consists of Chest X-ray images (anterior-posterior) selected from retrospective cohorts of pediatric patients of one to five years old from Guangzhou Women and Children's Medical Center, Guangzhou. In addition to Kaggle, the COVID-19 X-ray images are collected from the COVID-19 Chest X-ray dataset [9], organized by Dr. Joseph Paul Cohen of the University of Montreal. This dataset is a publicly open dataset of chest X-ray images of patients who are positive or suspected of COVID-19 or other viral and bacterial pneumonia. Both of these datasets include posterior-anterior chest images of patients with pneumonia.

## *4.2 Comparison of Different Algorithms*

Table 1 demonstrates the comparison among VGG16 transfer learning, SVM, SVM with clustering information, *k*NN-SVM, and *k*NN-SVM with clustering algorithms in terms of Accuracy, Precision, Recall, and F1-Score. Accuracy is calculated as the fraction of correct predictions in the test dataset. Precision is the ability of the

**Table 1** Performance analysis of different models for three classes (Normal, Pneumonia, and COVID-19) classification

| Algorithm | Accuracy | Precision | Recall | F1-score |
|---|---|---|---|---|
| VGG16 + Transfer learning | 0.75 | 0.73 | 0.69 | 0.66 |
| SVM | 0.86 | 0.88 | 0.83 | 0.85 |
| SVM + Clustering | 0.86 | 0.88 | 0.83 | 0.85 |
| *k*NN-SVM | 0.89 | 0.89 | 0.89 | 0.89 |
| *k*NN-SVM+Clustering | **0.92** | **0.92** | **0.91** | **0.91** |

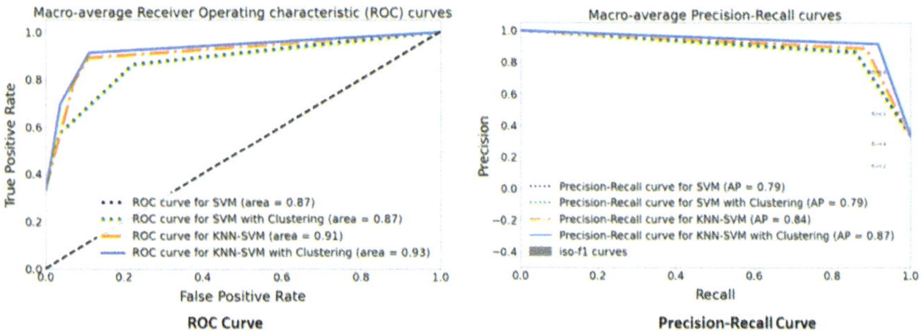


**Fig. 3** ROC and Precision-Recall curve for different algorithms

model to identify only the relevant instances. Recall is the ability of a model to find all the relevant cases. F1-score is the harmonic mean of Precision and Recall. Table 1 shows that the performance of *k*NN-SVM outperforms all other models. In addition, incorporating clustering information into classification tasks enhances the discriminating ability of the classifiers.

Figure 3 illustrates the Receiver Operating Characteristic (ROC) and Precision-Recall curves for all the algorithms executed in this study. The area under the ROC curve and Average precision are also mentioned in the legend of the figure. Results show that *k*NN-SVM with clustering information is superior to other algorithms (area under ROC and Average Precision (AP) for *k*NN-SVM are 0.93 and 0.87, respectively).

## 5 Conclusion and Future Works

In this research, we improved the performance of the transfer learning with the VGG16 model for COVID-19 pneumonia detection through a novel *k*NN-regularized Support Vector Machine (*k*NN-SVM). The *k*NN prioritizes the examples that are mostly surrounded by the examples from its own class to correctly classify while deciding decision boundary during training. This study shows that incorporating clustering information improves the performance of the classifiers. Results demonstrate that the machine learning algorithms are able to detect COVID-19-induced pneumonia from chest X-ray as accurate as radiologists. Extensive testing of the proposed method on other medical imaging classification problems remains an intriguing future direction.

## References


1. Ahmed, S.E.A.: Reconet: Multi-level preprocessing of chest x-rays for covid-19 detection using convolutional neural networks. MedRxiv (2020)
2. Al-Waisy, A.S.E.A.: Covid-chexnet: hybrid deep learning framework for identifying covid-19 virus in chest x-rays images. Soft Comput. 1–16 (2020)
3. Apostolopoulos, I.D., Mpesiana, T.A.: Covid-19: automatic detection from x-ray images utilizing transfer learning with convolutional neural networks. Phys. Eng. Sci. Med. **43**(2), 635–640 (2020)
4. Bassi, P.R., Attux, R.: A deep convolutional neural network for covid-19 detection using chest x-rays. Res. Biomed. Eng. 1–10 (2021)
5. Belkin, M. et al.: Manifold regularization: a geometric framework for learning from labeled and unlabeled examples. Mach. Learn. Res. **7**, 2399–2434 (2006)
6. Bernheim, A.E.A.: Chest CT findings in coronavirus disease-19 (covid-19): relationship to duration of infection. Radiology 200463 (2020)
7. Brunese, L., Mercaldo, F., Reginelli, A., Santone, A.: Explainable deep learning for pulmonary disease and coronavirus covid-19 detection from x-rays. Comput. Methods Prog. Biomed. **196**, 105608 (2020)
8. Cheng, H., Tan, P.N., Jin, R.: Efficient algorithm for localized support vector machine. IEEE Trans. Knowl. Data Eng. **22**(4), 537–549 (2010)
9. Cohen, J.P., Morrison, P., Dao, L.: Covid-19 image data collection (2020). arXiv:2003.11597. https://github.com/ieee8023/covid-chestxray-dataset
10. Cover, T., Hart, P.: Nearest neighbor pattern classification. IEEE Trans. Inf. Theor. **13**(1), 21–27 (1967)
11. Crammer, K., Singer, Y., Cristianini, N., Shawe-taylor, J., Williamson, B.: On the algorithmic implementation of multiclass kernel-based vector machines. Mach. Learn. Res. 265–292 (2001)
12. Das, N.N.E.A.: Automated deep transfer learning-based approach for detection of covid-19 infection in chest x-rays. IRBM (2020)
13. Deng, J., Dong, W., Socher, R., Li, L.J., Li, K., Fei-Fei, L.: Imagenet: a large-scale hierarchical image database. In: 2009 IEEE Conference on Computer Vision and Pattern Recognition, pp. 248–255. IEEE (2009)
14. Hable, R.: Universal consistency of localized versions of regularized kernel methods. J. Mach. Learn. Res. **14**(1), 153–186 (2013)
15. Hastie, T., Tibshirani, R., Friedman, J.: The Elements of Statistical Learning: Data Mining, Inference, and Prediction, 2nd edn. Springer (2009)
16. Hastie, T., Tibshirani, R.: Discriminant adaptive nearest neighbor classification. IEEE Trans. Pattern Anal. Mach. Intell. **18**(6), 607–616 (1996)
17. Jaiswal, A.K., Tiwari, P., Rathi, V.K., Qian, J., Pandey, H.M., Albuquerque, V.H.C.: Covidpen: a novel covid-19 detection model using chest x-rays and CT scans. Medrxiv (2020)
18. Kunapuli, G., Shavlik, J.: Mirror descent for metric learning: a unified approach. In: Proceedings of the ECML '12, pp. 859–874 (2012)

19. Ladicky, L., Torr, P.H.S.: Locally linear support vector machines. In: ICML, pp. 985–992 (2011)
20. Liu, Y.H. et al.: Face recognition using total margin-based adaptive fuzzy support vector machines. IEEE Trans. Neural Netw. **18**(1), 178–192 (2007)
21. Lv, D., Qi, W., Li, Y., Sun, L., Wang, Y.: A cascade network for detecting covid-19 using chest x-rays (2020). arXiv:2005.01468
22. Mei, X., Lee, H.C., Diao, K.y., Huang, M., Lin, B., Liu, C., Xie, Z., Ma, Y., Robson, P.M., Chung, M. et al.: Artificial intelligence–enabled rapid diagnosis of patients with covid-19. Nat. Med. **26**(8), 1224–1228 (2020)
23. Mooney, P.: Kaggle dataset: chest x-ray images (pneumonia) (2017). https://www.kaggle.com/paultimothymooney/chest-xray-pneumonia
24. Nakayama, H., Yun, Y.: Generating support vector machines using multi-objective optimization and goal programming. In: Jin, Y. (ed.), Multi-Objective Machine Learning, pp. 173–198. Springer (2006)
25. Ozturk, T., Talo, M., Yildirim, E.A., Baloglu, U.B., Yildirim, O., Acharya, U.R.: Automated detection of covid-19 cases using deep neural networks with x-ray images. Comput. Biol. Med. **121**, 103792 (2020)
26. Panwar, H., Gupta, P., Siddiqui, M.K., Morales-Menendez, R., Singh, V.: Application of deep learning for fast detection of covid-19 in x-rays using ncovnet. Chaos Solitons Fract. **138**, 109944 (2020)
27. Pham, T.D.: Classification of covid-19 chest x-rays with deep learning: new models or fine tuning? Health Inf. Sci. Syst. **9**(1), 1–11 (2021)
28. Rajpurkar, P., Irvin, J., Zhu, K., Yang, B., Mehta, H., Duan, T., Ding, D., Bagul, A., Langlotz, C., Shpanskaya, K., et al.: Chexnet: radiologist-level pneumonia detection on chest x-rays with deep learning (2017). arXiv:1711.05225
29. Salman, F.M., Abu-Naser, S.S., Alajrami, E., Abu-Nasser, B.S., Alashqar, B.A.: Covid-19 detection using artificial intelligence (2020)
30. Segata, N., Blanzieri, E., Bottou, L.: Fast and scalable local kernel machines. J. Mach. Learn. Res. 1883 (2009)
31. Simonyan, K., Zisserman, A.: Very deep convolutional networks for large-scale image recognition (2014). arXiv:1409.1556
32. Suykens, J.A.K., Vandewalle, J.: Least squares support vector machine classifiers. Neural Process. Lett. **9**(3), 293–300 (1999)
33. Tsochantaridis, I. et al.: Support vector machine learning for interdependent and structured output spaces. In: In ICML (2004)
34. Tsochantaridis, I., Joachims, T., Hofmann, T., Altun, Y.: Large margin methods for structured and interdependent output variables. Mach. Learn. Res. **6**, 1453–1484 (2005)
35. Waheed, A., Goyal, M., Gupta, D., Khanna, A., Al-Turjman, F., Pinheiro, P.R.: Covidgan: data augmentation using auxiliary classifier gan for improved covid-19 detection. IEEE Access **8**, 91916–91923 (2020)
36. Wang, Y., Yao, H., Zhao, S.: Auto-encoder based dimensionality reduction. Neurocomputing **184**, 232–242 (2016)
37. Weinberger, K.Q., Saul, L.K.: Distance metric learning for large margin nearest neighbor classification. J. Mach. Learn. Res. **10**, 207–244 (2009). http://dl.acm.org/citation.cfm?id=1577069.1577078
38. Xu, H., Caramanis, C., Mannor, S.: Robustness and regularization of support vector machines. J. Mach. Learn. Res. **10**, 1485–1510 (2009). http://dl.acm.org/citation.cfm?id=1577069.1755834
39. Xu, L., Schuurmans, D.: Unsupervised and semi-supervised multi-class support vector machines. In: National Conference on Artificial Intelligence. pp. 904–910. AAAI (2005). http://dl.acm.org/citation.cfm?id=1619410.1619478
40. Yoon, M., Yun, Y., Nakayama, H.: A role of total margin in support vector machines. In: Proceedings of the International Joint Conference on Neural Networks, vol. 3, pp. 2049–2053 (2003). https://doi.org/10.1109/IJCNN.2003.1223723

41. Zhang, H.E.A.: Svm-knn: discriminative nearest neighbor classification for visual category recognition. In: Proceedings of the 2006 IEEE Computer Society Conference on Computer Vision and Pattern Recognition, vol. 2. CVPR '06 (2006)
42. Zhu, J., Rosset, S., Hastie, T., Tibshirani, R.: 1-norm support vector machines. Technical report, Advances in Neural Information Processing Systems (2003)

# Quantum Simulation of Perfect State Transfer on Weighted Cubelike Graphs

**Jaideep Mulherkar, Rishikant Rajdeepak, and Sunitha VadivelMurugan**

**Abstract** A continuous-time quantum walk on a graph evolves according to the unitary operator $e^{-iAt}$, where $A$ is the adjacency matrix of the graph. Perfect state transfer (PST) in a quantum walk is the transfer of a quantum state from one node of a graph to another node with 100% fidelity. It can be shown that the adjacency matrix of a cubelike graph is a finite sum of tensor products of Pauli $X$ operators. We use this fact to construct an efficient quantum circuit for the quantum walk on cubelike graphs. In [5, 15], a characterization of integer weighted cubelike graphs is given that exhibit periodicity or PST at time $t = \pi/2$. We use our circuits to demonstrate PST or periodicity in these graphs on IBM's quantum computing platform [1, 10].

**Keywords** Continuous-time quantum walk · Perfect state transfer · Periodicity · Quantum circuits

## 1 Introduction

A quantum random walk is the quantum analogue of a classical random walk [12, 18, 19]. The study of classical random walks has led to many applications in science and engineering, such as in the study of randomized algorithms and a sampling approach called Markov chain Monte Carlo in computer science, in the study of social networks, in the behavior of stock prices in finance, in models of diffusion and study of polymers in Physics, and the motion of motile bacteria in biology. In [3, 7], the first models for quantum random walks were proposed. Since then, quantum walks have been a source of intense study. Researchers observed that there are some

J. Mulherkar · R. Rajdeepak (✉) · S. VadivelMurugan
Dhirubhai Ambani Institute of Information and Communication Technology, Gandhinagar 382007, India
e-mail: 201521006@daiict.ac.in

J. Mulherkar
e-mail: jaideep_mulherkar@daiict.ac.in

S. VadivelMurugan
e-mail: v_suni@daiict.ac.in



B. Rushi Kumar et al. (eds.), *Mathematics and Computing*, Springer Proceedings in Mathematics & Statistics 415, https://doi.org/10.1007/978-981-19-9307-7_10

startling differences between classical and quantum walks. For example, a quantum walk on a one-dimensional lattice spreads quadratically faster than a classical walk [16]. Quantum walks on cubelike graphs, such as the hypercubes, hit exponentially faster to the antipodal vertex as compared to classical counterparts [13].

Quantum walks on graphs are of two types: discrete and continuous. In the discrete case, a graph is associated with a Hilbert space of dimension $N \times \Delta$, where $N$ is the number of vertices, and $\Delta$ is the maximum degree of the graph. In the continuous case, a graph is associated with a Hilbert space of dimension $N$, and the evolution of the system is described by $e^{\iota tA}$, where $A$ is the adjacency matrix of the graph and $t$ is real. An essential feature of a quantum walk is a quantum state transfer from one vertex to another with high fidelity. When the transfer occurs with 100% fidelity, it is called perfect state transfer (PST). Some of the excellent survey papers on graph families that admit PST are [8, 9]. Among these graphs, cubelike graphs are the most famous ones that have been researched thoroughly for determining the existence and finding the pair of vertices admitting perfect state transfer in constant time [4, 6]. Notice that all cubelike graphs do not allow perfect state transfer. The study of PST on weighted graphs has been less studied. Recently, weighted abelian Cayley graphs have been characterized that exhibit PST [5].

In this paper, we look at the implementation of perfect state transfer on weighed cubelike graphs. Some of the efficient implementations of quantum walks are described in [2, 11, 14, 20, 21]. It can be shown that the adjacency matrix of a cubelike graph is the sum of the tensor products of Pauli $X$ operators. One then observes that generating efficient quantum circuits for quantum walks can then be done by quantum hamiltonian simulation techniques that have been described in [17]. We use quantum simulation techniques to verify the theoretical results of PST on weighted cubelike graphs.

## 2 Preliminaries

An undirected weighted graph $\Gamma$ consists of a triplet $(V, E, f)$, where $V$ is a non-empty set whose elements are called vertices; $E$ is a set of edges, where an edge is an unordered tuple of vertices, and $f : V \times V \to \mathbb{R}$ is a weight function that assigns non-zero real weights to edges. If $\Gamma$ is finite, then its adjacency matrix $A$ is defined by

$$A_{u,v} = f((u, v)), \qquad (u, v) \in V \times V.$$

The adjacency matrix $A$ is real and symmetric. A tuple $(u, u)$ is a loop if its weight is non-zero. If $f((u, u)) = 0$ for all $u \in V$, then the diagonal entries of $A$ are zero and the graph has no loops. A graph family of interest is a weighted cubelike graph which is defined as follows.

**Definition 1** Let $f$ be a real-valued function over a finite Boolean group $\mathbb{Z}_2^n$ of dimension $n > 0$. A cubelike graph, denoted by $Cay(\mathbb{Z}_2^n, f)$, is a graph with vertex-set $\mathbb{Z}_2^n$, and the weight of a pair $(u, v)$ of vertices is given by $f(u \oplus v)$, where $\oplus$ denotes the group addition, i.e., componentwise addition modulo 2. The adjacency matrix $A$ of $Cay(\mathbb{Z}_2^n, f)$ is given by

$$A_{u,v} = f(u \oplus v), \ u, v \in V.$$

An equivalent definition for an unweighted cubelike graph is given as follows: let $\Omega_f = \{u \in \mathbb{Z}_2^n : f(u) = 1\}$, then two vertices $u$ and $v$ are adjacent if $u \oplus v \in \Omega_f$. The cubelike graph, in this case, is denoted by $Cay(\mathbb{Z}_2^n, \Omega_f)$, see Figs. 1 and 2.

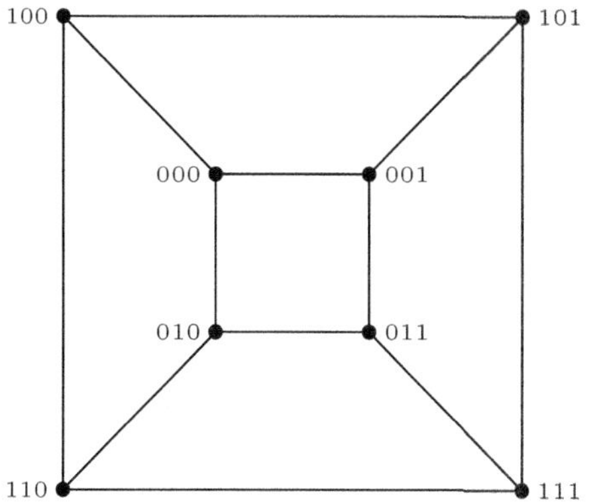


**Fig. 1** $Cay(\mathbb{Z}_2^3, \{001, 010, 100\})$

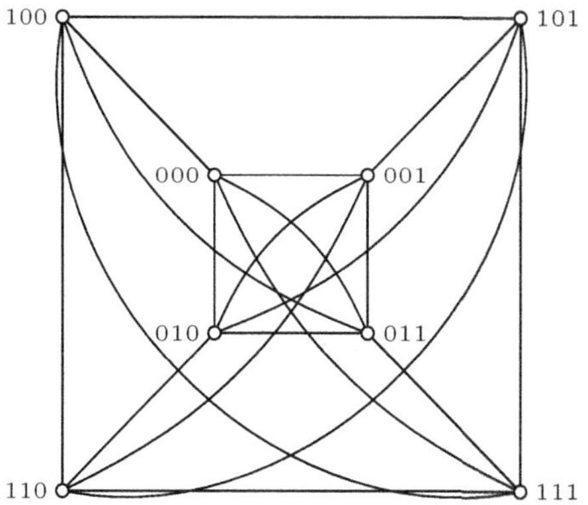


**Fig. 2** $Cay(\mathbb{Z}_2^3, \{001, 010, 011, 100, 111\})$

## 2.1 Continuous-Time Quantum Walk

Let $\Gamma$ be an undirected and weighted graph with or without loops and $A$ be the adjacency matrix. A quantum walk on $\Gamma$ is described by an evolution of the quantum system associated with the graph. Suppose the graph has $N$ vertices, then it is associated with a Hilbert space $\mathcal{H}_P \cong \mathbb{C}^N$ of dimension $N$, called the position space, and the computational basis is represented by

$$\{|v\rangle : v \text{ is a vertex in } \Gamma\}.$$

The continuous-time quantum walk (CTQW) on $\Gamma$ is described by the transition matrix $\mathcal{U}(t) = e^{-\iota t A}$, where $\iota = \sqrt{-1}$, that operates on the position space $\mathcal{H}_P$. In other words, if $|\psi(0)\rangle$ is the initial state of the quantum system associated with the graph, then the state of the system after time $t$ is given by

$$|\psi(t)\rangle = e^{-\iota t A} |\psi(0)\rangle .$$

**Definition 2** A graph is said to admit perfect state transfer (PST) if the quantum walker beginning at some vertex $u$ reaches a distinct vertex $v$ with probability 1, i.e., for some positive real $\tau$ and scalar $\lambda$,

$$|\langle v|e^{-\iota \tau A}|u\rangle| = |\lambda| = 1.$$

Alternatively, we say perfect state transfer occurs from the vertex $u$ to the vertex $v$. If $u = v$, we say the graph is periodic at $u$ with period $\tau$, and if the graph is periodic at every vertex with the same period $\tau$ then, the graph is periodic.

**Example 1** Consider the graph on the cycle of size 4, see Fig. 3, with the adjacency matrix $A$ given by

$$A = \begin{bmatrix} 0\,1\,1\,0 \\ 1\,0\,0\,1 \\ 1\,0\,0\,1 \\ 0\,1\,1\,0 \end{bmatrix}.$$

Then, the transition matrix at time $t = \pi/2$ is

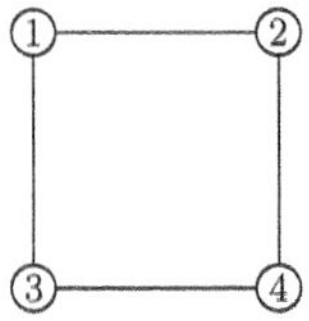


**Fig. 3** PST occurs between the pairs $\{1, 4\}$ and $\{2, 3\}$ with time $\frac{\pi}{2}$, and the graph is periodic with period $\pi$

$$\mathcal{U}(t=\frac{\pi}{2})=\begin{bmatrix}0&0&0&1\\0&0&1&0\\0&1&0&0\\1&0&0&0\end{bmatrix}.$$

Thus, perfect state transfer occurs between the pairs $\{1,4\}$ and $\{2,3\}$, both in time $\frac{\pi}{2}$. The graph is periodic with period $\pi$.

## 2.2 *Decomposition of the Adjacency Matrix of Weighted Cubelike Graph*

### 2.2.1 Group Representations

An $m$-degree representation of a finite group $G$ is a homomorphism $\rho$ from $G$ into the general linear group $GL(V)$ of an $m$-dimensional vector space $V$ over the field $\mathbb{F}$, where $\mathbb{F}$ is a complex or real field. Since $GL(V)$ is isomorphic to $GL_m(\mathbb{F})$, the general linear group of degree $m$ that consists of $m\times m$ invertible matrices, an equivalent definition for the group representation is the group homomorphism

$$\rho: G\to GL_m(\mathbb{F}).$$

The group algebra $\mathbb{C}[G]$ is an inner product space whose vectors are formal linear combinations of the group elements, i.e.,

$$\mathbb{C}[G]=\left\{\sum_{g\in G}\lambda_g g:\lambda_g\in\mathbb{C}\right\},$$

with the vector addition, the scalar multiplication, and the inner product defined by

$$\sum_{g\in G}\lambda_g g+\sum_{g\in G}\mu_g g=\sum_{g\in G}(\lambda_g+\mu_g)g,\qquad\text{(addition)},$$

$$\lambda\sum_{g\in G}\lambda_g g=\sum_{g\in G}(\lambda\lambda_g)g\qquad\text{(scalar multiplication)},$$

$$\left\langle\sum_{g\in G}\lambda_g g,\sum_{g\in G}\mu_g g\right\rangle=\sum_{g\in G}\lambda_g\bar{\mu}_g,\qquad\text{(inner product)}.$$

The regular representation on $G$, $\rho_{reg}:G\to GL(\mathbb{C}[G])$, is defined by;

$$\rho_{reg}(x)\left(\sum_{g\in G}\lambda_g g\right) = \sum_{g\in G}\lambda_g(xg) = \sum_{y\in G}\lambda_{x^{-1}y}y.$$

#### 2.2.2 The Decomposition

If $G = \mathbb{Z}_2^n$, then for $x \in \mathbb{Z}_2^n$ the regular representation acts on $\mathbb{Z}_2^n$ as

$$\rho_{reg}(x)y = x \oplus y = (x_1 \oplus y_1, \ldots, x_n \oplus y_n), \qquad y \in \mathbb{Z}_2^n.$$

Let $X$, $Y$, and $Z$ denote the three Pauli matrices that act on the computational basis $\{|0\rangle, |1\rangle\}$ of the two-dimensional Hilbert space $\mathbb{C}^2$ as

$$X|a\rangle = |a \oplus 1\rangle, \; Y|a\rangle = (-1)^a \iota |a \oplus 1\rangle, \; Z|a\rangle = (-1)^a |a\rangle, \; a \in \{0, 1\}.$$

The group element $y$ is also a vector in $\mathbb{C}[\mathbb{Z}_2^n]$ whose matrix representation is $|y\rangle = |y_1\rangle \otimes \cdots \otimes |y_n\rangle$. Hence, the action of $\rho_{reg}(x)$ over $y$ can be rewritten as

$$\begin{aligned}\rho_{reg}(x)y &= (X^{x_1}|y_1\rangle) \otimes \cdots \otimes (X^{x_n}y_n), \text{ where } X^{x_i}|y_i\rangle = |x_i \oplus y_i\rangle, \\ &= \left(X^{x_1} \otimes \cdots \otimes X^{x_n}\right)(|y_1\rangle \otimes \cdots \otimes |y_n\rangle).\end{aligned}$$

The adjacency matrix $A$ of $Cay(\mathbb{Z}_2^n, f)$ is decomposed by using the regular representation on $\mathbb{Z}_2^n$, viz., given $x, y \in \mathbb{Z}_2^n$, the value $\rho_{reg}(x)y = x \oplus y$ corresponds to the $(x, y)$-entry of $A$, so $A$ can be expressed as

$$A = \sum_{x\in\mathbb{Z}_2^n} f(x)\rho_{reg}(x). \tag{1}$$

Since $\rho_{reg}(x)$ commutes with $\rho_{reg}(y)$ for all $x, y \in \mathbb{Z}_2^n$, the evolution operator $\mathcal{U}(t) = e^{-\iota t A}$ is decomposed into

$$\mathcal{U}(t) = \prod_{x\in\mathbb{Z}_2^n} U(x, t), \qquad U(x, t) = e^{-\iota t f(x)\rho_{reg}(x)}. \tag{2}$$

### 2.3 *PST or Periodicity in Weighted Cubelike Graphs*

We simulate continuous-time quantum walk on $Cay(\mathbb{Z}_2^n, f)$ and verify the existence of perfect state transfer or periodicity as mentioned in the following theorem.

**Theorem 1** ([5, 15]) *Let $f : \mathbb{Z}_2^n \to \mathbb{Z}$ be an integer-valued function. For $x \in \mathbb{Z}_2^n$, define a subset $O_x = \{y \in \mathbb{Z}_2^n : \langle x|y\rangle \bmod 2 = 1\}$. Let $e_i$, $1 \le i \le n$, denote the $n$-tuple with entry 1 at position i and zero everywhere else. Let $\sigma \in \mathbb{Z}_2^n$ such that*

$$\sigma_i = 1 \textit{ only if } \sum_{y \in O_{e_i}} f(y) \textit{ mod } 2 = 1. \tag{3}$$

*Then,*

1. *if $\sigma$ is the identity element, i.e., $\sigma = \mathbf{0}$, then $Cay(\mathbb{Z}_2^n, f)$ is periodic with period $\frac{\pi}{2}$,*
2. *if $\sigma \neq \mathbf{0}$, then PST occurs between every pair $\{u, v\}$ satisfying $u \oplus v = \sigma$, with time $\tau = \frac{\pi}{2}$.*

**Note 1** Although PST or periodicity in weighted cubelike graph mentioned in [15] was done independently, it was only later that the authors realized that its generalized version, viz., PST on weighted abelian Cayley graph, has already been proved in another paper [5].

## 3 The Quantum Simulation

The idea to design a quantum circuit for CTQW on a cubelike graph has been taken from [17]; if the Hamiltonian is given by $A = Z_1 \otimes \cdots \otimes Z_n$, where $Z_i = Z$, then the phase shift applied to the system is $e^{-\iota t}$ if the parity of the $n$ qubits in the computational basis is even; otherwise, the phase shift applied is $e^{\iota t}$. Figure 4 illustrates the quantum circuit for $e^{-\iota t A}$, where $A = Z \otimes Z \otimes Z$.

### 3.1 *Quantum Circuits*

Let $x \in \mathbb{Z}_2^n$, then the regular representation $\rho_{reg}(x)$ is given by

$$\rho_{reg}(x) = \otimes_{i=1}^n X^{x_i} = H^{\otimes n} \left(\otimes_{i=1}^n Z^{x_i}\right) H^{\otimes n}, \text{ since } X = HZH.$$

Applying the changes to the operator $U(x, t)$ in Eq. 2, we get

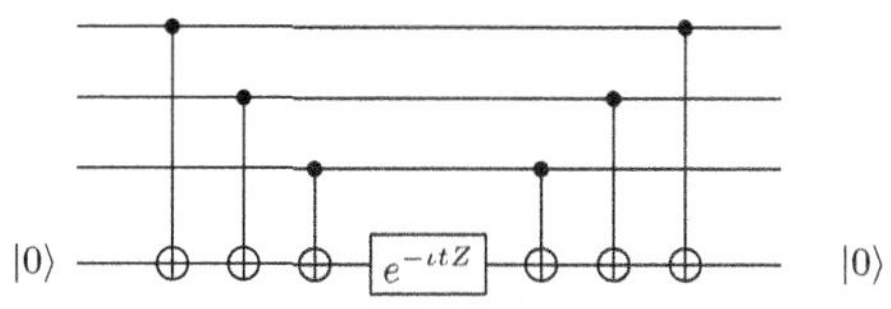


**Fig. 4** Quantum circuit to implement $e^{-\iota t A}$, where $A = Z \otimes Z \otimes Z$

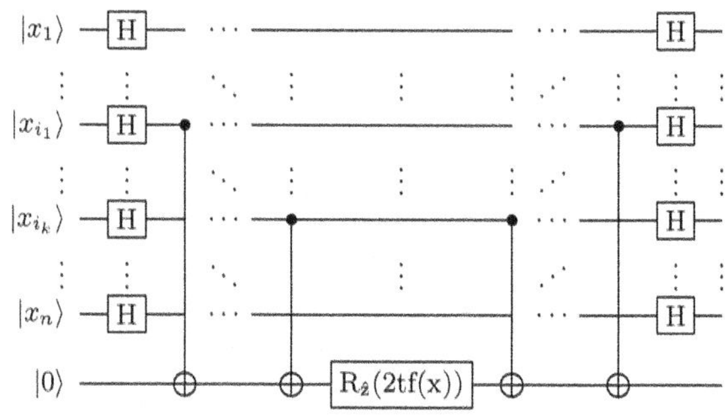


**Fig. 5** Quantum circuit for $U(x,t) = e^{-\iota t f(x)\rho_{reg}(x)}$

$$
\begin{aligned}
U(x,t) &= e^{-\iota t f(x)\rho_{reg}(x)} = e^{-\iota t f(x)[\otimes_{i=1}^{n} X^{x_i}]} \\
&= \sum_{l=0}^{\infty} \frac{(-\iota t f(x))^l}{l!} [\otimes_{i=1}^{n} X^{x_i}]^l \\
&= \sum_{l=0}^{\infty} \frac{(-\iota t f(x)))^{2l}}{(2l)!} I^{\otimes n} + \sum_{l=0}^{\infty} \frac{(-\iota t f(x))^{2l+1}}{(2l+1)!} [\otimes_{i=1}^{n} X^{x_i}] \\
&= H^{\otimes n} V(x,t) H^{\otimes n}, \qquad V(x,t) = e^{-\iota t f(x)[\otimes_{i=1}^{n} Z^{x_i}]}.
\end{aligned}
$$

We see that

$$
\begin{aligned}
\left(Z_1^{x_1} \otimes \cdots \otimes Z_n^{x_n}\right)|y\rangle &= (-1)^{x_1 y_1}|y_1\rangle \otimes \cdots \otimes (-1)^{x_n y_n}|y_n\rangle \\
&= (-1)^{\sum_{i=1}^{n} x_i y_i}|y_1\rangle \otimes \cdots \otimes |y_n\rangle \\
&= \begin{cases} |y\rangle, & \text{if } \langle x|y\rangle \bmod 2 = 0 \\ -|y\rangle, & \text{if } \langle x|y\rangle \bmod 2 = 1. \end{cases}
\end{aligned}
$$

This implies

$$
V(x,t)|y\rangle = \begin{cases} e^{-\iota t f(x) Z}|y\rangle & \text{if } \langle x|y\rangle \bmod 2 = 0 \\ e^{\iota t f(x) Z}|y\rangle & \text{if } \langle x|y\rangle \bmod 2 = 1. \end{cases}
$$

Thus, the action of the operator $V(x,t)$ is equivalent to the application of the rotation operator $R_{\hat{z}}(2tf(x))$ about the $\hat{z}$-axis if $\langle x|y\rangle$ is even, and $R_{\hat{z}}(-2tf(x))$ if $\langle x|y\rangle$ is odd. Hence, if $x$ has non-zero entries at positions $i_1,\ldots,i_k$, then the quantum circuit for the operator $e^{-\iota t f(x)\rho_{reg}(x)}$ is depicted by Fig. 5. Suppose elements in $\Omega_f = \{y : f(y) \neq 0\}$ are represented by $\Omega_f = \{x^{(1)},\ldots,x^{(\Delta)}\}$, where $\Delta$ is the cardinality of $\Omega_f$, then the quantum circuit for the continuous-time quantum walk is as shown in Fig. 6, where the initialized state, in general, is $|0\rangle^{\otimes n}$ along with an ancilla qubit with state $|0\rangle$.

**Remark 1** As seen in Fig. 6, the Hadamard gates $H$ applied at the end of $U(x^{(i)},t)$ and the beginning of $U(x^{(i+1)},t)$, $1 \leq i < \Delta$, are not required, because $H^2 = I$; thus, the actual number of $H$ gates required are $2n$. Secondly, the number of rotation

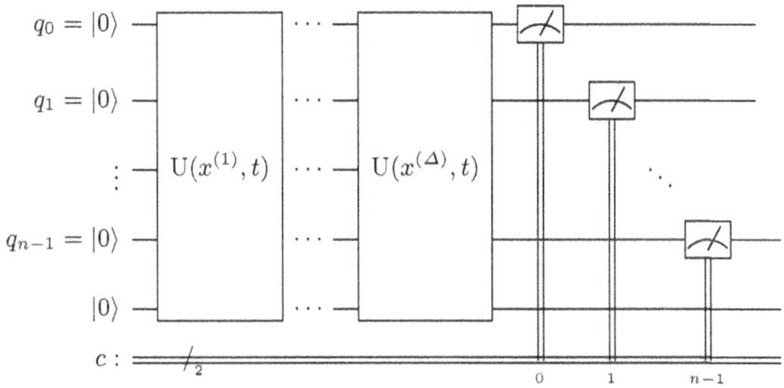


**Fig. 6** An illustration of CTQW quantum circuit on weighted cubelike graph

operators used is $\Delta$. Lastly, for each $x \in \Omega_f$, the number of CNOT gates applied is equal to the Hamming weight $wt(x)$ of $x$. Thus, the total number of CNOT gates used is $\sum_{x \in \Omega_f} wt(x)$.

## 3.2 Results

Recall that, if $u \oplus v = \sigma$, where $\sigma$ is given by Eq. 3 in Theorem 1, then $\{u, v\}$ is the PST pair. This partitions the vertex set into PST pairs. The graph shown in Fig. 1 admits PST between pairs {000, 111}, {001, 110}, {010, 101}, {011, 100}, and the other graph in Fig. 2 has PST pairs {000, 011}, {001, 010}, {100, 111}, {101, 110}. Since weighted cubelike graphs, as described in Theorem 1, are vertex-transitive, the study of PST between the pair $\{\mathbf{0}, \sigma\}$ is equivalent to any other pair. Therefore, every quantum circuit is initialized to state $|0\rangle^{\otimes n}$, see Figs. 7 and 8 which illustrate quantum circuits for the above graphs mentioned.

Suppose the weight function $f$ is defined by

$$f(001) = 4, \ f(011) = 8, \ \text{and} \ , \ f(101) = 3, \tag{4}$$

and zero on other elements, then the 3-tuple $\sigma$ is computed as (using Theorem 1)

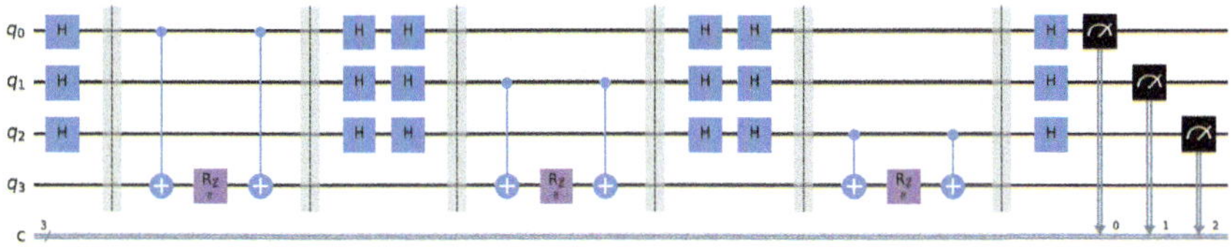


**Fig. 7** Quantum circuit for $Cay(\mathbb{Z}_2^3, \{001, 010, 100\})$

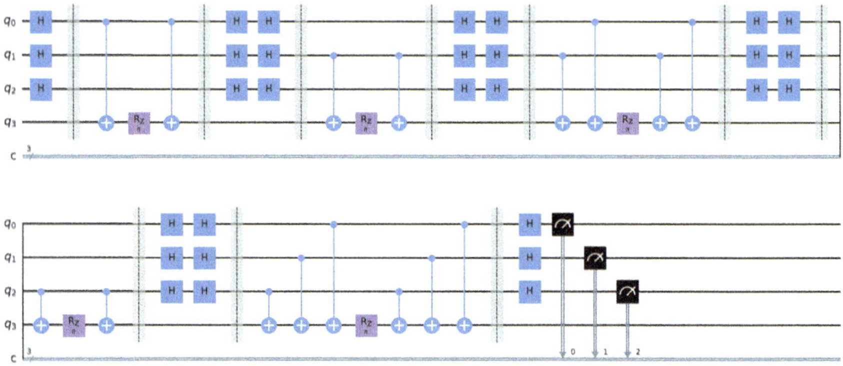


**Fig. 8** Quantum circuit for $Cay(\mathbb{Z}_2^3, \{001, 010, 011, 100, 111\})$

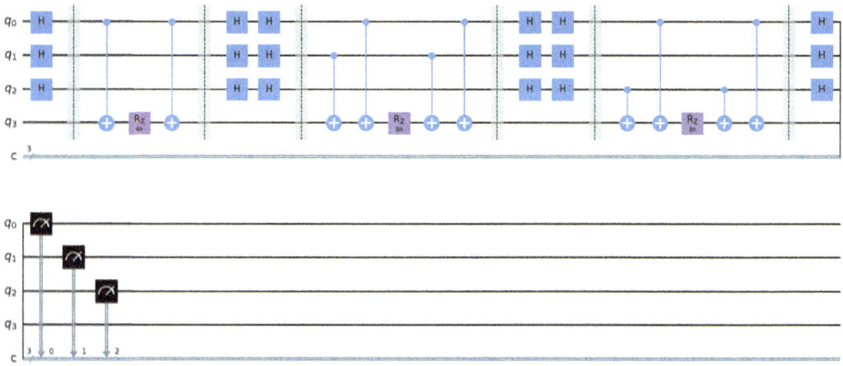


**Fig. 9** Quantum circuit for $Cay(\mathbb{Z}_2^3, \{f(001) = 4, f(011) = 8, f(101) = 3\})$

$$\begin{aligned}
O_{001} = \{001, 011, 101\} &\implies f(001) + f(011) + f(101) \bmod 2 = 1 \\
&\implies \sigma_1 = 1 \\
O_{010} = \{011\} &\implies f(011) \bmod 2 = 0 \\
&\implies \sigma_2 = 0 \\
O_{100} = \{101\} &\implies f(101) \bmod 2 = 1 \\
&\implies \sigma_3 = 1
\end{aligned}$$

Thus, $\sigma = 101$ and $\{000, 101\}$ is a PST pair. The same is obtained by simulating the quantum circuit shown in Fig. 9. On the other hand, if $f$ is defined by

$$f(010) = 4, \ f(011) = 7, \ f(100) = 8, \ f(101) = 2, \ f(110) = 5, \quad (5)$$

then $\sigma = 101$, and $\{000, 101\}$ is a PST pair.

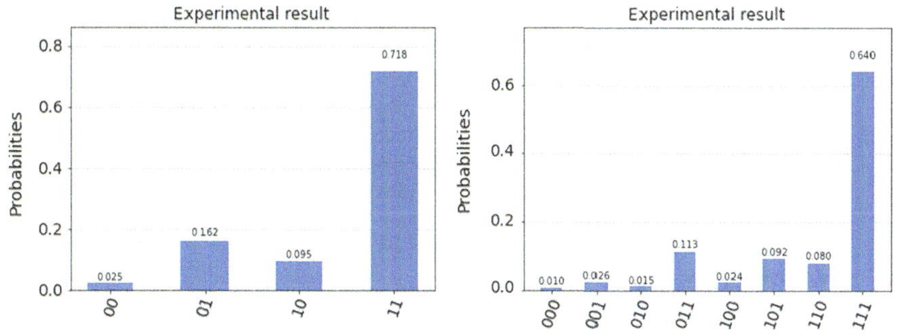


**Fig. 10** Experimented probability distribution of CTQW on $Cay(\mathbb{Z}_2^3, \{01, 10\})$ (left) and on $Cay(\mathbb{Z}_2^3, \{001, 010, 100\})$ (right) after time $\frac{\pi}{2}$

**Remark 2** Given a pair in a cubelike graph, we can assign weights to edges such that PST occurs between the given pair.

**Note 2** Quantum circuits displayed in Fig. 6 cannot be run on real quantum computers due to some technical issues such as quantum decoherence and state fidelity. We have, however, tested small graphs on the computer *ibmq_manila* as shown in Fig. 10.

## 4 Conclusion and Future Work

In this paper, we have experimentally tested perfect state transfer on IBM's quantum simulators and quantum computers on weighted cubelike graphs. We have used Hamiltonian simulation techniques to construct efficient circuits for continuous-time quantum random walks. We have verified the theoretical results of [5, 15] that PST or periodicity on integral weighted cubelike graphs occurs at time $t = \frac{\pi}{2}$, where weights are determined by Theorem 1. In the future, we plan to construct efficient quantum circuits for quantum walks on weighted abelian Cayley graphs.

## References


1. Abraham, H. et al.: Qiskit: An Open-Source Framework for Quantum Computing (2019). https://doi.org/10.5281/zenodo.2562110
2. Acasiete, F., Agostini, F., Moqadam, J., Portugal, R.: Implementation of quantum walks on ibm quantum computers. Quantum Inf. Process. **19**(426) (2020). https://doi.org/10.1007/s11128-020-02938-5
3. Aharonov, Y., Davidovich, L., Zagury, N.: Quantum random walks. Phys. Rev. A **48**, 1687–1690 (1993). https://doi.org/10.1103/PhysRevA.48.1687

4. Bernasconi, A., Godsil, C., Severini, P.: Quantum networks on cubelike graphs. Phys. Rev. A **78**, 052320 (2008). https://doi.org/10.1103/PhysRevA.78.052320
5. Cao, X., Feng, K., Tan, Y.Y.: Perfect state transfer on weighted abelian cayley graphs. Chin. Ann. Math. Ser. B **42**(4), 625–642 (2021). https://doi.org/10.1007/s11401-021-0283-4
6. Cheung, W., Godsil, C.: Perfect state transfer in cubelike graphs. Linear Algeb. Appl. **435**, 2468–2474 (2011). https://doi.org/10.1016/j.laa.2011.04.022
7. Farhi, E., Gutmann, S.: Quantum computation and decision trees. Phys. Rev. A **58**(2), 915–928 (1998). https://doi.org/10.1103/PhysRevA.58.915
8. Godsil, C.: State transfer on graphs. Disc. Math. **312**(1), 129–147 (2012). https://doi.org/10.1016/j.disc.2011.06.032
9. Godsil, C.: Periodic graphs. Electron. J. Comb. **18**(23) (2011). https://doi.org/10.37236/510
10. IBM: Quantum (2021). https://quantum-computing.ibm.com/
11. Ambrosiano, J., Adedoyin, A. et al.: Quantum Algorithm Implementations for Beginners (2020). arXiv:1804.03719
12. Kempe, J.: Quantum random walks: an introductory overview. Contemp. Phys. **44**(4), 307–327 (2003). https://doi.org/10.1080/00107151031000110776
13. Kempe, J.: Discrete quantum walks hit exponentially faster. Probab. Theory Relat. Fields **133**, 215–235 (2005). https://doi.org/10.1007/s00440-004-0423-2
14. Mulherkar, J., Rajdeepak, R., Sunitha, V.: Implementation of Hitting Times of Discrete Time Quantum Random Walks on Cubelike Graphs (2021). arXiv:2108.13769
15. Mulherkar, J., Rajdeepak, R., Sunitha, V.: Perfect State Transfer in Weighted Cubelike Graphs (2021). arXiv:2109.12607
16. Nayak, A., Vishwanath, A.: Quantum Walk on the Line (2000). arXiv:quant-ph/0010117
17. Nielsen, M.A., Chuang, I.L.: Quantum Computation and Quantum Information: 10th Anniversary Edition, 10th edn. Cambridge University Press, USA (2011)
18. Portugal, R.: Quantum Walks and Search Algorithms, 2 edn. Springer Publishing Company, Incorporated, Switzerland (2013). https://doi.org/10.1007/978-3-319-97813-0
19. Venegas-Andraca, S.E.: Quantum walks: a comprehensive review. Quantum Inf. Process. **11**(5), 1015–1016 (2012). https://doi.org/10.1007/s11128-012-0432-5
20. Wanzambi, E., Andersson, S.: Quantum Computing: Implementing Hitting Time for Coined Quantum Walks on Regular Graphs (2021). arXiv:2108.02723
21. Warat, P., Prapong, P., Unchalisa, T.: Implementation of quantum random walk on a real quantum computer. J. Phys.: Conf. Ser. **1719**, 012103 (2021). https://doi.org/10.1088/1742-6596/1719/1/012103

# Quadratically Sound Proof-of-Sequential-Work

**Souvik Sur and Dipanwita Roychowdhury**

**Abstract** Proof-of-sequential-work (PoSW) is a protocol which ensures that a prover must spend a specified number of sequential steps to evaluate a proof against some given statement, but can be efficiently verified. A crucial criterion for PoSW, known as soundness, is that a prover even with reasonable parallelism should not be able to compute the proof in steps much less than the specified amount. In particular, if a malicious prover skips $\alpha$ (known as soundness gap) fraction of computations to produce a proof, then the verifier should accept this proof with the probability $\leq (1-\alpha)^t$ using $t$ number of random challenges. While all the existing PoSWs [1, 4, 5] achieve soundness of $(1-\alpha)^t$, our proposed scheme gives a quadratic improvement of $(1-\alpha)^{2t}$. Our construction is based on linear hybrid cellular automata (LHCA), a widely used primitive in symmetric-key cryptography. Additionally, we show that our scheme is proven to be secure in the random oracle model.

**Keywords** Proofs-of-sequential-work · Cellular automata random oracle model · Soundness · Sequentiality

## 1 Introduction

A PoSW is a cryptographic protocol executed by a prover $\mathcal{P}$ and a verifier $\mathcal{V}$. Against an input $x$, $\mathcal{P}$ computes a commitment $\phi$ in $\Omega(N)$ sequential time. Then $\mathcal{V}$ asks $\mathcal{P}$ to provide proofs $\pi$ against $t$ number of challenges chosen uniformly at random. $\mathcal{V}$ accepts if and only if all the proofs $\pi$ validate the commitment $\phi$; rejects otherwise. To verify efficiently, $\mathcal{V}$ minimizes $t$ but keeps it sufficient to catch a malicious prover $\tilde{\mathcal{P}}$ skipping a fraction (say $\alpha$) of $N$. This fraction $\alpha$ is called the soundness of a

S. Sur (✉) · D. Roychowdhury
Department of Computer Science and Engineering, Indian Institute of Technology Kharagpur, Kharagpur 721302, WB, India
e-mail: souviksur@iitkgp.ac.in

D. Roychowdhury
e-mail: drc@cse.iitkgp.ac.in



B. Rushi Kumar et al. (eds.), *Mathematics and Computing*, Springer Proceedings in Mathematics & Statistics 415, https://doi.org/10.1007/978-981-19-9307-7_11

PoSW. Soundness ensures that every malicious prover $\tilde{\mathcal{P}}$ spending only $(1-\alpha)N$ sequential effort should be accepted with the probability $< (1-\alpha)^t$.

Mahmoody et al. introduced the first PoSW [5]. The prover finds a Merkle root of labels of a depth robust graph with $N$ vertices. The verifier asks for the labels of the parents of some of the vertices in order to verify the commitment. So, it takes $\Omega(N)$ time to compute the Merkle root.

Cohen and Pietrzak use a different graph in their PoSW [4]. They exploit an upward-directed binary tree with some additional edges. So, the labeling of the graph does not need a separate Merkle commitment as it is implicit in their graph.

Abusalah et al. added another dimension to PoSWs known as reversibility [1]. $\mathcal{P}$ undergoes $N$ reversible random permutations on an input $x$. $\mathcal{V}$ chooses a challenge permuted string. The proof is accepted only if the challenge string lies within the smallest number of reversed permutations.

Each of these schemes accepts any false proof with the probability at most $(1-\alpha)^t$ for a soundness gap of $0 \leq \alpha \leq 1$ and $t$ number of challenges chosen by the verifier. It is because these schemes allow $\mathcal{V}$ to verify only a single challenge in each of the $t$ rounds. The proposed PoSW in this paper scales down it to $(1-\alpha)^{2t}$ by checking two challenges simultaneously in each of the rounds.

## 1.1 Our Techniques

We propose a PoSW scheme based on linear hybrid cellular automata (LHCA). These automata fit in the current context, because they offer intrinsic randomness required for the soundness of a PoSW.

Briefly, our scheme works as follows. We use a $n$-bit linear hybrid cellular automaton where $n$ is the security parameter. The input $x \in \mathcal{X}$ passes through a hash function $h(\cdot)$ that produces a $n$-bit hash value. Both the LHCA and the hash function $h$ are public knowledge. The prover uses this hash output to initialize the state of the LHCA. The LHCA is then iterated $N \leq 2^n - 1$ times, where $N$ is the specified number of sequential steps and is again a public parameter. Under the assumption that the initializing state is a random $n$-bit vector (because it is a hash output), the sequence of the states can not be predicted unless computed. After each iteration, the prover needs to compute another hash (modeled as a random oracle $\mathsf{H}$) output with the inputs of the current state of the LHCA and the hash output computed in the last iteration. We call this output the label $\ell$ against the corresponding state. Using all the $N$ labels, the prover enumerates a Merkle tree with the root $\phi$ and announces it as the proof. The prover also stores the labels at the $d$ topmost levels of the Merkle tree.

During verification, the verifier gets the same initial state of the LHCA by calculating $h(x)$. The verifier then chooses $t$ integers uniformly at random within the range $[1, N]$. For each of these $t$ integers $\tau$, the verifier jumps onto the $\tau$-th state starting from the $h(x)$ and check the integrity of the labels of the $\tau$-th and either the $(\tau+1)$-th or the $(\tau-1)$-th states in the sequence. Therefore, for each $\tau$, two labels can be verified simultaneously. Essentially, the verifier computes the Merkle root $\ell_r$

using the labels supplied by the prover against those random challenges. If the computed root $\ell_r$ matches with the $\phi$ for all the $t$ challenges, the verifier accepts it; rejects otherwise. To jump on the $\tau$-th state, unlike the prover, the verifier multiplies the initial state vector $h(x)$ of the LHCA by $M^{\tau}$, where $M$ is the $n \times n$ transition matrix of the LHCA. This can be done in $\mathcal{O}(n^2 \log \tau)$ time with some precomputation. For $t$ challenges, this sums up to $\mathcal{O}(tn^2 \log \tau)$ time.

We show that our construction is correct and secure in the random oracle model. In particular, this scheme enforces a prover to compute a sequence of $N$ states querying the random oracle $\mathsf{H}$ at least $N$ times sequentially. Moreover, we show that depending upon the rule vector, an LHCA produces a sequence of random states during its evolution (See Theorem 1). This randomness is at the heart of our design of asymptotically difficult computations.

### *1.2 Organization of the Paper*

In Sect. 2, we present a succinct review of PoSW, random oracle, and cellular automata, and prove some results for use in later sections. We propose our PoSW scheme in Sect. 3. In Sect. 4, we establish the essential properties, correctness and soundness of the PoSW. Finally, Sect. 5 concludes the paper after highlighting an open problem in this context.

## 2 Preliminaries

We take $\mathcal{P}$ and $\mathcal{V}$ as the prover and the verifier, respectively. We denote three statistical security parameters with $w, t, n \in \mathbb{Z}^+$ and a time parameter $N \in \mathbb{Z}^+$. Let poly($n$) be some function $n^{\mathcal{O}(1)}$, and negl($n$) represents some function $n^{-\omega(1)}$. For some $x, z \in \{0, 1\}^*$, $x\|z$ implies concatenation of strings $x$ and $z$. The $i$-th bit of $x$ is represented by $x[i]$ and $x[i \dots j] = x[i]\| \dots x[j]$. We denote $|x|$ as the bit length of $x$.

If any randomized algorithm $\mathcal{A}$ outputs $y$ on an input $x$, we write $y \stackrel{R}{\leftarrow} \mathcal{A}(x)$. By $x \stackrel{\$}{\leftarrow} \mathcal{X}$, we mean that $x$ is sampled uniformly at random from $\mathcal{X}$. We consider $\mathcal{A}$ as efficient if it runs in probabilistic polynomial time (PPT). We assume $\mathsf{H}$:$\{0, 1\}^* \rightarrow \{0, 1\}^w$ is a random oracle. If an algorithm $\mathcal{A}$ queries the random oracle $\mathsf{H}$, it is denoted as $\mathcal{A}^{\mathsf{H}}$.

## 2.1 Proof of Sequential Work

**Definition 1** (*Proof of Sequential Work*) Assuming $\mathcal{X} \subseteq \{0,1\}^*$, a PoSW is a quadruple of Gen, $\mathsf{Solve}^{\mathsf{H}}$, $\mathsf{Open}^{\mathsf{H}}$, $\mathsf{Verify}^{\mathsf{H}}$ that implements a mapping $\mathcal{X} \rightarrow \{0,1\}^w$ is specified by four algorithms.

$\mathsf{Gen}(1^n, N) \rightarrow \mathbf{pp}$ is an algorithm that takes as input a security parameter $n$ and a targeted time bound $N$, and produces the public parameters **pp**.
$\mathsf{Solve}^{\mathsf{H}}(\mathbf{pp}, x) \rightarrow (\phi, \phi_{\mathcal{P}})$ takes an input $x \in \mathcal{X}$, and produces a proof $\phi \in \{0,1\}^w$ along with a $\phi_{\mathcal{P}} \in \{0,1\}^*$. $\mathcal{P}$ announces $\phi$, whereas stores $\phi_{\mathcal{P}}$ locally. Upon receiving $\phi$, $\mathcal{V}$ samples a set of random challenges $\gamma = \{\gamma_1, \gamma_2, \ldots, \gamma_t\}$, where each $\gamma_i \in [1, N]$.
$\mathsf{Open}^{\mathsf{H}}(\mathbf{pp}, x, \phi_{\mathcal{P}}, \gamma) \rightarrow \pi$ takes the challenge $\gamma$ and the $\phi_{\mathcal{P}}$ as the inputs, and sends the output $\pi \in \{0,1\}^*$ to $\mathcal{V}$.
$\mathsf{Verify}^{\mathsf{H}}(\mathbf{pp}, \pi, x, \gamma, \phi) \rightarrow \{0,1\}$ is an algorithm that takes an input $x$, a challenge $\gamma$, an output $\pi$, and a proof $\phi$, and either accepts (1) or rejects (0). The algorithm must be "significantly" faster than $\mathsf{Solve}^{\mathsf{H}}$. So, we require $\mathsf{Verify}^{\mathsf{H}}$ must run in $\mathrm{poly}(n, \log N)$ time.

The two desirable properties of a PoSW are now introduced.

**Definition 2** (*Correctness*) A PoSW is correct, if for all $n$, $N$, parameters **pp**, and $x \in \mathcal{X}$, we have

$$\Pr\left[\mathsf{Verify}^{\mathsf{H}}(\mathbf{pp}, \pi, x, \gamma, \phi) = 1 \;\middle|\; \begin{array}{l} \mathbf{pp} \leftarrow \mathsf{Gen}(1^n, N) \\ x \xleftarrow{\$} \mathcal{X} \\ (\phi, \phi_{\mathcal{P}}) = \mathsf{Solve}^{\mathsf{H}}(\mathbf{pp}, x) \\ \pi = \mathsf{Open}^{\mathsf{H}}(\mathbf{pp}, x, \phi_{\mathcal{P}}, \gamma) \end{array}\right] = 1.$$

$\mathcal{V}$ always accept a proof $\phi$ generated by $N$ sequential queries to H.

**Definition 3** (*Soundness*) A PoSW is sound if for all non-uniform algorithms $\tilde{\mathcal{P}}$ that run in parallel time $o(N)$, we have

$$\Pr\left[\begin{array}{l} \phi \neq \mathsf{Solve}^{\mathsf{H}}(\mathbf{pp}, x) \\ \mathsf{Verify}^{\mathsf{H}}(\mathbf{pp}, \pi, x, \gamma, \phi) = 1 \end{array} \;\middle|\; \begin{array}{l} \mathbf{pp} \leftarrow \mathsf{Gen}(1^n, N) \\ (x, \phi, \phi_{\mathcal{P}}) \leftarrow \tilde{\mathcal{P}}(1^n, 1^N, \mathbf{pp}) \\ \pi = \mathsf{Open}^{\mathsf{H}}(\mathbf{pp}, x, \phi_{\mathcal{P}}, \gamma) \end{array}\right] \leq negl(n).$$

All non-uniform parallel adversaries $\tilde{\mathcal{P}}$ should pass the verification at most negligible probability.

## 2.2 Cellular Automata

Cellular automata (CA, for short) is a model of computation. A one-dimensional CA can be visualized as a grid of cells that assume values at discrete time steps according to a set of functions working on the states of neighboring cells [3]. In this paper, we focus on one-dimensional binary cellular automaton. We call a CA binary when its cells can have only binary values, i.e., $\{0, 1\}$. Let $n$ be the number of cells in the CA. We call an ensemble of values at a time-step $t$ as a state of the CA and is denoted by an $n$-dimensional vector $s(t)$. Let, the $i$-th bit of the vector $s(t)$ be $b_i^t$. The cells are numbered $i = 0, 1, 2, \ldots, n-1$ in a left-to-right manner. The cells $b_i^t$ assume values from a set of boolean functions $r_i$, traditionally called rules. Two rules are used in this work.

Rule 90: $b_i^{t+1} = b_{i-1}^t \oplus b_{i+1}^t$.
Rule 150: $b_i^{t+1} = b_{i-1}^t \oplus b_i^t \oplus b_{i+1}^t$.

Note that the bits at the positions $i = 0$ and $i = n-1$ require the bits $b_{-1}^t$ and $b_n^t$ to be defined. We call a CA null-boundary CA if these boundary-bits $b_{-1}^t$ and $b_n^t$ are always assumed to be 0 for all $t$. Like the states, the ensemble of rules $r_0, \ldots, r_{n-1}$ can also be represented as bit-vector since there are only two rules (90 and 150). If $r_i = 0$, Rule 90 applies to the $i$-th cell of the CA. If $r_i = 1$, Rule 150 applies to the $i$-th cell of the CA. A rule $r_i$ is linear if it uses only linear operators, e.g., logical XOR $\oplus$. Rules 90 and 150 are linear as they use only $\oplus$ operator. A CA is called linear if it uses only linear rules. Further, a CA is called hybrid if all rules are not identical to each other. For example, the CA in Fig. 1 is a hybrid CA. The state transition of an 8-bit LHCA under a given rule vector is illustrated in Fig. 1.

An LHCA with rules 90 and 150, starting from an all-zero state $s(0) = 0^n$ continues to stay in that state. If the remaining $2^n - 1$ non-zero states occur in a single cycle, the LHCA is called a maximum-length LHCA. In order to derive the rule vector for a maximum-length LHCA of $n$ cells, we need a primitive polynomial of degree $n$ over $\mathbb{F}_2$. There exists a deterministic algorithm [2] to generate the rule vector for an $n$-cell maximum-length LHCA from this primitive polynomial.

As all rules are linear, an LHCA can also be characterized by an $n \times n$ linear map $M$ over $\mathbb{F}_2$, such that

$$M_{ij} = \begin{cases} 1, & \text{if the } b_i^{t+1} \text{ depends on } b_j^t \\ 0, & \text{otherwise} \end{cases}$$

| | | | | | | | | |
|---|---|---|---|---|---|---|---|---|
| State $s(t)$ at time $t$ | 0 | 1 | 0 | 1 | 1 | 1 | 0 | 0 |
| Rule vector | 1 | 0 | 1 | 0 | 1 | 1 | 0 | 0 |
| State $s(t+1)$ at time $t+1$ | 1 | 0 | 0 | 1 | 1 | 0 | 1 | 0 |

**Fig. 1** State change of an LHCA with rule vector $\langle 150, 90, 150, 90, 150, 150, 90, 90 \rangle$

This linear map $M$ is called the characteristic matrix of the LHCA. The $\tau$-th state $s(t+\tau)$ starting from the state $s(t)$ can be obtained as

$$s(t+\tau) = M^{\tau}s(t) \tag{1}$$

Thus, given an arbitrary $\tau < 2^n - 1$, the $\tau$-th state $s(\tau)$ can be found in $O(\log \tau)$-time via the characteristic matrix. Still, the states are updated using the rule vector as matrix exponentiation is much costlier than evaluating the rules of an LHCA. Given a rule vector, the characteristic matrix $M$ can be efficiently constructed by placing the rule vector along the principle diagonal of $M$, filling the super and the sub diagonals with 1s, and leaving all other entries as 0s.

We restate two theorems on LHCAs from [7].

**Theorem 1** *Let $\mathcal{L}$ be an $n$-cell LHCA governed by Rule 90 and Rule 150 only. If $\mathcal{L}$ is initialized by a random bit sequence (like the output of a well-designed hash function), then at any time $t$, the bits $b_i^{(t)}$ of $\mathcal{L}$ remain independent of one another with* $\Pr[b_i^t = 0] = \Pr[b_i^t = 1] = \frac{1}{2}$ *for all $i$.*

***Proof*** We proceed by induction on $t$. For $t = 0$, the result follows from the induction hypothesis that the initial state is a random bit sequence. For the inductive step, suppose that the bits of $\mathcal{L}$ are unbiased and independent at some time $t \geq 0$. It suffices to show that

$$\Pr[b_i^{t+1} = x \mid b_j^{t+1} = y] = \Pr[b_i^{t+1} = x] = \frac{1}{2} \tag{2}$$

for all $i, j$ with $i \neq j$, and for all $x, y \in \{0, 1\}$. For any $i$, if $r_i$ applies rule 90, then $b_i^{t+1} = b_{i-1}^t \oplus b_{i+1}^t$, and we say $support_i = \{i-1, i+1\}$. If $r_i$ applies rule 150, then $b_i^{t+1} = b_{i-1}^t \oplus b_i^t \oplus b_{i+1}^t$, and we say $support_i = \{i-1, i, i+1\}$. If $|i - j| > 2$, then $support_i \cap support_j = \emptyset$, and thus the bits $b_i^{t+1}$ and $b_j^{t+1}$ are independent. So assume that $j = i + 1$. Since $r_i = 0$ implies rule 90 and $r_i = 1$ denotes rule 150, we may write $b_i^{t+1} = b_{i-1}^t \oplus r_i b_i^t \oplus b_{i+1}^t$, and $b_j^{t+1} = b_{i+1}^{t+1} = b_i^t \oplus r_{i+1} b_{i+1}^t \oplus b_{i+2}^t$. The bit $b_i^{t+1}$ depends on $b_{i-1}^t$ but not on $b_{i+2}^t$. Moreover, the bit $b_j^{t+1}$ depends on $b_{i+2}^t$ but not on $b_{i-1}^t$. By induction hypothesis, $b_{i-1}^t$ and $b_{i+2}^t$ are independent, so Eq. (2) holds. The cases $j = i-1, i+2, i-2$ can be analogously handled. □

**Theorem 2** *If $\mathcal{L}$ is a null-boundary LHCA and the rule vector applies Rule* 150 *to both the boundary cells, then for all $t \geq 0$ and, for all $i, j$, the bits $b_i^{t+1}$ are independent of any single $b_j^t$ only.*

***Proof*** Let us call the bits $b_i^{t+1}$ at two boundaries $i \in \{0, n-1\}$ as the boundary bits and all the other bits at the positions $0 < i < n-1$ as the non-boundary bits. The non-boundary bits are dependent on multiple $b_j^t$s. If $r_i = 0$ (i.e., for rule 90), the non-boundary bit $b_i^{t+1} = b_{i-1}^t \oplus b_{i+1}^t$. Similarly, when $r_i = 1$ (i.e., for rule 150), $b_i^{t+1} = b_{i-1}^t \oplus b_i^t \oplus b_{i+1}^t$. On the contrary, if $r_0 = r_{n-1} = 0$, the boundary bits depend on a single $b_j^t$. Since $\mathcal{L}$ is a null-boundary CA, $b_{-1}^t = b_n^t = 0$ always. If $r_0 = 0$, then $b_0^{t+1} = b_1^t$, whereas if $r_{n-1} = 0$, then $b_{n-1}^{t+1} = b_{n-2}^t$. Taking $r_0 = r_{n-1} = 1$ makes $b_0^{t+1} = b_0^t \oplus b_1^t$ and $b_{n-1}^{t+1} = b_{n-2}^t \oplus b_{n-1}^t$. Now, by Theorem 1 at any $t \geq 0$, all $b_j^t$ are independent of each other. Therefore, $r_0 = r_{n-1} = 1$ makes each $b_i^{t+1}$ for $0 \leq i \leq n-1$ independent of any single $b_j^t$ only. □

The number of primitive polynomials of degree $n$ is $\phi(2^n - 1)/n$, where $\phi()$ denotes Euler's totient function. So, for any $n$, there exist a plenty of primitive polynomials satisfying Theorem 2. Therefore, finding such a rule vector would not be hard practically.

Theorem 1 implies that the bits of an LHCA remain independent of one another for all $t$, when it is initialized with a random state. Theorem. 2 suggests that there is no leakage of information between two successive states. It means, given a state $s(t)$ of an LHCA, the next state $s(t+1)$ appears as a pseudo-random bit-string which can not be guessed unless computed. Therefore, the sequence of states $s(0), s(1), \ldots, s(2^n - 1)$ looks like a pseudo-random sequence of bit-strings. These observations constitute the foundation of our PoSW scheme.

# 3 PoSW Based on Cellular Automata

In this section, we propose a PoSW scheme based on a maximum-length null-boundary LHCA $\mathcal{L}$. We denote the characteristic matrix of $\mathcal{L}$ by $M$. The number of cells of $\mathcal{L}$ is taken to be same as the security parameter $n \in \mathbb{Z}^+$. Moreover, the targeted sequential steps are taken as $\mathcal{O}(N)$. The four algorithms that specify our PoSW are now described.

## *3.1 The* $\mathsf{Gen}(1^n, N)$ *Algorithm*

This algorithm outputs the public parameters $\mathbf{pp} = \langle \mathcal{R}, \mathsf{H}, h, w, d, t \rangle$ having the following meanings.

1. $\mathcal{R}$ is the rule vector of a maximum-length, null-boundary LHCA $\mathcal{L}$.
2. We keep the size of $\mathcal{L}$ same as $n$.
3. As stated in Sect. 2.1 $w, t \in \mathbb{Z}^+$. Another integer $d < n$ is introduced.
4. We denote the set of all states of the LHCA by $\mathcal{S}$. We take $h : \mathcal{X} \rightarrow \mathcal{S}$ and $\mathsf{H} : \mathcal{S} \times \{0, 1\}^w \rightarrow \{0, 1\}^w$ to be two efficiently computable hash functions. $\mathsf{H}$ is modeled as a random oracle.

**Complexity**: Among the public parameters, only the rule vector needs to be computed. As already mentioned in Sect. 2, given a primitive polynomial of degree $n$, $\mathcal{R}$ is generated by a polynomial-time algorithm [2]. It is a randomized algorithm that succeeds with the probability $1/2$ if $n$ is even, and is deterministic if $n$ is odd.

The additional requirement of having Rule 150 at the boundary cells implies a constant expected number of searches for the appropriate rule vector. We can design $\mathsf{H}$ and $h$ from any good hash function. The parameter $t$ is used to confirm that the $\mathsf{Verify}$ algorithm succeeds with high probability for all outputs without violating the soundness.

### 3.2 The $\mathsf{Solve}^{\mathsf{H}}$ (pp, $x$) Algorithm

Without the loss of generality, we may consider $N$ to be the nearest smallest power of 2 greater than or equal to the specified number of steps. Essentially, $\mathsf{Solve}^{\mathsf{H}}$ equally partitions the sequence $\mathcal{S}$ into $2^d$ partitions (See Fig. 2). For each of these partitions, it computes the $\mathsf{H}$-sequence of length $N/2^d$ using $\ell_j \leftarrow \mathsf{H}(s(j)||\ell_{j-1})$ for $1 \leq j \leq N/2^d$. Subsequently for each partition, it finds the Merkle root $\ell_{r_{i-1}}$ of the Merkle tree $\{G_{i-1}\}_{1\leq i\leq 2^d}$ assuming $\ell_j$'s are sorted by their indices. Also, for each of the $i$-partitions, $\ell_1 \leftarrow \mathsf{H}(s(1)||\ell_{r_{i-1}})$ except the very first partition. The first one assumes $\ell_{r_0} = 0^w$. Finally, it computes the Merkle root $\ell_r$ of the tree $G$ from the roots of the trees $\{G_{i-1}\}_{1\leq i\leq 2^d}$. The algorithm $\mathsf{Solve}^{\mathsf{H}}$ is defined as follows.

1. Use $h$ to map the challenge $x \in \mathcal{X}$ to $s(0) \in \mathcal{S}$.
2. Initialize $salt \leftarrow 0^w$.
3. Repeat for $1 \leq i \leq 2^d$:
   (a) Assign $\tau \leftarrow \frac{(i-1)N}{2^d}$.
   (b) Initialize $\mathcal{L}$ with $s(\tau)$.
   (c) Compute $\ell_\tau \leftarrow \mathsf{H}(s(\tau)||salt)$.
   (d) Repeat for $1 \leq j \leq \frac{N}{2^d}$:
      i Update $\mathcal{L}$ to $s(\tau + j)$ from $s(\tau + j - 1)$.
      ii Compute the label $\ell_j \leftarrow \mathsf{H}(s(\tau + j - 1)||\ell_{j-1})$.
   (e) Compute the Merkle root $\ell_{r_{i-1}}$ of the tree $G_{i-1}$ from $\bigcup_{j=1}^{\frac{N}{2^d}} \ell_{j-1}$ (See Fig. 2).
   (f) Assign $salt \leftarrow \ell_{r_{i-1}}$.
4. Compute the Merkle root $\ell_r$ of the tree $G$ from the roots $\bigcup_{i=1}^{2^d} \ell_{r_{i-1}}$ of $\{G_{i-1}\}_{1\leq i\leq 2^d}$.
5. Announce the $\ell_r$ as $\phi$ and store all the labels of $G$ as $\phi_{\mathcal{P}}$.

**Complexity**: The effort spent by the prover to run $\mathsf{Solve}^{\mathsf{H}}$ (**pp**, $x$) is now deduced. We assume that each single call of the hash function $h$ takes $\mathcal{O}(1)$ time. Each state update $s(i) \rightarrow s(i+1)$ of $\mathcal{L}$ requires computing the next value of each of the $n$ cells. Depending on whether Rule 90 or Rule 150 is used for a cell, the update for that cell requires one or two two-input XOR calculations. For each transition, the number of XOR operations is in the range $[n, 2n]$. Although $\mathcal{S}$ has been partitioned into $2^d$ subsequences, each of these partitions takes the Merkle root of the last partition as an input to the very first $\mathsf{H}$-computation of its $\mathsf{H}$-sequence. Therefore, any malicious

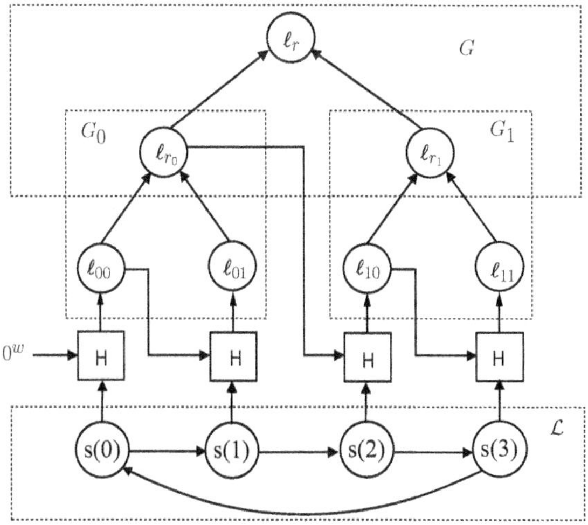


**Fig. 2** A schematic view of the PoSW scheme on an LHCA of size $n = 2$ and $d = 1$. The indices $i$s of the labels $\ell_i$s have been represented in binary format. When $N = 2^n - 1$, a dummy state $s(2^n - 1) = \{0\}^n$ has been added at the end of the cycle in order to make $N$ as a power of 2. In this example, it is $s(3)$

prover $\tilde{\mathcal{P}}$ having even $\mathcal{O}(N)$ parallel processors needs to compute the entire H-sequence sequentially spending $\mathcal{O}(N)$-time. The processor that computes the very first label $\ell$ in every $G_i$ in the sequence must wait until the Merkle root of the $G_{i-1}$ is computed. As there are $N$ iterations, the total effort of $\mathcal{P}$ is $\mathcal{O}(nN)$ along with $N$ sequential queries to the oracle H. When $N = \mathcal{O}(2^n)$, it becomes an exponential expression in the security parameter $n$.

### 3.2.1 Precompute and Jump

Before we proceed to describe the $\mathsf{Open}^{\mathsf{H}}$ and $\mathsf{Verify}^{\mathsf{H}}$ algorithm, let us define two procedures, Precompute and Jump.

Precompute($\cdot$) is a polynomial-time procedure that needs to be run only once unless the rule vector $\mathcal{R}$ is modified. Indeed, this procedure may be considered as a part of the Gen algorithm. Specifically, it uses the idea of fixed-base exponentiation [6] in order to reduce the running time of Verify by a factor of $n$. Given the rule vector $\mathcal{R}$, the transition matrix $M$ of $\mathcal{L}$ can be constructed efficiently as described in Sect. 2.2. The matrices $M^{2^i}$ for $i = 0, 1, 2, \ldots, n-1$ are precomputed and stored to be used in Jump.

Jump$(j, \tau) \rightarrow \{0, 1\}^n$ jumps from a state $s(j)$ to the state $s(j+\tau)$ in polynomial time using the precomputed matrices $M, M^2, \ldots, M^{2^{n-1}}$. First, it decomposes $\tau = 2^{i_1} + 2^{i_2} + \cdots + 2^{i_p}$ then computes $s(j+\tau) = (M^{2^{i_1}}(M^{2^{i_2}}(\cdots(M^{2^{i_p}}s(i))\cdots))$. The 1-bit positions $i_1, i_2, \ldots, i_p$ in $\tau$ can be identified in $\mathcal{O}(n)$ time. Since $N \leq 2^n - 1$, the computation of $s(j+\tau)$ needs at most $n$ matrix-vector multiplications; so, it requires $\mathcal{O}(n^2 \log \tau)$ time.

### 3.3 *The* $\mathsf{Open}^{\mathsf{H}}$ $(\mathrm{pp}, x, \phi_{\mathcal{P}}, \gamma)$

This algorithm is the standard Merkle commitment opening algorithm and works as follows:

1. Repeat for $0 \leq i \leq t-1$:
   (a) If $\gamma_i$ is even then assign $\tau \leftarrow (\gamma_i - 1)$; otherwise, $\tau \leftarrow \gamma_i$.
   (b) Enumerate $s(\tau) \leftarrow \mathsf{Jump}(0, \tau)$.
   (c) Find the set $\pi_i = \bigcup_{k \in T_i} \ell_k$, where $T_i \stackrel{\text{def}}{=} \bigcup_{j=1}^{\log N} \gamma_i[1 \ldots j-1] || \overline{\gamma_i[j]}$
2. Collect all $\pi = \bigcup_{i=1}^{t} \{\ell_{\gamma_{i-1}} \cup \ell_{\gamma_i} \cup \pi_i\}$

**Complexity**: The effort spent by the prover to open the commitments is now enumerated. For each $\gamma_i$, $\mathcal{P}$ needs to supply $\log N - 1$ number of labels, correspond to each levels of $G$ and $G_{\gamma_i}$. As $G$ is already stored in $\phi_{\mathcal{P}}$, $\mathcal{P}$ requires to figure out $G_{\gamma_i}$ only. So, $\mathcal{P}$ first finds out the partition of $\mathcal{S}$ containing the $\gamma_i$ and construct the Merkle tree $G_{\gamma_i}$. Then it sends the $\log(N/2^d)$ required labels. So, (s)he queries the oracle $\mathsf{H}$ for $\mathcal{O}(N/2^d - 1)$ times in each iteration. As there are $t$ iterations in total, it takes $\mathcal{O}(t(N/2^d - 1))$ sequential queries the oracle $\mathsf{H}$. Also a $\mathsf{Jump}(0, \tau)$ needs $\mathcal{O}(n^2 \log \tau)$ time yielding $\mathcal{O}(tn^2 \log \tau)$ time in $t$ iterations.

### 3.4 *The* $\mathsf{Verify}^{\mathsf{H}}$ $(\mathrm{pp}, x, \gamma, \pi, \phi)$ *Algorithm*

The $\mathsf{Verify}^{\mathsf{H}}$ algorithm runs the following steps:

1. Use $h$ to map the challenge $x \in \mathcal{X}$ to $s(0) \in \mathcal{S}$.
2. Repeat for $0 \leq i \leq t-1$:
   (a) If $\gamma_i$ is even then assign $\tau = \gamma_i - 1$; otherwise, $\tau = \gamma_i$.
   (b) Enumerate $s(\tau) \leftarrow \mathsf{Jump}(0, \tau)$.
   (c) Set a flag $f_i$ if and only if:
      i. $\ell_\tau \stackrel{?}{=} \mathsf{H}(s(\tau)||\ell_{\tau-1})$ and $\ell_{\tau+1} \stackrel{?}{=} \mathsf{H}(s(\tau+1)||\ell_\tau)$.
      ii. The Merkle root $\ell_r \stackrel{?}{=} \phi$, where $\ell_r$ can be computed recursively using $\ell_{\gamma_i[0 \ldots i]} = \mathsf{H}(\ell_{\gamma_i[0 \ldots i]||0} || \ell_{\gamma_i[0 \ldots i]||1})$.
3. Assign $f \leftarrow f_0 \wedge f_1 \wedge f_2 \wedge \ldots \wedge f_{t-1}$.
4. Accept if $f = 1$; reject otherwise.

**Complexity**: The effort needed for the verification is now evaluated. The initial query to $h$ should be done in $\mathcal{O}(1)$ time. For each $\gamma_i$, there is a single call for $\mathsf{Jump}$, update of $\mathcal{L}$, and Merkle root computation yielding $O(n^2 \log \tau)$, $O(n)$ time, and $\log N$ queries to $\mathsf{H}$, respectively. So, in $t$ iterations, the required effort is $t \log N$ queries to $\mathsf{H}$ plus $\mathcal{O}(tn^2 \log \tau)$ time; a polynomial in the security parameter $n$.

### 3.5 *Efficiency*

Here, we will discuss the efficiencies of both the prover $\mathcal{P}$ and the verifier $\mathcal{V}$ in terms of number of H-queries and the memory requirement for $\phi$ and $\phi_{\mathcal{P}}$.

**Prover's efficiency**: As already mentioned $\mathcal{P}$ needs to run $\mathsf{Solve}^{\mathsf{H}}$ and $\mathsf{Open}^{\mathsf{H}}$. So, $\mathcal{P}$ requires $\mathcal{O}(N + \frac{t(N-1)}{2^d}))$ number of H-queries. In general, for any $0 \leq \beta \leq 1$ and $d = n/\beta$, $\mathcal{P}$ requires $t.N^{\beta}$ sequential queries for $\mathsf{Open}^{\mathsf{H}}$ on top of the $N$ queries for $\mathsf{Solve}^{\mathsf{H}}$ to H and $N^{1-\beta}.w$ space to store $\phi_{\mathcal{P}}$.

**Verifier's efficiency**: $\mathcal{V}$ only executes $\mathsf{Verify}^{\mathsf{H}}$. So, $\mathcal{V}$ requires only $\mathcal{O}(t \log N)$ number of H-queries using $t.w. \log N$ space. Moreover, (s)he needs $t.w$ space for the random challenge $\gamma$.

## 4 Security of the Proposed PoSW

In this section, we establish two essential properties of PoSW. Throughout the analysis, we will assume that $\mathcal{P}$ evaluates $h(x)$ honestly as $h$ is called only once.

### 4.1 *Correctness*

According to Definition 2, any PoSW should always accept, a valid proof against the corresponding input from its domain. The following theorem establishes this correctness property of our PoSW scheme.

**Theorem 3** *The proposed* PoSW *is correct.*

***Proof*** In the algorithm $\mathsf{Solve}^{\mathsf{H}}$, since $h$ is a deterministic hash function, $s(0) \in \mathcal{S}$ is uniquely determined by the challenge $x \in \mathcal{X}$. Moreover, if $\phi$ is the correct evaluation of $\mathsf{Solve}^{\mathsf{H}}$, then $\forall i$, $f_i$ must be 1 results into $f = 1$ in $\mathsf{Verify}^{\mathsf{H}}$. So, $\mathcal{V}$ has to accept it. It therefore follows that

$$\Pr\left[\mathsf{Verify}^{\mathsf{H}}(\mathbf{pp}, x, \phi, \gamma, \pi) = 1 \;\middle|\; \begin{array}{l} \mathbf{pp} \leftarrow \mathsf{Gen}(1^n, N) \\ x \xleftarrow{\$} \mathcal{X} \\ (\phi, \phi_{\mathcal{P}}) = \mathsf{Solve}^{\mathsf{H}}(\mathbf{pp}, x) \\ \pi = \mathsf{Open}^{\mathsf{H}}(\mathbf{pp}, x, \phi_{\mathcal{P}}, \gamma) \end{array}\right] = 1.$$

□

## 4.2 Soundness

Here, we show that the PoSW is *sound*, means a dishonest prover $\tilde{\mathcal{P}}$ succeeds to convince $\mathcal{V}$ to accept a misleading proof with at most negligible probability.

**Theorem 4** *With parameters $t, w, N$ and a "soundness gap" $\alpha > 0$, even if $\tilde{\mathcal{P}}$ enumerates the sequence of states $S$ correctly but makes at most $(1-\alpha)N$ sequential queries to $\mathsf{H}$ after receiving $x$, and at most $q$ queries in total, then $\mathcal{V}$ will accept the proof $\phi$ with probability,*

$$\Pr[\tilde{\mathcal{P}}\ wins] < (1-\alpha)^{2t} + \frac{(n-1).w.q^2}{2^w}$$

***Proof*** Suppose $\tilde{\mathcal{P}}$ has made only $m = (1-\alpha)N$ queries to $\mathsf{H}$ to compute the correct labels $\ell_{i+1} = \mathsf{H}(s(i)||\ell_i)$. For each $\gamma_i$, $\mathcal{V}$ checks the correctness of the labels for both $\ell_{\gamma_i}$ and $\ell_{\gamma_{i+1}}$ by two ways. First, $\mathcal{V}$ checks if both $\ell_{\gamma_i} \stackrel{?}{=} \mathsf{H}(s(\gamma_i - 1)||\ell_{\gamma_i - 1})$ and $\ell_{\gamma_i + 1} \stackrel{?}{=} \mathsf{H}(s(\gamma_i)||\ell_{\gamma_i})$ are true. If yes, then if the corresponding $\ell_r \stackrel{?}{=} \phi$ or not. So on the $i$-th trial, i.e., for each $\gamma_i$, $\tilde{\mathcal{P}}$ passes the verification with the probability,

$$\begin{aligned}
\Pr[\tilde{\mathcal{P}}^{\gamma_i}\ wins] &= \Pr[\ell_r = \phi | \ell_{\gamma_i}, \ell_{\gamma_{i+1}} \text{ both are correct }] \\
&= \left(\frac{m}{N}\right)^{\frac{N}{2d}-2} \times \left(\frac{m-2i}{N-2i} \times \frac{m-2i-1}{N-2i-1}\right) \\
&< \left(\frac{m}{N}\right)^{\frac{N}{2d}}
\end{aligned} \tag{3}$$

Since $t << N$, we may assume none of the $\gamma_i$ and $\gamma_i \pm 1$ collide with each other, i.e., $\forall i, j, |\gamma_i - \gamma_j| \geq 3$. So, for all $i$, the events $[\tilde{\mathcal{P}}^{\gamma_i}\ win]$s are independent of each other. Therefore, $\tilde{\mathcal{P}}$ passes the verification with the probability,

$$\begin{aligned}
\Pr[\tilde{\mathcal{P}}\ wins] &= \prod_{i=0}^{t-1} \Pr[\tilde{\mathcal{P}}^{\gamma_i}\ win] \\
&< \prod_{i=0}^{t-1} \left(\frac{m}{N}\right)^2 \\
&= \left(\frac{m}{N}\right)^{2t} \\
&= (1-\alpha)^{2t}
\end{aligned} \tag{4}$$

Finally, $\tilde{\mathcal{P}}$ would have a minuscule advantage due to the collision and the sequentiality in random oracle $\mathsf{H}$ established in [4].

Therefore, $\tilde{\mathcal{P}}$ passes the verification with the probability,

$$\Pr[\tilde{\mathcal{P}}\ wins] < (1-\alpha)^{2t} + \frac{(n-1).w.q^2}{2^w}$$

□

## 5 Conclusion and Open Problem

This paper presents an idea of constructing a proof of sequential work based on linear hybrid cellular automata and hash function. Our scheme is proven to be correct and much more sound in the random oracle model. We have been able to link the security of our scheme with the inherent randomness of cellular automata and the sequentiality in the random oracle. It remains open to establish a verifiable delay function with similar constructions if we eliminate the random oracle from this and able to find some algebraic hardness assumptions from the same.

## References


1. Abusalah, H., Kamath, C., Klein, K., Pietrzak, K., Walter, M.: Reversible proofs of sequential work. In: Ishai, Y., Rijmen, V. (eds.) Advances in Cryptology—EUROCRYPT 2019—38th Annual International Conference on the Theory and Applications of Cryptographic Techniques, Darmstadt, Germany, 19–23 May 2019, Proceedings, Part II. Lecture Notes in Computer Science, vol. 11477, pp. 277–291. Springer (2019). https://doi.org/10.1007/978-3-030-17656-3_10
2. Cattell, K., Muzio, J.: Synthesis of one-dimensional linear hybrid cellular automata. IEEE Trans. CAD Integr. Circuits Syst. **15**, 325–335 (1996). https://doi.org/10.1109/43.489103
3. Chaudhuri, P.P., Chowdhury, D.R., Nandi, S., Chattopadhyay, S.: Additive Cellular Automata: Theory and Applications, vol. 1. Wiley (1997)
4. Cohen, B., Pietrzak, K.: Simple proofs of sequential work. In: Nielsen, J.B., Rijmen, V. (eds.) Advances in Cryptology—EUROCRYPT 2018—37th Annual International Conference on the Theory and Applications of Cryptographic Techniques, Tel Aviv, Israel, 29 Apr.–3 May 2018 Proceedings, Part II. Lecture Notes in Computer Science, vol. 10821, pp. 451–467. Springer (2018). https://doi.org/10.1007/978-3-319-78375-8_15
5. Mahmoody, M., Moran, T., Vadhan, S.P.: Publicly verifiable proofs of sequential work. In: Kleinberg, R.D. (ed.) Innovations in Theoretical Computer Science, ITCS '13, Berkeley, CA, USA, 9–12 Jan. 2013, pp. 373–388. ACM (2013). https://doi.org/10.1145/2422436.2422479
6. Menezes, A., van Oorschot, P.C., Vanstone, S.A.: Handbook of Applied Cryptography. CRC Press (1996). https://doi.org/10.1201/9781439821916
7. Sur, S., Das, A., Chowdhury, D.R.: Carrency: an energy-efficient proof-of-work scheme for crypto-currencies. In: Proceedings of the Seventh International Conference on Mathematics and Computing—ICMC 2021, Shibpur, India, pp. 23–38 (2021). https://doi.org/10.1007/978-981-16-6890-6_3

# On Some Properties of $K$-type Block Matrices in the Context of Complementarity Problem

**A. Dutta and A. K. Das**

**Abstract** In this article, we introduce $K$-type block matrices which include two new classes of block matrices, namely block triangular $K$-matrices and hidden block triangular $K$-matrices. We show that the solution of linear complementarity problem with $K$-type block matrices can be obtained by solving a linear programming problem. We show that block triangular $K$-matrices satisfy the least element property. We prove that hidden block triangular $K$-matrices are $Q_0$ and processable by Lemke's algorithm. The purpose of this article is to study properties of $K$-type block matrices in the context of the solution of linear complementarity problem.

**Keywords** $Z$-matrix · Hidden $Z$-matrix · Linear programming problem · Linear complementary problem · Semi-sublattice · $P$-matrix · $Q_0$-matrix

## 1 Introduction

The linear complementarity problem is a combination of linear and nonlinear systems of inequalities and equations. The problem may be stated as follows: Given $M \in R^{n\times n}$ and a vector $q \in R^n$, the linear complementarity problem, LCP$(M, q)$ is the problem of finding a solution $w \in R^n$ and $z \in R^n$ to the following system of linear equations and inequalities:

$$w - Mz = q, \quad w \geq 0, \; z \geq 0, \; w^T z = 0.$$

In complementarity theory, several matrix classes are considered due to the study of theoretical properties, applications, and its solution methods. For details see [11–13]. It is well known that the linear complementarity problem can be solved by a linear program if $M$ or its inverse is a $Z$-matrix, i.e., a real square matrix with non-

A. Dutta (✉)
Jadavpur University, Kolkata 700032, India
e-mail: aritradutta001@gmail.com

A. K. Das
Indian Statistical Institute, 203 B. T. Road, Kolkata 700 108, India



B. Rushi Kumar et al. (eds.), *Mathematics and Computing*, Springer Proceedings in Mathematics & Statistics 415, https://doi.org/10.1007/978-981-19-9307-7_12

positive off-diagonal elements. A number of authors have considered the special case of the linear complementarity problem under the restriction that $M$ is a $Z$-matrix. Chandrasekharan [16] considered $Z$-matrix solving a sequence of linear inequalities. Lemke's algorithm is a well-known technique for solving linear complementarity problem [1]. Mangasarian [8] showed that the following linear program:

$$\begin{aligned} \text{minimize} \;\; & p^T u \\ \text{subject to} \;\; & q + Mu \geq 0, \\ & u \geq 0 \end{aligned} \tag{1}$$

for an easily determined $p \in R^n$ solves the linear complementarity problem for a number of special cases especially when $M$ is a $Z$-matrix. Mangasarian [8] proved that least element of the polyhedral set $\{u : q + Mu \geq 0, u \geq 0\}$ in the sense of Cottle–Veinott can be obtained by a single linear program. It is well known that the quadratic programming problem

$$\begin{aligned} &\text{minimize } q^T u + \tfrac{1}{2} u^T M u \\ &\text{subject to } u \geq 0 \end{aligned}$$

can be formulated as a linear complementarity problem when $M$ is symmetric positive semidefinite. Mangasarian showed that this problem can be solved using a single linear program if $M$ is a $Z$-matrix. Hidden $Z$-matrices are the extension of $Z$-matrices. A matrix $M$ is said to be a hidden $Z$-matrix if $\exists$ two $Z$-matrices $X$ and $Y$ such that

1. $MX = Y$
2. $r^T X + s^T Y > 0$, for some $r, s \geq 0$.

For details, see [5, 6]. In this paper, we introduce block triangular $K$-matrix and hidden block triangular $K$-matrix. We call these two classes collectively $K$-type block matrix. We discuss the class of $K$-type block matrices in solution aspects for linear complementarity problem.

The paper is organized as follows. Section 2 presents some basic notations, definitions, and results. In sect. 3, we establish some results of these two matrix classes. We show that a linear complementarity problem with block triangular $K$-matrix and hidden block triangular $K$-matrix can be solved using linear programming problem.

## 2 Preliminaries

We denote the $n$-dimensional real space by $R^n$. $R^n_+$ denotes the nonnegative orthant of $R^n$. We consider vectors and matrices with real entries. Any vector $x \in R^n$ is a column vector and $x^T$ denotes the row transpose of $x$. $e$ denotes the vector of all 1. A matrix is said to be nonnegative or $M \geq 0$ if $m_{ij} \geq 0 \,\forall\, i, j$. A matrix is said to be positive if $m_{ij} > 0 \,\forall\, i, j$. Let $M$ and $N$ be two matrices with $M \geq N$, then $M - N \geq 0$. If $M$ is a matrix of order $n$, $\alpha \subseteq \{1, 2, \cdots, n\}$ and $\bar{\alpha} \subseteq \{1, 2, \cdots, n\} \setminus \alpha$, then $M_{\alpha\bar{\alpha}}$ denotes the submatrix of $M$ consisting of only the rows and columns of $M$ whose indices are in $\alpha$ and $\bar{\alpha}$, respectively. $M_{\alpha\alpha}$ is called a principal submatrix of M and $\det(M_{\alpha\alpha})$ is called a principal minor of $M$. Given a matrix $M \in R^{n\times n}$ and a vector $q \in R^n$, we define the feasible set $\text{FEA}(M, q) = \{z \in R^n : z \geq 0, q + Mz \geq 0\}$ and the solution set of LCP$(M, q)$ by $\text{SOL}(M, q) = \{z \in \text{FEA}(M, q) : z^T(q + Mz) = 0\}$.

We state the results of two-person matrix games in linear system with complementary conditions due to von Neumann [17] and Kaplansky [18]. The results say that there exist $\bar{x} \in R^m$, $\bar{y} \in R^n$ and $v \in R$ such that

$$\sum_{i=1}^{m} \bar{x}_i a_{ij} \leq v, \ \forall\, j = 1, 2, \ldots, n,$$
$$\sum_{j=1}^{n} \bar{y}_j a_{ij} \geq v, \ \forall\, i = 1, 2, \ldots, m.$$

The strategies $(\bar{x}, \bar{y})$ are said to be optimal strategies for player I and player II and $v$ is said to be the minimax value of game. We write $v(A)$ to denote the value of the game corresponding to the payoff matrix $A$. The value of the game, $v(A)$ is positive(nonnegative) if there exists a $0 \neq x \geq 0$ such that $Ax > 0$ $(Ax \geq 0)$. Similarly, $v(A)$ is negative(nonpositive) if there exists a $0 \neq y \geq 0$ such that $y^T A < 0$ $(y^T A \leq 0)$.

A matrix $M \in R^{n\times n}$ is said to be
- $PSD$-matrix if $x^T Mx \geq 0 \ \forall\, 0 \neq x \in R^n$.
- $P$ $(P_0)$-matrix if all its principal minors are positive (nonnegative).
- $S$-matrix [15] if there exists a vector $x > 0$ such that $Mx > 0$ and $\bar{S}$-matrix if all its principal submatrices are $S$-matrix.
- $Z$-matrix if off-diagonal elements are all non-positive and $K$ $(K_0)$-matrix if it is a $Z$-matrix as well as $P$ $(P_0)$-matrix.
- $Q$-matrix if for every $q$, LCP$(M, q)$ has at least one solution.
- $Q_0$-matrix if for $\text{FEA}(q, A) \neq \emptyset \Rightarrow \text{SOL}(q, A) \neq \emptyset$.

Now, we give some definitions, lemmas, and theorems which will be required for discussion in the next section.

**Lemma 1** ([1]) *If $A$ is a $P$-matrix, then $A^T$ is also $P$-matrix.*

**Lemma 2** *Let $A$ be a $P$-matrix. Then $v(A) > 0$.*

**Definition 1** [1] A subset $S$ of $R^n$ is called a meet semi-sublattice (under the componentwise ordering of $R^n$) if, for any two vectors $x$ and $y$ in $S$, their meet, the vector $z = \min(x, y)$ belongs to $S$.

**Definition 2** ([4]) The spectral radius $\sigma(M)$ of $M$ is defined as the maximum of the moduli $|\lambda|$ of all proper values $\lambda$ of $M$.

**Lemma 3** ([4]) *Let $M$ be a nonnegative matrix. Then there exists a proper value $p(M)$ of $M$, the Perron root of $M$, such that $p(M) \geq 0$ and $|\lambda| \leq p(M)$ for every proper value $\lambda$ of $M$. If $0 \leq M \leq N$, then $p(M) \leq p(N)$. Moreover, if $M$ is irreducible, the Perron–Frobenius root $p(M)$ is positive, simple and the corresponding proper value may be chosen positive. According to the Perron–Frobenius theorem, we have $\sigma(M) = p(M)$ for nonnegative matrices.*

**Definition 3** A matrix $W$ is said to have dominant principal diagonal if $|w_{ii}| > \sum_{k \neq i} |w_{ik}|$ for each $i$.

**Lemma 4** ([4]) *If $W$ is a matrix with dominant principal diagonal, then $\sigma(I - H^{-1}W) < 1$, where $H$ is the diagonal of $W$.*

**Theorem 1** ([4]) *The following four properties of a matrix are equivalent:*
*(i) All principal minors of $M$ are positive.*
*(ii) To every vector $x \neq 0$, there exists an index k such that $x_k y_k > 0$, where $y = Mx$.*
*(iii) To every vector $x \neq 0$, there exists a diagonal matrix $D_x$ with positive diagonal elements such that the inner product $(Mx, D_x x) > 0$.*
*(iv) To every vector $x \neq 0$, there exists a diagonal matrix $H_x \geq 0$ such that the inner product $(Mx, H_x x) > 0$.*
*(v) Every real proper value of $M$ as well as of each principal minor of $M$ is positive.*

**Lemma 5** ([1]) *If $F$ is a nonempty meet semi-sublattice that is closed and bounded below, then $F$ has a least element.*

**Lemma 6** ([8]) *If $z$ solves the linear program* min $p^T z$ *subject to* $Mz + q \geq 0, z \geq 0$ *and if the corresponding optimal dual variable $y$ satisfies $(I - M^T)y + p > 0$, then $z$ solves the linear complementarity problem LCP$(M, q)$.*

## 3 Main Results

In this paper, we introduce block triangular $K$-matrix and hidden block triangular $K$-matrix, which are defined as follows: A matrix $M \in R^{mn \times mn}$ is said to be a block triangular $K$-matrix if it is formed with block of $K$-matrices $M_{ij} \in R^{m \times m}$, either in upper triangular forms or in lower triangular forms. Here, $i$ and $j$ varry from 1 to $n$. For block upper triangular form of $M$, the blocks $M_{ij} = 0$ for $i < j$ and for block lower triangular form of $M$, the blocks $M_{ij} = 0$ for $i > j$.

Consider $M = \left[\begin{array}{cc|cc|cc} 1 & -1 & 0 & 0 & 0 & 0 \\ -1.5 & 2 & 0 & 0 & 0 & 0 \\ \hline 3 & -1 & 1 & -1 & 0 & 0 \\ -1 & 4 & -0.75 & 1 & 0 & 0 \\ \hline 1 & -1 & 1 & -0.5 & 5 & -1 \\ -0.5 & 1 & -0.5 & 1 & -10 & 6 \end{array}\right]$, which is a block triangular $K-$matrix.

The matrix $N \in R^{mn \times mn}$ is said to be hidden block triangular $K$-matrix if there exist two block triangular $K$-matrices $X$ and $Y$ such that $NX = Y$. $N$ is formed with block matrices either in upper triangular forms or in lower triangular forms. For block upper triangular form of $N$, the blocks $N_{ij} = 0$ for $i < j$ and $X$, $Y$ are formed with $K$ matrices in upper triangular form. Similarly, for block lower triangular form of $N$, the blocks $N_{ij} = 0$ for $i > j$ and $X, Y$ are formed with $K$ matrices in lower triangular form.

Consider $N = \left[\begin{array}{cc|cc} -1 & -1 & 0 & 0 \\ 5 & 4 & 0 & 0 \\ \hline -4.5 & -3 & 1 & 0.5 \\ 4 & 3.875 & -0.25 & 0.3125 \end{array}\right]$,

$X = \left[\begin{array}{cc|cc} 2 & -1 & 0 & 0 \\ -3 & 2 & 0 & 0 \\ \hline 3 & 0 & 4 & -1 \\ -2 & 1 & 0 & 4 \end{array}\right]$ and $Y = \left[\begin{array}{cc|cc} 1 & -1 & 0 & 0 \\ -2 & 3 & 0 & 0 \\ \hline 2 & -1 & 4 & 0 \\ 0 & 1 & -1 & 1 \end{array}\right]$, such that $NX = Y$.

Then $N$ is a hidden block triangular $K$-matrix.

**Theorem 2** *Let $M$ be a block triangular $K$-matrix. Then LCP$(M, q)$ is processable by Lemke's algorithm.*

***Proof*** Let $M$ be a block triangular $K$-matrix. Then $\exists\ z \in R^n$ such that $z_i(Mz)_i \leq 0\,\forall i \implies (z_1)_i(M_{11}z_1)_i \leq 0\,\forall i \implies z_1 = 0$, as $M_{11} \in K$; $(z_2)_i(M_{21}z_1 + M_{22}z_2)_i \leq 0\,\forall i \implies (z_2)_i(M_{22}z_2)_i \leq 0\,\forall i \implies z_2 = 0$, as $M_{22} \in K$. In similar way, $(z_n)_i (M_{n1}z_1 + M_{n2}z_2 + \cdots + M_{nn}z_n)_i \leq 0\,\forall i \implies (z_n)_i(M_{nn}z_n)_i \leq 0\,\forall i \implies z_n = 0$, as $M_{nn} \in K$ and $z_1 = z_2 = \cdots = z_{n-1} = 0$. Hence, $z = 0$. So $M$ is a $P$-matrix. Therefore, LCP$(M, q)$ is processable by Lemke's algorithm.

**Remark 1** ([3]) Let $M$ be a block triangular $K$-matrix. Then LCP$(M, q)$ is solvable by criss-cross method.

**Theorem 3** *If $M$ is a block triangular $K$-matrix and $q$ is an arbitrary vector, then the feasible region of LCP$(M, q)$ is a meet semi-sublattice.*

***Proof*** Let $F$ =FEA$(M, q)$. Let $x = \begin{bmatrix} x_1 \\ x_2 \\ \vdots \\ x_n \end{bmatrix}$, $y = \begin{bmatrix} y_1 \\ y_2 \\ \vdots \\ y_n \end{bmatrix} \in F$ are two feasible vectors. So $x \geq 0$, $y \geq 0$, $Mx + q \geq 0$, $My + q \geq 0$.

Let $z = \begin{bmatrix} z_1 \\ z_2 \\ \vdots \\ z_n \end{bmatrix} = \min(x, y)$. Then

$$Mx + q = \begin{bmatrix} M_{11}x_1 + q_1 \\ M_{21}x_1 + M_{22}x_2 + q_2 \\ \vdots \\ M_{n1}x_1 + M_{n2}x_2 + M_{n3}x_3 + \cdots + M_{nn}x_n + q_n \end{bmatrix} \geq 0.$$

$\implies x_1 \in$ FEA$(M_{11}, q_1)$, $x_2 \in$ FEA$(M_{22}, M_{21}x_1 + q_1), \cdots, x_n \in$ FEA$(M_{nn}, M_{n1}x_1 + M_{n2}x_2 + \cdots + M_{n(n-1)}x_{n-1} + q_n)$. In similar way, $My + q \geq 0 \implies y_1 \in$ FEA$(M_{11}, q_1)$, $y_2 \in$ FEA$(M_{22}, M_{21}x_1 + q_1), \ldots, y_n \in$ FEA$(M_{nn}, M_{n1}x_1 + M_{n2}x_2 + \cdots + M_{n(n-1)}x_{n-1} + q_n)$. Suppose $z = \min(x, y) \implies z_1 = \min(x_1, y_1)$,

$z_2 = \min(x_2, y_2), \ldots, z_n = \min(x_n, y_n)$. $M_{ij} \in K \implies z_1 \in \text{FEA}(M_{11}, q_1) \implies M_{11}z_1 + q_1 \geq 0$, $z_2 \in \text{FEA}(M_{22}, M_{21}z_1 + q_2) \implies M_{22}z_2 + M_{21}z_1 + q_2 \geq 0, \ldots,$ $z_n \in \text{FEA}(M_{nn}, M_{n1}z_1 + M_{n2}z_2 + \cdots + M_{n(n-1)}z_{n-1} + q_n) \implies M_{n1}z_1 + M_{n2}z_2 + \cdots + M_{n(n-1)}z_{n-1} + M_{nn}z_n + q_n \geq 0$. So $z = \min(x, y) \in \text{FEA}(M, q)$. Hence, the feasible region is a meet semi-sublattice.

Cottle et al. [1] showed that if $F$ is a nonempty meet semi-sublattice, that is closed and bounded below, then $F$ has a least element by Lemma 5. Now, we show that if the LCP$(M, q)$ is feasible, where $M$ is a block triangular $K$-matrix, then FEA$(M, q)$ contains a least element $u$.

**Theorem 4** *Let $M$ be a block triangular $K$-matrix and $q$ be an arbitrary vector. If the LCP$(M, q)$ is feasible, then FEA$(M, q)$ contains a least element $u$, where $u$ solves the LCP$(M, q)$.*

***Proof*** Let $F$ =FEA$(M, q)$. By Theorem 3, $F$ is a meet semi-sublattice. Let LCP$(M, q)$ be feasible. Then the set $F$ is obviously nonempty and bounded below by zero. Then the existence of the least element $l = \begin{bmatrix} l_1 \\ l_2 \\ \vdots \\ l_n \end{bmatrix}$ follows from Lemma 5.

That is $l = \begin{bmatrix} l_1 \\ l_2 \\ \vdots \\ l_n \end{bmatrix} \leq \begin{bmatrix} x_1 \\ x_2 \\ \vdots \\ x_n \end{bmatrix} = x \; \forall \, x \in F$ and $l \in F$.

Let $F_i = \text{FEA}(M_{ii}, M_{i(i-1)}z_{i-1} + M_{i(i-2)}z_{i-2} + \cdots + M_{i2}z_2 + M_{i1}z_1 + q_i)$.

Now, it is clear that $y_1 \in F_1, y_2 \in F_2, \ldots, y_n \in F_n$, where $y = \begin{bmatrix} y_1 \\ y_2 \\ \vdots \\ y_n \end{bmatrix} \in F$. As $M_{ii}$ are $Z$-matrices, $l_i$ is the least element of $F_i \; \forall \, i \in \{1, 2, \ldots, n\}$ and $l_i$ solves LCP$(M_{ii}, M_{i(i-1)}z_{i-1} + M_{i(i-2)}z_{i-2} + \cdots + M_{i2}z_2 + M_{i1}z_1 + q_i)$. So $l = \begin{bmatrix} l_1 \\ l_2 \\ \vdots \\ l_n \end{bmatrix}$ solves LCP$(M, q)$.

Mangasarian [8] showed that if $z$ solves the linear program, $\min p^T z$ subject to $Mz + q \geq 0, z \geq 0$ and if the corresponding optimal dual variable $y$ satisfies $(I - M^T)y + p > 0$, then $z$ solves the linear complementarity problem LCP$(M, q)$ by Lemma 6. Here, we show that if LCP$(M, q)$ with $M$, a block triangular $K$-matrix, has a solution which can be obtained by solving the linear program min $p^T x$ subject to $Mx + q \geq 0, x \geq 0$.

**Theorem 5** *The linear complementarity problem LCP$(M, q)$, where $M$ is a block triangular $K$-matrix, has a solution which can be obtained by solving the linear program* min $p^T x$ *subject to* $Mx + q \geq 0$, $x \geq 0$, *where* $p = r \geq 0$ *and* $Z_1$ *is a block triangular $K$-matrix with* $r^T Z_1 > 0$.

***Proof*** Let $M$ be a block triangular $K$-matrix. The linear program, min $p^T x$ subject to $Mx + q \geq 0$, $x \geq 0$ and the dual linear program, max $-q^T y$ subject to $-M^T y + p \geq 0$, $y \geq 0$ have solutions $x$ and $y$, respectively. $M$ can be written as $D - U$, where $D = \begin{bmatrix} D_{11} & 0 \cdots & 0 \\ D_{21} & D_{22} \cdots & 0 \\ \vdots & \vdots \;\; \vdots & \vdots \\ D_{n1} & D_{n2} \cdots & D_{nn} \end{bmatrix}$, $D_{ij}$'s are diagonal matrices with positive entries and

$U = \begin{bmatrix} U_{11} & 0 \cdots & 0 \\ U_{21} & U_{22} \cdots & 0 \\ \vdots & \vdots \;\; \vdots & \vdots \\ U_{n1} & U_{n2} \cdots & U_{nn} \end{bmatrix}$, $U_{ij}$'s are matrices with nonnegative entries. Consider $Z_1 = D - V$, a block triangular $K$-matrix and the matrix product $MZ_1 = D - W$, where

$V = \begin{bmatrix} V_{11} & 0 \cdots & 0 \\ V_{21} & V_{22} \cdots & 0 \\ \vdots & \vdots \;\; \vdots & \vdots \\ V_{n1} & V_{n2} \cdots & V_{nn} \end{bmatrix}$, $V_{ij}$'s are matrices with nonnegative entries and

$W = \begin{bmatrix} W_{11} & 0 \cdots & 0 \\ W_{21} & W_{22} \cdots & 0 \\ \vdots & \vdots \;\; \vdots & \vdots \\ W_{n1} & W_{n2} \cdots & W_{nn} \end{bmatrix}$, $W_{ij}$'s are matrices with nonnegtive entries. Since $Z_1$ is a block triangular $K$-matrix, it is a $P$-matrix. Hence, $v(Z_1) > 0$ and by Lemma 1 $v({Z_1}^T) > 0$. Let $r \geq 0$ be the value of $Z_1^T$, then $r^T Z_1 > 0$. Now, $0 < r^T Z_1 = p^T Z_1 = p^T Z_1 + y^T(-MZ_1 + D - W) = (p^T - y^T M)Z_1 + y^T(D - W) = (p^T - y^T M)(D - V) + y^T(D - W) \leq (p^T - y^T M + y^T)D$ as $p^T - y^T M \geq 0, y \geq 0, U \geq 0, V \geq 0$. This implies $(I - M^T)y + p > 0$, since $D_{ij}$'s are positive diagonal matrices. So, by Lemma 6, $x$ solves LCP$(M, q)$, which is a solution of min $p^t x$ subject to $Mx + q \geq 0$, $x \geq 0$.

**Corollary 1** *The solution of linear complementarity problem LCP$(M, q)$, with $M \in$ block triangular $K$-matrix can be obtained by solving the linear program* min $e^T x$ *subject to* $Mx + q \geq 0$, $x \geq 0$.

***Proof*** The identity matrix $I$ itself is a block triangular $K$-matrix. Then $e^T I > 0$.

**Theorem 6** *Let $M$ be a block triangular $K$-matrix. Then $M^{-1}$ exists and $M^{-1} \geq 0$.*

***Proof*** Assume that $Q = I - tM \geq 0, t > 0$. Let $p(Q)$ be the Perron-root of $Q$. Then we have $\det[(1 - p(Q))I - tM] = \det[Q - p(Q)I] = 0$. By Theorem 1,

$0 < p(Q) < 1$. Thus, the series $I + Q + Q^2 + \cdots$ converges to the matrix $(I - Q)^{-1} = (tM)^{-1} \geq 0$, since $Q^k \geq 0$ for $k = 1, 2, \cdots$. Therefore, $M^{-1}$ exists and $M^{-1} \geq 0$.

**Theorem 7** *Let $N$ be a block triangular $K$-matrix and $M$ ba a $Z$-matrix such that $M \leq N$. Then both $M^{-1}$ and $N^{-1}$ exist and $M^{-1} \geq N^{-1} \geq 0$.*

***Proof*** Let $N$ be a block triangular $K$-matrix and $M$ ba a $Z$-matrix such that $M \leq N$. Assume that $R = I - \alpha N \geq 0, \alpha > 0$. Then $S = I - \alpha M \geq R \geq 0$. Let $p(R)$ be a Perron root of $R$. Then we have $\det[(1 - p(R))I - \alpha N] = \det[R - p(R)I] = 0$. By Theorem 1, $0 < p(R) < 1$. Thus, the series $I + R + R^2 + \cdots$ converges to the matrix $(I - R)^{-1} = (\alpha N)^{-1}$. Since $S^k \geq R^k \geq 0$, for $k = 1, 2, \cdots$, the series $I + S + S^2 + \cdots$ converges to the matrix $(I - S)^{-1} = (\alpha M)^{-1}$. Therefore, $M^{-1}$ and $N^{-1}$ exist and $M^{-1} \geq N^{-1} \geq 0$.

**Corollary 2** *Assume that $M, N$ are block triangular $K$-matrices such that $M \leq N$. Then both $M^{-1}$ and $N^{-1}$ exist and $M^{-1} \geq N^{-1} \geq 0$.*

**Theorem 8** *Let $N$ be a hidden block triangular $K$-matrix. Then every diagonal block of $N$ is a hidden $Z$-matrix.*

***Proof*** Let $N$ be a hidden block triangular $K$-matrix with $NX = Y$, where $X$ and $Y$ are the block triangular $K$-matrices. Let

$$N = \begin{bmatrix} N_{11} & 0 \cdots & 0 \\ N_{21} & N_{22} \cdots & 0 \\ \vdots & \vdots \ \vdots & \vdots \\ N_{n1} & N_{n2} \cdots & N_{nn} \end{bmatrix}, X = \begin{bmatrix} X_{11} & 0 \cdots & 0 \\ X_{21} & X_{22} \cdots & 0 \\ \vdots & \vdots \ \vdots & \vdots \\ X_{n1} & X_{n2} \cdots & X_{nn} \end{bmatrix} \text{ and } Y = \begin{bmatrix} Y_{11} & 0 \cdots & 0 \\ Y_{21} & Y_{22} \cdots & 0 \\ \vdots & \vdots \ \vdots & \vdots \\ Y_{n1} & Y_{n2} \cdots & Y_{nn} \end{bmatrix}.$$

The block diagonal of $NX$ are $N_{ii}X_{ii}$ for $i \in \{1, 2, \cdots n\}$. So, $N_{ii}X_{ii} = Y_{ii}$ for $i \in \{1, 2, \cdots n\}$. $X_{ii}, Y_{ii}$ are $K$-matrices. Then $X_{ii}^T, Y_{ii}^T$ are also $K$-matrices. So $v(X_{ii}^T) > 0, v(Y_{ii}^T) > 0$. Let $r_i, s_i \in R^m{}_+$ such that $X_{ii}^T r_i + Y_{ii}^T s_i > 0 \implies r_i^T X_{ii} + s_i^T Y_{ii} > 0$. Hence, the block diagonals of $N$ are hidden $Z$-matrices.

**Theorem 9** *Let $N$ be a hidden block triangular $K$-matrix. Then every determinant of block diagonal matrices of $N$ is positive.*

***Proof*** Let $N$ be a hidden block triangular $K$-matrix with $NX = Y$, where $X$ and $Y$ are block triangular $K$-matrices. Let

$$N = \begin{bmatrix} N_{11} & 0 \cdots & 0 \\ N_{21} & N_{22} \cdots & 0 \\ \vdots & \vdots \ \vdots & \vdots \\ N_{n1} & N_{n2} \cdots & N_{nn} \end{bmatrix}, X = \begin{bmatrix} X_{11} & 0 \cdots & 0 \\ X_{21} & X_{22} \cdots & 0 \\ \vdots & \vdots \ \vdots & \vdots \\ X_{n1} & X_{n2} \cdots & X_{nn} \end{bmatrix} \text{ and } Y = \begin{bmatrix} Y_{11} & 0 \cdots & 0 \\ Y_{21} & Y_{22} \cdots & 0 \\ \vdots & \vdots \ \vdots & \vdots \\ Y_{n1} & Y_{n2} \cdots & Y_{nn} \end{bmatrix}.$$

The block diagonal of $NX$ are $N_{ii}X_{ii}$ for $i \in \{1, 2, \cdots n\}$. So, $N_{ii}X_{ii} = Y_{ii}$ for $i \in \{1, 2, \cdots n\}$. $X_{ii}, Y_{ii}$ are $K$-matrices. Then $\det(X_{ii}), \det(Y_{ii}) > 0 \ \forall\ i$. Hence, $\det(N_{ii}) > 0 \ \forall\ i$.

**Corollary 3** *Every block triangular $K$-matrix is a hidden block triangular $K$-matrix.*

***Proof*** Let $M$ be a block triangular $K$-matrix. Taking $X = I$, the identity matrix, it is clear that $M$ is a hidden block triangular $K$-matrices.

**Theorem 10** *The linear complementarity problem LCP($N, q$), where $N$ is a hidden block triangular $K$-matrix with $NX = Y$, $X$ and $Y$ are the block triangular $K$-matrices, has a solution which can be obtained by solving the linear program* min $p^T x$ *subject to* $Nx + q \geq 0$, $x \geq 0$, *where* $p = r + N^T s \geq 0$ *and* $r, s \geq 0$ *such that* $X^T r > 0$ *and* $Y^T s > 0$.

***Proof*** Let $N$ be a hidden block triangular $K$-matrix with $NX = Y$, where $X$ and $Y$ are the block triangular $K$-matrices. The linear program, min $p^T x$ subject to $Nx + q \geq 0, x \geq 0$ and the dual linear program, max $-q^T y$ subject to $-N^T y + p \geq 0,\ y \geq 0$ have solutions $x$ and $y$, respectively. $X$ can be written as $D - U$, where

$$D = \begin{bmatrix} D_{11} & 0 \cdots & 0 \\ D_{21} & D_{22} \cdots & 0 \\ \vdots & \vdots \; \vdots & \vdots \\ D_{n1} & D_{n2} \cdots & D_{nn} \end{bmatrix}$$, $D_{ij}$'s are diagonal matrices with positive entries and

$$U = \begin{bmatrix} U_{11} & 0 \cdots & 0 \\ U_{21} & U_{22} \cdots & 0 \\ \vdots & \vdots \; \vdots & \vdots \\ U_{n1} & U_{n2} \cdots & U_{nn} \end{bmatrix}$$, $U_{ij}$'s are matrices with nonnegative entries. $Y$ can be written as $D - V$. Then the matrix product $NX$ can be written as $D - V$, where

$$V = \begin{bmatrix} V_{11} & 0 \cdots & 0 \\ V_{21} & V_{22} \cdots & 0 \\ \vdots & \vdots \; \vdots & \vdots \\ V_{n1} & V_{n2} \cdots & V_{nn} \end{bmatrix}$$, $V_{ij}$'s are matrices with nonnegative entries. As $X, Y$ are block triangular $K$-matrices, so they are $P$-matrices. So, $v(X) > 0$, $v(Y) > 0$. Let $r \geq 0$ is the value of $X^T$ and $s \geq 0$ is the value of $Y^T$. Then $0 < r^T X + s^T Y = (r^T + s^T N)X = p^T X = p^T(D - U) = p^T(D - U) + y^T(-ND + NU + D - V)$, since $N(D - U) = D - V = (p^T - y^T N)(D - U) + y^T(D - V) \leq (y^T(I - N) + p^T)D$, since $-y^T N + p^T \geq 0, U \geq 0, V \geq 0, y \geq 0$. Now, $D_{ij}$'s are diagonal matrices with positive entries and $D$ is formed with the block matrices $D_{ij}$'s. Hence, $y^T(I - N) + p^T > 0$. By Lemma 6, $x$ solves the LCP($N, q$), which is a solution of min $p^T x$ subject to $Nx + q \geq 0$, $x \geq 0$.

**Lemma 7** *Let $N$ be a hidden block triangular $K$-matrix. Consider the LCP($\mathcal{N}, \bar{q}$), where* $\mathcal{N} = \begin{bmatrix} 0 & -N^T \\ N & 0 \end{bmatrix}$, $\bar{q} = \begin{bmatrix} r + N^T s \\ q \end{bmatrix}$ *and $r, s$ as mentioned in Theorem 10. If* $\begin{bmatrix} x \\ y \end{bmatrix} \in FEA(\mathcal{N}, \bar{q})$, *then* $(I - N^T)y + p > 0$, *where* $p = r + N^T s$.

***Proof*** Suppose $\begin{bmatrix} x \\ y \end{bmatrix} \in \text{FEA}(\mathcal{N}, \bar{q})$. Since $N$ is a hidden block triangular $K$-matrix, there exist two block triangular $K$-matrices $X$ and $Y$ such that $NX = Y$ and $r, s \geq 0$, $r^T X + s^T Y > 0$. Let $X = D - U$ and $Y = D - V$, where $U$ and $V$ are two square matrices with all nonnegative entries and $D$ is a block triangular diagonal matrix with positive entries as mentioned in Theorem 10. Then $0 < r^T X + s^T Y = r^T X + s^T NX = p^T(D - U) = p^T(D - U) + y^T(Y - NX) = p^T(D - U) + y^T(D - V - N(D - U)) = (-y^T N + p^T)(D - U) + y^T(D - V) \leq (y^T(I - N) + p^T)D$ since $\begin{bmatrix} x \\ y \end{bmatrix} \in \text{FEA}(\mathcal{N}, \bar{q})$, $U \geq 0, V \geq 0$. Since $D$ is a positive block triangular diagonal matrix, $(I - N^T)y + p > 0$.

**Theorem 11** *LCP($\mathcal{N}, \bar{q}$) has a solution iff LCP($N, q$) has a solution.*

***Proof*** Suppose LCP($\mathcal{N}, \bar{q}$) has a solution. Let $\bar{z} = \begin{bmatrix} x \\ y \end{bmatrix} \in \text{SOL}(\mathcal{N}, \bar{q})$. From the complementarity condition, it follows that $x^T(p - N^T y) + y^T(Nx + q) = 0$. Since $p - N^T y, Nx + q, x, y \geq 0$, and $x^T(p - N^T y) = 0, y^T(Nx + q) = 0$. By Lemma 7, it follows that $y + (p - N^T y) > 0$. This implies for all $i$ either $(p - N^T y)_i > 0$ or $y_i > 0$. Now if $(p - N^T y)_i > 0$, then $x_i = 0$. If $y_i > 0$ then $(q + Nx)_i = 0$. This implies $x_i(q + Nx)_i = 0 \; \forall \, i$. Therefore, $x$ solves LCP($N, q$).

Conversely, $x$ solves LCP($N, q$). Let $y = s$, where $s$ as mentioned in Theorem 10. Here, $p - N^T y = r + N^T s - N^T y = r + N^T s - N^T s = r \geq 0$. So $\bar{z} = \begin{bmatrix} x \\ s \end{bmatrix} \in \text{FEA}(\mathcal{N}, \bar{q})$. Further, $\mathcal{N}$ is $PSD$-matrix, which implies that $\mathcal{N} \in Q_0$. Therefore, $\bar{z}$ solves the LCP($\mathcal{N}, \bar{q}$).

**Theorem 12** *All hidden block triangular $K$-matrices are $Q_0$.*

***Proof*** Let $N$ be a hidden block triangular $K$-matrix. It is clear that feasibility of LCP($N, q$) implies the feasibility of LCP($\mathcal{N}, \bar{q}$). Note that $\mathcal{N} \in Q_0$. This implies that the feasible point of LCP($\mathcal{N}, \bar{q}$) is also a solution of LCP($\mathcal{N}, \bar{q}$). Hence, by Theorem 11, feasibility of LCP($N, q$) ensures the solvability of LCP($N, q$). Therefore, $N$ is a $Q_0$-matrix.

**Remark 2** Let $M = \begin{bmatrix} M_{11} & 0 \cdots & 0 \\ M_{21} & M_{22} \cdots & 0 \\ \vdots & \vdots \; \vdots & \vdots \\ M_{n1} & M_{n2} \cdots & M_{nn} \end{bmatrix}$, where $M_{ij} \in R^{m \times m}$ are $K$-matrices.

Let $z = \begin{bmatrix} z_1 \\ z_2 \\ \vdots \\ z_n \end{bmatrix}$ and $q = \begin{bmatrix} q_1 \\ q_2 \\ \vdots \\ q_n \end{bmatrix}$, where $z_i, q_i \in R^m$. Then

$$Mz + q = \begin{bmatrix} M_{11}z_1 + q_1 \\ M_{21}z_1 + M_{22}z_2 + q_2 \\ \vdots \\ M_{n1}z_1 + M_{n2}z_2 + M_{n3}z_3 + \cdots + M_{nn}z_n + q_n \end{bmatrix}.$$

First, we solve LCP$(M_{11}, q_1)$ and get the solution $w_1 = M_{11}z_1 + q_1, {w_1}^T z_1 = 0$. Then we solve LCP$(M_{22}, M_{21}z_1 + q_2)$ and get the solution $w_2 = M_{22}z_2 + M_{21}z_1 + q_2, {w_2}^T z_2 = 0$. Finally, we solve LCP$(M_{nn}, M_{n1}z_1 + M_{n2}z_2 + M_{n3}z_3 + \cdots + M_{n(n-1)}z_{n-1} + q_n)$ and get the solution $w_n = M_{nn}z_n + M_{n1}z_1 + M_{n2}z_2 + M_{n3}z_3 + \cdots + M_{n(n-1)}z_{n-1} + q_n, {w_n}^T z_n = 0$. So, $w = \begin{bmatrix} w_1 \\ w_2 \\ \vdots \\ w_n \end{bmatrix}$ and $z = \begin{bmatrix} z_1 \\ z_2 \\ \vdots \\ z_n \end{bmatrix}$ solve LCP$(M, q)$.

## 4 Conclusion

In this article, we introduce the class of block triangular $K$-matrix and the class of hidden block triangular $K$-matrix in the context of solution of linear complementarity problem. We call these two classes jointly $K$-type block matrices. We show that the linear complementarity problem with $K$-type block matrix is solvable by linear program. The linear complementarity problem with block triangular $K$-matrix is also processable by Lemke's algorithm as well as criss-cross method. We show that the hidden block triangular $K$-matrix is a $Q_0$-matrix.

**Acknowledgements** The author A. Dutta is thankful to the Department of Science and Technology, Govt. of India, INSPIRE Fellowship Scheme for financial support.

## References

1. Cottle, R.W., Pang, J.-S., Stone, R.E.: The linear complementarity problem. Soc. Ind. Appl. Math. (2009). 49.37.50.116
2. Das, A.K.: Properties of some matrix classes based on principal pivot transform. Ann. Oper. Res. **243**(1), 375–382 (2016). https://doi.org/10.1007/s10479-014-1622-6
3. Das, A.K., Jana, R.: Finiteness of criss-cross method in complementarity problem. In: International Conference on Mathematics and Computing. Springer, Singapore, 2017. Communications in Computer and Information Science, vol. 655. Springer, Singapore. https://doi.org/10.1007/978-981-10-4642-1_15

4. Fiedler, M., Ptak, V.: On matrices with non-positive off-diagonal elements and positive principal minors. Czechoslovak Math. J. **12.3**, 382–400 (1962). https://www.dml.cz/dmlcz/100526
5. Jana, R., Das, A.K., Dutta, A.: On hidden Z-matrix and interior point algorithm. Opsearch **56**(4), 1108–1116 (2019). https://doi.org/10.1007/s12597-019-00412-0
6. Jana, R., Dutta, A., Das, A.K.: More on hidden Z-matrices and linear complementarity problem. Linear Multilinear Algeb. **69**(6), 1151–1160 (2021). https://doi.org/10.1080/03081087.2019.1623857
7. Kaplansky, I.: A contribution to von Neumann's theory of games. Ann. Math. 474–479 (1945). DOI: doi.org/1969164
8. Mangasarian, O.L.: Linear complementarity problems solvable by a single linear program. Math. Programm. **10.1**, 263–270 (1976). https://doi.org/10.1007/BF01580671
9. Mohan, S.R., Neogy, S.K., Das, A.K.: A note on linear complementarity problems and multiple objective programming. Math. Programm. **100.2**, 339–344 (2004). https://doi.org/10.1007/s10107-003-0473-8
10. Neogy, S.K., Das, A.K.: On almost type classes of matrices with Q-property. Linear Multilinear Algeb. **53**(4), 243–257 (2005). https://doi.org/10.1080/03081080500092380
11. Neogy, S.K., Das, A.K.: Some properties of generalized positive subdefinite matrices. SIAM J. Matrix Anal. Appl. **27**(4), 988–995 (2006). https://doi.org/10.1137/040613585
12. Neogy, S.K., Das, A.K.: On singular N0-matrices and the class Q. Linear Algeb. Appl. **434**(3), 813–819 (2011). https://doi.org/10.1016/j.laa.2010.09.046
13. Neogy, S.K., Das, A.K.: On weak generalized positive subdefinite matrices and the linear complementarity problem. Linear and Multilinear Algeb. **61**(7), 945–953 (2013). https://doi.org/10.1080/03081087.2012.719507
14. Neogy, S.K. et al.: On a mixture class of stochastic game with ordered field property. Math. Programm. Game Theory Decis. Mak. 451–477 (2008). https://doi.org/10.1142/9789812813220_0025
15. Pang, Jong-Shi.: Hidden Z-matrices with positive principal minors. Linear Algeb. Appl. **23**, 201–215 (1979). https://doi.org/10.1016/0024-3795(79)90103-4
16. Pang, J.-S., Chandrasekaran, R.: Linear complementarity problems solvable by a polynomially bounded pivoting algorithm. Mathematical Programming Essays in Honor of George B. Dantzig Part II, pp. 13–27. Springer, Berlin, Heidelberg (1985). https://doi.org/10.1007/BFb0121072
17. Von Neumann, J.: A certain zero-sum two-person game equivalent to the optimal assignment problem. Contrib. Theory of Games **2.0**, 5–12 (1953). https://doi.org/10.1515/9781400881970
18. Kaplansky, I.: A contribution to von Neumann's theory of games. Ann. Math. 474–479 (1945)

# Families of Mordell Curves with Non-trivial Torsion and Rank of at Least Three

**Renz Jimwel S. Mina and Jerico B. Bacani**

**Abstract** In this study, we consider a particular type of elliptic curves called Mordell curves, and construct two infinite families of such curves with rank of at least three. We do this by using parametrizations due to Euler to obtain two rational points on these curves and obtain the third point from an elliptic curve of rank equal to two. We then show that the three points are of infinite order and are generally linearly independent.

**Keywords** Mordell curve · Elliptic cuve · Rank of elliptic curve

## 1 Introduction

Let $E$ be an elliptic curve over the field of rationals. It is known that the $\mathbb{Q}$-rational points on $E$ form an abelian group $E(\mathbb{Q})$, which is called the Mordell–Weil group. This group was proven by Mordell to be finitely generated, i.e., $E(\mathbb{Q}) \cong T \oplus \mathbb{Z}^r$, where the torsion group $T$ is the group of points with finite order and $r \geq 0$ is the rank of $E$. It is not known whether the rank is bounded. Twenty-eight is the largest known lower bound of rank of an elliptic curve over the rationals, and it was found by N. Elkies in 2006. Other works on elliptic curves with high ranks can be found in [11, 15, 16, 21, 22]. On the other hand, there are also lots of works done in constructing infinite families of elliptic curves with positive ranks. Some of these are found in [2, 6, 7, 12, 17].

In this paper, we will be dealing with a more specialized elliptic curve, called the Mordell curve, and is of the form

$$y^2 = x^3 + k. \tag{1}$$

R. J. S. Mina (✉) · J. B. Bacani
University of the Philippines Baguio, Baguio City 2600, Philippines
e-mail: rsmina1@up.edu.ph

J. B. Bacani
e-mail: jbbacani@up.edu.ph



B. Rushi Kumar et al. (eds.), *Mathematics and Computing*, Springer Proceedings in Mathematics & Statistics 415, https://doi.org/10.1007/978-981-19-9307-7_13

This curve has been extensively studied over the past years. Some of the works can be found in [4, 5, 9, 10, 14, 18–20]. On the other hand, the Mordell curve

$$y^2 = x^3 + k^2 \tag{2}$$

has not been extensively studied. There are some works on finding high-ranked Mordell curves of the form (2), but there are only few studies that are concerned with the infinite family of such curves. In 2004, Elkies and Rogers [3] found a particular elliptic curve of rank $r$ for every $r \leq 11$. In 2017, Izadi and Zargar [8] studied the twists of $y^2 = x^3 + 1$ which include the Mordell curve (2) and found infinite families of elliptic curves of rank of at least three parametrized by an elliptic curve of positive rank. Recently, in 2020, Choudhry and Zargar [1] constructed a parametrized family of Mordell curves with rank of at least three. From this family, they extracted a Mordell curve of rank 5.

Motivated by these works, we will also be constructing two infinite families of Mordell curves with rank of at least three. We do this by utilizing two parametrizations due to Euler. It is well known that a Mordell curve of the form (2) has a torsion group that is isomorphic to $\mathbb{Z}/3\mathbb{Z}$ whenever $k \in \mathbb{Z}$ [13]. On the other hand, if $k \notin \mathbb{Z}$, then by some rescaling, one can show that the torsion subgroup is still $\mathbb{Z}/3\mathbb{Z}$. So, this study will focus on proving that the rank of the two constructed families is three or more.

## 2 A Parametrized Family of Mordell Curves

Consider the following infinite family of Mordell curves

$$E_{(a,b)} : y^2 = x^3 + a^2b^2 \tag{3}$$

over the field $\mathbb{Q}(a, b)$. Forcing $a$ and $b$ to be the $x$-coordinates of two points on $E_{(a,b)}$, we get

$$a^3 + a^2b^2 = a^2(a + b^2) = u^2 \tag{4}$$
$$b^3 + a^2b^2 = b^2(b + a^2) = v^2 \tag{5}$$

for some rational functions $u$ and $v$. This means that $a + b^2$ and $b + a^2$ must be simultaneously squares. We use a parametrization due to Euler (and was compiled in [23]) for $a$ and $b$. That is, we let

$$a + b^2 = (b + p)^2 \tag{6}$$
$$b + a^2 = (a + q)^2 \tag{7}$$

for some variables $p$ and $q$. In this case, we get $a$ and $b$ to be

$$a = \frac{p(p+2q^2)}{1-4pq} \quad and \quad b = \frac{q(q+2p^2)}{1-4pq}. \tag{8}$$

Thus, we have the family of Mordell curves

$$E(p,q) : y^2 = x^3 + \frac{p^2q^2(p+2q^2)^2(q+2p^2)^2}{(1-4pq)^4} \tag{9}$$

with two rational points given by

$$P_1(p,q) = \left(\frac{p(p+2q^2)}{1-4pq}, \frac{p(p+2q^2)(q^2+p-2p^2q)}{(1-4pq)^2}\right) \tag{10}$$

$$P_2(p,q) = \left(\frac{q(q+2p^2)}{1-4pq}, \frac{q(q+2p^2)(p^2+q-2q^2p)}{(1-4pq)^2}\right). \tag{11}$$

We prove that these two points are independent. If we can find a specialization $(p,q) = (p_0, q_0)$ such that $P_1(p_0, q_0)$ and $P_2(p_0, q_0)$ are independent on the specialized curve over $\mathbb{Q}$ and are of infinite order, then by the specialization theorem of Silverman [24], the points are independent and hence the family of Mordell curves $E(p,q)$ has rank of at least two over $\mathbb{Q}$ for all but finitely many $(p,q)$. If we specialize at $(p,q) = (2,3)$ we obtain the curve

$$y^2 = x^3 + \frac{1742400}{279841} \tag{12}$$

with points $P_1(2,3) = \left(-\frac{40}{23}, \frac{520}{529}\right)$ and $P_2(2,3) = \left(-\frac{33}{23}, \frac{957}{529}\right)$. Using SAGE [25], we see that the two points are of infinite order. Moreover, the height pairing matrix of the two points has a nonzero determinant $\approx 7.18522262657344$ which implies that the points are independent. Hence, the family $E(p,q)$ has rank of at least two for all but finitely many $(p,q)$.

Now, we try to increase its rank by forcing the $x$-coordinate of the third point $P_3$ to be $x(P_3(p,q)) = -\frac{q(q+2p^2)}{1-4pq}$. Then, the corresponding $y$-coordinate is given by

$$y(P_3(p,q)) = \frac{q(q+2p^2)\sqrt{p^4+12p^3q^2+4p^2q^4-2p^2q+4pq^3-q^2}}{(1-4pq)^2}. \tag{13}$$

Note that $y(P_3(p,q))$ must be in $\mathbb{Q}(p,q)$. So, setting $q=1$, we obtain the quartic curve

$$u^2 = p^4 + 12p^3 + 2p^2 + 4p - 1. \tag{14}$$

Note that $(p,u) = (13/62, 761/3844)$ is on the curve. Thus, it is birationally equivalent to an elliptic curve in the Weierstrass equation given by

$$E' : g^2 = h^3 + 4104h - 112320 \tag{15}$$

with the transformation

$$p = \frac{g - 18h}{6h - 1872} \quad and \quad u = \frac{18p^2 + 108p + 6 - h}{18}. \tag{16}$$

Using SAGE, we see that $E'$ has rank equal to two, which has generators given by $(h, g) = (33, 243)$ and $(96, 1080)$. This implies that $E'$ has infinitely many rational points. Now, a point $(h, g) = (33, 243)$ on $E'$ corresponds to the $(p, u) = (13/62, 761/3844)$ on the quartic. Thus, we obtain three points given by

$$P_1(p) = \left(\tfrac{p(p+2)}{1-4p}, \tfrac{p(p+2)(-2p^2+p+1)}{(1-4p)^2}\right) \tag{17}$$

$$P_2(p) = \left(\tfrac{2p^2+1}{1-4p}, \tfrac{(2p^2+1)(p^2-2p+1)}{(1-4p)^2}\right) \tag{18}$$

$$P_3(p, h) = \left(-\tfrac{2p+1}{1-4p}, \tfrac{(2p^2+1)(18p^2+108p+6-h)}{18(1-4p^2)}\right). \tag{19}$$

If we specialize at $(h, g) = (33, 243)$ (with the corresponding value $p = 13/62$), we get the Mordell curve

$$E(13/62) : y^2 = x^3 + \frac{55474819252164}{147763360000} \tag{20}$$

with points

$$P_1(13/62) = \left(\tfrac{1781}{620}, \tfrac{7679672}{384400}\right) \tag{21}$$

$$P_2(13/62) = \left(\tfrac{4182}{620}, \tfrac{10040982}{384400}\right) \tag{22}$$

$$P_3(13/62, 33) = \left(-\tfrac{4182}{620}, \tfrac{3182502}{384400}\right). \tag{23}$$

Note that these points have infinite order and that the determinant of the corresponding height pairing matrix is the nonzero value $\approx$139.537172045240. Thus, these points are linearly independent. Using Silverman's specialization theorem, there exists an infinite family of Mordell curves of rank of at least three. We have proven our first main theorem.

**Theorem 1** *Let* $E(p, q) : y^2 = x^3 + \frac{p^2q^2(p+2q^2)^2(q+2p^2)^2}{(1-4pq)^4}$ *be defined over* $\mathbb{Q}(p, q)$. *Then the following holds:*

1. *The rank of* $E(p, q)$ *is at least two for all but finitely many pairs* $(p, q)$.
2. *There exists a subfamily of* $E(p, q)$ *whose rank is at least three parametrized by an elliptic curve of rank two.*

From this family, we can actually get Mordell curves of rank greater than three. For example, if $(h, g) = (33, 243)$, then we obtain the Mordell curve given by $y^2 = x^3 + \frac{55474819252164}{147763360000}$ with rank equal to 5.

## 3 Another Parametrized Family of Mordell Curves

Consider the following infinite family of Mordell curves

$$E_{(a,b,c)} : y^2 = x^3 + a^2b^2c^2 \tag{24}$$

over the field $\mathbb{Q}(a, b, c)$. Forcing $ab$, $ac$, and $bc$ to be the $x$-coordinates of three points on $E_{(a,b,c)}$, we get

$$a^3b^3 + a^2b^2c^2 = u^2 \tag{25}$$
$$a^3c^3 + a^2b^2c^2 = v^2 \tag{26}$$
$$b^3c^3 + a^2b^2c^2 = w^2, \tag{27}$$

for some rational functions $u$, $v$, and $w$. This implies that

$$ab + c^2 = u_1^2 \tag{28}$$
$$ac + b^2 = v_1^2 \tag{29}$$
$$bc + a^2 = w_1^2, \tag{30}$$

for some rational functions $u_1$, $v_1$, and $w_1$. We use a parametrization of $a$, $b$, and $c$ due to Euler [23] which is given by

$$\{a, b, c\} = \{s^2 + 8st, -8st + t^2, 4(s^2 + t^2)\}. \tag{31}$$

We then have the elliptic curve

$$E(s,t) : y^2 = x^3 + 16(s^2 + 8st)^2(t^2 - 8st)^2(s^2 + t^2)^2 \tag{32}$$

with three points given by

$$P_1(s,t) = ((s^2 + 8st)(t^2 - 8st), (t^2 - 8st)(s^2 + 8st)(4s^2 - st - 4t^2)) \tag{33}$$
$$P_2(s,t) = (4(s^2 + t^2)(s^2 + 8st), 4(s^2 + 8st)(s^2 + t^2)(2s^2 + 8st + t^2)) \tag{34}$$
$$P_3(s,t) = (4(s^2 + t^2)(t^2 - 8st), 4(t^2 - 8st)(s^2 + t^2)(s^2 - 8st + 2t^2)). \tag{35}$$

We show that two out of the three points are independent for all but finitely many pairs $(s, t)$. If we specialize at $(s, t) = (1, 4)$, we get the elliptic curve

$$E(1,4) : y^2 = x^3 + 1289097216 \tag{36}$$

with points

$$P_1(1,4) = (-528, 33792), \quad \textit{and} \quad P_2(1,2) = (2244, 112200). \tag{37}$$

Using SAGE, we see that the two points are of infinite order. Moreover, the calculated determinant of the height paring matrix of these two points is the nonzero value $\approx 5.71586499033672$, which implies that the points are independent. Hence, the family $E(s,t)$ has a rank of at least two for all but finitely many $(s,t)$.

Now, if we force the $x$-coordinate of the fourth point on $E(s,t)$ to be $x(P_4) = -4(s^2+t^2)(s^2+8st)$, then the corresponding $y$ coordinate is given by

$$y(P_4) = 4(s^2+t^2)(s^2+8st)\sqrt{-4s^4 - 32s^3t + 60s^2t^2 - 48st^3 + t^4}. \tag{38}$$

Note that $y(P_4)$ must be in $\mathbb{Q}(s,t)$. Setting $s=1$, we obtain the quartic curve over $\mathbb{Q}$ given by

$$u^2 = t^4 - 48t^3 + 60t^2 - 32t - 4. \tag{39}$$

Note that $(t,u) = (-37/41, 17605)$ is on the curve. Thus, it is birationally equivalent to an elliptic curve in the Weierstrass equation given by

$$E' : g^2 = h^3 + 28512h - 16236288, \tag{40}$$

with the transformation

$$t = \frac{2g + 144h - 24192}{12h - 57888}, \quad u = \frac{18t^2 - 432t + 180 - h}{18}. \tag{41}$$

Using SAGE, we see that the rank of $E'$ is two, meaning, it has infinitely many rational points with generators $(h,g) = (396, 7560)$ and $(504, 11232)$. If we consider $(h,g) = (396, 7560)$, the corresponding point on the quartic curve is $(t,u) = (-37/41, 17605)$. Thus, we obtain three points on $E(t)$, which are given by

$$P_1(t) = ((8t+1)(t^2-8t), (8t+1)(t^2-8t)(4-t-4t^2)) \tag{42}$$

$$P_2(t) = (4(t^2+1)(8t+1), 4(t^2+1)(8t+1)(t^2+8t+2)) \tag{43}$$

$$P_4(t,h) = \left(-4(t^2+1)(8t+1), 4(t^2+1)(8t+1)\left(\tfrac{18t^2-432t+180-h}{18}\right)\right), \tag{44}$$

where $h$ is the first coordinate of a rational point on $E'$. If we specialize at $(t,h) = (-37/41, 396)$, (knowing that $(h,g) = (396, 7560)$ corresponds to the point $(t,u) = (-37/41, 17605)$), we get the elliptic curve

$$E(-37/41) : y^2 = x^3 + \frac{1765180817543025000000}{13422659310152401} \tag{45}$$

with points

$$P_1(-37/41) = \left(-\tfrac{3443775}{68921}, -\tfrac{9522037875}{115856201}\right) \tag{46}$$

$$P_2(-37/41) = \left(-\tfrac{3111000}{68921}, \tfrac{23036955000}{115856201}\right) \tag{47}$$

$$P_4(-37/41, 396) = \left(\tfrac{3111000}{68921}, \tfrac{54769155000}{115856201}\right). \tag{48}$$

Note that all these points have infinite order and that the calculated determinant of the associated height pairing matrix is the nonzero value $\approx$156.789314658799. Thus, these points are linearly independent. Using the specialization theorem of Silverman, we have proven that there exists an infinite family of Mordell curves of the form $y^2 = x^3 + k^2$ with rank of at least three. We have proven our second main theorem.

**Theorem 2** *Let $E(s,t) : y^2 = x^3 + 16(s^2 + 8st)^2(t^2 - 8st)^2(s^2 + t^2)^2$ be defined over $\mathbb{Q}(s,t)$. Then the following holds:*

1. *The rank of $E(s,t)$ is at least two for all but finitely many pairs $(s,t)$.*
2. *There exists a subfamily of $E(s,t)$ whose rank is at least three parametrized by an elliptic curve of rank two.*

From this family, we can actually get Mordell curves of rank greater than three. For example, if $(h, g) = (801, -22815)$, then we obtain the Mordell curve $y^2 = x^3 + \frac{960282455448626099867025000000}{552284227795293031526 5024}$ with rank equal to 5.

## References


1. Choudhry, A., Zargar, A.S.: A parametrised family of Mordell curves with a rational point of order 3. Notes Number Theory Discret. Math. **26**(1), 40–44 (2020)
2. Dujella, A., Peral, J.C.: High rank elliptic curves with torsion $\mathbb{Z}/2\mathbb{Z} \times \mathbb{Z}/4\mathbb{Z}$ induced by Diophantine triples. LMS J. Comput. Math. **17**(1), 282–288 (2014)
3. Elkies, N.D., Rogers, N. F.: Elliptic curves $x^3 + y^3 = k$ of high rank. In: Buell, D. (ed.) Algorithmic Number Theory (ANTS-VI). Lecture Notes in Computer Science, vol. 3076, pp. 184–193. Springer, Berlin. (2004)
4. Ellison, W.J., Ellison, F., Pesek, J., Stahl, C.E., Stall, D.S.: The Diophantine equation $y^2 + k = x^3$. J. Number Theory. **4**(2), 107–117 (1972)
5. Gebel, J., Petho, A., Zimmer, H.G.: On Mordell's equation. Compos. Math. **110**(3), 335–367 (1998)
6. Izadi, F., Khoshnam, F., Moody, D.: Elliptic curves arising from Brahmagupta quadrilaterals. Bull. Aust. Math. Soc. **90**, 47–56 (2014)
7. Izadi, F., Nabardi, K.: A family of elliptic curves with rank $\geq 5$. Period. Math. Hung. **71**, 243–249 (2015)
8. Izadi, F., Zargar, A.S.: A note on twists of $y^2 = x^3 + 1$. Iran. J. Math. Sci. Inform. **12**(1), 27–34 (2017)
9. Kihara S.: On the rank of the elliptic curve $y^2 = x^3 + k$. Proc. Jpn. Acad. Ser. A Math. Sci. **63A**, 76–78 (1987)
10. Kihara S.: On the rank of the elliptic curve $y^2 = x^3 + k$ II. Proc. Jpn. Acad. Ser. A Math. Sci. **72A**, 228–229 (1996)
11. Kihara S.: On the rank of elliptic curves with three rational points of order 2. Proc. Jpn. Acad. Ser. A Math. Sci. **73**, 77–78 (1997)

12. Kihara S.: On an infinite family of elliptic curves with rank $\geq 14$ over $\mathbb{Q}$. Proc. Jpn. Acad. Ser. A Math. Sci. **73**(2), 32–32 (1997)
13. Knapp, A.: Elliptic Curves. Princeton University Press, Princeton (1992)
14. Ljunggren, W.: On the Diophantine equation $y^2 - k = x^3$. Acta Arith **8**(4), 451–463 (1963)
15. Mestre, J.F.: Construction d'une courbe elliptique de rang $\geq 12$,. C.R. Acad. Sci. Paris Ser. I **295**, 643–644 (1982)
16. Mestre, J.F.: Courbes elliptiques de rang $\geq 11$ sur $\mathbb{Q}(t)$. C.R. Acad. Sci. Paris Ser. I **313**, 139–142 (1991)
17. Moody, D., Sadek, M., Zargar, A.S.: Families of elliptic curve of rank $\geq 5$ over $\mathbb{Q}(t)$. Rocky Mountain J. Math. **49**(7), 2253–2266 (2019)
18. Mordell, L. J.: The Diophantine equation $y^2 + k = x^3$. Proc. Lond. Math. Soc. (2), **13**(1), 60–80 (1914)
19. Mordell, L. J.: A statement by Fermat. Proc. Lond. Math. Soc. (2) **18**(1) (1920)
20. Mordell, L. J.: The infinity of rational solutions of $y^2 = x^3 + k$. J. Lond. Math. Soc. (1) **411**(1), 523–525 (1966)
21. Nagao, K.: An example of elliptic curve over $\mathbb{Q}$ with rank $\geq 20$. Proc. Jpn. Acad. Ser. A Math. Sci. **69**, 291–293 (1993)
22. Nagao, K., Kouya, T.: An example of elliptic curve over $\mathbb{Q}$ with rank $\geq 21$. Proc. Jpn. Acad. Ser. A Math. Sci. **70**, 104–105 (1994)
23. Piezas III, T.: A collection of algebraic identities, https://sites.google.com/site/tpiezas/Home. Last accessed 04 March 2021
24. Silverman, J.H.: Heights and specialization map for families of abelian varieties. J. Reine Angew. Math. **342**, 197–211 (1983)
25. Stein, W.A., et al.: Sage Mathematics Software (Version 9.2). The Sage Development Team (2020). http://www.sagemath.org

# Applied Algebra and Analysis

# On the Genus of the Annihilator-Ideal Graph of Commutative Ring

**Selvakumar Krishnan and Karthik Shunmugaiah**

**Abstract** Let $R$ be a commutative ring with identity and $\mathbb{A}(R)$ be the set of all annihilator ideals of $R$. The annihilator-ideal graph of $R$, denoted by $A_I(R)$, is a simple graph with the vertex set $\mathbb{A}^*(R) := \mathbb{A}(R) \setminus \{(0)\}$, and two distinct vertices $I$ and $J$ are adjacent if and only if $ann(IJ) \neq ann(I) \cup ann(J)$. In this paper, we characterize all isomorphism classes of Artinian commutative rings whose $A_I(R)$ has genus one and crosscap one.

**Keywords** Annihilator-ideal graph · Planar · Genus · Crosscap

## 1 Introduction

In 2011, Behboodi et al. [3] introduced the annihilating ideal graph and found out more results on it. Let $\mathbb{A}(R)$ be the set of all ideals with non-zero annihilators. The *annihilating ideal graph* of a commutative ring, denoted by $\mathbb{AG}(R)$. The annihilating ideal graph with vertices $\mathbb{A}^*(R) := \mathbb{A}(R) \setminus \{(0)\}$, and two distinct vertices $I_1$ and $I_2$ are adjacent if and only if $I_1 I_2 = (0)$.

For any ring $R$, let $ann(x) = \{y \in R : xy = 0\}$ be the annihilator of $x \in R$. Badawi [2] introduced that the *annihilator graph* is the simple graph, it is denoted by $AG(R)$ whose vertex set is a set of all non-zero zero divisors of $R$ and two distinct vertices $z_1$ and $z_2$ are adjacent if and only if $ann(z_1 z_2) \neq ann(z_1) \cup ann(z_2)$.

In 2017, Salehifar et al. [9] introduced the *annihilator-ideal graph* $A_I(R)$ as the simple graph with vertex set $\mathbb{A}^*(R)$ and two distinct vertices $I$ and $J$ are adjacent if and only if $ann(IJ) \neq ann(I) \cup ann(J)$. It is seen that the annihilating-ideal graph is a subgraph of the annihilator-ideal graph $A_I(R)$.

By a graph $G = (V, E)$, we mean an undirected simple graph with vertex set $V$ and edge set $E$. Let $G$ be a simple graph, that is, no loops and no multiedges. A *complete bipartite graph* is a bipartite graph such that vertex set can be partitioned

S. Krishnan (✉) · K. Shunmugaiah
Manonmaniam Sundaranar University, Abishekapatti, Tirunelveli 627012, India
e-mail: selva_158@yahoo.co.in
URL: https://msuniv.irins.org/profile/174286



B. Rushi Kumar et al. (eds.), *Mathematics and Computing*, Springer Proceedings in Mathematics & Statistics 415, https://doi.org/10.1007/978-981-19-9307-7_14

into two disjoint sets of sizes $m$ and $n$ with every pair of vertices in the two sets adjacent. It is denoted by $K_{m,n}$. The *complete graph* on $n$ vertices, denoted $K_n$, is the graph in which every pair of distinct vertices is joined by an edge. A graph $G$ is said to be *unicyclic* if it contains a unique cycle. A graph $G$ is *planar* if a graph is drawn on a plane without edge crossing.

By a surface, we mean a connected two-dimension real manifold, i.e., a connected topological space such that each point has a neighborhood homeomorphic to an open disk. We denote $S_g$ for the surface formed by a connected sum of $g$ tori and $N_k$ for the one formed by a connected sum of $k$ projective planes. The *genus* $g(G)$ of a simple graph $G$ is the minimum $g$ such that $G$ can be embedded in $S_g$. Similarly, *crosscap number* $\overline{g}(G)$ is the minimum $k$ such that $G$ can be embedded in $N_k$. It is clear that $g(H) \leq g(G)$ and $\overline{g}(H) \leq \overline{g}(G)$ for any subgraph $H$ of $G$.

Note that the annihilating ideal graph is a subgraph of $A_I(R)$. In [7, 9], it has been shown that some basic properties of $A_I(R)$, the graphs $\mathbb{AG}(R)$ and $A_I(R)$ coincide and under which the $A_I(R)$ is a star graph. The following results are useful for further reference in this paper.

**Lemma 1** ([9, Lemma 2.4]) *If* $|V(A_I(R))| = 4$, *then* $A_I(R)$ *is isomorphic to* $C_4$ *or* $K_4$.

**Theorem 1** ([9, Theorem 3.12]) $A_I(R)$ *is a tree if and only if* $A_I(R)$ *is a star.*

**Theorem 2** *Let G be a connected graph. Then G is a split graph if and only if G contains no induced subgraph isomorphic to* $2K_2$, $C_4$, $C_5$.

## 2 Some Well-Known Graph of $A_I(R)$

In this section, we are identifying when the annihilator-ideal graph is isomorphic to some well-known graph. Also, we assume that $R$ is a commutative Artinian ring. Then $R \cong R_1 \times R_2 \times \cdots \times R_n$, where $R_i$ is an Artinian local ring for each $i$.

**Lemma 2** *Let R be a commutative ring with* $|Nil(R)^*| \geq 2$ *and* $A_N(R)$ *is the set of nilpotent ideal of R. Then the subgraph induced by* $A_N(R)$ *in* $A_I(R)$ *is complete.*

***Proof*** Suppose there are non-zero distinct ideals $J, K \in A_N(R)$ with $JK \neq (0)$. Suppose $ann(JK) = ann(J) \cup ann(K)$. Then $ann(JK) = ann(J)$ or $ann(JK) = ann(K)$. Without loss of generality, we consider $ann(JK) = ann(J)$. Since $K \in A_N(R)$, $K^{n_1} = (0)$, where $n_1 \geq 2$ is the nilpotency of $K$. If $JK^{\ell} \neq (0)$ for all $\ell$, $1 \leq \ell < n_1$, then $K^{n_1-1} \subseteq ann(JK) \setminus ann(J)$, a contradiction. Suppose, there exists a integer $m$, $1 \leq m < n_1$ such that $JK^m = (0)$. Hence, $K^{m-1} \subseteq ann(JK) \setminus ann(J)$, a contradiction. Thus, $J$ and $K$ are adjacent of $A_N(R)$.

**Theorem 3** *Let R be a commutative Artinian ring which is not a field. Then* $A_I(R)$ *is a tree if and only if* $R \cong F_1 \times F_2$, *where* $R_1$ *and* $R_2$ *are fields or R is a local ring with* $|\mathbb{A}^*(R)| \leq 2$.

***Proof*** Suppose $A_I(R)$ is tree. Since $R$ is Artinian, $R \cong R_1 \times R_2 \times \cdots \times R_n$, where every $(R_i, \mathfrak{m}_i)$ are Artinian local ring. If $n = 3$, then $R_1 \times (0) \times (0) - (0) \times R_2 \times (0) - (0) \times (0) \times R_3$ is a cycle in $A_I(R)$. Thus, $n \leq 2$.

Suppose $R \cong R_1 \times R_2$. If $R_1$ is not a field, then $R_1$ has a non-zero maximal ideal, say $\mathfrak{m}_1$. Let $n_1$ be the nilpotency of $\mathfrak{m}_1$. Let $I_1 = (0) \times R_2$, $I_2 = \mathfrak{m}_1 \times (0)$, $I_3 = \mathfrak{m}_1^{n_1-1} \times R_2$ and $I_4 = R_1 \times (0)$. Then $I_j \in \mathbb{A}^*(R)$, for each $j$. It is easy to see that $ann(I_3 I_4) = \mathfrak{m}_1 \times R_2$, $ann(I_3) \cup ann(I_4) \neq \mathfrak{m}_1 \times R_2$ and hence $I_3$ and $I_4$ are adjacent in $A_I(R)$. So, $(0) \times R_2 - \mathfrak{m}_1 \times (0) - \mathfrak{m}_1^{n_1-1} \times R_2 - R_1 \times (0) - (0) \times R_2$ is a cycle in $A_I(R)$. Therefore, we conclude that $R_1$ and $R_2$ are fields.

Finally, suppose $R$ is local ring. Then Lemma 2, $A_I(R)$ is complete and we have $|\mathbb{A}^*(R)| \leq 2$.

In view of Lemma 1 and Theorem 3, we have the following theorem.

**Theorem 4** *Let $R$ be a commutative Artinian ring which is not a field. Then $A_I(R)$ is star if and only if $R \cong R_1 \times R_2$, where $R_1$ and $R_2$ are fields or $R$ is a local ring with $|\mathbb{A}^*(R)| \leq 2$.*

**Theorem 5** *Let $R$ be a commutative Artinian ring which is not a field. Then $A_I(R)$ is a unicycle if and only if $R \cong R_1 \times R_2$, where $R_1$ is a field and $(R_2, \mathfrak{m}_2)$ is local ring with $\mathbb{A}^*(R_2) = \{\mathfrak{m}_2\}$, $\mathfrak{m}_2^2 = (0)$ or $|\mathbb{A}^*(R)| = 3$.*

***Proof*** First, suppose that $A_I(R)$ is unicycle graph and $R \cong \prod_{i=1}^n R_i$, where every $(R_i, \mathfrak{m}_i)$ is a Artinian local ring. Suppose $n \geq 3$. Let $I_1 = R_1 \times (0) \times (0) \times \cdots \times (0)$, $I_2 = (0) \times R_2 \times (0) \times \cdots \times (0)$, $I_3 = (0) \times (0) \times R_3 \times (0) \times \cdots \times (0)$, $J_1 = (0) \times R_2 \times R_3 \times (0) \times \cdots \times (0)$, $J_2 = R_1 \times (0) \times R_3 \times (0) \times \cdots \times (0)$ and $J_3 = R_1 \times R_2 \times (0) \times \cdots \times (0)$. Then $ann(J_1 J_2) = R_1 \times R_2 \times (0) \times R_4 \times \cdots \times R_n$, $ann(J_1) \cup ann(J_2) \neq R_1 \times R_2 \times (0) \times R_4 \times \cdots \times R_n$, $ann(J_1 J_3) = R_1 \times (0) \times R_3 \times \cdots \times R_n$, $ann(J_1) \cup ann(J_3) \neq R_1 \times (0) \times R_3 \times \cdots \times R_n$ and $ann(J_2 J_3) = (0) \times R_2 \times R_3 \times \cdots \times R_n$, $ann(J_2) \cup ann(J_3) \neq (0) \times R_2 \times R_3 \times \cdots \times R_n$. Hence, $I_1 - I_2 - I_3 - I_1$, as well as $J_1 - J_2 - J_3 - J_1$, are different cycles in $A_I(R)$, a contradiction. Thus, $n \leq 2$.

Assume that $n = 2$. Suppose $\mathfrak{m}_i \neq (0)$ and $n_i$ be the nilpotency of $\mathfrak{m}_i$ for $i = 1, 2$. Then $I_1 = \mathfrak{m}_1 \times (0)$, $I_2 = \mathfrak{m}_1^{n_1-1} \times R_2$, $I_3 = \mathfrak{m}_1^{n_1-1} \times \mathfrak{m}_2^{n_2-1}$, $I_4 = R_1 \times (0)$, $I_5 = (0) \times R_2$, for each $I_i \in \mathbb{A}^*(R)$ and $\mathfrak{m}_1 \times (0) - \mathfrak{m}_1^{n_1-1} \times R_2 - \mathfrak{m}_1^{n_1-1} \times \mathfrak{m}_2^{n_2-1} - \mathfrak{m}_1 \times (0)$ as well as $\mathfrak{m}_1^{n_1-1} \times \mathfrak{m}_2^{n_2-1} - R_1 \times (0) - (0) \times R_2 - \mathfrak{m}_1^{n_1-1} \times \mathfrak{m}_2^{n_2-1}$ are different cycles in $A_I(R)$, a contradiction. Thus any one of $R_i$ is a field. Consider $R_1$ is a field.

Suppose $R_2$ is not a field with $\mathfrak{m}_2 \neq (0)$. If $n_2 \geq 3$, then $R_1 \times (0) - (0) \times \mathfrak{m}_2 - (0) \times \mathfrak{m}_2^{n_2-1} - R_1 \times (0)$ and $R_1 \times (0) - (0) \times \mathfrak{m}_2 - R_1 \times \mathfrak{m}_2^{n_2-1} - (0) \times R_2 - R_1 \times (0)$ are two distinct cycles of $A_I(R)$, a contradiction. So, $n_2 = 2$. Suppose $R_2$ has non-zero proper ideal $I$ different from $\mathfrak{m}_2$. Then $R_1 \times (0) - (0) \times \mathfrak{m}_2 - (0) \times I - R_1 \times (0)$ and $R_1 \times (0) - (0) \times \mathfrak{m}_2 - R_1 \times I - (0) \times R_2 - R_1 \times (0)$ are two distinct cycles of $A_I(R)$, a contradiction. Hence, $\mathbb{A}^*(R_2) = \{\mathfrak{m}_2\}$.

Finally, if $R$ is local ring, then $A_I(R)$ is complete and by Lemma 2, we have $|\mathbb{A}^*(R)| = 3$. Conversely, $A_I(R) \cong C_4$ or $K_3$.

**Theorem 6** *Let $R$ be a commutative Artinian ring and $|\mathbb{A}^*(R)| \geq 2$. Then $A_I(R)$ is a split graph if and only if $R \cong F_1 \times F_2$, where each $F_i$ is a field or $R$ is a local ring but not field.*

***Proof*** Suppose that $A_I(R)$ is split graph, $R$ is Artinian, $R \cong \prod_{i=1}^{n} R_i$, where each $(R_i, \mathfrak{m}_i)$ is an Artinian local ring. If $n \geq 3$, then $R_1 \times (0) \times (0) \times \cdots \times (0) - (0) \times R_2 \times (0) \times \cdots \times (0) - R_1 \times (0) \times R_3 \times (0) \times \cdots \times (0) - (0) \times R_2 \times R_3 \times (0) \cdots \times (0) - R_1 \times (0) \times (0) \times \cdots \times (0)$ as induced cycle of length 4 in $A_I(R)$ and by Theorem 2, $A_I(R)$ is not a split graph, a contradiction. Hence, $n = 2$.

If $\mathfrak{m}_2 \neq (0)$ and has nilpotency $n_2 \geq 2$, then $R_1 \times (0) - (0) \times R_2 - R_1 \times \mathfrak{m}_2 - (0) \times \mathfrak{m}_2$ as induced cycle of length 4 in $A_I(R)$ and by Theorem 2, $A_I(R)$ is not a split graph, a contradiction. Hence, $R_1$ and $R_2$ are fields.

If $n = 1$, then $A_I(R)$ is complete and hence $A_I(R)$ is a split graph.

## 3 Genus of $A_I(R)$

In this section, we characterize when $A_I(R)$ is planar or genus one over $R$ is Artinian ring. The much needed results are given in the following section.

**Lemma 3** ([5, Theorem 4.4.5])

*(i) If $n \geq 3$, then*

$$g(K_n) = \left\lceil \frac{(n-3)(n-4)}{12} \right\rceil \quad \text{and} \quad \overline{g}(K_n) = \left\lceil \frac{(n-3)(n-4)}{6} \right\rceil.$$

*(ii) If $m, n \geq 2$, then*

$$g(K_{m,n}) = \left\lceil \frac{(m-2)(n-2)}{4} \right\rceil \quad \text{and} \quad \overline{g}(K_{m,n}) = \left\lceil \frac{(m-2)(n-2)}{2} \right\rceil.$$

**Lemma 4** ([5, Proposition 4.4.4]) *Let $G$ be a connected graph with $n \geq 3$ vertices and $q$ edges. If $G$ contains no cycle of length 3, then $g(G) \geq \lceil \frac{q}{4} - \frac{n}{2} + 1 \rceil$ and $\overline{g}(G) \geq \lceil \frac{q}{2} - n + 2 \rceil$.*

**Theorem 7** *If $(R, \mathfrak{m})$ is a local ring and there is an ideal $I$ of $R$ such that $I \neq \mathfrak{m}^i$ for every $i$, then $R$ has at least three distinct non-trivial ideals $J$, $K$, and $L$ such that $J$, $K$, and $L \neq \mathfrak{m}^i$ for every $i$.*

**Theorem 8** ([9, Theorem 4.1]) *Let $n = 3$. Then $A_I(R)$ is planar if and only if $R_1$, $R_2$, and $R_3$ are fields.*

**Lemma 5** ([9, Lemma 4.4]) *Let $n = 2$ such that $A_I(R)$ be planar. Then at least one of the rings $R_1$ or $R_2$ is field.*

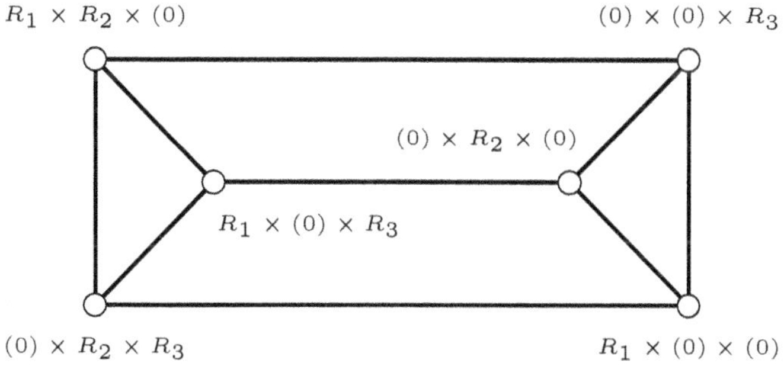


**Fig. 1** The graph $A_I(R_1 \times R_2 \times R_3)$

The following theorem we characterize when the annihilator-ideal graphs are planar.

**Theorem 9** *Let $R$ be an Artinian ring. $A_I(R)$ is planar if and only if $R$ is isomorphic to $R_1 \times R_2 \times R_3$, where each $R_i$ is a field, $R_1 \times R_2$ where $R_1$ is field and $R_2$ is local ring with exactly one proper ideal $\mathfrak{m}_2$, $R_1 \times R_2$, where $R_1$, $R_2$ is field or $R$ is local ring with $|\mathbb{A}^*(R)| \leq 4$.*

***Proof*** Suppose that $A_I(R)$ is planar and $R$ is Artinian, $R \cong \prod_{i=1}^{n} R_i$, where each $(R_i, \mathfrak{m}_i)$ is an Artinian local ring. If $n \geq 4$, then it contains a subgraph of $K_{3,3}$ with vertex set is $\{R_1 \times (0) \times (0) \times \cdots \times (0), (0) \times R_2 \times (0) \times \cdots \times (0), R_1 \times R_2 \times (0) \times \cdots \times (0)\} \subset \mathbb{A}^*(R)$ and $\{(0) \times (0) \times R_3 \times R_4 \times \cdots \times R_n, (0) \times (0) \times R_3 \times (0) \times \cdots \times (0), (0) \times (0) \times (0) \times R_4 \times \cdots \times R_n\} \subset \mathbb{A}^*(R)$. Thus, $n \leq 3$.

If $n = 3$, then by Theorem 8, we have $R_i$ are fields. If $n = 2$, then by Lemma 5, we have at least one of the rings $R_1$ or $R_2$ is field. Let $R_1$ be a field.

Suppose $\mathfrak{m}_2 \neq (0)$. Let $n_2$ be the nilpotency of $\mathfrak{m}_2$. If $n_2 \geq 3$, then the subgraph induced by the set $\{R_1 \times (0), R_1 \times \mathfrak{m}_2^{n_2-1}, R_1 \times \mathfrak{m}_2, (0) \times \mathfrak{m}_2^{n_2-1}, (0) \times \mathfrak{m}_2, (0) \times R_2\}$ contains $K_{3,3}$ as a subgraph of $A_I(R)$, a contradiction. Hence, $n_2 = 2$. Suppose $I$ is the non-zero proper ideal of $R_2$ different from $\mathfrak{m}_2$. Then the subgraph induced by the set $\{R_1 \times (0), R_1 \times I, R_1 \times \mathfrak{m}_2, (0) \times \mathfrak{m}_2, (0) \times I, (0) \times R_2\}$ contains $K_{3,3}$ as a subgraph $A_I(R)$, a contradiction. Thus, $\mathfrak{m}_2$ is the only one ideal in $R_2$.

Converse follows from Fig. 1 and $A_I(R)$ is $C_4$ or $K_m$, where $1 \leq m \leq 4$.

Now, we characterize all isomorphism classes of commutative Artinian ring $R$ whose $A_I(R)$ has genus one.

**Theorem 10** *Let $R$ be a commutative Artinian ring. Then $g(A_I(R)) = 1$ if and only if $R$ is isomorphic to one of the following ring:*

*(i)* *$R_1 \times R_2$, $\mathfrak{m}_i$ is the only non-zero proper ideal in $R_i$ and $n_i = 2$ for all $i = 1, 2$, where $n_i$ is the nilpotency of $\mathfrak{m}_i$;*
*(ii)* *$R_1 \times R_2$, $R_1$ is a field and $\mathfrak{m}_2$, $\mathfrak{m}_2^2$ are only ideals of $R_2$ and $n_2 = 3$, where $n_2$ is the nilpotency of $\mathfrak{m}_2$;*

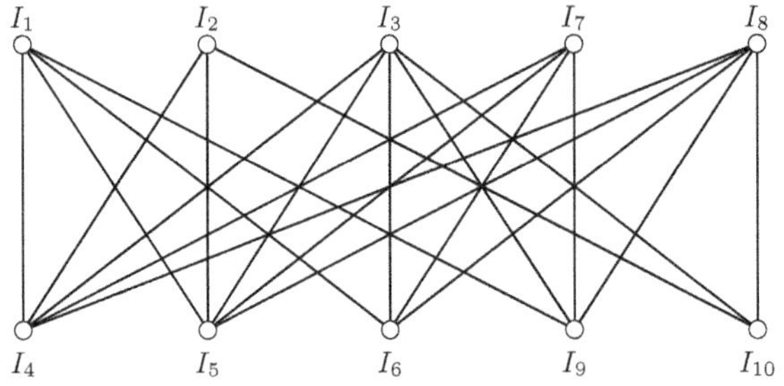


**Fig. 2** Subgraph $G$ of $A_I(R)$ induced by the set $\Omega_2$

*(iii)* $R_1 \times R_2$, *$R_1$ is a field and* $\mathfrak{m}_2$, $\mathfrak{m}_2^2$, $\mathfrak{m}_2^3$ *are only ideals of* $R_2$ *and* $n_2 = 4$, *where* $n_2$ *is the nilpotency of* $\mathfrak{m}_2$;
*(iv)* *$R$ is local ring with* $5 \leq |\mathbb{A}^*(R)| \leq 7$.

***Proof*** Assume that $g(A_I(R)) = 1$. Suppose that $n \geq 4$. Let $\Omega_1 = \{I_1, I_2, \dots, I_{13}\} \subset \mathbb{A}^*(R)$, where $I_1 = R_1 \times (0) \times \cdots \times (0)$, $I_2 = R_1 \times R_2 \times (0) \times \cdots \times (0)$, $I_3 = R_1 \times (0) \times R_3 \times (0) \times \cdots \times (0)$, $I_4 = R_1 \times R_2 \times (0) \times R_4 \times \cdots \times (0)$, $I_5 = R_1 \times R_2 \times R_3 \times (0) \times \cdots \times (0)$, $I_6 = (0) \times (0) \times R_3 \times R_4 \times (0) \times \cdots \times (0)$, $I_7 = (0) \times (0) \times (0) \times R_4 \times (0) \times \cdots \times (0)$, $I_8 = (0) \times (0) \times R_3 \times (0) \times \cdots \times (0)$, $I_9 = (0) \times R_2 \times R_3 \times (0) \times \cdots \times (0)$, $I_{10} = R_1 \times (0) \times (0) \times R_4 \times (0) \times \cdots \times (0)$, $I_{11} = (0) \times R_2 \times (0) \times R_4 \times (0) \times \cdots \times (0)$, $I_{12} = R_1 \times (0) \times R_3 \times R_4 \times (0) \times \cdots \times (0)$, $I_{13} = (0) \times R_2 \times R_3 \times R_4 \times (0) \times \cdots \times (0)$. Clearly, $I_1 I_6 = I_1 I_7 = I_1 I_8 = I_1 I_9 = I_1 I_{11} = I_1 I_{13} = I_2 I_6 = I_2 I_7 = I_2 I_8 = I_3 I_7 = I_3 I_{11} = I_4 I_8 = I_5 I_7 = I_7 I_8 = I_9 I_{10} = (0)$. Then $I_{12} \subset ann(I_2 I_9)$, $I_{12} \not\subset ann(I_2) \cup ann(I_9)$, $I_{12} \subset ann(I_2 I_{11})$, $I_{12} \not\subset ann(I_2) \cup ann(I_{11})$, $I_{12} \subset ann(I_2 I_{13})$, $I_{12} \not\subset ann(I_2) \cup ann(I_{13})$, $I_4 \subset ann(I_3 I_6)$, $I_4 \not\subset ann(I_3) \cup ann(I_6)$, $I_4 \subset ann(I_3 I_9)$, $I_4 \not\subset ann(I_3) \cup ann(I_9)$, $I_4 \subset ann(I_3 I_{13})$, $I_4 \not\subset ann(I_3) \cup ann(I_{13})$, $I_5 \subset ann(I_4 I_6)$, $I_5 \not\subset ann(I_4) \cup ann(I_6)$, $I_{12} \subset ann(I_4 I_9)$, $I_{12} \not\subset ann(I_4) \cup ann(I_9)$, $I_9 \subset ann(I_4 I_{12})$, $I_9 \not\subset ann(I_4) \cup ann(I_{12})$, $I_5 \subset ann(I_{11} I_{12})$, $I_5 \not\subset ann(I_{11}) \cup ann(I_{12})$, $I_3 \subset ann(I_4 I_{13})$, $I_3 \not\subset ann(I_4) \cup ann(I_{13})$, $I_4 \subset ann(I_5 I_6)$, $I_4 \not\subset ann(I_5) \cup ann(I_6)$, $I_{13} \subset ann(I_5 I_{10})$, $I_{13} \not\subset ann(I_5) \cup ann(I_{10})$, $I_{12} \subset ann(I_5 I_{11})$, $I_{12} \not\subset ann(I_5) \cup ann(I_{11})$, $I_{10} \subset ann(I_5 I_{13})$, $I_{10} \not\subset ann(I_5) \cup ann(I_{13})$. Then the induced subgraph $K_{5,5}$ is a subdivision of $\Omega_1$. By Lemma 3, $g(A_I(R)) \geq 3$ and thus $n \leq 3$.

Let $n = 3$. Then by Theorem 9, we assume that $R_1$ is not field and $n_1 \geq 2$ be the nilpotency of $\mathfrak{m}_1$. Consider the set $\Omega_2 = \{I_1, I_2, \dots, I_{10}\} \subset \mathbb{A}^*(R)$, where $I_1 = (0) \times (0) \times R_3$, $I_2 = (0) \times R_2 \times (0)$, $I_3 = (0) \times R_2 \times R_3$, $I_4 = \mathfrak{m}_1 \times (0) \times (0)$, $I_5 = R_1 \times (0) \times (0)$, $I_6 = \mathfrak{m}_1^{n_1-1} \times R_2 \times (0)$, $I_7 = \mathfrak{m}_1^{n_1-1} \times (0) \times R_3$, $I_8 = \mathfrak{m}_1^{n_1-1} \times R_2 \times R_3$, $I_9 = R_1 \times R_2 \times (0)$, $I_{10} = R_1 \times (0) \times R_3$. Then the subgraph $G$ is isomorphic to the graph in Fig. 2 induced by $\Omega_2$ and subgraph $G$ have no cycle of length 3, $|V(G)| = 10$ and $|E(G)| = 21$. Hence, by Theorem 4, $g(A_I(R)) \geq 2$. Hence, $n \leq 2$.

**Case 1.** Suppose $n = 2$ and each $R_i$ is not a field for $i = 1, 2$.
Then $\mathfrak{m}_i \neq (0)$ for $i = 1, 2$. Let $n_i$ be the nilpotency of $\mathfrak{m}_i$ for $i = 1, 2$. Suppose that $n_2 \geq 3$. Consider the set $\Omega_3 = \{I_1, I_2, \dots, I_9\} \subset \mathbb{A}^*(R)$, where $I_1 = \mathfrak{m}_1^{n_1-1} \times (0)$, $I_2 = R_1 \times (0)$, $I_3 = R_1 \times \mathfrak{m}_2^{n_2-1}$, $I_4 = R_1 \times \mathfrak{m}_2$, $I_5 = (0) \times R_2$, $I_6 = (0) \times \mathfrak{m}_2$,

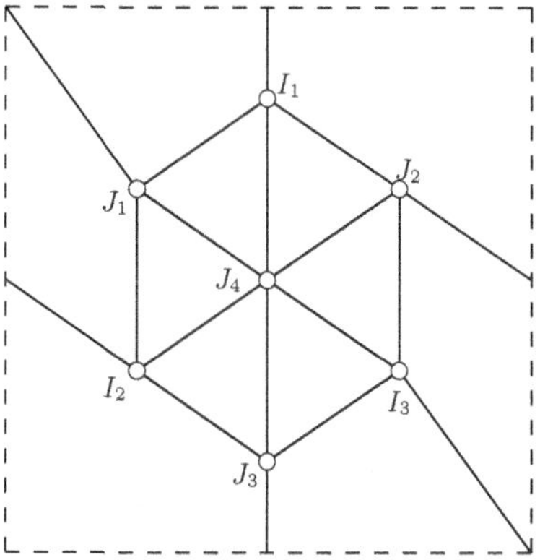


**Fig. 3** Embedding of $A_I(R_1 \times R_2)$ in $\mathbb{S}_1$

$I_7 = (0) \times \mathfrak{m}_2^{n_2-1}$, $I_8 = \mathfrak{m}_1 \times \mathfrak{m}_2$, $I_9 = \mathfrak{m}_1 \times \mathfrak{m}_2^{n_2-1}$. Then $K_{4,5}$ as a subgraph of $A_I(R)$ induced by $\Omega_3$ and by Lemma 3, $g(A_I(R)) \geq 2$, a contradiction. Hence, $n_i = 2$ for all $i = 1, 2$.

Suppose $R_2$ has non-zero proper ideal $I$ different from $\mathfrak{m}_2$. Since $\mathfrak{m}_2^2 = (0)$, $I^2 = (0)$. Consider the set $\Omega_4 = \{I_1, I_2, \ldots, I_9\} \subset \mathbb{A}^*(R)$, where $I_1 = \mathfrak{m}_1 \times (0)$, $I_2 = R_1 \times (0)$, $I_3 = R_1 \times I$, $I_4 = R_1 \times \mathfrak{m}_2$, $I_5 = (0) \times R_2$, $I_6 = (0) \times \mathfrak{m}_2$, $I_7 = (0) \times I$, $I_8 = \mathfrak{m}_1 \times \mathfrak{m}_2$, $I_9 = \mathfrak{m}_1 \times I$. Then $K_{4,5}$ as a subgraph of $A_I(R)$ induced by $\Omega_3$ and by Lemma 3, $g(A_I(R)) \geq 2$, a contradiction.

**Case 2.** Suppose $n = 2$ and $R_1$ is a field.

If $R_2$ is a field, then $A_I(R) \cong K_2$ is planar. Hence, $R_2$ is not a field. Let $n_2$ be the nilpotency of $\mathfrak{m}_2$. Suppose $n_2 \geq 5$. Consider the set $\Omega_5 = \{J_1, J_2, \ldots, J_9\} \subset \mathbb{A}^*(R)$, where $J_1 = R_1 \times (0)$, $J_2 = R_1 \times \mathfrak{m}_2$, $J_3 = R_1 \times \mathfrak{m}_2^{n_2-3}$, $J_4 = R_1 \times \mathfrak{m}_2^{n_2-2}$, $J_5 = (0) \times \mathfrak{m}_2$, $J_6 = (0) \times \mathfrak{m}_2^{n_2-3}$, $J_7 = (0) \times \mathfrak{m}_2^{n_2-2}$, $J_8 = (0) \times \mathfrak{m}_2^{n_2-1}$, $J_9 = (0) \times R_2$. It is easy to see that $K_{4,5}$ is a subgraph of $A_I(R)$, a contradiction. Thus, $n_2 \leq 4$.

Suppose $n_2 = 2$. If $|\mathbb{A}^*(R_2)| = 1$, then by Theorem 9, $A_I(R)$ is planar, a contradiction. Hence, $|\mathbb{A}^*(R_2)| \geq 2$ and by Theorem 1, there exist three distinct ideals $I$, $J, K \in \mathbb{A}^*(R_2)$ different from $\mathfrak{m}_2$. Consider the set $\Omega_6 = \{I_1, I_2, \ldots, I_9\} \subset \mathbb{A}^*(R)$, where $I_1 = R_1 \times (0)$, $I_2 = R_1 \times \mathfrak{m}_2$, $I_3 = R_1 \times I$, $I_4 = R_1 \times J$, $I_5 = (0) \times \mathfrak{m}_2$, $I_6 = (0) \times I$, $I_7 = (0) \times J$, $I_8 = (0) \times K$, $I_9 = (0) \times R_2$. Then $K_{4,5}$ as a subgraph of $A_I(R)$ and by Lemma 3, $g(A_I(R)) \geq 2$, a contradiction.

Suppose $n_2 = 3$ or $4$. If $I \in \mathbb{A}^*(R_2)$ with $I \neq \mathfrak{m}_2^i$ for all $i$, then by Theorem 1, there exist three distinct ideals $I, J, K \in \mathbb{A}^*(R_2)$ such that $I, J, K \neq \mathfrak{m}_2^i$ for all $i$. Consider the set $\Omega_7 = \{I_1, I_2, \ldots, I_{10}\} \subset \mathbb{A}^*(R)$, where $I_1 = R_1 \times (0)$, $I_2 = R_1 \times \mathfrak{m}_2^{n_2-1}$, $I_3 = R_1 \times I$, $I_4 = (0) \times \mathfrak{m}_2^{n_2-1}$, $I_5 = (0) \times \mathfrak{m}_2$, $I_6 = (0) \times I$, $I_7 = (0) \times J$, $I_8 = (0) \times K$, $I_9 = (0) \times R_2$, $I_{10} = R_1 \times \mathfrak{m}_2$. Then $K_{4,5}$ as a subgraph of $A_I(R)$ and by Lemma 3, $g(A_I(R)) \geq 2$ and by Lemmas 2 and 3, $g(A_I(R)) \geq 2$, a contradiction.

**Case 3.** $n = 1$. Then $A_I(R)$ is complete and so $5 \leq |V(A_I(R))| \leq 7$.

Converse follows from Figs. 3 and 4.

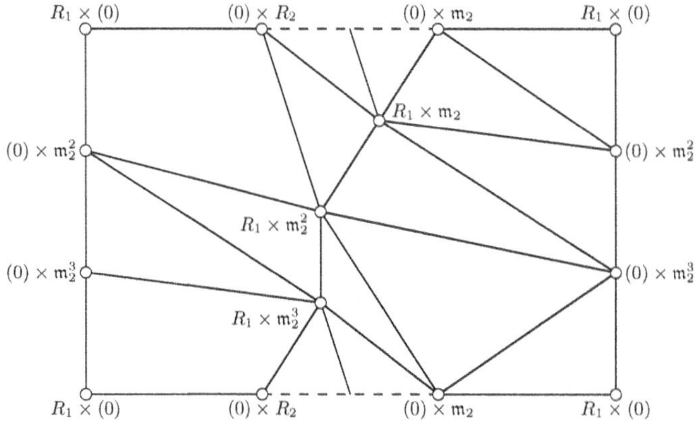


**Fig. 4** Embedding of $A_I(R_1 \times R_2)$ in $\mathbb{S}_1$,where $R_1$ is a fields

Now, we classify all commutative Artinian rings whose annihilator-ideal graph $A_I(R)$ is crosscap one.

**Theorem 11** *Let $R$ be a Artinian ring. Then $\overline{g}(A_I(R)) = 1$ if and only if $R$ is isomorphic to one of the following:*

*(i)* $R_1 \times R_2$, *$\mathfrak{m}_i$ is the only one non-zero proper ideal in $R_i$ and $n_i = 2$ for all $i = 1, 2$, where $n_i$ is the nilpotency of $\mathfrak{m}_i$;*

*(ii)* $R_1 \times R_2$, *$R_1$ is a field and $\mathfrak{m}_2$, $\mathfrak{m}_2^2$ are only ideals of $R_2$ and $n_2 = 3$, where $n_2$ is the nilpotency of $\mathfrak{m}_2$;*

*(iii) $R$ is local ring with $5 \leq |\mathbb{A}^*(R)| \leq 6$.*

***Proof*** Suppose that $\overline{g}(A_I(R)) = 1$ and $n \geq 4$. By Theorem 10, we have $K_{4,4}$ as a subgraph $A_I(R)$ and by Lemma 3, $\overline{g}(A_I(R)) \geq 2$. Thus, $n \leq 3$.

Suppose $n = 3$. Then by Theorem 9, $R_i$ is not a field for some $i$. Let $R_1$ is not a field and $n_1 \geq 2$ be the nilpotency of $\mathfrak{m}_1$. Consider the set $\Omega_2 = \{I_1, I_2, \ldots, I_{10}\} \subset \mathbb{A}^*(R)$, where $I_1 = (0) \times (0) \times R_3, I_2 = (0) \times R_2 \times (0), I_3 = (0) \times R_2 \times R_3, I_4 = \mathfrak{m}_1 \times (0) \times (0), I_5 = R_1 \times (0) \times (0), I_6 = \mathfrak{m}_1^{n_1-1} \times R_2 \times (0), I_7 = \mathfrak{m}_1^{n_1-1} \times (0) \times R_3, I_8 = \mathfrak{m}_1^{n_1-1} \times R_2 \times R_3, I_9 = R_1 \times R_2 \times (0), I_{10} = R_1 \times (0) \times R_3$. Then the subgraph $G$ is isomorphic to the graph in Fig. 2 induced by $\Omega_2$ and subgraph $G$ have no cycle of length 3, $|V(G)| = 10$ and $|E(G)| = 21$. Hence, by Theorem 4, $\overline{g}(A_I(R)) \geq 2$. Hence, $n \leq 2$.

Suppose $n = 2$ and each $R_i$ is not a field for $i = 1, 2$.

Then $\mathfrak{m}_i \neq (0)$ for $i = 1, 2$. Let $n_i$ be the nilpotency of $\mathfrak{m}_i$ for $i = 1, 2$. Suppose that $n_2 \geq 3$. Consider the set $\Omega_3 = \{I_1, I_2, \ldots, I_9\} \subset \mathbb{A}^*(R)$, where $I_1 = \mathfrak{m}_1^{n_1-1} \times (0), I_2 = R_1 \times (0), I_3 = R_1 \times \mathfrak{m}_2^{n_2-1}, I_4 = R_1 \times \mathfrak{m}_2, I_5 = (0) \times R_2, I_6 = (0) \times \mathfrak{m}_2, I_7 = (0) \times \mathfrak{m}_2^{n_2-1}, I_8 = \mathfrak{m}_1 \times \mathfrak{m}_2, I_9 = \mathfrak{m}_1 \times \mathfrak{m}_2^{n_2-1}$. Then the subgraph induced by $\Omega_3$ in $A_I(R)$ contains $K_{4,5}$ as a subgraph and by Lemma 3, $\overline{g}(A_I(R)) \geq 2$, a contradiction. Hence, $n_i = 2$ for all $i = 1, 2$ and $R_1 \times R_2$, $\mathfrak{m}_i$ is the only one non-zero proper ideal in $R_i$.

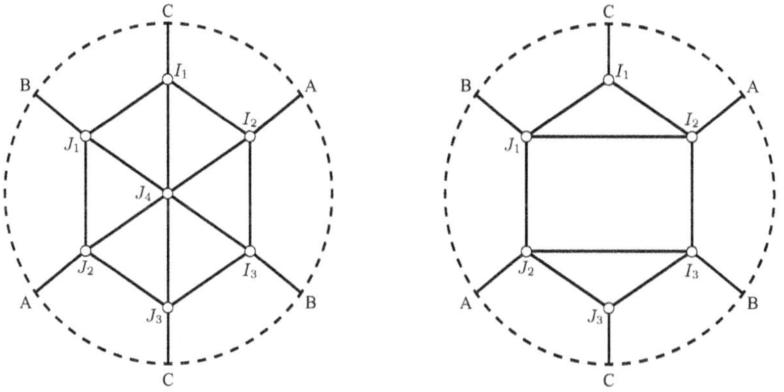


(a). The graph $A_I(R_1 \times R_2)$ in $\mathbb{N}_1$ (b). The graph $A_I(R_1 \times R_1)$ in $\mathbb{N}_1$, where $R_1$ is a fields

**Fig. 5** Embedding of annihilator-ideal graphs in $\mathbb{N}_1$

Suppose $n = 2$ and $R_i$ is not a field for some $i$.

Without loss of generality, we assume that $R_1$ is a field. If $R_2$ is a field, then $A_I(R) \cong K_2$ is planar. Hence, $R_2$ is not a field. Let $n_2$ be the nilpotency of $\mathfrak{m}_2$. Suppose $n_2 \geq 4$. Consider the set $\Omega_5 = \{J_1, J_2, \ldots, J_8\} \subset \mathbb{A}^*(R)$, where $J_1 = R_1 \times (0)$, $J_2 = R_1 \times \mathfrak{m}_2$, $J_3 = R_1 \times \mathfrak{m}_2^{n_2-2}$, $J_4 = R_1 \times \mathfrak{m}_2^{n_2-1}$, $J_5 = (0) \times \mathfrak{m}_2$, $J_6 = (0) \times \mathfrak{m}_2^{n_2-2}$, $J_7 = (0) \times \mathfrak{m}_2^{n_2-1}$ and $J_8 = (0) \times R_2$. Then the subgraph induced by $\Omega_5$ of $A_I(R)$ contains $K_{4,4}$ as a subgraph and by Lemma 3, $\overline{g}(A_I(R)) \geq 2$, a contradiction. Thus, $n_2 \leq 3$.

Suppose $n_2 = 2$. If $|\mathbb{A}^*(R_2)| = 1$, then by Theorem 9, $A_I(R)$ is planar, a contradiction. Hence, $|\mathbb{A}^*(R_2)| \geq 2$ and by Theorem 1, there exist three distinct ideals $I$, $J, K \in \mathbb{A}^*(R_2)$ different from $\mathfrak{m}_2$. Consider the set $\Omega_6 = \{I_1, I_2, \ldots, I_8\} \subset \mathbb{A}^*(R)$, where $I_1 = R_1 \times (0)$, $I_2 = R_1 \times \mathfrak{m}_2$, $I_3 = R_1 \times I$, $I_4 = R_1 \times J$, $I_5 = (0) \times \mathfrak{m}_2$, $I_6 = (0) \times I$, $I_7 = (0) \times J$, $I_8 = (0) \times K$. Then the subgraph induced by $\Omega_6$ in $A_I(R)$ contains $K_{4,4}$ as a subgraph and by Lemma 3, $\overline{g}(A_I(R)) \geq 2$, a contradiction.

Suppose $n_2 = 3$. If $I \in \mathbb{A}^*(R_2)$ with $I \neq \mathfrak{m}_2^i$ for all $i$, then by Theorem 1, there exist three distinct ideals $I$, $J$, $K \in \mathbb{A}^*(R_2)$ such that $I, J, K \neq \mathfrak{m}_2^i$ for all $i$. Consider the set $\Omega_7 = \{I_1, I_2, \ldots, I_8\} \subset \mathbb{A}^*(R)$, where $I_1 = R_1 \times (0)$, $I_2 = R_1 \times \mathfrak{m}_2^{n_2-1}$, $I_3 = (0) \times \mathfrak{m}_2^{n_2-1}$, $I_4 = (0) \times \mathfrak{m}_2$, $I_5 = (0) \times I$, $I_6 = (0) \times J$, $I_7 = (0) \times K$, $I_8 = (0) \times R_2$. Then $K_{3,5}$ as a subgraph of and by Lemma 3, $\overline{g}(A_I(R)) \geq 2$, a contradiction. Thus, $R$ is isomorphic to $R_1 \times R_2$, $R_1$ is a field and $\mathfrak{m}_2$, $\mathfrak{m}_2^2$ are only ideals of $R_2$. Finally, if $n = 1$, then $A_I(R)$ is complete graph and by Lemma 3, we have $5 \leq |\mathbb{A}^*(R)| \leq 6$.

Converse follows from Fig. 5.

# References


1. Atiyah, M.F., Macdonald, I.G.: Introduction to Commutative Algebra. Addison-Wesley Publishing Company, London (1969)
2. Badawi, A.: On the annihilator graph of a commutative ring. Commun. Algebra **42**(1), 108–121 (2014). https://doi.org/10.1080/00927872.2012.707262
3. Behboodi, M., Rakeei, Z.: The annihilating-ideal graph of commutative rings-I. J. Algebra Appl. **10**(4), 727–739 (2011). https://doi.org/10.1142/S0219498811004896
4. Behboodi, M., Rakeei, Z.: The annihilating-ideal graph of commutative rings-II. J. Algebra Appl. **10**(4), 741–753 (2011). https://doi.org/10.1142/S0219498811004902
5. Bojan, M., Carten, T.: Graphs on Surfaces. The Johns Hopkins University Press, Baltimore and London (1956)
6. Dummit, D.S., Foote, R.M.: Abstract Algebra, 2nd edn. Wiley, Singapore (2005)
7. Nikmehr, M. J., Hossini, S. M.: More on the annihilator-ideal graph of commutative ring. J. Algebra Appl. **10**, 1950160 (1–14) (2019)
8. Shaveisi, F., Nikandish, R.: The nil-graph of ideals of a commutative ring. Bull. Malays. Math. Sci. Soc. **39**, 3–11 (2016)
9. Salehifar, S., Khashayarmanesh, K., Afkhami, M.: On the annihilator-ideal graph of commutative rings. Ricerche Mat. **66**(2), 431–447 (2017)
10. White, A.T.: Graphs. Groups and Surfaces. North-Holland, Amsterdam (1973)

# The Radio $k$-chromatic Number for the Corona of Arbitrary Graph and $K_1$

**P. K. Niranjan**

**Abstract** For $k > 0$, an integer, and a connected simple graph $H$, a radio $k$-coloring of $H$ is a function $h$ which assigns every vertex of $H$ a non-negative integer so that for each duo of two separate vertices $x$ and $y$ of $H$, the absolute difference of color of the vertices is at least $1 + k - d(y, x)$. For a radio $k$-coloring $h$, the span $rc_k(h)$ of $h$ is the largest color allotted by it. The radio $k$-chromatic number, $rc_k(H)$, is $\min\{rc_k(h) : h \text{ is a radio } k\text{-coloring of } H\}$. In this manuscript, for the radio $k$-chromatic number of the corona of any arbitrary graph $H$ and $K_1$, we obtain an upper bound. Further, we derive a necessary condition for the lower bound to be exact. Furthermore, we corroborate that the given upper bound is sharp for radio $k$-chromatic number of $P_n \odot K_1$, where $P_n$ is a path with $n$ vertices and $n$ is odd.

**Keywords** Corona of graphs · Radio coloring · Graph operation · Radio number · Radio $k$-chromatic number · Span · Radio $k$-coloring

## 1 Introduction

Inspired by the frequency allocation problems radio $k$-coloring of graphs was defined. Detailed study of frequency assignment problem and graph theoretic formulation can be seen in Hale [5]. In 2001, Chartrand et al. [2] instigated a study of a graph coloring problem, namely, radio $k$-coloring of graphs, incentivized by frequency assignment to radio stations. For a connected simple graph $H$ and a natural number $k$, which is less than $diam(H)$, a radio $k$-coloring of $H$ is an allocation $h$ which assigns every vertex of $H$ some non-negative integer so that the difference between the color of every pair of distinct vertices $y$ and $x$ of $H$ is at least $1 + k - d(y, x)$. The span, $rc_k(h)$, of $h$ is the largest color utilized by $h$. The least of spans overall radio $k$-colorings of $H$ is known as the radio $k$-chromatic number of $H$ and is represented by $rc_k(H)$. An optimal radio $k$-coloring of $H$ is a radio $k$-coloring of $H$ whose span is $rc_k(H)$. Since our aim is to diminish the largest color utilized of radio $k$-coloring,

P. K. Niranjan (✉)
Department of Mathematics, RV College of Engineering, Mysuru Road, Bengaluru 560059, India
e-mail: niranjanpk704@gmail.com



B. Rushi Kumar et al. (eds.), *Mathematics and Computing*, Springer Proceedings in Mathematics & Statistics 415, https://doi.org/10.1007/978-981-19-9307-7_15

**Table 1** Special name for $rk$-coloring for some specific values of $k$

| $k$ | Name of the $rk$-coloring | $rk$-number |
|---|---|---|
| 1 | Proper coloring | Chromatic number ($\chi(H)$) |
| 2 | $L(2, 1)$-coloring | $L(2, 1)$-number ($\lambda(H)$) |
| $diam(H) - 2$ | Nearly antipodal coloring | Nearly antipodal number ($an'(H)$) |
| $diam(H) - 1$ | Antipodal coloring | Antipodal number ($an(H)$) |
| $diam(H)$ | Radio coloring | Radio number ($rn(H)$) |

we deem that every radio $k$-coloring assigns the color 0. Radio 1-coloring of $H$ is a usual vertex coloring of $H$ and $rc_1 = \chi(H)$. Radio 2-coloring of $H$ is an $L(2, 1)$-coloring of $H$ which was instigated by Griggs and Yeh [4]. In the studies of radio $k$-colorings, there are unique titles for some certain values of $k$, which are given in Table 1. Throughout this writing, we indicate a simple connected graph by a graph. In short, we refer to radio $k$-coloring as $rk$-coloring; radio $k$-chromatic number as $rk$-number; and radio number as $r$-number.

For the Cartesian product $H \Box G$ of arbitrary graphs, an upper bound for $rc_k(H \Box G)$ has been found by Kchikech et al. [8] as $\chi(G^k)(rc_k(H) + k - 1) - k$ and improved the same for $rk$-number of $P_n \Box P_n$ when $k \geq 2n - 3$. For $rn(P_n \Box K_{1,m})$, Ajayi and Adefokun [1] have obtained bounds. In [9], Kim et al. have deduced the $r$-number of $P_n \Box K_m$. Morris-Rivera et al. [16] have found $rn(C_n \Box C_n)$ of the toroidal grid. Saha and Panigrahi [21] have extended this result for the Cartesian product $C_n \Box C_m$ of two cycles, when $mn$ is even. In [11], Kola and Panigrahi have established a lower bound for $rc_k(H)$ of any graph $H$ and employing this bound, for prism graph $C_m \Box P_n$, they have established a lower bound for the $rk$-number. Furthermore, for $rn(C_m \Box P_2)$, when $m \equiv 2\ (mod\ 8)$ or $m \equiv 1\ (mod\ 4)$, they have ascertained that this bound is precisely the exact number. For several combinations of $m$ and $n$, Niranjan and Kola [18] have found $r$-number of $C_n \Box K_m$. Kola and Panigrahi [10] have found $r$-number of hypercube $Q_n$. The $L(2, 1)$-coloring of graphs is a special case of radio $k$-coloring. Several authors [6, 12, 14, 15, 19, 20] have studied $L(2, 1)$-coloring and parameters related to it for different operations of graphs.

Let $G$ and $H$ be graphs with set of vertices $\{x_1, x_2, \ldots, x_m\}$ and $\{y_1, y_2, \ldots, y_n\}$, respectively. The corona $H \odot G$ of $H$ and $G$ is the graph having the set of vertices $V(H) \bigcup \Big( \bigcup\limits_{i=1}^{n} \{y_i^j : 1 \leq j \leq m\} \Big)$ and the set of edges $E(H) \bigcup \Big( \bigcup\limits_{i=1}^{n} \{y_i y_i^j : 1 \leq j \leq m\} \Big) \bigcup \Big( \bigcup\limits_{i=1}^{n} \{y_i^j y_i^l : x_j x_l \in E(G)\} \Big)$. Identically, $H \odot G$ is the graph procured by taking one replicate of $H$ and for each vertex $y_i$ of $H$ consider one replicate of $G$, say $G_i$, and joining $y_i$ to each and every vertex of $G_i$ by a line. This is straightforward to observe that $H \odot G$ and $G \odot H$ are not isomorphic if $H$ and $G$ are not isomorphic, and $H \odot G$ is connected if and only if $H$ is connected. As well, it is straightforward to note that $diam(H \odot G) = diam(H) + 2$. In [17], Niranjan and Kola have determined the $r$-number for corona of the path $P_n$ and the cycle $C_m$, when $n$ is even, and obtained bounds for the $r$-number of $P_n \odot C_m$, for odd $n$. Further, similar results were obtained for $P_n \odot P_m$. Uma and Bhargavi [22] have obtained $rn(C_3 \odot P_n)$, $rn(C_4 \odot P_n)$ and $rn(C_5 \odot P_n)$ for $n \geq 2$. Das et al. [3] have obtained a procedure to get lower bound for $rc_k(H)$ as in Theorem 1.

**Theorem 1** ([3]) *If $h$ is an $rk$-coloring of a graph $H$, then*

$$rc_k(h) \geq |D_k| + 2\sum_{i=0}^{q}[|V_i|(q-i)] - 2q + \beta + \alpha - 1,$$

*where $V_i$'s and $D_k$ are described as pursues. If $k = 2q+1$ is odd, then $V_0 = V(C)$, where $C$ is a maximal clique in $H$. If $k = 2q$, then $V_0 = \{v\}$, where $v \in V(H)$. Define recursively $V_{i+1} = N(V_i)\backslash(V_0 \cup V_1 \cup V_2 \cup \cdots \cup V_i) for i = 0, 1, \ldots, q-1$. Let $D_k = V_0 \cup V_1 \cup V_2 \cup \cdots \cup V_q$. The vertices of $D_k$ receiving the smallest and the largest colors are in $V_\alpha$ and $V_\beta$, respectively.*

The following theorem is one of the main results in this article.

**Theorem 2** *If $k > 1$ and $H$ is a graph having $n$ vertices, then $rc_k(H \odot K_1) \leq 2rc_{k-2}(H) + k + 2n - 4$.*

## 2 Results

In this part of the manuscript, we first deduce an upper bound for the $r$-number of $H \odot K_1$. Later on, prove that the obtained upper bound is precisely exact for some classes of graphs. Furthermore, we give a necessary condition for the given bound to be the exact $rk$-number of the graph. The below definition proffers how considerably additional is the disparity betwixt any pair of successive colors utilized in an $rk$-coloring.

**Definition 1** Let $h$ be an $rk$-coloringof a graph $H$. Let $u_1, u_2, \ldots, u_m$ be the arrangement of elements of $V(H)$ so that $h(u_{j+1}) \geq h(u_j)$, $j = 1, 2, \ldots, m-1$. Define $\epsilon_j = h(u_j) - h(u_{j-1}) - (k + 1 - d(u_j, u_{j-1}))$, $2 \leq j \leq m$.

We refer the sums $\sum_{j=2}^{m} d(u_j, u_{j-1})$ and $\sum_{j=2}^{m} \epsilon_j$ as distance sum and epsilon sum, respectively. Lemma below gives the span of an $rk$-coloring as a function of $k$, a number of vertices $m$ in the graph, distance sum, and epsilon sum.

**Lemma 1** *Let $h$ be an $rk$-coloring of $H$ and let $u_1, u_2, u_3, \ldots, u_m$ be an arrangement of elements of $V(H)$ so that $h(u_{j+1}) \geq h(u_j)$, $1 \leq j \leq m-1$ and $\epsilon_j = h(u_j) - h(u_{j-1}) - (k + 1 - d(u_j, u_{j-1}))$, $2 \leq j \leq m$. Then*

$$rc_k(h) = (m-1)(k+1) - \sum_{j=2}^{m}[d(u_j, u_{j-1}) + \epsilon_j].$$

***Proof***

$$\begin{aligned} h(u_m) &= \sum_{j=2}^{m}[h(u_j) - h(u_{j-1})] + h(u_1) \\ &= \sum_{j=2}^{m}[1 + k - d(u_j, u_{j-1}) + \epsilon_j] + h(u_1) \\ &= (m-1)(1+k) - \sum_{j=2}^{m} d(u_j, u_{j-1}) + \sum_{j=2}^{m} \epsilon_j + h(u_1). \end{aligned}$$

Since $h(u_1) = 0$, $rc_k(h) = h(u_m) = (m-1)(k+1) - \sum_{j=2}^{m} d(u_j, u_{j-1}) + \sum_{j=2}^{m} \epsilon_j$.

Now, we prove the major outcome of the manuscript, which is Theorem 2. First, we consider an optimal radio $(k-2)$-coloring of $H$, using which we obtain a vertex ordering and a $rk$-coloring of $H \odot K_1$.

**Proof of Theorem** 2 Let $H$ be a graph, $n$ be the order of $H$, and $k > 1$. Let $h$ be an optimal radio $(k-2)$-coloring of $H$ and $v_1, v_2, \ldots, v_n$ be the associated vertex arrangement of $H$ such that $h(v_{i+1}) \geq h(v_i)$, $i = 1, 2, 3, \ldots, n-1$. Let $u_i$, $1 \leq i \leq n$, be the vertex of copy of $K_1$ corresponding to $v_i$. To get the required upper bound, we first order the vertices of $H \odot K_1$ depending on $n$ is odd or even, using which we generate a $rk$-coloring of $H \odot K_1$.

**Case I: $n = 2p$**

For $l = 1, 2, \ldots, p$, we choose $v_{2l-1}$ as $x_{2l-1}$; $u_{2l}$ as $x_{2l}$; $v_{2l}$ as $x_{n+2l}$ and $u_{2l-1}$ as $x_{n+2l-1}$. For $1 \leq l < j \leq 2n$, we have

$$d(x_j, x_l) = \begin{cases} d(v_{j'}, v_{l'}), & \text{if both } x_l \text{ and } x_j \text{ are on } H; \\ d(v_{j'}, v_{l'}) + 1, & \text{if only one of } x_l \text{ and } x_j \text{ is on } H; \\ d(v_{j'}, v_{l'}) + 2, & \text{if both } x_l \text{ and } x_j \text{ are on copies of } K_1, \end{cases}$$

where $j' \equiv j \ (mod \ n)$ and $l' \equiv l \ (mod \ n)$. So, by the choice of the vertices $x_l$, we have $d(x_{l+1}, x_l) = 1 + d(v_{l+1}, v_l)$ for $1 \leq l \leq n-1$ and $d(x_{l+1}, x_l) = 1 + d(v_{l-n}, v_{l+1-n})$ for $n+1 \leq l \leq 2n-1$, and $d(x_{n+1}, x_n) \geq 3$.

For $H \odot K_1$, we describe a coloring $g$ as pursues: $g(x_1) = h(v_1) = 0$; for $1 \leq l \leq n-1$, $g(x_{l+1}) = g(x_l) + [h(v_{l+1}) - h(v_l)] + 1$, $g(x_{n+l+1}) = g(x_{n+l}) + [h(v_{l+1}) -$

$h(v_l)] + 1$ and $g(x_{n+1}) = g(x_n) + k - 2$. Next, we show that $g$ is an $rk$-coloring of $H \odot K_1$. By the description of $g$, the condition for $rk$-coloring holds good for the pairs $x_l$ and $x_{l+1}$ for all $l$. For $1 \le l < j \le n$, $g(x_j) - g(x_l) = h(v_j) - h(v_l) + (j - l) \ge k - 2 + 1 - d(v_j, v_l) + (j - l) \ge k + 1 - d(v_j, v_l) - 2 + (j - l)$. If both $x_l$ and $x_j$ are on $H$ or on copies of $K_1$, then $j - l \ge 2$. So, $g(x_j) - g(x_l) \ge k + 1 - d(x_l, x_j)$. If only one of $x_l$ and $x_j$ is on $H$, then $g(x_j) - g(x_l) \ge 1 + k - (d(v_l, v_j) + 1) + (j - l) - 1 \ge k + 1 - d(x_l, x_j)$. Similarly, the result holds good for $n + 1 \le l < j \le 2n$. If $1 \le l \le n < j \le 2n$ and $j \ne l + n$, then clearly $g(x_j) - g(x_l) \ge k + 1 - d(x_l, x_j)$. If $1 \le l \le n$ and $j = n + l$, then $g(x_j) - g(x_l) \ge k$. Hence, $g$ is an $rk$-coloring of $H \odot K_1$.

Now, the span of $g$ is given by

$$\begin{aligned} rc_k(g) &= g(x_{2n}) = g(x_{2n}) - g(x_{n+1}) + g(x_{n+1}) - g(x_n) + g(x_n) - g(x_1) \\ &= \sum_{l=1}^{n-1}[g(x_{n+l+1}) - g(x_{n+l})] + g(x_{n+1}) - g(x_n) + \sum_{l=1}^{n-1}[g(x_{l+1}) - g(x_l)] \\ &= \sum_{l=1}^{n-1}[h(v_{l+1}) - h(v_l) + 1] + k - 2 + \sum_{l=1}^{n-1}[h(v_{l+1}) - h(v_{n+l}) + 1] \\ &= h(v_n) - h(v_1) + n - 1 + k - 2 + h(v_n) - h(v_1) + n - 1 \\ &= 2rc_{k-2}(H) + k + 2n - 4. \end{aligned}$$

Hence, $rc_k(H) \le 2rc_{k-2}(H) + k + 2n - 4$.

**Case II: $n = 2p + 1$**

For $l = 1, 2, \dots, p$, we choose $v_{2l-1}$ as $x_{2l-1}$; $u_{2l}$ as $x_{2l}$; $v_{2l}$ as $x_{n+2l}$ and $u_{2l-1}$ as $x_{n+2l-1}$. We choose $u_n$ as $x_n$ and $v_n$ as $x_{2n}$. For $1 \le l < j \le 2n$, we have

$$d(x_j, x_l) = \begin{cases} d(v_{j'}, v_{l'}), & \text{if both } x_l \text{ and } x_j \text{ are on } H; \\ d(v_{j'}, v_{l'}) + 1, & \text{if only one of } x_l \text{ and } x_j \text{ is on } H; \\ d(v_{j'}, v_{l'}) + 2, & \text{if both } x_l \text{ and } x_j \text{ are on copies of } K_1, \end{cases}$$

where $j' \equiv j \ (mod\ n)$ and $l' \equiv l \ (mod\ n)$. So, by the choice of the vertices $x_l$, we have $d(x_l, x_{l+1}) = 1 + d(v_{l+1}, v_l)$ for $1 \le l \le n - 2$ and $d(x_{l+1}, x_l) = 1 + d(v_{l+1-n}, v_{l-n})$ for $n + 1 \le l \le 2n - 2$, $d(x_{n-1}, x_n) = d(v_{n-1}, v_n) + 2$, $d(x_{2n-1}, x_{2n}) = d(v_{n-1}, v_n)$ and $d(x_n, x_{n+1}) \ge 3$.

Now, we describe a coloring $g$ of $H \odot K_1$ as pursues: $g(x_1) = h(v_1) = 0$; for $1 \le l \le n - 2$, $g(x_{l+1}) = g(x_l) + [h(v_{l+1}) - h(v_l)] + 1$, $g(x_{n+l+1}) = g(x_{n+l}) + [h(v_{l+1}) - h(v_l)] + 1$, $g(x_n) = g(x_{n-1}) + [h(v_n) - h(v_{n-1})]$, $g(x_{n+1}) = g(x_n) + k - 2$ and $g(x_{2n}) = g(x_{2n-1}) + [h(v_n) - h(v_{n-1})] + 2$. Analogous to Case I, here also we can demonstrate that $g$ is an $rk$-coloring of $H \odot K_1$.

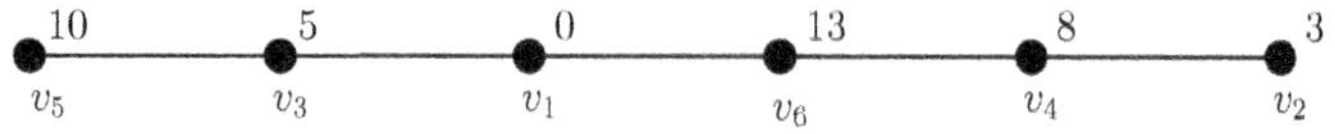


**Fig. 1** An optimal radio coloring (radio 5-coloring) of $P_6$

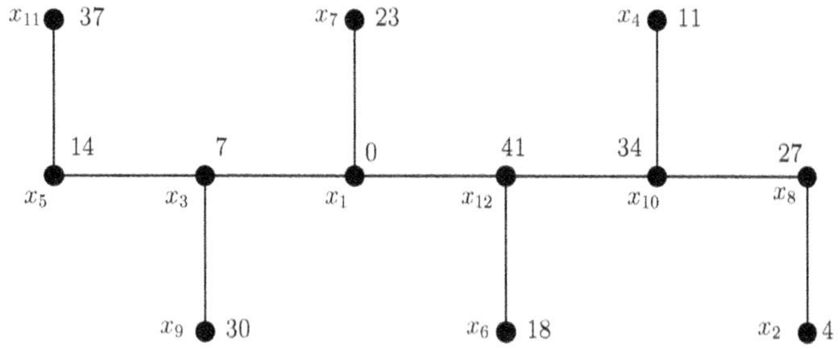


**Fig. 2** An arrangement of the vertices and a radio coloring (radio 7-coloring) of $P_6 \odot K_1$ obtained as in Theorem 2

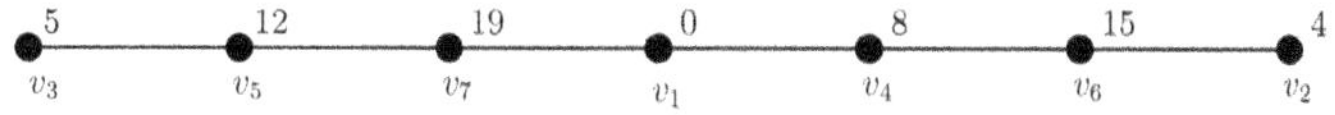


**Fig. 3** An optimal radio coloring (radio 6-coloring) of $P_7$

Now, the span of $g$ is given by $rc_k(h) = g(x_{2n}) = g(x_{2n}) - g(x_{n+1}) + g(x_{n+1}) - g(x_n) + g(x_n) - g(x_1) = h(v_n) - h(v_1) + n - 2 + k - 2 + h(v_n) - h(v_1) + n - 2 + 2 = 2rc_{k-2}(H) + k + 2n - 4$. Hence $rc_k(H) \leq 2rc_{k-2}(H) + k + 2n - 4$. □

**Example 1** In Fig. 1, an optimal radio coloring (radio 5-coloring) and hence a vertex ordering of $P_6$ is considered. Using this vertex ordering, a vertex ordering of $P_6 \odot K_1$ and hence a radio coloring (radio 7-coloring) of $P_6 \odot K_1$ is given in Fig. 2. In Fig. 3, an optimal radio coloring (radio 6-coloring) and hence a vertex ordering of $P_7$ is considered. Using this vertex ordering, a vertex ordering of $P_7 \odot K_1$ and hence a radio coloring (radio 8-coloring) of $P_7 \odot K_1$ is given in Fig. 4. In Fig. 5, an optimal radio 2-coloring of the cycle $C_6$ is considered and hence its vertices are arranged in non-decreasing order of their colors. Using this vertex arrangement, a vertex ordering of $C_6 \odot K_1$ and hence a radio 4-coloring of $C_6 \odot K_1$ is given in Fig. 6. In Fig. 7, an optimal radio 3-coloring of the cycle $C_6$ is considered and hence its vertices are arranged in non-decreasing order of their colors. Using this vertex arrangement, a vertex ordering of $C_6 \odot K_1$ and hence a radio 5-coloring of $C_6 \odot K_1$ is given in Fig. 8.

The next theorem equips an essential requirement for the inequality in Theorem 2 to be an equality.

**Theorem 3** *Let $k > 1$ and $H$ be a connected graph of order $n > 1$. If $rc_k(H \odot K_1) = 2rc_{k-2}(H) + k + 2n - 4$, then, in every optimal radio $(k-2)$-coloring of $H$, at least one pair of vertices of $H$ receiving the colors 1 and $rc_k(H)$ are adjacent.*

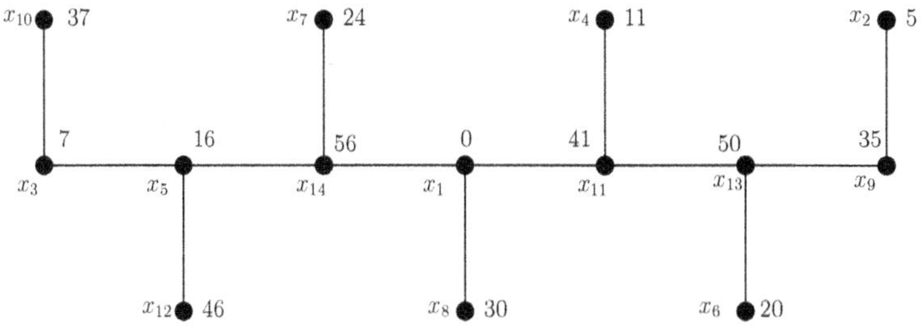


**Fig. 4** An arrangement of the vertices and a radio coloring (radio 8-coloring) of $P_7 \odot K_1$ obtained as in Theorem 2

**Fig. 5** An optimal radio 2-coloring of $C_6$

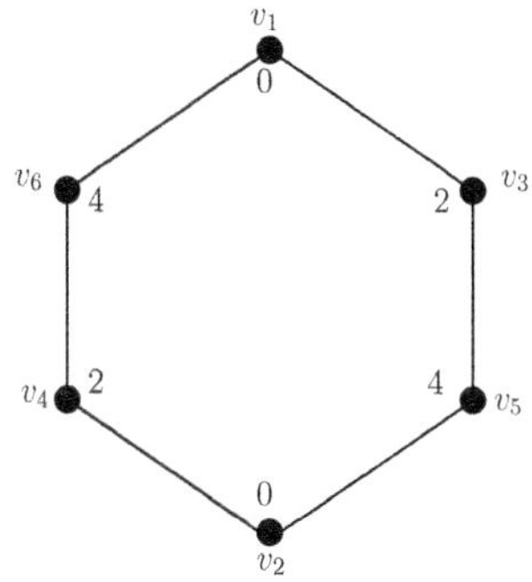


**Fig. 6** An arrangement of the vertices and a radio 4-coloring of $C_6 \odot K_1$ obtained as in Theorem 2

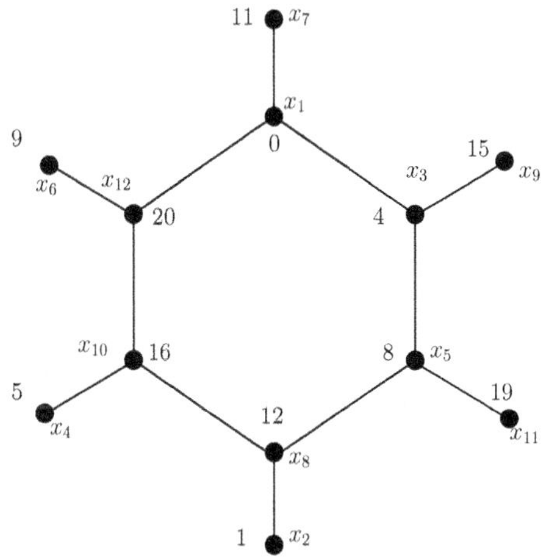

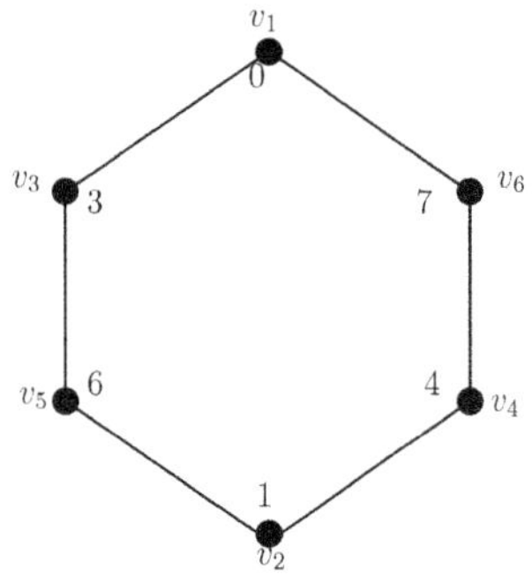


**Fig. 7** An optimal radio 3-coloring of $C_6$

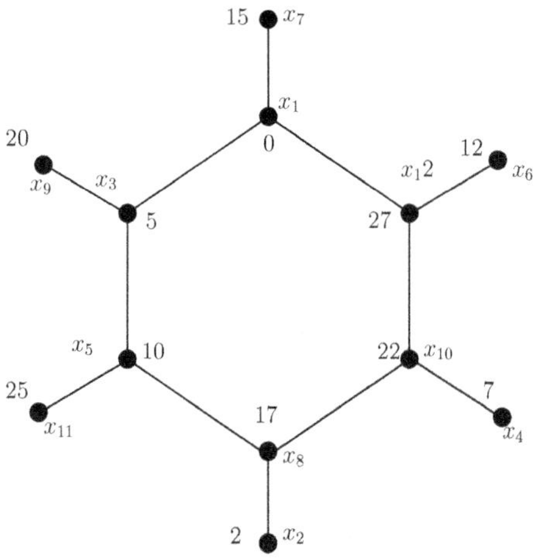


**Fig. 8** An arrangement of the vertices and a radio 5-coloring of $C_6 \odot K_1$ obtained as in Theorem 2

***Proof*** Suppose $h$ is an optimal radio $(k-2)$-coloring of $H$ such that none of the vertex pairs of $H$ receiving the colors 1 and $rc_k(H)$ are adjacent. We consider a vertex ordering $v_1, v_2, \ldots, v_n$ of $H$ such that $h(v_{i+1}) \geq h(v_i)$ for all $i$. We modify the coloring $g$ defined in the proof of Theorem 2 only for $x_{n+1}$ as follows, $g(x_{n+1}) = g(x_n) + k - 3$. Since $v_1$ and $v_n$ are not adjacent, with the above modification $h$ still remains as an $rk$-coloring of $H_1 \odot K_1$ and and $rc_k(g) = 2rc_{k-2}(H) + k + 2n - 5$, this is a repudiation to the attribute that $rc_k(H \odot K_1) = 2rc_{k-2}(H) + k + 2n - 4$.

For any path $P_n$ $(n \geq 2)$, Liu and Zhu [13] found that the $r$-number of $P_n$ is $\frac{n^2-2n+2}{2}$ for even $n$ and $\frac{n^2-2n+3}{2}$ for odd $n$. For any integer $k \geq n$, Kchikech et al. [7] have proved that $rc_k(P_n) = (n-1)k - \frac{1}{2}n(n-2) + 1$ for even $n$ and $rc_k(P_n) = (n-1)k - \frac{1}{2}(n-1)^2 + 2$ for odd $n$.

**Theorem 4** *For a path $P_n$ on even number of vertices and an odd integer $k$, $k > n$, $rc_k(P_n \odot K_1) = 2kn - n^2 - k$.*

$$rc_k(P_n \odot K_1) = \begin{cases} n^2 + n - 1, & if\ k = n + 1; \\ 2kn - n^2 - k, & if\ k > n + 1. \end{cases}$$

***Proof*** Let $n = 2q, k = 2p + 1 > n$, and $P_n : u_1u_2u_3 \ldots u_n$ be the path.

**Case I:** $k = n + 1$
From the outcome of Liu and Zhu [13] article, $rc_{n-1}(P_n) = \frac{n^2-2n+2}{2}$. Now, by Theorem 2, $rc_{n-1}(P_n \odot K_1) \leq n^2 + n - 1$. To get the lower bound, we utilize Theorem 1. We choose $V_0 = \{v_q, v_{q+1}\}$. By this choice of $V_0$, we get $|V_i| = 4$, $i = 1, 2, \ldots, q - 1$ and $|V_q| = 2$. Now, by Theorem 1, we get $rc_{n-1}(P_n \odot K_1) \geq n^2 + n - 1$. Hence, $rc_{n-1}(P_n \odot K_1) = n^2 + n - 1 = 2kn - n^2 - k$.

**Case II:** $k > n + 1$
From the outcome of Kchikech et al. [7] article, $rc_k(P_n) = k(n-1) - \frac{1}{2}(n-2)n + 1$ for even $n$ and $k \geq n$. Now, by Theorem 2, $rc_{n-1}(P_n \odot K_1) \leq 2kn - n^2 - k$. To get the lower bound, we choose $V_0 = \{v_q, v_{q+1}\}$. By this choice of $V_0$, we get $|V_i| = 4$, $i = 1, 2, \ldots, q - 1$ and $|V_q| = 2$. Now, by Theorem 1, we get $rc_{n-1}(P_n \odot K_1) \geq 2kn - n^2 - k$. Hence, $rc_{n-1}(P_n \odot K_1) = n^2 + n - 1 = 2kn - n^2 - k$.

## 3 Conclusion

For a graph, the problem of obtaining the $rk$-number is a non-trivial problem. In this manuscript, we obtained an upper bound for the $rk$-number of $H \odot K_1$, where $H$ is an arbitrary connected graph whose order is at least 2 and $k > 1$. Also, obtained a necessary condition for the upper bound to be exact. Further, we proved that the given upper bound is exact for $P_n \odot K_1$, for even $n$ and odd $k > n$. It is interesting to classify graphs for which the upper bound is exact.

## References


1. Ajayi, D.O., Adefokun, T.C.: On bounds of radio number of certain product graphs. J. Nigerian Math. Soc. **37**(2), 71–76 (2018)
2. Chartrand, G., Erwin, D., Harary, F., Zhang, P.: Radio labelings of graphs. Bull. Inst. Combin. Appl. **33**, 77–85 (2001)
3. Das, S., Ghosh, S.C., Nandi, S., Sen, S.: A lower bound technique for radio $k$-coloring. Discret. Math. **340**(5), 855–861 (2017)
4. Griggs, J.R., Yeh, R.K.: Labelling graphs with a condition at distance 2. SIAM J. Discret. Math. **5**(4), 586–595 (1992)
5. Hale, W.K.: Frequency assignment: theory and applications. Proc. IEEE **68**(12), 1497–1514 (1980)
6. Jacob, J., Laskar, R.C., Villalpando, J.J.: On the irreducible no-hole $L(2, 1)$-coloring of bipartite graphs and Cartesian products. J. Comb. Math. Comb. Comput. **78**, 49–64 (2011)

7. Kchikech, M., Khennoufa, R., Togni, O.: Linear and cyclic radio $k$-labelings of trees. Discussiones Mathematicae Graph Theory **27**(1), 105–123 (2007)
8. Kchikech, M., Khennoufa, R., Togni, O.: Radio $k$-labelings for Cartesian products of graphs. Discussiones Mathematicae Graph Theory **28**(1), 165–178 (2008)
9. Kim, B.M., Hwang, W., Song, B.C.: Radio number for the product of a path and a complete graph. J. Comb. Optim. **30**(1), 139–149 (2015)
10. Kola, S.R., Panigrahi, P.: An improved lower bound for the radio $k$-chromatic number of the hypercube $Q_n$. Comput. Math. Appl. **60**(7), 2131–2140 (2010)
11. Kola, S.R., Panigrahi, P.: A lower bound for radio $k$-chromatic number of an arbitrary graph. Contrib. Discret. Math. **10**(2) (2015)
12. Kuo, D., Yan, J.: On $L(2, 1)$-labelings of Cartesian products of paths and cycles. Discret. Math. **283**(1–3), 137–144 (2004)
13. Liu, D., Zhu, X.: Multilevel distance labelings for paths and cycles. SIAM J. Discret. Math. **19**(3), 610–621 (2005)
14. Mandal, N., Panigrahi, P.: L (2, 1)-colorings and irreducible no-hole colorings of cartesian product of graphs. Electron. Notes Discret. Math. **63**, 343–352 (2017)
15. Mandal, N., Panigrahi, P.: On irreducible no-hole l (2, 1)-coloring of cartesian product of trees with paths. AKCE Int. J. Graphs Comb. **17**(3), 1052–1058 (2020)
16. Morris-Rivera, M., Tomova, M., Wyels, C., Yeager, A.: The radio number of $C_n \square C_n$. Ars Comb. **120**, 7–21 (2015)
17. Niranjan, P.K., Kola, S.R.: On the radio number for corona of paths and cycles. AKCE Int. J. Graphs Comb. **17**(1), 269–275 (2020). https://doi.org/10.1016/j.akcej.2019.06.006
18. Niranjan, P.K., Kola, S.R.: The radio number for some classes of the cartesian products of complete graphs and cycles. J. Phys.: Conf. Ser. **1850**(1), 012014 (2021). https://doi.org/10.1088/1742-6596/1850/1/012014
19. Paul, S., Pal, M., Pal, A.: L (2, 1)-labeling of interval graphs. J. Appl. Math. Comput. **49**(1), 419–432 (2015)
20. Paul, S., Pal, M., Pal, A.: L (2, 1)-labeling of permutation and bipartite permutation graphs. Math. Comput. Sci. **9**(1), 113–123 (2015)
21. Saha, L., Panigrahi, P.: On the radio number of toroidal grids. Aust. J. Comb. **55**, 273–288 (2013)
22. Uma, J., Bhargavi, R.M.: Radio labeling on some corona graphs. In: AIP Conference Proceedings, vol. 2277, p. 100011. AIP Publishing LLC (2020)

# Some Parameters of Restricted Super Line Graphs

**Latha Devi Puli and K. Manjula**

**Abstract** The Restricted Super Line Graph of index $r$, $RL_r(G)$, of a graph $G$ has the $r$-element sets of $E(G)$ as its vertex set and two vertices being adjacent if exactly one edge of one set is adjacent to exactly one edge of the other. In this paper we characterize the complete graph $RL_r(G)$, totally disconnectedness, graph equations and connectedness of $RL_2(G)$.

**Keywords** Super line graphs · Restricted super line graphs · Dominating edge-pair

**AMS(2000)** 05C60 · 05C69 · 05C99

## 1 Introduction

All graphs considered here are finite, undirected and simple. We refer to [2, 5] for unexplained terminology and notations. For a given positive integer $r$ and a graph $G$ on at least $r$ edges, the *restricted super line graph* of index $r$ of $G$, denoted by $RL_r(G)$, is a simple graph whose vertex set consists of all possible $r$ - element subsets of $E(G)$ and two vertices $S = \{e_1, e_2, \ldots, e_r\}$ and $T = \{e'_1, e'_2, \ldots, e'_r\}$ in $V(RL_r(G))$ are adjacent if there exists exactly one pair of edges, say, $e_i, e'_j$ $(1 \leq i, j \leq r)$ that are adjacent in $G$. Restricted super line graph [3] is a modification of the super line graph $L_r(G)$ introduced by Bagga et al. [1]. The two graphs have the same vertex set and every edge of $RL_r(G)$ is an edge of $L_r(G)$. Therefore $RL_r(G)$ is a spanning subgraph of $L_r(G)$. If $r = 1$, the restricted super line graph $RL_r(G)$ coincides with the usual line graph.

For an edge $e$ of the graph $G$, the *neighborhood* $N(e)$ is the set of all edges of $G$ adjacent to $e$. i.e. if $e = uv$, then $N(e) = \{e' \in E - \{e\} : e'$is incident with $u$

L. D. Puli (✉)
Government First Grade College, Yelahanka, Bangalore, India
e-mail: drlathadevip@gmail.com

K. Manjula
Bangalore Institute of Technology, Bangalore, India



B. Rushi Kumar et al. (eds.), *Mathematics and Computing*, Springer Proceedings in Mathematics & Statistics 415, https://doi.org/10.1007/978-981-19-9307-7_16

or $v\}$. A pair of adjacent edges $a, b$ of $G$ is termed as a *dominating edge − pair* if $E(G) = N(a) \cup N(b)$. Throughout the paper $d(v)$ denotes the degree of a vertex $v$.

We recall the following results on $RL_r(G)$.

**Theorem 1** ([4]) *For any graph G,*

*(i)* $RL_r(G)$ *is a subgraph of* $L_r(G)$ *and*
*(ii)* $RL_r(G)) \cong L_r(G)$ *if and only if* $G \cong K_{1,2} \cup nK_2$ *or* $nK_2$

**Theorem 2** ([4]) *If G is a graph with q edges and no isolated vertices, then*

*(i)* $|V(RL_2(G)| = \binom{q}{2}$ *and*
*(ii) For a vertex* $S = \{e_i, e_j\}$ *of* $RL_2(G)$, $d(S) = \nu_{\{e_i e_j\}}\mu_{\{e_i e_j\}}$
*where* $\nu_{\{e_i e_j\}} = q - |N(e_i) \cup N(e_j)|$ *and* $\mu_{\{e_i e_j\}} = |N(e_i) \Delta N(e_j)|$

**Theorem 3** ([4]) *If H is a subgraph of G, then* $RL_r(H)$ *is a subgraph of* $RL_r(G)$.

**Theorem 4** ([4]) *The graph* $RL_2(C_n), RL_2(P_n)$ *is pancyclic for* $n > 5$ *(Fig. 1).*

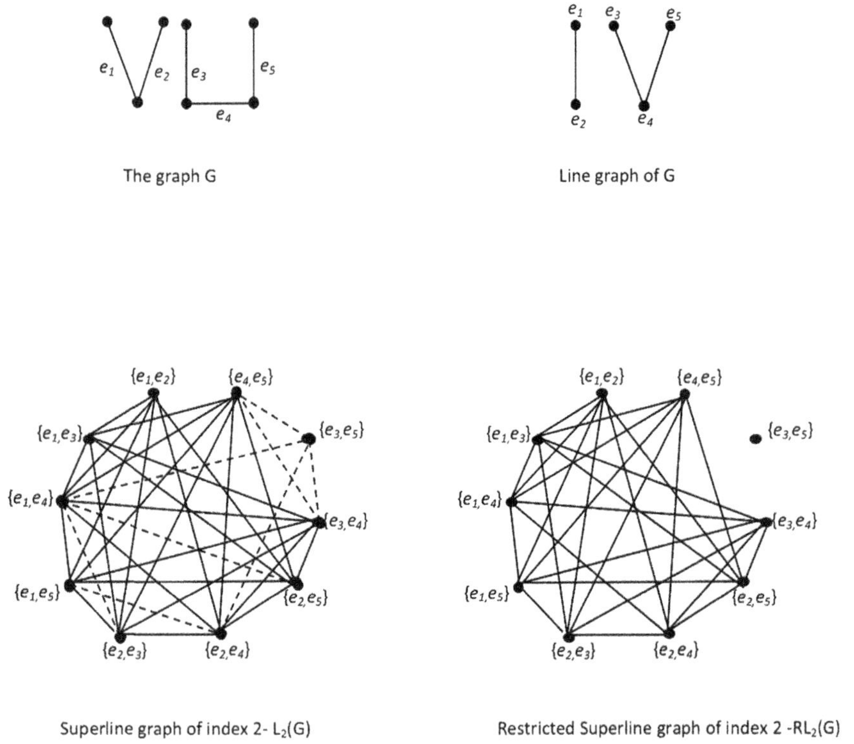


**Fig. 1** The graph $G$, its corresponding $L(G)$,$L_2(G)$ and $RL_2(G)$

## 2 Results on $RL_r(G)$

We note the following observations that are useful to understand the adjacency criteria in $RL_r(G)$.

**Observation 1** *The vertices* $S = \{e_1, e_2, \ldots, e_r\}$ *and* $T = \{e'_1, e'_2, \ldots, e'_r\}$ *are adjacent in* $RL_r(G)$ *only if the subgraph induced by the edge set* $\{e_1, e_2, \ldots, e_r, e'_1, e'_2, \ldots, e'_r\}$ *in* $G$ *has exactly one subgraph* $H$ *isomorphic to* $K_{1,2}$ *where one edge is in* $S$ *and the other in* $T$.

**Observation 2** *If the subgraph of* $G$ *induced by the edges of* $S$ *has vertex disjoint subgraphs* $H_1, H_2, \ldots, H_\alpha$ *in* $G$ *each isomorphic to* $K_{1,2}$, *then* $S$ *is adjacent to vertices of the form* $T_i = E(H_i) \cup X$ *where* $X \subseteq E(G) - \cup_{j=1}^{r} N(a_j)$, $|X| = r-2$, *for each* $i$, $1 \leq i \leq \alpha$.

**Observation 3** *If the subgraph of* $G$ *induced by the edges of* $S$ *is isomorphic to* $K_{1,2} \cup (r-2)K_2$, *then* $S$ *is adjacent to itself.*

**Theorem 5** *If* $G$ *is a graph with* $m$ *edges, then for any positive integer* $r$ *with* $r < m$.

*(i)* $|V(RL_r(G))| = \binom{m}{r}$

*(ii)* *For any vertex* $S = \{a_1, a_2, \ldots, a_r\}$ *of* $RL_r(G)$,

$$d(S) = \binom{|A|}{r-1}|B| + \alpha\binom{|A|}{r-2} - \lambda \text{ where}$$

$$A = E(G) - \cup_{i=1}^{r} N(a_i)$$
$$B = \{x/x \in N(a_i) \text{ for exactly one } i, 1 \leq i \leq r\}$$
$$\alpha = \text{Number of disjoint } K_{1,2}\text{'s in } S$$
$$\lambda = \begin{cases} 1 & \text{if the edges of } S \text{ induce exactly one } K_{1,2} \text{ in } G \\ 0 & \text{Otherwise} \end{cases}$$

***Proof*** (i) Follows directly by the definition of $RL_r(G)$.

(ii) Let $T$ be a vertex of $RL_r(G)$ which is adjacent to $S$. Then by Observation 1 the subgraph of $G$ induced by the edges of $S$ and $T$ has exactly one subgraph $H$ isomorphic to $K_{1,2}$, whose one edge is in $S$ and the other in $T$. Let the edges $a_l (1 \leq l \leq r)$ of $S$ and $b$ of $T$ form $H$. Then two cases arise:

**Case 1** $b \notin S$

By definition, the edge $b$ belongs to $B$ and the remaining $r-1$ elements of $T$ are not adjacent to any edge of $S$. so they belong to the set $A$.

**Case 2** $b \in S$

In this case, $H$ is contained in the subgraph induced by the edges of $S$ [$a_l$ and $b$ are not adjacent to any other edges $S$] and $T$ may or may not contain $a_l$. If $a_l$ is contained in $T$, then the other $r-2$ elements of $T$ belong to $A$.

Therefore,

$$d(S) = \binom{|A|}{r-1}|B| + \alpha\binom{|A|}{r-2} - \lambda$$

□

**Remark 1** For brevity, we have by the inclusion - exclusion principle that
$|A| = |E(G)| - (S_1 - S_2 + S_3 - \cdots + (-1)^r S_r)$
$|B| = S_1 - \binom{2}{1}S_2 + \binom{3}{2}S_3 - \cdots + (-1)^r\binom{r}{r-1}S_r$
where
$S_1 = \sum |N(a_i)|$
$S_2 = \sum |N(a_i) \cap N(a_j)|$
$S_3 = \sum |N(a_i) \cap N(a_j) \cap N(a_k)|$
$\vdots$
$S_r = \sum |N(a_1) \cap N(a_2) \cap \ldots \cap N(a_r)|$,

**Theorem 6** *For any integer $r \geq 2$, and a vertex $S = \{a_1, a_2, \ldots, a_r\}$ of $RL_r(G)$, $d(S) \neq 1$.*

***Proof*** **Claim**: If $S$ is not an isolated vertex then $d(S) > 1$.

Suppose to the contrary that $T = \{b_1, b_2, \ldots, b_r\}$ of $RL_r(G)$ is the only vertex adjacent to $S$. Then the subgraph of $G$ induced by the edges of $S$ and $T$ has exactly one subgraph isomorphic to $K_{1,2}$ induced by the edges $a_k$ and $b_l$ for some $k, l(1 \leq k, l \leq r)$. In relation to the edge $a_k$, the remaining $(r-1)$ edges of $S$ is classified according as they are adjacent to $a_k$ or not.

**Case 1** None of the $(r-1)$ edges in $S$ are adjacent to $a_k$.

**Sub case 1** Neither $a_k \in T$ nor $b_l \in S$.
Let $T'$ be the vertex obtained from $T$ by replacing $b_i$ $(i \neq l)$, with $a_k$. Then $T'$ exists (as $r > 2$). But $T \neq T'$, and $S$ is adjacent to $T'$ in $RL_2(G)$, a contradiction.

**Sub case 2** Let $a_k \in T$ and $b_l \notin S$.
Then $S'$ obtained from $S$ on replacing $a_k$ by $b_l$ is adjacent to $S$, a contradiction.

**Sub case 3** $a_k \notin T$and $b_l \in S$
Then the vertex $T'$ obtained on replacing $b_l$ of $T$ by $a_k$ is adjacent to $S$.

**Case 2** $a_k$ is adjacent to exactly one edge $a_p$ of $S$.

**Sub case 1** $a_p$ is not adjacent to any other edge of $S$.
Then $T'$ obtained from $T$ on replacing $b_l$ by $a_p$ is adjacent to $S$, a contradiction.

**Sub case 2** $a_p$ is adjacent to at least one more edge of $S$.
Then $T'$ obtained from $T$ on replacing $b_l$ by $a_k$ is adjacent to $S$, a contradiction.

**Case 3** $a_k$ is adjacent to more than one edge of $S$.

Let $e$ be an edge of $S$ not in the neighborhood of $a_k$. Such an edge does exist, otherwise $S$ will be an isolated vertex. $T'$ obtained from $T$ on replacing $b_i (i \neq l)$ with $e$ is adjacent to $S$, a contradiction.

Thus $d(S) > 1$

□

## 2.1 Completeness of $RL_r(G)$

The following theorem characterizes $RL_r(G)$ that are complete graphs.

**Theorem 7** *For any integer $r \geq 2$, $RL_r(G)$ is a complete graph if and only if $G \cong K_{1,2} \cup (r-1)K_2$.*

***Proof*** Suppose $RL_r(G)$ is a complete graph.

Let $S = \{a_1, a_2, \ldots, a_r\}$, $T = \{b_1, b_2, \ldots, b_r\}$ and $U = \{c_1, c_2, \ldots, c_r\}$ be any three vertices of $RL_r(G)$. Then as $S$, $T$ are adjacent, the subgraph of $G$ induced by the edges in $S$ and $T$ contains exactly one subgraph isomorphic to $K_{1,2}$ such that its one edge is in $S$ and the other in $T$. Again, as the vertices $T$ and $U$ are adjacent, the subgraph induced by the edges in $T$ and $U$ in $G$ contains exactly one $K_{1,2}$, whose one edge is in $T$ and the other in $U$. Let these two $K_{1,2}$s be formed by the edge-pairs $\{a_k, b_l\}$ and $\{b_m, c_j\}$. We claim that these $K_{1,2}$s are not distinct. If all the four edges $a_k$,$b_l$,$b_m$ and $c_j$ are distinct then $T$ is not adjacent to a vertex containing both $a_k$ and $c_j$, contradicting $RL_r(G)$ is complete. So, either $r < 2$ or $a_k = c_j$. But by hypothesis, $r \geq 2$. So $a_k = c_j$ and further if $b_l \neq b_m$ $T$ is not adjacent to any vertex containing $a_k$, in particular to $S$. This implies $b_l = b_m$ and therefore $G$ has exactly one $K_{1,2}$ and all other edges of $G$ are isolated edges. And, if the number of isolated edges exceeds $(r-1)$, then $RL_r(G)$ contains isolated vertices. Thus $G \cong K_{1,2} \cup (r-1)K_2$.

Conversely, suppose $G \cong K_{1,2} \cup (r-1)K_2$. Define:

$X = \{S \in V(RL_r(G))$ : $S$contains $r-2$ isolated edges and the two edges of $K_{1,2}\}$

*and*

$Y = \{S \in V(RL_r(G))$ : $S$ contains $r$ isolated edges$\}$

Then $X$ and $Y$ form a partition of $V(RL_r(G))$ with $|X| = r-1$ and $|Y| = 2$. For any two vertices $S$ and $T$ of $RL_r(G)$, the following three possibilities arise.

1. $S, T \in X$
2. $S, T \in Y$
3. $S \in X, T \in Y$.

In each of the cases, clearly the subgraph induced by edges of $S$ and $T$ contains a $K_{1,2}$ having one edge in $S$ and the other in $T$. Thus $S$ and $T$ are adjacent in $RL_r(G)$. So $RL_r(G)$ is a complete graph.

□

**Note**: Since for $r = 1$, $RL_r(G)$ is isomorphic to the line graph $L(G)$, in this case $RL_r(G)$ is complete if and only if $G$ isomorphic to $K_3$ or $K_{1,n}$.

## 3 Results on $RL_2(G)$

In this section, we obtain some graph equations and characterize totally disconnected $RL_2(G)$.

**Theorem 8** *If $G$ is an $(n, m)$ graph with no isolated vertices. Then*

1. *$RL_2(G) \cong G$ has no solution.*
2. *$RL_2(G) \cong L(G)$ if and only if $G \cong 3K_2$*
3. *$RL_2(G) \cong \overline{G}$ if and only if $G \cong K_3$*

***Proof*** 1. If $RL_2(G) \cong G$, then,

$$n = |V(G)| = |V(RL_2(G))| = \binom{m}{2}$$

Since $G$ has no isolated vertices we have $\binom{m}{2} = n \leq \sum_{i=1}^{n} d(v_i) = 2m$; $\frac{m(m-1)}{2} \leq 2m$ and therefore $m \leq 5$. Thus the feasible values of $(n, m)$ are $(3, 3)$, $(6, 4)$ and $(10, 5)$. The graphs associated with these parameters are shown in Fig. 2. The graphs $RL_2(K_3)$ and $RL_2(5K_2)$ are null graphs while in the other three cases, that is when $G$ is isomorphic to $P_2 \cup K_{1,3}$, $P_2 \cup P_4$ and $2P_3$, the corresponding restricted super line graph of index 2 are connected. Hence there exists no graph $G$ such that $RL_2(G)$ is isomorphic to $G$.

2. If $RL_2(G) \cong L(G)$, then $\binom{m}{2} = m \Rightarrow m = 3$. The graphs on three edges having no isolates and the corresponding line and restricted super line graphs are listed in Table 1. Clearly $RL_2(G) \cong L(G)$ if and only if $G \cong 3K_2$
3. Suppose $RL_2(G) \cong \overline{G}$. As in (1), the only possible pairs of $(n, m)$ are $(3, 3)$, $(6, 4)$ and $(10, 5)$ since $|V(G)| = |V(\overline{G})|$. Among the graphs in Fig. 2 we verify that $RL_2(G) \cong \overline{G}$ holds only when $G \cong K_3$.

□

**Theorem 9** *$RL_2(G)$ is totally disconnected if and only if $G$ is isomorphic to $K_{1,n}$, $K_3$, $nK_2$, $(n \geq 2)$ or a connected graph on four vertices.*

***Proof*** Sufficiency: Suppose $G \cong K_{1,n}$, $K_3$, $nK_2$ or any connected graph with four vertices as in Fig. 3. Let $S$ be a vertex of $RL_2(G)$. Then by Theorem 2, for a vertex $S =$

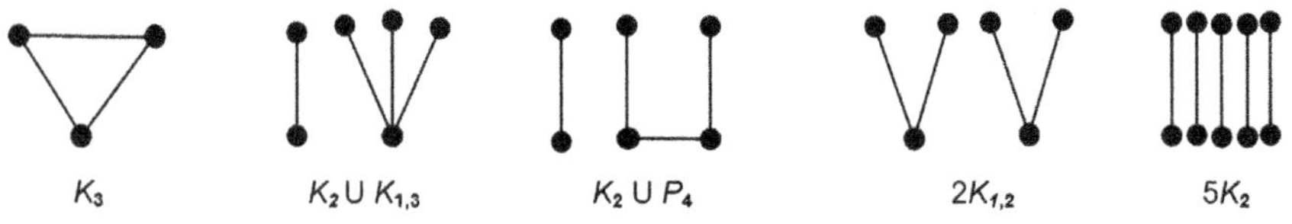


**Fig. 2** Graphs of Theorem 8

**Table 1** The graph $G$, $L(G)$, $RL_2(G)$ for case 2 of Theorem 8

| $G$ | $L(G)$ | $RL_2(G)$ |
|---|---|---|
| $3K_2$ | Null graph | Null graph |
| $K_{1,3}$ | $K_3$ | Null graph |
| $P_4$ | $K_{1,2}$ | Null graph |
| $K_{1,2} \cup K_2$ | $K_2 \cup K_1$ | $C_3$ |

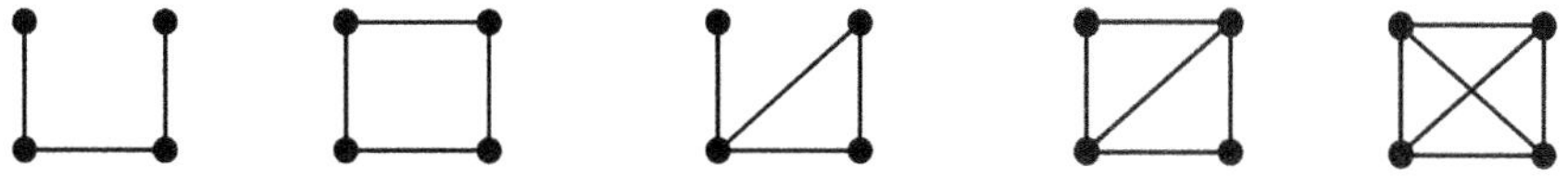

**Fig. 3** Connected spanning graphs of $K_4$ of Theorem 9

$\{e_i, e_j\}$ of $RL_2(G), d(S) = \nu_{\{e_i,e_j\}}.\mu_{\{e_i,e_j\}}$. It can be easily verified that $\nu_{\{e_i,e_j\}} = 0$ for every pair of edges $e_i, e_j$(dominating edge-pair) of $G \cong K_{1,n}$, $K_3$ and $\mu_{\{e_i,e_j\}} = 0$ for any $e_i, e_j$ of $G \cong nK_2$. If $G$ is a connected graph on four vertices,$\nu_{\{e_i,e_j\}} = 0$ for a pair of adjacent edges, and $\mu_{\{e_i,e_j\}} = 0$ for a pair of nonadjacent edges . Thus $d(S) = 0$ for every vertex $S$ and hence $RL_2(G)$ is totally disconnected.

Necessity: Suppose $RL_2(G)$ is totally disconnected. Then for any vertex $S = \{e_i, e_j\}$ of $RL_2(G)$, $d(S) = 0$. By Theorem 2, $\nu_{\{e_i,e_j\}}\mu_{\{e_i,e_j\}} = 0$ that is $|E(G) - (N(e_i) \cup N(e_j))||(N(e_i) \cup N(e_j)) - (N(e_i) \cap N(e_j))| = 0$.
Then $|E(G) - N(e_i) \cup N(e_j)| = 0$ or $|N(e_i) \cup N(e_j) - N(e_i) \cap N(e_j)| = 0$.
This implies either $E(G) - N(e_i) \cup N(e_j) = \phi$ or $N(e_i) \cup N(e_j) - N(e_i) \cap N(e_j) = \phi$.

**Case1.** For any $i, j$, $E(G) - N(e_i) \cup N(e_j) = \phi$.
This implies $E(G) = \phi$ or $E(G) = N(e_i) \cup N(e_j)$. But $E(G) \neq \phi$. So $E(G) = N(e_i) \cup N(e_j)$ holds for any two edges of $G$. This means every two edges of $G$ is a dominating edge-pair. Every pair of edges of $G$ is a dominating edge-pair if and only if $G$ is isomorphic to $K_3$ or $K_{1,n}$. Here $n \geq 2$ as $RL_2(G)$ has at least one vertex. So, $G$ has at least two edges. Thus $G$ is isomorphic to $K_3$ or $K_{1,n}$, $n \geq 2$.

**Case 2.** For any $i, j$, $N(e_i) \cup N(e_j) - N(e_i) \cap N(e_j) = \phi$.
Consequently either $N(e_i) \cup N(e_j) = \phi$ or $N(e_i) \cup N(e_j) = N(e_i) \cap N(e_j)$.

**Sub case 1.** For any $i, j$, $N(e_i) \cup N(e_j) = \phi$.
In this case, $N(e_i) = N(e_j) = \phi$. That is every edge of $G$ is an isolated edge. So, $G \cong nK_2$, $n \geq 2$

**Sub case 2.** For any $i, j$, $N(e_i) \cup N(e_j) = N(e_i) \cap N(e_j)$.
Here $N(e_i) = N(e_j)(\neq \phi)$. This implies $e_i$ and $e_j$ are a pair of nonadjacent edges that have non empty and equal neighborhoods. But there are no graphs existing in which every two edges are nonadjacent having non empty neighborhoods.

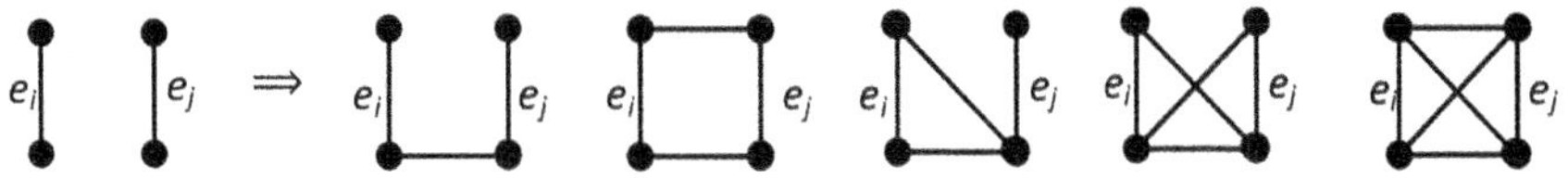


**Fig. 4** Graphs for which $N(e_i) = N(e_j)$

**Case 3** For some $i, j$, $N(e_i) \cup N(e_j) = E(G)$ and for some $N(e_i) = N(e_j)$.
In this case, for every pair of adjacent edges $e_i, e_j$ of $G$, $N(e_i) \cup N(e_j) = E(G)$ and for every pair of nonadjacent edges $e_i, e_j$ of $G$, $N(e_i) = N(e_j)$.
Let $e_i, e_j$ be non adjacent in $G$. Then the edge-pair $e_i, e_j$ induces a $2K_2$ (in $G$) which has 4 pendant vertices. For an edge $x$ of $G$, let $x \in N(e_i)$ then $x \in N(e_j)$ and therefore $x \in N(e_i) \cap N(e_j)$. Every edge $x$ joining these pendant vertices belongs to $N(e_i) \cap N(e_j)$. If $t$ is an edge of $G$ not joining these pendant vertices, then $t \notin N(e_i) \cap N(e_j)$. Thus the graphs $G$ for which $N(e_i) = N(e_j)$ for every pair of non adjacent edges are as shown in Fig. 4. Also it can be verified that for every pair of adjacent edges $e_i, e_j$ in these graphs, $N(e_i) \cup N(e_j) = E(G)$. Hence the result follows.

□

As a consequence we have the following.

**Theorem 10** *Let $G$ be a connected graph. A vertex $S = \{e_i, e_j\}$ of $RL_2(G)$ is an isolated vertex if and only if*

**(i)** *$e_i, e_j$ forms a dominating edge pair of $G$.*
**(ii)** *$e_i, e_j$ are any two non adjacent edges of a connected graph with four vertices (Fig. 5).*

We note that the diameter of a graph $G$ containing a dominating edge pair is at most three. So for a connected graph $G$, if $RL_2(G)$ has an isolated vertex then $diam(G) \leq 3$. However the converse is not true. The graph $H$ in Fig. 6 is of diameter three but $RL_2(G_1)$ contains no isolated vertex.

**Theorem 11** *For a disconnected graph $G$ having no isolates, $RL_2(G)$ has an isolated vertex if and only if $G \cong H \cup G_1$ where $H$ is a graph on 4 vertices without isolates and not isomorphic to $K_{1,3}$ and $G_1$ is the graph without isolates.*

***Proof*** Let $G$ be a disconnected graph having no isolates. Let $\{e_i, e_j\}$ be an isolated vertex of $RL_2(G)$. The subgraph of $G$ induced by $e_i, e_j$ can not be isomorphic to $K_{1,2}$ because for a disconnected graph $G$, $E(G) \neq N(e_i) \cup N(e_j)$. So, $e_i, e_j \in E(G)$ must be non adjacent and $N(e_i) = N(e_j)$. Therefore, by Theorem 10,$e_i, e_j$ may form a pair of nonadjacent edges or a pair of isolated edges in a connected graph with four vertices. Thus $e_i, e_j$ are any two edges of $H$.

Conversely, let $G \cong H \cup G_1$ where $H$ is a graph on 4 vertices not isomorphic to $K_{1,3}$. Then by Theorem 9, if $e_i, e_j$ are non adjacent edges of $H$, then $d(\{e_i, e_j\}) = 0$ as $\mu_{\{e_i, e_j\}} = 0$. Hence the result.

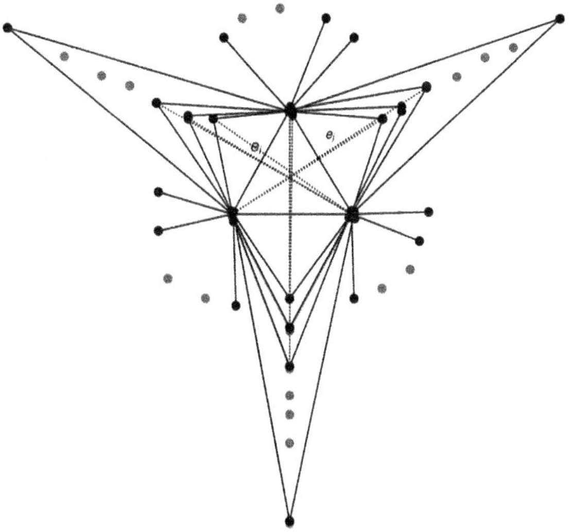


**Fig. 5** Graph in which the vertex $\{e_i, e_j\}$ is dominating edge-pair

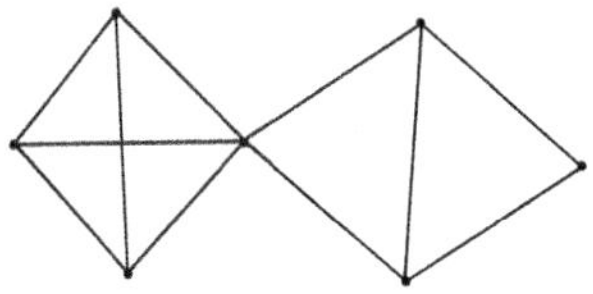

**Fig. 6** Graph $G_1$

□

**Conclusion:** Some results on $RL_r(G)$ were obtained. Graph equations on restricted super line graphs and super line graphs has been established. Characterization of $RL_2(G)$ for being null graph has been obtained. Continuing we want to characterize $RL_2(G)$ for being Hamiltonian and claw free.

**Acknowledgements** The authors wish to express their gratitude to the referees for their valuable inputs and also to Dr.B. Sooryanarayana for his helpful comments and suggestions.

## References


1. Bagga, J..S., Beineke, L..W., Varma, B..N.: Super Line graphs and their properties. In: Combinatorics, Graph Theory, Algorithms and Applications, vol. 1994, pp. 1–6. World Scientific Publishing, Beijing, New Jersey (1993)
2. Harary, F.: Graph Theory. Narosa Publishing House, New Delhi (1969)
3. Manjula, K., Sooryanarayana, B.: Restricted super line graphs. Far East J. Appl. Math. **I**(24), 23–37 (2006)
4. Manjula, K.: A study on Line graphs. Ph.D Thesis, Bangalore University (2004)
5. Devi Puli, L.: A study On derived graphs. Ph.D Thesis, Visvesvaraya Technological University (2015)

# Edge Constrained Eulerian Extensions

**Ghurumuruhan Ganesan**

**Abstract** In this paper, we study Eulerian extensions with edge constraints and use the probabilistic method to establish sufficient conditions for a given connected graph to be a subgraph of a Eulerian graph containing $m$ edges, for a given number $m$.

**Keywords** Eulerian extensions · Edge constraint · Probabilistic method

## 1 Introduction

In the Eulerian extension problem, a given graph is to be converted into a Eulerian graph by the addition of as few edges as possible and such problems have applications in routing and scheduling (Dorn et al. [3]). Boesch et al. [2] studied conditions under which a graph $G$ can be extended to a Eulerian graph and later Lesniak and Oellermann [6] presented a detailed survey on subgraphs and supergraphs of Eulerian graphs and multigraphs. For applications of Eulerian extensions to scheduling and parametric aspects, we refer to Höhn et al. [5] and Fomin and Golovach [4], respectively.

In this paper, we construct a Eulerian extension of graphs with a predetermined number of edges. Specifically, given a graph $G$ with maximum degree $\Delta$ and $b$ number of edges and given an integer $m > b$, we use the asymmetric probabilistic method to derive sufficient conditions for the existence of a Eulerian extension of $G$ with $m$ edges.

The paper is organized as follows: In Sect. 2, we state and prove our main result regarding Eulerian extensions with edge constraints.

G. Ganesan (✉)
IISER Bhopal, Bhopal, India
e-mail: gganesan82@gmail.com



B. Rushi Kumar et al. (eds.), *Mathematics and Computing*, Springer Proceedings in Mathematics & Statistics 415, https://doi.org/10.1007/978-981-19-9307-7_17

## 2 Edge Constrained Eulerian Extensions

Let $G = (V, E)$ be a graph with vertex set $V$ and edge set $E$. The vertices $u$ and $v$ are said to be adjacent in $G$ if the edge $(u, v)$ with endvertices $u$ and $v$ is present in $E$. We define $d_G(v)$ to be the degree of vertex $v$, i.e., the number of vertices adjacent to $v$ in $G$.

A sequence of vertices $\mathcal{W} := (u_1, u_2, \ldots, u_t)$ is said to be a *walk* if $u_i$ is adjacent to $u_{i+1}$ for each $1 \leq i \leq t-1$. If in addition the vertex $u_t$ is also adjacent to $u_1$, then $\mathcal{W}$ is said to be a *circuit*. We say that $\mathcal{W}$ is a Eulerian circuit if each edge of the graph $G$ occurs exactly once in $\mathcal{W}$. The graph $G$ is said to be a *Eulerian* graph if $G$ contains a Eulerian circuit.

Let $G$ be any graph. We say that a graph $H$ is a *Eulerian extension* of $G$ if $G$ and $H$ share the same vertex set, $G$ is a subgraph of $H$ and $H$ is Eulerian.

**Definition 1** For an integer $m \geq 1$, we say that a graph $G$ is *$m$-Eulerian extendable* if there exists a Eulerian extension $H$ of $G$ containing exactly $m$ edges.

We have the following result regarding $m$-Eulerian extendability. Throughout, constants do not depend on $n$.

**Theorem 1** *For every pair of constants $0 < \alpha, \beta < 1$ satisfying $\beta + 40\alpha^2 < \frac{1}{2}$ strictly, there exists a constant $N = N(\alpha, \beta) \geq 1$ such that the following holds for all $n \geq N$: Let $m$ be any integer satisfying*

$$2n \leq m \leq \alpha \cdot n^{\frac{3}{2}} \tag{2.1}$$

*and let $G \subset K_n$ be any connected graph containing $n$ vertices, $b$ edges and a maximum vertex degree $\Delta$. If*

$$\Delta \leq \beta \cdot n \text{ and } b \leq m - n, \tag{2.2}$$

*then $G$ is $m$-Eulerian extendable.*

To see the necessity of the bound $b \leq m - n$, we use the fact that a graph $H$ is Eulerian if and only if $H$ is connected and each vertex of $H$ has even degree (Theorem 1.2.26, pp. 27, West [7]). Therefore, to obtain a Eulerian extension of $G$, we only need to convert all odd-degree vertices into even-degree vertices.

Suppose that $n$ is even and all the vertices in $G$ have an odd degree. Because the degree of each vertex is at most $\frac{n}{2} - 1$ (see (2.2)), the sum of neighbourhood sizes of $2i - 1$ and $2i$ is at most $n - 2$. Therefore, for each $1 \leq i \leq \frac{n}{2}$, there exists a vertex $w_i$ neither adjacent to $2i - 1$ nor adjacent to $2i$ in the graph $G$. Adding the $n$ edges $\{(w_i, 2i-1), (w_i, 2i)\}_{1 \leq i \leq \frac{n}{2}}$ gives us a Eulerian extension of $G$.

In our proof of Theorem 1 below, we use the asymmetric probabilistic method for higher values of $m$ to obtain walks of predetermined lengths between pairs of odd-degree vertices and thereby construct the desired extension.

**Proof of Theorem** 1

As before, we use the fact that a graph $H$ is Eulerian if and only if $H$ is connected and each vertex of $H$ has an even degree. We assume that the vertex set of $G$ is $V := \{0, 1, 2, \ldots, n-1\}$ and also let $\mathcal{T}$ be the set of all odd-degree vertices in $G$ so that the number of odd degree vertices $\#\mathcal{T}$ is even. If there are vertices $u, v \in \mathcal{T}$ that are not adjacent to each other in $G$, then we mark the edge $(u, v)$ and also the endvertices $u$ and $v$. We then pick two new nonadjacent vertices $x$ and $y$ in $\mathcal{T} \setminus \{u, v\}$ and repeat the procedure. We continue this process until we reach one of the following two scenarios: Either the number of marked edges is $m - b$ in which case, we simply add the marked edges to $G$ and get the desired Eulerian extension $H$. Or, we are left with a set of marked edges of cardinality, say $l$ and a clique $\mathcal{C} := \{u_1, \ldots, u_{2z}\} \subset \mathcal{T}$ containing $2z$ unmarked vertices.

Let $G_0$ be the graph obtained by adding all the $l \leq \frac{n}{2}$ marked edges to $G$. If $\Delta_0$ and $b_0$ denote the maximum vertex degree and the number of edges in $G_0$, respectively, then

$$\Delta_0 \leq \Delta + 1 \text{ and } b_0 = b + l \leq b + n. \tag{2.3}$$

We now pair the vertices in $\mathcal{C}$ as $\{u_{2i-1}, u_{2i}\}_{1 \leq i \leq z}$ assuming that $z \geq 1$ (if not, we simply remove a marked edge $e$ from $G_0$ and label the endvertices of $e$ as $u_1$ and $u_2$). We use the probabilistic method to obtain $z$ edge-disjoint walks $\{\mathcal{W}_i\}_{1 \leq i \leq z}$ containing no edge of $G_0$ such that each walk $\mathcal{W}_i$ has $w$ edges and $u_{2i-1}$ and $u_{2i}$ as endvertices, where $w$ satisfies

$$b_0 + z \cdot w = m. \tag{2.4}$$

Adding the walks $\{\mathcal{W}_i\}_{1 \leq i \leq z}$ to $G_0$ would then give us the desired $m$-Eulerian extension.

In (2.4), we have assumed for simplicity that $w = \frac{m-b_0}{z}$ is an integer. If not, we write $m - b_0 = z \cdot w + r$ where $0 \leq r \leq w - 1$ and construct the $z - 1$ walks $\mathcal{W}_i$, $1 \leq i \leq z - 1$ each of length $w$ edges and the last walk $\mathcal{W}_z$ of length $w + r \leq 2w$. Again adding these walks to $G_0$ would give us the desired Eulerian extension with $m$ edges. For future use, we remark that the length $w$ of each walk added in the above process is bounded above by

$$w = \frac{m - b_0}{z} \leq m \leq \alpha \cdot n^{\frac{3}{2}}. \tag{2.5}$$

We begin with the pair of vertices $u_1$ and $u_2$. Let $\{X_i\}_{1 \leq i \leq w}$ be independent and identically distributed (i.i.d.) random variables uniformly distributed in the set $\{0, 1, \ldots, n-1\}$. Letting $\mathcal{S} := (u_1, X_1, \ldots, X_w, u_2)$, we would like to convert the sequence $\mathcal{S}$ into a walk $\mathcal{W}_1$ with endvertices $u_1$ and $u_2$ and containing no edge of $G_0$. The construction of $\mathcal{W}_1$ is split into two parts: In the first part, we collect the preliminary relevant properties of $\mathcal{S}$ and in the second part, we obtain the walk $\mathcal{W}_1$.

*Preliminary definitions and estimates*: An entry in $\mathcal{S}$ is defined to be a vertex and we define $(u_1, X_1)$, $(X_w, u_2)$ and $\{(X_i, X_{i+1})\}_{1 \le i \le w-1}$ to be the edges of $\mathcal{S}$. The neighbour set of a vertex $v$ in $\mathcal{S}$ the set of vertices $u$ such that either $(v, u)$ or $(u, v)$ appears as an edge of $\mathcal{S}$. The neighbour set of $v$ in the multigraph $G_0 \cup \mathcal{S}$ is the union of the neighbour set of $v$ in the graph $G_0$ and the neighbour set of $v$ in $\mathcal{S}$. The degree of a vertex $v$ in $G_0 \cup \mathcal{S}$ is defined to be the sum of the degree of $v$ in $G_0$ and the degree of $v$ in $\mathcal{S}$.

The three main ingredients used in the construction of the walk $\mathcal{W}_1$ are
(1) The degree of a vertex in the multigraph $G_0 \cup \mathcal{S}$,
(2) the number of "bad" vertices in $G_0 \cup \mathcal{S}$ and
(3) the number of "bad" edges in $\mathcal{S}$.
Below, we define and estimate each of the three quantities in that order.

We first estimate the degree of each vertex in the multigraph $G_0 \cup \mathcal{S}$. For any $0 \le v \le n-1$ and any $1 \le i \le w$, let $I_i = 1\!\!1(X_i = v)$ be the indicator function of the event that $X_i = v$. We have $\mathbb{P}(I_i = 1) = \frac{1}{n}$ and so if $D_v = \sum_{i=1}^{w} I_i$ denotes the number of times the entry $v$ appears in the sequence $(X_1, \ldots, X_w)$, then $\mathbb{E}D_v = \frac{w}{n}$ and so by the standard deviation estimate (A.30) in Appendix, we have

$$\mathbb{P}\left(D_v \ge \frac{2w}{n}\right) \le 2\exp\left(-\frac{w}{16n}\right). \tag{2.6}$$

If $\frac{w}{n} \ge 100 \log n$, then we get from (2.6) that $\mathbb{P}(D_v \ge \frac{2w}{n}) \le \frac{1}{n^2}$. Else, we use the Chernoff bound directly to get that

$$\mathbb{P}\left(D_v \ge 100 \log n\right) \le \frac{1}{n^2}. \tag{2.7}$$

Therefore setting $a_n := \max\left(\frac{2w}{n}, 100 \log n\right)$, we get that

$$\mathbb{P}\left(D_v \ge a_n\right) \le \frac{1}{n^2}. \tag{2.8}$$

If the event

$$E_{deg} := \bigcap_{0 \le v \le n-1} \{D_v \le a_n\} \tag{2.9}$$

occurs, then in $G_0 \cup \mathcal{S}$ each vertex has degree at most $\Delta_0 + 1 + a_n$, with the extra term 1 to account for the fact that vertices $X_1$ and $X_w$ are also adjacent to $u_1$ and $u_2$, respectively. By the union bound and (2.6), we therefore have

$$\mathbb{P}(E_{deg}) \ge 1 - \frac{1}{n}. \tag{2.10}$$

The next step is to estimate the number of "bad" vertices in $G_0 \cup \mathcal{S}$. Let $X_0 := u_1, X_{w+1} := u_2$ and for $0 \le i \le w-1$, say that vertex $X_i$ is *bad* if $X_i = X_{i+1}$

or $X_i = X_{i+2}$. For simplicity define $X_w$ to be bad always. If $J_i$ is the indicator function of the event that vertex $X_i$ is bad, then for $0 \leq i \leq w-1$, we have that

$$\frac{1}{n} \leq \mathbb{P}(J_i = 1) \leq \frac{2}{n}. \tag{2.11}$$

The term $N_{v,bad} := \sum_{i=0}^{w-1} J_i + 1$ denotes the total number of bad vertices in the sequence $\mathcal{S}$. To estimate $N_{v,bad}$, we split $N_{v,bad} - 1 = J(A) + J(B) + J(C)$, where

$$J(A) = J_1 + J_4 + \ldots,\ J(B) = J_2 + J_5 + \ldots \text{ and } J(C) = J_3 + J_6 + \ldots$$

so that each $J(u)$, $u \in \{A, B, C\}$ is a sum of i.i.d. random variables.

The term $J(A)$ contains at least $\frac{w}{3} - 1$ and at most $\frac{w}{3}$ random variables. As in the proof of (2.8), we use (2.11) and the standard deviation estimate (A.30) in Appendix to obtain that

$$\mathbb{P}\left(J(A) \geq \frac{a_n}{3}\right) \leq \frac{1}{n^2}$$

for all $n$ large. A similar estimate holds for $J(B)$ and $J(C)$ and so combining these estimates and using the union bound, we get that

$$\mathbb{P}\left(E_{v,bad}\right) \geq 1 - \frac{3}{n^2} \tag{2.12}$$

where $E_{v,bad} := \{N_{v,bad} \leq a_n + 1\}$ denotes the event that the number of bad vertices in $\mathcal{S}$ is at most $a_n + 1$.

The final estimate involves counting the number of bad edges in the sequence $\mathcal{S}$. For $0 \leq i \leq w$ say that $(X_i, X_{i+1})$ is a *bad edge* if one of the following two conditions hold:

$(d1)$ Either $\{X_i, X_{i+1}\}$ is an edge of $G_0$ or
$(d2)$ There exists $i+2 \leq j \leq w$ such that $\{X_i, X_{i+1}\} = \{X_j, X_{j+1}\}$.
To estimate the probability of occurrence of $(d1)$, let $e$ be an edge of $G_0$ with endvertices $u$ and $v$. We have that

$$\mathbb{P}\left(\{X_i, X_{i+1}\} = \{u, v\}\right) \leq \frac{2}{n^2}.$$

Similarly for any $i+2 \leq j \leq w$, the possibility $(d2)$ also occurs with probability at most $\frac{2}{n^2}$. Therefore if $L_i$ is the indicator function of the event that $(X_i, X_{i+1})$ is a bad edge, we have that

$$\mathbb{P}(L_i = 1) \leq \sum_{l=1}^{b_0} \frac{2}{n^2} + \sum_{j=i+2}^{w} \frac{2}{n^2} \leq \frac{2(b_0 + w)}{n^2}. \tag{2.13}$$

If $N_{e,bad} := \sum_{i=0}^{w} L_i$ denotes the total number of bad edges in $\mathcal{S}$, then from (2.13) and the fact that $L_0 \leq 1$ we have

$$\mathbb{E}N_{e,bad} \leq 1 + \frac{2(b_0 + w)w}{n^2} =: c_n. \tag{2.14}$$

Letting $E_{e,bad} := \{N_{e,bad} \leq K \cdot c_n\}$ denote the event that the number of bad edges in $\mathcal{S}$ is at most $K \cdot c_n$, for some large integer constant $K \geq 1$ to be determined later, we get from Markov inequality that

$$\mathbb{P}\left(E_{e,bad}\right) \geq 1 - \frac{1}{K}, \tag{2.15}$$

If $E_{valid}$ denotes the event that the first and last edges $(X_0, X_1)$ and $(X_w, X_{w+1})$ are valid edges not in $G_0$, then using the fact that the degree of any vertex in $G_0$ is at most $\frac{n}{2}$ (see (2.3) and (2.2) in the statement of the Theorem), we get that

$$\mathbb{P}(E_{valid}) \geq \left(\frac{1}{2} - \frac{1}{n}\right)^2. \tag{2.16}$$

Defining the joint event

$$E_{joint} := E_{valid} \bigcap E_{deg} \bigcap E_{v,bad} \bigcap E_{e,bad}$$

and using

$$\begin{aligned}\mathbb{P}\left(A \bigcap \bigcap_{i=1}^{l} B_i\right) &\geq \mathbb{P}(A) - \mathbb{P}\left(\bigcup_{i=1}^{l} B_i^c\right) \\ &\geq \mathbb{P}(A) - \sum_{i=1}^{l} \mathbb{P}\left(B_i^c\right) \end{aligned} \tag{2.17}$$

with $A = E_{valid}$ we get from (2.10), (2.12), (2.15) and (2.16) that

$$\mathbb{P}(E_{joint}) \geq \left(\frac{1}{2} - \frac{1}{n}\right)^2 - \frac{1}{K} - \frac{1}{n} - \frac{3}{n^2} \geq \frac{1}{21} \tag{2.18}$$

for all $n$ large, provided the constant $K = 5$, which we fix henceforth. This completes the preliminary estimates used in the construction of the walk $\mathcal{W}_1$.

*Construction of the walk* $\mathcal{W}_1$: Assuming that the event $E_{joint}$ occurs, we now convert $\mathcal{S}_0 := \mathcal{S}$ into a walk $\mathcal{W}_1$. We begin by "correcting" all bad vertices. Let $X_{i_1}$, $X_{i_2}, \ldots, X_{i_t}, i_1 < i_2 < \ldots < i_t$ be the set of all bad vertices. Thus for example either $X_{i_1} = X_{i_1+1}$ or $X_{i_1} = X_{i_1+2}$. Because the event $E_{deg}$ occurs, we get from the discussion following (2.9) that the degree of each vertex in $G_0 \cup \mathcal{S}_0$ is at most $\Delta_0 + a_n + 1$. From (2.3) and the first condition in (2.2), we get that

$$\Delta_0 \leq \Delta + 1 \leq \frac{n}{3} + 1 \tag{2.19}$$

and from the definition of $a_n$ prior to (2.8) and the upper bound $w \leq n^{\frac{3}{2}}$ in (2.5), we get that

$$a_n = \max\left(100 \log n, \frac{2w}{n}\right) \leq 100 \log n + \frac{2w}{n} \leq 100 \log n + 2\sqrt{n} \leq 3\sqrt{n} \tag{2.20}$$

for all $n$ large. Consequently, using $\beta < \frac{1}{2}$ strictly (see statement of Theorem 1),

$$\Delta_0 + a_n + 1 \leq \beta \cdot n + 1 + 3\sqrt{n} \leq \frac{n}{2} - 5 \tag{2.21}$$

for all $n$ large. From (2.21), we therefore get that there exists a vertex $v_1$ that is *not* a neighbour of $X_{i_1}$ in $G_0 \cup \mathcal{S}$. Similarly, the total number of neighbours of $v_1$ and $X_{i_1+3}$ in $G_0 \cup \mathcal{S}$ is at most

$$2\Delta_0 + 2a_n + 2 \leq 2\beta \cdot n + 2 + 6\sqrt{n} < n - 10 \tag{2.22}$$

for all $n$ large and so there exists a vertex $v_2 \neq X_{i_1}$ that is not a neighbour of $v_1$ and also not a neighbour of $X_{i_1+3}$ in $G_0 \cup \mathcal{S}$.

We now set $X^{(1)}_{i_1+1} = v_1$ and $X^{(1)}_{i_1+2} = v_2$ and $X^{(1)}_j = X_j$ for $j \neq i_1 + 1, i_1 + 2$ and call the resulting sequence $\mathcal{S}_1 := (X^{(1)}_1, \ldots, X^{(1)}_w)$. By construction, the degree of each vertex in the multigraph $G_0 \cup \mathcal{S}_1$ is at most $\Delta + a_n + 1 + 2$ and there are at most $t - 1$ bad vertices in $\mathcal{S}_1$. We now pick the bad vertex with the least index in $\mathcal{S}_1$ and repeat the above procedure with $\mathcal{S}_1$ to get a sequence $\mathcal{S}_2$ containing at most $t - 2$ bad vertices.

After $k \leq t$ iterations of the above procedure, the degree of each vertex in the multigraph $G_0 \cup \mathcal{S}_k$ would be at most

$$\Delta_0 + a_n + 1 + 2k \leq \Delta_0 + a_n + 1 + 2t \leq \Delta_0 + 3a_n + 3 \tag{2.23}$$

because the event $E_{joint} \subseteq E_{v,bad}$ occurs and so $t \leq a_n + 1$. Again using (2.19) and (2.20) and arguing as in (2.22), we get that the sum of the degrees of any two vertices in $G_0 \cup \mathcal{S}_t$ is at most $n - 10$ for all $n$ large. Thus, the above procedure indeed proceeds for $t$ iterations and by construction, the sequence $\mathcal{S}_t$ obtained at the end has no bad vertices.

We now perform an analogous procedure for correcting all bad edges in $\mathcal{S}_t$. For example if $(X_l, X_{l+1})$ is a bad edge in $\mathcal{S}_t$, then following an analogous argument as before we pick a vertex $Y_{l+1}$ that is neither adjacent to $X_l$ nor adjacent to $X_{l+2}$ in the sequence $\mathcal{S}_t$. We replace $X_{l+1}$ with $Y_{l+1}$ to get a new sequence $\mathcal{S}_{t+1}$. In the union $G_0 \cup \mathcal{S}_{t+1}$, the degree of each vertex is at most $\Delta_0 + 3a_n + 3 + 2$ (see (2.23)) and the number of bad edges is at most $r - 1$. At the end of $r \leq K \cdot c_n$ iterations, we obtain a multigraph $G_0 \cup \mathcal{S}_{t+r}$, where the degree of each vertex is at most

$$\Delta_0 + 3a_n + 3 + 2r \leq \Delta_0 + 3a_n + 3 + 2Kc_n,$$

since the event $E_{e,bad}$ occurs and therefore $N_{e,bad} \leq K \cdot c_n$ (see discussion preceding (2.15)). Substituting the expression for $c_n$ from (2.14) and using the second estimate for $a_n$ in (2.20), we get that $\Delta_0 + 3a_n + 3 + 2Kc_n$ is at most

$$\begin{aligned} &\Delta_0 + 300 \log n + \frac{6w}{n} + 3 + 2K + \frac{2K(b_0 + w)w}{n^2} \\ \leq\ &\Delta_0 + 301 \log n + \frac{6w}{n} + \frac{2K(b_0 + w)w}{n^2} \end{aligned} \tag{2.24}$$

for all $n$ large. Recalling that $u_1$ and $u_2$ are the endvertices of the starting sequence $\mathcal{S}_0$, we get that the final sequence $\mathcal{S}_{t+r}$ contains no bad edge and is therefore the desired walk $\mathcal{W}_1$ with endvertices $u_1$ and $u_2$. This completes the construction of the walk $\mathcal{W}_1$.

*Rest of the walks*: We now repeat the above procedure to construct the rest of the walks. We set $G_1 := G_0 \cup \mathcal{W}_1$ and argue as above to obtain a walk $\mathcal{W}_2$ with $w$ edges present in $\overline{G}_1$ and containing $u_3$ and $u_4$ as endvertices. Adding the walk $\mathcal{W}_2$ to $G_1$, we get a new graph $G_2$. In effect, to the graph $G_1$ containing $b_0 + w$ edges, we have added $w$ edges and by an argument analogous to (2.24), we have increased the degree of a vertex by at most

$$301 \log n + \frac{6w}{n} + \frac{2K(b_0 + 2w)w}{n^2},$$

in obtaining the graph $G_2$. We recall that (see the first paragraph of the proof) there are $z$ such walks to be created of which $z - 1$ have length $w$ and the final walk has length at most $2w$. Therefore after $z$ iterations, we get a graph $G_z$ with $m$ edges and whose maximum vertex degree $\Delta_z$ is at most

$$\begin{aligned} \Delta_z \leq \Delta_0 &+ \left(301 \log n + \frac{6w}{n}\right) \cdot (z - 1) + 301 \log n + \frac{12w}{n} \\ &+ 2K \sum_{k=1}^{z-1} \frac{(b_0 + k \cdot w)w}{n^2} + 2K \frac{(b_0 + (z - 1) \cdot w)2w}{n^2}. \end{aligned} \tag{2.25}$$

By construction $G_z$ is a Eulerian graph.

To verify the obtainability of $G_z$, we estimate $\Delta_z$ as follows. The term $z$ is no more than the size of a maximum clique in the original graph $G$ (see discussion prior to (2.3)) and since there are $m \leq n^{\frac{3}{2}}$ edges in $G$, the maximum size of a clique in $G$ is at most $n^{\frac{3}{4}}$. Therefore

$$z \leq n^{\frac{3}{4}}. \tag{2.26}$$

Also using (2.4) and (2.2), we get that $zw \leq m \leq \alpha \cdot n^{\frac{3}{2}}$ and so

$$\frac{wz}{n} \leq \alpha \cdot \sqrt{n} \leq \sqrt{n}. \tag{2.27}$$

Finally from (2.3), we have that $b_0 \leq b + n$ and so the second line in (2.25) is at most

$$\begin{aligned}\sum_{k=1}^{z} \frac{(b+n+k\cdot w)w}{n^2} &\leq \frac{z(b+n+zw)2w}{n^2} \\ &\leq \frac{2m(n+m)}{n^2} \\ &\leq 2\sqrt{n} + \frac{2m^2}{n^2} \\ &\leq \sqrt{n} + 2\alpha^2 \cdot n \end{aligned} \tag{2.28}$$

where the second inequality in (2.28) follows from the estimate $b + zw \leq b_0 + zw = m$ (see (2.4)), and the third and fourth estimates in (2.28) follow from the bound $m \leq \alpha \cdot n^{\frac{3}{2}}$ (see (2.1)).

Plugging (2.28), (2.27) and (2.26) into (2.25), we get that

$$\begin{aligned}\Delta_z &\leq \Delta_0 + 301 n^{\frac{3}{4}} \cdot \log n + \sqrt{n}\left(\frac{12}{10} + 2K\right) + 4K\alpha^2 \cdot n \\ &\leq (\beta + 4K\alpha^2)\cdot n + 1 + 301 n^{\frac{3}{4}} \cdot \log n + \sqrt{n}\left(\frac{12}{10} + 2K\right)\end{aligned} \tag{2.29}$$

for all $n$ large, where the second inequality in (2.29) is obtained by using $\Delta_0 \leq \Delta + 1 \leq \beta \cdot n + 1$ (see (2.3) and the first condition in (2.2)). From the statement of Theorem 1 and using $K = 5$, we have that

$$\beta + 4K\alpha^2 = \beta + 20\alpha^2 < \frac{1}{2}$$

strictly and so the degree of any vertex in $G_z$ is strictly less than $\frac{n}{2}$ and also, the sum of degrees of any two vertices in $G_z$ is at most

$$(2\beta + 40\alpha^2)\cdot n + 3 + 602 n^{\frac{3}{4}} \cdot \log n + \frac{12\sqrt{n}}{10} < n - 10$$

for all $n$ large. Thus, the graph $G_z$ can be obtained by the above probabilistic method as in the discussion following (2.21). □

# Appendix

Throughout, we use the following deviation estimate. Let $Z_i$, $1 \leq i \leq t$ be independent Bernoulli random variables satisfying

$$\mathbb{P}(Z_i = 1) = p_i = 1 - \mathbb{P}(Z_i = 0).$$

If $W_t = \sum_{i=1}^{t} Z_i$ and $\mu_t = \mathbb{E}W_t$, then for any $0 < \epsilon < \frac{1}{2}$ we have that

$$\mathbb{P}\left(|W_t - \mu_t| \geq \epsilon\mu_t\right) \leq 2\exp\left(-\frac{\epsilon^2}{4}\mu_t\right). \tag{A.30}$$

For a proof of (A.30), we refer to Corollary $A.1.14$, pp. 312, Alon and Spencer [1].

**Acknowledgements** I thank Profs. V. Raman, C. R. Subramanian and the referees for crucial comments that led to an improvement of the paper. I also thank IMSc for my fellowships.

# References

1. Alon, N., Spencer, J., et al.: The Probabilistic Method. Wiley Interscience (2008)
2. Boesch, F.T., Suffel, C., Tindell, R.: The spanning subgraph of Eulerian graphs. J. Graph Theory **1**, 79–84 (1977)
3. Dorn, F., Moser, H., Niedermeier, R., Weller, M.: Efficient algorithms for Eulerian extension and rural postman problem. SIAM J. Discret. Math. **27**, 75–94 (2013)
4. Fomin, F.V., Golovach, P.A.: Parameterized complexity of connected even/odd subgraph problems. In: Proceedings 29 th STACS, vol. 14, pp. 432–440 (2012)
5. Höhn, W., Jacobs, T., Megow, N.: On Eulerian extensions and their application to no-wait flowshop scheduling. J. Sched. **15**, 295–309 (2012)
6. Lesniak, L., Oellermann, O.R.: An Eulerian exposition. J. Graph Theory **10**(1986), 277–297 (1986)
7. West, D.B.: Introduction to Graph Theory. Prentice Hall, Hoboken (2001)

# Bounds of Some Energy-Like Invariants of Neighbourhood Corona of Graphs

**Chinglensana Phanjoubam and Sainkupar Mn. Mawiong**

**Abstract** For a finite simple graph $G$ with $n$ vertices, Laplacian-energy-like is defined as $LEL(G) = \sum_{j=1}^{n} \sqrt{\mu_j}$, where $\mu_j$ denotes Laplacian eigenvalue; and incidence energy $IE(G) = \sum_{j=1}^{n} \sqrt{q_j}$, where $q_j$ denotes signless Laplacian eigenvalue. In this paper, we give the bounds of some energy-like invariants, namely the Laplacian-energy-like and the incidence energy of the neighbourhood corona of two graphs. We observed that the bounds are sharp for the complete graph $K_n$.

**Keywords** Simple graph · Incidence energy · Laplacian-energy-like · Neighbourhood corona

## 1 Introduction

For a finite simple graph $G$ having $n$ vertices, its energy is defined [9] as $E(G) = \sum_{j=1}^{n} |\lambda_j|$, where $\lambda_j$ for $1 \leq j \leq n$ denotes the eigenvalue of the adjacency matrix of the graph $G$. This graph invariant was motivated by the molecular orbital theory of conjugated $\pi$-electron systems in chemistry and was introduced by Gutman [9] in mathematics, independent of chemical motivations. In [12], analogous to the energy of a graph, Laplacian energy is defined as $LE(G) = \sum_{j=1}^{n} \left| \mu_j - \frac{2m}{n} \right|$, where $\mu_j$ for $1 \leq j \leq n$ denotes the eigenvalue of the Laplacian matrix of the graph $G$

C. Phanjoubam (✉)
Department of Mathematics, North-Eastern Hill University, Shillong 793022, India
e-mail: phanjoubam17@gmail.com

S. Mn. Mawiong
Department of Basic Sciences and Social Sciences, North-Eastern Hill University, Shillong 793022, India



B. Rushi Kumar et al. (eds.), *Mathematics and Computing*, Springer Proceedings in Mathematics & Statistics 415, https://doi.org/10.1007/978-981-19-9307-7_18

having $m$ edges. Much beyond the applications in the molecular orbital theory of conjugated molecules, the Laplacian energy is found to have remarkable chemical applications [2]. Similar to the graph energy and Laplacian energy, Liu and Liu [17] proposed another graph invariant known as Laplacian-energy-like which is defined as $LEL(G) = \sum_{j=1}^{n} \sqrt{\mu_j}$. The chemical applications of the Laplacian-energy-like are much studied and are also described as a newly designed molecular descriptor [20]. Nikiforov introduced the idea of graph energy to a matrix as the sum of its singular values [18]. Nikiforov's idea inspired Jooyandeh et al. [15] to introduce the incidence energy of a graph which is defined as the sum of the singular values of the incidence matrix of the graph $G$. Moreover, note [10] that $IE(G) = \sum_{j=1}^{n} \sqrt{q_j}$, where $q_j$ for $1 \leq j \leq n$ denotes the eigenvalue of the signless Laplacian matrix of the graph $G$.

Laplacian-energy-like and incidence energy of a graph share many interesting relations. In particular, they coincide when the graph is bipartite and for more relations, we refer the readers to see [10]. The Laplacian-energy-like and the incidence energy are also much studied for some graph operations in recent years, particularly on regular graphs as well as semi-regular graphs. For some derived graphs of regular graphs, the bounds for the Laplacian-energy-like and the incidence energy have been determined in [1] and [3]. Motivated by such results, we give the bounds for the Laplacian-energy-like and the incidence energy of an important graph operation introduced recently in [8] called neighbourhood corona. The operation known as corona of graphs is defined in [7] and a variant called neighbourhood corona is defined recently in [8]. It is worth mentioning that the neighbourhood corona is quite useful in constructing a new family of expander graphs. Expander graphs are sparse but they are highly connected graphs. It has a wide number of applications in computer science, communication networks, complexity theory, etc. (see [16] for definitions and more details).

## 2 Preliminaries

We recall a few basic definitions and present some well-known results that are required in the subsequent sections. We consider only finite simple graph $G$ having $n$ number of vertices and $m$ number of edges throughout our paper. Let $V(G) = \{v_j : 1 \leq j \leq n\}$ be the vertex set of $G$. The adjacency matrix $A_G = (a_{jk})$ of the graph $G$ is an $n \times n$ matrix where $a_{jk}$ is defined as follows:

$$a_{jk} := \begin{cases} 1 & \text{if} v_j \text{is adjacent to } v_k \\ 0 & \text{otherwise.} \end{cases}$$

The degree of a vertex $v_j$ is denoted by $d_j$, $1 \leq j \leq n$. The graph $G$ is said to be regular with regularity $r$ if $d_j = r \; \forall \; j = 1, \ldots, n$. The diagonal matrix of the graph $G$ with $d_j$, $1 \leq j \leq n$ as its diagonal entries is denoted by $D_G$. The Laplacian matrix of the graph $G$ is defined as $L_G = D_G - A_G$. And the signless Laplacian matrix of the graph $G$ is defined as $Q_G = D_G + A_G$. Let $\mu_1 \leq \cdots \leq \mu_n$ be the eigenvalues of $L_G$, known as Laplacian eigenvalues. Let $q_1 \leq \cdots \leq q_n$ be the eigenvalues of $Q_G$, known as signless Laplacian eigenvalues. We remark that it is well known [4] that $\mu_1 = 0$ and $q_1 \geq 0$; and for an $r$-regular graph, $\mu_n \leq 2r$ and $q_n = 2r$.

We now recall the definition of neighbourhood corona, denoted by $G_1 \star G_2$, of two graphs $G_1$ and $G_2$ having $n_1$ and $n_2$ vertices respectively that are disjoint, introduced by Gopalapillai in [8].

**Definition 1** The neighbourhood corona $G_1 \star G_2$ is defined [8] as the graph obtained by taking one copy of $G_1$ and $n_1$ copies of $G_2$ and then joining every neighbour of the $j$th vertex of $G_1$ to every vertex in the $j$th copy of $G_2$ by an edge.

Liu and Zhou in [16] computed the Laplacian eigenvalues of the neighbourhood corona of a regular and any arbitrary graphs; and also the signless Laplacian eigenvalues of the neighbourhood corona of two regular graphs.

**Theorem 1** ([16]) *Let $G_1$ be a regular graph having $n_1 \geq 2$ vertices and regularity $r_1 \geq 1$, and $G_2$ be an arbitrary graph having $n_2 \geq 1$ vertices. Let the Laplacian eigenvalues of $G_1$ and $G_2$ be $0 = \mu_1 \leq \mu_2 \leq \cdots \leq \mu_{n_1}$ and $0 = \eta_1 \leq \eta_2 \leq \cdots \leq \eta_{n_2}$, respectively. Let $G = G_1 \star G_2$. Let*

$$\alpha_j, \bar{\alpha}_j = \frac{(n_2+1)r_1 + \mu_j \pm \sqrt{((n_2+1)r_1 + \mu_j)^2 - 4\mu_j((2n_2+1)r_1 - n_2\mu_j)}}{2}$$

*for each $j = 1, \ldots, n_1$. The Laplacian eigenvalues of $G$ are then given by*

$$\begin{bmatrix} r_1 + \eta_2 & \ldots & r_1 + \eta_{n_2} & \alpha_1 & \bar{\alpha}_1 & \ldots & \alpha_{n_1} & \bar{\alpha}_{n_1} \\ n_1 & \ldots & n_1 & 1 & 1 & \ldots & 1 & 1 \end{bmatrix}$$

*where the first row entries are the eigenvalues with their corresponding multiplicities listed in the second row.*

**Theorem 2** ([16]) *Let $G_1$ and $G_2$ be regular graphs having $n_1 \geq 2$ and $n_2 \geq 2$ vertices; and regularities $r_1 \geq 1$ and $r_2 \geq 1$, respectively. Let the signless Laplacian eigenvalues of $G_1$ and $G_2$ be $q_1 \leq \cdots \leq q_{n_1}$ and $\theta_1 \leq \cdots \leq \theta_{n_2}$, respectively. Let $G = G_1 \star G_2$. Then the signless Laplacian eigenvalues of $G$ consist of*
*(i) two eigenvalues which are the solutions of the following equation:*

$$x^2 - ((n_2+1)r_1 + 2r_2 + q_j)x + (2n_2r_1r_2 + (2n_2r_1 + 2r_2 + r_1)q_j - n_2{q_j}^2) = 0$$

*for each $j = 1, \ldots, n_1$;*
*(ii) $r_1 + \theta_j$, repeated $n_1$ times, for each $j = 1, \ldots, n_2 - 1$.*

**Definition 2** The first Zagreb index $M_1(G)$ of $G$ is defined [13] as

$$M_1(G) = \sum_{v_j \in V(G)} d_j{}^2.$$

**Remark 1** Notice that

$$\sum_{j=1}^{n} \mu_i{}^2 = tr(D - A)^2 = \sum_{j=1}^{n} (d_j{}^2 + d_j)$$
$$= M_1(G) + 2m.$$

So, for a regular graph $G$ with regularity $r$ we have $\sum_{i=1}^{n} \mu_j{}^2 = M_1(G) + rn.$

Notice that $M_1(G)$ satisfy the following well-known bounds.

**Lemma 1** ([6]) *Let $G$ be a graph having $n \geq 3$ vertices and $m$ edges. Then $M_1(G)$ satisfies the following:*

$$M_1(G) \leq m\left(\frac{2m}{n-1} + n - 2\right).$$

*The equality is true if and only if $G \cong S_n$ or $K_n$.*

**Lemma 2** ([13]) *Let $G$ be a graph having $n$ vertices and $m$ edges. Then $M_1(G)$ satisfies the following:*

$$M_1(G) \geq \frac{4m^2}{n}.$$

*The equality is true if and only if $G$ is isomorphic to a regular graph.*

**Lemma 3** ([5]) *Let $G$ be a graph having $n$ vertices and $m \geq 1$ edges. Let the Laplacian eigenvalues of $G$ be $0 = \mu_1 \leq \mu_2 \leq \cdots \leq \mu_n$. Then $\mu_2 = \mu_3 = \cdots = \mu_n$ if and only if $G \cong K_n$.*

**Lemma 4** ([3]) *Let $G$ be a graph having $n$ vertices and $m \geq 1$ edges. Let the signless Laplacian eigenvalues of $G$ be $q_1 \leq q_2 \cdots \leq q_n$. Then $q_1 = q_2 = \cdots = q_{n-1}$ if and only if $G \cong K_n$.*

**Lemma 5** (Ozeki's inequality) ([19]) *Let $\{x_j : 1 \leq j \leq n\}$ and $\{y_j : 1 \leq j \leq n\}$ be two sequences of numbers satisfying $C \geq x_j \geq c > 0$ and $D \geq y_j \geq d > 0$, $\forall\, j = 1, \ldots, n$. Then*

$$\left(\sum_{j=1}^{n} x_j y_j\right)^2 \geq \sum_{j=1}^{n} x_j{}^2 \sum_{j=1}^{n} y_j{}^2 - \frac{n^2}{4}(CD - cd)^2.$$

**Lemma 6** ([14]) *Let* $\{x_j : 1 \leq j \leq n\}$ *and* $\{y_j : 1 \leq j \leq n\}$ *be two sequences of numbers satisfying* $C \geq x_j \geq c \geq 0$ *and* $D \geq y_j \geq d \geq 0$, $\forall\ j = 1, \ldots, n$, *and* $CD \neq 0$. *Let* $x = \frac{c}{C}$ *and* $y = \frac{d}{D}$. *If* $(1+x)(1+y) \geq 2$, *then the above Ozeki's inequality holds.*

## 3 Bounds for the Laplacian-Energy-Like of Neighbourhood Corona

We consider the Laplacian-energy-like of the neighbourhood corona. We present its bounds in the following theorem and we also observe that the bounds are sharp for the complete graphs.

**Theorem 3** *Let* $G_1$ *be a regular graph having* $n_1 \geq 2$ *vertices,* $m_1$ *edges and regularity* $r_1 \geq 1$, *and* $G_2$ *be an arbitrary graph having* $n_2 \geq 1$ *vertices and* $m_2$ *edges. Let* $G = G_1 \star G_2$. *Let* $a = n_2\left(r_1 + \frac{M_1(G_1)}{n_1}\right)$ *and* $b = \frac{2m_2}{n_2 - 1} + r_1$. *Then we have the following:*

*(i)* $LEL(G) \leq n_1\sqrt{(n_2+2)r_1 + 2\sqrt{(2n_2+1)r_1{}^2 - a}} + n_1(n_2-1)\sqrt{b}$.
*The equality is true if and only if* $G_1 \cong K_{n_1}$ *and* $G_2 \cong K_{n_2}$.

*(ii)* $LEL(G) > n_1\sqrt{(n_2 - \frac{1}{2})r_1 + 2\sqrt{\left(2n_2 - \frac{1}{2}\right)r_1{}^2 - a}} + n_1(n_2-1)\sqrt{b - \frac{m_2}{2}}$.

***Proof*** Let the Laplacian eigenvalues of $G_1$ and $G_2$ be $0 = \mu_1 \leq \mu_2 \leq \cdots \leq \mu_{n_1}$ and $0 = \eta_1 \leq \eta_2 \leq \cdots \leq \eta_{n_2}$, respectively; and let

$$\alpha_j, \bar{\alpha}_j = \frac{(n_2+1)r_1 + \mu_j \pm \sqrt{((n_2+1)r_1+\mu_j)^2 - 4\mu_j((2n_2+1)r_1 - n_2\mu_j)}}{2}$$

for each $j = 1, \ldots, n_1$. Then from Theorem 1 and a simple computation, we have

$$\begin{aligned} LEL(G) &= \sum_{j=1}^{n_1}(\sqrt{\alpha_j} + \sqrt{\bar{\alpha}_j}) + n_1\sum_{j=2}^{n_2}\sqrt{r_1+\eta_j} \\ &= \sum_{j=1}^{n_1}\sqrt{(\sqrt{\alpha_j} + \sqrt{\bar{\alpha}_j})^2} + n_1\sum_{j=2}^{n_2}\sqrt{r_1+\eta_j} \\ &= \sum_{j=1}^{n_1}\sqrt{(n_2+1)r_1 + \mu_j + 2\sqrt{\mu_j(2n_2+1)r_1 - n_2\mu_j{}^2}} + n_1\sum_{j=2}^{n_2}\sqrt{r_1+\eta_j}. \end{aligned} \quad (1)$$

Note that $\sum_{j=1}^{n_1} \mu_j = r_1 n_1$, $\sum_{j=2}^{n_2} \eta_j = 2m_2$ and $\sum_{j=1}^{n_1} {\mu_j}^2 = r_1 n_1 + M_1(G_1)$. By the Cauchy-Schwarz inequality in (1), we get

$$LEL(G) \leq \sqrt{n_1 \sum_{j=1}^{n_1} \left( (n_2+1)r_1 + \mu_j + 2\sqrt{\mu_j(2n_2+1)r_1 - n_2{\mu_j}^2} \right)}$$
$$+ n_1 \sqrt{(n_2-1) \sum_{j=2}^{n_2} (r_1 + \eta_j)}$$
$$= \sqrt{{n_1}^2(n_2+2)r_1 + 2n_1 \sum_{i=1}^{n_1} \sqrt{\mu_i(2n_2+1)r_1 - n_2{\mu_i}^2}}$$
$$+ n_1\sqrt{(n_2-1)((n_2-1)r_1 + 2m_2)}$$

$$\leq \sqrt{{n_1}^2(n_2+2)r_1 + 2n_1 \sqrt{n_1 \sum_{j=1}^{n_1} \left(\mu_j(2n_2+1)r_1 - n_2{\mu_j}^2\right)}}$$
$$+ n_1(n_2-1)\sqrt{r_1 + \frac{2m_2}{n_2-1}}$$
$$= n_1 \sqrt{(n_2+2)r_1 + 2\sqrt{(2n_2+1){r_1}^2 - n_2\left(r_1 + \frac{M_1(G_1)}{n_1}\right)}}$$
$$+ n_1(n_2-1)\sqrt{\frac{2m_2}{n_2-1} + r_1}.$$

Notice that the above inequalities become equalities if and only if $\mu_2 = \cdots = \mu_{n_1}$ and $\eta_2 = \cdots = \eta_{n_2}$. Thus, Lemma 3 implies that the equality is true if and only if $G_1 \cong K_{n_1}$ and $G_2 \cong K_{n_2}$.

Next, let $x_j = \sqrt{(n_2+1)r_1 + \mu_j + 2\sqrt{\mu_j(2n_2+1)r_1 - n_2{\mu_j}^2}}$ and $y_j = 1$ for $j = 1, \ldots, n_1$. Choose $C = \sqrt{(n_2+3+2\sqrt{2})r_1}$, $c = \sqrt{(n_2+1)r_1}$, $d = 1$ and $D = 1$. Note that $0 \leq \mu_j \leq 2r_1\ \forall\ j = 1, \ldots, n_1$. Thus, $C \geq x_j \geq c > 0$ and $D \geq y_j \geq d > 0, \forall\ j = 1, \ldots, n_1$. Also, note that $(2+2\sqrt{2})r_1 < (C+c)^2$. So

$$(CD - cd)^2 = \frac{((2+2\sqrt{2})r_1)^2}{(C+c)^2} < 6r_1.$$

Thus by Ozeki's inequality, we have

$$\sum_{j=1}^{n_1} x_j > \sqrt{n_1 \sum_{j=1}^{n_1}\left((n_2+1)r_1 + \mu_j + 2\sqrt{\mu_j(2n_2+1)r_1 - n_2{\mu_j}^2}\right) - \frac{3{n_1}^2 r_1}{2}}$$
$$= \sqrt{{n_1}^2\left(n_2 - \frac{1}{2}\right) r_1 + 2n_1 \sum_{j=1}^{n_1}\sqrt{\mu_j(2n_2+1)r_1 - n_2{\mu_j}^2}}. \quad (2)$$

Again, let $x_j = \sqrt{\mu_j(2n_2+1)r_1 - n_2{\mu_j}^2}$ and $y_j = 1$ for $j = 1, \ldots, n_1$. Choose $C = \sqrt{2}r_1, c = 0, d = 1$ and $D = 1$. Since $0 \le \mu_j \le 2r_1$, we have $C \ge x_j \ge c \ge 0$ and $D \ge y_j \ge d \ge 0$, $\forall\ j = 1, \ldots, n_1$. Also, $CD \ne 0$. Note that $(CD - cd)^2 = 2{r_1}^2$. Since $\left(1 + \frac{c}{C}\right)\left(1 + \frac{d}{D}\right) = 2$, by Lemma 6, we have

$$\sum_{j=1}^{n_1}\left(\mu_j(2n_2+1)r_1 - n_2{\mu_j}^2\right) > \sqrt{n_1 \sum_{j=1}^{n_1}\left(\mu_j(2n_2+1)r_1 - n_2{\mu_j}^2\right) - \frac{1}{2}{n_1}^2{r_1}^2}$$
$$= n_1\sqrt{\left(2n_2 - \frac{1}{2}\right){r_1}^2 - n_2\left(r_1 + \frac{M_1(G_1)}{n_1}\right)}. \quad (3)$$

And let $x_j = \sqrt{r_1 + \eta_j}$ and $y_j = 1$, $j = 2, 3, \ldots, n_2$. We choose $C = \sqrt{r_1 + 2m_2}$, $c = \sqrt{r_1}$, $d = 1$ and $D = 1$. Since $0 \le \eta_j \le 2m_2$, we have $C \ge x_i \ge c > 0$ and $D \ge y_i \ge d > 0$, $\forall\ j = 2, 3, \ldots, n_2$. Note that $(CD - cd)^2 = \frac{4{m_2}^2}{(C+c)^2} < 2m_2$. Applying Ozeki's inequality, we have

$$\sum_{j=2}^{n_2}\sqrt{r_1 + \eta_j} > \sqrt{2m_2(n_2-1) + (n_2-1)^2\left(r_1 - \frac{m_2}{2}\right)}$$
$$= (n_2 - 1)\sqrt{\frac{2m_2}{n_2-1} + r_1 - \frac{m_2}{2}}. \quad (4)$$

From (1)–(4) we get $(ii)$.

The following corollary is now immediate by using the fact that $0 \le \eta_j \le 2r_2$ to obtain an inequality similar to (4) for a regular graph with regularity $r_2$ and from Lemmas 1 and 2.

**Corollary 1** *Let $G_1$ and $G_2$ be regular graphs having $n_1 \ge 2$ and $n_2 \ge 1$ vertices, $m_1$ and $m_2$ edges, and regularities $r_1 \ge 1$ and $r_2 \ge 1$, respectively. Let $G = G_1 \star G_2$ and let $a = \frac{r_2 n_2}{n_2 - 1} + r_1$. Then we have the following:*

*(i)* $LEL(G) \leq n_1\sqrt{(n_2+2)r_1 + 2\sqrt{(2n_2+1)r_1{}^2 - n_2 b}} + n_1(n_2-1)\sqrt{a}$ *where* $b = r_1 + m_1\left(\frac{r_1}{n_1-1} + \frac{n_1-2}{n_1}\right)$. *The equality is true if and only if* $G_1 \cong K_{n_1}$ *and* $G_2 \cong K_{n_2}$.

*(ii)* $$LEL(G) > n_1\sqrt{\left(n_2-\frac{1}{2}\right)r_1 + 2\sqrt{\left(n_2-\frac{1}{2}\right)r_1{}^2 - n_2 r_1}} + n_1(n_2-1)\sqrt{a-\frac{r_2}{2}}.$$

## 4 Bounds for the Incidence Energy of Neighbourhood Corona

We consider the incidence energy of neighbourhood corona, and we present its bounds in the following theorem. We observe similar to the case of Laplacian-energy-like that the bounds of the incidence energy of neighbourhood corona are sharp when the graphs are complete.

**Theorem 4** *Let* $G_1$ *and* $G_2$ *be regular graphs having* $n_1 \geq 2$ *and* $n_2 \geq 2$ *vertices, and regularities* $r_1 \geq 1$ *and* $r_2 \geq 1$, *respectively. Let* $G = G_1 \star G_2$ . *Let* $a = r_1 n_2(2r_1 + 2r_2 - 1) - \frac{n_2 M_1(G_1)}{n_1}$ *and* $b = r_1 + \frac{(n_2-2)r_2}{n_2-1}$. *Then we have the following:*

*(i)* $IE(G) \leq n_1\sqrt{(n_2+2)r_1 + 2r_2 + 2\sqrt{a + r_1(2r_2+1)}} + n_1(n_2-1)\sqrt{b}$.
*The equality is true if and only if* $G_1 \cong K_{n_1}$ *and* $G_2 \cong K_{n_2}$.

*(ii)* $IE(G) > n_1\sqrt{(n_2+1-4r_1)r_1 + 2r_2 + 2\sqrt{a + r_1\left(r_2+1-\frac{r_1}{2}\right)}} + n_1(n_2-1)\sqrt{b-\frac{r_2}{2}}$.

***Proof*** Let the signless Laplacian eigenvalues of $G_1$ and $G_2$ be respectively $0 \leq q_1 \leq q_2 \leq \cdots \leq q_{n_1}$ and $0 \leq \theta_1 \leq \theta_2 \leq \cdots \leq \theta_{n_2}$. Let

$$\beta_j, \bar{\beta}_j = (n_2+1)r_1 + 2r_2 + q_j \pm \sqrt{((n_2+1)r_1 + 2r_2 + q_j)^2 - 4(2n_2r_1r_2 + kq_j - n_2 q_j{}^2)}$$

for each $j = 1, \ldots, n_1$ and $k = 2n_2r_1 + 2r_2 + r_1$. Then from Theorem 2 and a simple computation, we have

$$
\begin{aligned}
IE(G) &= \sum_{j=1}^{n_1}\left(\sqrt{\frac{\beta_j}{2}}+\sqrt{\frac{\bar{\beta}_j}{2}}\right)+n_1\sum_{j=1}^{n_2-1}\sqrt{r_1+\theta_j}\\
&= \sum_{j=1}^{n_1}\sqrt{(n_2+1)r_1+2r_2+q_j+2\sqrt{2n_2r_1r_2+kq_j-n_2{q_j}^2}}\\
&\quad + n_1\sum_{j=1}^{n_2-1}\sqrt{r_1+\theta_j}. \qquad (5)
\end{aligned}
$$

Note that $\sum_{j=1}^{n_1} q_j = r_1n_1$, $\sum_{j=1}^{n_2-1}\theta_j = (n_2-2)r_2$ and $\sum_{j=1}^{n_1} {q_j}^2 = r_1n_1 + M_1(G_1)$. By the Cauchy-Schwarz inequality in (5) and a simple computation as in Theorem 3, we have

$$
\begin{aligned}
IE(G) &\le n_1\sqrt{(n_2+2)r_1+2r_2+2\sqrt{r_1n_2a+r_1(2r_2+1)-\frac{n_2M_1(G_1)}{n_1}}}\\
&\quad + n_1(n_2-1)\sqrt{r_1+\frac{(n_2-2)r_2}{n_2-1}}.
\end{aligned}
$$

Notice that the above inequality becomes an equality if and only if $q_1=\cdots=q_{n_1-1}$ and $\theta_1=\cdots=\theta_{n_2-1}$. Thus by Lemma 4, the equality is true if and only if $G_1 \cong K_{n_1}$ and $G_2 \cong K_{n_2}$.

Next, let $x_j = \sqrt{(n_2+1)r_1+2r_2+q_j+2\sqrt{2n_2r_1r_2+kq_j-n_2{q_j}^2}}$ and $y_j=1$ for $j=1,\ldots,n_1$. Choose $C=\sqrt{(n_2+3)r_1+2r_2+2\sqrt{2r_1(r_2n_2+2r_2+r_1)}}$, $c=\sqrt{(n_2+1)r_1+2r_2+2\sqrt{2r_1r_2n_2}}$, $d=1$ and $D=1$. Note that $0\le q_j\le 2r_1$ for $j=1,2,\ldots,n_1$. Thus $C\ge x_j\ge c>0$ and $D\ge y_j\ge d>0, \forall\, j=1,\ldots n_1$. Also note that

$$
(CD-cd)^2 = 4{r_1}^2\left(\frac{1}{C+c}+\frac{r_1+r_2}{(C+c)(e+f)}\right)^2 < 16{r_1}^2
$$

where $e=\sqrt{2r_1(r_2n_2+2r_2+r_1)}$ and $f=\sqrt{2r_1r_2n_2}$. Thus by Ozeki's inequality, we have

$$
\sum_{j=1}^{n_1} x_j > \sqrt{{n_1}^2((n_2+2-4r_1)r_1+2r_2)+2n_1\sum_{j=1}^{n_1}\sqrt{2n_2r_1r_2+kq_j-n_2{q_j}^2}}. \qquad (6)
$$

Again, let $x_j=\sqrt{2n_2r_1r_2+kq_j-n_2{q_j}^2}$ and $y_j=1$ for $j=1,\ldots,n_1$. Choose $C=\sqrt{2n_2r_1r_2+2r_1(2r_2+r_1)}$, $c=\sqrt{2n_2r_1r_2}$, $d=1$ and $D=1$. Since $0\le q_j\le 2r_1$, we have $C\ge x_j\ge c>0$ and $D\ge y_i\ge d>0, \forall\, j=1,\ldots,n_1$. Note that

$$(CD - cd)^2 = \frac{4{r_1}^2(2r_2 + r_1)^2}{(C + c)^2} < 2r_1(2r_2 + r_1).$$

Thus by Ozeki's inequality, we have

$$\sum_{j=1}^{n_1} x_j > n_1 \sqrt{r_1 n_2(2r_1 + 2r_2 - 1) + r_1\left(r_2 + 1 - \frac{r_1}{2}\right) - \frac{n_2 M_1(G_1)}{n_1}}. \qquad (7)$$

Finally, let $x_j = \sqrt{r_1 + \theta_j}$ and $y_j = 1, j = 1, 2, \ldots, n_2 - 1$. Choose $C = \sqrt{r_1 + 2r_2}$, $c = \sqrt{r_1}$, $d = 1$ and $D = 1$. Since $0 \le \theta_j \le 2r_2$, we have $C \ge x_j \ge c > 0$ and $D \ge y_j \ge d > 0$ for $j = 1, 2, \ldots, n_2 - 1$. Note that $(CD - cd)^2 = \frac{4{r_2}^2}{(C + c)^2} < 2r_2$. Thus by Ozeki's inequality, we have

$$\sum_{j=1}^{n_2-1} \sqrt{r_1 + \theta_j} > \sqrt{(n_2 - 2)r_2(n_2 - 1) + (n_2 - 1)^2\left(r_1 - \frac{r_2}{2}\right)}. \qquad (8)$$

From (5)–(8) we get the required result $(ii)$.

## 5 Conclusions

Neighbourhood corona is a recent graph operation which is used in the construction of graph expanders. In this paper, we have presented the bounds of the two energy-like invariants of neighbourhood corona of two graphs, namely $LEL$ and $IE$, which are important graph invariants with applications in molecular chemistry. We also obtained that the bounds for $LEL$ and $IE$ for neighbourhood corona of two graphs are sharp when all the graphs considered are the complete graph $K_n$.

**Acknowledgements** The second author acknowledges DST-SERB, India, grant no. MTR/2017/000403.

## References

1. Chen, X., Hou, Y., Li, J.: On two energy-like invariants of line graphs and related graph operations. J. Inequal. Appl. **51**, 1–15 (2016)
2. Consonni, V., Todeschini, R.: New spectral index for molecule description. MATCH Commun. Math. Comput. Chem. **60**, 3–14 (2008)
3. Cui, S.-Y., Tian, G.-X.: Some improved bounds on two energy-like invariants of some derived graphs. Open Math. **17**, 883–893 (2019)

4. Cvetković, D., Rowlinson, P., Simić, S.: An Introduction to the Theory of Graph Spectra, London Mathematical Society Student Texts, vol. 75. Cambridge University Press, Cambridge (2010)
5. Das, K.C.: A sharp upper bound for the number of spanning trees of a graph. Graphs Comb. **23**, 625–632 (2007)
6. De Caen, D.: An upper bound on the sum of squares of degrees in a graph. Discret. Math. **185**, 245–248 (1998)
7. Frucht, R., Harary, F.: On the corona of two graphs. Aequationes Math. **4**, 322–325 (1970)
8. Gopalapillai, I.: The spectrum of neighbourhood corona of graphs. Kragujevac J. Math. **35**, 493–500 (2011)
9. Gutman, I.: The energy of a graph. Ber. Math.-Stat. Sekt. Forschungszent. Graz. **103**, 1–22 (1978)
10. Gutman, I., Kiani, D., Mirzakhah, M.: On incidence energy of graphs. MATCH Commun. Math. Comput. Chem. **62**, 573–580 (2009)
11. Gutman, I., Polansky, O.E.: Mathematical Concepts in Organic Chemistry. Springer, Berlin (1986)
12. Gutman, I., Zhou, B.: Laplacian energy of a graph. Lin. Algebra Appl. **414**, 29–37 (2006)
13. Ilić, A., Stevanović, D.: On comparing Zagreb indices. MATCH Commun. Math. Comput. Chem. **62**, 681–687 (2009)
14. Izumino, S., Mori, H., Seo, Y.: On Ozeki's inequality. J. Inequal. Appl. **2**, 235–253 (1998)
15. Jooyandeh, M.R., Kiani, D., Mirzakhah, M.: Incidence energy of a graph. MATCH Commun. Math. Comput. Chem. **62**, 561–572 (2009)
16. Liu, X., Zhou, S.: Spectra of the neighbourhood corona of two graphs. Linear and Multilinear Algebra **62**, 1205–1219 (2014)
17. Liu, J., Liu, B.: A Laplacian-energy-like invariant of a graph. MATCH Commun. Math. Comput. Chem. **59**, 397–419 (2008)
18. Nikiforov, V.: The energy of graphs and matrices. J. Math. Anal. Appl. **326**, 1472–1475 (2007)
19. Ozeki, N.: On the estimation of the inequalities by the maximum, or minimum values. J. College Arts Sci. Chiba Univ. **5**, 199–203 (1968)
20. Stevanović, D., Ilić, A., Onisor, C., Diudea, M.: $LEL$-a newly designed molecular descriptor. Acta Chim. Slov. **56**, 410–417 (2009)
21. Wang, W., Luo, Y.: On Laplacian-energy-like invariant of a graph. Linear Algebra Appl. **437**, 713–721 (2012)

# Linear Recurrent Fractal Interpolation Function for Data Set with Gaussian Noise

**Mohit Kumar, Neelesh S. Upadhye, and A. K. B. Chand**

**Abstract** In this article, we use the linear recurrent fractal interpolation function approach to interpolate a data set with Gaussian noise on its ordinate. To investigate the variability at any intermediate point in the given noisy data set, we estimate the parameters of the probability distribution of the fractal function. In addition, we present a simulation study that experimentally confirms our theoretical findings.

**Keywords** Recurrent fractal interpolation function · Gaussian noise · Random noise · Normal distribution · Distribution of fractal function

## 1 Introduction

In 1986, Barnsley [1] introduced the concept of fractal interpolation based on the theory of iterated function systems (IFSs), and since then it has emerged as an important and powerful technique for modelling many natural phenomena. During the development of fractal interpolation theory, several researchers have generalized this concept in various ways and have developed different types of fractal interpolation functions (FIFs) such as recurrent FIF [2], alpha-FIF [3], vector-valued FIF [4], local FIF [5], random FIF [6–9], and many more (see, [10, 11]). Further, many of them have studied numerous analytical properties of these FIFs such as smoothness [12], stability [13], convexity [14], positivity [15], and shape preservation [16]. Presently, fractal interpolation has several applications in mathematics and various other applied science disciplines [17–19].

M. Kumar (✉) · N. S. Upadhye · A. K. B. Chand
Department of Mathematics, Indian Institute of Technology Madras, Chennai 600036, Tamil Nadu, India
e-mail: mohittripathi.5678@gmail.com

N. S. Upadhye
e-mail: neelesh@iitm.ac.in

A. K. B. Chand
e-mail: chand@iitm.ac.in

 
B. Rushi Kumar et al. (eds.), *Mathematics and Computing*, Springer Proceedings in Mathematics & Statistics 415, https://doi.org/10.1007/978-981-19-9307-7_19

During data collection process, we often encounter some random noise on data. However, dealing with noisy data is always a challenging task. In today's world, it is necessary to analyse the variability in the data before making any decisions based on it. Therefore, if we obtain noisy data with some fractal features, such as statistical self-similarity, and want to predict a missing or unknown value at any intermediate point in the given data. Then, we have to use this noisy data in such a way that the fractal properties are retained. In fact, dealing with noise and fractality together is more challenging. This article considers data set in $\mathbb{R}^2$ with Gaussian noise on the ordinate and applies the technique of recurrent fractal interpolation [2], which is useful for predicting variability at an interpolated value. The motivation behind using noise with a Gaussian distribution is that it is the most widely used distribution in statistics, and a significant number of theories for statistical tests of this distribution have already been developed in the literature.

The rest of the article is arranged as follows. In Sect. 2, we briefly discuss the theory of recurrent IFS and the construction of recurrent FIF. Section 3 describes the construction procedure of recurrent FIF for data set with Gaussian noise and estimates the parameters of probability distribution of this fractal function. In Sect. 4, we present a simulation study that validates our analytical findings. At the end, concluding observations are discussed in Sect. 5.

## 2 Preliminaries

For any $N \in \mathbb{N}$, let us denote $\mathbb{N}_N := \{1, 2, \dots, N\}$ and $\mathbb{N}_N^0 := \{0\} \bigcup \mathbb{N}_N$. Also, let $N \geq 2$ be an integer and $\Delta_y = \{(x_k, y_k) : k \in \mathbb{N}_N^0\}$ be a given data set in $\mathbb{R}^2$, where $x_0 < x_1 < \dots < x_N$. There are several functions passing through all the points $(x_k, y_k)$ of this data set, such as Lagrange's interpolation function, which is a unique $N$th degree polynomial passing through $N + 1$ given data points and various types of splines. However, polynomials or other smooth approximation functions may not be appropriate if the data points are derived from a curve or process that has fractal properties such as coastlines or electrocardiograms. In this context, fractal interpolation [1] is an effective approach.

In this article, we use recurrent fractal interpolation method [2] (a generalization of approach given in [1]), which is based on the theory of recurrent IFS (or RIFS).

### 2.1 Basics of RIFS

**Definition 1** ([20]) An iterated function system consists of a complete metric space $(X, d)$ together with a collection of continuous maps $W_k : X \to X$ for all $k \in \mathbb{N}_N$, and it is denoted by $\{X; W_k : k \in \mathbb{N}_N\}$.

If each $W_k$ is a contraction, then the IFS is referred to as a hyperbolic or contractive IFS.

**Definition 2** A recurrent IFS comprised of an IFS $\{X; W_k : k \in \mathbb{N}_N\}$ together with an $N \times N$ irreducible row-stochastic matrix $P = \left(p_{ij}\right)_{N\times N}$ satisfying

(i) $p_{ij} \in [0, 1],\ i, j \in \mathbb{N}_N$,

(ii) $\sum\limits_{j=1}^{N} p_{ij} = 1,\ i \in \mathbb{N}_N$,

(iii) for any $i, j \in \mathbb{N}_N$, there exist $k_1, k_2, \ldots, k_n \in \mathbb{N}_N$ with $k_1 = i$ and $k_n = j$ such that $p_{k_1k_2}p_{k_2k_3}\ldots p_{k_{n-1}k_n} > 0$.

We denote this RIFS by $\{X; P; W_k : k \in \mathbb{N}_N\}$.

Essentially, RIFS is a Markov chain with $N$ states and its $k$th state is represented by map $W_k$. Hence, $p_{ij}$ is the transition probability from state $i$ to state $j$. Moreover, item (iii) implies that the chain is irreducible, which means that every state in the chain is accessible from every other state. Here, the recurrent structure can also be given through an irreducible connection matrix $C = \left(c_{ij}\right)_{N\times N}$, where

$$c_{ij} = \begin{cases} 1 & \text{if } p_{ji} > 0, \\ 0 & \text{if } p_{ji} = 0. \end{cases} \tag{1}$$

Now, a brief construction of recurrent fractal by using RIFS is given as follows. For more detailed information, the reader can see Ref. [2]. Let us denote the product space

$$\tilde{\mathbb{H}}(X) := \underbrace{\mathbb{H}(X) \times \cdots \times \mathbb{H}(X)}_{N \text{ times}} = \mathbb{H}(X)^N,$$

where $\mathbb{H}(X)$ is the collection of non-empty compact subsets of $X$, which is complete with respect to the Hausdorff metric

$$h(A, B) = \max\{\max_{a\in A}\min_{b\in B} d(a, b), \max_{b\in B}\min_{a\in A} d(a, b)\},\ \ A, B \in \mathbb{H}(X).$$

Define a metric $\tilde{h}$ on $\tilde{\mathbb{H}}(X)$ by

$$\tilde{h}(\mathbf{A}, \mathbf{B}) := \max_{k\in\mathbb{N}_N} h(A_k, B_k),\ \text{ where } \mathbf{A} = (A_1, \ldots, A_N), \mathbf{B} = (B_1, \ldots, B_N).$$

Then $\left(\tilde{\mathbb{H}}(X), \tilde{h}\right)$ is a complete metric space. Next, we define a map

$$\mathbf{W} : \tilde{\mathbb{H}}(X) \to \tilde{\mathbb{H}}(X)$$

for all $(A_1, \ldots, A_N) \in \tilde{\mathbb{H}}(X)$ by

$$\mathbf{W}(A_1, \ldots, A_N) = \left( \bigcup_{j \in \Lambda(1)} W_1(A_j), \ldots, \bigcup_{j \in \Lambda(N)} W_N(A_j) \right),$$

where $\Lambda(i) = \{j : c_{ij} = 1\} \neq \emptyset$ for all $i \in \mathbb{N}_N$. In addition, $\mathbf{W}$ is a contraction on $\left(\tilde{\mathbb{H}}(X), \tilde{h}\right)$. Hence, there is a unique $\mathbf{G} = (G_1, \ldots, G_N) \in \tilde{\mathbb{H}}(X)$ such that $\mathbf{W}(\mathbf{G}) = \mathbf{G}$, and $G_i = \bigcup_{j \in \Lambda(i)} W_i(G_j),\ i \in \mathbb{N}_N$. This $\mathbf{G}$ is referred to as recurrent fractal or attractor or invariant set of the RIFS.

**Remark 1** We often call $G = \bigcup_{i \in \mathbb{N}_N} G_i$ as the attractor of the RIFS.

As one can see, RIFS is an extension of IFS that gives more complex local self-similar sets. Therefore, by employing the notion of RIFS, we can construct a more general FIF known as recurrent FIF. Here, we give a brief construction of RFIF for the data set $\Delta_y$. For detailed information, see Refs. [2, 21].

## 2.2 Construction of RFIF

Let us denote intervals $I := [x_0, x_N]$, and for all $k \in \mathbb{N}_N$, $I_k := [x_{k-1}, x_k]$, and $J_k := [x_{l(k)}, x_{r(k)}]$, where $l(k), r(k) \in \mathbb{N}_N^0$ with $l(k) < r(k)$. Now, define homeomorphisms $L_k : J_k \to I_k$ by

$$L_k(x) = a_k x + b_k = \left( \frac{x_k - x_{k-1}}{x_{r(k)} - x_{l(k)}} \right) x + \left( \frac{x_{r(k)} x_{k-1} - x_{l(k)} x_k}{x_{r(k)} - x_{l(k)}} \right), \quad k \in \mathbb{N}_N. \quad (2)$$

Here, $L_k$ satisfies $|L_k(x) - L_k(x^*)| \leq |a_k||x - x^*|,\ x, x^* \in J_k$. In addition, if $0 < |a_k| < 1$, i.e., $|x_k - x_{k-1}| < |x_{r(k)} - x_{l(k)}|$, then $L_k$ becomes contraction. Alternatively, to make $L_k$ a contraction, we take $I_k$ and $J_k$ so that the length of $I_k$ is less than the length of $J_k$. Now, we define continuous maps $M_k : J_k \times \mathbb{R} \to \mathbb{R}$ by

$$M_k(x, y) = c_k x + d_k y + e_k = \left( \frac{y_k - y_{k-1}}{x_{r(k)} - x_{l(k)}} - d_k \frac{y_{r(k)} - y_{l(k)}}{x_{r(k)} - x_{l(k)}} \right) x + d_k y +$$
$$\left( \frac{x_{r(k)} y_{k-1} - x_{l(k)} y_k}{x_{r(k)} - x_{l(k)}} - d_k \frac{x_{r(k)} y_{l(k)} - x_{l(k)} y_{r(k)}}{x_{r(k)} - x_{l(k)}} \right), \quad k \in \mathbb{N}_N. \quad (3)$$

Here, $M_k$ satisfies $|M_k(x, y) - M_k(x, y^*)| \leq |d_k||y - y^*|,\ x \in J_k$ and $y, y^* \in \mathbb{R}$. For $M_k$ to be a contraction on $y$-variable, we choose $d_k$ such that $0 \leq |d_k| < 1$.

Next, for all $k \in \mathbb{N}_N$, we consider $W_k : J_k \times \mathbb{R} \to I_k \times \mathbb{R}$ by

$$W_k(x, y) = (L_k(x), M_k(x, y)).$$

It is easy to verify that $W_k(x_{l(k)}, y_{l(k)}) = (x_{k-1}, y_{k-1})$ and $W_k(x_{r(k)}, y_{r(k)}) = (x_k, y_k)$. In addition, $W_k$ is a contraction with respect to some metric, equivalent to the

Euclidean metric in $\mathbb{R}^2$ and hence $\{I \times \mathbb{R}; W_k : k \in \mathbb{N}_N\}$ forms a contractive IFS. Also, we define a row-stochastic matrix $P = (p_{ij})_{N\times N}$ by

$$p_{ij} = \begin{cases} \frac{1}{n_i} & \text{if } I_i \subset J_j, \\ 0 & \text{if } I_i \not\subset J_j, \end{cases}$$

where $n_i$ denotes the number of $j$ such that $I_i \subset J_j$ for $i \in \mathbb{N}_N$. By appropriately selecting $J_k$'s, we can assume that $P$ is irreducible. Therefore, we construct RIFS $\{I \times \mathbb{R}; P; W_k : k \in \mathbb{N}_N\}$ associated with $\Delta_y$. Here, $C = (c_{ij})_{N\times N}$ can be obtained by using (1) such that

$$c_{ij} = \begin{cases} 1 & \text{if } I_j \subset J_i, \\ 0 & \text{if } I_j \not\subset J_i. \end{cases} \tag{4}$$

Let $\mathcal{C}(I)$ be the collection of real-valued continuous functions defined on $I$. Let us define a metric $d_\infty$ on $\mathcal{C}(I)$ by $d_\infty(f, g) :=\| f - g \|_\infty= \sup_{x\in I} |f(x) - g(x)|$. Then $(\mathcal{C}(I), d_\infty)$ is a complete metric space. Also, define $\mathcal{C}^*(I) := \{f \in \mathcal{C}(I) : f(x_k) = y_k,\ k \in \mathbb{N}_N^0\}$. Then $\mathcal{C}^*(I)$ is a closed subset of $(\mathcal{C}(I), d_\infty)$, and thus a complete metric space. Define an operator $T : \mathcal{C}^*(I) \to \mathcal{C}^*(I)$ by

$$Tf(x) := M_k\left(L_k^{-1}(x), f\left(L_k^{-1}(x)\right)\right),\ x \in I_k \text{ and } k \in \mathbb{N}_N.$$

It is easy to verify that $T$ is a contraction on $(\mathcal{C}^*(I), d_\infty)$, and hence there is a unique $f_y \in \mathcal{C}^*(I)$ such that

$$f_y(x) = M_k\left(L_k^{-1}(x), f_y\left(L_k^{-1}(x)\right)\right),\ x \in I_k \text{ and } k \in \mathbb{N}_N. \tag{5}$$

Such an $f_y$ is called RFIF associated with $\Delta_y$. Let $G := \{(x, f_y(x)) : x \in I\}$, and $G_k := \{(x, f_y(x)) : x \in I_k\}$ for all $k \in \mathbb{N}_N$. Then $G = \bigcup_{k\in\mathbb{N}_N} G_k$. Moreover,

$$\begin{aligned} G_k &= \{(x, f_y(x)) : x \in I_k\} = \{(x, M_k\left(L_k^{-1}(x), f_y\left(L_k^{-1}(x)\right)\right)) : x \in I_k\} \\ &= \{(L_k(x), M_k\left(x, f_y(x)\right)) : x \in J_k\} = \{W_k\left(x, f_y(x)\right) : x \in J_k\} \\ &= \bigcup_{j\in\Lambda(k)} W_k\left(G_j\right). \end{aligned}$$

Thus, $\mathbf{G} = (G_1, G_2, \ldots, G_N)$ is an attractor of the RIFS $\{I \times \mathbb{R}; P; W_k : k \in \mathbb{N}_N\}$ associated with the data set $\Delta_y$.

In the next section, we will construct linear RFIF for Gaussian noisy data set and discuss parameter estimation of distribution of this RFIF.

## 3 RFIF for Gaussian Noisy Data

Let us add Gaussian noise $\epsilon$ on $y$-variable, that is, $Y := y + \epsilon$. Therefore, for each $k \in \mathbb{N}_N^0$, $Y_k := y_k + \epsilon_k$, where $\epsilon_k \sim \mathcal{N}(0, \sigma_k^2)$ is a Gaussian noise. Let $\Delta_Y := \{(x_k, Y_k) : k \in \mathbb{N}_N^0\}$ be a data set with Gaussian noise. If we assume $\epsilon_k$'s are independent, then $Y_k$'s are as well. Here $Y_k \sim \mathcal{N}(y_k, \sigma_k^2)$. For $k \in \mathbb{N}_N$, we define $\mathcal{M}_k : J_k \times \mathbb{R} \to \mathbb{R}$ (a random analog of $M_k$) by $\mathcal{M}_k(x, Y) = C_k x + d_k Y + E_k$, where

$$\begin{aligned} C_k &= \frac{Y_k - Y_{k-1}}{x_{r(k)} - x_{l(k)}} - d_k \frac{Y_{r(k)} - Y_{l(k)}}{x_{r(k)} - x_{l(k)}}, \\ E_k &= \frac{x_{r(k)} Y_{k-1} - x_{l(k)} Y_k}{x_{r(k)} - x_{l(k)}} - d_k \frac{x_{r(k)} Y_{l(k)} - x_{l(k)} Y_{r(k)}}{x_{r(k)} - x_{l(k)}}. \end{aligned} \tag{6}$$

Define $\mathcal{W}_k : J_k \times \mathbb{R} \to I_k \times \mathbb{R}$ by $\mathcal{W}_k(x, Y) = (L_k(x), \mathcal{M}_k(x, Y))$, $k \in \mathbb{N}_N$ and construct RIFS $\{I \times \mathbb{R}; P; \mathcal{W}_k : k \in \mathbb{N}_N\}$ corresponds to the data set $\Delta_Y$, which is a random analog to the RIFS $\{I \times \mathbb{R}; P; W_k : k \in \mathbb{N}_N\}$ associated with $\Delta_y$. There is a unique RFIF $f_Y : I \to \mathbb{R}$ (a random analog of $f_y$ in (5)) such that

$$f_Y(x) = \mathcal{M}_k\left(L_k^{-1}(x), f_Y\left(L_k^{-1}(x)\right)\right), \quad x \in I_k, k \in \mathbb{N}_N. \tag{7}$$

To estimate the parameters of the probability distribution of RFIF $f_Y(x)$, we first have to write (7) in the explicit form.

### 3.1 *Explicit Expression of the RFIF*

It can be easily observed that the RIFS $\{I \times \mathbb{R}; P; \mathcal{W}_k : k \in \mathbb{N}_N\}$ is irreducible. Therefore, we can utilize all the maps $\mathcal{W}_k$'s (or $L_k$'s) of the RIFS over a sufficiently long period of time.

Let $t \in I_i$ be a given point. Then for any $x \in I_j$, we have a sequence $\{k_n\}_{n\in\mathbb{N}}$ in $\mathbb{N}_N$ such that $L_{k_1 k_2 \ldots k_n}(x) \to t$ as $n \to \infty$, where $L_{k_1 k_2 \ldots k_n}(x) = L_{k_1} \circ L_{k_2} \circ \cdots \circ L_{k_n}(x)$. In general, let $t \in I$, then there exists a sequence $\{k_n\}_{n\in\mathbb{N}}$, $k_n \in \mathbb{N}_N$, such that

$$\lim_{n\to\infty} L_{k_1 k_2 \ldots k_n}(x) = t, \text{ for any } x \in I. \tag{8}$$

As we know that $I_k = L_k(J_k) = \bigcup_{j\in\Lambda(k)} L_k(I_j)$. Therefore, $I = \bigcup_{k=1}^N I_k$ is the attractor of RIFS $\{I; C; L_k : k \in \mathbb{N}_N\}$. Hence (8) is verified. Now, by using induction in (7), we can easily get the following expression (for detailed information, see Ref. [9]).

$$\begin{aligned} &f_Y\left(L_{k_1 k_2 \ldots k_n}(x)\right) \\ &\quad = \left[C_{k_1} L_{k_2 k_3 \ldots k_n}(x) + d_{k_1} C_{k_2} L_{k_3 k_4 \ldots k_n}(x) + \cdots + d_{k_1} \ldots d_{k_{n-1}} C_{k_n} x\right] + \\ &d_{k_1} \ldots d_{k_n} f_Y(x) + \left[E_{k_1} + d_{k_1} E_{k_2} + \cdots + d_{k_1} \ldots d_{k_{n-1}} E_{k_n}\right], \quad x \in J_{k_n} \subset I. \end{aligned} \tag{9}$$

Define left shift operator $S(k_1k_2\dots k_n) := k_2k_3\dots k_n$ and its $j$-fold self-composition by $S^j(k_1k_2\dots k_n) := k_{j+1}k_{j+2}\dots k_n$ for $j \in \mathbb{N}_{n-1}$. Also, for simplicity of notation, we denote $D_j = \prod_{i=1}^{j} d_{k_i}$ and $t_j(x) = L_{S^j(k_1\dots k_n)}(x)$, for $j \in \mathbb{N}_n$. Thus (9) becomes

$$f_Y(L_{k_1\dots k_n}(x)) = D_n f_Y(x) + \sum_{j=1}^{n} D_{j-1}\left(C_{k_j}t_j(x) + E_{k_j}\right),\ x \in J_{k_n} \subset I. \quad (10)$$

Here we agreed that $L_{S^n(k_1\dots k_n)}(x) = x$ and $D_0 = \prod_{i=1}^{0} d_{k_i} = 1$. Since $f_Y$ is a continuous function and $\lim_{n\to\infty} D_n = 0$, therefore, as $n \to \infty$, we have from (8) and (10) that

$$f_Y(t) = \sum_{j=1}^{\infty} D_{j-1}\left(C_{k_j}t_j(x) + E_{k_j}\right),\ x \in I. \quad (11)$$

From (6), we get that

$$C_k z + E_k = \left(\frac{x_{r(k)} - z}{x_{r(k)} - x_{l(k)}}\right) Y_{k-1} + \left(\frac{z - x_{l(k)}}{x_{r(k)} - x_{l(k)}}\right) Y_k - d_k\left[\left(\frac{x_{r(k)} - z}{x_{r(k)} - x_{l(k)}}\right) Y_{l(k)} + \left(\frac{z - x_{l(k)}}{x_{r(k)} - x_{l(k)}}\right) Y_{r(k)}\right]. \quad (12)$$

Using (12) in (11), we get

$$\begin{aligned} f_Y(t) = &\sum_{j=1}^{\infty}\left[D_{j-1}\left(\frac{x_{r(k_j)} - t_j(x)}{x_{r(k_j)} - x_{l(k_j)}}\right)\right] Y_{k_j-1} \\ &+ \sum_{j=1}^{\infty}\left[D_{j-1}\left(\frac{t_j(x) - x_{l(k_j)}}{x_{r(k_j)} - x_{l(k_j)}}\right)\right] Y_{k_j} \\ &- \sum_{j=1}^{\infty}\left[D_j\left(\frac{x_{r(k_j)} - t_j(x)}{x_{r(k_j)} - x_{l(k_j)}}\right)\right] Y_{l(k_j)} \\ &- \sum_{j=1}^{\infty}\left[D_j\left(\frac{t_j(x) - x_{l(k_j)}}{x_{r(k_j)} - x_{l(k_j)}}\right)\right] Y_{r(k_j)}. \end{aligned} \quad (13)$$

As we know that $Y_{k_j-1}, Y_{k_j}, Y_{l(k_j)}, Y_{r(k_j)} \in \{Y_0, Y_1, \dots, Y_N\}$, since each $k_j \in \mathbb{N}_N$. So, we can separate the coefficients of $Y_i$ in (13) as follows

$$f_Y(t) = \sum_{i=0}^{N} \omega_i Y_i,\ t \in I, \quad (14)$$

where $\omega_i$ depends on the values of the sequence $\{k_j\}$ of $t$. As we can see from (14), for each $t \in I$, the RFIF $f_Y(t)$ is a random variable because $Y_i$'s are random variables.

### 3.2 Distribution of RFIF

It is known that the distribution of a linear combination of independent Gaussian random variables is Gaussian. Therefore, for each $t \in I$, we see from (14) that $f_Y(t)$ is a Gaussian random variable and its parameters are estimated as follows. Mean of $f_Y(t)$ is

$$\mathbb{E}\left[f_Y(t)\right] = \sum_{i=0}^{N} \omega_i \mathbb{E}[Y_i] = \sum_{i=0}^{N} \omega_i y_i.$$

If we assume that $y_k$ is a realization of $Y_k$, i.e., $\Delta_y$ is a realization of $\Delta_Y$, then from (14), we get that $f_y(t) = \sum_{i=0}^{N} \omega_i y_i$. Hence, $\mathbb{E}\left[f_Y(t)\right] = f_y(t)$. Variance of $f_Y(t)$ is

$$\mathbf{Var}\left[f_Y(t)\right] = \sum_{i=0}^{N} \omega_i^2 \mathbf{Var}[Y_i^2] = \sum_{i=0}^{N} \omega_i^2 \sigma_i^2.$$

Therefore, the probability distribution of $f_Y(t)$ is given by

$$f_Y(t) \sim \mathcal{N}\left(f_y(t), \sum_{i=0}^{N} \omega_i^2 \sigma_i^2\right). \tag{15}$$

## 4 Simulation

Let us consider the observed data set

$$\Delta_y = \{(0.5,\ 3.5), (1.3,\ 2.3), (1.9,\ 4.6), (2.4,\ 7.5), (3.5,\ 3.8), (4.3,\ 5.7), (5.2,\ 3.8)\}.$$

We can see from the above data set that $x_0 = 0.5$, $x_1 = 1.3$, $x_2 = 1.9$, $x_3 = 2.4$, $x_4 = 3.5$, $x_5 = 4.3$, $x_6 = 5.2$ and $y_0 = 3.5$, $y_1 = 2.3$, $y_2 = 4.6$, $y_3 = 7.5$, $y_4 = 3.8$, $y_5 = 5.7$, $y_6 = 3.8$. Therefore, $N = 6$, $I = [0.5,\ 5.2]$, and $I_1 = [0.5,\ 1.3]$, $I_2 = [1.3,\ 1.9]$, $I_3 = [1.9,\ 2.4]$, $I_4 = [2.4,\ 3.5]$, $I_5 = [3.5,\ 4.3]$, $I_6 = [4.3,\ 5.2]$. Let us consider $J_1 = J_2 = [1.9,\ 3.5]$, $J_3 = J_4 = [3.5,\ 5.2]$, $J_5 = J_6 = [0.5,\ 1.9]$ such that the row stochastic matrix $P$ is irreducible. Using (4), the recurrent structure is given by the connection matrix

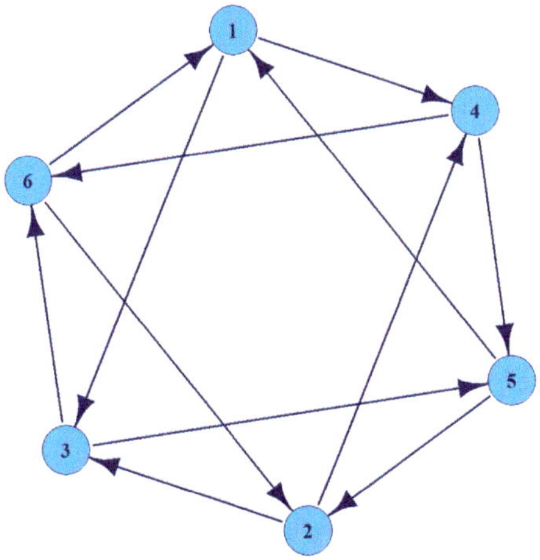


**Fig. 1** Directed graph of $C$

$$C = \begin{pmatrix} 0 & 0 & 1 & 1 & 0 & 0 \\ 0 & 0 & 1 & 1 & 0 & 0 \\ 1 & 1 & 0 & 0 & 0 & 0 \\ 1 & 1 & 0 & 0 & 0 & 0 \\ 0 & 0 & 0 & 0 & 1 & 1 \\ 0 & 0 & 0 & 0 & 1 & 1 \end{pmatrix}.$$

The irreducibility of $C$ can be confirmed by using its directed graph, as shown in Fig. 1, which is strongly connected and hence $C$ is irreducible.

Now, using (2), we can compute the parameters $a_k$ and $b_k$ of $L_k$ as given below.

$$(a_1, \dots, a_6) = (0.500,\ 0.375,\ 0.294,\ 0.647,\ 0.571,\ 0.643),$$
$$(b_1, \dots, b_6) = (-0.450,\ 0.587,\ 0.871,\ 0.135,\ 3.214,\ 3.979).$$

If we take vertical scaling vector $(d_1, \dots, d_6) = (0.3,\ 0.5, -0.3,\ 0.6, -0.5,\ 0.4)$. Then, by using (3), we can compute the parameters $c_k$ and $e_k$ of $M_k$ as given below.

$$(c_1, \dots, c_6) = (-0.600,\ 1.687,\ 1.706, -2.176,\ 1.750, -1.671),$$
$$(e_1, \dots, e_6) = (3.260, -3.206, -0.231,\ 12.838,\ 4.675,\ 5.136).$$

We may compute the values of RFIF $f_y$ using (5). The graph of $f_y$ (in blue color curve) is shown in Fig. 2. Here green color dots represent given initial data points $(x_k, y_k)$ of $\Delta_y$.

Now we add Gaussian noise on $y$-values of the initial data set $\Delta_y$, that is, for $k \in \mathbb{N}_6^0$, $Y_k = y_k + \epsilon_k$, where $\epsilon_k \sim \mathcal{N}(0, \sigma_k^2)$ with $(\sigma_0, \sigma_1, \dots, \sigma_6) = (1.8,\ 2.1,\ 1.7,\ 3.1,\ 2.2,\ 3.0,\ 1.9)$. Let us arbitrarily select a point $t = 3.133$ in the interval $I$. If we choose

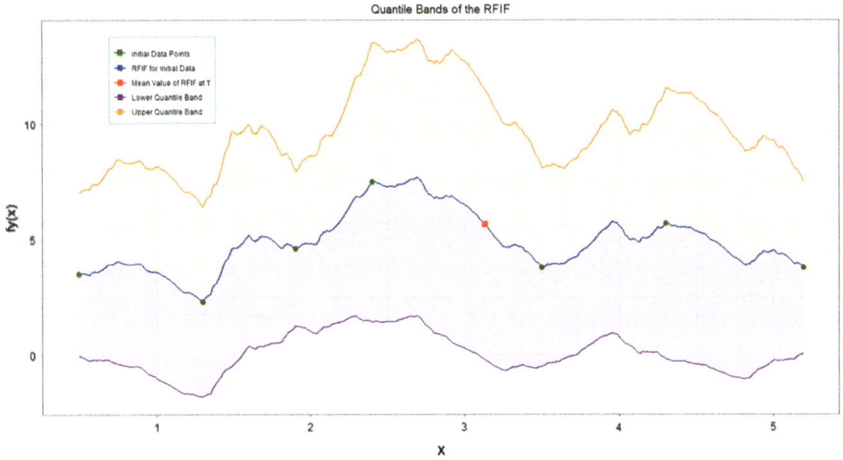


**Fig. 2** 95% quantile band of the RFIF $f_Y$

$x = 4.3$ in (8) with maximum tolerance error 0.001 then, we get a sequence (also known as code space) $\{4, 6, 1, 4, 6, 1, 3, 5, 1, 4\}$ of $t$. Therefore, parameters $\omega_i$ of (14) can be computed as follows:

$$(\omega_0, \omega_1, \ldots, \omega_6) = (-0.0672, 0.1567, -0.1157, 0.3716, 0.4324, 0.4196, -0.1975).$$

Hence, by using (15), we can estimate the distribution of RFIF $f_Y$ at the given point $t = 3.133$.

$$f_Y(3.133) \sim \mathcal{N}(5.6646, 4.1186). \tag{16}$$

The mean value of $f_Y(3.133)$, i.e. $\mathbb{E}[f_Y(3.133)] = f_y(3.133) = 5.6646$ is represented in Fig. 2 by a red color square. Moreover, blue curve in the same figure depicts the expected value of the RFIF $f_Y$. Further, for the given noisy data set $\Delta_Y$, we can obtain 95% quantile bands of the RFIF $f_Y$, which are shown in Fig. 2. This quantile band indicates that RFIF $f_Y$ will lie between the upper and lower quantile curves with a probability of 0.95.

Now, let us take 5000 random samples of $\Delta_Y$, which are deterministic data sets. For each observed sample we get a RFIF by using the same technique as we did for the initial data set $\Delta_y$. Therefore, we have 5000 random samples of RFIF $f_Y$ and hence we obtain 5000 observations of $f_Y(3.133)$. The histogram of these random observations of $f_Y(3.133)$ is shown in Fig. 3(i). In the same figure, we have fitted empirical probability density function (PDF) and the estimated Gaussian PDF of $f_Y(3.133)$ given in (16). As we can see that the estimated Gaussian PDF is very close to the empirical PDF, which implies that we have estimated the probability distribution of $f_Y(3.133)$ correctly. The same observation can also be seen from Fig. 3(ii), i.e., the estimated Gaussian cumulative distribution function (CDF) of $f_Y(3.133)$ is almost

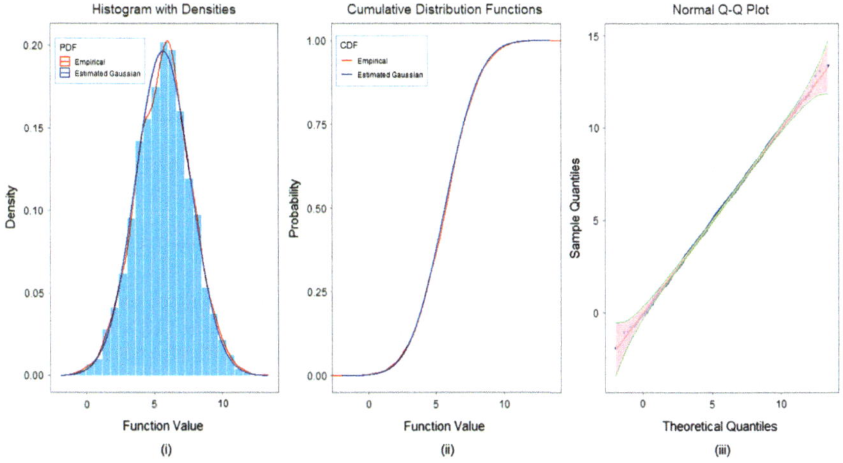


**Fig. 3** Histogram with empirical and estimated gaussian PDFs, empirical and estimated gaussian CDFs and normal Q-Q plot with 95% confidence bands for 5000 samples of $f_Y(3.133)$

similar to the empirical CDF of the observed samples of $f_Y(3.133)$. Normal quantile-quantile plot in Fig. 3(iii) depicts that random samples of $f_Y(3.133)$ are almost in a straight line, which implies that these samples are drawn from a normal or Gaussian population. The 95% confidence bands are also represented in the same figure, which shows that almost all the sample points lie between these lower and upper bands.

We can observe from the above simulation methods that the distribution of $f_Y(3.133)$ in (16) is valid. As we have chosen $t = 3.133$ arbitrarily in our simulation, therefore, we can say in general that for any given $t \in I$, RFIF $f_Y(t)$ is a Gaussian random variable, whose parameters are provided in (15) and may be estimated in the same manner as we did for $t = 3.133$.

## 5 Conclusion

For a data set containing Gaussian noise on the ordinate, the probability distribution of a linear recurrent fractal interpolation function at any given point is Gaussian. Therefore, if a data set includes Gaussian noise and is derived from a process that has fractal characteristics, then we can easily identify the variability at any intermediate point of the provided Gaussian noisy data set.

**Acknowledgements** The authors are thankful for the funding by Indian Institute of Technology Madras under the MHRD Project No.: SB20210848MAMHRD008558.

## References


1. Barnsley, M.F.: Fractal functions and interpolation. Constr. Approx. **2**(1), 303–329 (1986)
2. Barnsley, M.F., Elton, J.H., Hardin, D.P.: Recurrent iterated function systems. Constr. approx. **5**(1), 3–31 (1989)
3. Navascués, M.A.: Fractal polynomial interpolation. Z. Anal. Anwend. **24**(2), 401–418 (2005)
4. Massopust, P.R.: Vector-valued fractal interpolation functions and their box dimension. Aequat. Math. **42**(1), 1–22 (1991)
5. Massopust, P.R.: Local fractal interpolation on unbounded domains. Proc. Edinb. Math. Soc. **61**(1), 151–167 (2018)
6. Buzogány, E., Kolumbán, J., Soós, A.: Random fractal interpolation function using contraction method in probabilistic metric spaces. An. Univ. Bucuresti Mat. Inform. **51**(1), 13–24 (2002)
7. Luor, D.C.: Statistical properties of linear fractal interpolation functions for random data sets. Fractals **26**(1), 1–6 (2018)
8. Luor, D.C.: Fractal interpolation functions for random data sets. Chaos, Solitons Fractals **114**, 256–263 (2018)
9. Kumar, M., Upadhye, N.S., Chand, A.K.B.: Distribution of linear fractal interpolation function for random dataset with stable noise. Fractals **29**(4), 1–12 (2021)
10. Chand, A.K.B., Kapoor, G.P.: Generalized cubic spline fractal interpolation functions. SIAM J. Numer. Anal. **44**(2), 655–676 (2006)
11. Chand, A.K.B., Viswanathan, P.: A constructive approach to cubic Hermite fractal interpolation function and its constrained aspects. BIT Numer. Math. **53**, 841–865 (2013)
12. Navascués, M.A., Sebastián, M.V.: Smooth fractal interpolation. J. Inequal. Appl. **78734**, 1–20 (2006)
13. Chand, A.K.B., Kapoor, G.P.: Stability of affine coalescence hidden variable fractal interpolation functions. Nonlinear Anal. Theory Methods Appl. **68**(12), 3757–3770 (2008)
14. Viswanathan, P., Chand, A.K.B., Agarwal, R.P.: Preserving convexity through rational cubic spline fractal interpolation function. J. Comput. Appl. Math. **263**, 262–276 (2014)
15. Chand, A.K.B., Vijender, N., Agarwal, R.P.: Rational iterated function system for positive/monotonic shape preservation. Adv. Differ. Equ. **2014**(30), 1–19 (2014)
16. Chand, A.K.B., Vijender, N., Navascués, M.A.: Shape preservation of scientific data through rational fractal splines. Calcolo **51**, 329–362 (2014)
17. Banerjee, S., Easwaramoorthy, D., Gowrisankar, A.: Fractal Functions Dimensions and Signal Analysis. Springer International Publishing (2021)
18. Păcurar, C.M., Necula, B.R.: An analysis of COVID-19 spread based on fractal interpolation and fractal dimension. Chaos, Solitons Fractals **139**, 1–8 (2020)
19. Bajahzar, A., Guedri, H.: Reconstruction of fingerprint shape using fractal interpolation. Int. J. Adv. Comput. Sci. Appl. **10**(5), 103–114 (2019)
20. Barnsley, M.F., Demko, S.: Iterated function systems and the global construction of fractals. Proc. R. Soc. Lond. Ser. A **399**(1817), 243–275 (1985)
21. Barnsley, M.F.: Fractals Everywhere. 2nd edn., Morgan Kaufmann (2000)

# $C^1$-Rational Quadratic Trigonometric Spline Fractal Interpolation Functions

**Vijay and A. K. B. Chand**

**Abstract** Trigonometric interpolation has an essential role in geometric modeling of conic data. In this paper, a novel $C^1$-rational quadratic trigonometric spline fractal interpolation function with variable scaling and two families of shape parameters is introduced. We have investigated the convergence analysis of this fractal interpolant to a data-generating function in $C^3$ from the uniform error bound. When the conic data is positive and monotone, we have derived sufficient conditions based on the scaling functions and shape parameters so that the resultant trigonometric spline FIF preserves these fundamental shapes of conic data. The proposed results are verified by generating positive and monotonic trigonometric spline fractal interpolation functions.

**Keywords** Fractals · Iterated function system · Rational quadratic trigonometric fractal function · Positivity · Monotonicity

**AMS Classifications** 26A48 · 26C15 · 28A80 · 41A05 · 41A25 · 65D10

## 1 Introduction

Geometric modeling with splines has been a vital and fascinating area of research for the last six decades with applications ranging from animated films to simulated surgery. It attracts experts from numerical analysis, approximation theory, wavelets, classical and discrete geometry, engineering design, civil engineering, and computer science. This field is very active due to the continuous need for new techniques based on assumptions of data generating function and the nature of data. Researchers have tried various methods to tackle this problem with interpolating polynomials, splines, trigonometric splines, exponential splines, rational splines, etc. All these

Vijay (✉) · A. K. B. Chand
Department of Mathematics, Indian Institute of Technology Madras, Chennai 600036, India
e-mail: vijaysiwach975@gmail.com

A. K. B. Chand
e-mail: chand@iitm.ac.in



B. Rushi Kumar et al. (eds.), *Mathematics and Computing*, Springer Proceedings in Mathematics & Statistics 415, https://doi.org/10.1007/978-981-19-9307-7_20

non-recursive classical interpolants are either smooth or piecewise smooth and consequently, they can have non-differentiability at a few points. But if the data is generated from an irregular and non-differentiable function, it is not ideal to approximate them by these classical interpolants.

To unify the irregular objects and complex structures in nature and scientific phenomena, Mandelbrot [16] coined the term fractal in the literature. Fractal-based theory provides a robust framework to describe and analyze self-similar and scale-invariant patterns. Fractal-generating systems can model most of these complex phenomena by using simple self-referential rules with few parameters. Hutchinson [13] proposed the iterated function system (IFS) in 1981 to generate fractals through a common platform. Based on the structure of IFS, fractal interpolation functions (FIFs) were constructed by Barnsley [2] to fit non-smooth and irregular curves such as peaks of clouds, stalactite dangled roofs of caves, lightening, ECG curves, turbulence, profiles of mountain ranges, etc., from their data points [3]. But classical non-recursive interpolants and non-smooth fractal interpolants are not good enough to interpolate functions that have nowhere differentiability in their higher order derivatives. Using boundary conditions of fixed type, Barnsley and Harrington [4] developed the theory of $r$-times differentiable polynomial spline. Cubic spline FIFs with any type of boundary conditions were proposed in an elegant constructive way using moments and derivatives in [6, 9] respectively. Thus, fractal interpolation technique provides the flexibility of preferring a smooth or irregular model depending on the nature of the problem at hand. Other than modeling data with interpolant having a required degree of smoothness, preserving fundamental shapes like positivity, monotonicity, and convexity of data are crucial in data visualization. Chand and group proposed shape preserving interpolation and approximation using FIFs, see [7, 8, 15, 17–19].

The applications of trigonometric functions are familiar in the fields of medicines, harmonic motions, electronics, and automobile industries. Trigonometric interpolation functions are effective if our data is generated from a conic function. Several researchers have worked on shape-preserving trigonometric splines. Abbas [1] proposed a rational cubic trigonometric spline for the positivity-preserving feature of a prescribed positive data set. Han investigated quadratic and cubic trigonometric interpolating polynomials with shape parameters analogous to the quadratic and cubic B-spline curves, respectively [10, 11]. Ibraheem [14] constructed a $C^1$-rational cubic trigonometric spline to visualize positive data set. Bashir [5] proposed another cubic rational trigonometric spline to retain positivity, monotonicity, and constrained aspects of the given data set. Wang and Fan [20] introduced FIFs with variable scaling functions to approximate sophisticated curves with less self-similarity. Wang and Shan [21] investigated on smoothness, sensitivity, and stability of FIF with variable scaling functions. Smooth FIFs with variable scaling functions have been constructed in [19]. In this work, we have presented a novel $C^1$-rational trigonometric fractal interpolation function using variable scaling functions and studied its shape-preserving aspects.

A short description of our work is as follows. We construct a class of novel $C^1$-rational quadratic trigonometric spline (RQTS) fractal interpolation functions using variable scaling functions and two groups of shape parameters in Sect. 2 based on the theory of IFS and fractal function with variable scaling. In Sect. 3, we deduce the convergence analysis for our constructed fractal interpolant to a data-generating smooth function in $C^3$. In Sect. 4, we give bounds for the norm of variable scaling functions and shape parameters to obtain strictly positive RQTS FIFs for a strict positive data set. We obtain the bounds for shape parameters and scaling functions to get positive and monotone RQTS FIF for a positive and monotonic data in Sect. 5. Numerical examples of shape-preserving RQTS FIFs are given to support our theory inside Sects. 4 and 5. Conclusions of our work for this paper are summarized in Sect. 6.

## 2 Preliminaries and Construction of RQTS FIFs

Let us fix some notation for this paper: $I$ is denoted as a compact interval of $\mathbb{R}$. For $k \in \mathbb{N} \cup \{0\}$, $C^k(I)$ is the set of all $k$-times continuously differentiable real valued functions defined on $I$, and for $g \in C^k(I)$, $\|g\|_k = \max\{\|g^{(r)}\|_\infty r = 0, 1, 2, \ldots, k\}$. For any $j \in \mathbb{N}$, let $\mathbb{N}_j = \{1, 2, 3, \ldots, j\}$, and $\mathbb{N}_j^0 := \{0, 1, 2, 3, \ldots, j\}$.

Let $(\mathcal{X}, d_{\mathcal{X}})$ be a complete metric space. Consider a finite number of continuous functions $W_i : \mathcal{X} \to \mathcal{X}, i \in \mathbb{N}_{N-1}$. Then $\mathcal{I} := \{\mathcal{X}; W_i, i \in \mathbb{N}_{N-1}\}$ is called an IFS. If each $W_i$, $i \in \mathbb{N}_{N-1}$ is a contraction map with contractive factor $\alpha_i$, then $\mathcal{I}$ is known as an hyperbolic IFS. Let $\mathcal{H}(\mathcal{X})$ be a set of all compact subsets of $\mathcal{X}$ other than empty set. The Hausdorff metric $d_{\mathcal{H}(\mathcal{X})}$ on $\mathcal{H}(\mathcal{X})$ is defined by $d_{\mathcal{H}(\mathcal{X})}(A, B) = \max\{\mathcal{D}_B(A), \mathcal{D}_A(B)\}$, where $\mathcal{D}_B(A) = \max_{a \in A} \min_{b \in B} d_{\mathcal{X}}(a, b)$. From [3], it can be deduced that $\mathcal{H}(\mathcal{X})$ with Hausdorff metric is a complete metric space. Associated with the IFS $\mathcal{I}$, a Hutchinson map $W$ on $\mathcal{H}(\mathcal{X})$ is defined by $W(\mathcal{A}) = \bigcup_{i=1}^{N-1} W_i(\mathcal{A}), \forall \mathcal{A} \in \mathcal{H}(\mathcal{X})$. If our IFS $\mathcal{I}$ is *hyperbolic*, then easily we can prove that $W$ on $\mathcal{H}(\mathcal{X})$ is a contraction map with contractive factor $|\alpha|_\infty = \max\{|\alpha_i| : i \in \mathbb{N}_{N-1}\}$ [3]. Therefore, using the Banach fixed point theorem, the Hutchinson map defined above has a unique fixed point (say) $G$ such that for any initiator $\mathcal{A} \in \mathcal{H}(\mathcal{X})$, $\lim_{m \to \infty} W^{o(m)}(\mathcal{A}) = G$, where the limit is taken using Hausdorff metric. This fixed point $G$ is called the attractor or self-referential set or deterministic fractal corresponding to the IFS $\mathcal{I}$.

Let a set of interpolation points $\{(x_i, y_i) \in I \times \mathbb{R} : i \in \mathbb{N}_N\}$ be given, where $x_1 < x_2 < \cdots < x_N$ is a partition of $I = [x_1, x_N]$, and $\forall i \in \mathbb{N}_N$, $y_i \in [k_1, k_2] \subset \mathbb{R}$. Let $I_i := [x_i, x_{i+1}]$ and $F := I \times [k_1, k_2]$. Let $U_i : I \to I_i, i = 1, 2, \ldots, N-1$, be contractive homeomorphisms such that

$$\begin{aligned} &U_i(x_1) = x_i, \quad U_i(x_N) = x_{i+1}, \\ &|U_i(\mu_*) - U_i(\mu^*)| \le r|\mu_* - \mu^*|, \quad \forall \mu_*, \mu^* \in I, \end{aligned} \tag{2.1}$$

for some $0 \le r < 1$. For the values of ordinates, consider $N-1$ continuous functions $V_i : F \to \mathbb{R}$ satisfying

$$\begin{aligned} &V_i(x_1, y_1) = y_i, \quad V_i(x_N, y_N) = y_{i+1}, \\ &|V_i(\mu, \omega_*) - V_i(\mu, \omega^*)| \leq |\alpha_i||\omega_* - \omega^*|, \quad \forall \mu \in I, \ \omega_*, \omega^* \in [k_1, k_2], \end{aligned} \tag{2.2}$$

for some $-1 < \alpha_i < 1$, $i \in \mathbb{N}_{N-1}$. Now define mappings $W_i : F \to I_i \times \mathbb{R}$, $i = 1, 2, \dots, N-1$ by

$$W_i(x, y) = (U_i(x), V_i(x, y)), \forall (x, y) \in F.$$

Therefore, $\{F;\ W_i : i \in \mathbb{N}_{N-1}\}$ is an IFS for the data set $\{(x_i, y_i) : i \in \mathbb{N}_N\}$. For this IFS, the following vital result has been proved by Barnsley [2]:

**Theorem 1** *The IFS $\{F;\ W_i : i \in \mathbb{N}_{N-1}\}$ has a unique attractor $G$ which is the graph of a continuous function $\Upsilon : I \to \mathbb{R}$ satisfying $\Upsilon(x_i) = y_i$, $\forall i = 1, 2, \dots, N$. Additionally, if $C^*(I) := \{g \in C(I) : g(x_1) = y_1, g(x_N) = y_N\}$ is endowed with uniform metric and $T_\alpha : C^*(I) \to C^*(I)$ the Read-Bajraktarević (RB) operator is defined by $T_\alpha g(x) = V_i(U_i^{-1}(x), g(U_i^{-1}(x)))$, $x \in I_i$, $i \in \mathbb{N}_{N-1}$, then the function $\Upsilon$ is the unique fixed point possesses by $T_\alpha$.*

The above fixed point $\Upsilon$ of $T_\alpha$ is known as a fractal interpolation function which satisfies the following functional relation:

$$\Upsilon(x) = V_i(U_i^{-1}(x), \Upsilon(U_i^{-1}(x))), \quad \forall x \in I_i, \quad \forall i \in \mathbb{N}_{N-1}. \tag{2.3}$$

Most FIFs constructed for science and engineering problems are given by the maps

$$U_i(x) = a_i x + b_i, \quad V_i(x, y) = \alpha_i y + q_i(x), \quad i \in \mathbb{N}_{N-1}, \tag{2.4}$$

where $a_i$ and $b_i$ can be evaluated from (2.1), the free parameter $-1 < \alpha_i < 1$ is called vertical scaling factor of the map $W_i$, and $q_i \in C(I)$ such that the condition (2.2) holds. An IFS with variable scaling functions was presented in [20] by Wang and Fan using the iterated mappings

$$U_i(x) = a_i x + b_i, \quad V_i(x, y) = \alpha_i(x) y + q_i(x), \quad i \in \mathbb{N}_{N-1}, \tag{2.5}$$

where $\alpha_i(x)$ on $I$ is a Lipschitz function such that $\|\alpha_i\|_\infty = \sup\{|\alpha_i(x)| : x \in I\} < 1$. Assume that for $i \in \mathbb{N}_{N-1}$, the above maps $U_i(x)$ and $V_i(x, y)$ are satisfying (2.1)–(2.2). Then According to Theorem 1, the IFS $\{F\ (U_i(x), V_i(x, y)) : i \in \mathbb{N}_{N-1}\}$ defines a FIF $\Upsilon : I \to \mathbb{R}$ with variable scaling functions that satisfies the following recursive relation:

$$\Upsilon(x) = \alpha_i(U_i^{-1}(x))\Upsilon(L_i^{-1}(x)) + q_i(U_i^{-1}(x)), \quad \forall x \in I_i, \quad \forall i \in \mathbb{N}_{N-1}. \tag{2.6}$$

Barnsley–Harrington [4] developed conditions on parameters $\alpha_i$ and $q_i \in C^k(I)$ such that the IFS $\{I \times \mathbb{R};\ (U_i(x), V_i(x, y)) : I \in \mathbb{N}_{N-1}\}$ in (2.4) determines a FIF $\Upsilon \in C^k(I)$. This theorem has been extended in [19] to a FIF with variable scaling functions by developing conditions on $\alpha_i \in C^k(I)$ and $q_i \in C^k(I)$ such that the IFS

$\{I \times \mathbb{R};\ (U_i(x), V_i(x,y)) : i \in \mathbb{N}_{N-1}\}$ determines a FIF $\Upsilon \in C^k(I)$, where $U_i$ and $V_i$ are as defined in (2.5). Furthermore, for $n \in \mathbb{N}_k^0$, $\Upsilon$ satisfies

$$\Upsilon^{(n)}(U_i(x)) = a_i^{-n}\Big[\sum_{j=0}^{n}\binom{n}{j}\alpha_i^{(n-j)}(x)\Upsilon^{(j)}(x) + q_i^{(n)}(x)\Big],\quad i \in \mathbb{N}_{N-1}. \tag{2.7}$$

Based on the above theory, we construct a novel continuously differentiable RQTS FIF containing two shape parameters and a variable scaling function in each subinterval.

**Theorem 2** *Let $\{(x_i, y_i, d_i) \in \mathbb{R}^3 : i \in \mathbb{N}_N\}$ be a given Hermite data set with $x_1 < x_2 < \cdots < x_N$. Construct a rational IFS $\Omega = \{I \times \mathbb{R};\ (U_i(x), V_i(x,y)) : i \in \mathbb{N}_{N-1}\}$, where $U_i(x) = a_i x + b_i$, and $V_i(x,y) = \alpha_i(x)y + q_i(x)$ satisfying (2.1)–(2.2), and for each $i \in \mathbb{N}_{N-1}$, $\alpha_i \in C^1(I)$ satisfying $\|\alpha_i\|_1 < \frac{a_i}{2}$ and $q_i \in C^1(I)$ is of the form $q_i(x) = \frac{P_i^*(x)}{R_i^*(x)}$, where $P_i^*(x)$ is a quadratic trigonometric polynomial and $R_i^*(x) \neq 0$ (for all $x \in I$) is a preassigned quadratic trigonometric polynomial. Then, for a fixed choice of the rational IFS parameters, there exists a unique $C^1$-RQTS FIF $\Upsilon$ which satisfies $\Upsilon(x_i) = y_i$, $\Upsilon^{(1)}(x_i) = d_i$, $\forall i \in \mathbb{N}_N$.*

***Proof*** For $0 \le \theta = \frac{\pi}{2}\frac{x-x_1}{x_N-x_1} \le \frac{\pi}{2}$, let

$$q_i(x) = \frac{P_i^*(x)}{R_i^*(x)} = \frac{P_i(\theta)}{R_i(\theta)},\quad i \in \mathbb{N}_{N-1}, x \in I,$$

where $P_i^*(x) = P_i^*(x_1 + \frac{2}{\pi}\theta(x_N - x_1)) = P_i(\theta) = A_{i1}(1-\sin(\theta))^2 + A_{i2}(1-\sin(\theta))\sin(\theta) + A_{i3}(1-\cos(\theta))\cos(\theta) + A_{i4}(1-\cos(\theta))^2$, and $R_i^*(x) = R_i^*(x_1 + \frac{2}{\pi}\theta(x_N - x_1)) = R_i(\theta) = (1-\sin(\theta))^2 + \eta_i(1-\sin(\theta))\sin(\theta) + \beta_i(1-\cos(\theta))\cos(\theta) + (1-\cos(\theta))^2$.

The free parameters $\eta_i$ and $\beta_i$ to be chosen such that $\eta_i > 0$ and $\beta_i > 0$ to get a strictly positive denominator $R_i^*(x)$ of $q_i$. Consider $\mathcal{G} := \{g \in C^1(I)$: $g(x_1) = y_1$, $g(x_N) = y_N$, $g^{(1)}(x_1) = d_1$, and $g^{(1)}(x_N) = d_N\}$ be endowed with the metric induced by $C^1(I)$-norm on $I$. Define the RB operator $T_\alpha$ on $\mathcal{G}$ as

$$(T_\alpha g)(x) = \alpha_i(U^{-1}(x))g(U^{-1}(x)) + \frac{P_i^*(U^{-1}(x))}{R_i^*(U^{-1}(x))},\quad x \in I_i, i \in \mathbb{N}_{N-1}, \tag{2.8}$$

where $\alpha(x) = (\alpha_1(x), \alpha_2(x), \ldots, \alpha_{N-1}(x))$. For all $i = 1, 2, \ldots, N-1$, $\|\alpha_i\|_1 < \frac{a_i}{2} < 1$ implies $T_\alpha : \mathcal{G} \to \mathcal{G}$ is a contraction map on a complete metric space. Thus, there exists a unique fixed point say $\Upsilon \in \mathcal{G}$ corresponding to the IFS $\Omega$, which satisfies

$$\Upsilon(U_i(x)) = V_i(x, f(x)) = \alpha_i(x)\Upsilon(x) + q_i(x),\quad x \in I_i, i \in \mathbb{N}_{N-1}. \tag{2.9}$$

The join-up conditions $V_i(x_1, y_1) = y_i$, $V_i(x_N, y_N) = y_{i+1}$ are reduced to $\Upsilon(x_i) = y_i, \Upsilon(x_{i+1}) = y_{i+1}, i \in \mathbb{N}_{N-1}$, which are now interpolation and continuity conditions of $\Upsilon$ on $I$. Taking $x = x_1$ in (2.9), we get

$$\Upsilon(U_i(x_1)) = \alpha_i(x_1)\Upsilon(x_1) + A_{i1} \implies A_{i1} = y_i - \alpha_i(x_1)y_1.$$

Similarly, substituting $x = x_N$ in (2.9), we compute $A_{4i} = y_{i+1} - \alpha_i(x_N)y_N$.

Now, for $i \in \mathbb{N}_{N-1}$, by choosing $\|\alpha_i\|_1 < \frac{a_i}{2}, \alpha_i \in C^1(I)$ and $q_i \in C^1(I)$, we have $\Upsilon \in C^1(I)$ (see [19]), and $\Upsilon^{(1)}$ satisfies

$$a_i\Upsilon^{(1)}(U_i(x)) = \alpha_i(x)\Upsilon^{(1)}(x) + \alpha_i^{(1)}(x)\Upsilon(x) + q_i^{(1)}(x), \quad i \in \mathbb{N}_{N-1}, \ x \in I. \tag{2.10}$$

Assigning $x = x_1$ in (2.10), we have

$$a_i\Upsilon^{(1)}(U_i(x_1)) = \alpha_i(x_1)\Upsilon^{(1)}(x_1) + \alpha_i^{(1)}(x_1)\Upsilon(x_1) + q_i^{(1)}(x_1)$$
$$\implies A_{i2} = \eta_i(y_i - \alpha_i(x_1)y_1) + \tfrac{2(x_N - x_1)}{\pi}[a_i d_i - \alpha_i(x_1)d_1 - \alpha_i'(x_1)y_1].$$

Similarly, putting $x = x_N$ in (2.10), we obtain

$$A_{i3} = \beta_i(y_{i+1} - \alpha_i(x_N)y_N) + \tfrac{2(x_N - x_1)}{\pi}[\alpha_i(x_N)d_N + \alpha_i'(x_N)y_N - a_i d_{i+1}].$$

Therefore, the novel $C^1$-rational quadratic trigonometric spline FIF is given by

$$\Upsilon(U_i(x)) = \alpha_i(x)\Upsilon(x) + \frac{P_i^*(x)}{R_i^*(x)}, \quad i \in \mathbb{N}_{N-1}, \ x \in I, \tag{2.11}$$

where

$$\begin{aligned} P_i^*(x) = P_i(\theta) =& (y_i - \alpha_i(x_1)y_1)(1 - \sin(\theta))^2 + \big(\eta_i(y_i - \alpha_i(x_1)y_1) \\ &+ \frac{2(x_N - x_1)}{\pi}[a_i d_i - \alpha_i(x_1)d_1 - \alpha_i'(x_1)y_1]\big)(1 - \sin(\theta))\sin(\theta) \\ &+ \big(\beta_i(y_{i+1} - \alpha_i(x_N)y_N) + \frac{2(x_N - x_1)}{\pi}[\alpha_i(x_N)d_N + \\ &\alpha_i'(x_N)y_N - a_i d_{i+1}]\big)(1 - \cos(\theta))\cos(\theta) \\ &+ (y_{i+1} - \alpha_i(x_N)y_N)(1 - \cos(\theta))^2, \\ R_i^*(x) = R_i(\theta) =& (1 - \sin(\theta))^2 + \eta_i(1 - \sin(\theta))\sin(\theta) \\ &+ \beta_i(1 - \cos(\theta))\cos(\theta) + (1 - \cos(\theta))^2. \end{aligned} \tag{2.12}$$

This completes the existence of the proposed RQTS FIF in this result.

Now, let $S$ be the rational quadratic trigonometric spline function defined in [12]. For $x \in I_i = [x_i, x_{i+1}]$, let $S := S_i(x)$ such that

$$S_i(x) := \frac{P_i^{**}(\phi_i)}{R_i^{**}(\phi_i)}, \tag{2.13}$$

where

$$\begin{aligned}
P_i^{**}(\phi_i) :=& y_i(1-\sin(\phi_i))^2 + \left(\eta_i y_i + \frac{2h_i d_i}{\pi}\right)(1-\sin(\phi_i))\sin(\phi_i) \\
&+ \left(\beta_i y_{i+1} - \frac{2h_i d_{i+1}}{\pi}\right)(1-\cos(\phi_i))\cos(\phi_i) + y_{i+1}(1-\cos(\phi_i))^2, \\
R_i^{**}(\phi_i) :=& (1-\sin(\phi_i))^2 + \eta_i(1-\sin(\phi_i))\sin(\phi_i) \\
&+ \beta_i(1-\cos(\phi_i))\cos(\phi_i) + (1-\cos(\phi_i))^2, \\
\phi_i :=& \frac{\pi}{2}\left(\frac{x-x_i}{h_i}\right), \ x \in I_i, \text{ and } h_i := x_{i+1} - x_i.
\end{aligned} \tag{2.14}$$

**Remark 1** For all $i \in \mathbb{N}_{N-1}$ and for all $x \in I$, if we choose $\alpha_i(x) = 0$, then the rational quadratic trigonometric spline FIF $\Upsilon$ given in (2.11) modifies to the classical rational trigonometric spline interpolant $S$ given in [12].

**Remark 2** If derivative parameters $d_i$ are not given with the data $\{(x_i, y_i) : i \in \mathbb{N}_N\}$, then they must be determined either from the data $(x_i, y_i)$ or by any other appropriate methods. The arithmetic mean and the geometric mean methods are popular choices for calculating derivatives from data. Details for these methods are given in [8].

## 3 Convergence Analysis

Let us fix some notation for this section: $|y|_\infty := \max_{i\in\mathbb{N}_N} |y_i|$, $|d|_\infty := \max_{i\in\mathbb{N}_N} |d_i|$, $|\eta|_\infty := \max_{i\in\mathbb{N}_{N-1}} \eta_i$, $|\beta|_\infty := \max_{i\in\mathbb{N}_{N-1}} \beta_i$, $\xi_i := \min\{\eta_i, \beta_i\}$, $|\rho|_\infty := \max\{|\eta|_\infty, |\beta|_\infty\}$, $|\gamma|_\infty := \max_{i\in\mathbb{N}_{N-1}} \gamma_i$, $h := \max_{i\in\mathbb{N}_{N-1}} h_i$, $\alpha := (\alpha_1, \alpha_2, \dots, \alpha_{N-1})$, $\|\alpha\|_\infty := \max_{i\in\mathbb{N}_{N-1}} \|\alpha_i\|_\infty$, $\|\alpha\|_1 := \max_{i\in\mathbb{N}_{N-1}} \|\alpha_i\|_1$, and $\sigma := \min_{i\in\mathbb{N}_{N-1}} \sigma_i$, with

$$\sigma_i := \begin{cases} 1 & \text{if } \xi_i \geq 2 \\ \frac{\xi_i}{2} & \text{if } \xi_i < 2. \end{cases}$$

From [12], we know that for a data generating function $\Psi \in C^3(I)$, the classical RQTS function $S$ converges to $\Psi$ with order $O(h^3)$ as $h \to 0$. Here also we will show that after giving some restrictions on scaling functions, our RQTS FIF $\Upsilon$ is also converging to $\Psi$ with order $O(h^3)$ as $h \to 0$.

Note that the associated rational functions of RQTS FIF $q_i$ depend on the scaling factor $\alpha_i(x)$ and the shape parameters $\eta_i$ and $\beta_i$, and hence $q_i$ can be considered as a function of $\alpha_i$, $\eta_i$, $\beta_i$, and $\phi_i$. Thus, we can write $q_i(\alpha_i, \eta_i, \beta_i, \phi_i) = \frac{P_i(\alpha_i,\eta_i,\beta_i,\phi_i)}{R_i(\eta_i,\beta_i,\phi_i)}$, $i \in \mathbb{N}_{N-1}$, where $P_i$ and $R_i$ are as given in (2.12).

**Theorem 3** *For the original data generating function $\Psi \in C^3(I)$, let $\Upsilon$ be the RQTS FIF for data $\{(x_i, y_i, d_i) : i = 1, 2, \dots, N\}$, where $x_1 < x_2 < \dots < x_N$ and $S$ be the non-recursive counterpart of $\Upsilon$. Let the rational function $q_i$ required for $\Upsilon$ satisfying*

*$|q_i(\alpha_i(x), \eta_i, \beta_i, \phi_i) - q_i(0, \eta_i, \beta_i, \phi_i)| \leq \|\alpha\|_1 K_0$ for $\|\alpha_i\|_1 < \frac{a_i}{2}$ for all $i \in \mathbb{N}_{N-1}$, and for some real constant $K_0$, Then we have*

$$\|\Psi - \Upsilon\|_\infty \leq h^3 \|\Psi^{(3)}\|_\infty |\gamma|_\infty + \frac{\|\alpha\|_1(\|S\|_\infty + K_0)}{1 - \|\alpha\|_1},$$

*where $|\gamma|_\infty$ is some constant real number.*

***Proof*** Using triangle inequality for uniform norm, we have

$$\|\Psi - \Upsilon\|_\infty \leq \|\Psi - S\|_\infty + \|S - \Upsilon\|_\infty. \tag{3.15}$$

From [12], We know that for $x \in [x_i, x_{i+1}]$,

$$|\Psi(x) - S_i(x)| = |\Psi(x) - S(x)| \leq \|\Psi^{(3)}(\lambda)\| h_i^3 \gamma_i,$$

where $\gamma_i$ is some constant real number. Then, we obtain

$$\|\Psi - S\|_\infty \leq \|\Psi^{(3)}\|_\infty h^3 |\gamma|_\infty. \tag{3.16}$$

Now, for a given data set and $\|\alpha_i\|_1 < \frac{a_i}{2}$, $i \in \mathbb{N}_{N-1}$, the RQTS FIF $\Upsilon \in C^1(I)$ is the fixed point of the Read-Bajraktarević operator

$$(T_\alpha g)(x) = \alpha_i(U^{-1}(x))(g(U_i^{-1}(x)) + q_i(\alpha_i, \eta_i, \beta_i, \phi_i). \tag{3.17}$$

We know these interpolants $\Upsilon$ and $S$ are the fixed points of $T_\alpha$ with $\alpha \not\equiv 0$ and $\alpha = 0$ respectively.

Now

$$|T_\alpha \Upsilon(x) - T_\alpha S(x)| = |\{\alpha_i(U_i^{-1}(x))(\Upsilon(U_i^{-1}(x)) + q_i(\alpha_i, \eta_i, \beta_i, \phi_i)\} - \{\alpha_i(U_i^{-1}(x))(S(U_i^{-1}(x)) + q_i(\alpha_i, \eta_i, \beta_i, \phi_i)\}|,$$

$$\Longrightarrow |T_\alpha \Upsilon(x) - T_\alpha S(x)| \leq \|\alpha\|_\infty(\|\Upsilon - S\|_\infty) \leq \|\alpha\|_1(\|\Upsilon - S\|_\infty).$$

From the above inequality, we get

$$\|T_\alpha \Upsilon - T_\alpha S\|_\infty \leq \|\alpha\|_1 \|\Upsilon - S\|_\infty. \tag{3.18}$$

Let $x \in [x_i, x_{i+1}]$ and $\alpha \not\equiv 0$. Then, (3.17) implies

$$|T_\alpha S(x) - T_0 S(x)| = |\{\alpha_i(U_i^{-1})(S(U_i^{-1}(x)) + q_i(\alpha_i, \eta_i, \beta_i, \phi_i)\} - q_i(0, \eta_i, \beta_i, \phi_i)|,$$

$$\text{i.e. } |T_\alpha S(x) - T_0 S(x)| \leq \|\alpha\|_\infty \|S\|_\infty + \|\alpha\|_1 K_0 \leq \|\alpha\|_1(\|S\|_\infty + K_0).$$

$$\Longrightarrow \|T_\alpha S - T_0 S\|_\infty \leq \|\alpha\|_1(\|S\|_\infty + K_0). \tag{3.19}$$

Now using (3.18) and (3.19), we obtain

$$\begin{aligned}\|\Upsilon - S\|_\infty &= \|T_\alpha \Upsilon - T_0 S\|_\infty \le \|T_\alpha \Upsilon - T_\alpha S\|_\infty + \|T_\alpha S - T_0 S\|_\infty \\ &\le \|\alpha\|_1 \|\Upsilon - S\|_\infty + \|\alpha\|_1 (\|S\|_\infty + K_0).\end{aligned}$$

$$\Longrightarrow \|\Upsilon - S\|_\infty \le \frac{\|\alpha\|_1 (\|S\|_\infty + K_0)}{1 - \|\alpha\|_1}. \tag{3.20}$$

Now after putting (3.16) and (3.20) in the inequality (3.15), we can get our desired result.

Now to conclude about the convergence results, we will try to find upper bounds for $\|S\|_\infty$ and $K_0$. For $x \in I_i$, $S(x) = S_i(x)$, from (2.13) we have

$$|S_i(x)| \le \frac{\max\{|P_i^{**}(\phi_i)| : 0 \le \phi_i \le \frac{\pi}{2}\}}{\min\{|R_i^{**}(\phi_i)| : 0 \le \phi_i \le \frac{\pi}{2}\}}.$$

Now using the extremum calculations, we can easily get the following bounds

$$\begin{aligned}|P_i^{**}(\phi_i)| &\le |y|_\infty + \left(|\rho|_\infty |y|_\infty + \frac{2h}{\pi}|d|_\infty\right), \\ |R_i^{**}(\phi_i)| &\ge 1 + (\xi_i - 2)(\sin(\phi_i) + \cos(\phi_i) - 1) \ge \sigma_i.\end{aligned}$$

$$\Longrightarrow \|S\|_\infty \le \frac{(1 + |\rho|_\infty)|y|_\infty + \frac{2h}{\pi}|d|_\infty}{\min\{\sigma_i : i \in \mathbb{N}_{N-1}\}}.$$

Similarly, for $x \in I_i$ and $|I| := x_N - x_1$,

$$\begin{aligned}|q_i(\alpha_i, \eta_i, \beta_i, \phi_i) - q_i(0, \eta_i, \beta_i, \phi_i)| \le \frac{1}{\sigma_i}\Big\{&\|\alpha_i\|_\infty \big(\max\{|y_1|, |y_N|\}\big) \\ &+ \|\alpha_i\|_\infty \big(\max\{|\eta_i y_1|, |\beta_i y_N|\}\big) \\ &+ \frac{2|I|}{\pi}\|\alpha_i\|_\infty \big(\max\{|d_1|, |d_N|\}\big) \\ &+ \frac{2|I|}{\pi}\|\alpha_i^{(1)}\|_\infty \big(\max\{|y_1|, |y_N|\}\big)\Big\}.\end{aligned}$$

Thus, we can take

$$K_0 = \frac{(1 + |\rho|_\infty + \frac{2|I|}{\pi})(\max\{|y_1|, |y_N|\}) + \frac{2|I|}{\pi}\max\{|d_1|, |d_N|\}}{\min\{\sigma_i : i \in \mathbb{N}_{N-1}\}}.$$

**Convergence results**: In view of $\frac{a_i}{2} \leq \frac{h}{2|I|}$ and Theorem 3, it is found that the RQTS FIF $\Upsilon$ converges to $\Psi$ as $h \to 0$. Furthermore, if we choose our variable scaling functions such that $\|\alpha\|_1 < \min\left\{h^3, \frac{h}{2|I|}\right\}$, then Theorem 3 justifies that $\Upsilon$ converges to $\Psi \in C^3(I)$ with order $O(h^3)$ as $h \to 0$.

## 4 Positivity-Preserving RQTS FIF

In this section, we will restrict the associated shape parameters and variable scaling functions such that the $C^1$-RQTS FIF $\Upsilon$ satisfies $\Upsilon(x) > 0 \ \forall x \in I$, for a strictly positive data set.

**Theorem 4** *Let $\{(x_i, y_i)\, i = 1, 2, \ldots, N\}$ be a given set of strictly positive data satisfying $x_1 < x_2 < \cdots < x_N$, and $d_i$'s are chosen derivative values at knots $x_i$'s. For $i \in \mathbb{N}_{N-1}$, consider the iterated mappings $U_i$ and $V_i$ defined in (2.5) which are satisfying (2.1) and (2.2), respectively. Then the corresponding RQTS FIF $\Upsilon$ will be positive on $I$, if the non-negative variable scaling functions and shape parameters are chosen as*

$$\|\alpha_i\|_1 < \frac{a_i}{2}, \ \alpha_i(x_1) < \frac{y_i}{y_1}, \ \alpha_i(x_N) < \frac{y_{i+1}}{y_N},$$

$$\eta_i > \max\left\{0, \frac{-\frac{2|I|}{\pi}[a_i d_i - \alpha_i(x_1)d_1 - \alpha_i'(x_1)y_1]}{y_i - \alpha_i(x_1)y_1}\right\},$$

$$\beta_i > \max\left\{0, \frac{-\frac{2|I|}{\pi}[\alpha_i(x_N)d_N + \alpha_i'(x_N)y_N - a_i d_{i+1}]}{y_{i+1} - \alpha_i(x_N)y_N}\right\}, \quad \forall i \in \mathbb{N}_{N-1}.$$

***Proof*** According to the Theorem 2, for $\|\alpha_i\|_1 < \frac{a_i}{2}$ and for $\alpha_i(x), q_i(x) \in C^1(I)$, the RQTS FIF $\Upsilon \in C^1(I)$ and it satisfies the recursive formula

$$\Upsilon(U_i(x)) = \alpha_i(x)\Upsilon(x) + \frac{P_i(\theta)}{R_i(\theta)}, \quad i \in \mathbb{N}_{N-1}, \ x \in I.$$

Choosing $\eta_i > 0$ and $\beta_i > 0$, $R_i(\theta)$ becomes positive on $I$. Since $\Upsilon$ is the attractor of the IFS $\{F; (U_i(x), V_i(x, y)) : i \in \mathbb{N}_{N-1}\}$ and defined recursively, by the property of the attractor, to show $\Upsilon > 0$, it is sufficient to prove that for all $x \in I$, $\Upsilon(U_i(x)) > 0$ $\forall i \in \mathbb{N}_{N-1}$, whenever $\Upsilon(x) > 0$. Take $x \in I$, $\Upsilon(x) > 0$, and with these positive shape parameters and the non-negative variable scaling functions, the positivity of $\Upsilon(U_i(x))$ reduces to the positivity of $P_i(\theta)$ for all $\theta \in [0, 1]$. Now, if $A_{ij} > 0, \forall j \in \mathbb{N}_4$, then we have $P_i(\theta) > 0$. Thus,

$$A_{i1} > 0 \Longleftrightarrow \alpha_i(x_1) < \frac{y_i}{y_1},$$

$$A_{i4} > 0 \Longleftrightarrow \alpha_i(x_N) < \frac{y_{i+1}}{y_N},$$

$$A_{i2} > 0 \Longleftrightarrow \eta_i > \frac{-\frac{2|I|}{\pi}[a_i d_i - \alpha_i(x_1)d_1 - \alpha_i'(x_1)y_1]}{y_i - \alpha(x_1)y_1},$$

$$A_{i3} > 0 \Longleftrightarrow \beta_i > \frac{-\frac{2|I|}{\pi}[\alpha_i(x_N)d_N + \alpha_i'(x_N)y_N - a_i d_{i+1}]}{y_{i+1} - \alpha_i(x_N)y_N}.$$

Using the above conditions, we get the desired result.

**Remark 3** For all $i \in \mathbb{N}_{N-1}$ and $x \in I$, if we choose these variable scaling functions $\alpha_i(x) = 0$, then Theorem 4 gives sufficient conditions on shape parameters

$$\eta_i > \max\left\{0, \frac{-2h_i d_i}{\pi y_i}\right\} \text{ and } \beta_i > \max\left\{0, \frac{-2h_i d_{i+1}}{\pi y_{i+1}}\right\},$$

such that the RQTS function $S$ defined in [12] becomes positive for a prescribed positive data set $\{(x_i, y_i) : i \in \mathbb{N}_N\}$.

**Example 1** Consider strictly positive data set $\{(0.1, 1, -3), (0.3, 0.04, 2), (0.5, 0.6, 1), (0.7, 0.2, -3), (0.9, 4, 1)\}$. Following variable scaling functions and shape parameters are used in the construction of RQTS FIFs in Fig. 1a–f.

| Figure 1 | $\alpha$ | $\eta$ | $\beta$ |
|---|---|---|---|
| (a) | $(0, 0, 0, 0)$ | $(1, 1, 1, 3)$ | $(135, 1, 16, 1)$ |
| (b) | $(\frac{x}{9}, \frac{12}{100}, \frac{e^x}{20}, \frac{\sin(x)}{10})$ | $(1, 1, 1, 3)$ | $(135, 1, 16, 1)$ |
| (c) | $(\frac{x}{100}, \frac{x}{9}, \frac{e^{1-x}}{25}, \frac{1+\cos(x)}{20})$ | $(1, 1, 1, 3)$ | $(135, 1, 16, 1)$ |
| (d) | $(\frac{x}{100}, \frac{1-x}{90}, \frac{e^{1-x}}{25}, \frac{1+\cos(x)}{20})$ | $(1, 1, 1, 3)$ | $(135, 1, 16, 1)$ |
| (e) | $(\frac{x}{100}, \frac{x}{9}, \frac{e^x}{60}, \frac{1+\cos(x)}{20})$ | $(1, 1, 1, 3)$ | $(135, 1, 16, 1)$ |
| (f) | $(\frac{x}{100}, \frac{x}{9}, \frac{e^x}{60}, \frac{1+\cos(x)}{20})$ | $(100, 10, 1, 35)$ | $(135, 12, 16, 19)$ |

Figure 1a is the plot of the classical RQTS function defined in [12]. In Fig. 1a, we have used restricted shape parameters as described in Remark 3 to get a strictly positive RQTS function. In Fig. 1b, we do not restrict our scaling functions as prescribed by Theorem 4, and the corresponding RQTS FIF is not positive on $I = [0.1, 0.9]$. But when we restrict our shape parameters and scaling functions as prescribed in Theorem 4, then we get the positive RQTS FIFs plotted in Fig. 1c–f. To see the effects of variable scaling functions, we have plotted Fig. 1d with a different function $\alpha_2(x)$ from Fig. 1c, and by comparing these figures, we can observe the effects of $\alpha_2(x)$ on a positive RQTS FIF. Similarly, in Fig. 1e, we have used a different scaling function $\alpha_3(x)$ from Fig. 1c. Now, it is easy to observe the effects of $\alpha_3(x)$ on positive RQTS FIF. In Fig. 1f, we have used different shape parameters $\eta = (100, 10, 1, 35)$ and $\beta = (135, 12, 16, 19)$ with the same scaling functions as used for Fig. 1e to demonstrate the effects of shape parameters on a positive RQTS FIF.

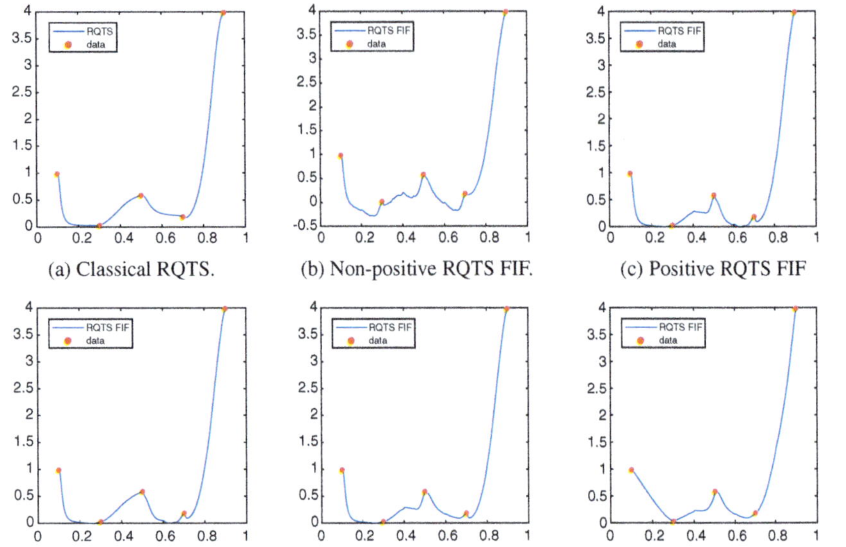


(a) Classical RQTS. (b) Non-positive RQTS FIF. (c) Positive RQTS FIF

(d) Effects of $\alpha_2(x)$ on Positive RQTS FIF (c) (e) Effects of $\alpha_3(x)$ on Positive RQTS FIF (c) (f) Effects of shape parameters on Positive RQTS FIF (e)

**Fig. 1** Positive or Non-positive RQTS FIFs

## 5 Monotonicity Preserving RQTS FIF

In this section, we will show that if scaling functions and shape parameters are restricted, we can obtain a positive and monotonically increasing $C^1$-RQTS FIF $\Upsilon$, when the data $\{(x_i, y_i, d_i) : i \in \mathbb{N}_N\}$ is positive and monotonically increasing, i.e., $0 < y_1 \leq y_2 \leq \cdots \leq y_N$ or $\Delta_i := \frac{y_{i+1}-y_i}{h_i} \geq 0$, $\forall i \in \mathbb{N}_{N-1}$. Note that if the data set is negative, one can add a suitable constant to the ordinates, and convert the data set to a positive one. The RQTS FIF can be shifted back by using equivalence of two dynamical systems. For a monotonically increasing interpolant, it is necessary that the derivatives parameters $d_i$ are non-negative $\forall i \in \mathbb{N}_N$. It is known from the calculus that a differentiable function $g$ is monotonically increasing on $I$ if and only if $g^{(1)}(x) \geq 0$ for all $x \in I$. From (2.7), we have

$$a_i \Upsilon^{(1)}(U_i(x)) = \alpha_i(x)\Upsilon^{(1)}(x) + \alpha_i^{(1)}(x)\Upsilon(x) + q_i^{(1)}(x), \quad i \in \mathbb{N}_{N-1}, \ \forall x \in I, \tag{5.21}$$

where $q_i^{(1)}(x) = \frac{\pi}{2|I|}\frac{\Gamma_i(\theta)}{R_i^2(\theta)}$,

$$\begin{aligned}\Gamma_i(\theta) &= M_{i1}^*(1-\sin(\theta))^3\cos(\theta) + M_{i2}^*(1-\sin(\theta))^2\sin(\theta)\cos(\theta)\\ &+ M_{i3}^*(1-\sin(\theta))(1-\cos(\theta))\sin(\theta) + M_{i4}^*(1-\sin(\theta))(1-\cos(\theta))\cos(\theta)\\ &+ M_{i5}^*(1-\sin(\theta))(1-\cos(\theta))((1-\cos(\theta)+\sin^2(\theta))\\ &+ M_{i6}^*(1-\sin(\theta))(1-\cos(\theta))((1-\cos(\theta)+\cos^2(\theta))\\ &+ M_{i7}^*((1-\sin(\theta))(1-\cos(\theta))^2\cos(\theta)) + M_{i8}^*((1-\sin(\theta))^2(1-\cos(\theta))\sin(\theta))\\ &+ M_{i9}^*((1-\cos(\theta))^2\sin(\theta)\cos(\theta)) + M_{i10}^*(1-\cos(\theta))^3\sin(\theta),\end{aligned}$$

$$\begin{aligned}&M_{i1}^* = A_{i2} - \eta_i A_{i1},\quad M_{i2}^* = M_{i1}^* + (A_{i3} - \beta_i A_{i1}),\\ &M_{i3}^* = \eta_i A_{i3} - \beta_i A_{i2} = M_{i4}^*,\quad M_{i5}^* = \eta_i A_{i4} - A_{i2},\\ &M_{i6}^* = A_{i3} - \beta_i A_{i1},\quad M_{i7}^* = 2(A_{i4} - A_{i1}) = M_{i8}^*,\\ &M_{i9}^* = M_{i10}^* + (\eta_i A_{i4} - A_{i2}),\quad M_{i10}^* = \beta_i A_{i4} - A_{i3}.\end{aligned}$$

Here we will use a similar argument as described in Theorem 4. Since $\Upsilon^{(1)}$ is defined recursively, to show $\Upsilon^{(1)} \geq 0$ on $I$, it's enough to show that $\Upsilon^{(1)}(U_i(x)) \geq 0$ for all $i \in \mathbb{N}_{N-1}$, whenever $\Upsilon^{(1)}(x) \geq 0$. Let $\Upsilon^{(1)}(x) \geq 0$ at grid points. Now choose our shape parameters $\eta_i > 0$ and $\beta_i > 0$, and monotonically increasing scaling functions $0 \leq \alpha_i(x) \in C^1(I)$ such that they satisfy the prescribed conditions in Theorem 4. Hence, $0 \leq \alpha_i(x)$, $0 \leq \alpha_i^{(1)}(x)$, and $0 < \Upsilon(x)$ for all $x \in I$, and $i \in \mathbb{N}_{N-1}$. Therefore, positivity of $\Upsilon^{(1)}(U_i(x))$ reduced to the positivity of $\Gamma_i(\theta)$ for all $\theta \in [0, 1]$. For the positivity of $\Gamma_i(\theta)$, it is sufficient to show that for all $j \in \mathbb{N}_{10}$, $M_{ij}^* \geq 0$, $\forall i \in \mathbb{N}_{N-1}$.

Now, observe that if $d_1 = 0$, the choice of scaling function $\alpha_i^{(1)}(x_1) \leq \frac{a_i d_i}{2y_1}$ gives us $M_{i1}^* \geq 0$ for all $i \in \mathbb{N}_{N-1}$. Similarly, if $d_N = 0$, the choice of scaling function $\alpha_i^{(1)}(x_N) \leq \frac{a_i d_{i+1}}{2y_N}$ gives us $M_{i10}^* \geq 0$ for all $i \in \mathbb{N}_{N-1}$. So assume that $d_1 \neq 0$ and $d_N \neq 0$. Then, for all $j \in \mathbb{N}_{10}$, $M_{ij}^* \geq 0$, if we choose

$$\begin{aligned}&\alpha_i(x_1) \leq \min\left\{\frac{a_i d_i}{2d_1}, \frac{y_i}{y_1}\right\},\quad \alpha_i(x_N) \leq \min\left\{\frac{a_i d_{i+1}}{2d_N}, \frac{y_{i+1}}{y_N}, \frac{y_{i+1}-y_i}{y_N}\right\},\\ &\alpha_i^{(1)}(x_1) \leq \frac{a_i d_i}{2y_1},\quad \alpha_i^{(1)}(x_N) \leq \frac{a_i d_{i+1}}{2y_N},\\ &\eta_i > \max\left\{0, \frac{\frac{4|I|}{\pi}[a_i d_i - \alpha_i(x_1)d_1 - \alpha_i^{(1)}(x_1)y_1]}{(y_{i+1} - \alpha_i(x_N)y_N) - (y_i - \alpha_i(x_1)y_1)}\right\},\\ &\beta_i > \max\left\{0, \frac{\frac{4|I|}{\pi}[a_i d_{i+1} - \alpha_i(x_N)d_N - \alpha_i^{(1)}(x_N)y_N]}{(y_{i+1} - \alpha_i(x_N)y_N) - (y_i - \alpha_i(x_1)y_1)}\right\},\quad i \in \mathbb{N}_{N-1}.\end{aligned} \tag{5.22}$$

The above results can be encapsulated in the following:

**Theorem 5** *Let $\{(x_i, y_i) : i = 1, 2, \ldots, N\}$ be a positive and monotonically increasing data set. Let $d_i$, $i \in \mathbb{N}_N$ be chosen so as to satisfy the necessary monotonic increasing condition. Then for $i \in \mathbb{N}_{N-1}$, the following conditions on non-negative monotonically increasing $\alpha_i(x)$ and shape parameters $\eta_i$ and $\beta_i$ are sufficient to preserve the properties of data by the RQTS FIF $\Upsilon$ on $I$:*

$$\|\alpha_i\|_1 < \frac{a_i}{2},\quad \alpha_i(x_1) \le \min\left\{\frac{a_i d_i}{2d_1}, \frac{y_i}{y_1}\right\},\quad \alpha_i(x_N) \le \min\left\{\frac{a_i d_{i+1}}{2d_N}, \frac{y_{i+1}-y_i}{y_N}\right\},$$

$$\alpha_i^{(1)}(x_1) \le \frac{a_i d_i}{2y_1},\quad \alpha_i^{(1)}(x_N) \le \frac{a_i d_{i+1}}{2y_N},$$

$$\eta_i > \max\left\{0, \frac{-\frac{2|I|}{\pi}[a_i d_i - \alpha_i(x_1)d_1 - \alpha_i^{(1)}(x_1)y_1]}{y_i - \alpha_i(x_1)y_1}, \frac{\frac{4|I|}{\pi}[a_i d_i - \alpha_i(x_1)d_1 - \alpha_i^{(1)}(x_1)y_1]}{(y_{i+1} - \alpha_i(x_N)y_N) - (y_i - \alpha_i(x_1)y_1)}\right\},$$

$$\beta_i > \max\left\{0, \frac{\frac{2|I|}{\pi}[a_i d_{i+1} - \alpha_i(x_N)d_N - \alpha_i^{(1)}(x_N)y_N]}{y_{i+1} - \alpha_i(x_N)y_N}, \frac{\frac{4|I|}{\pi}[a_i d_{i+1} - \alpha_i(x_N)d_N - \alpha_i^{(1)}(x_N)y_N]}{(y_{i+1} - \alpha_i(x_N)y_N) - (y_i - \alpha_i(x_1)y_1)}\right\}.$$

**Remark 4** If the data set is monotonically decreasing and positive, then we can choose a non-negative monotonically decreasing scaling function such that

$$\|\alpha_i\|_1 < \frac{a_i}{2},\quad \alpha_i(x_1) \le \min\left\{\frac{a_i d_i}{2d_1}, \frac{y_i - y_{i+1}}{y_1}\right\},\quad \alpha_i(x_N) \le \min\left\{\frac{a_i d_{i+1}}{2d_N}, \frac{y_{i+1}}{y_N}\right\},$$

$$\alpha_i^{(1)}(x_1) \ge \frac{a_i d_i}{2y_1},\quad \alpha_i^{(1)}(x_N) \ge \frac{a_i d_{i+1}}{2y_N},$$

and with the restrictions on the shape parameters as in Theorem 5, the resulting RQTS FIF will be monotonically decreasing.

**Remark 5** If $\Delta_i = 0$, then we take $\alpha_i = 0$ for the monotonicity of the FIF $\Upsilon$. Also in this case, $d_i = d_{i+1} = 0$. Consequently, $\Upsilon(U_i(x)) = y_i = y_{i+1}$, i.e., when $y_i = y_{i+1}$, our RQTS FIF $\Upsilon$ reduces to a constant on the interval $I_i = [x_i, x_{i+1}]$.

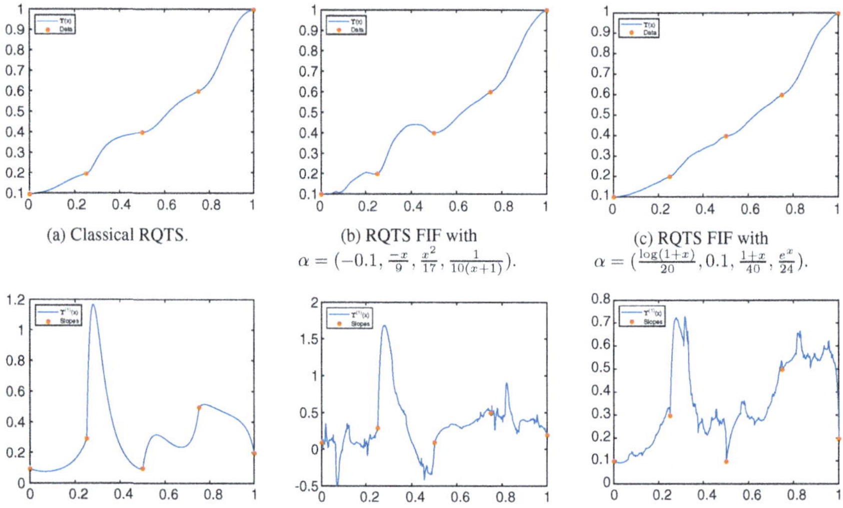


(a) Classical RQTS.

(b) RQTS FIF with $\alpha = (-0.1, \frac{-x}{9}, \frac{x^2}{17}, \frac{1}{10(x+1)})$.

(c) RQTS FIF with $\alpha = (\frac{\log(1+x)}{20}, 0.1, \frac{1+x}{40}, \frac{e^x}{24})$.

(d) First derivative of classical RQTS.

(e) First derivative of RQTS FIF with $\alpha = (-0.1, \frac{-x}{9}, \frac{x^2}{17}, \frac{1}{10(x+1)})$.

(f) First derivative of RQTS FIF with $\alpha = (\frac{\log(1+x)}{20}, 0.1, \frac{1+x}{40}, \frac{e^x}{24})$.

**Fig. 2** Monotonically increasing RQTS FIFs

**Example 2** For simplicity, we have taken a positive and monotonically increasing data set as $\{(0, 0.1, 0.1), (0.25, 0.2, 0.3), (0.5, 0.4, 0.1), (0.75, 0.6, 0.5), (1, 1, 0.2)\}$. For fixed shape parameters $\eta = (1, 2, 1, 1)$ and $\beta = (1, 11, 2, 1)$, Fig. 2a is the plot of classical RQTS, and Fig. 2b–c are generated by using scaling functions $(-0.1, \frac{-x}{9}, \frac{x^2}{17}, \frac{1}{10(x+1)})$ and $(\frac{\log(1+x)}{20}, 0.1, \frac{1+x}{40}, \frac{e^x}{24})$ respectively. For Fig. 2b, we do not restrict our parameters as prescribed by Theorem 5, and the corresponding RQTS FIF is not monotone in nature. When we restrict our parameters as prescribed by Theorem 5, then we get positive and monotone RQTS FIFs in Figs. 2a and c. Figure 2d–f are the plots of the first derivatives of the RQTS FIFs in Fig. 2a–c, respectively. Thus, it is easy to capture non-linearity associated with the derivatives of data generating conic function by using the proposed class of RQTS FIFs.

## 6 Conclusion

We have constructed a novel $C^1$-RQTS FIF with variable scaling functions to interpolate conic data that is partially self-similar in nature. The derivative of this RQTS FIF may not be differentiable at a dense subset of the given interpolation domain. The RQTS FIF can be reduced to its non-recursive classical rational quadratic trigonometric spline function when all scalings are zero. We have found sufficient conditions on scaling functions so that the proposed RQTS FIF has the same order of convergence as its non-recursive part. We have derived sufficient conditions on variable scaling functions and the shape parameters to retain fundamental shapes such as the positivity and monotonicity features of prescribed conic data.

## References

1. Abbas, M., Majid, A.A., Ali, Md.J.: Positivity preserving interpolation of positive data by cubic trigonometric spline. Mathematika **27**(1), 41–50 (2011)
2. Barnsley, M.F.: Fractal functions and interpolation. Constr. Approx. **2**(1), 303–329 (1986)
3. Barnsley, M.F.: Fractals Everywhere. Academic, Boston (1988)
4. Barnsley, M.F., Harrington, A.N.: The calculus of fractal interpolation functions. J. Approx. Theory **57**(1), 14–34 (1989)
5. Bashir, U., Ali, Md.J.: Data visualization using rational trigonometric spline. J. Appl. Math., Art. ID 531497 (2013)
6. Chand, A.K.B., Kapoor, G.P.: Generalized cubic spline fractal interpolation functions. SIAM J. Numer. Anal. **44**(2), 655–676 (2006)
7. Chand, A.K.B., Tyada, K.R.: Constrained shape preserving rational cubic fractal interpolation functions. Rocky Mt. J. Math. **48**(1), 75–105 (2018)
8. Chand, A.K.B., Vijender, N., Navascués, M.A.: Shape preservation of scientific data through rational fractal splines. Calcolo **51**(2), 329–362 (2014)
9. Chand, A.K.B., Viswanathan, P.: A constructive approach to cubic Hermite fractal interpolation function and its constrained aspects. BIT Numer. Math. **53**(4), 841–865 (2013)
10. Han, X.: Quadratic trigonometric polynomial curves with a shape parameter. Comput. Aided Geom. Design **19**(7), 503–512 (2002)

11. Han, X.: Cubic trigonometric polynomial curves with a shape parameter. Comput. Aided Geom. Design **21**(6), 535–548 (2004)
12. Hussain, M.Z., Saleem, S.: $C^1$-rational quadratic trigonometric spline. Egypt. Inf. J. **14**, 211–220 (2013)
13. Hutchinson, J.: Fractals and self-similarity. Indiana Univ. Math. J. **30**, 713–747 (1981)
14. Ibraheem, F., Hussain, M., Hussain, M.Z., Bhatti, A.A.: Positive data visualization using trigonometric function. J. Appl. Math., Art. ID 247120 (2012)
15. Katiyar, S.K., Chand, A.K.B., Saravana Kumar, G.: A new class of rational cubic spline fractal interpolation function and its constrained aspects. Appl. Math. Comput. **346**, 319–335 (2019)
16. Mandelbrot, B.: Fractals: Form, Chance and Dimension. W. H. Freeman, San Francisco (1977)
17. Viswanathan, P., Chand, A.K.B.: $\alpha$-fractal rational splines for constrained interpolation. Electron. Tran. Numer. Anal. **41**, 420–442 (2014)
18. Viswanathan, P., Chand, A.K.B., Navascués., M.A.: Fractal perturbation preserving fundamental shapes: bounds on the scale factors. J. Math. Anal. Appl. **419**(2), 804–817 (2014)
19. Viswanathan, P., Navascués, M.A., Chand, A.K.B.: Fractal polynomials and maps in approximation of continuous functions. Numer. Funct. Anal. Optim. **37**(1), 106–127 (2016)
20. Wang, H.Y., Fan, Z.L.: Analytical characteristics of fractal interpolation functions with function vertical scaling factors. Acta Math. Sinica **54**(1), 147–158 (2011)
21. Wang, H.Y., Shan, Y.J.: Fractal interpolation functions with variable parameters and their analytical properties. J. Approx. Theory **175**, 1–18 (2013)

# Cyclic Multivalued Iterated Function Systems

**R. Pasupathi, A. K. B. Chand, and M. A. Navascués**

**Abstract** IFS constitutes one of the powerful tools to generate fractal sets. Recently, a cyclic map is used in IFS to construct a new class of fractals. This paper is an effort to study multivalued IFSs with various types of cyclic multivalued maps such as cyclic multivalued $\phi$-contraction, cyclic multivalued Meir–Keeler contraction and cyclic multivalued contractive which are generalizations of contraction map, and the construction of fractals with the help of these IFSs have been established.

**Keywords** Iterated function systems · Cyclic Multivalued contractions · Fixed point · Fractal · Attractor

## 1 Introduction

Most of the systems resulting from the real-world phenomena or human artefacts are not of the regular classical Euclidean forms. Modelling or describing such complex structures proved to be a great challenge until fractal theory came into play. Fractal theory, introduced by Mandelbrot [22] proved to be one of the effective tools for capturing the complexity of the structure and for modelling a variety of phenomena in applied mathematics and engineering: approximation theory, geometric modelling, image processing, bio-engineering, signal processing, turbulence, etc. (see for instance [4–7, 14, 16, 24, 33]).

Hutchinson [13] introduced the concept of Iterated Function System (IFS) and Barnsley [2, 3] put the foundation of using IFS as a tool to generate fractal sets.

R. Pasupathi (✉) · A. K. B. Chand
Department of Mathematics, Indian Institute of Technology Madras, Chennai 600036, India
e-mail: pasupathi4074@gmail.com

A. K. B. Chand
e-mail: chand@iitm.ac.in

M. A. Navascués
Departmento de Matemática Aplicada, Escuela de Ingeniería y Arquitectura, Universidad de Zaragoza, 50018 Zaragoza, Spain
e-mail: manavas@unizar.es



B. Rushi Kumar et al. (eds.), *Mathematics and Computing*, Springer Proceedings in Mathematics & Statistics 415, https://doi.org/10.1007/978-981-19-9307-7_21

Over the years, IFS played a central role in generating many important fractal sets. An IFS is a finite collection of contraction mappings on a complete metric space $\mathcal{E}$. Let $\mathcal{C}(\mathcal{E})$ denote the set of all non-empty subsets of $\mathcal{E}$ which are compact. If $\mathcal{E}$ is given the Hausdorff metric, then $\mathcal{C}(\mathcal{E})$ becomes complete. The operator on $\mathcal{C}(\mathcal{E})$ induced by the given IFS turns out to be a contraction mapping, and hence there exists a unique set fixed point (say $G$) by the Banach contraction principle. We call $G$, the attractor of the IFS. Apart from the aforementioned application of IFS theory, this theory has remarkable applications in various fields like geometric modelling, pattern recognition, image processing, bio-medical engineering, etc. Due to its diverse applications, this area has drawn the attention of a lot of researchers in the last few decades. Many extensions have been done to this framework, say from a more generalized contraction, to a multivalued approach, to a countable and infinite set up, to different types of domain spaces, multifunction systems, etc. Some of the remarkable extensions are discussed below.

Hata [12] worked on IFS composed of $\phi$-contraction functions. The concept of the iterated multifunction system was introduced and studied in detail by Kunze et al. [19] and they further investigated the same with probability. Georgescu [10] worked on IFS consisting of generalized convex contractions in the framework of strong b-metric space. Iaona and Mihail [15] worked on IFS consisting of $\phi$-contractions . Maślanka and Strobin [23] investigated GIFS on $l_\infty$ sum of a metric space. IFS in a weak contraction setup was studied in detail by Okamura [29]. Lozinski worked on QIFS where the contractions act randomly with prescribed probability in the Hilbert space [17]. Fernau [9] introduced the concept of infinite IFS, which was further investigated by Secelean and many others (see for instance [35, 36]). Infinite IFS with a multivalued approach was investigated by Leśniak [20]. Also, he investigated homoclinic attractors in discontinuous IFSs in [21]. Samuel and Tetenov studied IFSs on uniform spaces [34]. Dumitru [8] studied generalized IFS containing Meir–Keeler functions. Pasupathi et al. [30] worked on the construction of fractals with IFS composed of cyclic contractions. Many other remarkable extensions in this theory can be found in [18, 25–28], and references therein.

This paper is devoted to the study of multivalued IFSs composed of different types of cyclic multivalued maps, and the existence of attractors of such maps is proven. Turning to the structure of our paper, in Sect. 2, the prerequisites are given. In Sect. 3, we discuss different types of cyclic generalized contractions and proved the existence of fixed points of the above functions by the iteration process. In Sect. 4, we construct fractal from the various types of new generalized multivalued IFSs consisting of cyclic multivalued $\phi$-contractions, cyclic multivalued Meir–Keeler contractions and cyclic multivalued contractive mappings.

# 2 Preliminary Facts

## 2.1 Hausdorff Metric

Let $(\mathcal{E}, \tau)$ be a metric space and let $\mathcal{C}(\mathcal{E})$ be the collection of non-empty subsets of $\mathcal{E}$ which are compact.

The metric $\omega : \mathcal{C}(\mathcal{E}) \times \mathcal{C}(\mathcal{E}) \to [0, \infty)$ is defined by

$$\omega(R, S) = \max\{\delta(R, S), \delta(S, R)\},$$

where

$$\delta(R, S) = \sup_{\alpha \in R} \tau(\alpha, S) \quad \text{and} \quad \tau(\alpha, S) = \inf_{\beta \in S} \tau(\alpha, \beta) \quad \forall\, R, S \in \mathcal{C}(\mathcal{E}).$$

We call $\omega$, the Hausdorff metric and the space $(\mathcal{C}(\mathcal{E}), \omega)$, the Hausdorff metric space.

The space $(\mathcal{C}(\mathcal{E}), \omega)$ is complete if $(\mathcal{E}, \tau)$ is complete and it is compact if $(\mathcal{E}, \tau)$ is compact.

**Lemma 1** ([3]) *If $(R_\lambda)_{\lambda \in \Lambda}$, $(S_\lambda)_{\lambda \in \Lambda}$ are finite collection of sets in $(\mathcal{C}(\mathcal{E}), \omega)$, then*

$$\omega\left(\bigcup_{\lambda \in \Lambda} R_\lambda\,,\ \bigcup_{\lambda \in \Lambda} S_\lambda\right) \leq \max_{\lambda \in \Lambda}\ \omega(R_\lambda, S_\lambda).$$

**Lemma 2** *Let $R, S \in \mathcal{C}(\mathcal{E})$ for some metric space $(\mathcal{E}, \tau)$. Then for any $\alpha \in R$, there exists $\beta \in S$ such that $\tau(\alpha, \beta) \leq \omega(R, S)$.*

***Proof*** Let $\alpha \in R$. Since $S$ is compact, there exists $\beta \in S$ which satisfies $\tau(\alpha, \beta) = \inf_{\gamma \in S} \tau(\alpha, \gamma)$. Thus $\tau(\alpha, \beta) \leq \delta(R, S) \leq \omega(R, S)$.

**Lemma 3** ([30]) *Suppose that $(\mathcal{E}, \tau)$ is a complete metric space. If $R$ is a closed subset of $\mathcal{E}$, then $\mathcal{C}(R)$ is also a closed subset of $\mathcal{C}(\mathcal{E})$ when equipped with the Hausdorff metric $\omega$.*

## 2.2 Iterated Function Systems

A map $g$ on a metric space $(\mathcal{E}, \tau)$ into itself is said to be a contraction if there exists a constant $0 \leq s < 1$, such that

$$\tau(g(\alpha), g(\beta)) \leq s\ \tau(\alpha, \beta) \quad \forall \alpha, \beta \in \mathcal{E}.$$

The constant $s$ is called a contractivity factor of $g$.

In 1922, Banach proved the following well-known result called the Banach contraction principle:

**Theorem 1** ([1]) *Let $(\mathcal{E}, \tau)$ be a metric space which is complete and let $g$ be a contraction map on $\mathcal{E}$. Then there is a unique point $\alpha^* \in \mathcal{E}$ obeying $g(\alpha^*) = \alpha^*$. And also, for each $\alpha \in \mathcal{E}$, the sequence $\{g^n(\alpha)\}_{m=1}^{\infty}$ converges to $\alpha^*$. That is $\lim_{n\to\infty} g^n(\alpha) = \alpha^*$ for all $\alpha \in \mathcal{E}$.*

A finite collection of contraction maps $(g_l)_{l=1}^{M}$, $M \in \mathcal{N}$ on a complete metric space $(\mathcal{E}, \tau)$ is called (hyperbolic) iterated function system (IFS), where $\mathcal{N}$ denotes the set of natural numbers.

We know that each $g_l$ induces a map on $\mathcal{C}(\mathcal{E})$, and consequently, we can define a set valued map (Hutchinson operator) on $\mathcal{C}(\mathcal{E})$ as $\mathcal{H}(R) = \bigcup_{l=1}^{M} g_l(R)$ (see Hutchinson [13]). He proved that every Hutchinson operator of a IFS has a unique invariant set $G$ (say) in $\mathcal{C}(\mathcal{E})$ by using the Banach contraction principle, such that

$$G = \mathcal{H}(G) = \bigcup_{l=1}^{M} g_l(G).$$

Moreover, $G = \lim_{n\to\infty} \mathcal{H}^n(S)$ for any $S \in \mathcal{C}(\mathcal{E})$. This set $G$ is called the attractor of the IFS. It is also called self-referential set (or) fractal.

## 2.3 Multivalued Maps

Now we discuss few basic concepts of multivalued maps. For the detailed exposition, the reader may consult [11].

Let $\mathcal{E}$ and $\mathcal{D}$ be two metric spaces. Consider a multivalued map $g : \mathcal{E} \to \mathcal{D}$. For the map $g$, denote $g^{-1}(R) := \{\alpha \in \mathcal{E} : g(\alpha) \subseteq R\}$ and $g_{+}^{-1}(R) = \{\alpha \in \mathcal{E} : g(\alpha) \cap R \neq \emptyset\}$.

**Definition 1** If $\mathcal{E} \subseteq \mathcal{D}$ and $g : \mathcal{E} \to \mathcal{D}$ is a multivalued map, then a point $\alpha \in \mathcal{E}$ is said to be a fixed point of $g$ if $\alpha \in g(\alpha)$. The collection of all fixed points of $g$ is identified by $Fix(g) = \{\alpha \in \mathcal{E} : \alpha \in g(\alpha)\}$.

**Definition 2** If $g : \mathcal{E} \to \mathcal{D}$ is a multivalued map and

1. if $g^{-1}(R)$ $(g_{+}^{-1}(R))$ is open in $\mathcal{E}$ for all open sets $R \subseteq \mathcal{D}$, then $g$ is said to be upper semicontinuous (lower semicontinuous), respectively,
2. if $g$ is both upper semicontinuous (u.s.c) and lower semicontinuous (l.s.c), then $g$ is said to be multivalued continuous.

**Proposition 1** ([11]) *If $g : (\mathcal{E}, \tau) \to (\mathcal{C}(\mathcal{E}), \omega)$ is a multivalued continuous map, then the induced map $g^* : (\mathcal{C}(\mathcal{E}), \omega) \to (\mathcal{C}(\mathcal{E}), \omega)$ defined by $g^*(R) = \bigcup_{\alpha\in R} g(\alpha)$ is a well-defined (single valued) continuous map.*

**Lemma 4** ([11]) *If $g : (\mathcal{E}, \tau) \to (\mathcal{C}(\mathcal{E}), \omega)$ is an u.s.c map and $R \in \mathcal{C}(\mathcal{E})$, then $g(R) \in \mathcal{C}(\mathcal{E})$.*

## 3 Cyclic Contractions and It's Fixed Point Principles

Kirk et al. [17] introduced various types of generalized contractions by using cyclic maps. We discuss the conclusion of Banach contraction principle of these contractions.

**Definition 3** Let $(\mathcal{E}, \tau)$ be a metric space and let $\{\Lambda_m\}_{m=1}^k$ be a set of non-empty subsets of $\mathcal{E}$. A map $g : \bigcup_{m=1}^k \Lambda_m \to \bigcup_{m=1}^k \Lambda_m$ is said to be a cyclic map on $\{\Lambda_m\}_{m=1}^k$ if it satisfies:

$$g(\Lambda_m) \subseteq \Lambda_{m+1}, \quad \text{for } m \in \mathcal{N}_k, \text{ where } \Lambda_{k+1} = \Lambda_1.$$

**Remark 1** If $\alpha^*$ is a fixed point of a cyclic map $g$ on $\{\Lambda_m\}_{m=1}^k$, then $\alpha^* \in \bigcap_{m=1}^k \Lambda_m$.

**Definition 4** A self-map $\phi$ on $[0, \infty)$ is said to be comparison function if $\phi$ is a right-continuous, non-decreasing and it satisfies $\phi(r) < r$ for any $r > 0$ and $\phi(0) = 0$.

**Remark 2** If $\phi$ is a comparison function, then $\lim_{n\to\infty} \phi^n(r) = 0$, for any $r \geq 0$.

Let us denote $\mathcal{N}_m$ to be the collection of first $m$ natural numbers. The following contractions are the generalizations of the Banach contraction in the cyclic form:

**Definition 5** A cyclic map $g : \bigcup_{m=1}^k \Lambda_m \to \bigcup_{m=1}^k \Lambda_m$ is said to be

1. cyclic contraction if we can find a constant $0 \leq s < 1$ such that $\tau(g(\alpha), g(\beta)) \leq s\, \tau(\alpha, \beta)$, $\forall\, \alpha \in \Lambda_m,\ \beta \in \Lambda_{m+1}$ for $m \in \mathcal{N}_k$.
2. cyclic $\phi$-contraction if $\tau(g(\alpha), g(\beta)) \leq \phi(\tau(\alpha, \beta))$, $\forall\, \alpha \in \Lambda_m,\ \beta \in \Lambda_{m+1}$ for $m \in \mathcal{N}_k$, where $\phi$ is a comparison function.
3. cyclic Meir–Keeler contraction if $\forall\, \mu > 0,\ \exists\, \nu > 0$ such that $\mu \leq \tau(\alpha, \beta) < \mu + \nu$ implies $\tau(g(\alpha), g(\beta)) < \mu, \forall \alpha \in \Lambda_m,\ \beta \in \Lambda_{m+1}$ for $m \in \mathcal{N}_k$.
4. cyclic contractive if $\tau(g(\alpha), g(\beta)) < \tau(\alpha, \beta)$, $\forall\, \alpha \in \Lambda_m,\ \beta \in \Lambda_{m+1}$ with $\alpha \neq \beta$, for $m \in \mathcal{N}_k$.

**Remark 3** 1. Every contraction map on a space $\mathcal{E}$ is a cyclic contraction on $\{\Lambda_m\}_{m=1}^k$ (Take $\Lambda_m = \mathcal{E}$ for each $m \in \mathcal{N}_k$) but the converse need not be true (see Example 1).
2. Every cyclic contraction with contractive factor $s$ is a cyclic $\phi$-contraction (where $\phi(t) = st$) and cyclic Meir–Keeler contraction (for every $\mu > 0$, choose $\nu = \frac{1-s}{s}\mu$).
3. Observe that every cyclic $\phi$-contraction and cyclic Meir–Keeler contraction is a cyclic contractive map .

**Example 1** Let $\Lambda_1 = [0, 2]$ and $\Lambda_2 = [1, 3]$. Define a mapping $g : \Lambda_1 \cup \Lambda_2 \to \Lambda_1 \cup \Lambda_2$ by

$$g(\alpha) = 17/8 - \alpha/8 \ \text{ if } \alpha \in [0, 2],$$

$$g(\alpha) = 15/8 \qquad \text{if } \alpha \in \left[2, \frac{11}{4}\right],$$

$$g(\alpha) = 37/8 - \alpha \ \text{ if } \alpha \in \left[\frac{11}{4}, 3\right].$$

Then the map $g$ is cyclic contraction and also continuous but not a contraction map.

**Example 2** Let $\Lambda_1 = [0, 1]$, $\Lambda_2 = [0, 3]$. For $m \geq 2$, define $g : \Lambda_1 \cup \Lambda_2 \to \Lambda_1 \cup \Lambda_2$ by

$$g(\alpha) = \alpha/m \ \text{ if } \alpha \in [0, 2],$$

$$g(\alpha) = 1/m \ \text{ if } \alpha \in (2, 3].$$

The map $g$ is non-continuous cyclic contraction.

**Proposition 2** ([17]) *Let $(\mathcal{E}, \tau)$ be a metric space which is complete and let $\{\Lambda_m\}_{m=1}^k$ be a set of subsets of $\mathcal{E}$ which are closed. If $g$ is a cyclic $\phi$-contraction on $\{\Lambda_m\}_{m=1}^k$, then there is a unique point $\alpha^* \in \bigcup_{m=1}^k \Lambda_m$ obeying fixed point condition $g(\alpha^*) = \alpha^*$.*

**Theorem 2** ([30]) *Let $(\mathcal{E}, \tau)$ be a metric space which is complete and let $\{\Lambda_m\}_{m=1}^k$ be a set of subsets of $\mathcal{E}$ which are closed. If $g$ is a cyclic $\phi$-contraction on $\{\Lambda_m\}_{m=1}^k$, then for any $\alpha \in \bigcup_{m=1}^k \Lambda_m$, $\lim_{n\to\infty} g^n(\alpha) = \alpha^*$, where $\alpha^*$ is the fixed point of $g$.*

**Proposition 3** *Let $(\mathcal{E}, \tau)$ be a metric space which is complete and let $\{\Lambda_m\}_{m=1}^k$ be a set of subsets of $\mathcal{E}$ which are closed. If $g$ is a cyclic Meir–Keeler contraction on $\{\Lambda_m\}_{m=1}^k$, then there is a unique point $\alpha^* \in \bigcup_{m=1}^k \Lambda_m$ satisfying $g(\alpha^*) = \alpha^*$ and for each $\alpha \in \bigcup_{m=1}^k \Lambda_m$, $\lim_{n\to\infty} g^n(\alpha) = \alpha^*$*

**Proposition 4** ([17]) *Let $(\mathcal{E}, \tau)$ be a metric space which is compact and let $\{\Lambda_m\}_{m=1}^k$ be a set of subsets of $\mathcal{E}$ which are closed. If $g$ is a cyclic contractive map on $\{\Lambda_m\}_{m=1}^k$, then there is a unique point $\alpha^* \in \bigcup_{m=1}^k \Lambda_m$ satisfying $g(\alpha^*) = \alpha^*$.*

**Theorem 3** ([32]) *Let $(\mathcal{E}, \tau)$ be a metric space which is compact and let $\{\Lambda_m\}_{m=1}^k$ be a set of subsets of $\mathcal{E}$ which are closed. If $g$ is a continuous cyclic contractive map on $\{\Lambda_m\}_{m=1}^k$, then for any $\alpha \in \bigcup_{m=1}^k \Lambda_m$, $\lim_{n\to\infty} g^n(\alpha) = \alpha^*$, where $\alpha^*$ is the fixed point of $g$.*

## 4 Cyclic Multivalued Iterated Function Systems

**Definition 6** Let $(\mathcal{E}, \tau)$ be a metric space and $\{\Lambda_m\}_{m=1}^k$ be a set of non-empty subsets of $\mathcal{E}$ . If a multivalued map $g : \bigcup_{m=1}^k \Lambda_m \to \bigcup_{m=1}^k \mathcal{C}(\Lambda_m)$ satisfies the following conditions:

1. $g(\alpha) \in \mathcal{C}(\Lambda_{m+1}), \forall \alpha \in \Lambda_m, \forall m \in \mathcal{N}_k$, where $\Lambda_{k+1} = \Lambda_1$,
2. $\omega(g(\alpha), g(\beta)) \leq \phi(\tau(\alpha, \beta)), \forall \alpha \in \Lambda_m, \quad \beta \in \Lambda_{m+1}$ for $m \in \mathcal{N}_k$,

where $\phi$ is a comparison function, then $g$ is said to be a cyclic multivalued $\phi$-contraction map on $\{\Lambda_m\}_{m=1}^k$. If $g$ satisfies only the first condition, then $g$ is said to be a cyclic multivalued map on $\{\Lambda_m\}_{m=1}^k$.

**Definition 7** A cyclic multivalued map $g$ on $\{\Lambda_m\}_{m=1}^k$ is said to be cyclic multivalued contractive if $g$ satisfies

$$\omega(g(\alpha), g(\beta)) < \tau(\alpha, \beta) \ \forall \ \alpha \in \Lambda_m, \quad \beta \in \Lambda_{m+1}, \text{ with } \alpha \neq \beta \ \text{ for } m \in \mathcal{N}_k. \quad (1)$$

**Definition 8** A cyclic multivalued map $g$ on $\{\Lambda_m\}_{m=1}^k$ is said to be cyclic multivalued Meir–Keeler contraction if $g$ satisfies $\forall \ \mu > 0, \ \exists \ \nu > 0$ such that

$$\mu \leq \tau(\alpha, \beta) < \mu + \nu \text{ implies } \omega(g(\alpha), g(\beta)) < \mu, \forall \alpha \in \Lambda_m, \ \beta \in \Lambda_{m+1}, \text{ for } m \in \mathcal{N}_k.$$

**Theorem 4** *Let $(\mathcal{E}, \tau)$ be a metric space and $\{\Lambda_m\}_{m=1}^k$ be a collection of subsets of $\mathcal{E}$. If $g$ is an u.s.c cyclic multivalued $\phi$-contraction on $\{\Lambda_m\}_{m=1}^k$, then the induced map $g^* : \bigcup_{m=1}^k \mathcal{C}(\Lambda_m) \to \bigcup_{m=1}^k \mathcal{C}(\Lambda_m)$ defined by $g^*(R) = \bigcup_{\alpha \in R} g(\alpha)$, for any $R \in \bigcup_{m=1}^k \mathcal{C}(\Lambda_m)$ is a (single valued) cyclic $\phi$-contraction on $\{\mathcal{C}(\Lambda_m)\}_{m=1}^k$.*

***Proof*** Let $R \in \mathcal{C}(\Lambda_m)$ for some $m \in \mathcal{N}_k$, this implies $g^*(R) \subseteq \Lambda_{m+1}$ and by Lemma 4, $g^*(R) \in \mathcal{C}(\Lambda_{m+1})$. Let $P \in \mathcal{C}(\Lambda_m)$ and $S \in \mathcal{C}(\Lambda_{m+1})$, for some $m \in \mathcal{N}_k$. Let $\gamma \in g^*(P)$, then there exists $\alpha \in P \subseteq \Lambda_m$ such that $\gamma \in g(\alpha)$. For $\alpha \in P$, there exists $\beta \in S \subseteq \Lambda_{m+1}$ such that $\tau(\alpha, \beta) \leq \omega(P, S)$.

Then, it is plain to see that

$$\tau(\gamma, g^*(S)) \leq \tau(\gamma, g(\beta)) \leq \omega(g(\alpha), g(\beta)) \leq \phi(\tau(\alpha, \beta)) \leq \phi(\omega(P, S)).$$

Since $\gamma$ is arbitrary in $g^*(P)$,

$$\delta(g^*(P), g^*(S)) \leq \phi(\omega(P, S)).$$

Similarly

$$\delta(g^*(S), g^*(P)) \leq \phi(\omega(S, P)).$$

Hence,

$$\omega(g^*(P), g^*(S)) \leq \phi(\omega(P, S)).$$

**Theorem 5** *Let $(\mathcal{E}, \tau)$ be a metric space and $\{\Lambda_m\}_{m=1}^k$ be a collection of subsets of $\mathcal{E}$. If $\{g_l\}_{l=1}^M$ is a collection of u.s.c cyclic multivalued $\phi_l$-contractions on $\{\Lambda_m\}_{m=1}^k$, then the Hutchinson map $\mathcal{H} : \bigcup_{m=1}^k \mathcal{C}(\Lambda_m) \to \bigcup_{m=1}^k \mathcal{C}(\Lambda_m)$ defined by $\mathcal{H}(R) := \bigcup_{l=1}^M g_l^*(R)$, for any $R \in \bigcup_{m=1}^k \mathcal{C}(\Lambda_m)$ is a cyclic $\phi$-contraction on $\{\mathcal{C}(\Lambda_m)\}_{m=1}^k$, where $\phi(r) := \max_{1 \leq l \leq M} \ \phi_l(r)$.*

***Proof*** Let $R \in \mathcal{C}(\Lambda_m)$ for some $m \in \mathcal{N}_k$. By Theorem 4, for each $l \in \mathcal{N}_M$, $g_l^*$ is a cyclic $\phi_l$-contraction. Therefore, $g_l^*(R) \in \mathcal{C}(\Lambda_{m+1})$ for all $l \in \mathcal{N}_M$. Hence $\mathcal{H}(R) = \bigcup_{l=1}^{M} g_l^*(R) \in \mathcal{C}(\Lambda_{m+1})$, and consequently, we have $\mathcal{H}(\mathcal{C}(\Lambda_m)) \subseteq \mathcal{C}(\Lambda_{m+1})$ for $m \in \mathcal{N}_k$.

Let $P \in \mathcal{C}(\Lambda_m)$ and $S \in \mathcal{C}(\Lambda_{m+1})$ for some $m \in \mathcal{N}_k$. Since $\{g_l^*\}_{l=1}^{M}$ are cyclic $\phi_l$-contractions, we have

$$\begin{aligned}\omega(\mathcal{H}(P), \mathcal{H}(S)) &= \omega\left(\bigcup_{l=1}^{M} g_l^*(P), \bigcup_{l=1}^{M} g_l^*(S)\right) \\ &\le \max_{1\le l\le M} \omega(g_l^*(P), g_l^*(S)) \\ &\le \max_{1\le l\le M} \phi_l(\omega(P, S)) \\ &\le \phi(\omega(P, S)).\end{aligned}$$

**Corollary 1** *Suppose that $(\mathcal{E}, \tau)$ is a complete metric space and $\{\Lambda_m\}_{m=1}^{k}$ is a collection of closed subsets of $\mathcal{E}$. If $\{g_l\}_{l=1}^{M}$ is a collection of u.s.c cyclic multivalued $\phi_l$-contractions on $\{\Lambda_m\}_{m=1}^{k}$, then the corresponding Hutchinson map $\mathcal{H}$ defined in the previous theorem has a unique invariant set G (say) and for each $R \in \bigcup_{m=1}^{k} \mathcal{C}(\Lambda_m)$, the sequence $(\mathcal{H}^n(R))_{n\ge 1}$ converges to G.*

***Proof*** By Lemma 3, for each $m \in \mathcal{N}_k$, $\mathcal{C}(\Lambda_m)$ is a closed non-empty subset of the complete metric space $(\mathcal{C}(\mathcal{E}), \omega)$.

By Theorem 5, $\mathcal{H}$ is a cyclic $\phi$-contraction on $\{\mathcal{C}(\Lambda_m)\}_{m=1}^{k}$. From Theorem 2, $\mathcal{H}$ has a unique set $G \in \bigcup_{m=1}^{k} \mathcal{C}(\Lambda_m)$, such that $\mathcal{H}(G) = G$ and for any $R \in \bigcup_{m=1}^{k} \mathcal{C}(\Lambda_m)$, $\lim_{n\to\infty} \mathcal{H}^n(R) = G$.

**Definition 9** A cyclic multivalued $\phi$-contraction IFS is a finite collection of u.s.c $\phi_l$-contraction maps $g_l : \bigcup_{m=1}^{k} \Lambda_m \to \bigcup_{m=1}^{k} \mathcal{C}(\Lambda_m)$, $l \in \mathcal{N}_M$ on a complete metric space $(\mathcal{E}, \tau)$, where $\{\Lambda_m\}_{m=1}^{k}$ are closed.

**Theorem 6** *Let $(\mathcal{E}, \tau)$ be a metric space and $\{\Lambda_m\}_{m=1}^{k}$ be a collection of subsets of $\mathcal{E}$. If $\{g_l\}_{l=1}^{M}$ is a collection of u.s.c cyclic multivalued Meir–Keeler contractions on $\{\Lambda_m\}_{m=1}^{k}$, then the Hutchinson map $\mathcal{H} : \bigcup_{m=1}^{k} \mathcal{C}(\Lambda_m) \to \bigcup_{m=1}^{k} \mathcal{C}(\Lambda_m)$ defined by $\mathcal{H}(R) := \bigcup_{l=1}^{M} g_l^*(R)$, for any $R \in \bigcup_{m=1}^{k} \mathcal{C}(\Lambda_m)$ is a cyclic Meir–Keeler contraction on $\{\mathcal{C}(\Lambda_m)\}_{m=1}^{k}$.*

***Proof*** Since $\{g_l\}_{l=1}^{M}$ are u.s.c cyclic multivalued maps on $\{\Lambda_m\}_{m=1}^{k}$, $\mathcal{H}$ is a well-defined cyclic map on $\{\mathcal{C}(\Lambda_m)\}_{m=1}^{k}$.

For a given $\mu > 0$, there exists $\nu_l > 0$, $l \in \mathcal{N}_M$, such that

$$\mu \le \tau(\alpha, \beta) < \mu + \nu_l \Rightarrow \omega(g_l(\alpha), g_l(\beta)) < \mu, \forall \alpha \in \Lambda_m,\ \beta \in \Lambda_{m+1}, \forall m \in \mathcal{N}_k.$$

Let $P \in \mathcal{C}(\Lambda_m)$ and $S \in \mathcal{C}(\Lambda_{m+1})$, for some $m \in \mathcal{N}_k$, such that $\mu \leq \omega(P, S) < \mu + \nu$, where $\nu = \min\{\nu_l : l \in \mathcal{N}_M\}$. Our claim is $\omega(\mathcal{H}(P), \mathcal{H}(S)) < \mu$.

Let $\gamma \in \mathcal{H}(P)$, then there exists $l \in \mathcal{N}_M$ and $\alpha \in P \subset \Lambda_m$ such that $\gamma \in g_l(\alpha)$. For this $\alpha \in P$, there exists $\beta \in S \subset \Lambda_{m+1}$, such that $\tau(\alpha, \beta) \leq \omega(P, S) < \mu + \nu$.

Case 1. If $\tau(\alpha, \beta) \geq \mu$, then $\mu \leq \tau(\alpha, \beta) < \mu + \nu$ and $\alpha \in \Lambda_m, \beta \in \Lambda_{m+1}$ implies

$$\delta(\gamma, \mathcal{H}(S)) \leq \omega(g_l(\alpha), g_l(\beta)) < \mu.$$

Case 2.: If $0 < \tau(\alpha, \beta) < \mu$,

$$\delta(\gamma, \mathcal{H}(S)) \leq \omega(g_l(\alpha), g_l(\beta)) < \tau(\alpha, \beta) < \mu.$$

By compactness of $\mathcal{H}(P)$,

$$\delta(\mathcal{H}(P), \mathcal{H}(S)) < \mu.$$

Similarly,

$$\delta(\mathcal{H}(S), \mathcal{H}(P)) < \mu.$$

Hence, $\mathcal{H}$ is a cyclic Meir–Keeler contraction on $\{\mathcal{C}(\Lambda_m)\}_{m=1}^k$.

**Corollary 2** *Suppose that $(\mathcal{E}, \tau)$ is a complete metric space and $\{\Lambda_m\}_{m=1}^k$ is a collection of closed subsets of $\mathcal{E}$. If $\{g_l\}_{l=1}^M$ is a collection of u.s.c cyclic multivalued Meir–Keeler contractions on $\{\Lambda_m\}_{m=1}^k$, then the corresponding Hutchinson map $\mathcal{H}$ has a unique invariant set $G$ (say) and for each $R \in \bigcup_{m=1}^k \mathcal{C}(\Lambda_m)$, the sequence $(\mathcal{H}^n(R))_{n\geq 1}$ converges to $G$.*

**Theorem 7** *Let $(\mathcal{E}, \tau)$ be a metric space and $\{\Lambda_m\}_{m=1}^k$ be a collection of subsets of $\mathcal{E}$. If $g$ is a multivalued continuous and cyclic multivalued contractive map on $\{\Lambda_m\}_{m=1}^k$, then the induced map $g^* : \bigcup_{m=1}^k \mathcal{C}(\Lambda_m) \to \bigcup_{m=1}^k \mathcal{C}(\Lambda_m)$ defined by $g^*(R) = \bigcup_{\alpha \in R} g(\alpha)$, for any $R \in \bigcup_{m=1}^k \mathcal{C}(\Lambda_m)$ is continuous cyclic contractive on $\{\mathcal{C}(\Lambda_m)\}_{m=1}^k$ when endowed with the Hausdorff metric $\omega$.*

***Proof*** By Proposition 1, $g^*$ is a (single valued) continuous map with respect to $\omega$. Since $g$ is a multivalued cyclic map on $\{\Lambda_m\}_{m=1}^k$, $g^*$ is a cyclic map on $\{\mathcal{C}(\Lambda_m)\}_{m=1}^k$. Let $P \in \mathcal{C}(\Lambda_m)$, $S \in \mathcal{C}(\Lambda_{m+1})$ with $P \neq S$ for some $m \in \mathcal{N}_k$. Let $\gamma \in g^*(P)$, then there exists $\alpha \in P \subseteq \Lambda_m$ such that $\gamma \in g(\alpha)$.

*Case 1.* If $\alpha \in S$, then $\tau(\gamma, g^*(S)) = 0 < \omega(P, S)$.

*Case 2.* If $\alpha \notin S$, then $\omega(g(\alpha), g(\beta)) < \tau(\alpha, \beta)$, $\forall\, \beta \in S \subseteq \Lambda_{m+1}$.
By compactness of $S$, there exists $\beta^* \in S \subseteq \Lambda_{m+1}$ satisfying $\tau(\alpha, \beta^*) = \inf_{\beta \in S} \tau(\alpha, \beta)$. Thus,

$$\tau(\gamma, g^*(S)) \leq \delta(g(\alpha), g(\beta^*)) \leq \omega(g(\alpha), g(\beta^*)) < \tau(\alpha, \beta^*)$$

$$= \inf_{\beta \in S} \tau(\alpha, \beta) \leq \omega(P, S).$$

Since $\gamma$ is an arbitrary element in $g^*(P)$ and $g^*(P)$ is compact,

$$\delta(g^*(P), g^*(S)) < \omega(P, S).$$

Similarly

$$\delta(g^*(S), g^*(P)) < \omega(S, P).$$

Hence, $g^*$ is a continuous cyclic contractive map on $\{\mathcal{C}(\Lambda_m)\}_{m=1}^k$.

**Lemma 5** *If $\{g_l\}_{l=1}^M$ are continuous maps on a Hausdorff metric space $(\mathcal{C}(\mathcal{E}), \omega)$ for any metric space $(\mathcal{E}, \tau)$, then the map $\mathcal{H} : \mathcal{C}(\mathcal{E}) \to \mathcal{C}(\mathcal{E})$ defined by $F(R) := \bigcup_{l=1}^M g_l(R)$ is also a continuous map.*

**Theorem 8** *Let $(\mathcal{E}, \tau)$ be a metric space and $\{\Lambda_m\}_{m=1}^k$ be a collection of subsets of $\mathcal{E}$. If $\{g_l\}_{l=1}^M$ is a collection of multivalued continuous and cyclic multivalued contractive functions on $\{\Lambda_m\}_{m=1}^k$, then the Hutchinson map $\mathcal{H} : \bigcup_{m=1}^k \mathcal{C}(\Lambda_m) \to \bigcup_{m=1}^k \mathcal{C}(\Lambda_m)$ defined by $\mathcal{H}(R) := \bigcup_{l=1}^M g_l^*(R)$, for any $R \in \bigcup_{m=1}^k \mathcal{C}(\Lambda_m)$ is continuous cyclic contractive on $\{\mathcal{C}(\Lambda_m)\}_{m=1}^k$ when equipped with the Hausdorff metric $\omega$.*

***Proof*** By Theorem 7 and Lemma 5, $\{g_l{}^*\}_{l=1}^M$ are cyclic contractive maps on $\{\mathcal{C}(\Lambda_m)\}_{m=1}^k$ and $\mathcal{H}$ is a continuous cyclic map. Let $P \in \mathcal{C}(\Lambda_m)$ and $S \in \mathcal{C}(\Lambda_{m+1})$, for any $m \in \mathcal{N}_k$, we have

$$\omega(\mathcal{H}(P), \mathcal{H}(S))) \leq \max_{1 \leq l \leq M} \omega(g_l^*(P), g_l^*(S)) < \max_{1 \leq l \leq M} \omega(P, S) = \omega(P, S).$$

Hence, the proof.

**Corollary 3** *Let $(\mathcal{E}, \tau)$ be a compact metric space and $\{\Lambda_m\}_{m=1}^k$ be a collection of closed subsets of $\mathcal{E}$. If $\{g_l\}_{l=1}^M$ is a collection of multivalued continuous and cyclic multivalued contractive functions on $\{\Lambda_m\}_{m=1}^k$, then the corresponding Hutchinson map $\mathcal{H}$ has a unique invariant set $G$ (say). Moreover, for any $R \in \bigcup_{m=1}^k \mathcal{C}(\Lambda_m)$, the sequence $(\mathcal{H}^n(R))_{n \geq 1}$ converges to $G$.*

***Proof*** By Theorems 8 and 3, we can conclude the proof.

## References


1. Banach, S.: Sur les operations dans les ensembles abstrait et leur application aux equations, integrals. Fundam. Math. **3**, 133–181 (1922)
2. Barnsley, M.F.: Fractal functions and interpolation. Constr. Approx. **2**, 303–329 (1986)
3. Barnsley, M.F.: Fractals Everywhere. Academic, Boston (1988)

4. Barnsley, M.F., Hurd, L.P.: Fractal Image Compression. AK Peters Ltd, Wellesley (1993)
5. Chand, A.K.B., Jha, S., Navascués, M.A.: Kantorovich-Bernstein $\alpha$-fractal function in $L^p$ spaces. Quaest. Math. **43**(2), 227–241 (2020)
6. Chand, A.K.B., Navascués, M.A., Viswanathan, P., Katiyar, S.K.: Fractal trigonometric polynomials for restricted range approximation. Fractals **24**(2), 1650022 (2016)
7. Chand, A.K.B., Vijender, N., Navascués, M.A.: Shape preservation of scientific data through rational fractal splines. Calcolo **51**(2), 329–362 (2014)
8. Dumitru, D.: Generalized iterated function systems containing Meir-Keeler functions. Ann. Univ. Bucureşti, Math. **LVIII**, 3–15 (2009)
9. Fernau, H.: Infinite iterated function systems. Math. Nachr. **170**, 79–91 (1994)
10. Georgescu, F.: Iterated function systems consisting of generalized convex contractions in the framework of complete strong b-metric spaces. Ann. Univ. Vest Timiş. Şer. Mat. Inform. *55*, 119–142 (2017)
11. Górniewicz, L.: Topological Fixed Point Theory of Multivalued Mappings. Kluwer, Dordrecht (1999)
12. Hata, M.: On some properties of set-dynamical systems. Proc. Jpn. Acad. **61**, 99–102 (1985)
13. Hutchinson, J.: Fractals and self-similarity. Indiana Univ. Math. J. **30**, 713–747 (1981)
14. Di Ieva, A., Grizzi, F., Jelinek, H., Pellionisz, A.J., Losa, G.A.: Fractals in the neurosciences, part I general principles and basic neurosciences. Neuroscientist **20**, 403–417 (2013)
15. Ioana, L., Mihail, A.: Iterated function systems consisting of $\phi$-contractions. Results Math. **72**, 2203–2225 (2017)
16. Jha, S., Chand, A.K.B., Navascués, M.A.: Approximation by shape preserving fractal functions with variable scaling. Calcolo **58**(8), 1–24 (2021)
17. Kirk, W.A., Srinivasan, P.S., Veeramani, P.: Fixed points for mappings satisfying cyclical contractive conditions. Fixed Point Theory **4**, 79–89 (2003)
18. Klimek, M., Kosek, M.: Generalized iterated function systems, multifunctions and Cantor sets. Ann. Polon. Math. **96**(1), 25–41 (2009)
19. Kunze, H., La Torre, D., Vrscay, E.: From iterated function systems to iterated multifunction systems. Comm. Appl. Nonlinear Anal. **15**, 1–13 (2008)
20. Leśniak, K.: Infinite iterated function systems: a multivalued approach. Bull. Pol. Acad. Sci. Math. **52**(1), 1–8 (2004)
21. Leśniak, K.: Homoclinic attractors in discontinuous iterated function systems. Chaos Solitons Fractals **81**, 146–149 (2015)
22. Mandelbrot, B.B.: The Fractal Geometry of Nature. Freeman, New York (1982)
23. Maślanka, Ł, Strobin, F.: On generalized iterated function systems defined on $l_\infty$-sum of a metric space. J. Math. Anal. Appl. **461**, 1795–1832 (2020)
24. Mazel, D.S., Hayes, M.H.: Using iterated function systems to model discrete sequences. U. IEEE Trans. Signal Process. **40**, 1724–1734 (1992)
25. Miculescu, R., Urziceanu, S.: The canonical projection associated with certain possibly infinite generalized iterated function systems as a fixed point. J. Fixed Point Theory Appl. **20**, 141 (2018)
26. Miculescu, R., Mihail, A., Urziceanu, S.: A new algorithm that generates the image of the attractor of a generalized iterated function system. Numer. Algorithms **83**, 1399–1413 (2020)
27. Mihail, A.: The shift space for recurrent iterated function systems. Rev Roum Math Pures Appl. **53**, 339–355 (2008)
28. Mihail, A., Miculescu, R.: Generalized IFSs on noncompact spaces. Fixed Point Theory Appl. 2010 (2010)
29. Okamura, K.: Self-similar measures for iterated function systems driven by weak contractions. Proc. Jpn. Acad. Ser. A Math. Sci. **94**, 31–35 (2018)
30. Pasupathi, R., Chand, A.K.B., Navascués, M.A.: Cyclic iterated function systems. J. Fixed Point Theory Appl. **22**(3), 1–17 (2020)
31. Pasupathi, R., Chand, A.K.B., Navascués, M.A.: Cyclic Meir-Keeler contraction and its fractals. Numer. Funct. Anal. Optim. **42**(9), 1053–1072 (2021)

32. Pasupathi, R., Chand, A.K.B., Navascués, M.A., Sebastián, M.V.: Cyclic generalized iterated function systems. Comp. Math. Methods **3**(6), 1–12 (2021)
33. Roy, A., Sujith, R.I.: Fractal dimension of premixed flames in intermittent turbulence. Combust. Flame **226**, 412–418 (2021)
34. Samuel, M., Tetenov, A.: On attractors of iterated function systems in uniform spaces. Sib Élektron Mat Izv **14**, 151–155 (2017)
35. Secelean, N.A.: Countable iterated function systems. Far East J. Dyn. Syst. **3**(2), 149–167 (2001)
36. Secelean, N.A.: The Existence of the Attractor of Countable Iterated Function Systems. Mediterr. J. Math. **9**, 61–79 (2012)

# On Almost Statistical Convergence of Weight $r$

**Ekrem Savaş**

**Abstract** In this paper, we introduce the opinion of $\tau$-almost statistical convergence of weight $r : \mathbb{R}^+ \to \mathbb{R}^+$ where $r(\xi_k) \to \infty$ for any sequence $(\xi_k)$ in $\mathbb{R}^+$ with $\xi_k \to \infty$. We also examine some relations.

**Keywords** Weight function $r$ · Statistical convergence · Almost convergence

## 1 Introduction

The following definition was given by Fast [1]: A sequence $\xi$ is said to be statistically convergent to the number $\gamma$ if for every $\omega > 0$

$$\lim_t \frac{1}{t}|\{k < t : |\xi_k - \gamma| \geq \omega\}| = 0.$$

In such case, we write $s - \lim \xi = \gamma$ or $\xi_k \to \gamma(s)$. Subsequently statistically convergent sequences have been considered in [5, 6].

Further the concept of $\tau$-almost statistical convergence was considered by Savas [5]. The goal of this note is to consider the idea of $\tau$-almost statistical convergence of weight $r$.

Lorentz [2] has expressed that

$$\hat{c} = \left\{\xi \in l_\infty : \lim_p \varphi_{p,q}(\xi) \text{ exists, uniformly in } q\right\}$$

where

$$\varphi_{p,q}(\xi) = \frac{\xi_q + \xi_{q+1} + \xi_{q+2} + \cdots + \xi_{q+p}}{p+1}.$$

E. Savaş (✉)
Department of Mathematics, Uşak University, Uşak, Turkey
e-mail: ekremsavas@yahoo.com



B. Rushi Kumar et al. (eds.), *Mathematics and Computing*, Springer Proceedings in Mathematics & Statistics 415, https://doi.org/10.1007/978-981-19-9307-7_22

**Definition 1** ([3]) A sequence $\xi = (\xi_k) \in m_\infty$ (the set of bounded sequences) is said to be strongly almost convergent to a number $\gamma$ if

$$\lim_{q\to\infty} \frac{1}{q} \sum_{k=1}^{q} \left|\xi_{k+p} - \gamma\right| = 0$$

uniformly in $p$.

By $\left[\hat{c}\right]$ we represent the space of all strongly almost convergent sequences. It has been observed that $c \subset [\hat{c}] \subset \hat{c} \subset m_\infty$ and the inclusions are strict.

**Definition 2** ([4]) Let $\tau = (\tau_q)$ be a non-decreasing sequence of positive numbers in such a way $\tau_{q+1} \leq \tau_q + 1$, $\tau_1 = 1$, $\tau_q \to \infty$ as $q \to \infty$. Let $I_q = \left[q - \tau_q + 1, q\right]$. The generalized de La Vallée–Poussin mean is interpreted such as

$$t_q(\xi) = \frac{1}{\tau_q} \sum_{k\in I_q} \xi_k.$$

A sequence $\xi = (\xi_k)$ is said to be $(V, \tau)$-summable to $\gamma$ if $t_q(\xi) \to \gamma$ as $q \to \infty$.

The opinion of $\tau$-statistical convergence was considered by Mursaleen [4]. Recall that a sequence $\xi = (\xi_k)$ is said to be $\tau$-statistically convergent if there is a complex number $\gamma$ in such a way

$$\lim_{q\to\infty} \frac{1}{\tau_q} \left|\{k \in I_q : |\xi_k - \gamma| \geq \omega\right| = 0.$$

The family of all $\tau$-statistically convergent sequences is represented by $\hat{S}_\tau$. Later the opinion of almost $\tau$-statistical convergence was studied by Savas [5].

For this paper we will consider function $r : \mathbb{R}^+ \to \mathbb{R}^+$ such that $r(\xi_k) \to \infty$ if $\xi_k \to \infty$. The family of all such functions will be represent by $\boldsymbol{R}$.

## 2 Main Results

**Definition 3** Let the sequence $\tau = (\tau_q)$ of real numbers be considered as above and let $r \in \mathbf{R}$. A sequence $\xi = (\xi_k)$ is said to be $\tau$-almost statistically convergent of weight $r$ if there is $\gamma$ in such a way

$$\lim_{q\to\infty} \frac{1}{r(\tau_q)} \left|\{k \in I_q : \left|\xi_{k+p} - \gamma\right| \geq \omega\right| = 0$$

uniformly in $p$. Consequently we write $\hat{S}_\tau^r - \lim \xi_k = \gamma$.

**Theorem 1** *Let $r \in \mathbf{R}$ and $\xi = (\xi_k)$, $\rho = (\rho_k)$ be sequences of complex numbers.*

(i) If $\hat{S}_\tau^r - \lim \xi_k = \xi_0$ and $\phi \in \mathbb{C}$, then $\hat{S}_\tau^r - \lim \phi\xi_k = \phi\xi_0$.
(ii) If $\hat{S}_\tau^r - \lim \xi_k = \xi_0$ and $\hat{S}_\tau^r - \lim \rho_k = \rho_0$, then $\hat{S}_\tau^r - \lim(\xi_k + \rho_k) = \xi_0 + \varrho_0$.

***Proof*** (i) For $\phi = 0$ the result is clear. Let $\phi \neq 0$. Now consider that

$$\frac{1}{r(\tau_q)} \left|\{k \in I_q : \left|\phi\xi_{k+p} - \phi\xi_0\right| \geq \omega\}\right| = \frac{1}{r(\tau_q)} \left|\left\{k \in I_q : \left|\xi_{k+p} - \xi_0\right| \geq \frac{\omega}{|\phi|}\right\}\right|.$$

(ii) Now

$$\frac{1}{r(\tau_q)} \left|\{k \in I_q : \left|\xi_{k+p} + \rho_{k+p} - (\xi_0 + \rho_0)\right| \geq \omega\}\right|$$
$$\leq \frac{1}{r(\tau_q)} \left|\left\{k \in I_q : \left|\xi_{k+p} - \xi_0\right| \geq \frac{\omega}{2}\right\}\right| + \frac{1}{r(\tau_q)} \left|\left\{k \in I_q : \left|\rho_{k+p} - \rho_0\right| \geq \frac{\omega}{2}\right\}\right|.$$

This completed proof.

**Definition 4** Let $\tau = \left(\tau_q\right)$ and $r \in \mathbf{R}$. Let $\varrho$ be a positive real number. A sequence $\xi = (\xi_k)$ is claimed to be strongly $\left(\hat{V}, \tau\right)$-almost summable of weight $r$ if there exists $\gamma$ such that

$$\lim_{q\to\infty} \frac{1}{r(\tau_q)} \sum_{k\in I_q} \left|\xi_{k+p} - \gamma\right|^{\varrho} = 0$$

uniformly in $p$. The family of all strongly $\left(\hat{V}, \tau\right)$-almost summable sequences of weight $r$ will be represent by $\left[\hat{V}_\varrho^r, \tau\right]$.

**Theorem 2** *Let $r_1, r_2 \in \mathbf{R}$ and there are $M > 0$ and $s \in \mathbb{N}$ in such a way $r_1(\tau_q)/r_2(\tau_q) \leq M$ for all $q \geq s$ then $\hat{S}_\tau^{r_1} \subseteq \hat{S}_\tau^{r_2}$.*

***Proof*** Note that, for all $p$

$$\frac{1}{r_2(\tau_q)} \left|\{k \in I_q : \left|\xi_{k+p} - \gamma\right| \geq \omega\}\right| = \frac{r_1(\tau_q)}{r_2(\tau_q)} \cdot \frac{1}{r_1(\tau_q)} \left|\{k \in I_q : \left|\xi_{k+p} - L\right| \geq \omega\}\right|$$
$$\leq M \cdot \frac{1}{r_1(\tau_q)} \left|\{k \in I_q : \left|\xi_{k+p} - L\right| \geq \omega\}\right|$$

for all $q \geq s$. If $\xi = (\xi_k) \in \hat{S}_\tau^{r_1}$ for all $\omega > 0$ and finally

$$\frac{1}{r_2(\tau_q)} \left|\{k \in I_q : \left|\xi_{k+p} - \gamma\right| \geq \varepsilon\}\right| = 0$$

uniformly in $p$ and so $\xi \in \hat{S}_\tau^{r_2}$. Therefore $\hat{S}_\tau^{r_1} \subseteq \hat{S}_\tau^{r_2}$.

**Corollary 1** *In especial let* $r \in \boldsymbol{R}$ *and there are* $T > 0$ *and a* $r \in \mathbb{N}$ *in such a way* $q/r(\tau_q) \leq T$ *for all* $q \geq s$ *then* $\hat{S}_\tau^r \subseteq \hat{S}_\tau$.

**Theorem 3** $\hat{S} \subseteq \hat{S}_\tau^r$ *if* $\liminf_{q\to\infty} \frac{r(\tau_q)}{q} > 0$.

***Proof*** For any $\omega > 0$, we have

$$\left\{k \leq t : \left|\xi_{k+p} - \gamma\right| \geq \omega\right\} \supseteq \left\{k \in I_q : \left|\xi_{k+p} - L\right| \geq \omega\right\}.$$

Therefore we get that for $p \in \mathbb{N}$

$$\begin{aligned}\frac{1}{t}\left|\left\{k \leq t : \left|\xi_{k+p} - \gamma\right| \geq \omega\right\}\right| &\geq \frac{1}{t}\left|\left\{k \in I_q : \left|\xi_{k+p} - \gamma\right| \geq \omega\right\}\right| \\ &\geq \frac{r(\tau_q)}{t} \cdot \frac{1}{r(\tau_q)}\left|\left\{k \in I_q : \left|\xi_{k+p} - \gamma\right| \geq \omega\right\}\right|.\end{aligned}$$

If $\xi \to \gamma(\hat{S})$ then $\frac{1}{t}\left|\{k \leq t : |\xi_k - \gamma| \geq \omega\}\right| \to 0$ as $t \to \infty$ and finally via above we write

$$\frac{1}{t}\left|\left\{k \leq t : \left|\xi_{k+p} - \gamma\right| \geq \varepsilon\right\}\right| \to 0$$

and so

$$\frac{1}{r(\tau_q)}\left|\left\{k \in I_q : \left|\xi_{k+p} - \gamma\right| \geq \varepsilon\right\}\right| \to 0$$

as $q \to \infty$. We get that $\xi \to \gamma(\hat{S}_\tau^r)$.

**Theorem 4** *Let* $r_1, r_2 \in \boldsymbol{R}$ *and there are* $T > 0$ *and a* $s \in \mathbb{N}$ *in such a way* $r_1(\tau_q)/r_2(\tau_q) \leq M$ *for all* $q \geq s$ *at that time* $\left[\hat{V}_\varrho^r, \tau\right] \subseteq \left[\hat{V}_\varrho^r, \tau\right]$.

***Proof*** The proof is easy and so is omitted.

**Corollary 2** *Let* $r \in \boldsymbol{R}$ *and there exist* $T > 0$ *and a* $s \in \mathbb{N}$ *in such a way* $q/r(\tau_q) \leq M$ *for all* $q \geq s$ *then* $\hat{S}_\tau^r \subseteq \hat{S}_\tau$.

**Theorem 5** *If* $0 < \varrho < \sigma < \infty$ *and* $r \in \boldsymbol{R}$ *at that time* $\left[\hat{V}_\sigma^r, \tau\right] \subset \left[\hat{V}_\varrho^r, \tau\right]$.

The proof is evident via Holder's inequality.

**Theorem 6** *Let* $r_1, r_2 \in \boldsymbol{R}$ *and there exist* $T > 0$ *and a* $s \in \mathbb{N}$ *in such a way* $r_1(\tau_q)/r_2(\tau_q) \leq T$ *for all* $q \geq s$ *and let* $0 < \varrho < \infty$. *If a sequence* $\xi = (\xi_k)$ *is strongly* $\left(\hat{V}, \tau\right)$*-almost summable of weight* $r_1$ *to* $\gamma$ *then it is* $\tau$*-almost statistically convergent of weight* $r_2$ *to* $\gamma$ *i.e* $\left[\hat{V}_\varrho^{r_1}, \tau\right] \subset \hat{S}_\tau^{r_2}$.

***Proof*** Let $\xi = (\xi_k) \in \left[\hat{V}_{\varrho}^{r_1}, \tau\right]$ and Let us take $\omega > 0$ and so

$$\sum_{k \in I_q} \left|\xi_{k+p} - \gamma\right|^{\varrho} = \sum_{\substack{k \in I_q \\ |\xi_{k+p}-\gamma| \geq \omega}} \left|\xi_{k+p} - \gamma\right|^{\varrho} + \sum_{\substack{k \in I_{\varrho} \\ |\xi_{k+q}-\gamma| < \omega}} \left|\xi_{k+p} - \gamma\right|^{\varrho}$$

$$\geq \sum_{\substack{k \in I_q \\ |\xi_{k+p}-\gamma| \geq \omega}} \left|\xi_{k+p} - \gamma\right|^{\varrho}$$

$$\geq \left|\{k \in I_q : \left|\xi_{k+p} - \gamma\right| \geq \omega\right| . \omega^{\varrho}.$$

Now it follows that

$$\frac{1}{r_1(\tau_q)} \sum_{k \in I_q} \left|\xi_{k+p} - \gamma\right|^{\varrho} \geq \frac{1}{r_1(\tau_q)} \left|\{k \in I_q : \left|\xi_{k+p} - \gamma\right| \geq \omega\right| . \varepsilon^{\varrho}$$

$$= \frac{r_2(\tau_q)}{r_1(\tau_q)} \cdot \frac{1}{r_2(\tau_q)} \left|\{k \in I_q : \left|\xi_{k+p} - \gamma\right| \geq \varepsilon\right| . \omega^{\varrho}$$

$$\geq \frac{1}{T} \cdot \frac{1}{r_2(\tau_q)} \left|\{k \in I_q : \left|\xi_{k+p} - \gamma\right| \geq \omega\right| . \omega^{\varrho}$$

for all $q \geq s$. If $\xi \to \gamma(\left[\hat{V}_{\varrho}^{r_1}, \tau\right])$ we get $\xi \to \gamma(\hat{S}_{\tau}^{r_2})$.

## References


1. Fast, H.: Sur la convergence statistique. Colloq. Math. **2**, 41–44 (1951)
2. Lorentz, G.G.: A contribution to the theory of divergent sequences. Acta Math. **80**, 167–190 (1948)
3. Maddox, I.J.: Spaces of strongly summable sequences. Quart. J. Math. **18**, 345–355 (1967)
4. Mursaleen, M.: $\lambda$-statistical convergence. Math. Slovaca **50**, 111–115 (2000)
5. Savas, E.: Strong almost convergence and almost $\tau$-statistical convergence. Hokkaido Math. J. **29**(3), 531–536 (2000)
6. Savas, R.: $\lambda-$ strongly bivariate summable functions of weight g. Suleyman Demirel Üniversitesi Fen Edebiyat Fakültesi Dergisi **15**(1), 80–89 (2020)

# Non-neighbor Topological Indices on Covid-19 Drugs with QSPR Analysis

**W. Tamilarasi and B. J. Balamurugan**

**Abstract** Coronavirus (COVID-19) is one of the recent infectious diseases caused by the virus SARS-CoV-2. The virus causes mild to severe respiratory problems which may lead to death in most cases. There is currently no precise or effective medication available to treat COVID-19 patients. Researchers and many pharmaceutical industries are working toward novel therapeutics and repurposed drugs for coronavirus. In this study, we consider some investigational antiviral drugs like Nitazoxanide, Imatinib, Famotidine, Galidesivir, and Artesunate that are used for the treatment of COVID-19. For this purpose, here we define various non-neighbor topological indices over the above aforesaid antiviral drugs to investigate the physicochemical properties associated with the indices. Further QSPR analysis was carried out between seven non-neighbor topological indices and eight physicochemical properties for the above drugs using the Linear regression method. The result obtained could aid in discovering new vaccines and drugs for COVID-19 disease.

**Keywords** Antiviral drugs · NN-polynomial · Non-neighbor topological indices · QSPR study · Linear regression

## 1 Introduction

The SARS-CoV-2 virus, which is the cause of COVID-19, spreads through respiratory droplets when an infected person sneezes or coughs. The outbreak of the disease was first reported in China and has spread worldwide. The person infected by this virus has symptoms like fever, cough, and shortness of breath. People with lung disease, diabetes, old age, and a compromised immune system are at higher risk of COVID-19. Although FDA has approved only one antiviral drug namely remdesivir

W. Tamilarasi · B. J. Balamurugan (✉)
Division of Mathematics, School of Advanced Sciences, Vellore Institute of Technology, Chennai Campus, Vandalur-Kelambakkam Road, Chennai 600127, Tamil Nadu, India
e-mail: balamurugan.bj@vit.ac.in

W. Tamilarasi
e-mail: tamilarasi.w2019@vitstudent.ac.in



B. Rushi Kumar et al. (eds.), *Mathematics and Computing*, Springer Proceedings in Mathematics & Statistics 415, https://doi.org/10.1007/978-981-19-9307-7_23

for the treatment of COVID-19, there are many investigational drugs like Nitazoxanide, Imatinib, Famotidine, Galidesivir, and Artesunate that are being tested. It has been recently reported [1] that the drug Nitazoxanide exhibits in vitro activity and antiviral effect against SARS-CoV-2. The drug Imatinib provide significant clinical impact for COVID-19 patient who is in critical condition [2]. Moreover, this drug will increase the endosomal pH level and reduce cell fusion of SARS-CoV-2 virus. It has also been reported [3] that Famotidine, another unique drug, is being tested currently for the treatment of COVID-19 based on its excellent ADMET properties. Galidesivir is a broad-spectrum antiviral drug that potentially fights against the coronavirus family: Ebola virus and some RNA viruses [4]. The drug Artesunate was considered an effective medicine for the treatment of COVID-19 because of its anti-inflammatory activity [5].

One branch of mathematical chemistry that examines the chemical characteristics of molecules is chemical graph theory. Topological index is an important tool in chemical graph theory which gives a mathematical measure of chemical graphs such as vertex degree, distance, eccentricity, spectrum, etc. Topological indices are mainly used in QSPR, QSAR, and QSTR studies that allow pharmacologists and chemists to use these data for drug discovery. Here we have considered non-neighbor topological indices which were first introduced by A. Rizwana et al. [6]. The indices considered for the study are non-neighbor First Zagreb index $\overline{M_1(G)}$ [6], non-neighbor Second Zagreb index $\overline{M_2(G)}$ [6], non-neighbor Harmonic index $\overline{H(G)}$ [6], non-neighbor Randić index $\overline{R_\alpha(G)}$ [7], non-neighbor Sum Connectivity index $\overline{SCI(G)}$ [8], non-neighbor ABC index $\overline{ABC(G)}$ [8], and non-neighbor Geometric Arithmetic index $\overline{GA(G)}$ [9].

The topological indices are computed by transforming the structure of a chemical compound into a molecular graph $G = (V, E)$ by representing the atoms as a vertex set $V$ and bonds between the atoms as an edge set $E$. The computation of topological indices directly by using the formula from the molecular graph is a tedious process. To overcome this there are several algebraic polynomials that can help to recover various topological indices. One of the polynomials is the Hosoya polynomial [10], which is used to deduce distance-based indices mainly the Weiner index. The indices which calculate the equidistant edges of the graph are found using Omega and Theta polynomials, whereas the indices which compute non-equidistant edges of the graph are deduced using Sadhana and PI polynomial [11]. M-polynomial [12] is used by many researchers to recover topological indices based on degree. Similar to M-polynomial, NM-polynomial [13] is also used to compute neighborhood degree sum-based topological indices. Here we have introduced a non-neighbor polynomial (NN-polynomial) to recover various non-neighbor topological indices.

Topological descriptor plays a significant role in mathematical chemistry, especially in QSPR/QSAR studies. Kirmani et al., in [14] carried out QSPR and QSAR analysis between topological index based on the degree and physicochemical properties of some COVID-19 drugs by using a multiple linear regression model. M.

C. Shanmukha et al., in [15] established QSPR analysis using a linear regression model for anticancer drugs. Havareh has designed QSPR model using degree-based, mostar type, and distance-based topological drugs that are used in the treatment of COVID-19 by curvilinear regression method [16]. Similarly, a QSPR model was designed using degree-based and neighborhood degree-based topological indices of novel drugs for the treatment of cancer by using the same curvilinear regression approach [17]. QSPR analysis using well-known degree-based topological indices for drugs used in the treatment of breast cancer was investigated [18] and was revealed that some of the indices possess high correlation values with the physicochemical properties of drugs. Zhong et al. [19] have investigated the QSPR analysis of valency-based topological indices of COVID-19 drugs and found that these indices serve as good predictive means in QSPR investigations. The above literature motivated us to carry out QSPR analysis between non-neighbor topological indices and physical and chemical properties for the aforesaid drugs using a linear regression model.

The Non-Neighbor polynomial of a graph G is defined as follows:

**Definition 1.1** The Non-Neighbor polynomial (NN-polynomial) of a graph $G$ is defined as

$NN(G; x, y) = \sum_{i \le j} e_{i,j} x^i y^j$, where $e_{i,j}, i, j \geq 1$, is a number of edges $uv$ of $G$ such that $\left\{\overline{d_G(u)}, \overline{d_G(v)}\right\} = \{i, j\}$ and $\overline{d_G(v)} = n - 1 - d_G(v)$ where $\overline{d_G(v)} =$ number of non-neighbors of the vertex $v \in G$ and $d_G(v) =$ degree of the vertex $v$.

The mathematical expression of the above indices is summarized in Table 1.

where $D_x = x\left(\frac{\partial(h(x,y))}{\partial x}\right)$, $D_y = y\left(\frac{\partial(h(x,y))}{\partial y}\right)$, $S_x = \int_0^x \frac{h(t,y)}{t} dt$, $S_y = \int_0^y \frac{h(x,t)}{t} dt$,

$J(h(x, y)) = h(x, x)$, $Q_k(h(x, y)) = x^k h(x, y)$For non-neighbor topological indices: $h(x, y) = NN(G; x, y)$.

## 2 Methodology and New Results

The edge partition technique is used to determine the NN-polynomial of the molecular graphs for the drugs Nitazoxanide (N), Imatinib (I), Famotidine (F), Galidesivir (G), and Artesunate (A). The aforementioned antiviral drugs' chemical structures are retrieved from "www.pubchem.ncbi.nlm.nih.gov." For each of the aforementioned medications, molecular graphs are created using their respective chemical structures. Using the formulae in Table 1, the various non-neighbor topological indices from the NN-polynomial are obtained.

**Table 1** Derivation formulae of topological indices from NN-polynomial

| Non-neighbor topological indices | Mathematical expression | $h(x, y) = NN(G; x, y)$ |
|---|---|---|
| Non-Neighbor First Zagreb index $\overline{M_1(G)}$ | $\sum_{uv\in E(G)} \left(\overline{d_G(u)} + \overline{d_G(v)}\right)$ | $\left(D_x + D_y\right)(h(x, y))\vert_{x=y=1}$ |
| Non-Neighbor Second Zagreb index $\overline{M_2(G)}$ | $\sum_{uv\in E(G)} \left(\overline{d_G(u)}.\overline{d_G(v)}\right)$ | $\left(D_x D_y\right)(h(x, y))\vert_{x=y=1}$ |
| Non-Neighbor Harmonic index $\overline{H(G)}$ | $\sum_{uv\in E(G)} \frac{2}{\overline{d_G(u)+d_G(v)}}$ | $(2S_x J)(h(x, y))\vert_{x=1}$ |
| Non-Neighbor Randić index $\overline{R_\alpha(G)}$ | $\sum_{uv\in E(G)} \left(\overline{d_G(u)}.\overline{d_G(v)}\right)^\alpha$ | $\left(D_x^k D_y^k\right)(h(x, y))\vert_{x=y=1}$ |
| Non-Neighbor Sum connectivity index $\overline{SCI(G)}$ | $\sum_{uv\in E(G)} \left(\overline{d_G(u)} + \overline{d_G(v)}\right)^\alpha$ | $\left(S_x^{1/2} J\right)(h(x, y))\vert_{x=1}$ |
| Non-Neighbor ABC index $\overline{ABC(G)}$ | $\sum_{uv\in E(G)} \sqrt{\frac{\overline{d_G(u)}+\overline{d_G(v)}-2}{\overline{d_G(u)}.\overline{d_G(v)}}}$ | $\left(D_x^{1/2} Q_{-2} J S_x^{1/2} S_y^{1/2}\right)(h(x, y))\vert_{x=1}$ |
| Non-Neighbor Geometric Arithmetic index $\overline{GA(G)}$ | $\sum_{uv\in E(G)} \frac{2\sqrt{\overline{d_G(u)}.\overline{d_G(v)}}}{d_u+d_v}$ | $\left(2S_x J D_x^{1/2} D_y^{1/2}\right)(h(x, y))\vert_{x=1}$ |

## 2.1 Computation of NN-polynomial and Topological Indices of Nitazoxanide

**Theorem 2.1.1** *The NN-polynomial of the chemical graph N of Nitazoxanide is*

$$NN(N; x, y) = 3x^{17}y^{17} + 10x^{17}y^{18} + 5x^{17}y^{19} + 4x^{18}y^{18}$$

**Proof** The chemical structure and chemical graph of Nitazoxanide are shown in Fig. 1a and 1b, respectively. Let $N$ be the chemical graph of Nitazoxanide with 21 vertices and 22 edges. Let $E_{i,j}$ denote the set of all edges $uv$ where $\{i, j\}$ denote the number of non-neighbor vertices at $u$ and $v$ respectively such that

$E_{i,j} = \left\{uv \in E(G) : \overline{d_G(u)} = i, \overline{d_G(v)} = j\right\}$ and $\left|E_{i,j}\right| = e_{i,j}$. From the molecular graph of Nitazoxanide we obtain,

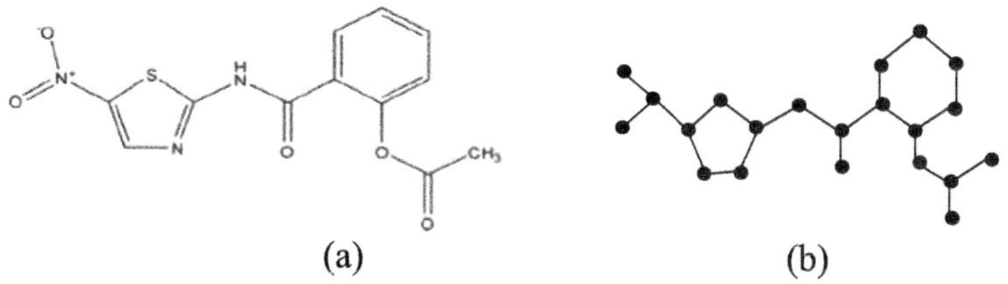


**Fig. 1** **a** Chemical structure and **b** Chemical graph of Nitazoxanide

$$e_{17,17} = 3, e_{17,18} = 10, e_{17,19} = 5, e_{18,18} = 4$$

Therefore, by Definition 1.1, we have

$$\begin{aligned} NN(N; x, y) &= \sum_{i \le j} e_{i,j} x^i y^j \\ &= e_{17,17} x^{17} y^{17} + e_{17,18} x^{17} y^{18} + e_{17,19} x^{17} y^{19} + e_{18,18} x^{18} y^{18} \\ &= 3x^{17} y^{17} + 10x^{17} y^{18} + 5x^{17} y^{19} + 4x^{18} y^{18} \end{aligned}$$

Hence the result.

In the following theorem, we recover some non-neighbor topological indices of Nitazoxanide from the NN-polynomial in Theorem 2.1.1 and by using the formula in Table 1.

**Theorem 2.1.2** *Let N be the molecular graph of Nitazoxanide, then we have*

$$\overline{M_1(N)} = 776, \overline{M_2(N)} = 6838, \overline{H(N)} = 1.2478, \overline{R_{-1/2}(N)} = 1.2485,$$

$$\overline{SCI(N)} = 3.7048, \overline{ABC(N)} = 7.2, \overline{GA(N)} = 21.9881.7$$

**Proof** Consider the NN-polynomial of Nitazoxanide from the above result. We have,

$$NN(N; x, y) = f(x, y) = 3x^{17} y^{17} + 10x^{17} y^{18} + 5x^{17} y^{19} + 4x^{18} y^{18}$$

Then

$$(D_x + D_y)(f(x, y)) = 102x^{17}y^{17} + 350x^{17}y^{18} + 180x^{17}y^{19} + 144x^{18}y^{18}$$
$$(D_x D_y)(f(x, y)) = 867x^{17}y^{17} + 3060x^{17}y^{18} + 1615x^{17}y^{19} + 1296x^{18}y^{18}$$
$$(2S_x J)(f(x, y)) = \frac{6}{34}x^{34} + \frac{20}{35}x^{35} + \frac{10}{36}x^{36} + \frac{8}{36}x^{36}$$
$$(D_x^k D_y^k)(f(x, y)) = 3(289)^{\alpha}x^{17}y^{17} + 10(306)^{\alpha}x^{17}y^{18} + 5(323)^{\alpha}x^{17}y^{19} + 4(324)^{\alpha}x^{18}y^{18}$$
$$\left(S_x^{1/2} J\right)(f(x, y)) = \frac{3}{\sqrt{34}}x^{34} + \frac{10}{\sqrt{35}}x^{35} + \frac{5}{\sqrt{36}}x^{36} + \frac{4}{\sqrt{36}}x^{36}$$
$$\left(D_x^{1/2} Q_{-2} J S_x^{1/2} S_y^{1/2}\right)(f(x, y)) = \frac{3\sqrt{32}}{17}x^{32} + \frac{10\sqrt{33}}{\sqrt{17\sqrt{18}}}x^{33} + \frac{5\sqrt{34}}{\sqrt{19}\sqrt{17}}x^{34} + \frac{4\sqrt{34}}{18}x^{34}$$
$$\left(2S_x J D_x^{1/2} D_y^{1/2}\right)(f(x, y)) = \frac{102}{34}x^{34} + \frac{20\sqrt{17\sqrt{18}}}{35}x^{35} + \frac{10\sqrt{19}\sqrt{17}}{36}x^{36} + \frac{144}{36}x^{36}$$

Now from Table 1, we get

$$\overline{M_1(N)} = (D_x + D_y)(f(x, y))|_{x=y=1} = 776$$
$$\overline{M_2(N)} = (D_x D_y)(f(x, y))|_{x=y=1} = 6838$$
$$\overline{H(N)} = (2S_x J)(f(x, y))|_{x=1} = 1.2478$$
$$\overline{R_{-1/2}(N)} = (D_x^k D_y^k)(f(x, y))|_{x=y=1} = 1.2485$$
$$\overline{SCI(N)} = \left(S_x^{1/2} J\right)(f(x, y))|_{x=1} = 3.7048$$
$$\overline{ABC(N)} = \left(D_x^{1/2} Q_{-2} J S_x^{1/2} S_y^{1/2}\right)(f(x, y))|_{x=1} = 7.2$$
$$\overline{GA(N)} = \left(2S_x J D_x^{1/2} D_y^{1/2}\right)(f(x, y))|_{x=1} = 21.9881$$

Hence the result.

## 2.2 *Computation of NN-polynomial and Topological Indices of Imatinib*

**Theorem 2.2.1** *The NN-polynomial of the chemical graph I of Imatinib is*

$$NN(I; x, y) = 3x^{33}y^{33} + 24x^{33}y^{34} + 3x^{33}y^{35} + 11x^{34}y^{34}$$

**Proof** The molecular graph of Imatinib ($I$) as shown in Fig. 2b contain 37 vertices and 41 edges. Let $E_{i,j}$ denote the set of all edges $uv$ where $\{i, j\}$ denote the number of non-neighbor vertices at $u$ and $v$ respectively such that $E_{i,j}. = \{uv \in E(G) : \overline{d_G(u)} = i, \overline{d_G(v)} = j\}$ and $|E_{i,j}| = e_{i,j}$. From the molecular graph of Imatinib, we get.

$$e_{33,33} = 3, e_{33,34} = 24, e_{33,35} = 3, e_{34,34} = 11$$

Therefore, by Definition 1.1, we have

$$\begin{aligned} NN(I; x, y) &= \sum_{i \leq j} e_{i,j}x^i y^j \\ &= e_{33,33}x^{33}y^{33} + e_{33,34}x^{33}y^{34} + e_{33,35}x^{33}y^{35} + e_{34,34}x^{34}y^{34} \\ &= 3x^{33}y^{33} + 24x^{33}y^{34} + 3x^{33}y^{35} + 11x^{34}y^{34} \end{aligned}$$

Hence the result.

Now, using the NN-polynomial from the previous theorem, we derive various non-neighbor topological indices for the chemical graph of imatinib in the following theorem. By applying the same methodology as in Theorem 2.1.2, we arrive at the following result.

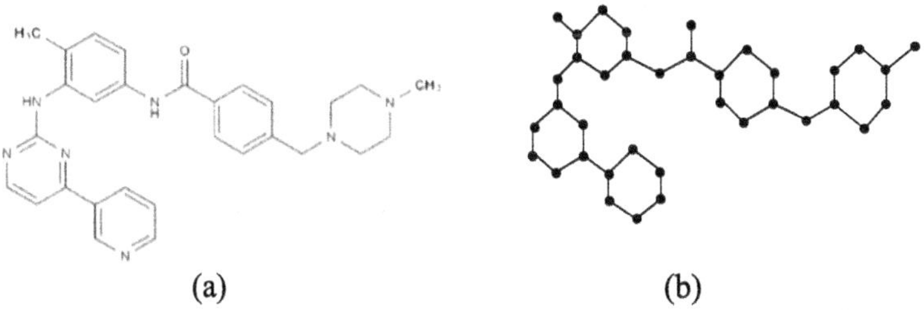


**Fig. 2** **a** Chemical structure and **b** Chemical graph of Imatinib

**Theorem 2.2.2** *Let I be the chemical graph of Imatinib, then*

$$\overline{M_1(I)} = 2758, \overline{M_2(I)} = 46376, \overline{H(I)} = 1.2190, \overline{R_{-1/2}(I)} = 1.219, \overline{SCI(I)} = 4.9989$$
$$\overline{ABC(I)} = 9.8491, \overline{GA(I)} = 40.996$$

## 2.3 Computation of NN-polynomial and Topological Indices of Famotidine

**Theorem 2.3.1** *The NN-polynomial of the chemical graph F of Famotidine is*

$$NN(F; x, y) = x^{15}y^{17} + 3x^{15}y^{18} + 9x^{16}y^{17} + 3x^{16}y^{18} + 4x^{17}y^{17}$$

**Proof** The chemical structure and chemical graph of Famotidine are shown in Fig. 3a and b, respectively. Let $F$ be the molecular graph of Famotidine with $|V(F)| = 20$ and $|E(F)| = 20$. Let $E_{i,j}$ denote the set of all edges $uv$ where $\{i, j\}$ denote the number of non-neighbor vertices at $u$ and $v$, respectively, such that $E_{i,j}. = \{uv \in E(G) : \overline{d_G(u)} = i, \overline{d_G(v)} = j\}$ and $|E_{i,j}| = e_{i,j}$. From the molecular graph of Famotidine, we obtain

$$e_{15,17} = 1, e_{15,18} = 3, e_{16,17} = 9, e_{16,18} = 3, e_{17,17} = 4$$

Therefore, by Definition 1.1, we have

$$\begin{aligned} NN(F; x, y) &= \sum_{i \le j} e_{i,j}x^i y^j \\ &= e_{15,17}x^{15}y^{17} + e_{15,18}x^{15}y^{18} + e_{16,17}x^{16}y^{17} + e_{16,18}x^{16}y^{18} + e_{17,17}x^{17}y^{17} \\ &= x^{15}y^{17} + 3x^{15}y^{18} + 9x^{16}y^{17} + 3x^{16}y^{18} + 4x^{17}y^{17} \end{aligned}$$

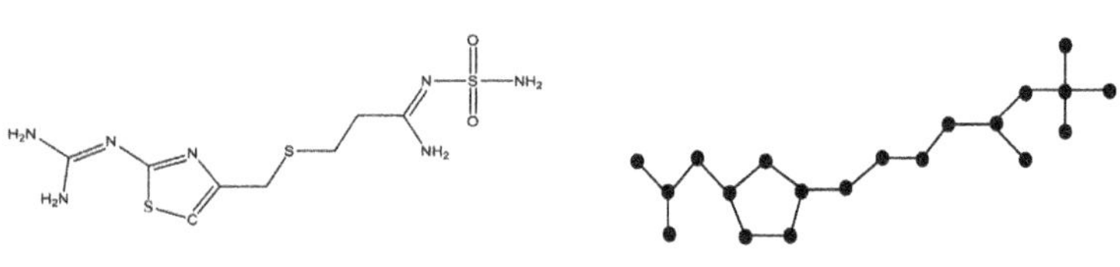


(a) (b)

**Fig. 3** **a** Chemical structure and **b** Chemical graph of Famotidine

Hence the result.

The NN-polynomial from the previous theorem is now used to generate a number of non-neighbor topological indices for the chemical graph of Famotidine in the following theorem by applying the same methodology as in Theorem 2.1.2, and the outcome is as follows:

**Theorem 2.3.2** *Let F be the chemical graph of Famotidine, then*

$$\overline{M_1(F)} = 666, \overline{M_2(F)} = 5533, \overline{H(I)} = 1.2015, \overline{R_{-1/2}(F)} = 1.2027, \overline{SCI(F)} = 3.4661$$
$$\overline{ABC(F)} = 7.7703, \overline{GA(F)} = 19.9761$$

## 2.4 Computation of NN-polynomial and Topological Indices of Galidesivir

**Theorem 2.4.1** *The NN-polynomial of the chemical graph* G *of Galidesivir is*

$$NN(\mathrm{G}; x, y) = 7x^{15}y^{15} + 7x^{15}y^{16} + 3x^{15}y^{17} + 3x^{16}y^{16} + x^{16}y^{17}$$

**Proof** From Fig. 4a and b, we see that the molecular graph of Galidesivir has 19 vertices and 21 edges. Let $E_{i,j}$ denote the set of all edges $uv$ where $\{i, j\}$ indicate the number of non-neighbor vertices at $u$ and $v$, respectively, such that $E_{i,j}. = \{uv \in E(G) : \overline{d_G(u)} = i, \overline{d_G(v)} = j\}$ and $|E_{i,j}| = e_{i,j}$. From the molecular graph of Galidesivir, we get

$$e_{15,15} = 7, e_{15,16} = 7, e_{15,17} = 3, e_{16,16} = 3, e_{16,17} = 1$$

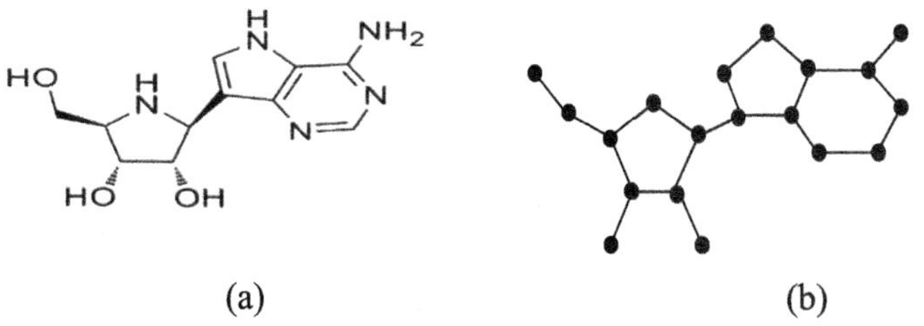


**Fig. 4** **a** Chemical structure and **b** Chemical graph of Galidesivir

Therefore, by Definition 1.1, we have

$$\begin{aligned} NN(G; x, y) &= \sum_{i \leq j} e_{i,j} x^i y^j \\ &= e_{15,15} x^{15} y^{15} + e_{15,16} x^{15} y^{16} + e_{15,17} x^{15} y^{17} + e_{16,16} x^{16} y^{16} + e_{16,17} x^{16} y^{17} \\ &= 7x^{15} y^{15} + 7x^{15} y^{16} + 3x^{15} y^{17} + 3x^{16} y^{16} + x^{16} y^{17} \end{aligned}$$

This completes the proof.

By using the NN-polynomial of the chemical graph of Galidesivir, the indices values are computed in Theorem 2.4.2 by following the same procedure as in Theorem 2.1.2.

**Theorem 2.4.2** *Let G be the chemical graph of Galidesivir, then*

$$\overline{M_1(G)} = 652, \overline{M_2(G)} = 5060, \overline{H(G)} = 1.3538, \overline{R_{-1/2}(G)} = 1.3543, \overline{SCI(G)} = 3.7699$$
$$\overline{ABC(G)} = 7.2958, \overline{GA(G)} = 20.9898$$

## 2.5 Computation of NN-polynomial and Topological Indices of Artesunate

**Theorem 2.5.1** *The NN-polynomial of the chemical graph A of Artesunate is*

$$NN(A; x, y) = 3x^{22}y^{23} + 4x^{22}y^{24} + x^{22}y^{25} + 3x^{23}y^{23} + 10x^{23}y^{24} + 5x^{23}y^{25} + 4x^{24}y^{24}$$

**Proof** The chemical structure and molecular graph of Artesunate are shown in Fig. 5a and b, respectively. Let $A$ be the molecular graph of Artesunate with 27 vertices and 30 edges. Let $E_{i,j}$ denote the set of all edges $uv$ where $\{i, j\}$ denote the number of non-neighbor vertices at $u$ and $v$, respectively, such that $E_{i,j}. = \{uv \in E(G) : \overline{d_G(u)} = i, \overline{d_G(v)} = j\}$ and $|E_{i,j}| = e_{i,j}$. From the molecular graph of Artesunate, we obtain

$$e_{22,23} = 3, e_{22,24} = 4, e_{22,25} = 1, e_{23,23} = 3, e_{23,24} = 10, e_{23,25} = 5, e_{24,24} = 4$$

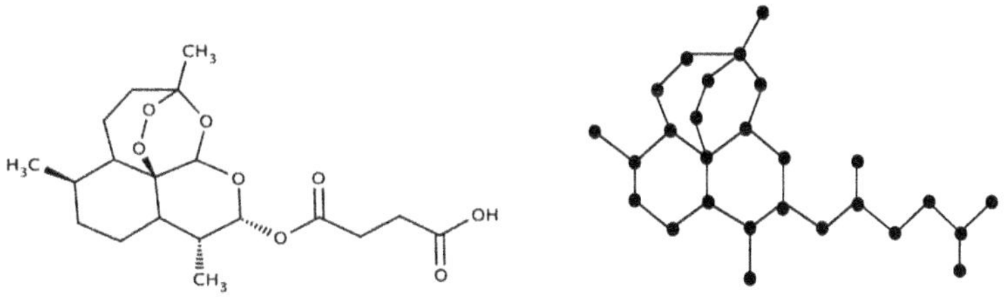


(a) (b)

**Fig. 5** **a** Chemical structure and **b** Chemical graph of Artesunate

Therefore, by Definition 1.1, we have

$$
\begin{aligned}
NN(A; x, y) &= \sum_{i \leq j} e_{i,j} x^i y^j \\
&= e_{22,23}x^{22}y^{23} + e_{22,24}x^{22}y^{24} + e_{22,25}x^{22}y^{25} + e_{23,23}x^{23}y^{23} + e_{23,24}x^{23}y^{24} + e_{23,25}x^{23}y^{25} +_{24,24} x^{24}y^{24} \\
&= 3x^{22}y^{23} + 4x^{22}y^{24} + x^{22}y^{25} + 3x^{23}y^{23} + 10x^{23}y^{24} + 5x^{23}y^{25} + 4x^{24}y^{24}
\end{aligned}
$$

Hence the result.

By using the NN-polynomial of the chemical graph of Artesunate, the indices values are computed in Theorem 2.5.2 by following the same procedure as in Theorem 2.1.2.

**Theorem 2.5.2** *Let A be the chemical graph of Artesunate, then*

$$
\overline{M_1(A)} = 1406, \overline{M_2(A)} = 16466, \overline{H(A)} = 1.2807, \overline{R_{-1/2}(A)} = 1.281, \overline{SCI(A)} = 4.3828
$$
$$
\overline{ABC(A)} = 8.5801, \overline{GA(A)} = 29.9864
$$

# 3 Non-neighbor Topological Indices in QSPR Studies

Here we define seven non-neighbor topological indices such as non-neighbor First Zagreb index $\overline{M_1(G)}$, non-neighbor Second Zagreb index $\overline{M_2(G)}$, non-neighbor Harmonic index $\overline{H(G)}$, non-neighbor Randić index $\overline{R_\alpha(G)}$, non-neighbor Sum connectivity index $\overline{SCI(G)}$, non-neighbor ABC index $\overline{ABC(G)}$, and non-neighbor Geometric Arithmetic index $\overline{GA(G)}$ for modeling eight Physical and chemical properties such as Boiling Point (BP), Enthalpy of Vaporization (EV), Flash Point (FP), Molar Refraction (MR), Polar Surface Area (PSA), Polarizability (P), Surface Tension (ST) and Molar Volume (MV) of five antiviral drugs Nitazoxanide, Imatinib, Famotidine, Galidesivir, and Artesunate. The values for these Physicochemical properties of drugs are obtained from ChemSpider. The above-mentioned non-neighbor topological indices and the experimental values of the physicochemical properties of drugs are represented in Table 2 and Table 3, respectively.

For the purpose of computation in the QSPR analysis study, a linear regression method is adopted. The linear regression method is represented by the following equation.

$$Y = A + B(X) \tag{1}$$

where $Y$ is the Physicochemical properties of drugs, $A$ is a constant, $B$ represents the regression coefficient and $X$ represent the topological index. Table 4 represents the value of $R^2$, the square of the correlation coefficient which is obtained by linear regression model between seven non-neighbor topological indices and eight Physicochemical characteristics of five aforesaid medications. The most significant maximum $R^2$ values in Table 4 are shown in bold, indicating that they represent the best predictor of the physicochemical characteristics of drugs. Table 5 shows the linear regression equation for the topological indices that fit the best and are the most predictable, along with the correlation coefficient value (R), F-Statistics, and P-value.

**Table 2** Non-Neighbor topological indices values of antiviral drugs

| Drugs | $\overline{M_1(G)}$ | $\overline{M_2(G)}$ | $\overline{H(G)}$ | $\overline{R_{-1/2}(G)}$ | $\overline{SCI(G)}$ | $\overline{ABC(G)}$ | $\overline{GA(G)}$ |
|---|---|---|---|---|---|---|---|
| Nitazoxanide | 776 | 6838 | 1.2478 | 1.2485 | 3.7048 | 7.2 | 21.9881 |
| Imatinib | 2758 | 46,376 | 1.2190 | 1.219 | 4.9989 | 9.8491 | 40.996 |
| Famotidine | 666 | 5533 | 1.2015 | 1.2027 | 3.4661 | 7.7703 | 19.9761 |
| Galidesivir | 652 | 5060 | 1.3538 | 1.3543 | 3.7699 | 7.2958 | 20.9898 |
| Artesunate | 1406 | 16,466 | 1.2807 | 1.281 | 4.3828 | 8.5801 | 29.9864 |

**Table 3** Physicochemical characteristics values of antiviral medications

| Drugs | Formula | BP | EV | FP | MR | PSA | P | ST | MV |
|---|---|---|---|---|---|---|---|---|---|
| Nitazoxanide | $C_{12}H_9N_3O_5S$ | 394 | – | – | 75.2 | 142 | 29.8 | 70.6 | 200.5 |
| Imatinib | $C_{29}H_{31}N_7O$ | 754.9 | – | – | 147.1 | 86 | 58.3 | 63.6 | 393 |
| Famotidine | $C_8H_{15}N_7O_2S_3$ | 662.4 | 97.4 | 354.4 | 79.1 | 238 | 31.3 | 97.3 | 183.6 |
| Galidesivir | $C_{11}H_{15}N_5O_3$ | 661.2 | 102.2 | 353.7 | 68.3 | 140 | 27.1 | 103.2 | 162.6 |
| Artesunate | $C_{19}H_{28}O_8$ | 507.1 | 85.1 | 175.6 | 92.2 | 101 | 36.6 | 51.4 | 292.2 |

**Table 4** $R^2$ value obtained by linear regression model between NN-topological indices and Physicochemical properties of the antiviral drugs

| Indices | BP | EV | FP | MR | PSA | P | ST | MV |
|---|---|---|---|---|---|---|---|---|
| $\overline{M_1(G)}$ | 0.223 | 0.934 | **0.999** | 0.976 | 0.470 | 0.977 | 0.373 | 0.960 |
| $\overline{M_2(G)}$ | 0.268 | 0.944 | 0.998 | **0.987** | 0.421 | **0.988** | 0.303 | 0.923 |
| $\overline{H(G)}$ | 0.011 | 0.062 | 0.001 | 0.211 | 0.093 | 0.209 | 0.065 | 0.142 |
| $\overline{R_{-1/2}(G)}$ | 0.012 | 0.064 | 0.001 | 0.217 | 0.089 | 0.215 | 0.068 | 0.147 |
| $\overline{SCI(G)}$ | 0.140 | 0.675 | 0.896 | 0.834 | **0.687** | 0.837 | **0.478** | 0.934 |
| $\overline{ABC(G)}$ | **0.293** | **0.990** | 0.864 | 0.927 | 0.320 | 0.928 | 0.349 | 0.936 |
| $\overline{GA(G)}$ | 0.171 | 0.811 | 0.992 | 0.931 | 0.557 | 0.932 | 0.454 | **0.979** |

**Table 5** The best linear regression model for the physical–chemical characteristics of antiviral drugs

| Linear model | R | F | P | Indicator |
|---|---|---|---|---|
| $BP = 20.469 + 70.702\ \overline{\mathbf{ABC}}$ | 0.541 | 1.245 | 0.345 | Not Significant |
| $EV = 201.432–13.515\ \overline{\mathbf{ABC}}$ | 0.995 | 102.603 | 0.062 | Not Significant |
| $FP = 511.408–0.238\ \overline{\mathbf{M_1}}$ | 0.999 | 2595.411 | 0.012 | Significant |
| $MR = 63.521 + 0.001\ \overline{\mathbf{M_2}}$ | 0.993 | 232.063 | 0.000 | Significant |
| $PSA = 462.203–78.928\ \overline{\mathbf{SCI}}$ | 0.8291 | 6.596 | 0.082 | Not Significant |
| $P = 25.179 + 0.0007\ \overline{\mathbf{M_2}}$ | 0.9939 | 243.005 | 0.000 | Significant |
| $ST = 177.57–24.689\ \overline{\mathbf{SCI}}$ | 0.6915 | 2.748 | 0.195 | Not Significant |
| $MV = -39.495 + 10.672\ \overline{\mathbf{GA}}$ | 0.989 | 140.167 | 0.001 | Significant |

# 4 Conclusion

Topological indices are used to identify various properties of the molecular structure of chemical compounds which are an essential part of drug discovery. Therefore, in this article, we have studied some investigational drugs for the treatment of COVID-19 like Nitazoxanide, Imatinib, Famotidine, Galidesivir, and Artesunate and established NN-polynomial expression from their molecular structures. Further, some non-neighbor topological indices are recovered from the NN-polynomial. The QSPR study shows a strong correlation value between various non-neighbor topological indices and physicochemical properties of the above drugs. In particular, non-neighbor First Zagreb index $\overline{M_1(G)}$ shows high significant correlation value (r = 0.999) with Flash point (FP), non-neighbor Second Zagreb index $\overline{M_2(G)}$ gives high positive significant correlation (r = 0.993) with Molar refraction (MR) and (r = 0.9939) with Polarizability (P). non-neighbor Geometric Arithmetic index $\overline{GA(G)}$ shows high significant correlation value (r = 0.989) with Molar volume (MV). The result of QSPR analysis will help chemists and pharmacists in new drug discovery for the treatment of Corona virus (COVID-19).

## 5 Data and Software Availability

In this article, the expression for NN-polynomial of molecular graphs of Nitazoxanide, Imatinib, Famotidine, Galidesivir, and Artesunate are obtained by using the edge partition method. The values of the physical and chemical properties of the above drugs are obtained from https://www.chemspider.com, an online chemical structure database. The chemical structure of antiviral drugs is obtained from www.pubchem.ncbi.nlm.nih.gov. Using SPSS Software, the linear regression analysis is performed (Statistical Package for the Social Sciences).

## References


1. Silva, M., Espejo, A., Pereyra, M.L., Lynch, M., Thompson, M., Taconelli, H., Baré, P., Pereson, M.J., Garbini, M., Crucci, P., Enriquez, D.: Efficacy of Nitazoxanide in reducing the viral load in COVID-19 patients. https://doi.org/10.1101/2021.03.03.21252509
2. Emadi, A., Chua, J.V., Talwani, R., Bentzen, S.M., Baddley, J.: Safety and efficacy of imatinib for hospitalized adults with COVID-19: a structured summary of a study protocol for a randomised controlled trial. Trials **21**, 897 (2020)
3. Malone, R.W., Tisdall, P., Fremont-Smith, P., Liu, Y., Huang, X.P., White, K.M., Miorin, L., Moreno, E., Alon, A., Delaforge, E., Hennecker, C.D., Wang, G., Potte, J., Blair, R.V., Roy, C.J., Smith, N., Hall, J.M., Tomera, K.M., Shapiro, G., Mittermaier, A., Kruse, A.C., García-Sastre, A., Roth, B.L., Glasspool-Malone, J., Ricke, D.O.: COVID-19: famotidine, histamine, mast cells, and mechanisms.Front. Pharmacol. **12**, 633680 (2021)
4. Ataei, M., Hosseinjani, H.: Molecular mechanisms of galidesivir as a potential antiviral treatment for COVID-19. J. Pharm. Care **8**(3), 150–151 (2020)
5. Uzun, T., Toptas, O.: Artesunate: could be an alternative drug to chloroquine in COVID-19 treatment? Chin. Med. **15**, 54 (2020)
6. Rizwana, A., Jeyakumar, G., Somasundaram, S.: Non-neighbor topological indices for hydrocarbons. Int. J. Sci. Eng. Sci. **1**(7), 16–19 (2017)
7. Roshini, G.R., Chandrakala, S.B., Indira, R., Sooryanarayana, B.: Non-neoghbor reduced randic and sum-connectivity index. Int. J. Math. Appl. **7**(3),127–133 (2019)
8. Chandrakala, S.B., Roshini, G.R., Sooryanarayana, B., Mihokova, M.: Non-neighbor sum-connectivity index and ABC index. Acta Univ. Matthiae Belii, Ser. Math. **27**, 43–58 (2019)
9. Roshini, G.R., Chandrakala, S.B., Vishu Kumar, M., Sooryanarayana, B.: Non-neighbor topological indices of honeycomb networks. Palest. J. Math. **10**, 52–58 (2021)
10. Haruo, H.: On some counting polynomials in chemistry. Discret. Appl. Math. **19**(1–3), 239–57 (1988)
11. Nadeem, M., Yousaf, A., Alolaiyan, H., Razaq, A.: Certain polynomials and related topological indices for the series of benzenoid graphs. Sci. Rep. (2019)
12. Deutsch, E., Klavzar, S.: M-polynomial and degree -based topological indices. Iran. J. Math. Chem. **6**(2), 93–102 (2015)
13. Mondal, S., Imran, M., De, N., Pal, A.: Neighborhood M-polynomial of titanium compounds. Arab. J. Chem. **14** (2021)
14. Kirmani, S.A.K, Ali, P., Azam, F.: Topological indices and QSPR/QSAR analysis of some antiviral drugs investigated for the treatment of COVID-19 patients. J. Quantum Chem. **121**(9), e26594 (2021)
15. Shanmukha, M.C., Basavarajappa, N.S, Shilpa, K.C, Usha, A.: Degree-based topological indices in anticancer drugs with QSPR analysis. Heliyon **6** (2020)

16. Havare, O.C.: Quantitative structure analysis of some molecules in drugs used in the treatment of COVID-19 with topological indices. Polycycl. Aromat. Compd. https://doi.org/10.1080/10406638.2021.1934045
17. Havare, O.C.: Topological indices and QSPR modeling of some novel drugs in the cancer treatment. Int. J. Quantum Chem. (2021)
18. Bokhary, S.A.U.H., Adnan, Siddiqui, M.K., Cancan, M.: On topological indices and QSPR analysis of drugs used for the treatment of Breast cancer. Polycycl. Aromat. Compd. (2021)
19. Zhong, J.F., Rauf, A., Naeem, M., Rahman, J., Aslam, A.: Quantitative structure-property relationship (QSPR) of valency based topological indices with Covid-19 drugs and application. Arab. J. Chem. **14** (2021)

# Some Results on Differential Polynomials of Meromorphic Functions Sharing Certain Values

**M. Tejuswini and N. Shilpa**

**Abstract** Off late, "Value Distribution Theory" concerning the differential polynomials of meromorphic functions is studied thoroughly. In this article, we consider a differential polynomial of a meromorphic function and its corresponding q-shift differential polynomial sharing the value 1, counted according to multiplicity and ignoring multiplicity to prove the uniqueness theorem. The concepts of normal families are employed to procure the main result, which in turn generalizes the existing result....

**Keywords** Normal Families · q-shift Differential Polynomials · Value Distribution · Meromorphic Functions · Uniqueness

## 1 Preliminaries

Throughout the article, the term "meromorphic function" means that the function has no other singularities other than poles in the whole complex plane $\mathbb{C}$. On the contrary, if the function is analytic everywhere in $\mathbb{C}$, then the function is called an "entire function". Let $\mathcal{F} = \{f : f\ is\ non\ constant\ meromorphicfunction\ in\ \mathbb{C}\}$. For $f,\ g \in \mathcal{F}$ and $a \in \mathbb{C} \cup \{\infty\}$, if the zeros of $f - a$ coincide in location and multiplicity with the zeros of $g - a$, then we say $f$ and $g$ share $a$ CM (counting multiplicities), if the coincidence happens only with location then $f$ and $g$ share $a$ IM (ignoring multiplicities) and $a$ is termed as the value point of $f$ and $g$. Now $f$ and $g$ share $\infty$ CM (IM) if $f$ and $g$ have same poles CM (IM). If $f - a$ has no zeros, then $a$ is the "Picard exceptional Value (PeV)" of $f$ [12]. Let $q \in \mathbb{Z}^+$, then the "counting function" $N_{(q}(r, \frac{1}{f-a})$ means that the $a$-points of $f$, CM whose multiplicities are not less than $q$ and the corresponding "reduced counting function" counted IM is given

M. Tejuswini (✉) · N. Shilpa
Department of Mathematics School of Engineering, Presidency University, Itagalpur, Bengaluru 560064, Karnataka, India
e-mail: 9tejuswini3@gmail.com

N. Shilpa
e-mail: shilpajaikumar@gmail.com



B. Rushi Kumar et al. (eds.), *Mathematics and Computing*, Springer Proceedings in Mathematics & Statistics 415, https://doi.org/10.1007/978-981-19-9307-7_24

by $\overline{N}_{(q}(r, \frac{1}{f-a})$ [9]. For the basic definitions and standard notations of Nevanlinna theory, the readers are referred to [10]. We use the notation $\rho(f), \mu(f), \Theta(a, f)$, and $\delta_k(a, f)$ to represent the terms "order of $f$", "lower order of $f$" "truncated defect of $f$", and the "deficiency of $f$", respectively, whose definitions can be seen in [3, 6].

**Definition 1** ([6]) Let $f \in \mathcal{F}$ be transcendental having infinitely many zeros then the "exponent of convergence of zeros" of $f$, is defined as

$$\lambda(f) = \limsup_{r \to \infty} \frac{log^{+} N(r, \frac{1}{f-a})}{log\, r}.$$

In 1981, the author duo Shibazaki–Yang gave the following result:

**Theorem 1** ([8]) *Let $f$ and $g$ be two entire functions of finite order. Suppose $f'$ and $g'$ share 1 value and if $\delta(0, f) > 0$, where 0 is the PeV of $g$, then either $f'g' \equiv 1$ or $f \equiv g$.*

For the meromorphic class of functions, in 2012, Banerjee–Majumder proved the following uniqueness result:

**Theorem 2** ([2]) *Let $f, g \in \mathcal{F}$. For $n \geq 13$, if $f^n(f-1)^2 f'$ and $g^n(g-1)^2 g'$ share 1 CM then, $f \equiv g$.*

The authors X. M. Li et.al., in 2014, gave a new perspective for the unicity theorems by considering a "differential polynomial" generated by a meromorphic function and its shift counterpart to prove the uniqueness result as stated below

**Theorem 3** ([6]) *Let $f \in \mathcal{F}$, $n, k \in \mathbb{Z}^{+}$ and the constant $\xi (\neq 0) \in \mathbb{C}$. If $[f^n(z)(f(z)-1)]^{(k)}$ and $[f^n(z+\xi)(f(z+\xi)-1)]^{(k)}$ share 1 value CM (1 IM) then $f(z+\xi) = f(z), \forall z \in \mathbb{C}$, whenever $\lambda\left(\frac{1}{f}\right) \notin [1, 2]$ and $\Theta(\infty, f) > \frac{2}{n}$ with $n > 3k + 11$ $(n > 9k + 20)$.*

This is the main motivation for the primary results of this paper stated below

**Theorem 4** *Let $f$ be a non constant meromorphic function and the constants $q, c \in \mathbb{C}$. For $m, n, k \in \mathbb{Z}^{+}$ satisfying $n > 3k + m + 10, k \geq 1$, if $[f^n(z)(f^m(z) - 1)]^{(k)}$ and $[f^n(qz+c)(f^m(qz+c)-1)]^{(k)}$ share 1 CM with $\lambda\left(\frac{1}{f}\right) \notin [1, 2]$ and $\Theta(\infty, f) > \frac{2}{n}$ then $f(qz+c) = f(z), \forall z \in \mathbb{C}$.*

**Theorem 5** *Let $f$ be a non constant meromorphic function and the constants $q, c \in \mathbb{C}$. For $m, n, k \in \mathbb{Z}^{+}$ satisfying $n > 9k + 4m + 16$, $k \geq 1$, if $[f^n(z)(f^m(z) - 1)]^{(k)}$ and $[f^n(qz+c)(f^m(qz+c)-1)]^{(k)}$ share 1 IM with $\lambda\left(\frac{1}{f}\right) \notin [1, 2]$ and $\Theta(\infty, f) > \frac{2}{n}$ then $f(qz+c) = f(z), \forall z \in \mathbb{C}$.*

## 2 Lemmas

**Lemma 1** ([7]) *Let $f \in \mathbb{F}$ and if*

$$F = \frac{\sum\limits_{p=0}^{k} a_p f^p}{\sum\limits_{q=0}^{l} b_q f^q}$$

*is in the most reduced form with $a_p$ and $b_q$ being constant coefficients and $a_k, b_l \neq 0$ then $T(r, F) = d\, T(r, f) + O(1)$, $d = max\,\{k, l\}$.*

**Lemma 2** ([5]) *Let $f, g \in \mathcal{F}$ and $k(\geq 1) \in \mathbb{Z}$. Suppose $f^{(k)}$ and $g^{(k)}$ share a polynomial $Q \neq 0$ CM and if*

$$\begin{aligned} \Delta_1 = (k+2)\Theta(\infty, f) + \Theta(0, f) + \delta_{k+1}(0, f) + 2\Theta(\infty, g) + \Theta(0, g) + \\ \delta_{k+1}(0, g) > k + 7, \end{aligned} \tag{1}$$

*and*

$$\begin{aligned} \Delta_2 = (k+2)\Theta(\infty, g) + \Theta(0, g) + \delta_{k+1}(0, g) + 2\Theta(\infty, f) + \Theta(0, f) + \\ \delta_{k+1}(0, f) > k + 7 \end{aligned} \tag{2}$$

*then either $f^{(k)}g^{(k)} = Q^2$ or $f \equiv g$.*

**Lemma 3** ([5]) *Let $f, g \in \mathcal{F}$ and $k(\geq 1) \in \mathbb{Z}$. Suppose $f^{(k)}$ and $g^{(k)}$ share a polynomial $Q \neq 0$ IM and if*

$$\begin{aligned} \Delta_3 = (3+2k)\Theta(\infty, f) + \Theta(0, f) + 2\delta_{k+1}(0, f) + (4+2k)\Theta(\infty, g) + \Theta(0, g) \\ +3\delta_{k+1}(0, g) > 4k + 13, \end{aligned} \tag{3}$$

*and*

$$\begin{aligned} \Delta_4 = (3+2k)\Theta(\infty, g) + \Theta(0, g) + 2\delta_{k+1}(0, g) + (4+2k)\Theta(\infty, f) + \Theta(0, f) \\ +3\delta_{k+1}(0, f) > 4k + 13 \end{aligned} \tag{4}$$

*then either $f^{(k)}g^{(k)} = Q^2$ or $f \equiv g$.*

**Lemma 4** ([4]) *Suppose $f \in \mathcal{F}$ has a spherical derivative which is bounded on $\mathbb{C}$, then $f$ is of order at most* 1.

**Lemma 5** ([6]) *For $f, g \in \mathcal{F}$ and constants $c_1, c_2, c \in \mathbb{C}\backslash\{0\}$ such that $(-1)^k (c_1c_2)^n(nc)^{2k} = 1$ if suppose $(f^n)^{(k)}(g^n)^{(k)} = 1$, then the functions take the form $f(z) = c_1e^{cz}$ and $g(z) = c_2e^{-cz}$ satisfying $n > 2k$ and $k \geq 1$.*

**Lemma 6** *Let $f,\ g \in \mathcal{F}$ and $n,\ k,\ m \in \mathbb{Z}^+$ with $n > 2k - m,\ k \geq 1$ if*

$$[f^n(z)(f^m(z) - 1)]^{(k)}[g^n(z)(g^m(z) - 1)]^{(k)} \equiv 1, \tag{5}$$

*then $\rho(f)$ and $\rho(g)$ is $\leq 2$.*

***Proof*** In case if $f$ and $g$ are "rational functions", then $\rho(f) = \rho(g) = 0$ and the lemma holds. Suppose $f$ and $g$ are "transcendental functions", we define two families of meromorphic functions say $F = \{f_w\}$ and $G = \{g_w\}$ where $f_w(z) = f(z + w)$ and $g_w(z) = g(z + w)$.

**Case 1:** Suppose one of the families say $F$ is normal on $\mathbb{C}$. By "Marty's theorem", the spherical derivative is bounded, i.e., for some $M > 0$, we have $f^{\#}(w) = f^{\#}_w(0) \leq M, \quad \forall\ w \in \mathbb{C}$ therefore by Lemma 4, $\rho(f(z))$ is maximum 2.

**Case 2:** Suppose one of the families say $F$ is not normal on $\mathbb{C}$. By "Marty's theorem", we have a sequence of functions $f_i(z)$ in $F$ such that $f_i(z) = f(w_i + z), z \in \{z : |z| < 1\}$ and $w_i$ is some infinite sequence in $\mathbb{C}$. Also as $|w_i| \to \infty$, we have

$$f^{\#}_i(0) = f^{\#}_i(w_i) \to \infty. \tag{6}$$

Now by Zalcman's lemma [11], we get:
(a) for $|z_i| < 1, z_i \to 0$,
(b) positive numbers $\rho_i$ such that $\rho_i \to 0^+$,
(c) sequence of functions $f_i(z_i + \rho_i\eta)$ of $f(w_i + z)$ where $f_i(z_i + \rho_i\eta) = f(w_i + z_i + \rho_i\eta)$ such that we have a spherically uniformly converging sequence $h_i(\eta)$ defined as

$$h_i(\eta) := \rho_i^{\frac{-k}{n}} f_i(z_i + \rho_i\eta), \tag{7}$$

which converges to $h(\eta)$ and satisfies the normalization $h^{\#}(\eta) \leq h^{\#}(0) = 1$. Therefore by Lemma 4 order of $h$ is at most 2. For $a_i = w_i + z_i$, the steps in the proof of "Zalcman's lemma" (refer[11]) gives the following results:

$$\rho_i = \frac{1}{f^{\#}_i(z_i)} = \frac{1}{f^{\#}(a_i)}, \tag{8}$$

$$f^{\#}(a_i) = f^{\#}_i(z_i) \geq f^{\#}_i(0) = f^{\#}(w_i). \tag{9}$$

Substituting (7) in (5), we get

$$\left[\rho_i^{\frac{k}{m+n}} h_i^{m+n}(\eta) - \rho_i^{\frac{k}{n}} h_i^n(\eta)\right]^{(k)} = \left[f_i^{m+n}(z_i + \rho_i\eta) - f_i^n(z_i + \rho_i\eta)\right]^{(k)}, \tag{10}$$

which converges spherically uniformly to

$$-[h^n(\eta)]^{(k)} \quad in \quad \mathbb{C} \setminus \tilde{h}^{-1}(\infty). \tag{11}$$

We claim that $[g^{m+n} - g^n]^{(k)}$ is non constant. If suppose $[g^{m+n} - g^n]^{(k)}$ is some constant then, $g^{m+n} - g^n = P_k$ where $P_k$ is a polynomial of degree $\leq k$ which in turn leads to $m + n \leq k$, a contradiction. Let us define

$$\hat{h}_i(\eta) = \rho_i^{\frac{-k}{n}} g_i(z_i + \rho_i \eta). \tag{12}$$

Substituting (12) in (5) and proceeding in the same manner as (10) and (11), we get the converging sequence $[\hat{h}^n(\eta)]^{(k)}$. By deducing according to formula of higher derivatives, we get

$$\left[\hat{h}^n(\eta)\right]^{(k)} \rightarrow -\tilde{h}(\eta). \tag{13}$$

By "Hurwitz's theorem", each zero and each pole of the non constant meromorphic function, $\tilde{h}(\eta)$ is of multiple $n$, so we write

$$\tilde{h}(\eta) = \hat{h}^n, \quad \forall\, \eta \in \mathbb{C}\backslash\{\tilde{h}^{-1}(\infty) \cup \hat{h}^{-1}(\infty)\}. \tag{14}$$

Using (13) and (14) in (5), we get

$$[\tilde{h}(\eta)]^{(k)}[\hat{h}^n(\eta)]^{(k)} \equiv 1. \tag{15}$$

With the assumption of (15) and that "order of $f$" is maximum 2, we have

$$\rho(f) = \rho(f^n) = \rho((f^n)^{(k)}) = \rho((g^n)^{(k)}) = \rho(g^n) = \rho(g) \leq 2, \tag{16}$$

$$\rho(h) = \rho(\tilde{h}) \leq 2. \tag{17}$$

The equations (15) and (17) together satisfy the conditions of Lemma 5, hence $h(z)$ and $\tilde{h}(z)$ are transcendental functions which takes the form

$$h(z) = c_1 e^{cz}, \qquad \tilde{h}(z) = c_2 e^{-cz}. \tag{18}$$

where the constants obey the condition $(-1)^k (c_1 c_2)^n (nc)^{2k} = 1$. Now

$$\frac{h_i'(\eta)}{h_i(\eta)} = \frac{\rho_i^{\frac{-k}{n}} f_i'(z_i + w_i + \rho_i \eta)}{\rho_i^{\frac{-k}{n}} f_i(z_i + w_i + \rho_i \eta)} = \frac{\rho_i f_i'(z_i + w_i + \rho_i \eta)}{f_i(z_i + w_i + \rho_i \eta)} \rightarrow \frac{h'(\eta)}{h(\eta)}. \tag{19}$$

Using Eq. (18) in (19), we get

$$\frac{h'(\eta)}{h(\eta)} = K. \tag{20}$$

Now,

$$\rho_i\left|\frac{f'(w_i+z_i)}{f(w_i+z_i)}\right| = \left(\frac{1+|f(w_i+z_i)|^2}{|f'(w_i+z_i)|}\right)\left(\left|\frac{f'(w_i+z_i)}{f(w_i+z_i)}\right|\right)$$
$$= \frac{1+|f(w_i+z_i)|^2}{f(w_i+z_i)} \to \left|\frac{h'(0)}{h(0)}\right|. \tag{21}$$

Again by using Eq. (18) in (21), we get

$$\left|\frac{h'(0)}{h(0)}\right| = |c|. \tag{22}$$

Therefore

$$\lim_{i\to\infty} f(w_i+z_i) \neq 0, \infty. \tag{23}$$

The results of (11), (18), and (23) gives

$$h_i(0) = \rho_i^{\frac{-k}{n}} f_i(z_i) = \rho_i^{\frac{-k}{n}} f(w_i+z_i) \to \infty. \tag{24}$$

Therefore

$$h_i(0) = h(0) = C, \tag{25}$$

which is a contradiction.

**Lemma 7** *Let $f \in \mathcal{F}$ be of finite order. For $n, m, k \in \mathbb{Z}^+$ with $n > 2k - m, k \geq 1$ and the constants $c, q \in \mathbb{C}$ if*

$$[f^n(z)(f^m(z)-1)]^{(k)}[f^n(qz+c)(f^m(qz+c)-1)]^{(k)} \equiv 1 \tag{26}$$

*then $\lambda\left(\frac{1}{f}\right) = \rho(f)$ and $\lambda\left(\frac{1}{f}\right) \in [1, 2]$.*

***Proof*** Let

$$G(z) = [f^n(z)(f^m(z)-1)]^{(k)}. \tag{27}$$

Therefore, (26) can be written as

$$G(z)G(qz+c) \equiv 1. \tag{28}$$

On the same lines of (16), one can get

$$\rho(f) = \rho(G). \tag{29}$$

Lets take up the emerging cases.

**Case 1:** In case if $f(z)$ is a "rational function", then $G(z)$ is also rational function. Suppose $qz_0 \in \mathbb{C}$ is a pole of $f(z)$, then by (26) and (27), $qz_0$ is a zero of $G(qz+c)$.

On the same lines, $qz_0 + c$ is a pole of $G(qz + c)$, $qz_0 + 2c$ is a zero of $G(qz + c)$ and so on. Proceeding same way, we find $G(z)$ has infinitely many poles and zeros which is a contradiction.

**Case 2:** Suppose $f(z)$ is a "transcendental meromorphic function", then by Lemma 6, we have $\rho(f)$ is maximum 2. Let's discuss the two subcases.

**Subcase 1:** Suppose $f(z)$ has no poles in $\mathbb{C}$, then by (26) and with the condition $n > 2k - m$, $f$ has no zeros in $\mathbb{C}$. If $\alpha(z)$ is a non constant polynomial, then $f$ can be written as

$$f(z) = e^{\alpha(z)}, \quad deg(\alpha) = \rho(f) \geq 1. \tag{30}$$

Using (30) in (26), we get

$$\begin{aligned}\left[f^{m+n}(z) - f^n(z)\right]^{(k)} &= \left[e^{(m+n)\alpha(z)} - e^{n\alpha(z)}\right]^{(k)} \\ &= R_k[\alpha'(z)]e^{n\alpha(z)},\end{aligned} \tag{31}$$

$$\begin{aligned}\left[f^{m+n}(qz + c) - f^n(qz + c)\right]^{(k)} &= \left[e^{(m+n)\alpha(qz+c)} - e^{n\alpha(qz+c)}\right]^{(k)} \\ &= R_k[\alpha'(qz + c)]e^{n\alpha(qz+c)},\end{aligned} \tag{32}$$

where

$$\begin{aligned}R_k[\alpha'(z)] = [(m + n)^k e^{m\alpha(z)} - n^k](\alpha'(z))^k + [(m + n)e^{m\alpha(z)} - n](\alpha(z))^{(k)} + \\ k[(m + n)^{k-1} e^{m\alpha(z)} - n^{k-1}]\alpha(z)^{(k-1)}\alpha(z)^{(k-2)},\end{aligned} \tag{33}$$

and

$$\begin{aligned}R_k[\alpha'(qz + c)] = q^k[[(m + n)^k e^{m\alpha(qz+c)} - n^k](\alpha'(qz + c))^k + [(m + n)e^{m\alpha(qz+c)} - \\ n](\alpha(qz + c))^{(k)} + k[(m + n)^{k-1} e^{m\alpha(qz+c)} - n^{k-1}]\alpha(qz + c)^{(k-1)}\alpha(qz + c)^{(k-2)}].\end{aligned} \tag{34}$$

Substituting (31) and (32) in (26), we get

$$R_k[\alpha'(z)]e^{n\alpha(z)} R_k[\alpha'(qz + c)]e^{n\alpha(qz+c)} \equiv 1.$$

Neglecting higher order terms of $R_k[\alpha'(z)]$ and $R_k[\alpha'(qz + c)]$, we get

$$\begin{aligned}\left[(m + n)^{2k} e^{m(\alpha(z)+\alpha(qz+c))} - n^{2k} - (m + n)^k n^k (e^{m\alpha(qz+c)} + e^{m\alpha(z)})\right] \times \\ \left[\alpha'(z)\alpha'(qz + c)\right]^k e^{n[\alpha(z)+\alpha(qz+c)]}\alpha'(z)\alpha'(qz + c) \equiv 1.\end{aligned} \tag{35}$$

Since $\alpha'(z)$ and $\alpha'(qz + c)$ are non zero constants, from (35), we see that $\alpha(z)$ is a constant which is a contradiction.

**Subcase 2:** Suppose $f$ has atleast one pole, say $z_1$, then as discussed in case 1, $G(z)$ has infinitely many poles and zeros. For sufficiently large positive integer $m$

$$n(qz_1 + 2mc, G) \geq m. \tag{36}$$

The "exponent of convergence of poles of $G$"

$$\lambda\left(\frac{1}{G}\right) = \limsup_{qz_1+2mc\to\infty} \frac{log^+ n(qz_1 + 2mc, G)}{log(qz_1 + 2mc)} \geq \limsup_{m\to\infty} \frac{logm}{log(qz_1 + 2mc)} \geq 1. \tag{37}$$

**Subcase 2.1:** Suppose that

$$\lambda\left(\frac{1}{G}\right) < \rho(f). \tag{38}$$

With the assumption that $n > 2k - m$ and using Eqs. (26) to (28) in Lemma 2.4 of [6], we write

$$\lambda(f) \leq \lambda\left(\frac{1}{G}\right). \tag{39}$$

Combining Eqs. (38) and (39), we get

$$\lambda(f) < \rho(f). \tag{40}$$

Let $a_1, a_2 \ldots a_m, \ldots$ and $b_1, b_2 \ldots b_q, \ldots$ be the non zero zeros and poles of $f(z)$ each counted according to its multiplicity, respectively, then the cannonical products of zeros and poles of $f(z)$ can be written as

$$h_1(z) = z^{n_0} \prod_{m=1}^{\infty} E_{n_m}\left(\frac{z}{a_m}\right) \quad and \tag{41}$$

$$h_2(z) = z^{\hat{n_0}} \prod_{q=1}^{\infty} \hat{E}_{\hat{n}_q}\left(\frac{z}{b_q}\right), \tag{42}$$

where $n_0, \hat{n_0} \geq 0$ are integers. Let $g(z)$ be an "entire function", we rewrite $f(z)$ in the form

$$f(z) = \frac{h_1(z)}{h_2(z)} e^{g(z)}. \tag{43}$$

$$Define \quad h(z) = \frac{h_1(z)}{h_2(z)},$$

Hence

$$f(z) = h(z)e^{g(z)}. \tag{44}$$

From Theorem 4.3.6 of [1], we write (45) and (46) as follows:

$$\lambda(h_1) = \rho(h_1) = \lambda(f), \tag{45}$$

$$\lambda(h_2) = \rho(h_2) = \lambda(\frac{1}{f}). \tag{46}$$

Equations (45) and (46) concludes that

$$N(r, f) \leq N(r, G). \tag{47}$$

Equations (47) and (40) gives that

$$\lambda\left(\frac{1}{f}\right) \leq \lambda\left(\frac{1}{G}\right) < \lambda(f) < \rho(f). \tag{48}$$

Using Eqs. (46) and (48) for (44), we can write

$$\rho(h) < \rho(f) \quad and \quad \rho(f) = \rho(e^g) < \infty.$$

Therefore, $g$ is a non constant polynomial with

$$\rho(f) = \rho(e^g) = deg(g).$$

Replacing (44) in the left inequality of (26) individually, we have

$$\begin{aligned}[f^{m+n}(z) - f^n(z)]^{(k)} &= \left[h^{m+n}(z)e^{(m+n)g(z)} - h^n(z)e^{ng(z)}\right]^{(k)} \\ &= K_1(z)e^{(m+n)g(z)} + K_2(z)e^{ng(z)},\end{aligned} \tag{49}$$

where

$$\begin{aligned}K_1(z) &= (h^{m+n}(z))^{(k)} + k(m+n)g'(z)(h^{m+n}(z))^{(k-1)} \\ &+\sum_{j=2}^{k}\left\{\binom{k}{j}\left((m+n)g'(z)\right)^j + H_{j-1}\left((m+n)g'(z)\right)\right\}\left(h^{m+n}(z)\right)^{(k-j)}, \\ K_2(z) &= -[(h^n(z))^{(k)} + kng'(z)(h^n(z))^{(k-1)} \\ &+\sum_{j=2}^{k}\left\{\binom{k}{j}\left(ng'(z)\right)^j + H_{j-1}\left(ng'(z)\right)\right\}\left(h^n(z)\right)^{(k-j)}],\end{aligned} \tag{50}$$

and

$$\left[f^{m+n}(qz+c) - f^{n}(qz+c)\right]^{(k)} = \big[h^{m+n}(qz+c)e^{(m+n)g(qz+c)} - h^{n}(qz+c)e^{ng(qz+c)}\big]^{(k)}$$

$$= K_1(qz+c)e^{(m+n)g(qz+c)} + K_2(qz+c)e^{ng(qz+c)}, \quad (51)$$

where

$$K_1(qz+c) = q^k[(h^{m+n}(qz+c))^{(k)} + k(m+n)g'(qz+c)(h^{m+n}(qz+c))^{(k-1)}$$
$$+\sum_{j=2}^{k}\left\{\binom{k}{j}((m+n)g'(qz+c))^j + H_{j-1}\left((m+n)g'(qz+c)\right)\right\}\left(h^{m+n}(qz+c)\right)^{(k-j)}],$$
$$K_2(qz+c) = -q^k[(h^n(qz+c))^{(k)} + k\,n\,g'(qz+c)(h^n(qz+c))^{(k-1)}$$
$$+\sum_{j=2}^{k}\left\{\binom{k}{j}\left(ng'(qz+c)\right)^j + H_{j-1}\left(n\,g'(qz+c)\right)\right\}\left(h^n(qz+c)\right)^{(k-j)}]. \quad (52)$$

Here $H_{j-1}(v)$ is a polynomial of degree $\leq j-1$. Using (49) and (51) in (26), we get

$$[K_1(z)e^{(m+n)g(z)} + K_2(z)e^{ng(z)}][K_1(qz+c)e^{(m+n)g(qz+c)} + K_2(qz+c)e^{ng(qz+c)}] \equiv 1.$$

which gives

$$K_1(z)K_1(qz+c)e^{(m+n)[g(z)+g(qz+c)]} + K_2(z)K_2(qz+c)e^{n[g(z)+g(qz+c)]}+$$
$$K_1(z)K_2(qz+c)e^{[(m+n)g(z)+ng(qz+c)]} + K_1(qz+c)K_2(z)e^{[(m+n)g(qz+c)+ng(z)]} \equiv 1, \quad (53)$$

Arguing in a similar way as in the subcase 1 of Lemma 7, we see that $g(z)$ is a constant polynomial, a contradiction to the lemma statement.
**Subcase 2.2:** suppose that

$$\lambda\left(\frac{1}{G}\right) = \rho(f). \quad (54)$$

$$N(r, G) = N(r, (f^{m+n} - f^n)^{(k)})$$
$$\leq N(r, f^{m+n} - f^n) + k\overline{N}(r, f)$$
$$N(r, G) \leq (m+n+k)N(r, f). \quad (55)$$

Using Eqs. (30), (48), (54), and Lemma 6, we get

$$\lambda\left(\frac{1}{f}\right) = \rho(f) \in [1, 2], \quad (56)$$

which proves the lemma.

## 3 Proof of Theorems

**Theorem** 4: Let

$$F_1 = f^n(z)(f^m(z) - 1), \tag{57}$$

$$G_1 = f^n(qz + c)(f^m(qz + c) - 1). \tag{58}$$

Using the condition $n > 3k + m + 10$ and Lemma 1 for $F_1$ defined in (57), we have

$$\begin{aligned}\Theta(\infty, F_1) &= 1 - \limsup_{r\to\infty} \frac{\overline{N}(r, F_1)}{T(r, F_1)} \\ &\geq 1 - \limsup_{r\to\infty} \frac{T(r, f)}{(m+n)T(r, f) + O(1)} \geq 1 - \frac{1}{m+n}.\end{aligned} \tag{59}$$

$$\begin{aligned}\Theta(0, F_1) &= 1 - \limsup_{r\to\infty} \frac{\overline{N}\left(r, \frac{1}{F_1}\right)}{T(r, F_1)} \\ &\geq 1 - \limsup_{r\to\infty} \frac{T(r, f)}{(m+n)T(r, f) + O(1)} \geq 1 - \frac{2}{m+n}.\end{aligned} \tag{60}$$

$$\begin{aligned}\delta_{k+1}(0, F_1) &= 1 - \limsup_{r\to\infty} N_{k+1}\left(r, \frac{1}{F_1}\right) \\ &\geq 1 - \frac{(m+k+1)\overline{N}\left(r, \frac{1}{f}\right)}{m+n} \geq 1 - \frac{m+k+1}{m+n}.\end{aligned} \tag{61}$$

Similarly, for $G_1$, we obtain

$$\Theta(\infty, G_1) \geq 1 - \frac{1}{m+n}. \tag{62}$$

$$\Theta(0, G_1) \geq 1 - \frac{2}{m+n}. \tag{63}$$

$$\delta_{k+1}(0, G_1) \geq 1 - \frac{m+k+1}{m+n}. \tag{64}$$

Substituting (59)–(64) in (1) and (2), we get

$$\Delta_1 = k + 8 - \frac{3k + 2m + 10}{m + n}. \tag{65}$$

$$\Delta_2 = k + 8 - \frac{3k + 2m + 10}{m + n}. \tag{66}$$

For $n > 3k + m + 10$, we get $\Delta_1 > k + 7$ and $\Delta_2 > k + 7$. Applying Lemma 2, for $F_1^{(k)}$ and $G_1^{(k)}$ sharing 1 CM, we get $F_1^{(k)} G_1^{(k)} \equiv 1$ or $F_1 \equiv G_1$. We look into the following two cases:

**Case 1:** Suppose that $F_1^{(k)} G_1^{(k)} \equiv 1$, then by Lemma 7, we have $\lambda \left( \frac{1}{f} \right) = \rho(f) \in [1, 2]$ which contradicts the statement of the Theorem 4.

**Case 2:** Suppose that $F_1 \equiv G_1$, then by (57) and (58), we write

$$f^n(z)(f^m(z) - 1) = f^n(qz + c)(f^m(qz + c) - 1). \tag{67}$$

Let

$$\phi(z) = \frac{f(z)}{f(qz + c)}. \tag{68}$$

We take up the subsequent two possibilities:

**Subcase 1:** Suppose that the non constant $\phi$ is a "meromorphic function", then using (68), (67) can be rewritten as

$$f^n(qz + c) - f^n(z) = f^{m+n}(qz + c) - f^{m+n}(z).$$
$$f^m(qz + c) = \frac{1 - \phi^n(z)}{1 - \phi^{m+n}(z)}. \tag{69}$$

Using Eqs. (68), (69), and Lemma 1, we write

$$T(r, f^m(z)) = T(r, \phi^m(z) f^m(qz + c))$$
$$mT(r, f) \leq (m + n)T(r, \phi) + S(r, \phi) \;\; (or) \;\; T(r, f) \leq nT(r, \phi) + S(r, \phi). \tag{70}$$

Now using "Second Fundamental Theorem" of Nevanlinna, the reduced counting function will be

$$\overline{N}(r, f) \geq (n - 2)T(r, \phi) + S(r, \phi). \tag{71}$$

Using Eqs. (70) and (71), the truncated defect

$$\begin{aligned}\Theta(\infty, f) &= 1 - \limsup_{r\to\infty} \frac{\overline{N}(r, f)}{T(r, f)} \\ &\le 1 - \frac{n-2}{n} \le \frac{2}{n},\end{aligned} \tag{72}$$

which is a contradiction.

**Subcase 2:** Suppose that $\phi$ is a constant and if $\phi^{m+n} \not\equiv 1$ then the right equality of Eq. (69) becomes a constant and hence $f^m(qz + c)$ becomes a constant which is wrong. Therefore, $\phi^{m+n} \equiv 1$. Using (68), (67) can be rewritten as

$$(1 - \phi^{m+n}(z))f^m(qz + c) = 1 - \phi^n(z). \tag{73}$$

Therefore

$$\phi^{m+n} \equiv \phi^n \equiv \phi \equiv 1. \tag{74}$$

Equation (74) in (68) gives the complete proof of the Theorem. ■

**Theorem** 5: Substituting the values (59)–(64) in (3) and (4), we get

$$\Delta_3 = 4k + 14 - \frac{9k + 5m + 16}{m + n}. \tag{75}$$

$$\Delta_4 = 4k + 14 - \frac{9k + 5m + 16}{m + n}. \tag{76}$$

For $n > 9k + 4m + 16$, we have $\Delta_3 > 4k + 13$ and $\Delta_4 > 4k + 13$. With the assumption that $F_1^{(k)}$ and $G_1^{(k)}$ share 1 IM, applying Lemma 3, we get $F_1^{(k)} G_1^{(k)} \equiv 1$ or $F_1 \equiv G_1$. Following a similar methodology as in the proof of Theorem 4, we arrive at the desired result. ■

**Corollary 1** *If $m = 1$ and $q = 1$ in the Theorem 4, as a special case, we get the results of Theorem 3.*

## References


1. Ash, R.: Complex Variables. Academic, New York (1971)
2. Banerjee, A., Majumder, S.: Certain non-linear differential polynomials sharing 1-points with finite weight. Thai J. Math. **10**(2), 321–336 (2012)
3. Banerjee, A., Sahoo, P.: Uniqueness and Weighted value sharing of differential polynomials of meromorphic functions. Acta Univ. Sapientiae Math. **2**(3), 181–196 (2011)
4. Chang, J.M., Zalcman, L.: Meromorphic functions that share a set with their derivatives. J. Math. Anal. Appl. **138**, 1020–1028 (2008)

5. Li, X.M., Yi, H.X.: Uniqueness of meromorphic functions whose certain non-linear differential polynomials share a polynomial. J. Comput. Math. Appl. **62**, 539–550 (2011)
6. Li, X.M., Yi, H.X., Shi, Y.: Value sharing of certain differential polynomials and their shifts of meromorphic functions. J. Comput. Methods Funct. Theory **14**, 63–84 (2014)
7. Mokhonko, A.Z.: On the Nevanlinna characteristics of some meromorphic functions. J. Th. Funct. Funct. Anal. Appl. **14**, 83–87 (1971)
8. Shibazaki, K., Yang, C.C.: Unicity theorems for entire functions of finite order. Mem. Natl. Defense Acad. Jpn. **21**(3), 67–71 (1981)
9. Shilpa, N., Achala, L.N.: Uniqueness of meromorphic functions of a certain non linear differential polynomials. Int. Elec. J. Pure Appl. Math. **10**(1), 23–39 (2016)
10. Yang, C.C., Yi, H.X.: Uniqueness Theory of Meromorphic Functions. Kluwer Academic Publishers, China (2003)
11. Zalcman, L.: A Heuristic princle in complex function theory. J. Am. Math. Monthly **82**, 813–817 (1975)
12. Zhang, Q.C.: Meromorphic functions sharing three values. Indian J. Pure Appl. Math. **30**, 667–682 (1999)

# A Subclass of Pseudo-Type Meromorphic Bi-Univalent Functions of Complex Order Associated with Linear Operator

**Asha Thomas, Thomas Rosy, and G. Murugusundaramoorthy**

**Abstract** In this article, we construct a new subclass of pseudo-type meromorphic bi-univalent function class of complex order, associated with linear operator and investigate the initial coefficient estimates $|b_0|$, $|b_1|$ and $|b_2|$. Furthermore, several new or known outcomes of our result are mentioned.

**Keywords** Analytic · Univalent · Meromorphic functions · Pseudo-type functions · Bi-univalent · Coefficient bounds

## 1 Introduction and Definitions

Let $\mathcal{A}$ denote class of analytic functions

$$f(\xi) = \xi + \sum_{n=2}^{\infty} a_n \xi^n, \qquad (1)$$

in $\mathbb{U} = \{\xi : |\xi| < 1\}$. Also, let $\mathcal{S}$ consist of functions in $\mathcal{A}$ which are univalent and normalized by $f(0) = f'(0) - 1 = 0$ in $\mathbb{U}$.

For $f_1, f_2 \in \mathcal{A}$, $f_1$ is subordinate to $f_2$, denoted by $f_1(\xi) \prec f_2(\xi)$, if there exists $\varpi$ defined on $\mathbb{U}$ with $\varpi(0) = 0$ and $|\varpi(z)| < 1$ satisfies $f_1(\xi) = f_2(\varpi(\xi))$. Ma and Minda [8] consolidated different subclasses of starlike and convex functions where either

$$\frac{\xi f'(\xi)}{f(\xi)} \quad \text{or} \quad 1 + \frac{\xi f''(\xi)}{f'(\xi)}$$

is subordinate to a more general function and are denoted by $\mathfrak{S}^*_\Sigma(\varphi)$ and $\mathfrak{K}_\Sigma(\varphi)$, respectively. In this article, $\varphi$ is assumed to be an analytic function in the unit disk

A. Thomas (✉) · T. Rosy
Department of Mathematics, Madras Christian College, Tambaram, Chennai 600059, Tamilnadu, India
e-mail: ashasarah.shiju@gmail.com

G. Murugusundaramoorthy
School of Advanced Sciences, VIT University, Vellore 632014, Tamilnadu, India

 
B. Rushi Kumar et al. (eds.), *Mathematics and Computing*, Springer Proceedings in Mathematics & Statistics 415, https://doi.org/10.1007/978-981-19-9307-7_25

$\mathbb{U}$, which satisfies $\varphi(0) = 1$ and $\varphi'(0) > 0$ and with respect to the real axis $\varphi(\mathbb{U})$ is symmetric. This function is written as

$$\varphi(\xi) = 1 + \beta_1\xi + \beta_2\xi^2 + \beta_3\xi^3 + \cdots, (\beta_1 > 0). \tag{1.2}$$

Setting $\varphi(\xi)$ as

$$\varphi(\xi) = \left(\frac{1+\xi}{1-\xi}\right)^{\delta} = 1 + 2\delta\xi + 2\delta^2\xi^2 + \frac{4\delta^2 + 2\delta}{3}\xi^3 + \cdots, \quad 0 < \delta \leq 1 \tag{1.3}$$

we have $\beta_1 = 2\delta$, $\beta_2 = 2\delta^2$, $\beta_3 = \frac{4\delta^2+2\delta}{3}$.

If

$$\varphi(\xi) = \frac{1 + (1 - 2\omega)\xi}{1 - \xi} = 1 + 2(1 - \omega)\xi + 2(1 - \omega)\xi^2 + \cdots, \quad (0 \leq \omega < 1) \tag{1.4}$$

then $\beta_1 = \beta_2 = \beta_3 = 2(1 - \omega)$.

Let $\Sigma'$ denote the class of all meromorphic univalent functions $\mathsf{g}$ of the form

$$\mathsf{g}(\xi) = \xi + b_0 + \sum_{n=1}^{\infty} \frac{b_n}{\xi^n}, \tag{1.5}$$

defined on $\mathbb{U}^* = \{\xi : 1 < |\xi| < \infty\}$. Since $\mathsf{g} \in \Sigma'$ is univalent its inverse $\mathsf{g}^{-1} = \upsilon$ exists and satisfies

$$\mathsf{g}^{-1}(\mathsf{g}(\xi)) = \xi \text{ and } \mathsf{g}^{-1}(\mathsf{g}(w)) = w \text{ for } \xi \in \mathbb{U}^*, M > 0\,, M < |w| < \infty$$

where

$$\mathsf{g}^{-1}(w) = \upsilon(w) = w + \sum_{n=0}^{\infty} \frac{C_n}{w^n}, \quad M < |w| < \infty \tag{1.6}$$

Now $\mathsf{g} \in \Sigma'$ is meromorphic bi-univalent if $\mathsf{g}^{-1} \in \Sigma'$, akin to the bi-univalent analytic functions [5]. Let the class of all meromorphic bi-univalent functions be denoted by $\mathfrak{M}_{\Sigma'}$. In literature, the coefficient bounds of meromorphic univalent functions were studied extensively, the bound $|b_2| \leq \frac{2}{3}$ for meromorphic univalent functions $\mathsf{g} \in \Sigma'$ with $b_0 = 0$ was estimated by Schiffer [13] and Duren [3] proved that $|b_n| \leq \frac{2}{(n+1)}$ on the coefficient of meromorphic univalent functions $\mathsf{g} \in \Sigma'$ with $b_k(0) = 0$ for $1 \leq k < \frac{n}{2}$. For the coefficient of $\upsilon \in \mathfrak{M}_{\Sigma'}$, Springer [15] proved $|C_3| \leq 1$; $|C_3 + \frac{1}{2}C_1^2| \leq \frac{1}{2}$ and conjectured $|C_{2n-1}| \leq \frac{(2n-1)!}{n!(n-1)!}$, (n=1,2,...).

Springer's conjecture was proved true for $n = 3, 4, 5$ by Kubota [7] and the coefficient bounds $C_{2n-1}$, $1 \leq n \leq 7$ for the inverse meromorphic univalent functions in $\mathbb{U}^*$ was obtained by Schober [12] and also proved the sharpness. The coefficient bounds for a class consisting of inverses of meromorphic starlike univalent functions of order $\delta$ in $\mathbb{U}^*$ was estimated [6, 16].

For $\mathsf{g} \in \Sigma'$ as in (1.5), a linear differential operator is defined as follows [10, 14]:

$$F_\zeta^0 \mathsf{g}(\xi) = \mathsf{g}(\xi),$$

$$F_\zeta^1 \mathsf{g}(\xi) = (1-\zeta)\mathsf{g}(\xi) + \zeta\xi \mathsf{g}'(\xi) = F_\zeta \mathsf{g}(\xi) \qquad (\zeta \geq 0) \tag{1.7}$$

$$F_\zeta^\nu \mathsf{g}(\xi) = F_\zeta(F_\zeta^{\nu-1}\mathsf{g}(\xi)) \qquad (\nu \in \mathfrak{N} = \{1, 2, 3, \ldots\}) \tag{1.8}$$

Then from (1.7) and (1.8), we get

$$F_\zeta^\nu \mathsf{g}(\xi) = \xi + (1-\zeta)^\nu b_0 + \sum_{n=1}^{\infty}[1-(n+1)\zeta]^\nu b_n \xi^{-n} \qquad (\nu \in \mathfrak{N} = \{0, 1, 2, 3, \ldots\}). \tag{1.9}$$

A new subclass $\mu$ - pseudo starlike function of order $\vartheta$ $(0 \leq \vartheta < 1)$ satisfying the analytic condition

$$Re\left(\frac{\xi(f'(\xi))^\mu}{f(\xi)}\right) > \vartheta,\ \xi \in \mathbb{U},\ 1 \leq \mu \in \mathbb{R} \tag{1.10}$$

and denoted by $\mathcal{L}_\mu(\vartheta)$ was defined by Babalola [1] and he remarked that for $\mu > 1$, the classes of $\mu-$ pseudo starlike functions represent the analytic starlike functions. Also, when $\mu = 1$, we have the class of starlike functions of order $\vartheta$ (1-pseudo starlike functions of order $\vartheta$) and for $\mu = 2$, we get the class of functions, which is a product combination of geometric expressions for bounded turning and starlike functions.

Motivated by the earlier works [2, 4, 9, 10, 17, 18], a new subclass of pseudo-type meromorphic bi-univalent functions class, denoted by $\mathfrak{P}_{\Sigma'}^\gamma(\eta, \mu, \varphi, \zeta, \nu)$ where $\gamma \in \mathbb{C}\backslash\{0\}$ is introduced and the coefficient bounds $|b_0|$, $|b_1|$ and $|b_2|$ are determined when associated with the linear operator as defined in (1.9). Several outcomes of the new results are discussed.

**Definition 1** For $0 < \eta \leq 1$ and $\mu \geq 1$, a function $\mathsf{g}(\xi) \in \Sigma'$ given by (1.5) is said to be in the class $\mathfrak{P}_{\Sigma'}^\gamma(\eta, \mu, \varphi, \zeta, \nu)$ if the following conditions are satisfied:

$$1 + \frac{1}{\gamma}\left[(1-\eta)\left(\frac{F_\zeta^\nu \mathsf{g}(\xi)}{\xi}\right)^\mu + \eta\left(\frac{\xi(F_\zeta^\nu \mathsf{g}'(\xi))^\mu}{F_\zeta^\nu \mathsf{g}(\xi)}\right) - 1\right] \prec \varphi(\xi) \tag{1.11}$$

and

$$1 + \frac{1}{\gamma}\left[(1-\eta)\left(\frac{F_\zeta^\nu \upsilon(w)}{w}\right)^\mu + \eta\left(\frac{w(F_\zeta^\nu \upsilon'(w))^\mu}{F_\zeta^\nu \upsilon(w)}\right) - 1\right] \prec \varphi(w) \tag{1.12}$$

where $\xi, w \in \mathbb{U}^*$, $\gamma \in \mathbb{C}\backslash\{0\}$ and the function $\upsilon$ is given by (1.6).

By suitably specializing the parameter $\eta$, we obtain new subclasses of $\mathfrak{P}^{\gamma}_{\Sigma'}(\eta, \mu, \varphi, \zeta, \nu)$ as illustrated in the following Examples.

**Example 1** For $\eta = 1$, $\mathrm{g} \in \Sigma'$ is in the class $\mathfrak{P}^{\gamma}_{\Sigma'}(1, \mu, \varphi, \zeta, \nu) \equiv \mathfrak{P}^{\gamma}_{\Sigma'}(\mu, \varphi, \zeta, \nu)$ if the following conditions hold:

$$1 + \frac{1}{\gamma}\left(\frac{\xi(F^{\nu}_{\zeta}\mathrm{g}'(\xi))^{\mu}}{F^{\nu}_{\zeta}\mathrm{g}(\xi)} - 1\right) \prec \varphi(\xi) \quad \text{and} \quad 1 + \frac{1}{\gamma}\left(\frac{w(F^{\nu}_{\zeta}\upsilon'(w))^{\mu}}{F^{\nu}_{\zeta}\upsilon(w)} - 1\right) \prec \varphi(w)$$

where $\xi, w \in \mathbb{U}^*$, $\mu \geq 1$, $\gamma \in \mathbb{C}\backslash\{0\}$ and the function $\upsilon$ is given by (1.6).

**Remark 1** We note that $\mathfrak{P}^{\gamma}_{\Sigma'}(1, 1, \varphi, \zeta, \nu) \equiv \mathfrak{S}^{\gamma}_{\Sigma'}(\varphi)$

**Example 2** For $\eta = 1$ and $\gamma = 1$, $\mathrm{g} \in \Sigma'$ given by (1.5) is in the class $\mathfrak{P}^{1}_{\Sigma'}(1, \mu, \varphi, \zeta, \nu) \equiv \mathfrak{P}_{\Sigma'}(\mu, \varphi, \zeta, \nu)$ if the following conditions hold:

$$\frac{\xi(F^{\nu}_{\zeta}\mathrm{g}'(\xi))^{\mu}}{F^{\nu}_{\zeta}\mathrm{g}(\xi)} \prec \varphi(\xi) \quad \text{and} \quad \frac{w(F^{\nu}_{\zeta}\upsilon'(w))^{\mu}}{F^{\nu}_{\zeta}\upsilon(w)} \prec \varphi(w)$$

where $\xi, w \in \mathbb{U}^*$, $\mu \geq 1$ and $\upsilon$ is given by (1.6).

**Example 3** For $\eta = 0$, $\mathrm{g} \in \Sigma'$ given by (1.5) is in the class $\mathfrak{P}^{\gamma}_{\Sigma'}(1, \mu, \varphi, \zeta, \nu) \equiv \mathfrak{R}^{\gamma}_{\Sigma'}(\mu, \varphi, \zeta, \nu)$ if the following conditions hold:

$$1 + \frac{1}{\gamma}\left[\left(\frac{F^{\nu}_{\zeta}\mathrm{g}(\xi)}{\xi}\right)^{\mu} - 1\right] \prec \varphi(\xi) \quad \text{and} \quad 1 + \frac{1}{\gamma}\left[\left(\frac{F^{\nu}_{\zeta}\upsilon(w)}{w}\right)^{\mu} - 1\right] \prec \varphi(w)$$

where $\xi, w \in \mathbb{U}^*$, $\mu \geq 1$ and the function $\upsilon$ is given by (1.6).

## 2 Coefficient Bounds

The bounds $|b_0|$, $|b_1|$ and $|b_2|$ for functions in the class $\mathfrak{P}^{\gamma}_{\Sigma'}(\eta, \mu, \varphi, \zeta, \nu)$ are estimated.

**Lemma 1** ([11]) *If $\Phi \in \mathfrak{P}$, the class of all functions with $\Re(\Phi(\xi)) > 0$, $(\xi \in \mathbb{U})$ then*

$$|c_k| \leq 2, \text{ for each } k,$$

*where*

$$\Phi(\xi) = 1 + c_1\xi + c_2\xi^2 + \cdots \text{ for } \xi \in \mathbb{U}.$$

The functions $p(\xi), q(w) \in \mathfrak{P}$ is defined by

$$p(\xi) = \frac{1 + r(\xi)}{1 - r(\xi)} = 1 + \frac{p_1}{\xi} + \frac{p_2}{\xi^2} + \cdots$$

and

$$q(w) = \frac{1 + s(w)}{1 - s(w)} = 1 + \frac{q_1}{w} + \frac{q_2}{w^2} + \cdots .$$

we obtain

$$r(\xi) = \frac{p(\xi) - 1}{p(\xi) + 1} = \frac{1}{2}\left[\frac{p_1}{\xi} + \left(p_2 - \frac{p_1^2}{2}\right)\frac{1}{\xi^2} + \cdots\right]$$

and

$$s(w) = \frac{q(w) - 1}{q(w) + 1} = \frac{1}{2}\left[\frac{q_1}{w} + \left(q_2 - \frac{q_1^2}{2}\right)\frac{1}{w^2} + \cdots\right].$$

Also, observe that for $p(\xi), q(w) \in \mathfrak{P}$, for each i

$$|p_i| \leq 2 \text{ and } |q_i| \leq 2 \quad \text{for each } i.$$

**Theorem 1** *Let g of the form (1.5) be in $\mathfrak{P}^{\gamma}_{\Sigma'}(\eta, \mu, \varphi, \zeta, \nu)$. Then*

$$|b_0| \leq \frac{|\gamma||\beta_1|}{|\mu - \mu\eta - \eta||(1 - \zeta)^{\nu}|}, \tag{2.1}$$

$$|b_1| \leq \frac{|\gamma|}{2|\mu - \eta - 2\mu\eta||(1 - 2\zeta)^{\nu}|}\left(4|(\beta_1 - \beta_2)^2| + 4|\beta_1^2| + 8|\beta_1(\beta_1 - \beta_2)| + \frac{|\mu(\mu - 1)(1 - \eta) + 2\eta|^2|\gamma|^2|\beta_1|^4}{|\mu - \mu\eta - \eta|^4}\right)^{\frac{1}{2}} \tag{2.2}$$

*and*

$$|b_2| \leq \frac{|\gamma|}{2|\mu - \eta - 3\mu\eta||(1 - 3\zeta)^{\nu}|}\left(2|\beta_1| + 4|\beta_2 - \beta_1| + 2|\beta_1 - 2\beta_2 + \beta_3| + \frac{|\mu(\mu - 1)(\mu - 2)(1 - \eta) - 6\eta||\gamma|^2|\beta_1|^3}{3|\eta|^3}\right) \tag{2.3}$$

*with* $0 < \eta \leq 1, \mu \geq 1, \gamma \in \mathbb{C}\backslash\{0\}$ *and* $\xi, w \in \mathbb{U}^*$.

***Proof*** From (1.11) and (1.12)

$$1+\frac{1}{\gamma}\left[(1-\eta)\left(\frac{F_{\zeta}^{\nu}g(\xi)}{\xi}\right)^{\mu}+\eta\left(\frac{\xi(F_{\zeta}^{\nu}g'(\xi))^{\mu}}{F_{\zeta}^{\nu}g(\xi)}\right)-1\right]=\varphi(r(\xi)) \tag{2.4}$$

and

$$1+\frac{1}{\gamma}\left[(1-\eta)\left(\frac{F_{\zeta}^{\nu}\upsilon(w)}{w}\right)^{\mu}+\eta\left(\frac{w(F_{\zeta}^{\nu}\upsilon'(w))^{\mu}}{F_{\zeta}^{\nu}\upsilon(w)}\right)-1\right]=\varphi(s(w)). \tag{2.5}$$

Using (1.5), (1.6), (1.11), and (1.12), we have

$$\begin{aligned}&1+\frac{1}{\gamma}\left[(1-\eta)\left(\frac{F_{\zeta}^{\nu}g(\xi)}{\xi}\right)^{\mu}+\eta\left(\frac{\xi(F_{\zeta}^{\nu}g'(\xi))^{\mu}}{F_{\zeta}^{\nu}g(\xi)}\right)-1\right]\\&=1+\beta_1p_1\frac{1}{2\xi}+\left[\frac{1}{2}\beta_1\left(p_2-\frac{p_1^2}{2}\right)+\frac{1}{4}\beta_2p_1^2\right]\frac{1}{\xi^2}\\&+\left[\frac{\beta_1}{2}\left(p_3-p_1p_2+\frac{p_1^3}{4}\right)+\frac{\beta_2}{2}\left(p_1p_2-\frac{p_1^3}{2}\right)+\beta_3\frac{p_1^3}{8}\right]\frac{1}{\xi^3}\cdots\end{aligned} \tag{2.6}$$

and

$$\begin{aligned}&1+\frac{1}{\gamma}\left[(1-\eta)\left(\frac{F_{\zeta}^{\nu}\upsilon(w)}{w}\right)^{\mu}+\eta\left(\frac{w(F_{\zeta}^{\nu}\upsilon'(w))^{\mu}}{F_{\zeta}^{\nu}\upsilon(w)}\right)-1\right]\\&=1+\beta_1q_1\frac{1}{2w}+\left[\frac{1}{2}\beta_1\left(q_2-\frac{q_1^2}{2}\right)+\frac{1}{4}\beta_2q_1^2\right]\frac{1}{w^2}\\&+\left[\frac{\beta_1}{2}\left(q_3-q_1q_2+\frac{q_1^3}{4}\right)+\frac{\beta_2}{2}\left(q_1q_2-\frac{q_1^3}{2}\right)+\beta_3\frac{q_1^3}{8}\right]\frac{1}{w^3}\cdots\quad.\end{aligned} \tag{2.7}$$

Equating the coefficients of $\xi^{-1}, \xi^{-2}, \xi^{-3}, \ldots$ and $w^{-1}, w^{-2}, w^{-3}, \ldots$ in (2.6) and (2.7), we get

$$\frac{(\mu-\mu\eta-\eta)(1-\zeta)^{\nu}}{\gamma}b_0=\frac{1}{2}\beta_1p_1, \tag{2.8}$$

$$\frac{1}{2\gamma}\left[\left(\mu(\mu-1)(1-\eta)+2\eta\right)(1-\zeta)^{2\nu}b_0^2+2(\mu-\eta-2\eta\mu)(1-2\zeta)^{\nu}b_1\right]=\frac{1}{2}\beta_1\left(p_2-\frac{p_1^2}{2}\right)+\frac{1}{4}\beta_2p_1^2. \tag{2.9}$$

$$\frac{1}{6\gamma}\Big[\big(\mu(\mu-1)(\mu-2)(1-\eta)-6\eta\big)(1-\zeta)^{3\nu}b_0^3+6\big(\mu(\mu-1)(1-\eta)+2\eta+\eta\mu\big)(1-\zeta)^{\nu}(1-2\zeta)^{\nu}b_0b_1$$
$$+6(\mu-\eta-3\eta\mu)(1-3\zeta)^{\nu}b_2\Big]=\left[\frac{\beta_1}{2}\left(p_3-p_1p_2+\frac{p_1^3}{4}\right)+\frac{\beta_2}{2}\left(p_1p_2-\frac{p_1^3}{2}\right)+\beta_3\frac{p_1^3}{8}\right], \tag{2.10}$$

$$\frac{-(\mu-\mu\eta-\eta)}{\gamma}(1-\zeta)^{\nu}b_0=\frac{1}{2}\beta_1q_1, \tag{2.11}$$

$$\frac{1}{2\gamma}\Big[\big(\mu(\mu-1)(1-\eta)+2\eta\big)(1-\zeta)^{2\nu}b_0^2+2(\eta-\mu+2\eta\mu)(1-2\zeta)^{\nu}b_1\Big]=\frac{1}{2}\beta_1\left(q_2-\frac{q_1^2}{2}\right)+\frac{1}{4}\beta_2q_1^2 \tag{2.12}$$

and

$$\frac{1}{6\gamma}\big[(6\eta-\mu(\mu-1)(\mu-2)(1-\eta)(1-\zeta)^{3\nu})b_0^3$$
$$+6\big(\mu(\mu-1)(1-\eta)-\mu(1-\eta)+3\eta+3\eta\mu\big)(1-\zeta)^{\nu}(1-2\zeta)^{\nu}b_0b_1+6(\eta-\mu+3\eta\mu)(1-3\zeta)^{\nu}b_2\big]$$
$$=\left[\frac{\beta_1}{2}\left(q_3-q_1q_2+\frac{q_1^3}{4}\right)+\frac{\beta_2}{2}\left(q_1q_2-\frac{q_1^3}{2}\right)+\beta_3\frac{q_1^3}{8}\right]\quad . \tag{2.13}$$

From (2.8) and (2.11), we get

$$p_1=-q_1 \tag{2.14}$$

and

$$b_0^2=\frac{\gamma^2\beta_1^2}{8(\mu-\mu\eta-\eta)^2(1-\zeta)^{2\nu}}(p_1^2+q_1^2). \tag{2.15}$$

Applying Lemma 1 for the coefficients $p_1$ and $q_1$, we have

$$|b_0|\leq\frac{|\gamma||\beta_1|}{|\mu-\mu\eta-\eta||(1-\zeta)^{\nu}|}.$$

$|b_1|$ is determined using (2.9), (2.12), (2.14) and (2.15),

$$2(\mu-\eta-2\eta\mu)^2(1-2\zeta)^{2\nu}\frac{b_1^2}{\gamma^2}+[\mu(\mu-1)(1-\eta)+2\eta]^2(1-\zeta)^{4\nu}\frac{b_0^4}{2\gamma^2}$$
$$=(\beta_1-\beta_2)^2\frac{p_1^4}{8}+\frac{\beta_1^2}{4}(p_2^2+q_2^2)+\beta_1(\beta_2-\beta_1)\frac{(p_1^2p_2+q_1^2q_2)}{4}. \tag{2.16}$$

By Lemma 1 and using (2.15), we get

$$|b_1|^2 \leq \frac{|\gamma^2|}{4|\mu - \eta - 2\eta\mu|^2|(1-2\zeta)^{2\nu}|} \times$$
$$\left(4|(\beta_1 - \beta_2)^2| + 4|\beta_1|^2 + 8|\beta_1(\beta_1 - \beta_2)| + \frac{|\mu(\mu-1)(1-\eta) + 2\eta|^2|\gamma|^2|\beta_1|^4}{|\mu - \mu\eta - \eta|^4}\right).$$

$$\implies |b_1| \leq \frac{|\gamma|}{2|\mu - \eta - 2\eta\mu||(1-2\zeta)^{\nu}|} \times$$
$$\sqrt{4|(\beta_1 - \beta_2)^2| + 4|\beta_1|^2 + 8|\beta_1(\beta_1 - \beta_2)| + \frac{|\mu(\mu-1)(1-\eta) + 2\eta|^2|\gamma|^2|\beta_1|^4}{|\mu - \mu\eta - \eta|^4}}.$$

$|b_2|$ is determined using (2.10) and (2.13) with $p_1 = -q_1$,

$$\frac{1}{\gamma} b_0 b_1 = \frac{\beta_1[p_3 + q_3] + (\beta_2 - B_1)p_1[p_2 - q_2]}{2[2\mu(\mu-1)(1-\eta) - (1-\eta)\mu + 5\eta + 4\eta\mu](1-\zeta)^{\nu}(1-2\zeta)^{\nu}}. \tag{2.17}$$

Subtracting (2.13) from (2.10) and using $p_1 = -q_1$, we have

$$\begin{aligned} &2(\mu - \eta - 3\eta\mu)(1-3\zeta)^{\nu}\frac{b_2}{\gamma} \\ &= -(\mu - \eta - 3\mu\eta)(1-\zeta)^{\nu}(1-2\zeta)^{\nu}\frac{b_0 b_1}{\gamma} - [\mu(\mu-1)(\mu-2)(1-\eta) - 6\eta](1-\zeta)^{3\nu}\frac{b_0^3}{3\gamma} + \frac{\beta_1}{2}(p_3 - q_3) \\ &\quad + \frac{\beta_2 - \beta_1}{2}(p_2 + q_2)p_1 + \frac{\beta_1 - 2\beta_2 + \beta_3}{4}p_1^3. \end{aligned} \tag{2.18}$$

Substituting for $\frac{b_0 b_1}{\gamma}$ and $\frac{b_0^3}{\gamma}$ in (2.18), further computation yields,

$$\begin{aligned} \frac{b_2}{\gamma} = &\frac{-\beta_1}{2(\mu - \eta - 3\eta\mu)(1-3\zeta)^{\nu}}\left(\frac{\mu - 3\eta - 4\eta\mu - \mu(\mu-1)(1-\eta)}{2\mu(\mu-1)(1-\eta) - \mu + 5\eta + 5\eta\mu}p_3\right. \\ &\left. + \frac{2\eta + \eta\mu + \mu(\mu-1)(1-\eta)}{2\mu(\mu-1)(1-\eta) - \mu + 5\eta + 5\eta\mu}q_3\right) \\ &- \frac{(\beta_2 - \beta_1)p_1}{2(\mu - \eta - 3\eta\mu)(1-3\zeta)^{\nu}}\left(\frac{\mu - 3\eta - 4\eta\mu - \mu(\mu-1)(1-\eta)}{2\mu(\mu-1)(1-\eta) - \mu + 5\eta + 5\eta\mu}p_2\right. \\ &\left. - \frac{2\eta + \eta\mu + \mu(\mu-1)(1-\eta)}{2\mu(\mu-1)(1-\eta) - \mu + 5\eta + 5\eta\mu}q_2\right) \\ &+ \frac{\beta_1 - 2\beta_2 + \beta_3}{8(\mu - \eta - 3\eta\mu)(1-3\zeta)^{\nu}}p_1^3 - \frac{(\mu(\mu-1)(\mu-2)(1-\eta) - 6\eta)\gamma^2\beta_1^3}{48(\mu - \eta - 3\eta\mu)(1-3\zeta)^{\nu}\eta^3}p_1^3. \end{aligned}$$

Applying Lemma 1 in the above equation yields

$$|b_2| \leq \frac{|\gamma|}{2|\mu-\eta-3\eta\mu||(1-3\zeta)^{\nu}|}\times \left(2|\beta_1|+4|\beta_2-\beta_1|+2|\beta_1-2\beta_2+\beta_3| + \frac{|\mu(\mu-1)(\mu-2)(1-\eta)-6\eta||\gamma|^2|\beta_1|^3}{3|\eta|^3}\right). \tag{2.19}$$

By taking $\eta = 1$, we get the results mentioned below.

**Theorem 2** *Let* $g$ *of the form (1.5) be in the class* $\mathfrak{P}^{\gamma}_{\Sigma'}(\mu,\varphi,\zeta,\nu)$. *Then*

$$|b_0| \leq \frac{|\gamma|\,|\beta_1|}{|(1-\zeta)^{\nu}|}, \tag{2.20}$$

$$|b_1| \leq \frac{|\gamma|}{|1+\mu||(1-2\zeta)^{\nu}|}\sqrt{|(\beta_1-\beta_2)^2|+|\beta_1^2|+2|\beta_1(\beta_1-\beta_2)|+|\gamma|^2\,|\beta_1|^4} \tag{2.21}$$

*and*

$$|b_2| \leq \frac{|\gamma|}{|1+2\mu||(1-3\zeta)^{\nu}|}\left(|\beta_1|+2|\beta_2-\beta_1|+|\beta_1-2\beta_2+\beta_3|+|\gamma|^2\,|\beta_1|^3\right) \tag{2.22}$$

*where* $\gamma \in \mathbb{C}\backslash\{0\}$, $\mu \geq 1$ *and* $\xi, w \in \mathbb{U}^*$.

By taking $\eta = 1$ and $\gamma = 1$, we state the following results.

**Theorem 3** *Let* $g$ *of the form (1.5) be in the class* $\mathfrak{P}_{\Sigma'}(\mu,\varphi,\zeta,\nu)$. *Then*

$$|b_0| \leq \frac{|\beta_1|}{|(1-\zeta)^{\nu}|},$$

$$|b_1| \leq \frac{1}{|1+\mu||(1-2\zeta)^{\nu}|}\sqrt{|(\beta_1-\beta_2)^2|+|\beta_1^2|+2|\beta_1(\beta_1-\beta_2)|+|\beta_1|^4}$$

*and*

$$|b_2| \leq \frac{1}{|1+2\mu||(1-3\zeta)^{\nu}|}\left(|\beta_1|+2|\beta_2-\beta_1|+|\beta_1-2\beta_2+\beta_3|+|\beta_1|^3\right)$$

*where* $\mu \geq 1$, $\xi, w \in \mathbb{U}^*$.

# 3 Corollaries and Concluding Remarks

For $\mathsf{g}$ of the form (1.5) and $\mathsf{g} \in \mathfrak{P}^{\gamma}_{\Sigma'}\left(\eta, \mu, \left(\frac{1+\xi}{1-\xi}\right)^{\delta}, \zeta, \nu\right) \equiv \mathfrak{P}^{\gamma}_{\Sigma'}(\eta, \mu, \delta, \zeta, \nu)$ setting $\beta_1 = 2\delta, \quad \beta_2 = 2\delta^2$ and $\beta_3 = \frac{4\delta^2+2\delta}{3}$, and similarly,
for $\mathsf{g} \in \mathfrak{P}^{\gamma}_{\Sigma'}\left(\eta, \mu, \frac{1+(1-2\omega)\xi}{1-\xi}, \zeta, \nu\right) \equiv \mathfrak{P}^{\gamma}_{\Sigma'}(\eta, \mu, \omega, \zeta, \nu)$ setting $\beta_1 = \beta_2 = \beta_3 = 2(1-\omega)$ analogous results of Theorems 1, 2 and 3 can be derived.

**Corollary 1** *Let* $\mathsf{g}$ *of the form (1.5) be in* $\mathfrak{P}^{\gamma}_{\Sigma'}(\eta, \mu, \delta, \zeta, \nu)$ *. Then*

$$|b_0| \leq \frac{2|\gamma|\delta}{|\mu - \mu\eta - \eta||(1-\zeta)^{\nu}|}, \tag{3.1}$$

$$|b_1| \leq \frac{2|\gamma|\delta}{|\mu - \eta - 2\eta\mu||(1-2\zeta)^{\nu}|}\sqrt{(\delta-2)^2 + \frac{|\mu(\mu-1)(1-\eta)+2\eta|^2|\gamma^2|}{|\mu-\mu\eta-\eta|^4}\delta^2} \tag{3.2}$$

*and*

$$|b_2| \leq \frac{2|\gamma|\delta}{|\mu - \eta - 3\eta\mu||(1-3\zeta)^{\nu}|}\left(3 - 2\delta + \left(\frac{4-6\delta+2\delta^2}{3}\right)\right. \\ \left. + \frac{2|\gamma|^2\delta^2|\mu(\mu-1)(\mu-2)(1-\eta)-6\eta|}{3|\eta|^3}\right) \tag{3.3}$$

*where* $\gamma \in \mathbb{C}\backslash\{0\}, 0 < \eta \leq 1, \mu \geq 1$ *and* $\xi, w \in \mathbb{U}^*$.

**Corollary 2** *Let* $\mathsf{g}$ *of the form (1.5) be in* $\mathfrak{P}^{\gamma}_{\Sigma'}(\eta, \mu, \omega, \zeta, \nu)$*. Then*

$$|b_0| \leq \frac{2|\gamma|(1-\omega)}{|\mu - \mu\eta - \eta||(1-\zeta)^{\nu}|}, \tag{3.4}$$

$$|b_1| \leq \frac{2|\gamma|(1-\omega)}{|\mu - \eta - 2\eta\mu||(1-2\zeta)^{\nu}|}\sqrt{1 + \frac{|\mu(\mu-1)(1-\eta)+2\eta|^2|\gamma^2|}{|\mu-\mu\eta-\eta|^4}(1-\omega)^2} \tag{3.5}$$

*and*

$$|b_2| \leq \frac{2|\gamma|(1-\omega)}{|\mu - \eta - 3\eta\mu||(1-3\zeta)^{\nu}|}\left(1 + \frac{2|\gamma|^2(1-\omega)^2|\mu(\mu-1)(\mu-2)(1-\eta)-6\eta|}{3|\eta|^3}\right) \tag{3.6}$$

*where* $\gamma \in \mathbb{C}\backslash\{0\}, 0 < \eta \leq 1, \mu \geq 1$ *and* $\xi, w \in \mathbb{U}^*$.

**Concluding Remarks:** We remark that, when $\eta = 1$ and $\mu = 1$, the coefficient bounds $b_0, b_1$ and $b_2$ for functions in the class $\mathfrak{S}^{\gamma}_{\Sigma'}(\varphi, \zeta, \nu)$, can be obtained which gives us the results discussed in Theorems of [9]. Also, the bounds for the function $\mathfrak{g}$ given by (1.5) in the subclass $\mathfrak{S}^{\gamma}_{\Sigma'}(\varphi, \zeta, \nu)$ can be determined by taking $\varphi(\xi)$ as given in (1.3) and (1.4), respectively.

## References


1. Babalola, K.O.: On $\lambda-$ pseudo starlike functions. J. Class. Anal. **3**(2), 137–147 (2013)
2. Deniz, E.: Certain subclasses of bi-univalent functions satisfying subordinate conditions. J. Class. Anal. **2**(1), 49–60 (2013)
3. Duren, P.L.: Coefficients of meromorphic schlicht functions. Proc. Amer. Math. Soc. **28**, 169–172 (1971)
4. Janani, T., Murugusundaramoorthy, G.: Coefficient estimates of meromorphic bi- starlike functions of complex order. Int. J. Anal. Appl. **4**(1), 68–77 (2014)
5. Janani, T., Murugusundaramoorthy, G., Vijaya, K.: New subclass of pseudo- type meromorphic bi-univalent functions of complex order. Novi Sad J. Math. Soc. **48**, 93–102 (2018)
6. Kapoor, G.P., Mishra, A.K.: Coefficient estimates for inverses of starlike functions of positive order. J. Math. Anal. Appl. **329**(2), 922–934 (2007)
7. Kubota, Y.: Coefficients of meromorphic univalent functions. Kodai Math. Semin. Rep. **28**(2–3), 253–261 (1977)
8. Ma, W.C., Minda, D.: A unified treatment of some special classes of functions. In: Proceedings of the Conference on Complex Analysis, Tianjin, pp. 157–169 (1992)
9. Murugusundaramoorthy, G., Janani, T., Cho, N.E.: Coefficient estimates of Mocanu type meromorphic bi-univalent functions of complex order. Proc. Jangjeon Math. Soc. **19**, 691–700 (2016)
10. Naik, A., Panigrahi, T., Murugusundaramoorthy, G.: Coefficient estimate for class of meromorphic bi-bazilevic type functions associated with linear operator by convolution. Jordan J. Math. Stat. **14**(2), 287–305 (2021)
11. Pommerenke, C.: Univalent Functions. Vandenhoeck & Ruprecht, Göttingen (1975)
12. Schober, G.: Coefficients of inverses of meromorphic univalent functions. Proc. Amer. Math. Soc. **67**(1), 111–116 (1977)
13. Schiffer, M.: On an extremum problem of conformal representation. Bulletin de la Société Mathématique de France **66**, 48–55 (1938)
14. Shaba, T.G., Khan, M.G., Ahmad, B.: Coefficients bounds for certain new subclasses of meromorphic bi-univalent functions associated with Al-Oboudi differential operator palestine. J. Math. **9**(2), 1–11 (2020)
15. Springer, G.: The coefficient problem for schlicht mappings of the exterior of the unit circle. Trans. Amer. Math. Soc. **70**, 421–450 (1951)
16. Srivastava, H.M., Mishra, A.K., Kund, S.N.: Coefficient estimates for the inverses of starlike functions represented by symmetric gap series. Panamerican Math. J. **21**(4), 105–123 (2011)
17. Srivastava, H.M., Joshi, S.B., Joshi, S.S., Pawar, H.: Coefficient estimates for certain subclasses of meromorphically bi-univalent functions. Palestine J. Math. **5**(Special Issue: 1), 250–258 (2016)
18. Xu, Q.H., Lv, C.B., Srivastava, H.M.: Coefficient estimates for the inverses of a certain general class of spirallike functions. Appl. Math. Comput. **219**, 7000–7011 (2013)

# Bi-Starlike Function of Complex Order Involving Double Zeta Functions in Shell Shaped Region

**V. Malathi and K. Vijaya**

**Abstract** In the contemporary paper concerning double zeta functions, we demarcated two new-fangled subclasses of bi-starlike and bi-convex function of complex order in the open unit disc linked with shell-shaped region and acquired Taylor–Maclaurin coefficients $|a_2|$ and $|a_3|$ of functions in these classes. Furthermore, we determine the Fekete–Szegö inequalities and the significance of the results which are new and are also piercing out as corollaries.

## 1 Introduction and Definitions

Let $\mathfrak{A}$ signify the class of functions of the form

$$f(t) = t + \sum_{n=2}^{\infty} a_n t^n \tag{1}$$

which are analytic in the open unit disc $\mathbb{D} = \{t : |t| < 1\}$ and normalized by $f(0) = 0$ and $f'(0) = 1$. Additionally, let $\mathfrak{S}$ symbolize the class of all functions in $\mathfrak{A}$ which are univalent in $\mathbb{D}$. Some of the substantial and well-investigated subclasses of $\mathfrak{S}$ comprise (for example) the class of starlike $\mathfrak{S}^*(\varrho)$ and convex functions $\mathfrak{K}(\varrho)$ of order $\varrho (0 \leq \varrho < 1)$ in $\mathbb{D}$, respectively. The convolution or Hadamard product of two functions $f, h \in \mathfrak{A}$ is defined as

$$(f * h)(t) = t + \sum_{n=2}^{\infty} a_n b_n t^n, \tag{2}$$

V. Malathi · K. Vijaya (✉)
Department of Mathematics, School of Advanced Sciences, Vellore Institute of Technology, Vellore 632014, Tamilnadu, India
e-mail: kvijaya@vit.ac.in



B. Rushi Kumar et al. (eds.), *Mathematics and Computing*, Springer Proceedings in Mathematics & Statistics 415, https://doi.org/10.1007/978-981-19-9307-7_26

where $f$ is given by (1) and $h(t) = t + \sum_{n=2}^{\infty} b_n t^n$.

For $\mathbf{h}_1, \mathbf{h}_2 \in \mathfrak{A}$ and $\mathbf{h}_1$, is subordinate to $\mathbf{h}_2$, is written $\mathbf{h}_1 \prec \mathbf{h}_2$, provided there is an analytic function $\varpi$ definite on $\mathbb{D}$ with $\varpi(0) = 0$ and $|\varpi(z)| < 1$ sustaining $\mathbf{h}_1(z) = \mathbf{h}_2(\varpi(z))$.

The study of operators plays a central role in the geometric function theory and its correlated fields. In recent years, there has been collective importance in problems concerning evaluations of various families of the Hurwitz–Lerch zeta function [10]. These functions ascend naturally in many branches of analytic function theory and have plentiful applications in mathematics [1].

We recall Hurwitz–Lerch Zeta function [23], assumed as

$$\mathfrak{H}(t, \ell, s) := \sum_{n=0}^{\infty} \frac{t^n}{(n+\ell)^s} \tag{3}$$

($\ell \in \mathbb{C} \setminus \mathbb{Z}_0^-; s \in \mathbb{C}; \Re(s) > 1$ and $|t| < 1$ where, as usual, $\mathbb{Z}_0^- := \mathbb{Z} \setminus \mathbb{N}$, ($\mathbb{Z} := \{0, \pm 1, \pm 2, \pm 3, \ldots\}$). It is clear that $\mathfrak{H}(2\pi i\lambda, s, \ell)$ is an ordinary Lerch zeta function and note that

$$\mathfrak{H}(t, 1, \ell) = \ell^{-1} {}_2F_1(\ell, 1; \ell + 1, t),$$

where ${}_2F_1$ is the Gaussian hypergeometric function. Recent investigations on Hurwitz–Lerch Zeta function can be found in [6, 15], and also the references stated therein. The double zeta function of Barnes [2] is defined by

$$\zeta(x, \ell, \tau) = \sum_{n=0}^{\infty} \sum_{j=0}^{\infty} (j + \ell + \tau)^{-x},$$

where $\ell \neq 0$ and $\tau \in \mathbb{C} \setminus \{0\}$ with $|\arg(\tau)| < \pi$. Bin-Saad [3] posed a generalized form of double zeta function as

$$\zeta_\tau^\kappa(t, s, \ell) = \sum_{n=0}^{\infty} (\kappa)_n \mathfrak{H}(t, s, \ell + n\tau) \frac{t^n}{n!}$$

where $\tau \in \mathbb{C} \setminus \{0\}$; $\kappa \in \mathbb{C} \setminus \mathbb{Z}_0^-$; $\ell \in \mathbb{C} \setminus \{-(j + \tau n)\}$, $n, j \in \mathbb{N}_0 := \mathbb{N} \cup \{0\}$, $|s| < 1$; $|t| < 1$ and $(\kappa)_n$ is the Pochhammer symbol defined by

$$(\kappa)_n = \begin{cases} 1, & n = 0 \\ \kappa(\kappa+1)(\kappa+2)\ldots(\kappa+n-1), & n \in \mathbb{N}. \end{cases} \tag{4}$$

Hurwitz–Lerch zeta function in [18] is given by

$$\Lambda_n(t, \ell, s) = \frac{\mathfrak{H}(t, s, \ell + n\tau)}{\mathfrak{H}(t, \ell, s)}, \quad n \in \mathbb{N}_0. \tag{5}$$

It is clear that $\Lambda_0(t, \ell, s) = 1$. Now consider the function

$$\Upsilon_\kappa(t, \ell, s) = \sum_{n=0}^{\infty} \frac{(\kappa)_n}{n!} \Lambda_n(t, \ell, s) t^n, \tag{6}$$

which implies

$$t\Upsilon_\kappa(t, \ell, s) = z + \sum_{n=2}^{\infty} \frac{(\kappa)_{n-1}}{(n-1)!} \Lambda_{n-1}(t, \ell, s) t^n.$$

Thus,

$$t\Upsilon_\kappa(t, \ell, s) * (z\Upsilon_\kappa(t, \ell, s))^{-1} = \frac{t}{(1-t)^\delta} = t + \sum_{n=2}^{\infty} \frac{(\delta)_{n-1}}{(n-1)!} t^n, \ \delta > -1$$

poses a linear operator

$$\mathfrak{I}_\kappa^\delta(t, \ell, s) f(t) = (t\Upsilon_\kappa(t, \ell, s))^{-1} * f(t) = t + \sum_{n=2}^{\infty} \frac{(\delta)_{n-1}}{(\kappa)_{n-1} \Lambda_{n-1}(t, \ell, s)} a_n t^n \tag{7}$$

where $\kappa \in \mathbb{C} \setminus \mathbb{Z}_0^-$; $\tau \in \mathbb{C} \setminus \{0\}$; $\ell \in \mathbb{C} \setminus \{-(\jmath + \tau n)\}, n, \jmath \in \mathbb{N}_0$, $|s| < 1$; $|t| < 1$ and $\Lambda_n(t, a, s)$ is defined in (5). It is clear that $I_\kappa^\delta(t, \ell, s) f(t) \in \mathfrak{A}$.

$$\mathfrak{I}_\kappa^\delta f(t) = \mathfrak{I}_\kappa^\delta(t, \ell, s) f(t) = t + \sum_{n=2}^{\infty} \mathfrak{w}_n \, a_n \, t^n, \tag{8}$$

where

$$\mathfrak{w}_n = \frac{(\delta)_{n-1}}{(\kappa)_{n-1} \Lambda_{n-1}(t, \ell, s)}, \tag{9}$$

$\kappa \in \mathbb{C} \setminus \mathbb{Z}_0^-$; $\ell \in \mathbb{C} \setminus \{-(\jmath + \tau n)\}, n, \jmath \in \mathbb{N}_0, \tau \in \mathbb{C} \setminus \{0\}$; $|s| < 1$; $|t| < 1$ and $\Lambda_n(t, \ell, s)$ is defined in (5) .

It is well recognized that each $f \in \mathfrak{S}$ has an inverse $f^{-1}$ demarcated by

$$f^{-1}(f(t)) = t \quad (z \in \mathbb{D})$$
$$\text{and} \quad f(f^{-1}(w)) = w \quad (|w| < r_0(f); r_0(f) \geq 1/4)$$

where

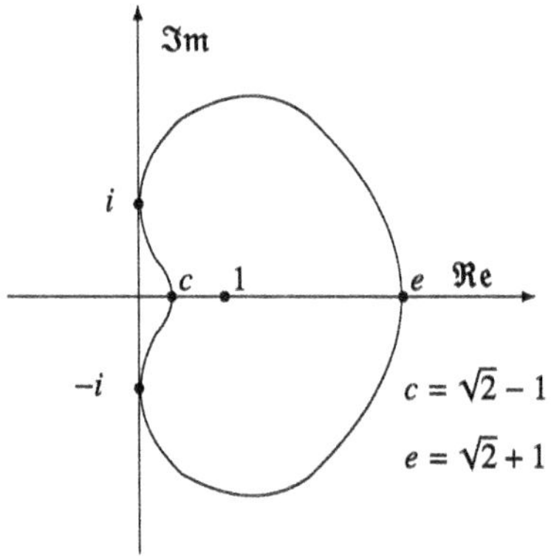


**Fig. 1** The boundary of the set $\wp(\mathbb{D})$

$$f^{-1}(w) = g(w) = w - a_2w^2 + (2a_2^2 - a_3)w^3 - (5a_2^3 - 5a_2a_3 + a_4)w^4 + \cdots . \tag{10}$$

A function $f \in \mathfrak{A}$ is said to be bi-univalent in $\mathbb{D}$ if both $f$ and $f^{-1}$ are univalent in $\mathbb{D}$. Let $\Sigma$ signify the class of bi-univalent functions in $\mathbb{D}$ given by (1). Formerly, Brannan and Taha [5] presented certain subclasses of $\Sigma$, specifically bi-starlike functions $\mathfrak{S}^*_\Sigma(\varrho)$ of order $\varrho (0 < \varrho \leq 1)$ and bi-convex function $\mathfrak{K}_\Sigma(\varrho)$ of order $\varrho$. For each $f \in \mathfrak{S}^*_\Sigma(\varrho)$ and $f \in \mathfrak{K}_\Sigma(\varrho)$, non-sharp estimates on Taylor–Maclaurin coefficients $|a_2|$ and $|a_3|$ were established [5, 28] and succeeding coefficients:

$$|a_n| \qquad (n \in \mathbb{N} \setminus \{1, 2\}; \quad \mathbb{N} := \{1, 2, 3, \cdots\})$$

is still an open problem (see [4, 5, 13, 16, 28]). Numerous researchers (see [12, 22, 24]) have familiarized and inspected many interesting subclasses of $\Sigma$ and they have originate non-sharp approximations on the coefficients $|a_2|$ and $|a_3|$.

Making use of the above subordination, Lately in [20] Raina and Sokol, defined a

$$\mathrm{S}^*(\wp) = \left\{ f \in \mathfrak{A} : \frac{tf'(t)}{f(t)} \prec t + \sqrt{1 + t^2} =: \wp(t) \right\}$$

where the branch of the square root is chosen to be the principal one, that is $\wp(0) = 1$. The function $\wp(t) := t + \sqrt{1 + t^2}$ maps the unit disc $\mathbb{D}$ onto a shell-shaped region on the right half plane, and it is analytic and univalent on $\mathbb{D}$. The range $\wp(\mathbb{D})$ is symmetric regarding the real axis and $\wp(z)$ is a function with positive real part in $\mathbb{D}$, with $\wp(0) = \wp'(0) = 1$. Besides, it is a starlike domain with respect to the point $\wp(0) = 1$ (see Fig. 1) also see [21].

Inspired by the aforementioned works, we define a subclass of bi-univalent functions, namely $\Sigma$ as follows.

Inspired by the work of Silverman and Silvia [26] (also see [27]) and a recent study by Srivastava et al. [25], and by the earlier work of Deniz [9] and Huo Tang et al. [12], in the present paper we introduce two new-fangled subclasses given in Definitions 1 and 2 comprising the linear operator $\mathfrak{I}^\delta_\kappa$ and determine estimates of $|a_2|$

and $|a_3|$. Some associated classes are also well-thought-out, and linking to earlier recognized results are stated as corollaries.

**Definition 1** Let $f \in \Sigma$ given by (1) and let $f \in \mathfrak{S}^{\delta,\kappa}_{\Sigma,\wp}(\vartheta, \lambda)$ if

$$1 + \frac{1}{\vartheta}\left(\frac{t(\mathfrak{I}^{\delta}_{\kappa} f(t))'}{\mathfrak{I}^{\delta}_{\kappa} f(t)} + \left(\frac{1+e^{i\lambda}}{2}\right)\frac{t^2(\mathfrak{I}^{\delta}_{\kappa} f(t))''}{\mathfrak{I}^{\delta}_{\kappa} f(t)} - 1\right) \prec \wp(t) \tag{11}$$

and

$$1 + \frac{1}{\vartheta}\left(\frac{w(\mathfrak{I}^{\delta}_{\kappa} g(w))'}{\mathfrak{I}^{\delta}_{\kappa} g(w)} + \left(\frac{1+e^{i\lambda}}{2}\right)\frac{w^2(\mathfrak{I}^{\delta}_{\kappa} g(w))''}{\mathfrak{I}^{\delta}_{\kappa} g(w)} - 1\right) \prec \wp(w) \tag{12}$$

where $\vartheta \in \mathbb{C}\backslash\{0\}$ $\lambda \in (-\pi, \pi]$, $t, w \in \mathbb{D}$ and $g$ is given by (10).

**Definition 2** Let $f$ given by (1) and so $f \in \mathfrak{K}^{\delta,\kappa}_{\Sigma,\wp}(\vartheta, \lambda)$ if

$$1 + \frac{1}{\vartheta}\left(\frac{[t(\mathfrak{I}^{\delta}_{\kappa} f(t))' + \left(\frac{1+e^{i\lambda}}{2}\right) t^2(\mathfrak{I}^{\delta}_{\kappa} f(t))'']'}{(\mathfrak{I}^{\delta}_{\kappa} f(t))'} - 1\right) \prec \wp(t) \tag{13}$$

and

$$1 + \frac{1}{\vartheta}\left(\frac{[w(\mathfrak{I}^{\delta}_{\kappa} g(w))' + \left(\frac{1+e^{i\lambda}}{2}\right) w^2(\mathfrak{I}^{\delta}_{\kappa} g(w))'']'}{(\mathfrak{I}^{\delta}_{\kappa} g(w))'} - 1\right) \prec \wp(w) \tag{14}$$

where $\vartheta \in \mathbb{C}\backslash\{0\}$ $\lambda \in (-\pi, \pi]$, $t, w \in \mathbb{D}$ and $g$ is given by (10).

**Remark 1** By fixing $\lambda = 0$, and $f \in \Sigma$ given by (1), we note that $\mathfrak{S}^{\delta,\kappa}_{\Sigma,\wp}(\vartheta, 1) \equiv \mathfrak{S}^{\delta,\kappa}_{\Sigma,\wp}(\vartheta)$ and $\mathfrak{K}^{\delta,\kappa}_{\Sigma,\wp}(\vartheta, 1) \equiv \mathfrak{K}^{\delta,\kappa}_{\Sigma,\wp}(\vartheta)$.

1. Let $f \in \mathfrak{S}^{\delta,\kappa}_{\Sigma,\wp}(\vartheta)$ if

$$\left[1 + \frac{1}{\vartheta}\left(\frac{t(\mathfrak{I}^{\delta}_{\kappa} f(t))'}{\mathfrak{I}^{\delta}_{\kappa} f(t)} - 1\right)\right] \prec \wp(t) \text{ and } \left[1 + \frac{1}{\vartheta}\left(\frac{w(\mathfrak{I}^{\delta}_{\kappa} g(w))'}{\mathfrak{I}^{\delta}_{\kappa} g(w)} - 1\right)\right] \prec \wp(w)$$

2. Also, $f \in \mathfrak{K}^{\delta,\kappa}_{\Sigma,\wp}(\vartheta)$ if

$$\left[1 + \frac{1}{\vartheta}\left(\frac{t(\mathfrak{I}^{\delta}_{\kappa} f(t))''}{(\mathfrak{I}^{\delta}_{\kappa} f(t))'}\right)\right] \prec \wp(t) \text{ and } \left[1 + \frac{1}{\vartheta}\left(\frac{w(\mathfrak{I}^{\delta}_{\kappa} g(w))''}{(\mathfrak{I}^{\delta}_{\kappa} g(w))'}\right)\right] \prec \wp(w),$$

where $\vartheta \in \mathbb{C}\backslash\{0\}$ $t, w \in \mathbb{D}$ and $g$ is given by (10).

**Remark 2** Assuming $\vartheta = 1$, and for $f \in \Sigma$ given by (1), we two new classes as below

$\mathfrak{S}^{\delta,\kappa}_{\Sigma,\wp}(1, \lambda) \equiv \mathfrak{S}^{\delta,\kappa}_{\Sigma,\wp}(\lambda)$ and $\mathfrak{K}^{\delta,\kappa}_{\Sigma,\wp}(1, \lambda) \equiv \mathfrak{K}^{\delta,\kappa}_{\Sigma,\wp}(\lambda)$.

1. Let $f \in \mathfrak{S}^{\delta,\kappa}_{\Sigma,\wp}(\lambda)$ if

$$\left(\frac{t(\mathfrak{I}^{\delta}_{\kappa}f(t))'}{\mathfrak{I}^{\delta}_{\kappa}f(t)}+\left(\frac{1+e^{i\lambda}}{2}\right)\frac{t^2(\mathfrak{I}^{\delta}_{\kappa}f(t))''}{\mathfrak{I}^{\delta}_{\kappa}f(t)}\right)\prec \wp(t)$$

and

$$\left(\frac{w(\mathfrak{I}^{\delta}_{\kappa}g(w))'}{\mathfrak{I}^{\delta}_{\kappa}g(w)}+\left(\frac{1+e^{i\lambda}}{2}\right)\frac{w^2(\mathfrak{I}^{\delta}_{\kappa}g(w))''}{\mathfrak{I}^{\delta}_{\kappa}g(w)}\right)\prec \wp(w).$$

2. Let $f \in \mathfrak{K}^{\delta,\kappa}_{\Sigma,\wp}(\lambda)$ if

$$\left(\frac{[t(\mathfrak{I}^{\delta}_{\kappa}f(t))'+\left(\frac{1+e^{i\lambda}}{2}\right)t^2(\mathfrak{I}^{\delta}_{\kappa}f(t))'']'}{(\mathfrak{I}^{\delta}_{\kappa}f(t))'}\right)\prec \wp(z)$$

and

$$\left(\frac{[w(\mathfrak{I}^{\delta}_{\kappa}g(w))'+\left(\frac{1+e^{i\lambda}}{2}\right)w^2(\mathfrak{I}^{\delta}_{\kappa}g(w))'']'}{(\mathfrak{I}^{\delta}_{\kappa}g(w))'}\right)\prec \wp(w),$$

where $\lambda \in (-\pi,\pi]$, $t, w \in \mathbb{D}$ and $g$ as given by (10).

## 2 Coefficient Estimates for $f \in \mathfrak{S}^{\delta,\kappa}_{\Sigma,\wp}(\vartheta,\lambda)$ and $f \in \mathfrak{K}^{\delta,\kappa}_{\Sigma,\wp}(\vartheta,\lambda)$

For notational simplicity, in the sequel, we let

$$\mathfrak{w}_2=\frac{(\delta)_1}{(\kappa)_1\Lambda_1(t,\ell,s)}, \tag{15}$$

$$\mathfrak{w}_3=\frac{(\delta)_2}{(\kappa)_2\Lambda_2(t,\ell,s)} \tag{16}$$

where $\tau \in \mathbb{C}\setminus\{0\}$; $\kappa \in \mathbb{C}\setminus\mathbb{Z}_0^-$; $\ell \in \mathbb{C}\setminus\{-(\jmath+\tau n)\}$, $n, \jmath \in \mathbb{N}_0$, $|s|<1$; $|t|<1$ and $\Lambda_n(t,s,a)$ is defined in (5). Also

$$\wp(t) := t+\sqrt{1+t^2} = 1+t+\frac{1}{2}t^2-\frac{1}{8}t^4+\cdots. \tag{17}$$

For deriving our main results, we need the following lemma.

**Lemma 1** ([17]) *If $h \in \mathfrak{P}$, then $|c_k| \leq 2$ for each $k$, where $\mathfrak{P}$ is the family of all functions $h$ analytic in $\mathbb{D}$ for which $\Re(h(z)) > 0$ and*

$$h(z) = 1 + c_1 t + c_2 t^2 + \cdots \quad \textit{for}\ z \in \mathbb{D}.$$

Let $p(t)$ and $q(t)$ by

$$p(t) := \frac{1 + u(t)}{1 - u(t)} = 1 + p_1 z + p_2 t^2 + \cdots$$

and

$$q(t) := \frac{1 + v(t)}{1 - v(t)} = 1 + q_1 z + q_2 t^2 + \cdots .$$

It follows that

$$u(t) := \frac{p(t) - 1}{p(t) + 1} = \frac{1}{2}\left[p_1 z + \left(p_2 - \frac{p_1^2}{2}\right)t^2 + \cdots\right]$$

and

$$v(t) := \frac{q(t) - 1}{q(t) + 1} = \frac{1}{2}\left[q_1 z + \left(q_2 - \frac{q_1^2}{2}\right)t^2 + \cdots\right].$$

Then $p(t)$ and $q(t)$ are analytic in $\mathbb{D}$ with $p(0) = 1 = q(0)$.

Since $u, v : \mathbb{D} \to \mathbb{D}$, the functions $p(t)$ and $q(t)$ have a positive real part in $\mathbb{D}$, for each $i$

$$|p_i| \leq 2 \quad \text{and} \quad |q_i| \leq 2. \tag{18}$$

**Theorem 1** *Let $f$ be assumed as in* (1) *and $f \in \mathfrak{S}_{\Sigma,\wp}^{\delta,\kappa}(\vartheta, \lambda)$. Then*

$$|a_2| \leq \frac{\sqrt{2}\,|\vartheta|}{\sqrt{2\left|\vartheta[(5 + 3e^{i\lambda})\mathfrak{w}_3 - (2 + e^{i\lambda})\mathfrak{w}_2^2] + (2 + e^{i\lambda})^2\mathfrak{w}_2^2\right|}}. \tag{19}$$

*and*

$$|a_3| \leq \frac{|\vartheta|^2}{|2 + e^{i\lambda}|^2\mathfrak{w}_2^2} + \frac{|\vartheta|}{|5 + 3e^{i\lambda}|\mathfrak{w}_3}. \tag{20}$$

***Proof*** It follows from (11) and (12) that

$$1 + \frac{1}{\vartheta}\left(\frac{t(\mathfrak{I}_\kappa^\delta f(t))'}{\mathfrak{I}_\kappa^\delta f(t)} + \left(\frac{1 + e^{i\lambda}}{2}\right)\frac{t^2(\mathfrak{I}_\kappa^\delta f(t))''}{\mathfrak{I}_\kappa^\delta f(t)} - 1\right) = \wp(u(t)) \tag{21}$$

and

$$1+\frac{1}{\vartheta}\left(\frac{w(\mathfrak{I}_{\kappa}^{\delta}g(w))'}{\mathfrak{I}_{\kappa}^{\delta}g(w)}+\left(\frac{1+e^{i\lambda}}{2}\right)\frac{w^2(\mathfrak{I}_{\kappa}^{\delta}g(w))''}{\mathfrak{I}_{\kappa}^{\delta}g(w)}-1\right)=\wp(v(w)), \tag{22}$$

where

$$\begin{aligned}\wp(u(t)) &= \sqrt{1+\left(\frac{p(t)-1}{p(t)+1}\right)^2}+\frac{p(t)-1}{p(t)+1}\\ &= 1+\frac{p_1}{2}t+\left(\frac{p_2}{2}-\frac{p_1^2}{8}\right)t^2+\left(\frac{p_3}{2}-\frac{p_1p_2}{4}\right)t^3+\cdots.\end{aligned} \tag{23}$$

and similarly we get

$$\wp(v(w))=1+\frac{q_1}{2}w+\left(\frac{q_2}{2}-\frac{q_1^2}{8}\right)w^2+\left(\frac{q_3}{2}-\frac{q_1q_2}{4}\right)w^3+\cdots. \tag{24}$$

Now, equating the coefficients in (21) and (22), we get

$$\frac{1}{\vartheta}(2+e^{i\lambda})\mathfrak{w}_2a_2=\frac{1}{2}p_1, \tag{25}$$

$$\frac{1}{\vartheta}\left[(5+3e^{i\lambda})\mathfrak{w}_3a_3-(2+e^{i\lambda})\mathfrak{w}_2^2a_2^2\right]=\frac{1}{2}\left(p_2-\frac{p_1^2}{2}\right)+\frac{1}{8}p_1^2, \tag{26}$$

$$-\frac{1}{\vartheta}(2+e^{i\lambda})\mathfrak{w}_2a_2=\frac{1}{2}q_1, \tag{27}$$

and

$$\frac{1}{\vartheta}\left([2(5+3e^{i\lambda})\mathfrak{w}_3-(2+e^{i\lambda})\mathfrak{w}_2^2]a_2^2-(5+3e^{i\lambda})\mathfrak{w}_3a_3\right)=\frac{1}{2}\left(q_2-\frac{q_1^2}{2}\right)+\frac{1}{8}q_1^2. \tag{28}$$

From (25) and (27), we get

$$p_1=-q_1 \tag{29}$$

and

$$8(2+e^{i\lambda})^2\mathfrak{w}_2^2a_2^2=\vartheta^2(p_1^2+q_1^2). \tag{30}$$

Now from (26), (28) and (30), we obtain

$$\left(2\{2\vartheta[(5+3e^{i\lambda})\mathfrak{w}_3-(2+e^{i\lambda})\mathfrak{w}_2^2]+(2+e^{i\lambda})^2\mathfrak{w}_2^2\}\right)a_2^2=\vartheta^2(p_2+q_2). \tag{31}$$

Applying Lemma 1 and using (18), we have

$$|a_2| \leq \frac{\sqrt{2}\,|\vartheta|}{\sqrt{2\left|\vartheta[(5+3e^{i\lambda})\mathfrak{w}_3 - (2+e^{i\lambda})\mathfrak{w}_2^2] + (2+e^{i\lambda})^2\mathfrak{w}_2^2\right|}}.$$

Next, by subtracting (26) from (28) and using (29), we get

$$\frac{2}{\vartheta}(5+3e^{i\lambda})\mathfrak{w}_3(a_3 - a_2^2) = \frac{1}{4}(p_2 - q_2).$$

Upon substituting the value of $a_2^2$ from (30), we get

$$a_3 = \frac{\vartheta^2(p_1^2+q_1^2)}{8(2+e^{i\lambda})^2\Psi^2} + \frac{\vartheta(p_2-q_2)}{4(5+3e^{i\lambda})\mathfrak{w}_3}.$$

Applying Lemma 1 and using (18), we get

$$|a_3| \leq \frac{|\vartheta|^2}{|2+e^{i\lambda}|^2\mathfrak{w}_2^2} + \frac{|\vartheta|}{|5+3e^{i\lambda}|\mathfrak{w}_3}.$$

□

**Theorem 2** *Let* $f \in \mathfrak{K}_{\Sigma,\wp}^{\delta,\kappa}(\vartheta,\lambda)$, $\vartheta \in \mathbb{C}\backslash\{0\}$ *and* $\lambda \in (-\pi,\pi]$. *Then*

$$|a_2| \leq \frac{|\vartheta|}{\sqrt{|\vartheta[3(5+3e^{i\lambda})\mathfrak{w}_3 - 4(2+e^{i\lambda})\mathfrak{w}_2^2] + 2(2+e^{i\lambda})^2\mathfrak{w}_2^2|}} \tag{32}$$

*and*

$$|a_3| \leq \frac{|\vartheta|^2}{4|2+e^{i\lambda}|^2\mathfrak{w}_2^2} + \frac{|\vartheta|}{3|5+3e^{i\lambda}|\mathfrak{w}_3}. \tag{33}$$

***Proof*** From (13) and (14) equivalently we have

$$1 + \frac{1}{\vartheta}\left(\frac{[t(\mathfrak{J}_\kappa^\delta f(t))' + \left(\frac{1+e^{i\lambda}}{2}\right)t^2(\mathfrak{J}_\kappa^\delta f(t))'']'}{(\mathfrak{J}_\kappa^\delta f(t))'} - 1\right) = \wp(u(t)) \tag{34}$$

and

$$1 + \frac{1}{\vartheta}\left(\frac{[w(\mathfrak{J}_\kappa^\delta g(w))' + \left(\frac{1+e^{i\lambda}}{2}\right)w^2(\mathfrak{J}_\kappa^\delta g(w))'']'}{(\mathfrak{J}_\kappa^\delta g(w))'} - 1\right) = \wp(v(w)), \tag{35}$$

and on lines similar to the proof of Theorem 1, from (34) and (35), we get

$$\frac{2}{\vartheta}(2+e^{i\lambda})\mathfrak{w}_2 a_2 = \frac{1}{2}p_1, \tag{36}$$

$$\frac{1}{\vartheta}[3(5+3e^{i\lambda})\mathfrak{w}_3 a_3 - 4(2+e^{i\lambda})\mathfrak{w}_2^2 a_2^2] = \frac{1}{2}\left(p_2 - \frac{p_1^2}{2}\right) + \frac{1}{8}p_1^2, \tag{37}$$

and

$$-\frac{2}{\vartheta}(2+e^{i\lambda})\mathfrak{w}_2 a_2 = \frac{1}{2}q_1, \tag{38}$$

$$\frac{1}{\vartheta}[3(5+3e^{i\lambda})(2a_2^2 - a_3)\mathfrak{w}_3 - 4(2+e^{i\lambda})\mathfrak{w}_2^2 a_2^2] = \frac{1}{2}\left(q_2 - \frac{q_1^2}{2}\right) + \frac{1}{8}q_1^2. \tag{39}$$

From (36) and (38), we get

$$p_1 = -q_1 \tag{40}$$

and

$$32(2+e^{i\lambda})^2\mathfrak{w}_2^2 a_2^2 = \vartheta^2(p_1^2 + q_1^2). \tag{41}$$

Now from (37), (39) and (41), we obtain

$$a_2^2 = \frac{\vartheta^2(p_2+q_2)}{4[\vartheta[3(5+3e^{i\lambda})\mathfrak{w}_3 - 4(2+e^{i\lambda})\mathfrak{w}_2^2] + 2(2+e^{i\lambda})^2\mathfrak{w}_2^2]}. \tag{42}$$

Applying Lemma 1 and by (18), we get the desired inequality given in (32). Now by subtracting (37) from (39), and using (40), we get

$$\frac{6}{\vartheta}(5+3e^{i\lambda})(a_3 - a_2^2)\mathfrak{w}_3 = \frac{1}{2}(p_2 - q_2).$$

Upon substituting the value of $a_2^2$ given (41), the above equation leads to

$$a_3 = \frac{\vartheta(p_2 - q_2)}{12(5+3e^{i\lambda})\mathfrak{w}_3} + \frac{\vartheta^2(p_1^2+q_1^2)}{32(2+e^{i\lambda})^2\mathfrak{w}_2^2}. \tag{43}$$

Applying Lemma 1 and by (18), we get the preferred estimate in (33). □

Fixing $\lambda = \pi$ in Theorems 1 and 2, we can state the following:

**Corollary 1** *Let $f$ be given by* (1) *and $f \in \mathfrak{S}_{\Sigma,\wp}^{\delta,\kappa}(\vartheta)$. Then*

$$|a_2| \leq \frac{\sqrt{2}\,|\vartheta|}{\sqrt{2|\vartheta|(2\mathfrak{w}_3 - \mathfrak{w}_2^2) + \mathfrak{w}_2^2}} \quad \text{and} \quad |a_3| \leq \frac{|\vartheta|^2}{\mathfrak{w}_2^2} + \frac{|\vartheta|}{2\mathfrak{w}_3}.$$

**Corollary 2** *Let $f$ be given by* (1) *and $f \in \mathfrak{K}_{\Sigma,\wp}^{\delta,\kappa}(\vartheta)$. Then*

$$|a_2| \leq \frac{|\vartheta|}{\sqrt{2|\vartheta|(3\mathfrak{w}_3 - 2\mathfrak{w}_2^2) + 2\mathfrak{w}_2^2}} \quad \text{and} \quad |a_3| \leq \frac{|\vartheta|^2}{4\mathfrak{w}_2^2} + \frac{|\vartheta|}{6\mathfrak{w}_3}.$$

Taking $\vartheta = 1$ in Theorems 1 and 2, we state the following results:

**Corollary 3** *Let $f$ be given by* (1) *and $f \in \mathfrak{S}^{\delta,\kappa}_{\Sigma,\wp}(\lambda)$. Then*

$$|a_2| \leq \frac{\sqrt{2}}{\sqrt{2|(5+3e^{i\lambda})\mathfrak{w}_3 - (2+e^{i\lambda})\mathfrak{w}_2^2| + |2+e^{i\lambda}|^2\mathfrak{w}_2^2}}$$

*and*

$$|a_3| \leq \frac{1}{|2+e^{i\lambda}|^2\mathfrak{w}_2^2} + \frac{1}{|5+3e^{i\lambda}|\mathfrak{w}_3}.$$

**Corollary 4** *Let $f$ be given by* (1) *and $f \in \mathfrak{K}^{\delta,\kappa}_{\Sigma,\wp}(\lambda)$. Then*

$$|a_2| \leq \frac{1}{\sqrt{\{[3|5+3e^{i\lambda}|\mathfrak{w}_3 - 4\left|2+e^{i\lambda}\right|\mathfrak{w}_2^2] + 2|2+e^{i\lambda}|^2\mathfrak{w}_2^2\}}}$$

*and*

$$|a_3| \leq \frac{1}{4\left|2+e^{i\lambda}\right|^2\mathfrak{w}_2^2} + \frac{1}{3\left|5+3e^{i\lambda}\right|\mathfrak{w}_3}.$$

## 3 Fekete–Szegö Inequality for $f \in \mathfrak{S}^{\delta,\kappa}_{\Sigma,\wp}(\vartheta,\lambda)$

In this section, due to Fekete–Szegö [11], we prove the following result:

**Theorem 3** *Let $f \in \mathfrak{S}^{\delta,\kappa}_{\Sigma,\wp}(\vartheta,\lambda)$ and $\aleph \in \mathbb{R}$. Then*

$$|\, a_3 - \aleph a_2^2 \,| \leq \begin{cases} \frac{\vartheta}{3|5+3e^{i\lambda}|\mathfrak{w}_3}, & 0 \leq |\, \hbar(\aleph,\vartheta) \,| \leq \frac{\vartheta}{3|5+3e^{i\lambda}|\mathfrak{w}_3} \\ 2|\vartheta||\hbar(\aleph,\vartheta)|, & |\hbar(\aleph,\vartheta)| \geq \frac{\vartheta}{3|5+3e^{i\lambda}|\mathfrak{w}_3}. \end{cases}$$

*where*

$$\hbar(\aleph,\vartheta) = \frac{\vartheta^2(1-\aleph)}{4[\vartheta[3(5+3e^{i\lambda})\mathfrak{w}_3 - 4(2+e^{i\lambda})\mathfrak{w}_2^2] + 2(2+e^{i\lambda})^2\mathfrak{w}_2^2]}.$$

***Proof*** From (42) and (43)

$$\begin{aligned} a_3 - \aleph a_2^2 &= \frac{(1-\aleph)\vartheta^2(p_2+q_2)}{4[\vartheta[3(5+3e^{i\lambda})\mathfrak{w}_3 - 4(2+e^{i\lambda})\mathfrak{w}_2^2] + 2(2+e^{i\lambda})^2\mathfrak{w}_2^2]} + \frac{\vartheta(p_2-q_2)}{12(5+3e^{i\lambda})\mathfrak{w}_3} \\ &= \left[\hbar(\aleph,\vartheta) + \frac{\vartheta}{12(5+3e^{i\lambda})\mathfrak{w}_3}\right] p_2 + \left[\hbar(\aleph,\vartheta) - \frac{\vartheta}{12(5+3e^{i\lambda})\mathfrak{w}_3}\right] q_2 \end{aligned}$$

where

$$\hbar(\aleph, \vartheta) = \frac{\vartheta^2(1-\aleph)}{4[\vartheta[3(5+3e^{i\lambda})\mathfrak{w}_3 - 4(2+e^{i\lambda})\mathfrak{w}_2^2] + 2(2+e^{i\lambda})^2\mathfrak{w}_2^2]}.$$

Thus, by applying Lemma 1, we get

$$| a_3 - \aleph a_2^2 | \leq \begin{cases} \frac{\vartheta}{3|5+3e^{i\lambda}|\mathfrak{w}_3}, & 0 \leq | \hbar(\aleph, \vartheta) | \leq \frac{\vartheta}{3|5+3e^{i\lambda}|\mathfrak{w}_3} \\ 2|\vartheta||\hbar(\aleph, \vartheta)|, & |\hbar(\aleph, \vartheta)| \geq \frac{\vartheta}{3|5+3e^{i\lambda}|\mathfrak{w}_3}. \end{cases}$$

In particular, by taking $\aleph = 1$, we get

$$| a_3 - a_2^2 | \leq \frac{\vartheta}{3|5+3e^{i\lambda}|\mathfrak{w}_3}$$

□

## 4 Concluding Remarks

Lately, various subclasses of starlike functions were introduced, see [7, 8, 14], by subordinating (or fixing) some particular functions such as functions linked with Bell numbers, shell-like curve connected with Fibonacci numbers, functions associated with conic domains and rational functions. Instead of $\wp$ in (17), one can determine new results for the subclasses introduced in this paper.

**Acknowledgements** If you want to include acknowledgments of assistance and the like at the end of an individual chapter please use the `acknowledgement` environment—it will automatically be rendered in line with the preferred layout.

## References

1. Aleksandar, I.: The Riemann Zeta-function: Theory and Applications. Wiley, New York (1985)
2. Barnes, E.W.: The theory of the double gamma function. Philos. Trans. Roy. Soc. A **196**, 265–387 (1901)
3. Bin-Saad, M.G.: Hypergeometric seires assotiated with the Hurwitz-Lerch zeta function. Acta Math. Univ. Comenianae **LXXVIII**(2), 269–286 (2009)
4. Brannan, D.A., Clunie, J.G. (eds.): Aspects of Contemporary Complex Analysis (Proceedings of the NATO Advanced Study Institute held at the University of Durham, Durham; July 1–20, 1979). Academic, New York (1980)
5. Brannan, D.A., Taha, T.S.: On some classes of bi-univalent functions. Studia Univ. Babeś-Bolyai Math. **31**(2), 70–77 (1986)
6. Choi, J., Srivastava, H.M.: Certain families of series associated with the Hurwitz-Lerch Zeta function. Appl. Math. Comput. **170**, 399–409 (2005)

7. Cho, N.E., Kumar, S., Kumar, V., Ravichandran, V., Serivasatava, H.M.: Starlike functions related to the Bell numbers. Symmetry **11**, Article ID: 219 (2019)
8. Dzoik, J., Raina, R.K., Sokół, J.: On certain subclasses of starlike functions related to a shell-like curve connected with Fibonacci numbers. Math. Comput. Model. **57**, 1203–1211 (2013)
9. Deniz, E.: Certain subclasses of bi-univalent functions satisfying subordinate conditions. J. Class. Anal. **2**(1), 49–60 (2013)
10. Erdelyi, A., Magnus, W., Oberhettinger, F., Tricomi, F.G.: Higher Transcendental Functions, vol. I. McGraw-Hill, New York (1953)
11. Fekete, M., Szegö, G.: Eine Bemerkung über ungerade schlichte Funktionen. J. London. Math. Soc. **8**, 85–89 (1933)
12. Tang, H., Deng, G.-T., Li, S.-H.: Coefficient estimates for new subclasses of Ma-Minda bi-univalent functions. J. Inequal. Appl. **2013**, 317 (2013)
13. Lewin, M.: On a coefficient problem for bi-univalent functions. Proc. Amer. Math. Soc. **18**, 63–68 (1967)
14. Kanas, S., Răducanu, D.: Some classes of analytic functions related to conic domains. Math. Slovaca **64**, 1183–1196 (2014)
15. Murugusundaramoorthy, G.: Subordination results for spirallike functions associated with Hurwitz-Lerch zeta function. Integral Trans. Spec. Funct. **23**(2), 97–103 (2012)
16. Netanyahu, E.: The minimal distance of the image boundary from the origin and the second coefficient of a univalent function in $|z|<1$. Arch. Rational Mech. Anal. **32**, 100–112 (1969)
17. Pommerenke, C.: Univalent Functions. Vandenhoeck & Ruprecht, Göttingen (1975)
18. Rabhaw, I., Darus, M.: On operator defined by double zeta functions. TAMKANG J. MATH. **42**(2), 163–174 (2011)
19. Ruscheweyh, S.: New criteria for univalent functions. Proc. Amer. Math. Soc. **49**, 109–115 (1975)
20. Raina, R.K., Sokól, J.: On Coefficient estimates for a certain class of starlike functions. Hacettepe. J. Math. Statist. **44**, 1427–1433 (2015)
21. Raina, R.K., Sokól, J.: On Coefficient estimates for a certain class of starlike functions. Hacettepe. J. Math. Stat. **44**, 1427–1433 (2015)
22. Srivastava, H.M., Mishra, A.K., Gochhayat, P.: Certain subclasses of analytic and bi-univalent functions. Appl. Math. Lett. **23**, 1188–1192 (2010)
23. Srivastava, H.M., Choi, J.: Series Associated with the Zeta and Related Functions. Kluwer Academic Publishers, Dordrecht (2001)
24. Srivastava, H.M., Murugusundaramoorthy, G., Magesh, N.: Certain subclasses of bi-univalent functions associated with the Hohlov operator. Global J. Math. Anal. **1**(2), 67–73 (2013)
25. Srivastava, H.M., Raducanu, D., Zaprawa, P.A.: Certain subclass of analytic functions defined by means of differential subordination. Filomat. **30**(14), 3743–3757 (2016)
26. Silverman, H.M., Silvia, E.: Characterizations for subclasses of univalent functions. Math. Japon. **50**, 103–109 (1999)
27. Silverman, H.: A class of bounded starlike functions. Int. J. Math. Math. Sci. **17**, 249–252 (1994)
28. Taha, T.S.: Topics in Univalent Function Theory, Ph.D. Thesis, University of London (1981)

# Fuzzy Rule-Based Expert System for Multi Assets Portfolio Optimization

**Garima Bisht and Sanjay Kumar**

**Abstract** Portfolio optimization has always been a topic of wide interest for investors. They always want to maximize their return for a given level of risk or minimize the risk for a given level of return. Modern Portfolio Theory (MPT) helps investors in portfolio selection but doesn't consider the uncertainty and complexity associated with the real market. Thus, to deal with the uncertainty of the real market, we use fuzzy logic in portfolio selection. In this paper, we have found the results with Statistical method (using Lagrange's multipliers method) and then by using Fuzzy logic toolbox of MATLAB (Triangular membership function and Gaussian membership function). The results obtained by both the methods are then compared. This study also examines the testing data sets.

**Keywords** Lagrange's multiplier · Fuzzy expert system · MATLAB · Triangular and Gaussian membership function

## 1 Introduction

In the investment world, there exist different motives of investors, but the most prominent among them is to get the highest return at a given level of risk or to get minimum risk at a given level of return. The financial market despite its benefits and rewards is the most complex industry which requires critical analysis to evaluate risk and return. The application of fuzzy set theory in real estate investment, especially on the allocation of assets in investment portfolios, has been a relatively explored area. So, here we make use of fuzzy logic in portfolio selection which considers the uncertainty of the investment world and thus gives a closer result to the financial market as compared to the Statistical method.

Investment portfolio theories guide the investors that how should an individual investor or financial institution allocate their money and other capital assets within an investing portfolio. An investing portfolio has long-term goals which are independent

G. Bisht (✉) · S. Kumar
Department of Mathematics, Statistics and Computer Science, G. B. Pant, University of Agriculture and Technology, Pantnagar 263145, Uttarakhand, India
e-mail: garimabisht98@gmail.com

 
B. Rushi Kumar et al. (eds.), *Mathematics and Computing*, Springer Proceedings in Mathematics & Statistics 415, https://doi.org/10.1007/978-981-19-9307-7_27

of the day-to-day fluctuations of the financial market. Investment portfolio theories help the investors to calculate the expected return and risk associated with the allocation of assets. An investor will face the trade-off between expected / anticipated return and risk, subject to various constraints on account that the market imperfections can't be ignored [1]. Each investor is different, having different financial goals, different levels of risk tolerance, and personal preferences which are often defined as the objectives and constraints. Objectives can be the type of return being expected, while constraints may include factors such as time horizon, etc. Thus, it is really a balancing act between risk and return with each investor having a unique requirement as well as financial outlook [2].

An expert system is a computer program that emulates human expertise like decision-making, the ability to solve complex problems. Imprecision, incompleteness, and vagueness are the main characteristics of the information expressed using natural language. Management of uncertainty due to linguistic representation was one of the major issues of conventional expert systems. Hence, there comes the concept of fuzzy set theory for the management of uncertainty due to linguistic representation of information [3]. A fuzzy rule-based system which is commonly known as Mamdani fuzzy inference system (FIS) was developed [4]. A fuzzy rule-based expert system is a collection of fuzzy membership functions and rules of the form "If x is low and y is high then z is medium". Here, *x, y, and z* are input and output variables, and low, high, and medium are fuzzy sets defined for *x, y, and z,* respectively. Later an FIS in which the conclusion part of the fuzzy rule was constituted by a weighted linear combination of crisp input rather than a fuzzy set was given [5].

In the past, many researchers have developed different types of methods to predict the ambiguity of real market, thus making an optimal portfolio selection. The ambiguity of a financial market is traditionally dealt with the probabilistic methods. A number of experimental studies showed the restrictions of probabilistic approaches in depicting the uncertainty of the financial markets, but the integrated use of fuzzy methods, quantitative analysis, qualitative analysis, the expert's knowledge and the manager's individual opinions can be effective for portfolio selection problem [6]. Efficient portfolios designed by the Markowitz model did achieve better than any domestic individual security. By capitalizing inefficient portfolios, the portfolios located on the efficient frontier, the depositors afford to get maximum return on savings by taking a certain given level of risk, maximum Sharpe ratio or minimum risk [7].

By utilizing a different perspective, a new definition of the risk for the random fuzzy portfolio selection was given [8] which considers the portfolio selection problem when the returns of the securities contain both randomness and fuzziness. Then two fuzzy mathematical programming models were developed considering the expert's knowledge of the classical quadratic programming approach of Modern Portfolio Theory (MPT) through the fuzzy set theory in obtaining portfolio return optimization involving direct real estate investment [9].

Then researchers integrated fuzzy set theory with genetic algorithms to develop a methodology for effective stock selection. A stock scoring mechanism using fundamental variables and applied fuzzy membership functions was developed to re-scale the scores properly. The scores were then used to obtain the relative rankings of stocks and the genetic algorithm was employed for the optimization of model parameters and feature selection for input variables [10]. Many researchers revised the MPT, they established that many inherent flaws of the MPT theory had marred the efficacy of the theory. Among all other things of MPT, its simplistic assumptions and direct correlation of risks and returns were identified as significant flaws [11].

Later, the difficulties were examined in optimizing the diffuse limited value with reverence to the structures of the parametric representation of diffuse figures as a convex constraint function [12]. A new method for the selection of the portfolio of new product development under uncertainty and inaccuracy [13] also takes into account the time-related effects of the project completion time, when the competition becomes relevant and when the product becomes obsolete and is no longer of any interest. An alternative solution to quadratic programming in the portfolio allocation situation [14] was given by describing the Markowitz mapping model in two factors: quadratic efficiency function and quadratic programming configuration. They focused on a universal approach to numerical resolution of non-QP portfolio allocation models and considered procedures that had been applied correctly to machine learning and large-scale optimization. Later, the uncertain variables were introduced to describe security performance and the return of contextual factors in portfolio selection models [15]. They estimated security performance by expert evaluation rather than historical data.

In this paper, Mamdani fuzzy rule-based expert systems are developed for analyzing the different portfolios with three assets for their risk and return using triangular and Gaussian membership functions. Performances of the fuzzy rule-based systems are also tested using testing data and conventional methodology of portfolio optimization.

The rest of the paper is structured as follows. Numerical method is discussed in Sect. 2. Problem formulation is presented in Sect. 3. Fuzzy rule-based model is developed in Sect. 4 for evaluating expected return and risk with triangular and gaussian fuzzy sets. Results are discussed in Sect. 5. At last, conclusions are given in Sect. 6.

## 2 Numerical Method

### *2.1 Lagrange's Multiplier Method*

Considering n assets portfolio, we need to minimize the risk given by variance and the constraints are considered:

1. Sum of weights of assets is equal to 1.
2. $\mathrm{E}(r_P) = \sum_{i=1}^{n} wi E(r)$

Let the n assets i = 1, 2, 3 … n be represented by column vector w = $[w_1\ w_2 \ldots w_n]^T$ and the returns on assets are represented by column vector R = $[r_1\ r_2 \ldots r_n]^T$

We consider a column vector e = $[1\ 1\ 1 \ldots 1]^T$ and the covariance matrix as

$$C = \begin{bmatrix} \sigma_{11} & \cdots & \sigma_{1n} \\ \vdots & \ddots & \vdots \\ \sigma_{n1} & \cdots & \sigma_{nn} \end{bmatrix}$$

then the optimization condition and constraints are written as

min. $\sigma 2 = w^T C w$

subject to, $w^T \mathrm{e} = 1$

$$w^T \mu = \mathrm{R}$$

defining a Lagrange's function

$$L\left(w^T, \lambda_1, \lambda_2\right) = \left(w^T C w - \sigma 2\right) + \lambda_1\left(1 - w^T \mathrm{e}\right) + \lambda_2\left(R - w^T \mu\right) \tag{1}$$

and differentiating Eq. (1) partially w.r.t $w^T, \lambda_1, \lambda_2$

$$\frac{\partial L}{\partial w^T} = wC - \lambda_1 e - \lambda_2 \mu \tag{2}$$

$$\frac{\partial L}{\partial \lambda_1} = 1 - w^T \mathrm{e} \tag{3}$$

$$\frac{\partial L}{\partial \lambda_2} = \mathrm{R} - w^T \mu \tag{4}$$

After putting all of them to zero, we get

$$w = \lambda_1 \mathrm{e} C - 1 + \lambda_2 \mu C - 1 \tag{5}$$

$$1 = \mathrm{e}\lambda_1 \mathrm{e}^{\mathrm{T}} C - 1 + \lambda_2 \mu^{\mathrm{T}} C - 1 \tag{6}$$

$$R = \lambda_1 \mu \mathrm{e}^T C - 1 + \lambda_2 \mu^T C - 1\mu \tag{7}$$

putting, $a = \mathrm{e}^{\mathrm{T}} C^{-1} \mathrm{e}$, $b = \mu^{\mathrm{T}} C^{-1} \mathrm{e}$, $c = \mu^{\mathrm{T}} C^{-1} \mu$

we have

$$1 = a\lambda_1 + b\lambda_2 \tag{8}$$

$$\mathrm{R} = b\lambda_1 + c\lambda_2 \tag{9}$$

$$\begin{bmatrix} 1 \\ R \end{bmatrix} = \begin{bmatrix} a & b \\ b & c \end{bmatrix} \begin{bmatrix} \lambda_1 \\ \lambda_2 \end{bmatrix} \tag{10}$$

$$\begin{aligned} \sigma 2 &= w^T C w \\ &= \lambda_1^2 c + 2\lambda_1\lambda_2 b + \lambda_2^2 a \\ &= [\lambda_1 \lambda_2] \begin{bmatrix} a & b \\ b & c \end{bmatrix} \begin{bmatrix} \lambda_1 \\ \lambda_2 \end{bmatrix} \end{aligned} \tag{11}$$

putting the value of $\begin{bmatrix} \lambda_1 \\ \lambda_2 \end{bmatrix}$ from Eqs. (10) to (11), we have

$$\sigma 2 = \frac{\left[\boldsymbol{c}\boldsymbol{R}^2 - 2\boldsymbol{b}\boldsymbol{R} + \boldsymbol{a}\right]}{\left(\boldsymbol{a}\boldsymbol{c} - \boldsymbol{b}^2\right)}$$

## 3 Problem Formulation

Let *A*, *B*, and *C* be the three assets with expected returns of 5, 10, and 15% and standard deviations of 10, 20, and 30%. The coefficient of correlations between assets return are taken as $\rho_{AB} = 0$, $\rho_{BC} = 0.5$, and $\rho_{AC} = 0.5$, .Let $X_{\mathrm{A}}$, $X_{\mathrm{B}}$, and $X_{\mathrm{C}}$ denote the number of units of assets *A*, *B*, and *C*, respectively, in the portfolio.

Lagrange's method is used to obtain training data on return and portfolio variance (Table 1) for a fuzzy rule-based expert system.

## 4 Fuzzy Rule-Based Model

We have developed fuzzy rule-based expert systems for evaluating expected returns and risk in different portfolios. Table 1 is used to define the universe of discourse for different input and output variables.

A fuzzy expert system is an expert system that uses fuzzy logic instead of crisp two-valued logic and is a collection of membership functions and rules that are used to reason about data. Fuzzy expert systems are oriented numerical processing unlike conventional expert systems, which use mainly symbolic reasoning. Fuzzy rule-based expert systems are generally called fuzzy inference systems (FIS) or fuzz logic controllers. As shown in Fig. 1, the following are the main steps in the development of a fuzzy inference system:

**Table 1** Risk and return in different portfolios using statistical model

| $X_A$ | $X_B$ | $X_C$ | $\mu$ | $\sigma$ |
|---|---|---|---|---|
| 100 | 0 | 0 | 5 | 10 |
| 90 | 10 | 0 | 5.5 | 9.21 |
| 80 | 20 | 0 | 6 | 8.94 |
| 70 | 30 | 0 | 6.5 | 9.21 |
| 63 | 34 | 3 | 7 | 9.92 |
| 57 | 36 | 7 | 7.5 | 10.76 |
| 52 | 36 | 12 | 8 | 11.68 |
| 46 | 38 | 16 | 8.5 | 12.66 |
| 41 | 38 | 21 | 9 | 13.7 |
| 35 | 40 | 25 | 9.5 | 14.79 |
| 29 | 42 | 29 | 10 | 15.9 |
| 24 | 42 | 34 | 10.5 | 17.04 |
| 18 | 44 | 38 | 11 | 18.20 |
| 13 | 44 | 43 | 11.5 | 19.38 |
| 7 | 46 | 47 | 12 | 20.57 |
| 1 | 48 | 51 | 12.5 | 21.78 |
| 0 | 40 | 60 | 13 | 23.06 |
| 0 | 30 | 70 | 13.5 | 24.55 |
| 0 | 20 | 80 | 14 | 26.23 |
| 0 | 10 | 90 | 14.5 | 28.05 |
| 0 | 0 | 100 | 15 | 30 |

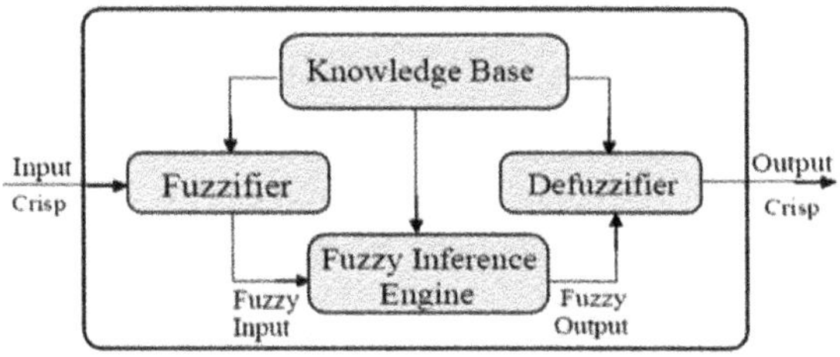


**Fig. 1** Block diagram of fuzzy inference system

1. Fuzzification
2. Fuzzy Inference
3. Defuzzification.

In this fuzzy rule-based model, no. of units of assets $X_A$, $X_B$, and $X_C$, expected return ($\mu$), and risk ($\sigma$) are fuzzified using fuzzy sets $A_i$, $B_i$, $C_i$, $D_i$, and $E_i$. Figures 2 and 3 show the graphical representation of these fuzzy sets.

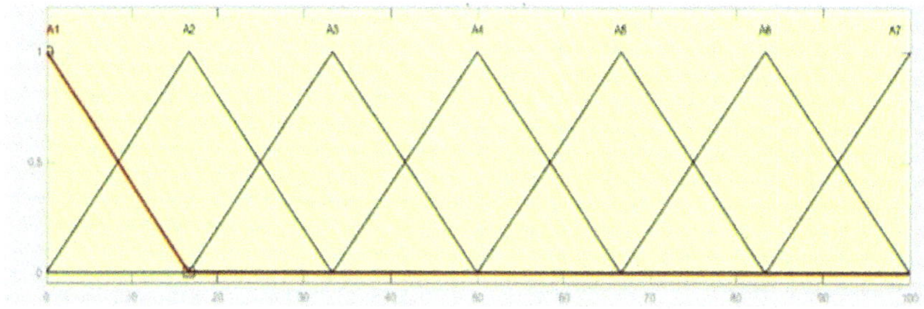

**Fig. 2** Triangular fuzzy sets

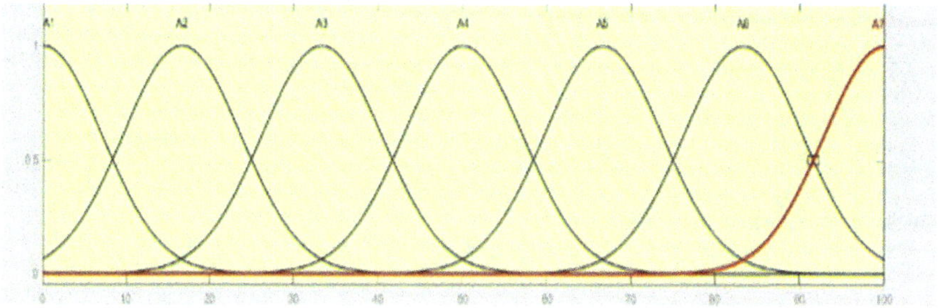

**Fig. 3** Gaussian fuzzy sets

**Table 2** Rule base for fuzzy rule-based expert system for expected return

| | |
|---|---|
| $(A_1, B_2, C_6, D_6)$ | $(A_2, B_2, C_6, D_6)$ |
| $(A_1, B_2, C_6, D_7)$ | $(A_1, B_3, C_5, D_6)$ |
| $(A_1, B_2, C_7, D_6)$ | $(A_1, B_3, C_5, D_7)$ |
| $(A_1, B_2, C_7, D_7)$ | $(A_1, B_3, C_6, D_6)$ |
| $(A_1, B_3, C_6, D_6)$ | $(A_1, B_3, C_6, D_7)$ |

## *4.1 Rule Base for Fuzzy Rule-Based Expert System*

Fuzzy rule "If $X_A$ is $A_i$ and $X_B$ is $B_i$ and $X_C$ is $C_i$ then $\mu$ is $D_i$" is abbreviated as ($A_i$, $B_i$, $C_i$: $D_i$). Table 2 shows some of the rules used in the fuzzy rule-based model.

## *4.2 Inferencing*

In this process, the membership grades are calculated and inferencing of different rules is done by using min–max methods. The rule base view is shown in Fig. 4.

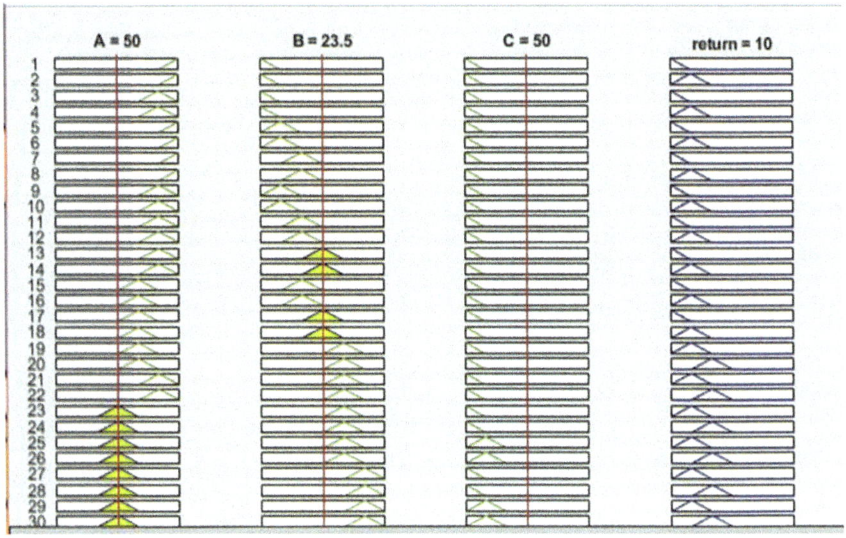


**Fig. 4** Rule base view for portfolio return

## 4.3 *Defuzzification*

Defuzzification is the process in which fuzzy output is converted into crisp output. The centroid defuzzification method is used to find a point representing the center of gravity. It is calculated by using the following equation:

$$ZCOA = \frac{\int \mu A(z).z.dz}{\int \mu A(z).dz} \tag{12}$$

## 4.4 *Testing Data Set*

We take the following random data for $X_A$, $X_B$, and $X_C$ (Table 3) and compute expected return and risk associated with different portfolios to verify the performance of developed all fuzzy rule-based expert systems.

**Table 3** Testing data set

| S.No | $X_A$ | $X_B$ | $X_C$ |
|---|---|---|---|
| 1 | 95 | 3 | 2 |
| 2 | 90 | 5 | 5 |
| 3 | 88 | 10 | 2 |
| 4 | 85 | 10 | 5 |
| 5 | 68 | 32 | 0 |
| 6 | 62 | 34 | 4 |
| 7 | 63 | 30 | 7 |
| 8 | 70 | 10 | 20 |
| 9 | 60 | 30 | 10 |
| 10 | 57 | 34 | 9 |
| 11 | 55 | 35 | 10 |
| 12 | 54 | 36 | 10 |
| 13 | 52 | 36 | 12 |
| 14 | 45 | 30 | 25 |
| 15 | 28 | 45 | 27 |
| 16 | 27 | 42 | 31 |
| 17 | 13 | 30 | 57 |
| 18 | 4 | 46 | 50 |
| 19 | 0 | 42 | 58 |
| 20 | 2 | 3 | 95 |
| 21 | 0 | 4 | 96 |

## 5 Results

Tables 4 and 5 show the expected return and risk associated with portfolios which are computed using fuzzy rule-based expert systems with triangular and Gaussian fuzzy sets.

Tables 6 and 7 show the comparison between the Statistical, Triangular, and Gaussian methods for portfolio return and risk in the testing data.

Now by using fuzzy rule-based expert systems (Tables 4 and 5), we have the following observations:

- When the assets are taken in the ratio 80:20:0 the return by the statistical method is 6%, whereas the returns obtained by triangular and Gaussian methods are 6.43% and 6.44%.
- When the assets are taken in the ratio 0:20:80 then the risk in the portfolio by triangular and Gaussian methods are 25% and 24.9% which is less than the risk obtained by the statistical method which is 26.23%.

**Table 4** Expected return in different portfolios using statistical and Mamdani FIS with triangular and gaussian membership functions

| Assets | | | Expected return | | |
|---|---|---|---|---|---|
| $X_A$ | $X_B$ | $X_C$ | Conventional | Triangular | Gaussian |
| 100 | 0 | 0 | 5 | 6.36 | 6.36 |
| 90 | 10 | 0 | 5.5 | 6.41 | 6.41 |
| 80 | 20 | 0 | 6 | 6.43 | 6.44 |
| 70 | 30 | 0 | 6.5 | 6.37 | 7.23 |
| 63 | 34 | 3 | 7 | 7.48 | 7.54 |
| 57 | 36 | 7 | 7.5 | 8.18 | 8.22 |
| 52 | 36 | 12 | 8 | 8.32 | 8.36 |
| 46 | 38 | 16 | 8.5 | 8.34 | 8.43 |
| 41 | 38 | 21 | 9 | 8.90 | 8.88 |
| 35 | 40 | 25 | 9.5 | 9.78 | 9.85 |
| 29 | 42 | 29 | 10 | 10.4 | 10.3 |
| 24 | 42 | 34 | 10.5 | 10.6 | 10.8 |
| 18 | 44 | 38 | 11 | 10.8 | 10.8 |
| 13 | 44 | 43 | 11.5 | 11 | 11 |
| 7 | 46 | 47 | 12 | 11.3 | 11.4 |
| 1 | 48 | 51 | 12.5 | 11.7 | 11.6 |
| 0 | 40 | 60 | 13 | 11.7 | 11.9 |
| 0 | 30 | 70 | 13.5 | 13.6 | 13.5 |
| 0 | 20 | 80 | 14 | 13.6 | 13.6 |
| 0 | 10 | 90 | 14.5 | 13.6 | 13.6 |
| 0 | 0 | 100 | 15 | 13.6 | 13.6 |

From Fig. 5, when the combinations of assets are in the ratio 0:30:70, the return from both the statistical and Gaussian methods is 13.5%, whereas the risk obtained from the statistical method is 24.55% and the risk obtained from the Gaussian method is 23.7%. Hence, by using fuzzy methods, we can considerably reduce the risk for the same amount of return or increase the return for the same amount of risk level.

- From Table 8, it is observed that the RMSE value in computing expected return using a fuzzy rule-based model with triangular membership function is 0.690365, while with Gaussian membership function, it is 0.700040. Even though there is a very small difference observed in RMSE, we conclude that the triangular membership function outperforms the Gaussian membership functions in the fuzzy rule-based model for computing expected return in portfolios.
- Since RMSE value in computing risk associated with different portfolios using fuzzy rule-based models with the Gaussian membership function is greater than the triangular membership function, it can be concluded that the triangular

**Table 5** Portfolio variance in different portfolios using statistical and Mamdani FIS with triangular and gaussian membership functions

| Assets | | | Portfolio Variance | | |
|---|---|---|---|---|---|
| $X_A$ | $X_B$ | $X_C$ | Conventional | Triangular | Gaussian |
| 100 | 0 | 0 | 10 | 11.4 | 11.4 |
| 90 | 10 | 0 | 9.21 | 11.5 | 11.5 |
| 80 | 20 | 0 | 8.94 | 11.6 | 11.6 |
| 70 | 30 | 0 | 9.21 | 11.5 | 11.5 |
| 63 | 34 | 3 | 9.92 | 12.4 | 12.3 |
| 57 | 36 | 7 | 10.76 | 13.3 | 13.3 |
| 52 | 36 | 12 | 11.68 | 13.3 | 13.4 |
| 46 | 38 | 16 | 12.66 | 13.2 | 13.4 |
| 41 | 38 | 21 | 13.7 | 14.6 | 14.5 |
| 35 | 40 | 25 | 14.79 | 15.6 | 15.9 |
| 29 | 42 | 29 | 15 | 16.6 | 16.5 |
| 24 | 42 | 34 | 17.04 | 17.6 | 17.8 |
| 18 | 44 | 38 | 18.20 | 19 | 18.3 |
| 13 | 44 | 43 | 19.38 | 19.6 | 19.7 |
| 7 | 46 | 47 | 20.57 | 20.8 | 20.9 |
| 1 | 48 | 51 | 21.78 | 22.6 | 22.3 |
| 0 | 40 | 60 | 23.06 | 22.8 | 22.5 |
| 0 | 30 | 70 | 24.55 | 24.6 | 23.7 |
| 0 | 20 | 80 | 26.23 | 25 | 24.9 |
| 0 | 10 | 90 | 28.05 | 25.8 | 26 |
| 0 | 0 | 100 | 30 | 27.1 | 27.1 |

membership function is better than the Gaussian membership function for a fuzzy rule-based model for computing risk.

From Table 9, it can be concluded that in computing expected return associated with different portfolios for testing data, the triangular membership function outperforms the Gaussian membership function in fuzzy rule-based expert systems.

## 6 Conclusions

In the present study, we suggest the use of fuzzy logic in multi asset portfolio optimization problems. The main reason for using fuzzy logic in the study of multi assets portfolio optimization is to develop comprehensive models of asset portfolio optimization for investors pursuing either aggressive or conservative strategies. Fuzzy

**Table 6** Expected return in different portfolios using statistical and Mamdani FIS with triangular and gaussian membership functions for the testing data

| Assets | | | Expected return | | |
|---|---|---|---|---|---|
| $X_A$ | $X_B$ | $X_C$ | Conventional | Triangular | Gaussian |
| 95 | 3 | 2 | 5.35 | 6.41 | 6.41 |
| 90 | 5 | 5 | 5.75 | 6.41 | 6.41 |
| 88 | 10 | 2 | 5.7 | 6.39 | 6.38 |
| 85 | 10 | 5 | 6 | 6.39 | 6.38 |
| 68 | 32 | 0 | 6.6 | 6.43 | 6.44 |
| 62 | 34 | 4 | 7.1 | 7.58 | 7.61 |
| 63 | 30 | 7 | 7.2 | 7.46 | 7.6 |
| 70 | 10 | 20 | 7.5 | 7.84 | 7.83 |
| 60 | 30 | 10 | 7.5 | 6.53 | 6.59 |
| 57 | 34 | 9 | 7.6 | 8.17 | 8.21 |
| 55 | 35 | 10 | 7.75 | 8.24 | 8.29 |
| 54 | 36 | 10 | 7.8 | 8.28 | 8.32 |
| 52 | 36 | 12 | 8 | 8.32 | 8.36 |
| 45 | 30 | 25 | 9 | 8.34 | 8.41 |
| 28 | 45 | 27 | 9.95 | 10.5 | 10.4 |
| 27 | 42 | 31 | 10.2 | 10.6 | 10.5 |
| 13 | 30 | 57 | 12.2 | 13.5 | 13.2 |
| 4 | 46 | 50 | 12.3 | 11.7 | 11.5 |
| 0 | 42 | 58 | 12.9 | 11.7 | 11.6 |
| 2 | 3 | 95 | 14.65 | 13.6 | 13.6 |
| 0 | 4 | 96 | 14.8 | 13.6 | 13.6 |

rule-based models with different types of membership functions (triangular and Gaussian) are developed to analyze the risk and return in three assets problem with known individual risk and return. The results obtained by fuzzy logic for the portfolio parameter, e.g., portfolio variance and expected return are a little bit different from the result obtained by the conventional method, but this result seems very near to reality. The variation may also be due to the reason that in the conventional method, non-stochastic uncertainty is left without any reason. But in the present study, by the use of fuzzy logic, a large amount of uncertainty is dealt with in the form of a rule base to give a better result.

The expected return from different portfolios computed using fuzzy rule-based models is better than conventional statistical models. Also, in particular, for fuzzy methods, we have seen by RMSE values that the triangular membership functions give better results than the Gaussian membership functions for both training and testing data.

**Table 7** Portfolio variance in different portfolios using statistical and Mamdani FIS with triangular and gaussian membership functions for the testing data

| Assets | | | Portfolio variance | | |
|---|---|---|---|---|---|
| $X_A$ | $X_B$ | $X_C$ | Conventional | Triangular | Gaussian |
| 95 | 3 | 2 | 9.85 | 11.5 | 11.5 |
| 90 | 5 | 5 | 9.96 | 11.5 | 11.5 |
| 88 | 10 | 2 | 9.39 | 11.5 | 11.5 |
| 85 | 10 | 5 | 9.70 | 11.5 | 11.5 |
| 68 | 32 | 0 | 10.3 | 11.6 | 11.6 |
| 62 | 34 | 4 | 10.08 | 12.6 | 12.4 |
| 63 | 30 | 7 | 10.29 | 12.5 | 12.4 |
| 70 | 10 | 20 | 11.95 | 11.8 | 11.9 |
| 60 | 30 | 10 | 10.81 | 13 | 12.9 |
| 57 | 34 | 9 | 10.94 | 13.4 | 13.4 |
| 55 | 35 | 10 | 11.21 | 13.4 | 13.4 |
| 54 | 36 | 10 | 11.30 | 13.3 | 13.3 |
| 52 | 36 | 12 | 11.68 | 13.3 | 13.4 |
| 45 | 30 | 25 | 13.83 | 13.4 | 13.6 |
| 28 | 45 | 27 | 15.81 | 16.8 | 16.7 |
| 27 | 42 | 31 | 16.35 | 17 | 16.9 |
| 13 | 30 | 57 | 21.32 | 24.6 | 23.5 |
| 4 | 46 | 50 | 21.3 | 22 | 21.9 |
| 0 | 42 | 58 | 22.79 | 22.1 | 21.3 |
| 2 | 3 | 95 | 28.90 | 27 | 27 |
| 0 | 4 | 96 | 29.20 | 26.9 | 26.9 |

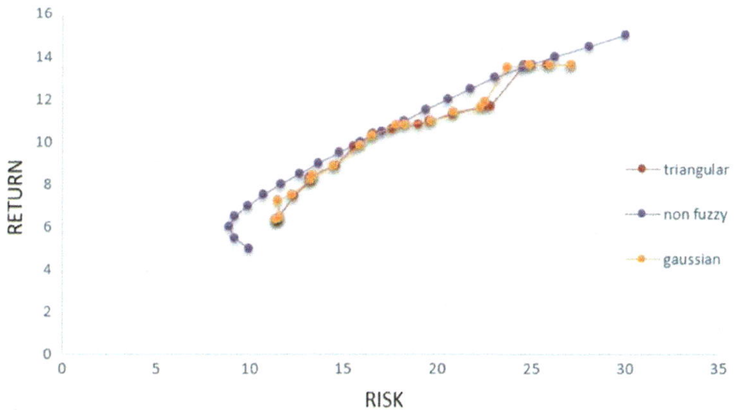


**Fig. 5** Graph between return and risk for non-fuzzy, triangular, and gaussian method

**Table 8** RMSE in expected return and portfolio variance (Training data)

| Membership function type | RMSE | |
|---|---|---|
| | Expected return | Portfolio variance |
| Triangular | 0.690365 | 1.599639 |
| Gaussian | 0.700040 | 1.601557 |

**Table 9** RMSE in expected return and portfolio variance (Testing data)

| Membership function type | RMSE | |
|---|---|---|
| | Expected return | Portfolio variance |
| Triangular | 0.735987 | 1.828523 |
| Gaussian | 0.741154 | 1.743934 |

Fuzzy set theory is a convenient method for portfolio optimization. Using fuzzy data instead of crisp data has the advantage of reducing uncertainty. The financial system has a number of uncertainties that can never be eliminated completely and hence cannot be neglected. This approach gives a new dimension to study in the field of finance.

## References


1. Markowitz, H.: Portfolio selection. J. Finance. **7**, 77–91 (1952)
2. Allen, J., Bhattacharya, S., Smarandache, F.: Fuzziness and funds allocation in portfolio optimization. Int. J. Soc. Econ. **30**, 619–632 (2003)
3. Zadeh, L.A.: Fuzzy sets. Inf. Control **8**, 338–353 (1965)
4. Mamdani, E.H., Assilian, S.: An experiment in linguistic synthesis with a fuzzy logic controller. Int. J. Man. Mach. Stud. **7**, 1–13 (1975)
5. Takagi, T., Sugeno, M.: Fuzzy identification of systems and its applications to modeling and control. IEEE Trans. Syst. Man Cybern. SMC-15, 116–132 (1985)
6. Wang, S., Zhu, S.: On fuzzy portfolio selection problems. Fuzzy Optim. Decis. Mak. **1**, 361–377 (2002)
7. Ivanova, M., Dospatliev, L.: Application of Markowitz portfolio optimization on bulgarian stock market from 2013 to 2016. Int. J. Pure Appl. Math. **117** (2018)
8. Huang, X.: Portfolio selection with a new definition of risk. Eur. J. Oper. Res. **186**, 351–357 (2008)
9. Hasuike, T., Katagiri, H., Ishii, H.: Portfolio selection problems with random fuzzy variable returns. Fuzzy Sets Syst. **160**, 2579–2596 (2009)
10. Huang, X.: Mean-risk model for uncertain portfolio selection. Fuzzy Optim. Decis. Mak. **10**, 71–89 (2011)
11. Omisore, I.: The modern portfolio theory as an investment decision tool. J. Account. Tax. **4** (2012)
12. Fard, O.S., Ramezanzadeh, M.: On fuzzy portfolio selection problems: a parametric representation approach. Complexity **2017** (2017)

13. Fernandez, E., Gomez-Santillan, C., Rangel-Valdez, N., Cruz-Reyes, L., Balderas, F.: An interval-based evolutionary approach to portfolio optimization of new product development projects. Math. Probl. Eng. **2019** (2019)
14. Perrin, S., Roncalli, T.: Machine learning optimization algorithms and portfolio allocation. SSRN Electron. J. (2019)
15. Lv, L., Zhang, B., Peng, J., Ralescu, D.A.: Uncertain portfolio selection with borrowing constraint and background risk. Math. Probl. Eng. **2020** (2020)

# Stability Analysis of Additive Time-Varying T–S Fuzzy System Using Augmented Lyapunov Functional

**Bhuvaneshwari Ganesan and Manivannan Annamalai**

**Abstract** This article discusses the stability analysis problem of Takagi–Sugeno (T–S) fuzzy system with additive time-varying delay components. To find a stability region and to stabilize the system, a state feedback control scheme is considered. A Lyapunov–Krasovskii functional is constructed to obtain less conservative results by utilizing the integral inequality based on non-orthogonal polynomials and the conditions are derived as linear matrix inequality form. The stability conditions are obtained for the system involving two delay components and the proposed result is validated through numerical examples.

**Keywords** Additive time-varying delays · T–S fuzzy system · Stability · Linear matrix inequality

## 1 Introduction

In real world, there exist delays in physical systems inherently. Avoiding these delays when modeling physical system into mathematical model gives only the approximated results. In order to get more accurate results, the time delays must be included in mathematical models. Time-delay systems are fundamental mathematical representations of real-world events such as chemical engineering system, power system, biological system, and so on. The presence of delay causes the system to be unstable and gives poor performance. As a result, substantial research has been focused on analysis and synthesis challenges of time-delayed systems. Researchers have been more focused on determining the stability of systems of various kinds, such as neutral system [4], stochastic system [10], fuzzy system [11], singular system [14], and hybrid system [15].

The majority of work focused on determining the maximum upper bound for delayed system and analyzing its stability. It has been accomplished through the appli-

B. Ganesan · M. Annamalai (✉)
Division of Mathematics, Vellore Institute of Technology, Vandalur-Kelambakkam Road,,
Chennai 600127, Tamil Nadu, India
e-mail: manivannanmku@gmail.com

 
B. Rushi Kumar et al. (eds.), *Mathematics and Computing*, Springer Proceedings in Mathematics & Statistics 415, https://doi.org/10.1007/978-981-19-9307-7_28

cation of Lyapunov stability theory by developing appropriate Lyapunov–Krasovskii functional (LKF). The construction of proper LKF ensures to get less conservative results in analyzing stability of the system. There are various types of LKF which have been used in the literature such as discretized LKF [5], polynomial-type LKF [6], augmented LKF [7], relaxed LKF [18], etc.

Takagi and Sugeno first introduced the concept of fuzzy IF-THEN rules for nonlinear systems to make it into linear subsystems by employing input–output data. Another primary role of T–S fuzzy system is that the control and stability conditions can be expressed as linear matrix inequality (LMI). This methodology is used in nonlinear systems, which has wide applications in many practical problems. Discrete-time [16] and continuous-time [13] systems are two types of time-varying T–S fuzzy systems. These systems addressed the problem with time delays such as constant delay, discrete delay, distributed delay, and additive time-varying delays. In order to handle system with such delays, various control methodologies have been employed to stabilize the system, such as state feedback control, sliding mode control, fuzzy logic control, and adaptive control.

Many researchers have investigated the stability of nonlinear system with additive time-varying delays. A new stability results have been studied for the nonlinear system with additive time-varying delays via new augmented LKFs in [2]. In [8], stability problem of a system involves two additive time-varying delays which have been investigated by using a quadratic function negative-determination lemma. Stabilization problem of switched T–S fuzzy system has been investigated with additive time-varying delays and robust stabilization is also investigated in [1]. In [20], a stability and stabilization problem via new LKFs has been studied for additive time-varying delayed T–S fuzzy system. In [21], a local stability and stabilization problem has been investigated for nonlinear systems with parameter uncertainty and two additive time-varying delays via T–S fuzzy model.

In this paper, a stability and stabilization problem for T–S fuzzy system with additive time-varying delays has been considered. A state feedback controller involves state with additive time-varying delays which is employed to stabilize the system. LKFs are considered in an augmented form and an integral inequality based on non-orthogonal polynomials has been applied to get less conservative results. Furthermore, the stability conditions have been obtained in the form of LMI. Finally, the advantages of proposed method have been validated through numerical example.

## 2 Problem Formulations

Consider the delayed T–S fuzzy model with additive time-varying delays as follows: Fuzzy Plant Rule $i(i = 1, 2, \ldots, p)$ : IF $s_1$ is $w_{i1}$, and, ..., and $s_q$ is $w_{iq}$ THEN

$$\begin{cases} \dot{x}(t) = A_i x(t) + B_i x(t - \hbar_1(t) - \hbar_2(t)) + C_i u(t), \\ x(t) = \phi(t),\ t \in [-\bar{\hbar}, 0],\ t \geq 0, \end{cases} \tag{1}$$

where $x(t) \in \mathbb{R}^n$ represents the state vector and $u(t) \in \mathbb{R}^n$ is control input; $s_m$, $w_{im}$ $(m = 1, \ldots, q)$ represents the premise variables and associated fuzzy sets, respectively; $p$ denotes the number of IF-THEN rules; $A_i$, $B_i$ and $C_i$ are appropriate dimensional known matrices. $\hbar_1(t)$, $\hbar_2(t)$ are two additive positive time-varying bounded delays satisfying the following conditions:

$$0 \le \hbar_1(t) \le \hbar_1, \quad \dot{\hbar}_1(t) \le \mu_1 < 1, \quad 0 \le \hbar_2(t) \le \hbar_2, \quad \dot{\hbar}_2(t) \le \mu_2 < 1, \tag{2}$$

and $\bar{\hbar} = \hbar_1 + \hbar_2$. $\phi(t)$ denotes initial condition and it is continuously differentiable function on $[-\bar{\hbar}, 0]$. $\hbar_1$ and $\hbar_2$ are constant and positive scalars which represent the upper bound of two additive time-varying delays.

By adopting standard fuzzy inference, the overall fuzziness of the design can be denoted as follows:

$$\begin{cases} \dot{x}(t) = \sum_{i=1}^{p} \zeta_i(s(t))\Big[A_i x(t) + B_i x(t - \hbar_1(t) - \hbar_2(t)) + C_i u(t)\Big], \\ x(t) = \phi(t), \ t \in [-\bar{\hbar}, 0], \ t \ge 0, \end{cases} \tag{3}$$

where $s(t) = [s_1(t), \ldots, s_q(t)]$ and

$$\zeta_i(s(t)) = \frac{\psi_i(s(t))}{\sum_{i=1}^{p} \psi_i(s(t))} \ge 0, \text{ and } \psi_i(s(t)) = \prod_{m=1}^{q} w_{im}(s_m(t))$$

with $w_{im}(s_m(t))$ representing the grade membership of $s_m(t)$ in $w_{im}$. It is clear to see that

$$\psi_i(s(t)) > 0, \quad \forall i = 1, \ldots, p, \quad \sum_{i=1}^{p} \psi_i(s(t)) > 0, \quad \text{for any } s(t).$$

Hence $\zeta_i(s(t))$ satisfy, $\quad \zeta_i(s(t)) \ge 0, \quad \forall i = 1, \ldots, p, \quad \sum_{i=1}^{p} \zeta_i(s(t)) = 1$, for any $s(t)$.

Now, to stabilize the delayed T–S fuzzy system, consider the state feedback control design with additive time delay as follows:

**Controller rule**: IF $s_1$ is $w_{i1}$ and , ...,and $s_q$ is $w_{iq}$, THEN

$$u(t) = K_{ai} x(t) + K_{bi} x(t - \hbar_1(t) - \hbar_2(t)),$$

where $K_{ai}$ and $K_{bi}$ are unknown control gain matrices. Therefore, the complete fuzzy control rule is inferred as

$$u(t) = \sum_{i=1}^{p} \zeta_i(s(t))[K_{ai} x(t) + K_{bi} x(t - \hbar_1(t) - \hbar_2(t))]. \tag{4}$$

By adopting (4) in (3), the closed-loop system can be obtained as follows:

$$\begin{cases} \dot{x}(t) = \sum_{i=1}^{p}\sum_{l=1}^{p} \zeta_i(s(t))\zeta_l(s(t))\Big[A_i x(t) + B_i x(t - \hbar_1(t) - \hbar_2(t)) \\ \qquad\qquad + C_i\big(K_{al}x(t) + K_{bl}x(t - \hbar_1(t) - \hbar_2(t))\big)\Big], \\ x(t) = \phi(t),\ t \in [-\bar{\hbar}, 0],\ t \geq 0. \end{cases} \tag{5}$$

The major goal of this paper is to establish stability of additive time-varying delayed T–S fuzzy system (5). Besides that, the problem deals with finding the control gain matrices $K_{al}$ and $K_{bl}$ and to stabilize the system (5). Some important lemmas are introduced before deriving the main results as follows.

Most existing results for delayed systems have been used in memoryless controller design of the form $u(t) = Kx(t)$. The controller considered in this paper contains state vector, also a state with two additive time-varying delays of the form $u(t) = K_a x(t) + K_b x(t - \hbar_1(t) - \hbar_2(t))$.

## 2.1 Preliminaries

This section provides some lemmas that can be used in the main result to obtain stability criteria of the delay-dependent T–S fuzzy system.

**Lemma 1** ([19]) *For two scalars a and b with $b > a$, a vector $z : [a, b] \to \mathbb{R}^n$, and $n \times n$ real matrices $R > 0$, $H_i (i = 1, 2)$ and $Y_j (j = 1, 2, 3)$ satisfying*

$$\Theta := \begin{bmatrix} Y_1 & Y_2 & H_1 \\ * & Y_3 & H_2 \\ * & * & R \end{bmatrix} \geq 0, \text{ the following inequality holds:}$$

$$\int_a^b \dot{z}^T(s)R\dot{z}(s)ds \geq \frac{1}{b-a}\chi_1^T R\chi_1 + \chi_2^T\Big(H_1 + H_1^T - \frac{b-a}{3}Y_1\Big)\chi_2 + \chi_3^T\Big[15(H_2 + H_2^T) - 20(b-a)Y_3\Big]\chi_3 + 20\chi_3^T H_2^T L_2\chi_1.$$

*Where* $\chi_1 := z(b) - z(a)$, $\chi_2 := z(b) + z(a) - (2/(b-a))\int_a^b z(s)ds$,

$$\chi_3 := \frac{4}{b-a}\int_a^b z(s)ds - \frac{8}{(b-a)^2}\int_a^b\int_\theta^b z(s)dsd\theta.$$

**Lemma 2** ([17]) *For any constant positive symmetric matrix $L \in \mathbb{R}^{m\times m}$, scalar $\kappa > 0$, vector function $z : [0, \kappa] \to \mathbb{R}^m$ such that the integration concerned is well defined, then*

$$\kappa\int_0^\kappa z^T(s)Lz(s)ds \geq \Big(\int_0^\kappa z(s)ds\Big)^T L\Big(\int_0^\kappa z(s)ds\Big).$$

# 3 Main Results

In this section, the stability criteria conditions are derived by choosing suitable LKFs and using the above-mentioned lemmas. Now, the following notations are given to understand the main results:

$$e_i = \left[0_{n\times(i-1)n}\ I_n\ 0_{n\times(15-i)n}\right]\ (i = 1, \ldots, 15),$$

$$\xi^T(t) = \Big[x^T(t)\ \ x^T(t-\bar{\hbar})\ \ x^T(t-\hbar_1)\ \ x^T(t-\hbar_2)\ \ x^T(t-\hbar_1(t))\ \ x^T(t-\hbar_2(t))$$
$$x^T(t-\hbar(t))\ \ x^T(t-\hbar_1(t)-\hbar_2(t))\ \ \dot{x}^T(t)\ \ \frac{1}{\hbar_2-\hbar_1}\int_{t-\hbar_2}^{t-\hbar_1} x^T(s)ds\ \ \int_{t-\hbar_1}^{t} x^T(s)ds$$
$$\int_{t-\hbar_2}^{t} x^T(s)ds\ \ \frac{1}{(\hbar_2-\hbar_1)^2}\int_{t-\hbar_2}^{t-\hbar_1}\int_{\theta}^{t-\hbar_1} x^T(s)dsd\theta\ \ \frac{1}{\hbar_2^2}\int_{t-\bar{\hbar}}^{t-\hbar_1}\int_{\theta}^{t-\hbar_1} x^T(s)dsd\theta$$
$$\frac{1}{\hbar_1^2}\int_{t-\bar{\hbar}}^{t-\hbar_2}\int_{\theta}^{t-\hbar_2} x^T(s)dsd\theta\Big].$$

**Theorem 1** *For given scalars and control gain matrices* $\hbar_1 > 0, \hbar_2 > 0, \mu_1, \mu_2, K_{al}, K_{bl}$, *the system (5) with additive time-varying delays* $\hbar_1(t), \hbar_2(t)$ *satisfying condition (2) is asymptotically stable if there exist positive definite symmetric matrices* $P, Q_i, R_i, S_i (i = 1, 2, 3), T_i (i = 1, 2)$ *and any matrices* $L_i, Z_i (i = 1, 2, 3)$ *such that the following LMI is satisfied:*

$$\Omega_{i,l} = \begin{bmatrix}
\varphi_{1il}^1 & 0 & \varphi_1^3 & \varphi_1^4 & 0 & 0 & 0 & \varphi_{1il}^8 & \varphi_{1il}^9 & 0 & 0 & 0 & 0 & \varphi_1^{14} & \varphi_1^{15} \\
* & \varphi_2^2 & 0 & 0 & 0 & 0 & 0 & 0 & 0 & 0 & 0 & 0 & 0 & 0 & 0 \\
* & * & \varphi_3^3 & \varphi_3^4 & 0 & 0 & 0 & 0 & 0 & \varphi_3^{10} & 0 & 0 & \varphi_3^{13} & \varphi_3^{14} & \varphi_3^{15} \\
* & * & * & \varphi_4^4 & 0 & 0 & 0 & 0 & 0 & \varphi_4^{10} & 0 & 0 & \varphi_4^{13} & \varphi_4^{14} & \varphi_4^{15} \\
* & * & * & * & \varphi_5^5 & 0 & 0 & 0 & 0 & 0 & 0 & 0 & 0 & 0 & 0 \\
* & * & * & * & * & \varphi_6^6 & 0 & 0 & 0 & 0 & 0 & 0 & 0 & 0 & 0 \\
* & * & * & * & * & * & \varphi_7^7 & 0 & 0 & 0 & 0 & 0 & 0 & 0 & 0 \\
* & * & * & * & * & * & * & \varphi_8^8 & \varphi_{8il}^9 & 0 & 0 & 0 & 0 & 0 & 0 \\
* & * & * & * & * & * & * & * & \varphi_9^9 & 0 & 0 & 0 & 0 & \varphi_9^{14} & \varphi_9^{15} \\
* & * & * & * & * & * & * & * & * & \varphi_{10}^{10} & 0 & 0 & \varphi_{10}^{13} & 0 & 0 \\
* & * & * & * & * & * & * & * & * & * & \varphi_{11}^{11} & 0 & 0 & 0 & 0 \\
* & * & * & * & * & * & * & * & * & * & * & \varphi_{12}^{12} & 0 & 0 & 0 \\
* & * & * & * & * & * & * & * & * & * & * & * & \varphi_{13}^{13} & 0 & 0 \\
* & * & * & * & * & * & * & * & * & * & * & * & * & \varphi_{14}^{14} & \varphi_{14}^{15} \\
* & * & * & * & * & * & * & * & * & * & * & * & * & * & \varphi_{15}^{15}
\end{bmatrix} < 0, \quad (6)$$

*where*

$\varphi_{1il}^{1} = Q_2 + Q_3 + \hbar_1 T_1 + \hbar_2 T_2 - S_1 + 2\beta N A_i + 2\beta NCK_{al},\ \varphi_1^3 = \frac{\hbar_2^2}{2} P_{12} + S_1,\ \varphi_1^4 = \frac{\hbar_1^2}{2} P_{13},$
$\varphi_{1il}^{8} = \beta N B_i + \beta NCK_{bl},\ \varphi_{1il}^{9} = P_{11} + A_i^T N^T + K_{al}^T C^T N^T - \beta N,\ \varphi_1^{14} = -P_{12}\hbar_2^2,$
$\varphi_1^{15} = -P_{13}\hbar_1^2,\ \varphi_2^2 = -R_2 - R_3,\ \varphi_3^3 = (\hbar_2 - \hbar_1)R_1 + R_2 - S_1 - \frac{1}{\hbar_2 - \hbar_1} S_2$
$- (L_1 + L_1^T - \frac{\hbar_2 - \hbar_1}{3} Z_1),\ \varphi_3^4 = \frac{1}{\hbar_2 - \hbar_1} S_2 - (L_1 + L_1^T - \frac{\hbar_2 - \hbar_1}{3} Z_1),$
$\varphi_3^{10} = 2(L_1 + L_1^T - \frac{\hbar_2 - \hbar_1}{3} Z_1) - 80L_2^T,\ \varphi_3^{13} = 160L_2^T,\ \varphi_3^{14} = \frac{\hbar_2^2\hbar_1^2}{2} P_{14}^T,\ \varphi_3^{15} = \frac{\hbar_2^4}{2} P_{15},$
$\varphi_4^4 = -(\hbar_2 - \hbar_1)R_1 + R_3 - \frac{1}{\hbar_2 - \hbar_1} S_2 - (L_1 + L_1^T - \frac{\hbar_2 - \hbar_1}{3} Z_1),$
$\varphi_4^{10} = 2(L_1 + L_1^T - \frac{\hbar_2 - \hbar_1}{3} Z_1) + 80L_2^T,\ \varphi_4^{13} = -160L_2^T,\ \varphi_4^{14} = \frac{\hbar_1^4}{2} P_{15}^T,\ \varphi_4^{15} = \frac{\hbar_1^2\hbar_2^2}{2} P_{16}^T,$
$\varphi_5^5 = (1-\mu_1)Q_1 - (1-\mu_1)Q_2 + (1-\mu_1)S_3,\ \varphi_6^6 = -(1-\mu_2)Q_3,\ \varphi_7^7 = -(1-\mu_1-\mu_2)Q_1,$
$\varphi_8^8 = -(1-\mu_1-\mu_2)S_3,\ \varphi_{8il}^9 = B_i^T N^T + K_{bl}^T C^T N^T,\ \varphi_9^9 = \hbar_1^2 S_1 + (\hbar_2 - \hbar_1)S_2 - 2N,$
$\varphi_9^{14} = \hbar_1^2 P_{12},\ \varphi_9^{15} = \hbar_2^2 P_{13},\ \varphi_{10}^{10} = -4(L_1 + L_1^T - \frac{\hbar_2 - \hbar_1}{3} Z_1) - 16\big[15(L_2 + L_2^T) - 20(\hbar_2 - \hbar_1)Z_3\big],$
$\varphi_{10}^{13} = 32\big[15(L_2 + L_2^T) - 20(\hbar_2 - \hbar_1)Z_3\big],\ \varphi_{11}^{11} = \frac{-1}{\hbar_1} T_1,\ \varphi_{12}^{12} = \frac{-1}{\hbar_2} T_2,$
$\varphi_{13}^{13} = -64\big[15(L_2 + L_2^T) - 20(\hbar_2 - \hbar_1)Z_3\big],\ \varphi_{14}^{14} = -2\hbar_1^2\hbar_2^2 P_{14},\ \varphi_{14}^{15} = -\hbar_2^4 P_{15} - \hbar_1^4 P_{15},$
$\varphi_{15}^{15} = -2\hbar_1^2\hbar_2^2 P_{16}.$

***Proof*** Construct the LKF in the following form:

$$V(x_t) = \sum_{\upsilon=1}^{5} V_\upsilon(x_t),$$

where

$$\begin{aligned}
V_1(x_t) =& \eta^T(t) P \eta(t),\\
V_2(x_t) =& \int_{t-\hbar(t)}^{t-\hbar_1(t)} x^T(s) Q_1 x(s) ds + \int_{t-\hbar_1(t)}^{t} x^T(s) Q_2 x(s) ds + \int_{t-\hbar_2(t)}^{t} x^T(s) Q_3 x(s) ds,\\
V_3(x_t) =& (\hbar_2 - \hbar_1) \int_{t-\hbar_2}^{t-\hbar_1} x^T(s) R_1 x(s) ds + \int_{t-\bar{\hbar}}^{t-\hbar_1} x^T(s) R_2 x(s) ds + \int_{t-\bar{\hbar}}^{t-\hbar_2} x^T(s) R_3 x(s) ds,\\
V_4(x_t) =& \int_{-\hbar_1}^{0} \int_{t+\theta}^{t} x^T(s) T_1 x(s) ds d\theta + \int_{-\hbar_2}^{0} \int_{t+\theta}^{t} x^T(s) T_2 x(s) ds d\theta,\\
V_5(x_t) =& \hbar_1 \int_{-\hbar_1}^{0} \int_{t+\theta}^{t} \dot{x}^T(s) S_1 \dot{x}(s) ds d\theta + \int_{-\hbar_2}^{-\hbar_1} \int_{t+\theta}^{t} \dot{x}^T(s) S_2 \dot{x}(s) ds d\theta\\
&+ \int_{t-\hbar_1(t)-\hbar_2(t)}^{t-\hbar_1(t)} x^T(s) S_3 x(s) ds,
\end{aligned}$$

with $\eta = col\Big\{x(t), \int_{t-\bar{\hbar}}^{t-\hbar_1} \int_{\theta}^{t-\hbar_1} x(s) ds d\theta, \int_{t-\bar{\hbar}}^{t-\hbar_2} \int_{\theta}^{t-\hbar_2} x(s) ds d\theta\Big\}.$

The derivative of $V(x_t)$ is derived as follows:

$$\dot{V}_1(x_t)) = 2\eta^T(t)P\dot{\eta}(t),$$
$$= 2\xi^T(t)\left\{\begin{bmatrix} e_1 \\ \hbar_1^2 e_{14} \\ \hbar_2^2 e_{15}\end{bmatrix}^T \begin{bmatrix} P_{11} & P_{12} & P_{13} \\ * & P_{14} & P_{15} \\ * & * & P_{16}\end{bmatrix}\begin{bmatrix} e_9 \\ \frac{\hbar_2^2}{2}e_3 - \hbar_2^2 e_{14} \\ \frac{\hbar_1^2}{2}e_4 - \hbar_1^2 e_{15}\end{bmatrix}\right\}\xi(t) = \xi^T(t)\Upsilon_1\xi(t), \quad (7)$$

$$\dot{V}_2(x_t) \le \xi^T(t)\Big\{e_1^T[Q_2 + Q_3]e_1 + e_5^T[(1-\mu_1)Q_1 - (1-\mu_1)Q_2]e_5 - (1-\mu_1-\mu_2)e_7^T Q_1 e_7$$
$$- (1-\mu_2)e_6^T Q_3 e_6\Big\}\xi(t) = \xi^T(t)\Upsilon_2\xi(t), \quad (8)$$

$$\dot{V}_3(x_t) = \xi^T(t)\Big\{e_2^T[-R_2 - R_3]e_2 + e_3^T[(\hbar_2 - \hbar_1)R_1 + R_2]e_3 + e_4^T[-(\hbar_2 - \hbar_1)R_1$$
$$+ R_3]e_4\Big\}\xi(t) = \xi^T(t)\Upsilon_3\xi(t), \quad (9)$$

$$\dot{V}_4(x_t) \le \xi^T(t)\Big\{e_1^T[\hbar_1 T_1 + \hbar_2 T_2]e_1 - \frac{1}{\hbar_1}e_{11}^T T_1 e_{11} - \frac{1}{\hbar_2}e_{12}^T T_2 e_{12}\Big\}\xi(t) = \xi^T(t)\Upsilon_4\xi(t), \quad (10)$$

$$\dot{V}_5(x_t) \le \xi^T(t)\Big\{\hbar_1^2 e_9^T S_1 e_9 - [e_1 - e_3]^T S_1[e_1 - e_3] + (\hbar_2 - \hbar_1)e_9^T S_2 e_9 + (1-\mu_1)e_5^T S_3 e_5$$
$$- (1-\mu_1-\mu_2)e_8^T S_3 e_8\Big\}\xi(t) - \int_{t-\hbar_2}^{t-\hbar_1}\dot{x}^T(s)S_2\dot{x}(s)ds$$
$$= \xi^T(t)\Upsilon_5\xi(t) - \int_{t-\hbar_2}^{t-\hbar_1}\dot{x}^T(s)S_2\dot{x}(s)ds. \quad (11)$$

applying Lemma 1 in the integral $-\int_{t-\hbar_2}^{t-\hbar_1}\dot{x}^T(s)S_2\dot{x}(s)ds$ yields

$$-\int_{t-\hbar_2}^{t-\hbar_1}\dot{x}^T(s)S_2\dot{x}(s)ds \le \xi^T(t)\Big\{\frac{-1}{\hbar_2 - \hbar_1}[e_3 - e_4]^T S_2[e_3 - e_4] - [e_3 + e_4 - 2e_{10}]^T$$
$$\times \Big(L_1 + L_1^T - \frac{\hbar_2 - \hbar_1}{3}Z_1\Big)[e_3 + e_4 - 2e_{10}]$$
$$- [4e_{10} - 8e_{13}]^T\big(15(L_2 + L_2^T) - 20(\hbar_2 - \hbar_1)Z_3\big)$$
$$\times [4e_{10} - 8e_{13}]\Big\}\xi(t) = \xi^T(t)\Upsilon_6\xi(t). \quad (12)$$

The following equation is obtained from the system (5) for any matrix $N$ and any scalar $\beta$

$$0 = [e_9 + \beta e_1]2N\Big\{\sum_{i=1}^{p}\sum_{l=1}^{p}\zeta_i(s(t))\zeta_l(s(t))\Big[A_i e_1 + B_i e_8 + C_i(K_{al}e_1 + K_{bl}e_8\Big] - e_9\Big\}$$
$$= \xi^T(t)\Upsilon_7\xi(t). \quad (13)$$

From (7) to (13), the upper bound of $\dot{V}(x_t)$ is obtained as

$$\dot{V}(x_t) \leq \sum_{i=1}^{p}\sum_{l=1}^{p} \zeta_i(s(t))\zeta_l(s(t))\xi^T(t)\left\{\sum_{a=1}^{7} \Upsilon_a\right\}\xi(t) = \sum_{i=1}^{p}\sum_{l=1}^{p} \zeta_i(s(t))\zeta_l(s(t))\xi^T(t)\Omega_{i,l}\xi(t), \tag{14}$$

where $\xi(t)$ is given in the main results and $\Omega_{i,l}$ is given in (6). If the LMI (6) hold then the condition defined in (14) is satisfied. Thus the system (5) is asymptotically stable, this completes the proof.

**Remark 1** In the derivative of $V_5(x(t))$ there exists single integral term $\int_{t-\hbar_2}^{t-\hbar_1} \dot{x}^T(s)S_2\dot{x}(s)ds$ in which integral inequality based on non-orthogonal polynomials has been applied. This integral inequality helps to derive a less conservative result.

**Theorem 2** *For given scalars* $\hbar_1 > 0, \hbar_2 > 0, \mu_1, \mu_2$ *and unknown control gain matrices* $K_{al}, K_{bl}$*, the system (5) with additive time delays* $\hbar_1(t), \hbar_2(t)$ *satisfying condition (2) is asymptotically stable if there exist positive definite symmetric matrices* $\check{P}, \check{Q}_i, \check{R}_i, \check{S}_i (i = 1, 2, 3), \check{T}_i (i = 1, 2)$ *and any matrices* $\check{L}_i, \check{Z}_i (i = 1, 2, 3)$ *such that the following LMI is satisfied:*

$$\check{\Omega}_{i,l} = \begin{bmatrix}
\check{\varphi}_{1il}^1 & 0 & \check{\varphi}_1^3 & \check{\varphi}_1^4 & 0 & 0 & 0 & \check{\varphi}_{1il}^8 & \check{\varphi}_{1il}^9 & 0 & 0 & 0 & 0 & \check{\varphi}_1^{14} & \check{\varphi}_1^{15} \\
* & \check{\varphi}_2^2 & 0 & 0 & 0 & 0 & 0 & 0 & 0 & 0 & 0 & 0 & 0 & 0 & 0 \\
* & * & \check{\varphi}_3^3 & \check{\varphi}_3^4 & 0 & 0 & 0 & 0 & 0 & \check{\varphi}_3^{10} & 0 & 0 & \check{\varphi}_3^{13} & \check{\varphi}_3^{14} & \check{\varphi}_3^{15} \\
* & * & * & \check{\varphi}_4^4 & 0 & 0 & 0 & 0 & 0 & \check{\varphi}_4^{10} & 0 & 0 & \check{\varphi}_4^{13} & \check{\varphi}_4^{14} & \check{\varphi}_4^{15} \\
* & * & * & * & \check{\varphi}_5^5 & 0 & 0 & 0 & 0 & 0 & 0 & 0 & 0 & 0 & 0 \\
* & * & * & * & * & \check{\varphi}_6^6 & 0 & 0 & 0 & 0 & 0 & 0 & 0 & 0 & 0 \\
* & * & * & * & * & * & \check{\varphi}_7^7 & 0 & 0 & 0 & 0 & 0 & 0 & 0 & 0 \\
* & * & * & * & * & * & * & \check{\varphi}_8^8 & \check{\varphi}_{8il}^9 & 0 & 0 & 0 & 0 & 0 & 0 \\
* & * & * & * & * & * & * & * & \check{\varphi}_9^9 & 0 & 0 & 0 & 0 & \check{\varphi}_9^{14} & \check{\varphi}_9^{15} \\
* & * & * & * & * & * & * & * & * & \check{\varphi}_{10}^{10} & 0 & 0 & \check{\varphi}_{10}^{13} & 0 & 0 \\
* & * & * & * & * & * & * & * & * & * & \check{\varphi}_{11}^{11} & 0 & 0 & 0 & 0 \\
* & * & * & * & * & * & * & * & * & * & * & \check{\varphi}_{12}^{12} & 0 & 0 & 0 \\
* & * & * & * & * & * & * & * & * & * & * & * & \check{\varphi}_{13}^{13} & 0 & 0 \\
* & * & * & * & * & * & * & * & * & * & * & * & * & \check{\varphi}_{14}^{14} & \check{\varphi}_{14}^{15} \\
* & * & * & * & * & * & * & * & * & * & * & * & * & * & \check{\varphi}_{15}^{15}
\end{bmatrix} < 0, \tag{15}$$

*where*

$$\check{\varphi}_{1il}^1 = \check{Q}_2 + \check{Q}_3 + \hbar_1\check{T}_1 + \hbar_2\check{T}_2 - \check{S}_1 + 2\beta A_i\check{N} + 2\beta CF_{al},\ \check{\varphi}_1^3 = \frac{\hbar_2^2}{2}\check{P}_{12} + \check{S}_1,\ \check{\varphi}_1^4 = \frac{\hbar_1^2}{2}\check{P}_{13},$$

$$\check{\varphi}_{1il}^8 = \beta B_i\check{N} + \beta CF_{bl},\ \check{\varphi}_{1il}^9 = \check{P}_{11} + \check{N}^T A_i^T + F_{al}^T C^T - \beta\check{N},\ \check{\varphi}_1^{14} = -\check{P}_{12}\hbar_2^2,\ \check{\varphi}_1^{15} = -\check{P}_{13}\hbar_1^2,$$

$$\check{\varphi}_2^2 = -\check{R}_2 - \check{R}_3,\ \check{\varphi}_3^3 = (\hbar_2 - \hbar_1)\check{R}_1 + \check{R}_2 - \check{S}_1 - \frac{1}{\hbar_2 - \hbar_1}\check{S}_2 - (\check{L}_1 + \check{L}_1^T - \frac{\hbar_2 - \hbar_1}{3}\check{Z}_1),$$

$$\check{\varphi}_3^4 = \frac{1}{\hbar_2 - \hbar_1}\check{S}_2 - (\check{L}_1 + \check{L}_1^T - \frac{\hbar_2 - \hbar_1}{3}\check{Z}_1),\ \check{\varphi}_3^{10} = 2(\check{L}_1 + \check{L}_1^T - \frac{\hbar_2 - \hbar_1}{3}\check{Z}_1) - 80\check{L}_2^T,$$

$$\check{\varphi}_3^{13} = 160\check{L}_2^T,\ \check{\varphi}_3^{14} = \frac{\hbar_2^2\hbar_1^2}{2}\check{P}_{14}^T,\ \check{\varphi}_3^{15} = \frac{\hbar_2^4}{2}\check{P}_{15},\ \check{\varphi}_4^4 = -(\hbar_2 - \hbar_1)\check{R}_1 + \check{R}_3 - \frac{1}{\hbar_2 - \hbar_1}\check{S}_2$$

$$-(\check{L}_1+\check{L}_1^T-\frac{\hbar_2-\hbar_1}{3}\check{Z}_1),\ \check{\varphi}_4^{10}=2(\check{L}_1+\check{L}_1^T-\frac{\hbar_2-\hbar_1}{3}\check{Z}_1)+80\check{L}_2^T,\ \check{\varphi}_4^{13}=-160\check{L}_2^T,$$

$$\check{\varphi}_4^{14}=\frac{\hbar_1^4}{2}\check{P}_{15}^T,\ \check{\varphi}_4^{15}=\frac{\hbar_1^2\hbar_2^2}{2}\check{P}_{16}^T,\ \check{\varphi}_5^5=(1-\mu_1)\check{Q}_1-(1-\mu_1)\check{Q}_2+(1-\mu_1)\check{S}_3,\ \check{\varphi}_6^6=-(1-\mu_2)\check{Q}_3,$$

$$\check{\varphi}_7^7=-(1-\mu_1-\mu_2)\check{Q}_1,\ \check{\varphi}_8^8=-(1-\mu_1-\mu_2)\check{S}_3,\ \check{\varphi}_{8il}^9=\check{N}^TB_i^T+F_{bl}^TC^T,$$

$$\check{\varphi}_9^9=\hbar_1^2\check{S}_1+(\hbar_2-\hbar_1)S_2-2\check{N},\ \check{\varphi}_9^{14}=\hbar_1^2\check{P}_{12},\ \check{\varphi}_9^{15}=\hbar_2^2\check{P}_{13},\ \check{\varphi}_{10}^{10}=-4(\check{L}_1+\check{L}_1^T-\frac{\hbar_2-\hbar_1}{3}\check{Z}_1)$$

$$-16\big[15(\check{L}_2+\check{L}_2^T)-20(\hbar_2-\hbar_1)\check{Z}_3\big],\ \check{\varphi}_{10}^{13}=32\big[15(\check{L}_2+\check{L}_2^T)-20(\hbar_2-\hbar_1)\check{Z}_3\big],\ \check{\varphi}_{11}^{11}=\frac{-1}{\hbar_1}\check{T}_1,$$

$$\check{\varphi}_{12}^{12}=\frac{-1}{\hbar_2}\check{T}_2,\ \check{\varphi}_{13}^{13}=-64\big[15(\check{L}_2+\check{L}_2^T)-20(\hbar_2-\hbar_1)\check{Z}_3\big],\ \check{\varphi}_{14}^{14}=-2\hbar_1^2\hbar_2^2\check{P}_{14},$$

$$\check{\varphi}_{14}^{15}=-\hbar_2^4\check{P}_{15}-\hbar_1^4\check{P}_{15},\ \check{\varphi}_{15}^{15}=-2\hbar_1^2\hbar_2^2\check{P}_{16}.$$

*Then the control gain matrices can be constructed as* $K_{al}=F_{al}\check{N}^{-1}$, $K_{bl}=F_{bl}\check{N}^{-1}$.

***Proof*** Let us now consider $K_{al}\check{N}=F_{al}$, $K_{bl}\check{N}=F_{bl}$ and $\Gamma=col\{\check{N},\check{N},\check{N},\check{N},\check{N},\check{N},\check{N},\check{N},\check{N},\check{N},\check{N},\check{N},\check{N},\check{N},\check{N}\}$ where $\check{N}=N^{-1}$. Let us now consider the other matrices as $\check{P}=\check{N}P\check{N}$, $\check{Q}_i=\check{N}Q_i\check{N}$, $\check{R}_i=\check{N}R_i\check{N}$, $\check{S}_i=\check{N}S_i\check{N}$, $\check{T}_i=\check{N}T_i\check{N}$, $\check{L}_i=\check{N}L_i\check{N}$, $\check{Z}_i=\check{N}Z_i\check{N}$. Pre- and post-multiplication of $\Gamma^T$ and $\Gamma$ in LMI (6) leads to LMI (15). The proof is complete.

## 4 Numerical Examples

**Example 1** Consider the delayed system (5) with parameters

$$A_1=\begin{bmatrix}-2.1 & 0.1\\-0.2 & -0.9\end{bmatrix},\ B_1=\begin{bmatrix}-1.1 & 0.1\\-0.8 & -0.9\end{bmatrix},\ C_1=\begin{bmatrix}0.14 & 0\\0.1 & 1.15\end{bmatrix},$$
$$A_2=\begin{bmatrix}-1.9 & 0\\-0.2 & -1.1\end{bmatrix},\ B_2=\begin{bmatrix}-0.9 & 0\\-1.1 & -1.2\end{bmatrix},\ C_2=\begin{bmatrix}0.13 & -0.1\\0 & 0.12\end{bmatrix}.$$

Membership function is chosen in the form that $\zeta_1(t)=\dfrac{1}{1+e^{-2x_1(t)}}$ and $\zeta_2(t)=1-\zeta_1(t)$. Moreover, let $\mu_1=0.1$, $\mu_2=0.1$, $\hbar_1=0.1$, $\beta=0.1$ and solving the LMIs in Theorem 2, the obtained maximum upper bound $\hbar_2$ is 3.2562. Also, the control gain matrices corresponding to Theorem 2 are obtained as

$$K_{a1}=\begin{bmatrix}-197.6648 & -197.9296\\19.2615 & 18.8824\end{bmatrix},\ K_{a2}=\begin{bmatrix}-197.6648 & -197.9296\\19.2615 & 18.8824\end{bmatrix},$$
$$K_{b1}=\begin{bmatrix}9.9122 & 2.6148\\-0.1506 & 0.4711\end{bmatrix},\ K_{b2}=\begin{bmatrix}9.9122 & 2.6148\\-0.1506 & 0.4711\end{bmatrix}.$$

The state response of the closed-loop system is obtained by assuming $\hbar_1(t)=0.4+0.1\sin t$, $\hbar_2(t)=0.8\sin t$ under initial condition $x(0)=[2\ \ -2]^T$. The state

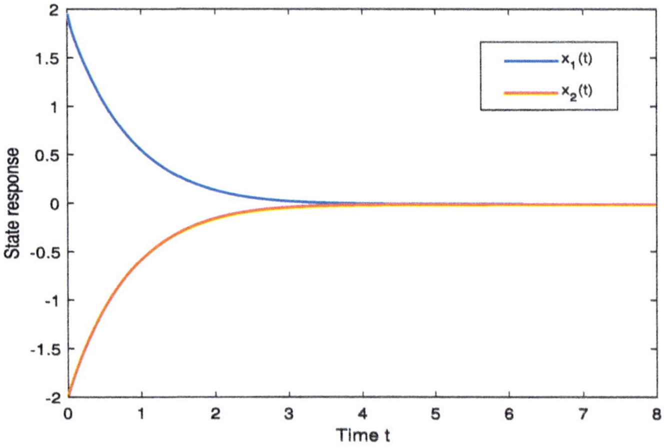


**Fig. 1** State trajectories with $\hbar_1(t) = 0.4 + 0.1 \sin t$, $\hbar_2(t) = 0.8 \sin t$ (Example (1))

trajectory of the closed-loop system (5) under the obtained control gain matrices is expressed in Fig. 1. This implies that the additive time-varying delayed T–S fuzzy system converge to origin under the proposed controller.

**Example 2** Consider the delayed system (5) with $C = 0$ gives

$$\dot{x}(t) = \sum_{i=1}^{p} \zeta_i(s(t))\Big[A_i x(t) + B_i x(t - \hbar_1(t) - \hbar_2(t))\Big], \tag{16}$$

and the parameters are as follows:

$$A_1 = \begin{bmatrix} -2 & 0 \\ 0 & -0.9 \end{bmatrix}, \quad B_1 = \begin{bmatrix} -1 & 0 \\ -1 & -1 \end{bmatrix}.$$

Consider the LMIs in Theorem 2 with $C = 0$, for different values of $\hbar_1$ and $\mu_1 = 0.1$, $\mu_2 = 0.1$ the maximum allowable upper bound $\hbar_2$ is calculated and tabulated in Table 1, and for different values of $\hbar_2$ and $\mu_1 = 0.1$, $\mu_2 = 0.1$ the allowable upper bound $\hbar_1$ is calculated and tabulated in Table 2. When compared with the existing results, the acquired results, as shown in the table, are less conservative. Moreover, for the proposed T–S fuzzy system, the delay-dependent conditions obtained increase the delay bound.

**Table 1** The obtained MAUBs $\hbar_2$ under $\mu_1 = 0.1,\ \mu_2 = 0.1$

| Methods | $\hbar_1 = 1.0$ | $\hbar_1 = 1.1$ | $\hbar_1 = 1.2$ | $\hbar_1 = 1.5$ |
|---|---|---|---|---|
| [12] | 1.198 | 1.027 | 0.980 | 0.610 |
| [3] | 0.9999 | 1.0770 | 0.9725 | 0.6807 |
| [9] | 1.2136 | 1.1136 | 1.0137 | 0.7137 |
| Theorem 2 | 1.7231 | 1.6953 | 1.5135 | 1.2356 |

**Table 2** The obtained MAUBs $\hbar_1$ under $\mu_1 = 0.1,\ \mu_2 = 0.1$

| Methods | $\hbar_2 = 0.3$ | $\hbar_2 = 0.4$ | $\hbar_2 = 0.5$ |
|---|---|---|---|
| [12] | 1.708 | 1.645 | 1.574 |
| [3] | 1.8804 | 1.7798 | 1.6759 |
| [9] | 1.9137 | 1.8137 | 1.7136 |
| Theorem 2 | 2.4135 | 2.3651 | 2.2355 |

## 5 Conclusion

The stability problem of T–S fuzzy system has been studied with two additive time-varying delays. A state feedback control design has been considered to stabilize the system. The control design takes the form of a state with additive time delays. In order to get less conservative results, augmented-type LKFs are constructed and an integral inequality based on non-orthogonal polynomials has been employed. The conservative results in the form of linear matrix inequalities have been obtained. Two numerical examples have been given to illustrate the improvement and efficacy of the proposed method.

## References


1. Ahmida, F., Tissir, E.H.: Stabilization of switched T-S fuzzy systems with additive time-varying delays. In: Proceedings of the Mediterranean Conference on Information & Communication Technologies 2015, pp. 401–408. Springer, Cham (2016)
2. Chen, W., Gao, F., Liu, G.: New results on delay-dependent stability for nonlinear systems with two additive time-varying delays. Eur. J. Control **58**, 123–130 (2021)
3. Ding, L., He, Y., Wu, M., Wang, Q.: New augmented Lyapunov-Krasovskii functional for stability analysis of systems with additive time–varying delays. Asian J. Control **20**(4), 1663–1670 (2018)
4. Han, Q.L.: A descriptor system approach to robust stability of uncertain neutral systems with discrete and distributed delays. Automatica **40**(10), 1791–1796 (2004)
5. Han, Q.L., Gu, K.: Stability of linear systems with time-varying delay: a generalized discretized Lyapunov functional approach. Asian J. Control **3**(3), 170–180 (2001)
6. Huang, Y.B., He, Y., An, J., Wu, M.: Polynomial-type Lyapunov-Krasovskii functional and Jacobi-Bessel inequality: further results on stability analysis of time-delay systems. IEEE Trans. Autom. Control **66**(6), 2905–2912 (2020)

7. Kwon, O.M., Park, M.J., Lee, S.M., Park, J.H.: Augmented Lyapunov-Krasovskii functional approaches to robust stability criteria for uncertain Takagi-Sugeno fuzzy systems with time-varying delays. Fuzzy Sets Syst. **201**, 1–19 (2012)
8. Liu, M., He, Y., Jiang, L.: A binary quadratic function negative-determination lemma and its application to stability analysis of systems with two additive time-varying delay components. IET Control Theory & Appl. **15**(17), 2221–2231 (2021)
9. Liu, M., He, Y., Wu, M., Shen, J.: Stability analysis of systems with two additive time-varying delay components via an improved delay interconnection Lyapunov-Krasovskii functional. J. Frankl. Inst. **356**(6), 3457–3473 (2019)
10. Muralisankar, S., Manivannan, A., Balasubramaniam, P.: Robust stability criteria for uncertain neutral type stochastic system with Takagi-Sugeno fuzzy model and Markovian jumping parameters. Commun. Nonlinear Sci. Numer. Simul. **17**(10), 3876–3893 (2012)
11. Lian, Z., He, Y., Zhang, C.K., Wu, M.: Stability and stabilization of T-S fuzzy systems with time-varying delays via delay-product-type functional method. IEEE Trans. Cybern. **50**(6), 2580–2589 (2019)
12. Tang, H., Han, Y., Xiao, X., Yu, H.: Improved stability criterion for linear systems with two additive time-varying delay. In: 2016 35th Chinese Control Conference (CCC), pp. 1637–1641. IEEE (2016)
13. Wang, L., Lam, H.K.: A new approach to stability and stabilization analysis for continuous-time Takagi-Sugeno fuzzy systems with time delay. IEEE Trans. Fuzzy Syst. **26**(4), 2460–2465 (2017)
14. Wang, Y., Xia, Y., Shen, H., Zhou, P.: SMC design for robust stabilization of nonlinear Markovian jump singular systems. IEEE Trans. Autom. Control **63**(1), 219–224 (2017)
15. Wu, L., Ho, D.W.: Sliding mode control of singular stochastic hybrid systems. Automatica **46**(4), 779–783 (2010)
16. Wu, L., Su, X., Shi, P., Qiu, J.: A new approach to stability analysis and stabilization of discrete-time T-S fuzzy time-varying delay systems. IEEE Trans. Syst. Man Cybernetics Part B (Cybernetics) **41**(1), 273–286 (2010)
17. Yoneyama, J.: Robust sampled-data stabilization of uncertain fuzzy systems via input delay approach. Inf. Sci. **198**, 169–176 (2012)
18. Zhang, B., Lam, J., Xu, S.: Stability analysis of distributed delay neural networks based on relaxed Lyapunov-Krasovskii functionals. IEEE Trans. Neural Netw. Learn. Syst. **26**(7), 1480–1492 (2014)
19. Zhang, X.M., Lin, W.J., Han, Q.L., He, Y., Wu, M.: Global asymptotic stability for delayed neural networks using an integral inequality based on nonorthogonal polynomials. IEEE Trans. Neural Netw. Learn. Syst. **29**(9), 4487–4493 (2017)
20. Zhao, T., Huang, M., Dian, S.: Stability and stabilization of T-S fuzzy systems with two additive time-varying delays. Inf. Sci. **494**, 174–192 (2019)
21. Zhao, T., Chen, C., Dian, S.: Local stability and stabilization of uncertain nonlinear systems with two additive time-varying delays. Commun. Nonlinear Sci. Num. Simul. **83**, 105097 (2020)

# Fractional Calculus and Integral Equations

# Solution of Fractional Differential Equations by Using Conformable Fractional Differential Transform Method with Adomian Polynomials

**R. S. Teppawar, R. N. Ingle, and R. A. Muneshwar**

**Abstract** The conformable fractional differential transform method (CFDTM) with Adomian polynomials are used in this work to solve fractional differential equations (FDEs). We shall solve nonlinear and singular Lane–Emden equations by employing this innovative approach. We first compute the differential transform (DT) of the nonlinear term in the conformable fractional sense, but in our novel technique, we substitute such nonlinear terms with recurrence relations in their Adomian polynomial of index $m$. The components of the dependent variable are eventually replaced by FDT of the same index. Because Adomian polynomials may be used for analytic nonlinear function, the CFDTM becomes much more helpful and significant. Furthermore, we compute the solution for nonlinear FDEs by using CFDTM with Adomian polynomials and these solutions are correlated with solutions calculated by using FDT method. The solutions are analyzed numerically and graphically by using Python software and the outcomes show that this technique is very effective and simple.

**Keywords** Fractional differential equation · Conformable fractional differential transform method · Adomian polynomials · Singular Lane–Emden equation

## 1 Introduction and Preliminaries

Fractional calculus has grown more relevant in mathematical study in recent decades. There is no standard form for fractional derivative definition. However, the most widely employed definitions are found in [7, 10]. Recently, several authors [2, 8] proposed a new limit concept for fractional derivatives, from which he deduced various results of fractional derivatives.

R. S. Teppawar (✉) · R. A. Muneshwar
P.G Department of Mathematics, N.E.S. Science College, Nanded 431602, MH, India
e-mail: rajeshteppawar@gmail.com

R. N. Ingle
Department of Mathematics, Bahirji Smarak Mahavidyalaya, Basmathnagar, Hingoli 431512, MH, India



B. Rushi Kumar et al. (eds.), *Mathematics and Computing*, Springer Proceedings in Mathematics & Statistics 415, https://doi.org/10.1007/978-981-19-9307-7_29

**Table 1** $\beta$-fractional derivative of functions

| No. | Function $\phi$ | $T_\beta(\phi(\vartheta))$ |
|---|---|---|
| 1 | $e^{c\vartheta}$ | $e^{c\vartheta}c\vartheta^{1-\beta}$ |
| 2 | $\sin c\vartheta$ | $c\vartheta^{1-\beta}\cos(c\vartheta)$ |
| 3 | $\cos c\vartheta$ | $-c\vartheta^{1-\beta}\sin(c\vartheta)$ |
| 4 | $a^{\vartheta}$ | $(a^{\vartheta}\log a)\vartheta^{1-\beta}$ |
| 5 | $\frac{1}{\beta}\vartheta^{\beta}$ | 1 |
| 6 | 1 | 0 |

**Definition 1** ([8]) Let $\psi\colon [0,\infty) \to \mathbb{R}$ be a function and $\forall \beta \in (0,1)$, then conformable fractional derivative of $\psi$ of order $\beta$ is defined as

$$T_\beta(\psi)(\vartheta) = \lim_{\varepsilon\to 0}\frac{\psi\left(\vartheta + \varepsilon\vartheta^{1-\beta}\right) - \psi(\vartheta)}{\varepsilon}, \quad \vartheta > 0. \tag{1.1}$$

**Definition 2** ([8]) The most useful result is that

$$T_{n\beta}(\psi)(\vartheta) = \vartheta^{\lceil\beta\rceil-\beta}\psi^{\lceil\beta\rceil}(\vartheta), \tag{1.2}$$

where $\beta \in (s, s+1]$ and $\psi$ is a $(s+1)$-differentiable function at $\vartheta > 0$.

**Theorem 1** ([8]) *If $\Phi$ and $\Psi$ are $\beta$-differentiable functions at $\vartheta > 0$, then*

1. $T_\beta(\sigma\Phi + \sigma\Psi)(\vartheta) = \sigma T_\beta(\Phi)(\vartheta) + \sigma T_\beta(\Psi)(\vartheta), \quad \forall \ \sigma \in \mathbb{R}.$
2. $T_\beta(\Phi\Psi)(\vartheta) = \Phi T_\beta(\Psi)(\vartheta) + \Psi T_\beta(\Phi)(\vartheta).$
3. $T_\beta\left(\frac{\Phi}{\Psi}\right)(\vartheta) = \left(\frac{\Psi(\vartheta)T_\beta\Phi(\vartheta) - \Phi(\vartheta)T_\beta\Psi(\vartheta)}{\Psi(\vartheta)^2}\right).$
4. $T_\beta\Phi(\vartheta) = \vartheta^{1-\beta}\frac{d\Phi(\vartheta)}{d\vartheta}.$

Conformable fractional derivative of some functions in table form.

If $0 < \beta \leq 1$ and $c \in \mathbb{R}$, then following table we have (Table 1).

Fractional differential equations (FDEs) are used to simulate a wide range of physical events, and they may be solved using a variety of transform methods [3, 6, 9, 11]. For this answer, Emrah Ünal and Ahmet Gökdoan [6] created the CFDTM. For both linear and nonlinear CFDEs, the CFDTM provides a recursive approach for determining the series solution. The difficulty with this strategy is that obtaining the differential transform of a nonlinear function will be difficult to calculate. We offer a more powerful strategy for employing the CFDTM to solve nonlinear FDEs in this study. Instead of using a nonlinear function, Adomian polynomials are used, and the dependent components are substituted with their analogous DT component. We suggested a technique combining the DTM with the ADM. This approach has the benefit of being able to combine two powerful strategies for producing an approximate series solution.

The following are some fundamental definitions of the CFDTM utilized in this paper:

**Definition 3** ([6]) If $\phi(\vartheta)$ is infinitely $\beta$-differentiable function with $\beta \in (0, 1]$, then CFDTM of $\phi(\vartheta)$ is defined as

$$\Phi_\beta(\vartheta) = \frac{1}{\beta^l l!}\left[(T_\beta^{\vartheta_0}\phi)^{(l)}(\vartheta)\right], \tag{1.3}$$

where $(T_\beta^{\vartheta_0}\phi)^{(l)}(\vartheta)$ signifies the use of the fractional derivative $l$ many times.

**Definition 4** ([6]) If $\Phi_\beta(l)$ be the CFDTM of $\phi(\vartheta)$ then inverse CFDTM of $\Phi(l)$ is defined as

$$\phi(\vartheta) = \sum_{l=0}^{\infty} \Phi_\beta(l)\vartheta^{\beta l}.$$

The CFDT of initial circumstances is defined as

$$\Phi_\beta(l) = \begin{cases} \frac{1}{(\beta l)!}\left[\frac{d^{(\beta l)}y(\vartheta)}{d\vartheta^{(\beta l)}}\right]_{\zeta=\zeta_0} & \text{for } \beta l \in \mathbb{Z}^+ \\ 0 & \text{for } \beta l \notin \mathbb{Z}^+, \end{cases}$$

where $\Phi_\beta(l)$ is the fractional differential transform of $y(\vartheta)$.

Some fundamental properties of the CFDTM can be found in [6]. Let $y(\vartheta)$, $x(\vartheta)$ and $z(\vartheta)$ be functions of time $\vartheta$ and $Y(l)$, $X(l)$ and $Z(l)$ are their corresponding FDT with order $\beta$. If $c$ and $d$ are constants then the following holds.

**Theorem 2** *If* $y(\vartheta) = cx(\vartheta) \pm dz(\vartheta)$, *then* $Y_\beta(l) = c\ X(l) \pm d\ Z_\beta(l)$.

**Theorem 3** *If* $y(\vartheta) = x(\vartheta)z(\vartheta)$, *then* $Y_\beta(l) = \sum_{r=0}^{l} X_\beta(r)Z_\beta(l-r)$.

**Theorem 4** *If* $y(\vartheta) = x(\vartheta)z(\vartheta)$, *then* $Y_\alpha(l) = \sum_{r=0}^{l} X_\alpha(r)Z_\alpha(l-r)$.

**Theorem 5** *If* $y(\vartheta) = \vartheta^r$ *then* $Y_\beta(l) = \delta(l - \frac{r}{\beta})$ *where* $\delta(l) = \begin{cases} 1 & for\ l = 0 \\ 0 & for\ l \neq 0 \end{cases}$.

**Theorem 6** *If* $\phi(\vartheta) = T_\beta^{\vartheta_0}(y(\vartheta))$, *for* $0 < \beta \leq 1$, *then* $\Phi_\beta(l) = \beta(l+1)Y_\beta(l+1)$.

**Theorem 7** *If* $\phi(\vartheta) = T_\beta^{\vartheta_0}(y(\vartheta))$, *for* $s < \beta \leq s+1$, *then*

$$\Phi_\beta(l) = Y_\beta\left(l + \frac{\beta}{\alpha}\right) = \frac{\Gamma(l\alpha + \beta + 1)}{\Gamma(l\alpha + \beta - s)}.$$

## 2 Modified Conformable Fractional Differential Transform Method (CFDTM)

**Case I:** If $T_{\beta}^{t_0} y = \phi(y)$ where $\phi(y)$ is nonlinear function of $y$. In this case we will derive some consequential relation of our expected algorithm, for this we consider the nonlinear function $\phi(y)$ which is approximated by

$$\phi(y) = \sum_{\eta=0}^{\infty} A_{\eta}, \tag{2.1}$$

where the $A_{\eta}$ are defined [1] as

$$A_{\eta} = \frac{1}{\eta!}\left[\frac{d^{\eta}}{d\sigma^{\eta}}\left[\phi\left(\sum_{i=0}^{\infty}\sigma^{i} y_i\right)\right]\right]_{\sigma=0}, \eta = 0, 1, \ldots$$

Adomian polynomials of $\phi(y)$ are organized as

$$\begin{aligned}
A_0 &= \phi\left(y_0\right) \\
A_1 &= y_1\phi^{(1)}\left(y_0\right), \\
A_2 &= y_2\phi^{(1)}\left(y_0\right) + \frac{1}{2!}y_1^2\phi^{(2)}\left(y_0\right) \\
A_3 &= y_3\phi^{(1)}\left(y_0\right) + y_1 y_2\phi^{(2)}\left(y_0\right) + \frac{1}{3!}y_1^3\phi^{(3)}\left(y_0\right) \\
&\vdots
\end{aligned} \tag{2.2}$$

The DT components of $\phi(y)$ are computed using the characteristics of CFDTM and may be indicted as follows:

$$\begin{aligned}
\Phi(0) &= \phi(y(0)) \\
&= \phi(Y_{\beta}(0)) \\
\Phi(1) &= \left.\frac{d}{d\vartheta}\phi(y(\vartheta))\right|_{\vartheta=0} \\
&= y'(0)\phi^{(1)}(y(0)) \\
&= Y_{\beta}(1)\phi^{(1)}(Y_{\beta}(0)), \\
\Phi(2) &= \frac{1}{2!}\left(y''(0)\phi^{(1)}(y(0)) + \left(y'(0)\right)^2\phi^{(2)}(y(0))\right) \\
&= Y_{\beta}(2)\phi^{(1)}(Y_{\beta}(0)) + \frac{1}{2!}(Y_{\beta}(1))^2\phi^{(2)}(Y_{\beta}(0)), \\
\Phi(3) &= Y_{\beta}(3)\phi^{(1)}(Y_{\beta}(0)) + Y_{\beta}(1)Y_{\beta}(2)\phi^{(2)}(Y_{\beta}(0)) + \frac{1}{3!}(Y_{\beta}(1))^3\phi^{(3)}(Y_{\beta}(0)) \\
&\vdots
\end{aligned} \tag{2.3}$$

**Case II:** If $T_\beta y = \psi(y^{(\beta)})$, where $\psi(y^{(\beta)})$ is nonlinear function of $y^{(\beta)}$. Here Adomian polynomials of the nonlinear function $\psi\left(y^{(\beta)}\right)$ are given as follows:

$$
\begin{aligned}
A_0 &= \psi\left(y_0^{(\beta)}\right)\\
A_1 &= y_1^{(\beta)}\psi^{(1)}\left(y_0^{(\beta)}\right),\\
A_2 &= y_2^{(\beta)}\psi^{(1)}\left(y_0^{(\beta)}\right) + \frac{1}{2!}\left(y_1^{(\alpha)}\right)^2\psi^{(2)}\left(y_0^{(\beta)}\right),\\
A_3 &= y_3^{(\beta)}\psi^{(1)}\left(y_0^{(\beta)}\right) + y_1^{(\beta)}y_2^{(\beta)}\psi^{(2)}\left(y_0^{(\beta)}\right) + \frac{1}{3!}\left(y_1^{(\beta)}\right)^3\psi^{(\beta)}\left(y_0\right) \qquad (2.4)\\
A_4 &= y_4^{(\beta)}\psi^{(1)}\left(y_0^{(\beta)}\right) + \left(y_1^{(\beta)}y_3^{(\beta)} + \frac{1}{2!}\left(y_2^{(\beta)}\right)^2\right)\psi^{(2)}\left(y_0^{(\beta)}\right)\\
&\quad + \frac{1}{2!}\left(y_1^{(\beta)}\right)^2 y_2^{(\beta)}\psi^{(\beta)}\left(y_0^{(\beta)}\right) + \frac{1}{4!}\left(y_1^{(\beta)}\right)^4\psi^{(4)}\left(y_0^{(\alpha)}\right)\\
&\vdots
\end{aligned}
$$

The fractional power series expansion of order $\alpha$ of nonlinear function $\psi\left(y^{(\beta)}\right)$ is given as

$$
\begin{aligned}
\psi\left(y^{(\beta)}\right) &= \sum_{l=0}^{\infty}\Psi(l)\vartheta^{l\alpha}\\
&= \psi\left(\sum_{l=0}^{\infty}Y_\beta(l)\vartheta^{l\alpha}\right),
\end{aligned}
$$

where $Y_\beta(k)$ denotes the CFDT of $y^{(\beta)}$. We may then use the characteristics of the fractional differential transform to arrive at a solution:

$$
\begin{aligned}
\Psi(0) &= \psi\left(y^{(\beta)}(\vartheta)\right)\Big|_{\vartheta=0}\\
&= \psi\left(Y_\beta(0)\right)\\
&= \psi\left(\Gamma(\beta+1)Y_\beta\left(\frac{\beta}{\alpha}\right)\right)\\
\Psi(1) &= Y_\beta(1)\psi^{(1)}\left(Y_\beta(0)\right)\\
&= \frac{\Gamma(\alpha+\beta+1)}{\Gamma(\alpha+\beta-m)}Y_\beta\left(1+\frac{\beta}{\alpha}\right)\psi^{(1)}\left(\Gamma(\beta+1)Y_\beta\left(\frac{\beta}{\alpha}\right)\right)\\
\Psi(2) &= Y_\beta(2)\psi^{(1)}\left(Y_\beta(0)\right) + \frac{1}{2!}\left(Y_\beta(1)\right)^2\psi^{(2)}\left(Y_\beta(0)\right).
\end{aligned}
$$

Similarly,

$$
\begin{aligned}
&=\frac{\Gamma\left(2\alpha+\beta+1\right)}{\Gamma\left(2\alpha+\beta-m\right)}Y_{\beta}\left(2+\frac{\beta}{\alpha}\right)\psi^{(1)}\left(\Gamma(\beta+1)Y_{\beta}\left(\frac{\beta}{\alpha}\right)\right)\\
&+\frac{1}{2!}\left(\frac{\Gamma\left(\alpha+\beta+1\right)}{\Gamma\left(\alpha+\beta-m\right)}Y_{\beta}\left(1+\frac{\beta}{\alpha}\right)\right)^{2}\psi^{(2)}\left(\Gamma(\beta+1)Y_{\beta}\left(\frac{\beta}{\alpha}\right)\right)\\
\Psi(3)&=\frac{\Gamma\left(3\alpha+\beta+1\right)}{\Gamma\left(3\alpha+\beta-m\right)}Y_{\beta}\left(3+\frac{\beta}{\alpha}\right)\psi^{(1)}\left(\Gamma(\beta+1)Y_{\beta}\left(\frac{\beta}{\alpha}\right)\right)\\
&+\frac{\Gamma\left(\alpha+\beta+1\right)}{\Gamma\left(\alpha+\beta-m\right)}Y_{\beta}\left(1+\frac{\beta}{\alpha}\right)\frac{\Gamma\left(\beta+1+\frac{2}{\alpha}\right)}{\Gamma\left(1+\frac{2}{\alpha}\right)}Y_{\beta}\left(2+\frac{\beta}{\alpha}\right)\psi^{(2)}\left(\Gamma(\beta+1)Y_{\beta}\left(\frac{\beta}{\alpha}\right)\right)\\
&+\frac{1}{3!}\left(\frac{\Gamma\left(\alpha+\beta+1\right)}{\Gamma\left(\alpha+\beta-m\right)}Y_{\beta}\left(1+\frac{\beta}{\alpha}\right)\right)^{3}\psi^{(3)}\left(\Gamma(\beta+1)Y_{\beta}\left(\frac{\beta}{\alpha}\right)\right)\\
\Psi(4)&=\frac{\Gamma\left(4\alpha+\beta+1\right)}{\Gamma\left(4\alpha+\beta-m\right)}Y_{\beta}\left(4+\frac{\beta}{\alpha}\right)\psi^{(1)}\left(\Gamma(\beta+1)Y_{\beta}\left(\frac{\beta}{\alpha}\right)\right)\\
&+\left(\frac{\Gamma\left(\alpha+\beta+1\right)}{\Gamma\left(\alpha+\beta-m\right)}Y_{\beta}\left(1+\frac{\beta}{\alpha}\right)\frac{\Gamma\left(3\alpha+\beta+1\right)}{\Gamma\left(3\alpha+\beta-m\right)}Y_{\beta}\left(3+\frac{\beta}{\alpha}\right)\right.\\
&\left.+\frac{1}{2!}\left(\frac{\Gamma\left(\alpha+\beta+1\right)}{\Gamma\left(\alpha+\beta-m\right)}Y_{\beta}\left(1+\frac{\beta}{\alpha}\right)\right)^{2}\right)\psi^{(2)}\left(\Gamma(\beta+1)Y_{\beta}\left(\frac{\beta}{\alpha}\right)\right)\\
&+\frac{1}{2!}\left(\frac{\Gamma\left(\alpha+\beta+1\right)}{\Gamma\left(\alpha+\beta-m\right)}Y_{\beta}\left(1+\frac{\beta}{\alpha}\right)\right)^{2}\left(\frac{\Gamma\left(2\alpha+\beta+1\right)}{\Gamma\left(2\alpha+\beta-m\right)}Y_{\alpha}\left(2+\frac{\beta}{\alpha}\right)\right)\\
&\times\psi^{(3)}\left(\Gamma(\beta+1)Y_{\beta}\left(\frac{\beta}{\alpha}\right)\right)\\
&+\frac{1}{4!}\left(\frac{\Gamma\left(\alpha+\beta+1\right)}{\Gamma\left(\alpha+\beta-m\right)}Y_{\beta}\left(1+\frac{\beta}{\alpha}\right)\right)^{4}\psi^{(4)}\left(\Gamma(\beta+1)Y_{\beta}\left(\frac{\beta}{\alpha}\right)\right)\\
&\vdots
\end{aligned}
\tag{2.5}
$$

**Case III:** If $T_{\beta}y=\varphi\left(y, y^{(\beta)}\right)$, where $\varphi\left(y, y^{(\beta)}\right)$.

Consider the nonlinear CFDE as

$$T_{\beta}y=\varphi\left(y, y^{(\beta)}\right),$$

where $\varphi\left(y, y^{(\beta)}\right)$ denotes a nonlinear function. Here $\varphi\left(y, y^{(\beta)}\right)$ is analytic in $y$, and with regard to the provided circumstances differential transform, its Adomian polynomials are analytic. Now, by comparing Eqs. (2.2) with (2.3) and Eqs. (2.4) with (2.5), withal by superseding each $y_l$ and $T_{\gamma}y_l$ in the $A_l$ by $Y_{\gamma}(l)$ and $\frac{\Gamma(l\alpha+\gamma+1)}{\Gamma(l\alpha+\gamma-m)}Y_{\gamma}(l+\gamma/\alpha)$, respectively, the formulae for $\widetilde{A}_l$ are obtained as follows:

$$\frac{\Gamma\left(l\alpha+\gamma+1\right)}{\Gamma\left(l\alpha+\gamma-m\right)}Y_{\gamma}(l+\gamma/\alpha)=\widetilde{A}_l. \tag{2.6}$$

In particular, if $\gamma=\alpha$, then Eq. (2.6) becomes

$$\gamma(l+1)Y_{\gamma}(l+1)=\widetilde{A}_l. \tag{2.7}$$

The following are some of the advantages and benefits of employing this approach or methodology to evaluate the FDT of nonlinear terms. When we compare this method to Caputo, conformable, classical approach, and many other algorithms proposed by various authors in [4–6, 12], we can see that it is superior. This strategy required less computational effort, according to our findings. Because this method is based on the algebraic recurrence relation, it does not require integration. This enables for simple series solution calculation while also allowing this approach to calculate additional series solution terms as needed.

# 3 Applications of CFDTM

In this part, we'll look at how the proposed CFDTM works with Adomian polynomials and solve some fractional differential equations with different types of nonlinearity.

**Example 1** Consider the nonlinear FDE

$$T_\beta y + e^y = 0, \quad 0 < \beta \le 1, \quad y(0) = 0. \tag{3.1}$$

We apply Theorem 6 and Eq. (2.7) to Eq. (3.1), then we get the recurrence relations shown below:

$$\begin{cases} \beta(l+1)Y_\beta(l+1) = -\widetilde{A}_l, \quad l = 0, 1, 2, \ldots \\ Y_\beta(0) = 0, \end{cases} \tag{3.2}$$

where the $\widetilde{A}_l$ are obtained from the $A_l$ of the $e^y$ as follows:

$$\left|\begin{array}{l|l} A_0 = e^{y_0} & \widetilde{A}_0 = e^{Y_\beta(0)} \\ A_1 = y_1 e^{y_0} & \widetilde{A}_1 = Y_\beta(1) e^{Y(0)} \\ A_2 = \left(y_2 + \frac{(y_1)^2}{2}\right) e^{y_0} & \widetilde{A}_2 = \left(Y_\beta(2) + \frac{(Y_\beta(1))^2}{2}\right) e^{Y_\beta(0)} \\ A_3 = \left(y_3 + y_1 y_2 + \frac{(y_1)^3}{3!}\right) e^{y_0} & \widetilde{A}_3 = \left(Y_\beta(3) + Y_\beta(1) Y_\beta(2) + \frac{(Y_\beta(1))^3}{3!}\right) e^{Y_\beta(0)} \\ \vdots & \vdots \end{array}\right|$$

Now by using these values of $\widetilde{A}_l$ in Eq. (3.2), we obtained the following differential transform components: $Y_\beta(1) = \frac{-1}{\beta}, \quad Y_\beta(2) = \frac{1}{2\beta^2}, \quad Y_\beta(3) = -\frac{1}{3\beta^3}, \ldots$. Therefore approximate solution of (3.1) is as follows:

$$y(\vartheta) = -\frac{\vartheta}{\beta} + \frac{\vartheta^{2\beta}}{2\beta^2} - \frac{\vartheta^{3\beta}}{3\beta^3} + \frac{\vartheta^{4\beta}}{4\beta^4} - \frac{\vartheta^{5\beta}}{5\beta^5} + \frac{\vartheta^{6\beta}}{6\beta^6} - \cdots$$

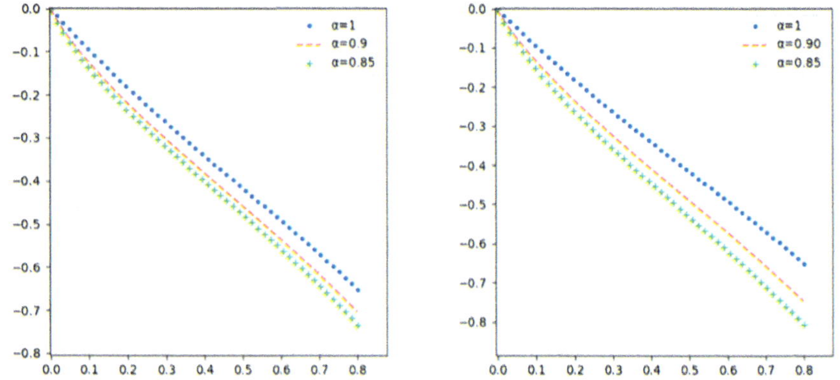


(a) Graph of solution of 3.1 for different value of $\alpha$ by FDTM.

(b) Graph of solution of 3.1 for different value of $\alpha$ by CFDTM.

**Fig. 1** Comparison of the fourth iteration approximate solutions of CFDTM with the FDTM

If $\beta \to 1$ then we get the Taylor series of the precise $y(\vartheta) = -(1 + \ln \vartheta)$. $y(\vartheta) = -\vartheta + \frac{\vartheta^2}{2} - \frac{\vartheta^3}{3} + \frac{\vartheta^4}{4} - \frac{\vartheta^5}{5} + \frac{\vartheta^6}{6} - \frac{\vartheta^7}{7} + \cdots$ (Fig. 1).

**Example 2** Consider the fractional Riccati equation

$$T_\beta y = 1 - y^2, \quad 0 < \beta \leq 1, \quad y(0) = 0. \tag{3.3}$$

We apply Theorem 6 and Eq. (2.7) to Eq. (3.3), then we get the following recurrence relations:

$$\begin{cases} \beta(l+1)Y_\beta(l+1) = \delta(l) - \widetilde{A}_l, & \forall\, l \\ y(0) = 0, \end{cases} \tag{3.4}$$

where the $\widetilde{A}_l$ of $y^2$ are as follows:

$$\begin{array}{|l|l|} A_0 = {y_0}^2 & \widetilde{A}_0 = (Y_\beta(0))^2 \\ A_1 = 2y_1y_0 & \widetilde{A}_1 = 2Y_\beta(1)Y_\beta(0) \\ A_2 = {y_1}^2 + 2y_0y_2 & \widetilde{A}_2 = (Y_\beta(1))^2 + 2Y_\beta(0)Y_\beta(2) \\ A_3 = 2y_0y_3 + 2y_1y_2 & \widetilde{A}_3 = 2Y_\beta(0)Y_\beta(3) + 2Y_\beta(1)Y_\beta(2) \\ \vdots & \vdots \end{array}$$

Now by using these values of $\widetilde{A}_l$ in Eq. (3.4), we obtained the following differential transform components:

$$Y_\beta(1) = \frac{1}{\beta}, \quad Y_\beta(2) = 0, \quad Y_\beta(3) = -\frac{1}{3\beta^3}, \quad Y_\beta(4) = 0, \quad Y_\beta(5) = \frac{2}{15\beta^{15}}, \ldots$$

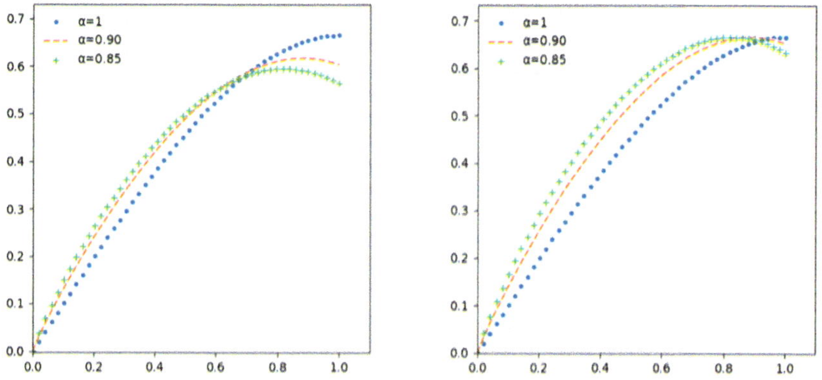


(a) Graph of solution of 3.4 for different value of $\beta$ by FDTM.

(b) Graph of solution of 3.4 for different value of $\beta$ by CFDTM.

**Fig. 2** Comparison of the fourth approximate solutions of (3.4), by CFDTM with the FDTM

Therefore approximate solution of (3.3) is as follows:

$$y(\vartheta) = \frac{1}{\beta} - \frac{\vartheta^{3\beta}}{3\beta^3} + \frac{2\vartheta^{5\beta}}{15\beta^5} - \cdots .$$

As $\beta \to 1$, then we have (Fig. 2).

$$y(\vartheta) = \vartheta - \frac{\vartheta^3}{3} + \frac{2\vartheta^5}{15} - \frac{17\vartheta^7}{315} + \cdots .$$

**Example 3** Consider the nonlinear FDE

$$T_\beta y = \sec y, \quad 0 < \beta \le 1, \quad y(0) = 0. \tag{3.5}$$

We apply Theorem 6 and Eq. (2.7) to Eq. (3.5), then we get

$$\begin{cases} \beta(l+1)Y_\beta(l+1) = \tilde{A}_l, \quad k = 0, 1, 2, \dots \\ Y_\beta(0) = 0, \end{cases} \tag{3.6}$$

where the $\widetilde{A}_l$ for the sec $y$ is as follows:

$$\left| \begin{array}{l} A_0 = \sec(y_0) \\ A_1 = \sec(y_0)\tan(y_0)y_1 \end{array} \right| \begin{array}{l} \widetilde{A}_0 = \sec(Y_\beta(0)) \\ \widetilde{A}_1 = \sec(Y_\beta(0))\tan(Y_\beta(0))Y_\beta(1) \end{array} \right|$$

$$A_2 = \tfrac{1}{2}\sec(y_0)(\sec(y_0))^2(y_1)^2 + (\tan(y_0))^2(y_1)^2 + 2\tan(y_0)y_2$$

$$\widetilde{A}_2 = \tfrac{1}{2}\sec(Y_\beta(0))(\sec(Y_\beta(0)))^2(Y_\beta(1))^2 + (\tan(Y_\beta(0)))^2(Y_\beta(1))^2 + 2\tan(Y_\beta(0))Y_\beta(2)$$

$$A_3 = \frac{1}{6}\sec(y_0)(\sec(y_0))^2(y_1)^2(5\tan(y_0)(y_1)^3 + 6y_1y_2) + \tan(y_0)((\tan(y_0))^2(y_1)^3 \\ +6\tan(y_0)y_1y_2 + 6y_3)$$

$$\widetilde{A}_3 = \frac{1}{6}\sec(Y_\beta(0))(\sec(Y_\beta(0)))^2(Y_\beta(1))^2(5\tan(Y_\beta(0))(Y_\beta(1))^3 + 6Y_\beta(1)Y_\alpha(2)) \\ +\tan(Y_\beta(0))((\tan(Y_\beta(0)))^2(Y_\beta(1))^3 + 6\tan(Y_\alpha(0))Y_\beta(1)Y_\beta(2) + 6Y_\beta(3))$$

$$\vdots$$

Now by using these values of $\widetilde{A}_l$ in Eq. 3.6, we obtained the following conformable differential transform components: $Y_\beta(1) = \frac{1}{\beta}$, $Y_\beta(2) = 0$, $Y_\beta(3) = \frac{1}{6\beta^3}$, $Y_\beta(4) = 0, \ldots$.

If we take the series solution's limit as $\beta \to 1$, we obtain (Fig. 3)

$$y(\vartheta) = \vartheta + 0.166667\vartheta^3 + 0.075\vartheta^5 + 0.0446429\vartheta^7 + \cdots.$$

**Example 4** Consider the nonlinear FDE

$$T_\beta y + \frac{2}{x}y' + e^y = 0, \quad 1 < \beta \le 2, \tag{3.7}$$

subjected to $y(0) = 0, \quad y'(0) = 0$.

We apply Theorem 7, Eq. (2.6) and $\alpha = \frac{\beta}{2}$, then we get the following recurrence relations:

$$\begin{cases} \dfrac{\Gamma\left(\beta\frac{(l+1)}{2}+1\right)}{\Gamma\left(\beta\frac{(l+1)}{2}-1\right)}Y_\beta(l+1) + 2(l+1)Y_\beta(l+1) + \widetilde{A}_{l-1} = 0, \quad l = 1, 2, \ldots \\ Y_\beta(0) = 0, \quad Y_\beta(1) = 0, \end{cases} \tag{3.8}$$

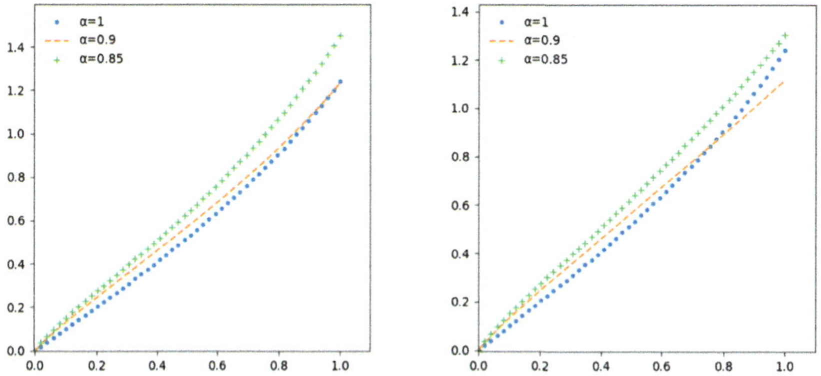


(a) Graph of solution of 3.5 for different value of $\beta$ by FDTM.

(b) Graph of solution of 3.5 for different value of $\beta$ by CFDTM.

**Fig. 3** Comparison of the fourth approximate solutions of CFDTM with the FDTM

where the $\widetilde{A}_l$ of $e^y$ is as follows:

$$\begin{array}{l|l} A_0 = e^{y_0} & \widetilde{A}_0 = e^{Y_\beta(0)} \\ A_1 = y_1 e^{y_0} & \widetilde{A}_1 = Y_\alpha(1) e^{Y(0)} \\ A_2 = \left(y_2 + \frac{(y_1)^2}{2}\right) e^{y_0} & \widetilde{A}_2 = \left(Y_\beta(2) + \frac{(Y_\beta(1))^2}{2}\right) e^{Y_\beta(0)} \\ A_3 = \left(y_3 + y_1 y_2 + \frac{(y_1)^3}{3!}\right) e^{y_0} & \widetilde{A}_3 = \left(Y_\beta(3) + Y_\beta(1) Y_\beta(2) + \frac{(Y_\beta(1))^3}{3!}\right) e^{Y_\beta(0)} \\ \vdots & \vdots \end{array}$$

Now by using these values of $\widetilde{A}_l$ in Eq. (3.8), we obtained the following differential transform components:

$$Y_\beta(1) = 0, \quad Y_\beta(2) = -\frac{\Gamma(\beta-1)}{\Gamma(\beta+1) + 4\Gamma(\beta-1)}, \quad Y_\beta(3) = 0,$$
$$Y_\beta(4) = \frac{\Gamma(\beta-1)}{\Gamma(\beta+1) + 4\Gamma(\beta-1)} \frac{\Gamma(2\beta-1)}{\Gamma(2\beta+1) + 8\Gamma(2\beta-1)} \cdots$$

As $\beta \to 2$, we obtain (Fig. 4)

$$y(\vartheta) = -\frac{1}{3 \times 2!}\vartheta^2 + \frac{1}{5 \times 4!}\vartheta^4 - \frac{8}{21 \times 6!}\vartheta^6 + \cdots$$

**Example 5** Consider the nonlinear FDE

$$T_\beta y + \frac{2}{x} y' + y^2 = 0, \quad 1 < \beta \leq 2, \tag{3.9}$$

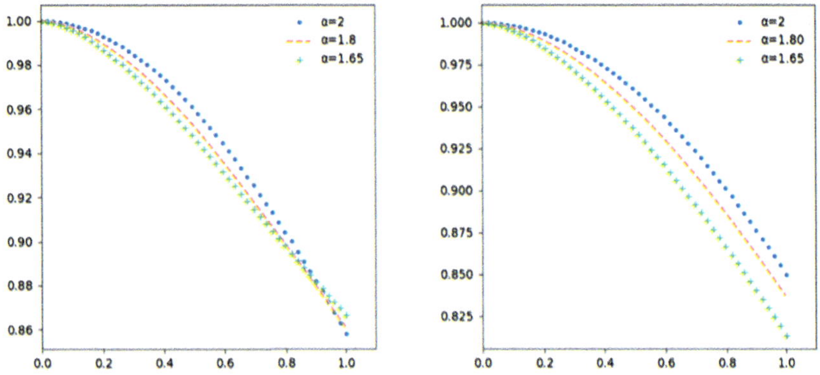


(a) Graph of solution of 3.7 for different value of $\beta$ by FDTM.

(b) Graph of solution of 3.7 for different value of $\beta$ by CFDTM.

**Fig. 4** Comparison of the fourth approximate solutions of CFDTM with the FDTM

subjected to the

$$y(0) = 1, \quad y'(0) = 0.$$

We apply Theorem 7, Eq. (2.7) and $\alpha = \frac{\beta}{2}$, then we get the following recurrence relations:

$$\begin{cases} \frac{\Gamma\left(\beta\frac{(l+1)}{2}+1\right)}{\Gamma\left(\beta\frac{(l+1)}{2}-1\right)} Y_\beta(l+1) + 2(l+1)Y_\beta(l+1) + \widetilde{A}_{l-1} = 0, \quad l = 1, 2, \ldots \\ Y_\beta(0) = 1, \quad Y_\beta(1) = 0, \end{cases} \tag{3.10}$$

where the $\widetilde{A}_l$ of $y^2$ are as follows:

$$\begin{array}{|l|l|} A_0 = {y_0}^2 & \widetilde{A}_0 = (Y_\beta(0))^2 \\ A_1 = 2y_1y_0 & \widetilde{A}_1 = 2Y_\beta(1)Y_\beta(0) \\ A_2 = {y_1}^2 + 2y_0y_2 & \widetilde{A}_2 = (Y_\beta(1))^2 + 2Y_\beta(0)Y_\beta(2) \\ A_3 = 2y_0y_3 + 2y_1y_2 & \widetilde{A}_3 = 2Y_\beta(0)Y_\beta(3) + 2Y_\alpha(1)Y_\beta(2) \\ \vdots & \vdots \end{array}$$

Now by using these values of $\widetilde{A}_l$ in Eq. (3.10), we obtained the following differential transform components:

$$Y_\beta(0) = 1, \quad Y_\beta(1) = 0, \quad Y_\beta(2) = -\frac{\Gamma(\beta-1)}{\Gamma(\beta+1)+4\Gamma(\beta-1)}, \quad Y_\beta(3) = 0,$$

$$Y_\beta(4) = \frac{2\Gamma(\beta-1)}{\Gamma(\beta+1)+4\Gamma(\beta-1)} \frac{\Gamma(2\beta-1)}{\Gamma(2\beta+1)+8\Gamma(2\beta-1)}, \cdots$$

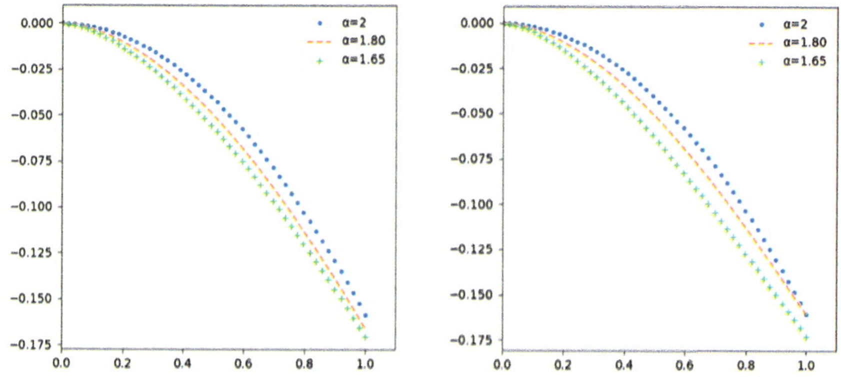


(a) Graph of solution of 3.9 for different value of $\beta$ by FDTM.

(b) Graph of solution of 3.9 for different value of $\beta$ by CFDTM.

**Fig. 5** Comparison of the fourth approximate solutions of CFDTM with the FDTM

As $\beta \to 2$, then we have (Fig. 5)

$$y(\vartheta) = 1 - \frac{1}{6}\vartheta^2 + \frac{2}{120}\vartheta^4 - \cdots$$

## 4 Conclusion

In this paper, we discussed the concepts of redesign of the conformable fractional differential transform method with Adomian polynomials for finding nonlinear fractional differential equations. By using this new approach, we solved some nonlinear and singular Lane–Emden equations. A nonlinear fractional differential equation has been put ahead and examined by adopting CFDTM with Adomian polynomials. In the Caputo and conformable fractional touch, we firstly determine the DT of the nonlinear term, but, in this new technique, we replace such calculation by recurrence relation in its Adomian polynomial. The dependent components are eventually substituted for the same index's equivalent FDT. As Adomian polynomials are suited in any nonlinear analytic function, this emphasizes the usefulness and applicability of the CFDTM. A suggested approach is a combination of the DTM and ADM. This method's strength is that it effectively combines these two powerful strategies for generating approximate series solutions. Furthermore, the solutions computed by using CFDTM with Adomian polynomials for some fractional order are correlated with solutions obtained for the same fractional order by adopting FDTM. The solution was analyzed graphically by using Python software.

**Acknowledgements** The corresponding author is supported by Council of Scientific and Industrial Research, Human Resource Development Group, Junior Research Fellowship Research Project Grant, Government of India (No. 08/581(0006)/2019-EMR-I) and the authors are thankful to referees for their fruitful discussions and helpful suggestions on this paper.

## References

1. Adomian, G.: Solving Frontier Problems of Physics: The Decomposition Method. Kluwer Academic Publishers, Boston (1994)
2. Almeida, R., Guzowska, M., Odzijewicz, T.: A Remark on Local Fractional Calculus and Ordinary Derivatives. arXiv:1612.00214
3. A. Arikoglu, I. Ozkol, Solution of fractional differential equations by using differential transform method. Chaos Soliton Fract. **34**, 1473–1481 (2007)
4. Daftardar-Gejji, V., Jafari, H.: Adomian decomposition: a tool for solving a system of fractional differential equations. J. Math. Anal. Appl. **301**(2), 508–518 (2005)
5. Elsaid, A.: Fractional differential transform method combined with the Adomian polynomials. Apll. Math. Comput. **218**, 6899–6911 (2012)
6. ünal, E., Gökdoğan, A.: Solution of conformable fractional ordinary differential equations via differential transform method. Optik, 264–273 (2017)
7. Podlubny I.: Fractional Differential Equations. Academic, USA (1999)
8. Khalil, R., Al Horani, M., Yusuf, A., Sababhed, M.: A new definition of fractional derivative. J. Comput. Appl. Math. **264**, 65–70 (2014)
9. Marasi, H.R., Sharifi, N., Piri, H.: Modified differental transform method for singular lane Emden equations in integer and fractional order. TWMS J. App. Eng. Math. **5**(1), 124–131 (2015)
10. Millar, K.S.: An Introduction to Fractional Calculus and Fractional Differential Equations. Wiley, New York (1993)
11. Teppawar, R.S., Ingle, R.N., Thorat, S.N.: Some results and applicatios on conformable fractional Kamal transform. J. Math. Comput. Sci. **11**(5), 6581–6598 (2021)
12. Zhou, J.K.: Differential Transformation and Its Applications for Electrical Circuits. Huazhong University Press, Wuhan, China (1986). In Chinese

# Generalized Results on Existence & Uniqueness with Wronskian and Abel Formula for $\alpha$-Fractional Differential Equations

**R. A. Muneshwar, K. L. Bondar, V. D. Mathpati, and Y. H. Shirole**

**Abstract** R. A. Muneshwar et al. has proposed a new $\alpha$-fractional derivative notion based on the limit. This topic will be continued in this article, and some conclusions on existence and uniqueness theorems for linear $\alpha$-fractional differential equations will be discussed. Moreover, we derived the Wronskian determinant formula and Abel's formula for $\alpha$-fractional differential equations. In addition, we provide applications of the obtained results.

**Keywords** Fractional derivative · Existence and uniqueness theorems · Abels formula

## 1 Introduction and Preliminaries

The idea of fractional derivation has gained prominence in mathematical study during the last few decades. There is no known method for obtaining an exact solution to fractional differential equations [12, 13, 17], however there are approximate and numerical solutions. For defining the fractional derivative, there is no standard form. However, the Riemann-Liouville and Caputo definitions of fractional derivatives are the most often utilised. Some writers have recently suggested a revised definition of the fractional derivative [11]. Later, many authors studied this new theory, which can be found in [1–3, 10]. Several investigations on this theory and the application of fractional differential equations based on Hadamard, Riemann-Liouville, and Caputo derivatives have been found in the literature [5, 6, 14, 15, 17].

R. A. Muneshwar (✉) · V. D. Mathpati · Y. H. Shirole
P.G Department of Mathematics, N.E.S. Science College, Nanded 431602, MH, India
e-mail: muneshwarrajesh10@gmail.com

K. L. Bondar
Government Vidarbha Institute of Science and Humanities, Amravati, MH, India



B. Rushi Kumar et al. (eds.), *Mathematics and Computing*, Springer Proceedings in Mathematics & Statistics 415, https://doi.org/10.1007/978-981-19-9307-7_30

## 2 $\alpha$-Fractional Derivative

Muneshwar et al. [16] introduced the concept of $\alpha$-fractional derivative and integral by doing some appropriate modification in the traditional definition of a derivative, which is

**Definition 1** (*Modified $\alpha$-Fractional Derivative* [16]) If $\vartheta > 0$ with $\forall \alpha \in (0, 1]$ and $\Psi : [0, \infty) \to \mathbb{R}$ then modified $\alpha$-fractional derivative of order $\alpha$ is given by

$$T_\alpha(\Psi(\vartheta)) = \lim_{\mu \to 0} \frac{\Psi(\vartheta e^{\mu\vartheta^{1-\alpha}}) - \Psi(\vartheta)}{\mu}$$

**Remark** Throughout this paper it is not explicitly mention that the underlying elements $\xi$ & $\xi_0$ is satisfies $\xi$ & $\xi_0 \geq a > 0$ and $I = (a, b)$.

By using this definition we deduce the following results which can be found in [16].

**Theorem 1** ([16]) *If $\Psi$ is a $\alpha$-differential function at $\vartheta > 0$, then*

$$T_\alpha \Psi(\vartheta) = \vartheta^{2-\alpha} \frac{d\Phi(\vartheta)}{d\vartheta}$$

**Definition 2** ([16]) If $\gamma \in [0, \vartheta)$ then new $\alpha$-fractional integral of $\Phi$ is defined by

$$I_\alpha^\gamma \Phi(\vartheta) = \int_\gamma^\vartheta \frac{\Phi(\mu)}{\mu^{2-\alpha}} d\mu,$$

if integral exists.

Following results is obtained by using the Definition 1.

**Theorem 2** ([16]) *If $\Phi$ and $\Psi$ are $\alpha$-differentiable functions at point $\vartheta > 0$ then*

1. $T_\alpha(\beta\Phi) = \beta T_\alpha(\Phi), \quad \forall \ \beta \in \mathbb{R}.$
2. $T_\alpha(\Phi + \Psi)(\vartheta) = T_\alpha(\Phi)(\vartheta) + T_\alpha(\Psi)(\vartheta).$
3. $T_\alpha(\Phi\Psi)(\vartheta) = \Phi T_\alpha(\Psi)(\vartheta) + \Psi T_\alpha(\Phi)(\vartheta).$
4. $T_\alpha\left(\frac{\Phi}{\Psi}\right)(\vartheta) = \left(\frac{\Psi(\vartheta)T_\alpha\Phi(\vartheta) - \Phi(\vartheta)T_\alpha\Psi(\vartheta)}{\Psi(\vartheta)^2}\right).$
5. $T_\alpha\left(\frac{1}{\Psi}\right) = -\frac{T_\alpha\Psi}{\Psi^2}.$
6. $T_\alpha(\Phi \circ \Psi)(\vartheta) = T_\alpha\Phi(\Psi(\vartheta))T_\alpha\Psi(\vartheta).$

**Theorem 3** ([16]) *If $0 < \alpha \leq 1$ and $c \in \mathbb{R}$ then, by using Theorem 1, we have*

1. $T_\alpha(\vartheta^n) = n\vartheta^{n+1-\alpha}$
2. $T_\alpha\left(\frac{1}{\alpha-1}t^{\alpha-1}\right) = 1$
3. $T_\alpha(\beta) = 0, \quad \forall \ \beta \in \mathbb{R}$
4. $T_\alpha(e^{ct}) = e^{ct}ct^{2-\alpha}$

5. $T_\alpha(\sin ct) = ct^{2-\alpha}\cos(ct)$
6. $T_\alpha(\cos ct) = -ct^{2-\alpha}\sin(ct)$
7. $T_\alpha(logt) = t^{1-\alpha}$
8. $T_\alpha(a^t) = (a^t loga)t^{2-\alpha}$
9. $T_\alpha(\tan t) = t^{2-\alpha}\sec^2 t$
10. $T_\alpha(\cot t) = -t^{2-\alpha}\csc^2 t$.

## 3 Existence and Uniqueness Theorems

Consider a fractional differential equation of order $n\alpha$,

$$ {}^nD_\alpha\gamma + p_{n-1}(\xi)^{n-1}D_\alpha\gamma + \cdots + p_2(\xi)^2D_\alpha\gamma + p_1(\xi)D_\alpha\gamma + p_0(\xi)\gamma = 0, \quad (3.1)$$

where $D_\alpha\gamma = D_\alpha D_\alpha \ldots D_\alpha\gamma$. Corresponding, non-homogeneous case is as follows.

$$ {}^nD_\alpha\gamma + p_{n-1}(\xi)^{n-1}D_\alpha\gamma + \cdots + p_2(\xi)^2D_\alpha\gamma + p_1(\xi)D_\alpha\gamma + p_0(\xi)\gamma = f(\xi). \quad (3.2)$$

We define an nth order differential operator as follows.

$$L_\alpha[\gamma] = {}^nD_\alpha\gamma + p_{n-1}(\xi)^{n-1}D_\alpha\gamma + \cdots + p_2(\xi)^2D_\alpha\gamma + p_1(\xi)D_\alpha\gamma + p_0(\xi)\gamma = 0. \quad (3.3)$$

**Theorem 4** *Let $\xi^{\alpha-2}p(\xi), \xi^{\alpha-2}q(\xi) \in C(I)$ are continuous functions defined on $I$. If $\gamma$ is the $\alpha$-differentiable and $\alpha \in (0, 1]$, then the IVP*

$$D_\alpha\gamma + p(\xi)\gamma = q(\xi) \quad (3.4)$$

$$\gamma(\xi_0) = \gamma_0,$$

*has a only one solution on $I$, where $\xi_0 \in I$.*

***Proof*** Consider the IVP as,

$$D_\alpha\gamma + p(\xi)\gamma = q(\xi)$$
$$\Rightarrow \xi^{2-\alpha}\gamma' + p(\xi)\gamma = q(\xi)$$

$$\Rightarrow \gamma' + \xi^{\alpha-2}p(\xi)\gamma = \xi^{\alpha-2}q(\xi). \quad (3.5)$$

Proof is follows from classical theories of existence and uniqueness.

**Theorem 5** *Let $\xi^{\alpha-2}p_{n-1}(\xi), \ldots, \xi^{\alpha-2}p_1(\xi), \xi^{\alpha-2}p_0(\xi), \xi^{\alpha-2}q(\xi) \in C(I)$, are continuos functions defined on $I$. If $\gamma$ be $n$ times $\alpha$-differentiable function and $\alpha \in (0, 1]$, then a solution $\gamma(\xi)$ of the IVP*

$$
{}^{n}D_{\alpha}\gamma + p_{n-1}(\xi)^{n-1}D_{\alpha}\gamma + \cdots + p_{2}(\xi)^{2}D_{\alpha}\gamma + p_{1}(\xi)D_{\alpha}\gamma + p_{0}(\xi)\gamma = q(\xi), \tag{3.6}
$$
$$
\gamma(\xi_0) = \gamma_0,\, D_{\alpha}\gamma(\xi_0) = \gamma_1, \ldots, {}^{n-1}D_{\alpha}\gamma(\xi_0) = \gamma_{n-1}, \quad a < \xi_0 \leq b
$$

*is exists on I and it is unique, for* $\xi_0 \in I$.

***Proof*** We begin by changing the variables as follows:

$$
\nu_1 = \gamma,\, \nu_2 = D_{\alpha}\gamma,\, \nu_3 = {}^{2}D_{\alpha}\gamma, \ldots, \nu_n = {}^{n}D_{\alpha}\gamma.
$$

Therefore, we have

$$
\begin{aligned}
D_{\alpha}\nu_1 &= \nu_2 \\
D_{\alpha}\nu_2 &= \nu_3 \\
&\vdots \\
D_{\alpha}\nu_{n-1} &= \nu_{n-1}
\end{aligned}
$$

$D_{\alpha}\nu_n = -p_{n-1}\nu_n - \cdots - p_2\nu_3 - p_1\nu_2 - p_0\nu_1 + q(\xi)$.

Now, in matrix form, we can rewrite the supplied IVP as follows:

$$
D_{\alpha}\begin{bmatrix} \nu_1 \\ \nu_2 \\ \vdots \\ \nu_{n-1} \\ \nu_n \end{bmatrix} + \begin{bmatrix} 0 & -1 & 0 & 0 & \cdots & 0 \\ 0 & 0 & -1 & 0 & \cdots & 0 \\ \vdots & \vdots & \vdots & \vdots & \cdots & \vdots \\ 0 & 0 & 0 & 0 & \cdots & -1 \\ p_0 & p_1 & p_2 & p_3 & \cdots & p_{n-1} \end{bmatrix} \begin{bmatrix} \nu_1 \\ \nu_2 \\ \vdots \\ \nu_{n-1} \\ \nu_n \end{bmatrix} = \begin{bmatrix} 0 \\ 0 \\ \vdots \\ 0 \\ q(t) \end{bmatrix}
$$

$$
D_{\alpha}\vartheta(\xi) + P(\xi)\vartheta(\xi) = Q(\xi),
$$
$$
\vartheta'(\xi) + \xi^{\alpha-2}P(\xi)\vartheta(\xi) = Q(\xi)\xi^{\alpha-2}. \tag{3.7}
$$

As $\xi_0 \in I$, then proof is follows from classical theories of existence and uniqueness [8].

**Theorem 6** *If* $\{\gamma_i\}_{i=1}^{2}$ *be set of n times* $\alpha$*-differentiable and* $c_1, c_2$ *are arbitrary numbers then* $L_{\alpha}$ *is linear.*

***Proof*** If $\{\gamma_i\}_{i=1}^{2}$ be set of $n$ times $\alpha$-differentiable and $c_1, c_2 \in R$ then by the Definition 1, we have,

$$
\begin{aligned}
L_{\alpha}[c_1\gamma_1 + c_2\gamma_2] &= {}^{n}D_{\alpha}(c_1\gamma_1 + c_2\gamma_2) + p_{n-1}(t)^{n-1}D_{\alpha}(c_1\gamma_1 + c_2\gamma_2) + \cdots \\
&\quad + p_1(\xi)D_{\alpha}(c_1\gamma_1 + c_2\gamma_2) + p_0(\xi)(c_1\gamma_1 + c_2\gamma_2) \\
&= c_1({}^{n}D_{\alpha}\gamma_1 + p_{n-1}(\xi)^{n-1}D_{\alpha}\gamma_1 + \cdots + p_0(\xi)\gamma_1) \\
&\quad + c_2({}^{n}D_{\alpha}\gamma_2 + p_{n-1}(\xi)^{n-1}D_{\alpha}\gamma_2 + \cdots + p_0(\xi)\gamma_2).
\end{aligned}
$$

We have

$$L_\alpha[c_1\gamma_1 + c_2\gamma_2] = c_1 L_\alpha[\gamma_1] + c_2 L_\alpha[\gamma_2].$$

**Theorem 7** *If $\gamma_1(\xi), \gamma_2(\xi), \ldots, \gamma_n(\xi)$ be the solutions of equation $L_\alpha[\gamma] = 0$ then there linear combination*

$$\gamma = c_1\gamma_1 + c_2\gamma_2 + \cdots + c_n\gamma_n \tag{3.9}$$

*is also solution of $L_\alpha[\gamma] = 0$, where $c_k$, $k = 1, \ldots, n$, are arbitrary constants.*

***Proof*** Let $\{\gamma_i\}_{i=1}^n$ be set of solutions of $L_\alpha[\gamma] = 0$ and let us consider,

$$\gamma = c_1\gamma_1 + c_2\gamma_2 + \cdots + c_n\gamma_n,$$

where $c_k$, for $k = 1, \ldots, n$, are arbitrary constants then by Theorem 6, we have,

$$L_\alpha(\gamma) = c_1 L_\alpha(\gamma_1) + c_2 L_\alpha(\gamma_2) + \cdots + c_n L_\alpha(\gamma_n) = 0.$$

Hence, we are through.

**Definition 3** Let $\gamma_1(\xi), \gamma_2(\xi), \ldots, \gamma_n(\xi)$ are at least $(n-1)$ times $\alpha$-differentiable functions. If $\alpha \in (0, 1]$, then determinant

$$W_\alpha(\gamma_1, \gamma_2, \ldots, \gamma_n)(\xi_0) = \begin{vmatrix} \gamma_1 & \gamma_2 & \cdots & \gamma_n \\ D_\alpha\gamma_1 & D_\alpha\gamma_2 & \cdots & D_\alpha\gamma_n \\ \vdots & \vdots & \vdots & \vdots \\ {}^{n-1}D_\alpha\gamma_1 & {}^{n-1}D_\alpha\gamma_2 & \cdots & {}^{n-1}D_\alpha\gamma_n \end{vmatrix} \tag{3.10}$$

is called $\alpha$-Wronskian of the functions $\gamma_1(\xi), \gamma_2(\xi), \ldots, \gamma_n(\xi)$.

**Definition 4** Any set $\{\gamma_i\}_{i=1}^n$ of $n$ solutions of $L_\alpha(\gamma) = 0$, is called as a fundamental set if any solution $\gamma$ satisfies the Eq. (3.9).

**Theorem 8** *Let $\{\gamma_i\}_{i=1}^n$ be set of $n$ solutions of $L_\alpha(\gamma) = 0$. If there exists $\xi_0 \in I$, such that*

$$W_\alpha(\gamma_1, \gamma_2, \ldots, \gamma_n)(\xi_0) \neq 0, \tag{3.11}$$

*then $\{\gamma_i\}_{i=1}^n$ is a fundamental set of solutions.*

***Proof*** As $\{\gamma_i\}_{i=1}^n$ be set of $n$ solutions of $L_\alpha(\gamma) = 0$, on $I$, then

$$\gamma = \sum_{i=1}^n c_i\gamma_i$$

is also a solution of $L_\alpha[\gamma] = 0$. Now it is sufficient to find the constants, $c_k$, for $1 \le k \le n$. The system of linear equations may be written as follows:

$$\begin{aligned} c_1\gamma_1(\xi_0) + c_2\gamma_2(\xi_0) + \cdots + c_n\gamma_n(\xi_0) &= \gamma(\xi_0) \\ c_1 D_\alpha\gamma_1(\xi_0) + c_2 D_\alpha\gamma_2(\xi_0) + \cdots + c_n D_\alpha\gamma_n(\xi_0) &= D_\alpha\gamma(\xi_0) \\ \vdots \quad \vdots \quad &\vdots \\ c_1{}^{n-1}D_\alpha\gamma_1(\xi_0) + c_2{}^{n-1}D_\alpha\gamma_2(\xi_0) + \cdots + c_n{}^{n-1}D_\alpha\gamma_n(\xi_0). &= {}^{n-1}D_\alpha\gamma(\xi_0) \end{aligned} \tag{3.12}$$

By using Cramer's rule, we obtain

$$c_k = \frac{W_\alpha^k(\xi_0)}{W_\alpha(\xi_0)}, \quad 1 \le k \le n. \tag{3.13}$$

As $W_\alpha(\xi_0) \neq 0$, then it follows that the constants $c_1, c_2, \ldots, c_n$, are exist.

**Theorem 9** *Let* $\xi^{\alpha-2}p_{n-1}(\xi), \ldots, \xi^{\alpha-2}p_1(\xi), \xi^{\alpha-2}p_0(\xi) \in C(I)$ *are continuos functions and* $\{\gamma_i\}_{i=1}^n$ *be set of solutions of* $L_\alpha[\gamma] = 0$ *on* $I$. *If* $\xi_0 \ge a > 0$ *then the set* $\{\gamma_i\}_{i=1}^n$ *be the fundamental set of solutions of* $L_\alpha[\gamma] = 0$.

***Proof*** Let $\xi^{\alpha-2}p_{n-1}(\xi), \ldots, \xi^{\alpha-2}p_1(\xi), \xi^{\alpha-2}p_0(\xi) \in C(I)$ are continuos functions and $\{\gamma_i\}_{i=1}^n$ be set of solutions of $L_\alpha[\gamma] = 0$ on $I$ & $\xi_0 \in I$. Now we consider the following $n$ IVPs,

$$\begin{aligned} L_\alpha[\gamma] = 0, \gamma(\xi_0) = 1, D_\alpha\gamma(\xi_0) = 0, \ldots, n-1D_\alpha\gamma(\xi_0) = 0 \\ L_\alpha[\gamma] = 0, \gamma(\xi_0) = 0, D_\alpha\gamma(\xi_0) = 1, \ldots, n-1D_\alpha\gamma(\xi_0) = 0 \\ \vdots \\ L_\alpha[\gamma] = 0, \gamma(\xi_0) = 0, D_\alpha\gamma(\xi_0) = 1, \ldots, n-1D_\alpha\gamma(\xi_0) = 1. \end{aligned}$$

From [8], it gives that there is the solution $\gamma_k$ of $k$th IVP, $\forall\, k$. As

$$W_\alpha(\xi) = \begin{vmatrix} 1\,0 \cdots 0 \\ 0\,1 \cdots 0 \\ 0\,0 \cdots 0 \\ 0\,0 \cdots 1 \end{vmatrix} = 1 \neq 0.$$

Then by the Theorem 8, the set $\{\gamma_i\}_{i=1}^n$ be the fundamental set.

**Theorem 10** *If* $\{\gamma_i\}_{i=1}^n$ *be set of* $n$ *solutions of equation* $L_\alpha[\gamma] = 0$ *on* $I$ *and* $a \ge a_0 > 0$ *then following are holds*

1. $W_\alpha(\xi)$ *is satisfies the differential equation*

$$D_\alpha W + P_{n-1}(\xi)W = 0.$$

2. *If* $\xi_0 \in I$ *then*

$$W_\alpha(\xi) = W_\alpha(\xi_0)e^{-\int_{\xi_0}^{\xi} \nu^{\alpha-2}(P_{n-1}(\nu))}d\nu.$$

*Moreover, if* $W_\alpha(\xi_0) \neq 0$ *then* $W_\alpha(\xi) \neq 0$, *for all* $\xi \in I$.

***Proof*** Let $\{\gamma_i\}_{i=1}^{n}$ be set of $n$ solutions of equation $L_\alpha[\gamma] = 0$ on $I$. Now we introduce new variables

$$\nu_1 = \gamma, \nu_2 = D_\alpha\gamma, \nu_3 = {}^2D_\alpha\gamma, \ldots, \nu_n = {}^nD_\alpha\gamma. \tag{3.14}$$

then, as shown below, we may rewrite these differential equations in matrix form.

$$D_\alpha \begin{bmatrix} \nu_1 \\ \nu_2 \\ \vdots \\ \nu_{n-1} \\ \nu_n \end{bmatrix} = \begin{bmatrix} 0 & 1 & 0 & 0 & \cdots & 0 \\ 0 & 0 & 1 & 0 & \cdots & 0 \\ \vdots & \vdots & \vdots & \vdots & \cdots & \vdots \\ 0 & 0 & 0 & 0 & \cdots & 1 \\ -p_0 & -p_1 & -p_2 & -p_3 & \cdots & -p_{n-1} \end{bmatrix} \begin{bmatrix} \nu_1 \\ \nu_2 \\ \vdots \\ \nu_{n-1} \\ \nu_n \end{bmatrix} \tag{3.15}$$

$$\Rightarrow D_\alpha\vartheta(\xi) = P(\xi)\vartheta(\xi)$$

$$\Rightarrow D_\alpha W_\alpha(\xi) = (p_{11} + p_{22} + \cdots + p_{nn})W_\alpha(\xi).$$

$$\Rightarrow \frac{D_\alpha W_\alpha(\xi)}{W_\alpha(\xi)} = -p_{n-1}(\xi)$$

$$\Rightarrow ln(W_\alpha(\xi)) - ln(W_\alpha(\xi_0)) = -\nu^{\alpha-2}(p_{n-1}(\xi))$$

$$\Rightarrow W_\alpha(\xi) = W_\alpha(\xi_0)e^{-\int_{\xi_0}^{\xi} \nu^{\alpha-2}(p_{n-1}(\nu))d\nu}. \tag{3.16}$$

If $W_\alpha(\xi_0) \neq 0$ then by Eq. 3.16, we have, $W_\alpha(\xi) \neq 0$, for all $\xi \in I$.

**Theorem 11** *Let* $\xi^{\alpha-2}p_{n-1}(\xi), \ldots, \xi^{\alpha-2}p_1(\xi), \xi^{\alpha-2}p_0(\xi) \in C(I)$ *are continuos functions. If* $\{\gamma_i\}_{i=1}^{n}$ *is a fundamental set of solutions of* $L_\alpha[\gamma] = 0$, *on* $I$ *then* $W_\alpha(\xi) \neq 0$, *for all* $\xi \in I$.

***Proof*** If $\xi^{\alpha-2}p_{n-1}(t), \ldots, \xi^{\alpha-2}p_1(\xi), \xi^{\alpha-2}p_0(\xi) \in C(I)$ and suppose that $\xi_0$ be any point in $I$ and $\xi_0 \geq a > 0$, then by Theorem 5, $\exists$ a unique solution $\gamma(\xi)$ of the IVP,

$$L_\alpha[\gamma] = 0, \gamma(\xi_0) = 1, D_\alpha\gamma(\xi_0) = 0, \ldots, {}^{n-1}D_\alpha\gamma(\xi_0) = 0. \tag{3.17}$$

As $\{\gamma_i\}_{i=1}^{n}$ is a fundamental set of solutions then $\exists$ unique constants $c_1, c_2, \ldots, c_n$, such that

$$
\begin{aligned}
c_1\gamma_1(\xi) + c_2\gamma_2(\xi) + \cdots + c_n\gamma_n(\xi) &= \gamma(\xi) \\
c_1 D_\alpha\gamma_1(\xi) + c_2 D_\alpha\gamma_2(\xi) + \cdots + c_n D_\alpha\gamma_n(\xi) &= D_\alpha\gamma(\xi) \\
\vdots \quad \vdots \quad \vdots & \\
{c_1}^{n-1} D_\alpha\gamma_1(\xi) + {c_2}^{n-2} D_\alpha\gamma_2(\xi) + \cdots + c_n D_\alpha\gamma_n(\xi) &= D_\alpha\gamma(\xi),
\end{aligned}
\tag{3.18}
$$

for all $\xi \in I$. In particular, if $\xi = \xi_0$, then we obtain the system

$$
\begin{aligned}
c_1\gamma_1(\xi_0) + c_2\gamma_2(\xi_0) + \cdots + c_n\gamma_n(\xi_0) &= \gamma(\xi_0) \\
c_1 D_\alpha\gamma_1(\xi_0) + c_2 D_\alpha\gamma_2(\xi_0) + \cdots + c_n D_\alpha\gamma_n(\xi_0) &= D_\alpha\gamma(\xi_0) \\
\vdots \quad \vdots \quad \vdots & \\
{c_1}^{n-1} D_\alpha\gamma_1(\xi_0) + {c_2}^{n-1} D_\alpha\gamma_2(\xi_0) + \cdots + {c_n}^{n-1} D_\alpha\gamma_n(\xi_0) &= D_\alpha\gamma(\xi_0).
\end{aligned}
\tag{3.19}
$$

This system has a unique solution

$$
c_k = \frac{W_\alpha^k}{W_\alpha(\xi_0)}, \quad 1 \le k \le n. \tag{3.20}
$$

Here, for every $k$

$$
W_\alpha^k = \begin{bmatrix}
\gamma_1(\xi_0) & \cdots & \gamma_{k-1}(\xi_0) & 1 & \gamma_{k+1}(\xi_0) & \cdots & \gamma_n(\xi_0) \\
D_\alpha\gamma_1(\xi_0) & \cdots & D_\alpha\gamma_{k-1}(\xi_0) & 0 & D_\alpha\gamma_{k+1}(\xi_0) & \cdots & D_\alpha\gamma_n(\xi_0) \\
\vdots & \vdots & \vdots & \vdots & \vdots & \vdots & \vdots \\
{}^{n-1}D_\alpha\gamma_1(\xi_0) & \cdots & {}^{n-1}D_\alpha\gamma_{k-1}(\xi_0) & 0 & {}^{n-1}D_\alpha\gamma_{k+1}(\xi_0) & \cdots & {}^{n-1}D_\alpha\gamma_n(\xi_0)
\end{bmatrix}
$$

As $W_\alpha(\xi_0) \neq 0$ then $c_1, c_2, \ldots, c_n$, are exist and by Theorem 10, we conclude that $W_\alpha(\xi) \neq 0$, for all $\xi \in I$.

**Theorem 12** *If* $\xi^{\alpha-2}p_{n-1}(\xi), \ldots, \xi^{\alpha-2}p_1(\xi), \xi^{\alpha-2}p_0(\xi) \in C(I)$ *are continuos functions then the set* $\{\gamma_i\}_{i=1}^n$ *is a fundamental set of solutions of equation* $L_\alpha[\gamma] = 0$ *on* $I$ *iff the set of solutions* $\{\gamma_i\}_{i=1}^n$ *is linearly independent.*

***Proof*** Suppose that $\{\gamma_i\}_{i=1}^n$ is a fundamental set then by Theorem 11, it follows that $\exists\, \xi_0 \in I$ such that $W_\alpha(\xi_0) \neq 0$. Assume that

$$
c_1\gamma_1(\xi) + c_2\gamma_2(\xi) + \cdots + c_n\gamma_n(\xi) = 0, \tag{3.21}
$$

for all $\xi \in I$. By repeated $\alpha$-differentiation of the Eq. 3.21, we can find unique constants $c_1, c_2, \ldots, c_n$, such that

$$c_1\gamma_1(\xi) + c_2\gamma_2(\xi) + \cdots + c_n\gamma_n(\xi) \quad = 0$$

$$c_1 D_\alpha\gamma_1(\xi) + c_2 D_\alpha\gamma_2(\xi) + \cdots + c_n D_\alpha\gamma_n(\xi) \quad = 0$$

$$\vdots \quad \vdots \quad \vdots$$

$$c_1{}^{n-1}D_\alpha\gamma_1(\xi) + c_2{}^{n-1}D_\alpha\gamma_2(\xi) + \cdots + c_n{}^{n-1}D_\alpha\gamma_n(\xi) = 0.$$

By Cramer's rule, we have, $c_i = 0,\ \forall\ i$. Thus, set of solutions $\{\gamma_i\}_{i=1}^n$ is linearly independent.

Conversely, suppose $\{\gamma_i\}_{i=1}^n$ is a linearly independent set.

**Claim**: $\{\gamma_i\}_{i=1}^n$ is a fundamental set of solutions.

If possible, $\{\gamma_i\}_{i=1}^n$, is not a fundamental set of solutions then, by Theorem 8, we get $W_\alpha(\xi) = 0$, for all $\xi \in I$. If we choose any $\xi_0 \in I$ then $W_\alpha(\xi_0) = 0$. But as $W_\alpha(\xi_0) \neq 0$ then the matrix

$$\begin{bmatrix} \gamma_1(\xi_0) & \gamma_2(\xi_0) & \cdots & \gamma_n(\xi_0) \\ D_\alpha\gamma_1(\xi_0) & D_\alpha\gamma_2(\xi_0) & \cdots & D_\alpha\gamma_n(\xi_0) \\ \vdots & \vdots & \vdots & \vdots \\ {}^{n-1}D_\alpha\gamma_1(\xi_0) & {}^{n-1}D_\alpha\gamma_2(\xi_0) & \cdots & {}^{n-1}D_\alpha\gamma_n(\xi_0) \end{bmatrix} \tag{3.22}$$

is not invertible which means that there exist $c_1, c_2, \ldots, c_n$ with $c_1^2 + c_2^2 + \cdots + c_n^2 \neq 0$, such that

$$c_1\gamma_1(\xi_0) + c_2\gamma_2(\xi_0) + \cdots + c_n\gamma_n(\xi_0) = 0$$
$$c_1 D_\alpha\gamma_1(\xi_0) + c_2 D_\alpha\gamma_2(\xi_0) + \cdots + c_n D_\alpha\gamma_n(\xi_0) = 0$$
$$\vdots \quad \vdots \quad \vdots$$
$$c_1{}^{n-1}D_\alpha\gamma_1(\xi_0) + c_2{}^{n-1}D_\alpha\gamma_2(\xi_0) + \cdots + c_n{}^{n-1}D_\alpha\gamma_n(\xi_0) = 0.$$

Now, let

$$\gamma(\xi) = c_1\gamma_1(\xi) + c_2\gamma_2(\xi) + \cdots + c_n\gamma_n(\xi), \tag{3.23}$$

for all $\xi \in I$. Then $\gamma(\xi)$ is the solution of the differential equation and

$$\gamma(\xi_0) = D_\alpha\gamma(\xi_0) = \cdots = {}^{n-1}D_\alpha\gamma(\xi_0) = 0.$$

However, the IVP's solution is the zero function. According to Theorem 5,

$$c_1\gamma_1(\xi) + c_2\gamma_2(\xi) + \cdots + c_n\gamma_n(\xi) = 0,$$

for all $\xi \in I$ with constants $c_i^s$ not all equal to zero then $\{\gamma_i\}_{i=1}^n$ are linearly dependent set, a contradiction.

**Theorem 13** *If $\{\gamma_i\}_{i=1}^n$ be LI solutions of the equation $L_\alpha[\gamma]=0$ on $I$ and $a \geq \xi_0 > 0$, then the general solution of the equation is*

$$\gamma = c_1\gamma_1 + c_2\gamma_2 + \cdots + c_n\gamma_n$$

*where $c_k$, $\forall\, k$.*

***Proof*** Let $\{\gamma_i\}_{i=1}^n$ be the set of LI solutions of $L_\alpha[\gamma]=0$ on $I$. Particular solution at any $\xi=\xi_0$ is obtained by using initial conditions as follows,

$$\gamma(\xi_0)=\lambda_0,\ D_\alpha\gamma(\xi_0)=\lambda_1,\ldots,{}^{n-1}D_\alpha\gamma(\xi_0)=\lambda_{n-1}, \tag{3.24}$$

where $\xi_0 \in I$ and $\lambda_0, \lambda_1, \lambda_2, \ldots, \lambda_{n-1}$ are arbitrary constants. If we choose constants $c_1, c_2, \ldots, c_n$, which satisfy the conditions (3.24), then proof is completed. To do so, we can use the set of equations below.

$$\begin{aligned}
c_1\gamma_1(\xi_0)+c_2\gamma_2(\xi_0)+\cdots+c_n\gamma_n(\xi_0)&=\lambda_0\\
c_1D_\alpha\gamma_1(\xi_0)+c_2D_\alpha\gamma_2(\xi_0)+\cdots+c_nD_\alpha\gamma_n(\xi_0)&=\lambda_1\\
\vdots \quad \vdots \quad &\vdots\\
c_1{}^{n-1}D_\alpha\gamma_1(\xi_0)+c_2{}^{n-1}D_\alpha\gamma_2(\xi_0)+\cdots+c_n{}^{n-1}D_\alpha\gamma_n(\xi_0)&=\lambda_{n-1}.
\end{aligned} \tag{3.25}$$

Above system of equation can be written as follows,

$$\begin{bmatrix} \gamma_1(\xi_0) & \gamma_2(\xi_0) & \cdots & \gamma_n(\xi_0)\\ D_\alpha\gamma_1(\xi_0) & D_\alpha\gamma_2(\xi_0) & \cdots & D_\alpha\gamma_n(\xi_0)\\ \vdots & \vdots & \vdots & \vdots\\ {}^{n-1}D_\alpha\gamma_1(\xi_0) & {}^{n-1}D_\alpha\gamma_2(\xi_0) & \cdots & {}^{n-1}D_\alpha\gamma_n(\xi_0)\end{bmatrix}\begin{bmatrix} c_1\\ c_2\\ \vdots\\ c_n\end{bmatrix}=\begin{bmatrix}\lambda_0\\ \lambda_1\\ \vdots\\ \lambda_{n-1}\end{bmatrix}. \tag{3.26}$$

As $\{\gamma_i\}_{i=1}^n$ are linearly independent solutions of $L_\alpha[\gamma]=0$, then $W_\alpha(\xi_0)\neq 0$ and hence system (3.25) has a only one solution.

**Theorem 14** *Let $\gamma_p$ be PI of $L_\alpha[\gamma]=q(\xi)$. If $\{\gamma_i\}_{i=1}^n$, be a fundamental set of solutions of $L_\alpha[\gamma]=0$, then the general solution of $L_\alpha[\gamma]=q(\xi)$ is*

$$\gamma = c_1\gamma_1 + c_2\gamma_2 + \cdots + c_n\gamma(n) + \gamma_p, \tag{3.27}$$

*where $c_k$, for $k=1,\ldots,n$, are arbitrary constants.*

***Proof*** Let $\Upsilon(\xi)$ and $\gamma_p(\xi)$ be the general solution and PI of $L_\alpha[\gamma]=q(\xi)$. If $u(\xi)=\Upsilon(\xi)-\gamma_p(\xi)$, then we have

$$L_\alpha[u]=L_\alpha[\Upsilon(\xi)-\gamma_p(\xi)]=L_\alpha[\Upsilon(\xi)]-L_\alpha[\gamma_p(\xi)]=q(\xi)-q(\xi)=0. \tag{3.28}$$

This implies that $u(\xi)$ is a solution of $L_\alpha[\gamma] = 0$. Therefore by Theorem 7, we have,

$$u(\xi) = \sum_{i=1}^{n} \{c_i \gamma_i(\xi)\}$$

And so

$$\Upsilon(\xi) - \gamma_p(\xi) = \sum_{i=1}^{n} \{c_i \gamma_i(\xi)\}$$

$$\Rightarrow \Upsilon(\xi) = \sum_{i=1}^{n} \{c_i \gamma_i(\xi)\} + \gamma_p(\xi) \tag{3.29}$$

be the general solution of $L_\alpha[\gamma] = q(\xi)$.

**Theorem 15** *Let $\xi \in (a, b)$ and consider the fractional differential equation,*

$$D_\alpha D_\alpha \gamma + P(\xi) D_\alpha \gamma + Q(\xi)\gamma = 0. \tag{3.30}$$

*where $P(\xi)$, $Q(\xi)$ be continuous functions defined on $(a, b)$. If $\gamma_1, \gamma_2$ be two linearly independent solutions of* (3.30) *existing on $(a, b)$ then $W_\alpha[\gamma_1, \gamma_2] = e^{-I_\alpha(P)}$*

***Proof*** We apply the operator $D_\alpha$ on $W_\alpha$ to get

$$\begin{aligned} D_\alpha(W_\alpha[\gamma_1, \gamma_2]) &= D_\alpha(\gamma_1 D_\alpha \gamma_2 - \gamma_2 D_\alpha \gamma_1) \\ &= D_\alpha \gamma_1 D_\alpha \gamma_2 + \gamma_1 D_\alpha D_\alpha \gamma_2 - D_\alpha \gamma_2 D_\alpha \gamma_1. \end{aligned}$$

But, $\gamma_1$ and $\gamma_2$ satisfies (3.30). Hence

$$D_\alpha D_\alpha \gamma_1 = -P(\xi) D_\alpha \gamma_1 - Q(\xi)\gamma_1$$

and

$$D_\alpha D_\alpha \gamma_2 = -P(\xi) D_\alpha \gamma_2 - Q(\xi)\gamma_2.$$

Therefore

$$\begin{aligned} D_\alpha(W_\alpha[\gamma_1, \gamma_2]) &= -P(\xi)(\gamma_1 D_\alpha \gamma_2 - \gamma_2 D_\alpha \gamma_1) \\ &= -P(\xi) W_\alpha[\gamma_1, \gamma_2]. \end{aligned}$$

Thus

$$\frac{D_\alpha(W_\alpha[\gamma_1, \gamma_2])}{W_\alpha[\gamma_1, \gamma_2]} = -P(\xi).$$

Consequently,

$$W_\alpha[\gamma_1, \gamma_2] = e^{I_\alpha(-P(\xi))} \tag{3.31}$$

This completes the proof.

**Theorem 16** *Let $P_1(\xi), P_2(\xi), \ldots, P_n(\xi)$ are the continuous functions in*

$$ {}^{n}D_{\alpha} + P_1(x)^{n-1}D_{\alpha} + \cdots + P_n(\xi)D_{\alpha} = 0, \qquad (*)$$

*where $\xi \in (a, b)$. If $\{\gamma_i\}_{i=1}^{n}$, be the set of n linearly independent solutions of above equation existing on $(a, b)$ containing a point $\xi_0$ then $W(\gamma_1, \gamma_2, \ldots, \gamma_n) = ce^{I_{\alpha}(-P_1(\xi))}$*

***Proof*** Let

$$W_{\alpha}(\gamma_1, \gamma_2, \ldots, \gamma_n) = \begin{vmatrix} \gamma_1 & \gamma_2 & \cdots & \gamma_n \\ D_{\alpha}\gamma_1 & D_{\alpha}\gamma_2 & \cdots & D_{\alpha}\gamma_n \\ \vdots & \vdots & \vdots & \vdots \\ {}^{n-1}D_{\alpha}\gamma_1 & {}^{n-1}D_{\alpha}\gamma_2 & \cdots & {}^{n-1}D_{\alpha}\gamma_n \end{vmatrix}.$$

We apply the operator $D_{\alpha}$ on $W_{\alpha}$ is a sum of n determinants.

$$D_{\alpha}W_{\alpha} = A_1 + A_2 + \cdots + A_k, \ldots, A_n,$$

where $A_k$ differs from $W_{\alpha}$ only in the $k$th rows and $k$th row of $A_k$ is obtained by applying $D_{\alpha}$ on the $k$th row of $W_{\alpha}$. Thus

$$D_{\alpha}(W_{\alpha}(\xi)) = \begin{vmatrix} D_{\alpha}\gamma_1 & D_{\alpha}\gamma_2 & \cdots & D_{\alpha}\gamma_n \\ D_{\alpha}\gamma_1 & D_{\alpha}\gamma_2 & \cdots & D_{\alpha}\gamma_n \\ {}^{2}D_{\alpha}\gamma_1 & {}^{2}D_{\alpha}\gamma_2 & \cdots & {}^{2}D_{\alpha}\gamma_n \\ \vdots & \vdots & \vdots & \vdots \\ {}^{n-1}D_{\alpha}\gamma_1 & {}^{n-1}D_{\alpha}y_2 & \cdots & {}^{n-1}D_{\alpha}\gamma_n \end{vmatrix} + \begin{vmatrix} D_{\alpha}\gamma_1 & D_{\alpha}\gamma_2 & \cdots & D_{\alpha}\gamma_n \\ {}^{2}D_{\alpha}\gamma_1 & {}^{2}D_{\alpha}\gamma_2 & \cdots & {}^{2}D_{\alpha}\gamma_n \\ {}^{2}D_{\alpha}\gamma_1 & {}^{2}D_{\alpha}\gamma_2 & \cdots & {}^{2}D_{\alpha}\gamma_n \\ \vdots & \vdots & \vdots & \vdots \\ {}^{n-1}D_{\alpha}\gamma_1 & {}^{n-1}D_{\alpha}\gamma_2 & \cdots & {}^{n-1}D_{\alpha}\gamma_n \end{vmatrix}$$

$$+ \cdots + \begin{vmatrix} \gamma_1 & \gamma_2 & \cdots & \gamma_n \\ D_{\alpha}\gamma_1 & D_{\alpha}\gamma_2 & \cdots & D_{\alpha}\gamma_n \\ {}^{2}D_{\alpha}\gamma_1 & {}^{2}D_{\alpha}\gamma_2 & \cdots & {}^{2}D_{\alpha}\gamma_n \\ \vdots & \vdots & \vdots & \vdots \\ {}^{n}D_{\alpha}\gamma_1 & {}^{n}D_{\alpha}\gamma_2 & \cdots & {}^{n}D_{\alpha}\gamma_n \end{vmatrix}.$$

The first $n-1$ determinants $A_1, A_2, \ldots, A_{n-1}$ are all zero, since they each have two identical rows. Thus

$$D_{\alpha}(W_{\alpha}(\xi)) = \begin{vmatrix} \gamma_1 & \gamma_2 & \cdots & \gamma_n \\ D_{\alpha}\gamma_1 & D_{\alpha}\gamma_2 & \cdots & D_{\alpha}\gamma_n \\ {}^{2}D_{\alpha}\gamma_1 & {}^{2}D_{\alpha}\gamma_2 & \cdots & {}^{2}D_{\alpha}\gamma_n \\ \vdots & \vdots & \vdots & \vdots \\ {}^{n-1}D_{\alpha}\gamma_1 & {}^{n-1}D_{\alpha}\gamma_2 & \cdots & {}^{n-1}D_{\alpha}\gamma_n \\ {}^{n}D_{\alpha}\gamma_1 & {}^{n}D_{\alpha}\gamma_2 & \cdots & {}^{n}D_{\alpha}\gamma_n \end{vmatrix} \qquad (**)$$

Now, as $\gamma_1, \gamma_2, \ldots, \gamma_n$ are solution of equation (*), we have

$${}^{n}D_{\alpha}\gamma_k + {}^{n-1}D_{\alpha}\gamma_k + \cdots + D_{\alpha}\gamma_k = 0 \quad \forall\, k = 1, 2, \ldots, n.$$

$$\Rightarrow {}^{n}D_{\alpha}\gamma_k = -{}^{n-1}D_{\alpha}\gamma_k - \cdots - D_{\alpha}\gamma_k = 0 \quad \forall\, k = 1, \ldots., n.$$

Putting $k = 1, 2, \ldots, n$ in above equation, we get ${}^{n}D_{\alpha}\gamma_1, {}^{n}D_{\alpha}\gamma_2, \ldots, {}^{n}D_{\alpha}\gamma_n$ and substitute these values in Eq. (**), we have

$$D_{\alpha}(W_{\alpha}(\xi)) = \begin{vmatrix} \gamma_1 & \cdots & \gamma_n \\ D_{\alpha}\gamma_1 & \cdots & D_{\alpha}\gamma_n \\ {}^{2}D_{\alpha}\gamma_1 & \cdots & {}^{2}D_{\alpha}\gamma_n \\ \vdots & \vdots & \vdots \\ {}^{n-2}D_{\alpha}\gamma_1 & \cdots & {}^{n-2}D_{\alpha}\gamma_n \\ -P_1{}^{n-1}D_{\alpha}\gamma_1 - \cdots - P_n D_{\alpha}\gamma_1 & \cdots & -P_1{}^{n-1}D_{\alpha}\gamma_n - \cdots - P_n D_{\alpha}\gamma_n \end{vmatrix}.$$

Now we multiplying first row by $P_n$, the second row by $P_{n-1}$,..., the $(n-1)$th row by $P_2$ and adding these to the last row, we have

$$D_{\alpha}(W_{\alpha}(\xi)) = \begin{vmatrix} \gamma_1 & \gamma_2 & \cdots & \gamma_n \\ D_{\alpha}\gamma_1 & D_{\alpha}\gamma_2 & \cdots & D_{\alpha}\gamma_n \\ {}^{2}D_{\alpha}\gamma_1 & {}^{2}D_{\alpha}\gamma_2 & \cdots & {}^{2}D_{\alpha}\gamma_n \\ \vdots & \vdots & \vdots & \vdots \\ {}^{n-2}D_{\alpha}\gamma_1 & {}^{n-2}D_{\alpha}\gamma_2 & \cdots & {}^{n-2}D_{\alpha}\gamma_n \\ -P_1{}^{n-1}D_{\alpha}\gamma_1 & -P_1{}^{n-1}D_{\alpha}\gamma_2 & \cdots & -P_1{}^{n-1}D_{\alpha}\gamma_n \end{vmatrix}.$$

Thus

$$D_{\alpha}(W_{\alpha}(\xi)) = -P_1 W_{\alpha}(\xi)$$

$$\Rightarrow \frac{D_{\alpha}(W_{\alpha}(x))}{W_{\alpha}(\xi)} = -P_1$$

$$\Rightarrow lnW - lnC = I_{\alpha}(-P_1(\xi))$$

$$\Rightarrow W_{\alpha}(\xi) = ce^{-\int_{\xi_0}^{\xi} P_1(\xi)}d\xi$$

$$\Rightarrow W_{\alpha}(\gamma_1, \gamma_2, \ldots, \gamma_n)(\xi) = ce^{-\int_{\xi_0}^{\xi} P_1(\xi)d\xi}.$$

# 4 Abel's Formula

We'll start with fractional differential equations.

$$D_\alpha \gamma + a(\xi)\gamma = b(\xi), \tag{4.1}$$

where, $0 < \alpha \leq 1$.

Multiply (4.1) by $e^{I(a)}$, we get

$$e^{I(a)} D_\alpha \gamma + e^{I(a)} a(\xi)\gamma = e^{I(a)} b(\xi).$$

As a result, using the results on conformable fractional derivatives from [4], we have

$$D_\alpha(e^{I(a)}\gamma) = e^{I(a)} b(\xi).$$

Hence

$$\gamma = e^{-I(a)} I_\alpha(e^{I(a)} b(\xi)), \tag{4.2}$$

is a solution of (4.1).

Now, think about $\gamma_1$ to be an solution of (3.30). Our goal is to seek out a second solution $\gamma_2$ of Eq. (3.30).

From (3.31), we have $W_\alpha[\gamma_1, \gamma_2] = e^{I_\alpha(-P(\xi))}$, from which we get

$$\gamma_1 D_\alpha \gamma_2 - \gamma_2 D_\alpha \gamma_1 = e^{I_\alpha(-P(\xi))}$$

and so

$$D_\alpha \gamma_2 - \gamma_2 \frac{D_\alpha \gamma_1}{\gamma_1} = \frac{e^{I_\alpha(-P(\xi))}}{\gamma_1}. \tag{4.3}$$

Equation (4.3) is a fractional linear equation, with $a(\xi) = \frac{D_\alpha \gamma_1}{\gamma_1}$, and $b(\xi) = \frac{e^{I_\alpha(-P(\xi))}}{\gamma_1}$. Hence, using the fact

$$I_\alpha\left(\frac{D_\alpha \gamma_1}{\gamma_1}\right) = ln\gamma_1,$$

and by using formula (4.2), we get

$$\gamma_2 = \gamma_1 I_\alpha\left(\frac{e^{-I_\alpha(P)}}{\gamma_1^2}\right) \tag{4.4}$$

## 5 Applications

**Example 1** Consider the fractional differential equation

$$D_{\frac{1}{2}} D_{\frac{1}{2}} \gamma + \xi D_{\frac{1}{2}} \gamma = 0.$$

Clearly, $\gamma_1 = 1$ is a solution of such equation. Using formula (4.4) and the definition of $I_{\frac{1}{2}}(\Phi)$ to get

$$\gamma_2 = \gamma_1 I_{\frac{1}{2}} \left( \frac{e^{-I_\alpha(P)}}{\gamma_1^2} \right) = I_{\frac{1}{2}}(e^{-2\sqrt{\xi}}).$$

Clearly, $\gamma_2$ satisfies the above equation.

**Example 2** Consider the fractional differential equation

$$D_{3/2} D_{3/2} \gamma + \frac{1}{2} \tan \sqrt{\xi} D_{3/2} \gamma = 0.$$

Clearly, $\gamma_1 = 1$ is a solution of such equation, noting that $D_{2/3} 1 = 0$. Hence using formula (4.4), to get

$$\gamma_2 = \gamma_1 I_{3/2} \left( \frac{e^{-I_\alpha(P)}}{\gamma_1^2} \right) = I_{3/2} \left( e^{-I_{3/2}(\frac{1}{2} \tan \sqrt{\xi})} \right).$$

We get $\gamma_2 = I_{3/2}(\cos \sqrt{\xi})$. Using integral again we see that such $\gamma_2$ is a solution of the equation.

**Example 3** Consider the fractional differential equation

$$D_{3/2} D_{3/2} \gamma + \frac{1}{2} \cot \sqrt{\xi} D_{3/2} \gamma = 0.$$

Clearly, $\gamma_1 = 1$ is a solution of such equation. Using formula (4.4) and the definition of $I_\alpha(\Phi)$ to get

$$\gamma_2 = \gamma_1 I_{3/2} \left( \frac{e^{-I_{3/2}(P)}}{\gamma_1^2} \right) = I_{3/2}(e^{-\xi}).$$

$$\gamma_2 = I_{3/2}(\sin^{-1}(\sqrt{\xi})).$$

# 6 Conclusion

In this study, we provide some results on linear $\alpha$-fractional differential equations' existence and uniqueness theorems. It has been revealed that the outcomes of this research are identical to those of the standard instance. In addition, the Abel's formula and Wronskian determinant of $\alpha$-fractional differential equations were developed. In addition, we provide applications of the obtained results.

# References


1. Abdeljawad, T.: On conformable fractional calculus. J. Comput. Appl. Math. **279**, 57–66 (2015). https://doi.org/10.1016/j.cam.2014.10.016
2. Almeida, R., Guzowska, M., Odzijewicz, T.: A remark on local fractional calculus and ordinary derivatives 1–13 (2007) arXiv:1612.00214
3. Abdeljawad, T., AL Horani, M., Khalil, R.: Conformable fractional semigroups of operators. J. Semigroup Theory Appl. **2015** (2015)
4. Anderson, D., Avery, R.: Fractional-order boundary value problem with Sturm-Liouville boundary conditions. Electron. J. Differ. Eqs. **2015**(29), 1–10 (2015)
5. Bonilla, B., Rivero, M., Trujillo, J.: Linear differential equations of fractional order. In: Advances in Fractional Calculus, pp. 77–91 (2007)
6. Bonilla, B., Rivero, M., Trujillo, J.: On systems of linear fractional differential equations with constant coefficients. Appl. Math. Comput. **187**(1), 68–78 (2007). https://doi.org/10.1016/j.amc.2006.08.104
7. Błasik, M.: Numerical scheme for a two-term sequential fractional differential equation. Sci. Res. Inst. Math. Comput. Sci. **10**, 17–29 (2011)
8. Finan, M.: A Second Course in Elementary Ordinary Differential Equations. Arkansas Tech University (2013)
9. Hammad, M., Khalil, R.: Abel's formula and Wronskian for conformable fractional differential equations. Int. J. Differ. Eqs. Appl. **13**(2), 177–183 (2014)
10. Khalil, R.: Fractional Fourier series with applications. Am. J. Comput. Appl. Math. **4**(6), 187–191 (2014)
11. Khalil, R., Al Horani, M., Yousef, A., Sababheh, M.: A new definition of fractional derivative. J. Comput. Appl. Math. **264**, 65–70 (2014). https://doi.org/10.1016/j.cam.2014.01.002
12. Kilbas, A., Srivastava, H., Trujillo, J.: Theory and applications of fractional differential equations, vol. 204. Elsevier Science Limited, London (2006)
13. Klimek, M.: Sequential fractional differential equations with Hadamard derivative. Commun. Nonlinear Sci. Numer. Simul. **16**(12), 4689–4697 (2011). https://doi.org/10.1016/j.cnsns.2011.01.018
14. Loghmani, G., Javanmardi, S.: Numerical methods for sequential fractional differential equations for Caputo operator. Bull. Malays. Math. Sci. Soc. **35**(2), 315–323 (2011)
15. Miller, K., Ross, B.: An Introduction to the Fractional Calculus and Fractional Differential Equations. A Wiley-Interscience Publication, New York (1993)
16. Muneshwar, R.A., Bondar, K.L., Shirole, Y.H.: Solution of linear and non-linear partial differential equation of fractional order. Proyecc. J. Math. **40**(5), 1175–1190 (2021)
17. Podlubny, I.: Fractional Differential Equations: An Introduction to Fractional Derivatives, Fractional Differential Equations, to Methods of Their Solution and Some of Their Applications. Academic press, London (1998)

# Method of Directly Defining the Inverse Mapping for Nonlinear Ordinary and Partial Fractional-Order Differential Equations

**Dulashini Karunarathna and Mangalagama Dewasurendra**

**Abstract** We extend the Method of Directly Defining the inverse Mapping (MDDiM) to determine approximate solutions for fractional-order ordinary and partial differential equations. The Riccati, Abel, and time-fractional Rosenau-Hyman equations were solved here. The MDDiM was utilized for the first time to solve fractional-order ordinary and partial differential equations. By considering the sum of the initial three terms of the series solution, we were able to get approximate solutions for the fractional Riccati ordinary differential equation and the time-fractional Rosenau-Hyman equation. We also used the fourth-order series solution to get an approximate solution for the Abel differential equation. By determining the ideal option of the convergence control value for quick convergence, as well as alternative fractional orders on solutions, we were able to achieve solution graphs and minimum errors.

**Keywords** Fractional-order differential equation · Method of directly defining the inverse mapping · Riccati differential equation · Abel differential equation · Rosenau-Hyman equation

## 1 Introduction

The Method of Directly Defining inverse Mapping (MDDiM) has been used to tackle mathematical and real-world problems involving nonlinear ordinary and partial differential equations [1, 3, 4, 8, 10, 11]. We used our innovative method so-called MDDiM to solve nonlinear fractional-order ordinary and partial equations in this study.

We used the MDDiM that we extended to solve fractional-order Riccati equations in the first case. The fractional-order ordinary Riccati differential equations are a type of nonlinear differential equation that may be used for a wide range of problems in

D. Karunarathna (✉) · M. Dewasurendra
Department of Mathematics, University of Peradeniya, Peradeniya 20400, Sri Lanka
e-mail: dulashinik@sci.pdn.ac.lk; dulashinikarunarathna111@gmail.com

M. Dewasurendra
e-mail: mangalagama.dewasurendra@sci.pdn.ac.lk



B. Rushi Kumar et al. (eds.), *Mathematics and Computing*, Springer Proceedings in Mathematics & Statistics 415, https://doi.org/10.1007/978-981-19-9307-7_31

electrical networks, chemical physics, engineering, acoustics, and material science [9]. The MDDiM was used for the first time to solve fractional-order ordinary and partial differential equations. We were able to produce estimated three-term solutions, solution graphs, and minimum error of the solution by determining the best choice of $h$ for fast convergence, and different fractional order $\alpha$ on the solution.

In the second example, we solved a fractional-order Abel differential equation using MDDiM. This equation has a long history in many areas of pure and applied mathematics [6]. Here, we obtained approximate solutions by getting the sum of the initial four terms of the series solution. The solution graphs and minimum error values were determined by choosing the best value of $h$ for fast convergence, and different $\alpha$ values.

The time-fractional Rosenau-Hyman equation, which was discovered as a mathematical model to research pattern creation of nonlinear dispersion in liquid drops, was solved within the third example. The MDDiM was utilized for the first time to solve a fractional partial differential equation. By summing the first three terms of the series solution, we were able to get approximate answers. For validation, we compared the MDDiM solution to the solution found in the literature. The primary goal of this research is to demonstrate that the MDDiM may be used to derive analytical solutions to fractional-order differential equations.

## 2 Methodology

**Extension of MDDiM for partial differential equations of fractional order**

We use the differential equation of this type as an example,

$$\mathcal{N}[D_t^{\alpha} u(x,t)] - g(x,t) = 0 \tag{1}$$

where $\mathcal{N}$ is the fractional nonlinear operator, $D_t^{\alpha}$ denote the fractional derivative, $x$ and $t$ are independent variables, $g$ is a known function, and $u$ is an unknown function. To apply the MDDiM, the higher order deformation equation is constructed as

$$(1-q)L[\varphi(x,t;q) - u_0(x,t)] = qh\Big(\mathcal{N}[D_t^{\alpha}\varphi(x,t;q)] - g(x,t)\Big), \quad q \in [0,1] \tag{2}$$

where $L$ is an auxiliary linear operator, and $h \neq 0$ is an auxiliary parameter.

When $q = 0$ and $q = 1$, Eq. (2) can be reduced to

$$\varphi(x,t;0) = u_0(x,t), \quad \varphi(x,t;1) = u(x,t) \tag{3}$$

respectively [2]. Then, the MDDiM solution $\varphi(x,t;q)$ will vary from the $u_0(x,t)$ to the solution $u(x,t)$.

Taylor series expansion of $\varphi(x, t; q)$ is given as

$$\varphi(x, t; q) = u_0(x, t) + \sum_{m=1}^{\infty} u_m(x, t)q^m \tag{4}$$

where

$$u_m(x, t) = \frac{1}{m!}\frac{\partial^m \varphi(x, t; q)}{\partial q^m}|_{q=0}. \tag{5}$$

we obtain the $m$th-order deformation equation by differentiating the Eq. (2) $m$-times with respect to $q$ and dividing it by $m!$ and finally considering $q = 0$,

$$L[u_m(x, t) - \chi_m u_{m-1}(x, t)] = hR_m[\overrightarrow{u}_{m-1}(x, t)] \tag{6}$$

where

$$\chi_m = \begin{cases} 1, \text{ when } & m > 1, \\ 0, & \text{otherwise.} \end{cases} \tag{7}$$

$$R_m[\overrightarrow{u}_{m-1}(x, t)] = \frac{1}{(m-1)!}\Big(\frac{\partial^{m-1}}{\partial q^{m-1}}\Big(\mathcal{N}\Big[D_t^{\alpha}\varphi(x, t; q)\Big] - g(x, t)\Big)\Big)|_{q=0}. \tag{8}$$

Define the solution space function as

$$S = \sum_{k=0}^{+\infty} \psi_k(x, t). \tag{9}$$

Define the approximate solution space and the space for the initial guess respectively as

$$S^* = \sum_{k=0}^{\mu} \psi_k(x, t) \tag{10}$$

and

$$\hat{S} = \sum_{k=\mu+1}^{+\infty} \psi_k(x, t) \tag{11}$$

so that $S = \hat{S} \cup S^*$. Finally, we directly define the inverse mapping $\mathscr{J}$. We obtain the final version of the higher order deformation equation for the MDDiM:

$$u_m(x,t) = \chi_m u_{m-1}(x,t) + h\mathscr{J}[R_m[u_{m-1}(x,t)]] + \sum_{n=1}^{\mu} a_{m,n}\varphi_n(x,t). \tag{12}$$

We define the square residual error function $E(h)$ to find the error of the MDDiM solution and $h$ values which give the optimal errors,

$$E(h) = \int_{\Omega} (\mathcal{N}[D_t^{\alpha} u(x,t)] - g(x,t))^2 dx\, dt. \tag{13}$$

## 3 Example 01: MDDiM Solutions for Fractional Riccati Differential Equation

Consider the fractional-order ordinary Riccati differential equation [9],

$$\frac{d^{\alpha} y}{dx^{\alpha}} = -y^2 + 1, \quad 0 < \alpha \leq 1, \tag{14}$$

with initial condition $y(0) = 0$. Many scholars are interested in Riccati differential equations. The variational iteration approach was used to generate approximate solutions to the ordinary Riccati differential equation for certain of them. In this study, we use MDDiM to solve the ordinary Riccati differential equation.

Consider an $n$th-order nonlinear differential equation $\mathcal{N}[D_x^{\alpha} y(x)] - f(x) = 0$, and the deformation Eq. (12) of MDDiM for this example:

$$y_k(x) = \chi_k y_{k-1}(x) + h\mathscr{J}[R_{m-1}[y(x)]] + a_{k,0} + a_{k,1}x \quad \text{for} \quad k \geq 1. \tag{15}$$

By considering

$$\mathcal{N}[D_x^{\alpha} y(x)] - f(x) = \frac{d^{\alpha} y}{dx^{\alpha}} + y^2 - 1 \tag{16}$$

with initial condition $y(0) = 0$, we came up with an initial prediction of $y_0(x) = x$. We have a lot of freedom in MDDiM to create an inverse linear mapping directly. For this example, we chose

$$\mathscr{J}[x^k] = \frac{x^{k+1}}{Ak^3 + 1}. \tag{17}$$

Here, $A$ is an arbitrary constant.

### 3.1 Results and Discussion

The sum of the first three term solution can be written as

$$y(x) = y_0(x) + y_1(x) + y_2(x). \tag{18}$$

Taking $A = 100$, we obtained terms for Eq. (18) by considering different values of $\alpha$, and MDDiM solution when $\alpha = 0.9$ included for Eq. (19).

$$y(x) = x + \frac{2hx^2}{101} + h\left(0.955579096\, x^{\frac{11}{10}} + \frac{x^3}{801} - x\right) + \cdots \tag{19}$$

Here, the optimal value of $h$ is determined by minimizing the residual error of the sum of three term solution, and it was $h = -1.0727831$. Then, we can say that 3rd order MDDiM solution is accurate enough with the squared residual error $7.190345 \times 10^{-5}$. For various $\alpha$, we also got MDDiM solutions and convergence control parameter values. Squared residual errors of MDDiM solutions and convergence control parameter values with various $\alpha$ values are shown in Table 1. Figure 1 shows the graph of MDDiM solution versus $x$ for various $\alpha$ values.

We utilized MDDiM to solve the fractional Riccati differential equation in the first case. When a minimum error was reached, we obtained approximate answers

**Table 1** Squared residual errors and convergence control parameter values for different $\alpha$

| $\alpha$ values | Square residual error E(h) | $h$ values |
|---|---|---|
| 0.9 | $7.190345 \times 10^{-5}$ | −1.072783 |
| 0.8 | $2.789923 \times 10^{-5}$ | −0.016020 |
| 0.7 | $3.101696 \times 10^{-5}$ | −0.270407 |

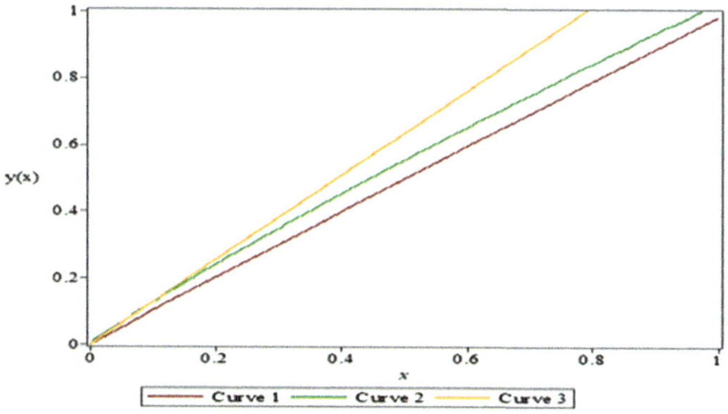


**Fig. 1** MDDiM solution graphs for different values of $\alpha$. Curve 1: $\alpha = 0.9$; Curve 2: $\alpha = 0.8$; Curve 3: $\alpha = 0.7$

by examining only the first three terms. The comparability of the MDDiM solutions with the fractional variation iteration approach [9] was investigated to assess the correctness of the MDDiM solutions. The MDDiM solutions would be acceptable to us. MDDiM may therefore be used to solve ordinary and partial fractional-order differential equations.

## 4 Example 02: MDDiM Solutions for Abel Differential Equation

To solve an Abel differential equation, we use the MDDiM, which is a semi-analytical approach. The Abel differential equations appear in N. H. Abel's work on elliptic function theory. The Riccati equation is a natural generalization of the first kind of Abel's differential equations.

Consider the nonlinear fractional-order Abel differential equation [6],

$$\frac{d^{\alpha}y}{dx^{\alpha}} = -y^3 \sin x - xy^2 + x^2 y - x^3, \quad 0 < \alpha \leq 1, \tag{20}$$

with initial conditions $y(0) = 0$. Many studies have been performed on solutions of the Abel differential equations. Some of them such as Optimal Homotopy Analysis Method [12] have been used to solve fractional Abel differential equation. In this study, we solve the fractional Abel differential equation using MDDiM.

Consider the fractional-order nonlinear differential equation $\mathcal{N}[D_x^{\alpha} y(x)] - f(x) = 0$, and the higher order deformation equation of MDDiM given by

$$y_k(x) = \chi_k y_{k-1}(x) + h \mathscr{J}[R_{m-1}[y(x)]] + \sum_{n=1}^{\mu} a_{k,n}\phi_n \quad \text{for} \quad k \geq 1. \tag{21}$$

Here, $\mathscr{J}$ is the inverse linear mapping, and $\mathcal{N}$ is the fractional nonlinear operator. The convergence control parameter, $h$, needs to be found. By applying MDDiM [11] to Eq. (21), we obtained higher order deformation equation (20):

$$y_k = \chi_k y_{k-1} + h \mathscr{J}[R_{m-1}[y(x)]] + a_{k,0} + a_{k,1}x \quad \text{for} \quad k \geq 1. \tag{22}$$

By considering $\mathcal{N}[D_x^{\alpha} y(x)] - f(x) = \frac{d^{\alpha}y}{dx^{\alpha}} + y^3 \sin x + xy^2 - x^2 y + x^3$, with initial condition $y(0) = 0$, we obtained an initial guess as $y_0(x) = 0$. In this method, we have great freedom to choose an inverse linear mapping [8]. For this example, we chose the inverse mapping as $\mathscr{J}[x^k] = \frac{x^{k-1}}{Ak+1}$. Here, A is an arbitrary constant, and Maple 16 package was used to attain the following results.

**Table 2** Squared residual errors and convergence control parameter values for different $\alpha$

| $\alpha$ values | Square residual error E(h) | $h$ values |
|---|---|---|
| 0.4 | $6.688762 \times 10^{-21}$ | −0.623904 |
| 0.50 | $2.173158 \times 10^{-21}$ | −0.974734 |
| 0.60 | $1.756642 \times 10^{-24}$ | −0.684494 |
| 0.70 | $2.702231 \times 10^{-24}$ | −0.498490 |
| 0.80 | $5.410933 \times 10^{-24}$ | −0.285775 |
| 0.90 | $1.250342 \times 10^{-23}$ | −0.323124 |
| 0.98 | $7.573952 \times 10^{-23}$ | −0.170613 |

## 4.1 Results and Discussion

The first four term solution can be written as

$$y(x) = y_0(x) + y_1(x) + y_2(x) + y_3(x). \tag{23}$$

Taking $A = 100$, we obtained terms for Eq. (23) by considering different values of $\alpha$, and MDDiM solution when $\alpha = 0.98$ included for Eq. (5):

$$y(x) = \frac{hx^2}{2} + h^2\left(0.2454090594x^{\frac{1}{50}} + \frac{x^3}{20}\right) + \cdots \tag{24}$$

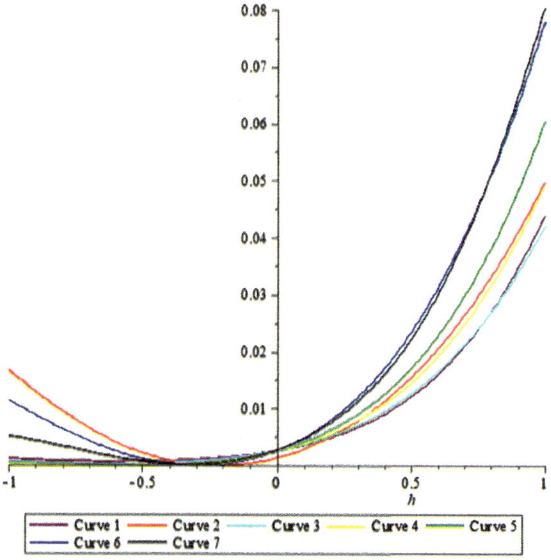


**Fig. 2** Squared residual error curves of MDDiM solution for different values of $\alpha$ with fixed $x = 0.5$.
Curve 1: $\alpha = 0.40$;
Curve 2: $\alpha = 0.98$;
Curve 3: $\alpha = 0.50$;
Curve 4: $\alpha = 0.60$;
Curve 5: $\alpha = 0.70$;
Curve 6: $\alpha = 0.80$;
Curve 7: $\alpha = 0.90$

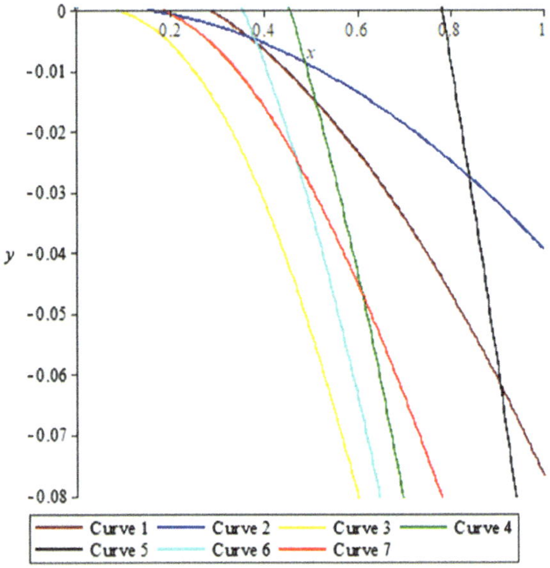


**Fig. 3** MDDiM solution for different values of $\alpha$. Curve 1: $\alpha = 0.40$; Curve 2: $\alpha = 0.50$; Curve 3: $\alpha = 0.60$; Curve 4: $\alpha = 0.70$; Curve 5: $\alpha = 0.90$; Curve 6: $\alpha = 0.50$; Curve 7: $\alpha = 0.98$

Here, values of $h$ are made out by minimizing the residual error of the obtained MDDiM solution and it was $h = -0.1706131$. The corresponding MDDiM approximation shows to be accurate enough with the squared residual error $7.5739522 \times 10^{-23}$. Table 2 represents squared residual errors of MDDiM solutions and convergence control parameter values with various $\alpha$ values. Figure 3 represents the graph of the approximation MDDiM solutions for different values of $\alpha$ (Fig. 2).

## 5 Example 03: MDDiM Solutions for Time-Fractional Rosenau-Hyman Equation

We use the MDDiM semi-analytical approach to solve the time-fractional Rosenau-Hyman equation. The Rosenau-Hyman equation is a mathematical model for studying pattern generation in liquid droplets with nonlinear dispersion. This is the first time a fractional-order partial differential equation has been studied using the MDDiM.

Consider the time-fractional Rosenau-Hyman equation [5],

$$\frac{\partial^\alpha u}{\partial t^\alpha} = u\frac{\partial^3 u}{\partial x^3} + u\frac{\partial u}{\partial x} + 3\frac{\partial u}{\partial x}\frac{\partial^2 u}{\partial x^2}, \tag{25}$$

$t > 0$ and $0 < \alpha \leq 1$ subject to the initial condition $u(x, 0) = -\frac{8c}{3}\cos^2\left(\frac{x}{4}\right)$. The exact solution to this problem is given as follows [5]:

$$u(x,t) = -\frac{8c}{3}\cos^2\left(\frac{x-ct}{4}\right), \tag{26}$$

where $c$ is an arbitrary constant. The nonlinear operator and higher order deformation equation are defined separately:

$$N[D_t^{\alpha}u(x,t)] - f(x,t) = \frac{\partial^{\alpha}u}{\partial t^{\alpha}} - u\frac{\partial^3 u}{\partial x^3} - u\frac{\partial u}{\partial x} - 3\frac{\partial u}{\partial x}\frac{\partial^2 u}{\partial x^2}. \tag{27}$$

$$u_k(x,t) = \chi_k u_{k-1}(x,t) + h\,\mathscr{J}[R_{m-1}[u(x,t)]] + a_{k,0} + a_{k,1}x \quad \text{for} \quad k \geq 1. \tag{28}$$

By considering a nonlinear operator with a given initial condition $u(x,0) = -\frac{8c}{3}\cos^2\left(\frac{x}{4}\right)$, we implemented initial guess as $u_0(x,t) = -\frac{8c}{3}\cos^2\left(\frac{x}{4}\right)$. Using the freedom of directly defining the inverse mapping [8], we chose

$$\mathscr{J} = \frac{x^{k+1}}{Ak+0.8} \tag{29}$$

where $A$ is an arbitrary constant.

## 5.1 Results and Discussion

We choose an approximation solution by examining the sum of three term solution with $A = 1$ :

$$u_1(x,t) = \frac{5}{4}ht\Big(\frac{8}{3}c^2\cos^2\left(\frac{x}{4}\right)\sin\left(\frac{x}{4}\right) - 4c\cos\left(\frac{x}{4}\right)\sin\left(\frac{x}{4}\right)\Big(-\frac{1}{3}c\cos^2\left(\frac{x}{4}\right)\Big)\Big). \tag{30}$$

$$u_2(x,t) = u_0(x,t) + 2u_1(x,t) + \frac{5h^2c^2}{12}\Big(\frac{5}{2}t\sin\left(\frac{x}{2}\right) + \frac{ct^{\alpha+1}}{0.8+\alpha}\cos\left(\frac{x}{2}\right)\Big). \tag{31}$$

MDDiM solutions and residual errors were obtained and displayed by establishing the proper suited values for $h$. Because we were able to obtain enough accurate residual error as shown in Table 3, we just analyzed three terms of the series solution. Figure 4 displays the MDDiM solution of $u(x,t)$ versus time $t$ for various $\alpha$ values.

**Table 3** Squared residual errors and convergence control parameter values for different $\alpha$

| $\alpha$ values | $E(h)$ values | $h$ values |
|---|---|---|
| 0.25 | $1.93078226 \times 10^{-5}$ | −0.93171114 |
| 0.50 | $7.34478224 \times 10^{-6}$ | −0.93984369 |
| 0.75 | $2.80931060 \times 10^{-4}$ | −0.92866480 |

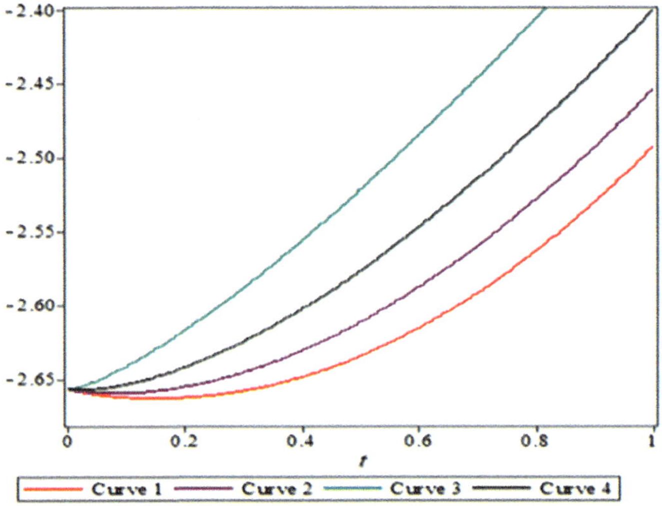


**Fig. 4** MDDiM solution plot for different $\alpha$ when $x = \frac{\pi}{13}$, $c = 1$ and $h = -1.02$. Curve 1: $\alpha = 1$; Curve 2: $\alpha = 0.75$; Curve 3: $\alpha = 0.25$; Curve 4: $\alpha = 0.50$

## 6 Conclusions

We have expanded the Method of Directly Defining inverse Mapping in this work to solve nonlinear fractional-order ordinary and partial differential equations with applications in science and engineering. These examples show how to use extended MDDiM to solve fractional-order ordinary and partial differential equations. All computations related to the aforementioned examples are worked out in this paper using the Maple 16 package.

We used MDDiM to solve the fractional Riccati differential equation and the Abel differential equation in the first and second examples, respectively. By examining only the first three terms and four terms where the smallest error occurs, we were able to derive approximate answers. Comparisons using the fractional variation iteration technique and Homotopy Analysis Method solutions were done to validate the MDDiM solutions of the Riccati differential problem and the Abel differential equation.

We used extended MDDiM to solve the time-fractional Rosenau-Hyman equation in the third problem. In the approximation series solutions, we achieved approximate solutions with less complicated terms, reducing calculation time. The MDDiM is particularly good at handling solutions of a class of nonlinear partial differential equations of fractional order, according to our findings. Comparisons with q-HAM solutions [5] for different fractional orders were made to validate the MDDiM solutions. The MDDiM solutions are something we could agree on. We also found the MDDiM solutions with the lowest errors and the optimal choice of $h$ for quick convergence, as well as the impact of various fractional order $\alpha$ on the solution (see Fig. 4). In addition, this unique technique can be utilized to examine increasingly complex models in the future.

## References

1. Baxter, M., Dewasurendra, M., Vajravelu, K.: A method of directly defining the inverse mapping for solutions of coupled systems of nonlinear differential equations. Numer. Algorithm **77**, 1199–1211 (2017)
2. Dewasurendra, M., Baxter, M., Vajravelu, K.: A method of directly defining the inverse mapping for solutions of non-linear coupled systems arising in convection heat transfer in a second grade fluid. Appl. Math. Comput., Elsevier **339**, 758–767 (2018)
3. Dewasurendra, M., Vajravelu, K.: On the method of inverse mapping for solutions of coupled systems of nonlinear differential equations arising in nanofluid flow. Heat Mass Transf., Appl. Math. Nonlinear Sci. **3**, 1–4 (2018)
4. Dewasurendra, M., Zhang, Y., Vajravelu, K.: A method of directly defining the inverse mapping (MDDiM) for solutions of non-linear coupled systems arising in SIR and SIS epidemic models. Commun. Numer. Anal. **2**, 64–77 (2019)
5. Iyiola, O.S., Ojo, G.O., Mmaduabuchi, O.: The fractional Rosenau-Hyman model and its approximate solution. Alex. Eng. J. **55**, 1655–1659 (2016)
6. Jafari, H., Sayevand, K., Tajadodi, H., Baleanu, D.: Homotopy analysis method for solving Abel differential equation of fractional order. Cent. Eur. J. Phys. **11**, 1523–1527 (2013)
7. Liao, S.: Proposed Homotopy Analysis Techniques for the Solutions of Nonlinear Problems, Ph.D. thesis, Shanghai Jiao Tong University (1992)
8. Liao, S., Zhao, Y.: On the method of directly defining the inverse mapping for nonlinear differential equations. Numer. Algorithm **72**, 989–1020 (2016)
9. Merdan, M.: On the solutions fractional Riccati differential equation with modified Riemann-Liouville derivative. Int. J. Differ. Eqs. **2012**, 1–17 (2012)
10. Sahabandu, C.W., Dewasurendra, M., Juman, Z.A.M.S., Vajravelu, K., Chamkha, A.J.: Semi-analytical method for propagation of harmonic waves in nonlinear magneto-thermo-elasticity. Comput. Math. Appl. **105**, 107–111 (2022)
11. Sahabandu, C.W., Karunarathna, D., Sewvandi, P., Juman, Z.A.M.S., Dewasurendra, M., Vajravelu, K.: A Method of Directly Defining the inverse Mapping for a nonlinear partial differential equation and for systems of nonlinear partial differential equations. Comput. Appl. Math. **40**, 1–16 (2021)
12. Van Gorder, R.A., Vajravelu, K.: On the selection of auxiliary functions, operators, and convergence control parameters in the application of the homotopy analysis method to nonlinear differential equations: a general approach. Commun. Nonlinear Sci. Numer. Simul. **14**, 4078–4089 (2009)

# Existence Results for Nonlocal Impulsive Fractional Neutral Functional Integro Differential Equations with Bounded Delay

**M. Latha Maheswari and R. Nandhini**

**Abstract** We consider the impulsive neutral fractional functional integro-differential equation under bounded delay with nonlocal condition in the Banach spaces. The existence condition for the solution of this problem is studied using Darbo Sadovskii's fixed point theorem with Hausdorff's measure of noncompactness.

**Keywords** Nonlocal condition · Fractional integro-differential equation · Neutral functional integro-differential equation

## 1 Introduction

Motivated by the paper of Suresh [12], we provide the impulsive neutral fractional functional integro—differential equation under bounded delay with nonlocal condition in Banach space

$$ {}^{c}D^{\beta}\left[y(\delta)+f\left(\delta, y(\delta), y_{\delta}\right)\right]=A(\delta) y(\delta)+\int_{0}^{\delta} G(\delta, r) g\left(r, y(r), y_{r}\right) d r, \quad (1) $$

$$ y_{0}=\Phi+h_{l}(y), \quad (2) $$

$$ \left.\Delta y\right|_{\delta=\delta_{k}}=\mathcal{I}_{k}\left(y\left(\delta_{k}^{-}\right)\right), \quad (3) $$

where $\delta \in [0, a]$ in Eq. (1), $A = A(\delta)$ is the bounded linear operator defined on D(A) which is dense in the Banach space $Y$ and for $y \in C([0, a]; Y)$, $\|A(\delta)\| \leq m$, and $y_\delta$ : $[-\mu, 0] \to Y$ defined by $y_\delta\ (\theta) = y\ \ (\delta\ +\ \theta\ )$ for $\theta \in [-\mu, 0]$; $g, f:\ [0, a] \times Y \times C([-\mu, 0]; Y) \to Y$, $G : [0, a] \times [0, a] \to (0, +\infty)$, $h_l : C([0, a]; Y) \to C([-\mu, 0]; Y)$ and $0 < \delta_1 < \delta_2 < \cdots < \delta_p < a$, $\mathcal{I}_j : Y \to Y$, $j = 1, 2, \ldots p$,

M. Latha Maheswari (✉) · R. Nandhini
PSG College of Arts and Science, Coimbatore 641014, TamilNadu, India
e-mail: lathamahespsg@gmail.com

R. Nandhini
e-mail: nanmathpsg@gmail.com



B. Rushi Kumar et al. (eds.), *Mathematics and Computing*, Springer Proceedings in Mathematics & Statistics 415, https://doi.org/10.1007/978-981-19-9307-7_32

are suitable functions, $y(\delta_k^+) = lim_{t\to 0^+} y(\delta_k + t)$ denote the right limit and $y(\delta_k^-) = lim_{t\to 0^-} y(\delta_k + t)$ denote the left limit of $y(\delta)$ at $\delta = \delta_k$, $\triangle y\big|_{\delta=\delta_k} = y(\delta_k^+) - y(\delta_k^-)$, $a, m, \mu, p > 0$ and $0 < \beta < 1$ are suitable constants.

## 2 Preliminaries

Here $Y$ denotes the Banach space associated by the norm $\|.\|$. $C([m, n], Y)$ is a Banach space which has all continuous Y—valued functions on $[m, n]$ associated with the norm

$$\|y\|_{[m,n]} = sup_{y\in[m,n]}\{\|y(r)\|\} \quad \forall\ y \in C([m, n], Y).$$

**Proposition 1** ([5]) *For $\beta_1, \beta_2 > 0$ and $f$ as a suitable functions we have,*

*(i)* ${}^cD_{0+}^{\beta_1} f(y) = I_{0+}^{1-\beta_1} Df(y) = I_{0+}^{1-\beta_1} f'(y), 0 < \beta_1 < 1.$
*(ii)* ${}^cD_{0+}^{\beta_1}\,{}^cD_{0+}^{\beta_2} f(y) \neq {}^cD_{0+}^{\beta_1+\beta_2} f(y).$
*(iii)* ${}^cD_{0+}^{\beta_1}\,{}^cD_{0+}^{\beta_2} f(y) \neq {}^cD_{0+}^{\beta_2}\,{}^cD_{0+}^{\beta_1} f(y).$

From the above conditions, it is clear that the differential operator does not satisfy the semigroup and commutative properties.

For convenience, we assume ${}^cD_{0+}^{\beta}$ as ${}^cD^{\beta}$.

**Lemma 1** ([6]) *Let $\xi_V(.)$ denote Hausdorff's measure of noncompactness and the bounded sets $E_1,\ E_2 \subset V$ (real Banach space) meet the following properties.*

*(1)* *$E_1$ is pre-compact iff $\xi_V(E_1) = 0$.*
*(2)* *$\xi_V(E_1) = \xi_V(\bar{E_1}) = \xi_V(conv\ E_1)$ where $conv\ E_1$ denote the convex hull of $E_1$.*
*(3)* *$\xi_V(E_1) \leq \xi_V(E_2)$ where $E_1 \subseteq E_2$.*
*(4)* *$\xi_V(E_1 + E_2) \leq \xi_V(E_1) + \xi_V(E_2)$ where $E_1 + E_2 = \{y + z;\ y \in E_1, z \in E_2\}$.*
*(5)* *$\xi_V(E_1 \cup E_2) \leq max\{\xi_V(E_1), \xi_V(E_2)\}$.*
*(6)* *$\xi_V(\lambda\ E_1) = |\lambda|\xi_V(E_1)$ for any $\lambda \in R$.*
*(7)* *For any $E_1 \subseteq D(T)$, Banach space $Z$ and constant $k > 0$, the condition $\xi_Z(TE_1) \leq k\xi_V(E_1)$, holds whenever the map $T : D(T) \subseteq V \to Z$ is Lipschitz continuous.*
*(8)* *$\xi_V(E_1) = inf\{dY(E_1, E_2) : E_2 \subseteq V$ be precompact$\}$ $= inf\{dY(E_1, E_2) : E_2 \subseteq V$ be finite valued$\}$ where $dY(E_1, E_2)$ indicate the non symmetric (or symmetric) Hausdorff distance between $E_1$ and $E_2$ in $V$.*
*(9)* *If $\{U_n\}_{n=1}^{+\infty}$ is a $\downarrow$ sequence of bounded closed non empty subsets of $V$ and $lim_{n\to+\infty}\xi_V(U_n) = 0$, then $\cap_{n=1}^{+\infty} U_n$ is non empty and compact in $V$.*

**Lemma 2** ([12]) *If $U \subseteq C([0, a]; Y)$ is bounded, then*

$$\xi(U(\delta)) \leq \xi_c(U),$$

*$\forall\,\delta \in [0, a]$, where $U(\delta) = \{w(\delta);\ w \in U\} \subseteq Y$. In addition, if $U$ is equicontinuous on $[0, a]$, then $\xi U(\delta)$ is continuous on $[0, a]$ and*

$$\xi_c(U) = sup\{\xi(U(\delta)), \delta \in [0, a]\}.$$

**Lemma 3** ([12]) *If* $\{w_n\}_{n=1}^{\infty} \subset L^1((a, b); Y)$ *is uniformly integrable, then* $\xi\left(\{w_n(\delta)\}_{n=1}^{\infty}\right)$ *is measurable and*

$$\xi\left(\left\{\int_a^t w_n(r)dr\right\}_{n=1}^{\infty}\right) \leq \varphi \int_a^t \xi\left(\{w_n(r)\}_{n=1}^{\infty}\right) dr,$$

*where* $\varphi = 1$ *if* $\{w_n\}$ *is equicontinuous and* $\varphi = 2$ *if* $\{w_n\}$ *is not equicontinuous.*

**Lemma 4** ([12]) *If* $U \subseteq C([0, a]; Y)$ *is bounded and equicontinuous, then* $\xi(U(r))$ *is continuous and*

$$\xi\left(\int_0^{\delta} U(r)dr\right) \leq \int_0^{\delta} \xi(U(r))dr,$$

$\forall\ \delta \in [0, a]$ *where,*

$$\int_0^{\delta} U(r)dr = \left\{\int_0^{\delta} w(r)dr : w \in U\right\}.$$

## 3 Existence Theorem

The integral equation of (1)–(3) is defined as,

$$\begin{aligned} y(\delta) &= \left[\Phi(0) + h_l(y)(0) + f\left(0, \Phi(0) + h_l(y)(0), \Phi + h_l(y)\right)\right] - f\left(\delta, y(\delta), y_\delta\right) \\ &+ \frac{1}{\Gamma(\beta)} \sum_{0<\delta_i<\delta} \int_{\delta_{i-1}}^{\delta_i} (\delta_i - r)^{\beta-1} A(r)y(r)dr + \frac{1}{\Gamma(\beta)} \int_{\delta_i}^{\delta} (\delta - r)^{\beta-1} A(r)y(r)dr \\ &+ \frac{1}{\Gamma(\beta)} \sum_{0<\delta_i<\delta} \int_{\delta_{i-1}}^{\delta_i} (\delta_i - r)^{\beta-1} \int_0^r G(r, h)g\left(h, y(h), y_h\right) dhdr \\ &+ \frac{1}{\Gamma(\beta)} \int_{\delta_i}^{\delta} (\delta - r)^{\beta-1} \int_0^r G(r, h)g\left(h, y(h), y_h\right) dhdr + \sum_{0<\delta_i<\delta} \mathcal{I}_i\left(y(\delta_i^-)\right), \end{aligned} \quad (4)$$

where $0 \leq \delta \leq a$.

From the ideology of Hausdorff's measure of noncompactness and its applications in Banach Spaces, we consider the following hypotheses:

(H1) $g : [0, a] \times Y \times C\left([-\mu, 0]; Y\right) \to Y$ satisfies the cartheodory—type condition.
i.e. $g(., y, \Phi) : [0, a] \to Y$ is measurable, $\forall\ (y, \Phi) \in Y \times C([-\mu, 0]; Y)$ and
$g(\delta, .) : Y \times C\left([-\mu, 0]; Y\right) \to Y$ is continuous, for a.e. $\delta \in [0, a]$.

(H2) $\exists$ an integrable function $\alpha : [0, a] \to [0, +\infty)$ and a non-decreasing function which is continuous and denoted by $\Omega : [0, +\infty) \to [0, +\infty) \ni$,

$$\|g(\delta, y, \Phi)\| \leq \alpha(\delta)\Omega\left(\|y\| + \|\Phi\|_{[-\mu,0]}\right)$$

$\forall\, (\delta, y, \Phi) \in [0, a] \times Y \times C([-\mu, 0]; Y)$.

(H3) $\exists$ an integrable function $\tau : [0, a] \to [0, +\infty) \ni$,

$$\xi\left(g(\delta, H_1, H_2)\right) \leq \tau(\delta)\left(\xi(H_1)\right) + sup_{-\mu\leq\theta\leq 0}\ \xi\left(H_2(\theta)\right),$$

for a.e. $\delta \in [0, a]$ and any bounded subset $H_1 \subset Y$ and $H_2 \subset C([-\mu, 0]; Y)$, where $H_2(\theta) = \{v(\theta) : v \in H_2\}$.

(H4) $\exists$ a $0 < \gamma < 1$, $\ni f$ is $Y_\gamma$—valued, $(A)^\gamma f(.)$ is continuous and $\exists\ c_1 > 0$, $c_2 > 0$ and $\mathcal{L}_f > 0 \ni$,

$$\|(A)^\gamma f(\delta, y, \Phi)\| \leq c_1\left(\|y\| + \|\Phi\|_{[-\mu,0]}\right) + c_2,$$

and

$$\left\|(A)^\gamma f(\delta, y_1, \Phi_1) - (A)^\gamma f(\delta, y_2, \Phi_2)\right\| \leq \mathcal{L}_f\left(\|y_1 - y_2\| + \|\Phi_1 - \Phi_2\|_{[-\mu,0]}\right),$$

$\forall\, \delta \in [0, a]$, $y, y_1, y_2 \in Y$ and $\Phi, \Phi_1, \Phi_2 \in C([-\mu, 0] : Y)$.

(H5) $h_l : C([0, a]; Y) \to C([-\mu, 0]; Y)$ is Lipschitz continuous satisfying the following criteria:
$\exists$ a $\mathcal{L}_h > 0 \ni$,

$$\|h_l(y_1) - h_l(y_2)\|_{[-\mu,0]} \leq \mathcal{L}_h\, \|y_1 - y_2\|_{[0,a]},$$

$\forall\, y_1, y_2 \in C([0, a], Y)$.

(H6) $h_l$ is bounded uniformly. (i.e.) $\exists$ a $N_\epsilon > 0 \ni$

$$\|h_l(y)\|_{[-\alpha,0]} \leq N_\epsilon,$$

$\forall\, y \in C([0, a]; Y)$.

(H7) $\exists$ constants $d_j > 0 \ni$,

$$\|\mathcal{I}_j(y)\| \leq d_j,\ j = 1, 2, \ldots p,$$

where $d = max\{d_j\}$, $j = 1, 2, \ldots, p$.

(H8) $\mathcal{I}_j : Y \to Y$ is continuous and $\exists$ constants $l_j \ni$,

$$\left\|\mathcal{I}_j(y_1) - \mathcal{I}_j(y_2)\right\| \leq l_j\, \|y_1 - y_2\|,$$

$j = 1, 2, \ldots p\ \forall\, y_1, y_2 \in Y$.

(H9) For each $\delta \in [0, a]$, $G(\delta, .)$ is measurable on $[0, \delta]$ and

$$G(\delta) = ess\ sup\{|G(\delta, r)| : 0 \leq r \leq \delta\}$$

is bounded on $[0, a]$,

(H10) The map $\delta \mapsto G_\delta$ is continuous from $[0, a]$ to $\mathcal{L}^\infty([0, a]; R^+)$, here $G_\delta(r) = G(\delta, r)$.

(H11) The following holds
$m\left[2c_1 + \frac{a^\beta}{\Gamma(\beta+1)}\right] + \frac{Ga^{\beta+1}}{\Gamma(\beta+1)} \int_0^a \alpha(r)dr\ lim\ inf_{k\to\infty} \frac{\Omega(2k)}{k} < 1$,
where $G = sup_{0\leq\delta\leq a} G(\delta)$.

**Theorem 1** *Consider the hypotheses $H_1 - H_{11}$ holds, then $\forall\ \Phi \in C([-\mu, 0]; Y)$, Eqs. (1)–(3) has atleast a solution provided,*

$$\mathcal{L}_0 + \frac{4akl_a}{\Gamma(\beta)} \int_0^a \tau(\delta)d\delta < 1.$$

***Proof*** Consider the mapping
$\Lambda : C([-\mu, a]; Y) \to C([-\mu, a]; Y)$ defined as $\Lambda = \Lambda_1 + \Lambda_2$, where

$$\Lambda_1 y(\delta) = \begin{cases} \Phi(\delta) + h_l(y)(\delta), & \text{for } \delta \in [-\mu, 0], \\ \Phi(0) + h_l(y)(0) + f(0, \Phi(0) + h_l(y)(0), \Phi + h_l(y)) - f(\delta, y(\delta), y_\delta) & \\ \quad + \frac{1}{\Gamma(\beta)} \sum_{0<\delta_i<\delta} \int_{\delta_{i-1}}^{\delta_i} (\delta_i - r)^{\beta-1} A(r)y(r)dr & \\ \quad + \frac{1}{\Gamma(\beta)} \int_{\delta_i}^{t} (\delta - r)^{\beta-1} A(r)y(r)dr + \sum_{0<\delta_i<\delta} \mathcal{I}_k\left(y(\delta_k^-)\right), & \text{for } \delta \in [0, a], \end{cases}$$

and

$$\Lambda_2 y(\delta) = \begin{cases} 0, & \text{for } \delta \in [-\mu, 0], \\ \frac{1}{\Gamma(\beta)} \sum_{0<\delta_i<\delta} \int_{\delta_{i-1}}^{\delta_i} (\delta_i - r)^{\beta-1} \int_0^r G(r, h)g(h, y(h), y_h)\, dhdr & \\ \quad + \frac{1}{\Gamma(\beta)} \int_{\delta_i}^{\delta} (\delta - r)^{\beta-1} \int_0^r G(r, h)g(h, y(h), y_h)\, dhdr, & \text{for } \delta \in [0, a]. \end{cases}$$

It is easy to see that $\Lambda$ is well defined in $C([-\mu, 0]; Y)$. Furthermore, $\Lambda$ is continuous by the usual methodology involving the hypotheses (H1)–(H6). We show that the fixed point of $\Lambda$ is the solution of (1)–(3).

First to prove $\Lambda(B_k) \subset B_k$, for $k \in N_\epsilon$. Contrarily suppose $\forall\ k \in N_\epsilon$, $\exists\ y^k \in B_k$ and $\delta^k \in [0, a] \ni$

$$\left\|\Lambda y^k(\delta^k)\right\| > k,$$

If $\delta^k \in [-\mu, 0]$ then,

$$\begin{aligned} k < \left\|\Lambda y^k(\delta^k)\right\| &\leq \left\|\Lambda_1 y^k(\delta^k)\right\| + \left\|\Lambda_2 y^k(\delta^k)\right\| \\ &\leq \left\|\Phi(\delta^k) + h_l(y^k)(\delta^k)\right\| \\ &\leq \|\Phi\|_{[-\mu,0]} + N_\epsilon. \end{aligned} \tag{5}$$

If $\delta^k \in [0, a]$ then,

$$k < \left\|\Lambda y^k(\delta^k)\right\| \leq \left\|\Lambda_1 y^k(\delta^k)\right\| + \left\|\Lambda_2 y^k(\delta^k)\right\|. \tag{6}$$

Consider $\left\| \Lambda_1 y^k(\delta^k) \right\|$,

$$\begin{aligned}
\left\| \Lambda_1 y^k(\delta^k) \right\| &\leq \|\Phi(0)\| + \left\| h_l(y^k)(0) \right\| + \left\| f\left(0, \Phi(0) + h_l(y^k)(0), \Phi + h_l(y^k)\right) \right\| \\
&+ \left\| g\left(\delta^k, y^k(\delta^k), y^k{}_{\delta^k}\right) \right\| \\
&+ \frac{1}{\Gamma(\beta)} \sum_{0<\delta^k{}_i<\delta^k} \int_{\delta^k{}_{i-1}}^{\delta^k{}_i} (\delta^k{}_i - r)^{\beta-1} \|A(r)\| \left\| y^k(r) \right\| dr \\
&+ \frac{1}{\Gamma(\beta)} \int_{\delta^k{}_i}^{\delta^k} (\delta^k - r)^{\beta-1} \|A(r)\| \left\| y^k(r) \right\| dr \\
&+ \sum_{0<\delta^k{}_i<\delta^k} \left\| \mathcal{I}_i \left( y^k(\delta^{k-}{}_i) \right) \right\| \\
&\leq \|\Phi(0)\| + N_\epsilon + 2m \left[ c_1 \left( \|\Phi(0)\| + N_\epsilon + k \right) + c_2 \right] \\
&+ \frac{a^\beta mk}{\Gamma(\beta+1)} + d
\end{aligned} \tag{7}$$

Similarly, consider $\left\| \Lambda_2 y^k(\delta^k) \right\|$,

$$\begin{aligned}
\left\| \Lambda_2 y^k(\delta^k) \right\| &\leq \frac{1}{\Gamma(\beta)} \sum_{0<\delta_i^k<\delta^k} \int_{\delta^k{}_{i-1}}^{\delta^k{}_i} (\delta^k{}_i - r)^{\beta-1} \int_0^r \left\| G(r,h) g\left(h, y^k(h), y_h^k\right) \right\| dh dr \\
&+ \frac{1}{\Gamma(\beta)} \int_{\delta^k{}_i}^{\delta^k} (\delta^k - r)^{\beta-1} \int_0^r \left\| G(r,h) g\left(h, y^k(h), y_h^k\right) \right\| dh dr \\
&\leq \frac{a^{\beta+1}}{\Gamma(\beta+1)} G \Omega(2k) \int_0^\delta \alpha(r) dr
\end{aligned} \tag{8}$$

Substituting (7) and (8) in (6), we get

$$\begin{aligned}
k &< \left\| \Lambda y^k(\delta^k) \right\| \\
&\leq \|\Phi(0)\| + N_\epsilon + 2m \left[ c_1 \left( \|\Phi(0)\| + N_\epsilon \right) + c_2 \right] + d \\
&+ km \left[ 2c_1 + \frac{a^\beta}{\Gamma(\beta+1)} \right] \frac{a^{(\beta+1)}}{\Gamma(\beta+1)} G \int_0^\delta \alpha(r) dr \Omega(2k).
\end{aligned} \tag{9}$$

Let the right hand side of Eq. (9) be represented as $\mathcal{L}_k$ then,

$$k < \left\| \Lambda y^k(\delta^k) \right\| \leq \mathcal{L}_k. \tag{10}$$

Hence, from (5) and (9), we obtain

$$k < max\left( \|\Phi\|_{[-\mu,0]} + N_\epsilon,\ \mathcal{L}_k \right).$$

Divide the above equation by $k$, and take $lim\ inf_{k\to\infty}$, we have

$$1 < 2mc_1 + \frac{a^{\beta}}{\Gamma(\beta+1)}m + \frac{a^{(\beta+1)}m}{\Gamma(\beta+1)}G\int_0^{\delta}\alpha(r)dr lim\ inf_{k\to\infty}\frac{\Omega(2k)}{k}$$

which contradicts $H_{11}$.

$\therefore \exists$ a $k \in N_{\epsilon}$, $\ni \Lambda(B_k) \subset B_k$. Now, we restrict $\Lambda$ on such $B_k$.

Next to justify that $\Lambda$ is a $\xi_v$—contraction.

For,

$$\begin{aligned} \|\Lambda_1 y_1(\delta) - \Lambda_1 y_2(\delta)\| &= \|h_l(y_1)(\delta) - h_l(y_2)(\delta)\| \\ &\le \mathcal{L}_h \|y_1 - y_2\|_{[0,a],} \text{ for } \delta \in [-\mu, 0], \end{aligned} \tag{11}$$

and

$$\begin{aligned} \|\Lambda_1 y_1(\delta) - \Lambda_1 y_2(\delta)\| \le\ & \mathcal{L}_h \|y_1 - y_2\|_{[0,a]} \\ & + m\mathcal{L}_f [2\mathcal{L}_h \|y_1 - y_2\|] + m\left[\mathcal{L}_f\|y_1 - y_2\| + \mathcal{L}_h\|y_1 - y_2\|\right] \\ & + 2\frac{a^{\beta}}{\Gamma(\beta+1)}m \|y_1 - y_2\| + \mathcal{L}_I \|y_1 - y_2\| \\ \le\ & \mathcal{L}_0 \|y_1 - y_2\|, \text{ for } \delta \in [0, a]. \end{aligned} \tag{12}$$

From (11) and (12) it follows that

$$\|\Lambda_1 y_1 - \Lambda_1 y_2\| \le \mathcal{L}_0 \|y_1 - y_2\|_{[-\mu,a]}.$$

$\therefore \Lambda_1$ is Lipschitzian with Lipschitz constant $\mathcal{L}_0$.

By applying Lemmas 2–4, $\forall$ bounded subset $U \subset C([-\mu, a]; Y)$ and any $\epsilon > 0$, we can consider $\{y_n\}_{n=0}^{\infty} \subset U \ni$

$$\begin{aligned} \xi_v(U) &\le 2\xi_c\left(\{y_n\}_{n=0}^{\infty}\right) + \epsilon \\ &\le \frac{4Gal_a}{\Gamma(\beta)}\xi_v(U)\int_0^a \tau(r)dr + \epsilon. \end{aligned}$$

Since $\epsilon > 0$ is arbitrary, from the theorem, we obtain

$$\begin{aligned} \xi_v(\Lambda U) &= \xi_v(\Lambda_1 U) + \xi_v(\Lambda_2 U) \\ &\le \left(\mathcal{L}_0 + \frac{4Gal_a}{\Gamma(\beta)} \in \delta_0^{\omega}\tau(r)dr\right)\xi_v(w). \end{aligned}$$

We conclude that $\Lambda$ is a $\xi_v$—contraction. Hence from the Darbo Sadovskii's fixed point theorem, any fixed point $y$ of $\Lambda$ is an integral equation of (1)–(3). □

## 4 Conclusion

In this work, using Darbo Sadovskii's fixed point theorem with Hausdorff's measures of noncompactness, the existence condition for the solution of the impulsive neutral fractional functional integro—differential equations under bounded delay with nonlocal condition in Banach spaces is studied. In future, we shall study about the stability of solution of the above problem with interval impulse condition.

## References

1. Kilbas, A.A., Srivastava, H.M., Trujillo, J.J.: Theory and Applications of Fractional Differential Equations. Elsevier Publishers (2006)
2. Anguraj, A., Lathamaheswari, M., Chang, Y.K.: Solutions of a fractional functional integrodifferential equations with interval impulses and infinite delay. Nonlinear Stud. (2015)
3. Anguraj, A., Lathamaheswari, M.: Existence of solutions for fractional impulsive neutral functional infinite delay integro-differential equations with nonlocal conditions. J. Nonlinear Sci. Its Appl. **5**(4), 271–280 (2012)
4. Bainov, D.D., Simeonov, P.S: Impulsive Differential Equations: Periodic Solutions and Applications. Longman Scientific and Technical, Harlow (1993)
5. Balachandran, K., Kiruthika, S., Trujillo, J.J.: Existence results for fractional impulsive integrodifferential equations in Banach space. Commun. Nonlinear Sci. Numer. Simulat. (2010)
6. Banas, J., Goebel, K.: Measure of noncompactness in Banach spaces. In: Lecture Notes in Pure and Applied Mathematics, vol. 60. Dekker, New York (1980)
7. Benchohra, M., Ntouyas, S.K.: Nonlocal Cauchy problems for neutral functional differential and integrodifferential inclusions. J. Math. Anal. Appl. **258**, 573–590 (2001)
8. Fan, Z.: Impulsive problems for semilinear differential equations with nonlocal conditions. Nonlinear Anal. **72**, 1104–1109 (2010)
9. Hernandez, E., Rabello, M., Henrquez, H.R.: Existence of solutions for impulsive partial neutral functional differential equations. J. Math. Anal. Appl. **331**, 1135–1158 (2007)
10. Lakshmikantham, V., Bainov, D.D., Simeonov, P.S.: Theory of Impulsive Differential Equations. World Scientific, Singapore (1989)
11. Samoilenko, A.M., Perestyuk, N.A.: Impulsive Differential Equations. World Scientific, Singapore (1995)
12. Suresh, M.L., Gunasekar, T., Paul Samuel, F.: Existence results for nonlocal impulsive neutral functional integro-differential equations. Int. J. Pure Appl. Math. **116**(23), 337–345 (2017)
13. Xue, X.: Existence of solutions for semilinear nonlocal Cauchy problems in Banach spaces. Electron. J. Differ. Eqs. **64**, 1–7 (2005)

# An Application of Conformable Fractional Differential Transform Method for Smoking Epidemic Model

G. Tamil Preethi, N. Magesh, and N. B. Gatti

**Abstract** The nonlinear differential system of tobacco smoking model is proposed and discussed in the current work by making use of conformable fractional differential transform method. The fractional approximation of Taylor's power series expansion is used to provide the result. The logical approach of the fractional order of differential transform method is discussed through stimulation technique. Maple software is used for computational tasks in order to find additional iteration.

**Keywords** Tobacco smoking model · Conformable fractional differential transform method · $\zeta$-differentiable

## 1 Introduction

Infectious disease modelling has been used to investigate the mechanisms of spread of disease, time of an epidemic and example techniques for controlling a lethal disease [12, 34]. Daniel Bernoulli, a physics specialist who devised a mathematical formula for sickness in 1760, produced the first mathematical model of disease dissemination. To protect the method of inoculating against smallpox [17], a model was developed. In the twentieth century, William Hamer [16] and Ronald Ross [26] used mass action regulation to provide an alternative to armed conflict. The rise of compartmental models was recognised in the 1920s. The epidemic models of Kermack-McKendrick [21] and Reed-Frost (1928) both elaborates the interaction prevailing among inclined, inflamed and a community of healthy people. The Kermack-McKendrick epidemiological model had a lot of success in predicting the outcome of outbreaks that were quite similar to those found in numerous historical pandemics [7, 21].

G. T. Preethi · N. Magesh (✉)
Post-Graduate and Research Department of Mathematics, Government Arts College for Men, Krishnagiri 635001, Tamilnadu, India
e-mail: nmagi_2000@yahoo.co.in

N. B. Gatti
Department of Mathematics, Government Science College, Chitradurga 577501, Karnataka, India



B. Rushi Kumar et al. (eds.), *Mathematics and Computing*, Springer Proceedings in Mathematics & Statistics 415, https://doi.org/10.1007/978-981-19-9307-7_33

Stochastic and deterministic models are two types of epidemic models. Stochastic refers to the state of being or having a random variable. A stochastic version is to predict probability distributions of capacity outcomes over time using random version in one or many inputs. The stochastic utilisation of chance changes in the threat of inhalation, disease and other contamination processes. Stochastic approaches can be used to determine statistical agent-level illness distribution for smaller or to large populations [14, 15]. Deterministic or compartmental mathematical models are commonly utilised in dealing with large populations, contingent upon tuberculosis. In a deterministic variant, peoples are divided into compartments, where each simulating a different level of the scourge. The costs of switching classes are technically defined as derivatives, hence the model is based on Differential Equations. When creating styles, it is necessary to assume that the count of population in a compartment varies with time where the epidemic technique is predictable. Alternatively, using the simplest history that was used to develop the model, the fluctuations in population of a compartment can be computed [7, 28].

The tobacco industry switched its marketing efforts to primitive and emerging countries in Africa, the former Soviet Union, Asia, the Middle East and Latin America as the cost of smoking reduced in the traditional markets of North America and Western Europe. Due to the typically shaky regulatory climate in certain nations, it was also recommended that the company target populations there [9]. If current trends continue, tobacco use will kill around 11 million people every year in some part of the world by 2020, with 70% of these fatalities occurring in developing and underdeveloped countries. According to the WHO, smoking causes 275 million deaths among children and adolescents each year, with more than 15 million people expected to die from smoking-related diseases by 2030 [34]. High blood pressure, discoloured enamel, bad breath and coughing are the most common short-term smoking adverse effects. Long-term smoking is now the leading cause of morbidity, oropharyngeal, esophageal cancer, corneal ulcer, heart disease and gum disease. As a result, smoking is a major health issue that affects people all over the world. Smoking spreads through social interaction in a similar way to many infectious diseases [5, 8, 31, 35–37]. Mathematical modelling has been widely utilised to study the effects of smoking. In 2000, Castillo-Garsow et al. began working with a basic mathematical model for tobacco use, recovery and relapse. He divided the entire population into three categories: smokers ($\mathcal{S}$), potential smokers ($\mathcal{P}$) and quitters ($\mathcal{Q}$). As a new phenomenia, many researcher are much interested in proceedings of smoking mathematical model with different classes like snuffing class, regular and irregular smokers, temporarily quit smokers and permently quit smokers, etc.

Guillaume de l'Hopital, a French Mathematician who came out with the idea of fractional calculus and his first textbook on infinitesimal calculus gives the ideas of differential calculus and its applications. His name is strongly connected with l'Hopital's rule for calculating limits involving indeterminate forms. A ridiculous idea for conformable fractional derivative was proposed by Khalil et al. [19]. The definitions and properties of conformable derivatives are illustrated in [24], where in establishing all the properties of fractional derivation, offered the chain rule definition, which was applied to the base of the series.

It was first introduced by Zhou in 1986 that a semi analytical-numerical technique called Differential Transform Method (DTM) could be used in order to solve differential equation (linear and nonlinear) problems that could be used in electric circuits. For differential equations, this approach generates an analytical polynomial solution. It is not the same as the traditional high-order Taylor series approach, which requires symbolic computations of the data functions necessary derivatives. It takes a lengthy time to compute higher orders using the Taylor series method. The DTM is a method for getting analytical Taylor series solutions to differential equations using an iterative process [23, 25, 32]. Acan provided a new conformable fractional reduced differential transform method as well as a conformable variation iteration methodology based on the novel specified fractional derivative [2]. It is apparent that more research and explanations on this conformable fractional derivative approach are possible. In this paper we analysed CFDTM for different orders of $\zeta$. For more details and basic properties one can refer [4, 6, 18, 23, 29].

The epidemiology purposed is to comprehend and to control the spread of disease (if possible). We divided the overall population into five categories in this study. At time $'t'$, number of population consist of (1) $\mathfrak{p}(t)$, smokers who might be interested (potential smokers), (2) $\mathfrak{o}(t)$, smokers who only smoke on occasion (occasional smokers), (3) $\mathfrak{s}(t)$, smokers, (4) $\mathfrak{q}(t)$, smokers who have temporarily stopped smoking (temporarily quit smokers), (5) $\mathfrak{l}(t)$, smokers who have made a permanent decision to stop smoking (permanently quit smokers). The suggested smoking model is proposed in [3, 33] with initial conditions

$$
\begin{aligned}
\frac{d\,\mathfrak{p}(t)}{dt} &= \chi - \alpha\, \mathfrak{p}(t)\, \mathfrak{s}(t) - \vartheta\, \mathfrak{p}(t), \\
\frac{d\,\mathfrak{o}(t)}{dt} &= \alpha\, \mathfrak{p}(t)\, \mathfrak{s}(t) - (\lambda_1 + \vartheta)\, \mathfrak{o}(t), \\
\frac{d\,\mathfrak{s}(t)}{dt} &= \lambda_1\, \mathfrak{o}(t) + \lambda_2\, \mathfrak{s}(t)\, \mathfrak{q}(t) - (\vartheta + \mu)\, \mathfrak{s}(t), \\
\frac{d\,\mathfrak{q}(t)}{dt} &= -\lambda_2\, \mathfrak{s}(t)\, \mathfrak{q}(t) - \vartheta\, \mathfrak{q}(t) + \mu(1 - \nu)\, \mathfrak{s}(t), \\
\frac{d\,\mathfrak{l}(t)}{dt} &= \nu\, \mu\, \mathfrak{s}(t) - \vartheta\, \mathfrak{l}(t).
\end{aligned}
\tag{1}
$$

## 2 Conformable Fractional Differential Transform Method

We describe the fundamental definition and properties of the conformable fractional one-dimensional differential transform method incorporated and analysed by Ünal and Gökdoğan, H. M. Srivastava and Hatira Günerhan in order to expand the analytical and continuous function $f(t)$ in terms of a fractional power series [30, 32].

**Definition 1** ([30, 32]) If we assume that $f(t)$ is an infinitely $\zeta-$differentiable function for some $\zeta \in (0,\ 1]$, then the conformable fractional differential transform of $f(t)$ is

$$F_\zeta(p) = \frac{1}{\zeta^p p!}\left[\left(T_\zeta^{t_0} f\right)^{(p)}(t)\right]_{t=t_0}, \tag{2}$$

where $\left(T_\zeta^{t_0} f\right)^{(p)}(t)$ signifies the fractional derivative's $p$th iteration $\left(T_\zeta^{t_0} f\right)(t)$ for a function $f\ :\ (t_0,\ \infty) \to \mathbb{R}$ given by

$$\left(T_\zeta^{t_0} f\right)(t) := \lim_{\varepsilon \to 0}\left\{\frac{f\left(t + \varepsilon\,(t - t_0)^{1-\zeta}\right) - f(t)}{\varepsilon}\right\}, \quad t > t_0 \geq 0; \quad 0 < \zeta \leq 1. \tag{3}$$

**Definition 2** ([30, 32]) If $F_\zeta(p)$ indicates the conformable fractional differential transform of the function $f(t)$ established by Definition 1, then $F_\zeta(p)$'s inverse fractional differential transform is characterised by

$$f(t) = \sum_{p=0}^{\infty} F_\zeta(p)(t - t_0)^{\zeta\,p} = \sum_{p=0}^{\infty} \frac{1}{\zeta^p p!}\left[\left(T_\zeta^{t_0} f\right)^{(p)}(t)\right]_{t=t_0}(t - t_0)^{\zeta\,p}. \tag{4}$$

Definition 3 follows from the application of Definitions 1 and 2.

**Definition 3** ([30, 32]) For integer-order derivatives, the conformable fractional differential transform (CFDT) of the initial conditions is defined as follows:

$$F_\zeta(p) = \begin{cases} \frac{1}{(\zeta\,p)!}\left[\frac{d^{\zeta p}\{f(t)\}}{dt^{\zeta p}}\right]_{t=t_0}; & \zeta p \in \mathbb{N} \\ 0; & \zeta p \notin \mathbb{N}, \end{cases} \quad \text{for} \quad p = 0,\ 1, \ldots,\ \left[\frac{n}{\zeta}\right] - 1, \tag{5}$$

where $n$ is the order of the associated fractional differential equation and $\mathbb{N}$ is the set of positive integers. For properties one can refer [23, 24, 32].

## 3 Conformable Fractional Tobacco Smoking Model Mathematical Modelling

The following conformable fractional differential system provides a conformable fractional model of the genuine development of an infectious outbreak in a large population.

$$\begin{cases} T_\zeta\ \mathfrak{p}(t) = \chi - \alpha\ \mathfrak{p}(t)\ \mathfrak{s}(t) - \vartheta\ \mathfrak{p}(t), \\ T_\zeta\ \mathfrak{o}(t) = \alpha\ \mathfrak{p}(t)\ \mathfrak{s}(t) - (\lambda_1 + \vartheta)\ \mathfrak{o}(t), \\ T_\zeta\ \mathfrak{s}(t) = \lambda_1\ \mathfrak{o}(t) + \lambda_2\ \mathfrak{s}(t)\ \mathfrak{q}(t) - (\vartheta + \mu)\ \mathfrak{s}(t), \\ T_\zeta\ \mathfrak{q}(t) = -\lambda_2\ \mathfrak{s}(t)\ \mathfrak{q}(t) - \vartheta\ \mathfrak{q}(t) + \mu\ (1 - \nu)\ \mathfrak{s}(t), \\ T_\zeta\ \mathfrak{l}(t) = \ \nu\ \mu\ \mathfrak{s}(t) - \vartheta\ \mathfrak{l}(t), \end{cases} \tag{6}$$

with the initial conditions $\mathfrak{p}(0) = \mathfrak{p}_0$, $\mathfrak{o}(0) = \mathfrak{o}_0$, $\mathfrak{s}(0) = \mathfrak{s}_0$, $\mathfrak{q}(0) = \mathfrak{q}_0$, $\mathfrak{l}(0) = \mathfrak{l}_0$.

## 3.1 Application of Conformable Fractional Differential Transform Method

By using the properties of discussed in Sect. 2, Eq. (6) can be revised as follows:

$$\begin{aligned} &\zeta(p+1)\mathcal{P}_\zeta(p+1) = \chi\ \delta(p) - \alpha \sum_{r=0}^{p} \mathcal{P}_\zeta(r)\ \mathcal{S}_\zeta(p-r) - \vartheta\ \mathcal{P}_\zeta(p), \\ &\zeta(p+1)\mathcal{O}_\zeta(p+1) = \alpha \sum_{r=0}^{p} \mathcal{S}_\zeta(r)\ \mathcal{P}_\zeta(p-r) - (\lambda_1 + \vartheta)\ \mathcal{O}_\zeta(p), \\ &\zeta(p+1)\mathcal{S}_\zeta(p+1) = \ \lambda_1\ \mathcal{O}_\zeta(p) + \lambda_2 \sum_{r=0}^{p} \mathcal{S}_\zeta(r)\ \mathcal{Q}_\zeta(p-r) - (\vartheta + \mu)\ \mathcal{S}_\zeta(p), \\ &\zeta(p+1)\mathcal{Q}_\zeta(p+1) = -\lambda_2 \sum_{r=0}^{p} \mathcal{Q}_\zeta(r)\ \mathcal{S}_\zeta(p-r) - \vartheta\ \mathcal{Q}_\zeta(p) + \ \mu\ (1-\nu)\ \mathcal{S}_\zeta(p), \\ &\zeta(p+1)\mathcal{L}_\zeta(p+1) = \ \nu\ \mu\ \mathcal{S}_\zeta(p) - \vartheta\ \mathcal{L}_\zeta(p). \end{aligned} \tag{7}$$

With commencing values $\mathfrak{p}_{(0)} = \mathcal{P}(0) = 40$, $\mathfrak{o}_{(0)} = \mathcal{O}(0) = 10$, $\mathfrak{s}_{(0)} = \mathcal{S}(0) = 20$, $\mathfrak{q}_{(0)} = \mathcal{Q}(0) = 10$, $\mathfrak{l}_{(0)} = \mathcal{L}(0) = 5$ and parameters values from the above table applying in (7), we obtain the series for the classical order $\zeta = 1$, upto certain order by inverse DTM as follows (Table 1):

$$\begin{aligned} &\mathfrak{p}\ (t) = \sum_{p=0}^{\infty} \mathcal{S}_\zeta(p) t^{p\zeta} = 40 - 113.0\,t + 207.1690000\,t^2 + \cdots \\ &\mathfrak{o}\ (t) = \sum_{p=0}^{\infty} \mathcal{O}_\zeta(p) t^{p\zeta} = 10 + 111.4800000\,t - 207.2424800\,t^2 + \cdots \\ &\mathfrak{s}\ (t) = \sum_{p=0}^{\infty} \mathcal{S}_\zeta(p) t^{p\zeta} = 20 - 16.48000000\,t + 7.284480000\,t^2 + \cdots \\ &\mathfrak{q}\ (t) = \sum_{p=0}^{\infty} \mathcal{Q}_\zeta(p) t^{p\zeta} = 10 + 15.0\,t - 7.136000000\,t^2 + \cdots \\ &\mathfrak{l}\ (t) = \sum_{p=0}^{\infty} \mathcal{L}_\zeta(p) t^{p\zeta} = 5 + 1.850000000\,t - 0.6129500000\,t^2 + \cdots \end{aligned} \tag{8}$$

**Table 1** Parameter values and its descriptions

| Parameters | Descriptions | Values |
|---|---|---|
| $\chi$ | The pace of recruitment in $\mathfrak{s}$ | 1 |
| $\alpha$ | Contact rate between $\mathfrak{s}$ and $\mathfrak{p}$ | 0.14 |
| $\vartheta$ | Death rate due to natural causes | 0.05 |
| $\lambda_1$ | Percentage of occasional change to regular smokers | 0.002 |
| $\lambda_2$ | Smokers and temporary quitters relapse to smoking with higher contact rate | 0.0025 |
| $\mu$ | Quitting smokers rate | 0.8 |
| $\nu$ | A small percentage of smokers who has given up smoking for good | 0.1 |

## 4 Result and Discussion

An important and productive way to recognise epidemological problems is by graphical methods (Figs. 1, 2, 3, 4, 5 Tables 2, 3, 3, 4, 5, 6).

As the time increases the potential smokers learn the habit of smoking very fastly where an occasional smokers initially feel hard to quit smoking but gradually, they try to reduce their smoking habit. As a third part, smokers who have the habit of smoking will have the passion of smoking, sometimes they try to quit smoking but they actually

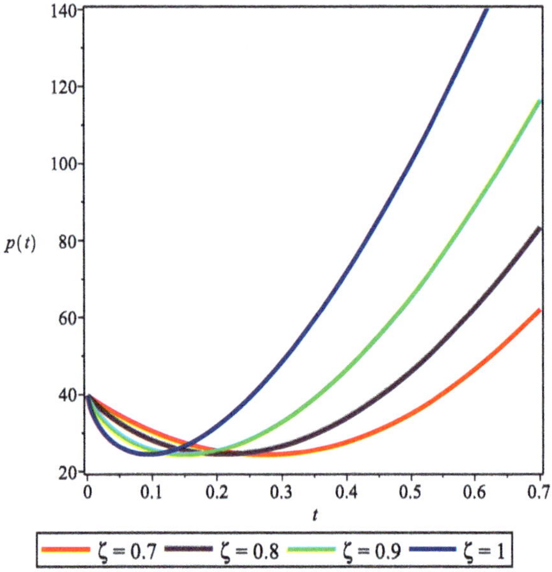


**Fig. 1** Plots of CFDTM for potential smokers $p(t)$ at different values of $\zeta$

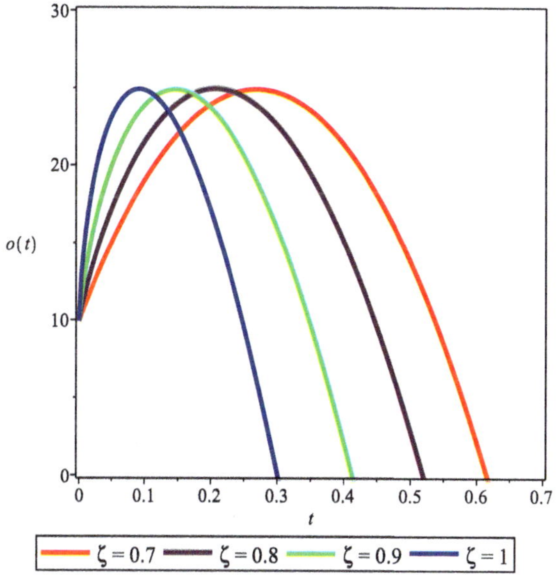


**Fig. 2** Plots of CFDTM for occasional smokers $o(t)$ at different values of order $\zeta$

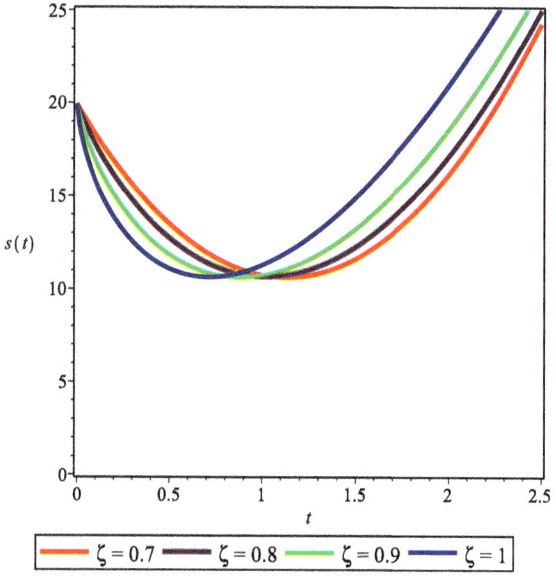


**Fig. 3** Plots by CFDTM for $s(t)$ for different values of order $\zeta$

**Fig. 4** Plots by CFDTM for $q(t)$ for different values of order $\zeta$

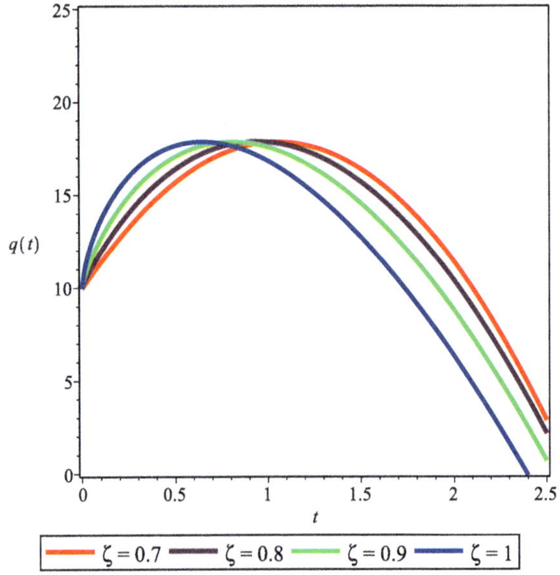


**Fig. 5** Plots of CFDTM solution for quit smokers $l(t)$ at different values of order $\zeta$

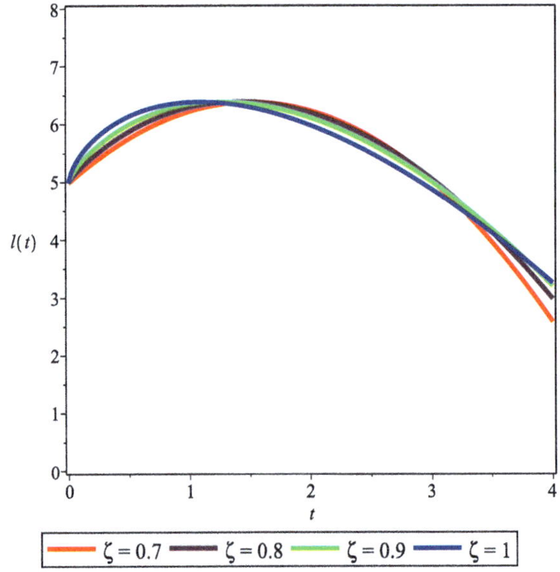

**Table 2** Numerical solutions by CFDTM for potential smokers $p(t)$ for different values of order $\zeta$

| $p(t)$ | $\zeta = 0.8$ | $\zeta = 0.9$ | $\zeta = 1$ |
|---|---|---|---|
| $t = 0$ | 40 | 40 | 40 |
| $t = 0.1$ | 25.74439 | 28.24708 | 30.77169 |
| $t = 0.2$ | 25.67126 | 24.61943 | 25.68676 |
| $t = 0.3$ | 33.24419 | 26.79993 | 24.74521 |
| $t = 0.4$ | 46.85714 | 34.11128 | 27.94704 |
| $t = 0.5$ | 65.65487 | 46.16542 | 35.29225 |
| $t = 0.6$ | 89.08462 | 62.69775 | 46.78084 |
| $t = 0.7$ | 116.75113 | 83.51082 | 62.41281 |
| $t = 0.8$ | 148.35345 | 108.44879 | 82.18816 |
| $t = 0.9$ | 183.65234 | 137.38393 | 106.10689 |
| $t = 1.0$ | 222.45156 | 170.20864 | 134.16900 |

**Table 3** Numerical solutions by CFDTM for occasional smokers $o(t)$ at different values of order $\zeta$

| $o(t)$ | $\zeta = 0.8$ | $\zeta = 0.9$ | $\zeta = 1$ |
|---|---|---|---|
| $t = 0$ | 10 | 10 | 10 |
| $t = 0.05$ | 20.00159 | 17.19205 | 15.05589 |
| $t = 0.10$ | 23.95158 | 21.53886 | 19.07557 |
| $t = 0.15$ | 24.98654 | 24.04824 | 22.05904 |
| $t = 0.20$ | 23.79569 | 24.97880 | 24.00630 |
| $t = 0.25$ | 20.73104 | 24.47119 | 24.91734 |
| $t = 0.30$ | 16.01388 | 22.61818 | 24.79217 |
| $t = 0.35$ | 9.79929 | 19.48730 | 23.63079 |
| $t = 0.40$ | 4.20349 | 15.13090 | 21.43320 |
| $t = 0.45$ | 1.68228 | 9.59122 | 18.19939 |
| $t = 0.50$ | 0.28400 | 2.90347 | 13.92938 |

don't. People those who temporarily stopped smoking have more chances of quitting the habit of smoking permanently. It can be seen that the result of the epidemic system of Eq. (6) fully agrees with the result obtained by the conformable fractional differential transform method by tabular values and in given figures for different fractional order $\zeta$.

**Table 4** Numerical solutions by CFDTM for $s(t)$ for different values of order $\zeta$

| $s(t)$ | $\zeta = 0.8$ | $\zeta = 0.9$ | $\zeta = 1$ |
|---|---|---|---|
| $t = 0$ | 20 | 20 | 20 |
| $t = 0.1$ | 17.02102 | 17.83730 | 18.42484 |
| $t = 0.2$ | 15.18220 | 16.19461 | 16.99537 |
| $t = 0.3$ | 13.79554 | 14.83357 | 15.71160 |
| $t = 0.4$ | 12.73006 | 13.70102 | 14.57351 |
| $t = 0.5$ | 11.92306 | 12.76993 | 13.58112 |
| $t = 0.6$ | 11.33692 | 12.02332 | 12.73441 |
| $t = 0.7$ | 10.94620 | 11.44928 | 12.03339 |
| $t = 0.8$ | 10.73243 | 11.03887 | 11.47806 |
| $t = 0.9$ | 10.68144 | 10.78505 | 11.06842 |
| $t = 1.0$ | 10.78200 | 10.68207 | 10.80448 |

**Table 5** Numerical solutions by CFDTM for $q(t)$ for different values of order $\zeta$

| $q(t)$ | $\zeta = 0.8$ | $\zeta = 0.9$ | $\zeta = 1$ |
|---|---|---|---|
| $t = 0$ | 10 | 10 | 10 |
| $t = 0.1$ | 12.69159 | 11.95858 | 11.42864 |
| $t = 0.2$ | 14.32495 | 13.42918 | 12.71456 |
| $t = 0.3$ | 15.53215 | 14.63096 | 13.85776 |
| $t = 0.4$ | 16.43465 | 15.61330 | 14.85824 |
| $t = 0.5$ | 17.09091 | 16.40147 | 15.71600 |
| $t = 0.6$ | 17.53613 | 17.01138 | 16.43104 |
| $t = 0.7$ | 17.79415 | 17.45426 | 17.00336 |
| $t = 0.8$ | 17.88236 | 17.73855 | 17.43296 |
| $t = 0.9$ | 17.81410 | 17.87091 | 17.71984 |
| $t = 1.0$ | 17.60000 | 17.85679 | 17.86400 |

## 5 Conclusion

We wish to discuss the fractional order equations which are more approximate for biological modelling. The proposed model depends on time and rapid growth of smoking habits among populations. In the present work, we calculated the estimated solutions by an analytical method called conformable fractional differential transform method for the current trending epidemiological smoking model. We illustrated the competence and validity of the suggested model based on smoking. The obtained results are shown as fractional order series. The importance and logical approach of the fractional order differential transform method is explained by the tabular values and pictorial depiction. The manifest of considered model shows the productivity and well-structured behaviour of fractional order. Following our results, the current

**Table 6** Numerical solutions by CFDTM solution for quit smokers $l(t)$ at different values of order $\varsigma$

| $l(t)$ | $\varsigma = 0.8$ | $\varsigma = 0.9$ | $\varsigma = 1$ |
|---|---|---|---|
| $t = 0$ | 5 | 5 | 5 |
| $t = 0.1$ | 5.34244 | 5.24678 | 5.17887 |
| $t = 0.2$ | 5.56519 | 5.44113 | 5.34548 |
| $t = 0.3$ | 5.74310 | 5.60891 | 5.49983 |
| $t = 0.4$ | 5.88996 | 5.75569 | 5.64192 |
| $t = 0.5$ | 6.01224 | 5.88423 | 5.77176 |
| $t = 0.6$ | 6.11380 | 5.99624 | 5.88933 |
| $t = 0.7$ | 6.19718 | 6.09292 | 5.99465 |
| $t = 0.8$ | 6.26425 | 6.17514 | 6.08771 |
| $t = 0.9$ | 6.31641 | 6.24359 | 6.16851 |
| $t = 1.0$ | 6.35476 | 6.29882 | 6.23705 |

situation will continue for approximately years together. Our estimates give a number of the order may increase, but to bring under control there are many precautions and rehabilitation centres to overcome the smoking habits. Our graphical representation gives us the clear picture that smoking habit can gradually decrease.

## References


1. Abdeljawad, T.: On conformable fractional calculus. J. Comput. Appl. Math. **279**, 57–66 (2015)
2. Acan, O., Firat, O., Keskin, Y.: Conformable variational iteration method, conformable fractional reduced differential transform method and conformable homotopy analysis method for non-linear fractional partial differential equations. Waves Random Complex Media **30**(2), 250–268 (2020)
3. Abdullah, M., Ahmad, A., Raza, N., Farman, M., Ahmad, M.O.: Approximate solution and analysis of smoking epidemic model with Caputo fractional derivatives. Int. J. Appl. Comput. Math. **4**(112), 1–16 (2018)
4. Alkhudhari, Z., Al-Sheikh, S., Al-Tuwairqi, S.: Global dynamics of mathematical model on smoking. ISRN Appl. Math. **847075**, 1–7 (2014)
5. Alrabaiah, H., Zeb, A., Alzahrani, E., Shah, K.: Dynamical analysis of fractional-order tobacco smoking model containing snuffing class. Alex. Eng. J. **60**, 3669–3678 (2021)
6. Alzahrani, E., Zeb, A.: Stability analysis and prevention strategies of tobacco smoking model. Bound. Value Probl. **3**, 1–13 (2020)
7. Brauer, F., Castillo-Chavez, C.: Mathematical Models in Population Biology and Epidemiology. Springer, New York (2001)
8. Brownlee, J.: Certain considerations on the causation and course of epidemics. Proc. R. Soc. Med. **2**, 243–258 (1909)
9. Brownlee, J.: The mathematical theory of random migration and epidemic distribution. Proc. R. Soc. Edinb. **31**, 262–289 (1912)
10. Castillo-Garsow, C., Jordan-Salivia, G., Rodriguez Herrera, A.: Mathematical models for dynamics of tobacco use, recovery and relapse. Technical Report Series BU-1505-M, Cornell University (2000)

11. Choi, H., Jung, I., Kang, Y.: Giving up smoking dynamic on adolescent nicotine dependence a mathematical modeling approach. In: KSIAM: Spring Conference, p. 2011. Daejeon, Korea (2011)
12. Daley, D.J., Gani, J.: Epidemic Modeling: An Introduction. Cambridge University Press, New York (2005)
13. Ertürk, V.S., Zaman, G., Momani, S.: A numeric-analytic method for approximating a giving up smoking model containing fractional derivatives. Comput. Math. Appl. **64**(10), 3065–3074 (2012)
14. Nakamura, G.M., Monteiro, A.C.P., Cardoso, G.C., Martinez, A.S.: Efficient method for comprehensive computation of agent-level epidemic dissemination in networks. Sci. Rep. **7**(1), 40885 (2017)
15. Nakamura, G.M., Cardoso, G.C., Martinez, A.S.: Improved susceptible-infectious-susceptible epidemic equations based on uncertainties and autocorrelation functions. R. Soc. Open Sci. **7**(2), 191504 (2020)
16. Hamer, W.: Epidemiology Old and New. Kegan Paul, London (1928)
17. Hethcote, H.W.: The mathematics of infectious diseases. Soc. Ind. Appl. Math. **42**, 599–653 (2000)
18. Christopher, A.J., Prakash, A., Magesh, N., Tamil Preethi, G.: Techniques, Certain Efficient, to solve the unreported cases of 2019—nCoV epidemic model. Italin. J. Pure App. Math. 2021 (2019)
19. Khalil, R., Horani, M.A., Yousef, A., Sababheh, M.: A new definition of fractional derivative. J. Comput. Appl. Math. **264**, 65–70 (2014)
20. Kendall, D.G.: Deterministic and stochastic epidemics in closed populations. In: Proceedings of the Third Berkeley Symposium on Mathematical Statistics and Probability, Contributions to Biology and Problems of Health, vol. 4, pp. 149–165 (1956)
21. Kermack, W.O., McKendrick, A.G.: Contributions to the mathematical theory of epidemics, part 1. Proc. R. Soc. Edinb., Sect. A, Math. **115**, 700–721 (1927)
22. Ham, O.K.: Stages and processes of smoking cessation among adolescents. West. J. Nursing Res. **29**(3), 301–315 (2007)
23. Hatami, M., Ganji, D.D., Sheikholeslami, M.: Differential Transformation Method for Mechanical Engineering. Academic Press is an imprint of Elsevier (2016)
24. Mamat, M., Syouri, S., Alghrouz, I.M., Sulaiman, I.M., Sufahani, S.F.: Conformable fractional differential transform method for solving fractional derivatives. Int. J. Adv. Sci. Technol. **29**(7), 1734–1743 (2020)
25. Odibat, Z.M., Bertelle, C., Aziz-Alaoui, M.A., Duchamp, G.H.: A multi-step differential transform method and application to non-chaotic or chaotic systems. Comput. Math. Appl. **59**, 1462–1472 (2010)
26. Ross, R.: The Prevention of Malaria. Dutton (1910)
27. Ross, R., Hudson, H.: An application of the theory of probabilities to the study of a priori pathometry. -Part II. Proc. R. Soc. London. Ser. A, Contain. Pap. Math. Phys. Character **93**, 212–225 (1917)
28. Sharomi, O., Gumel, A.B.: Curtailing smoking dynamics: a mathematical modeling approach. Appl. Math. Comput. **195**(2), 475–499 (2008)
29. Singh, J., Kumar, D., Qurashi, M.A., Baleanu, D.: A new fractional model for giving up smoking dynamics. Adv. Differ. Eqs. **88**, 1–16 (2017)
30. Srivastava, H.M., Günerhan, H.: Analytical and approximate solutions of fractional-order susceptible-infected-recovered epidemic model of childhood disease. Math. Meth. Appl. Sci. 1–7 (2018)
31. Swartz, J.B.: Use of a multistage model to predict time trends in smoking induced lung cancer. J. Epidemiol. Community Health **46**(3), 311–315 (1992)
32. Ünal, E., Gökdoğan, A.: Solution of conformable fractional ordinary differential equations via differential transform method. Optik Internat. J. Light Electron Opt. **218**, 264–273 (2017)
33. Veeresha, P., Prakasha, D.G., Baskonus, H.M.: Solving smoking epidemic model of fractional order using a modefied homotopy analysis transform method. Math. Sci. **13**, 115–128 (2019)

34. World Health Organization: "Prevention from Smoking". 26 Jan 2013. https://www.who.int/news-room/fact-sheets/detail/tobacco
35. Zaman, G.: Qualitative behavior of giving up smoking models. Bull. Malays. Math. Sci. Soc. **34**(2), 403–415 (2011)
36. Zeb, A., Zaman, G., Momani, S.: Square-root dynamics of a giving up smoking model. Appl. Math. Model. **37**(7), 5326–5334 (2013)
37. Zeb, A., Zaman, G., Jung, I.H., Khan, M.: Optimal campaign strategies in fractional-order smoking dynamics. Zeitschrift fur Naturforschung A **69**(5–6), 225–231 (2014)

# Solvability of Infinite System of Volterra Integral Equations in the Tempered Spaces

**Rahul and N. K. Mahato**

**Abstract** In this paper, we discuss the existence solution of an infinite system of Volterra integral equations with $n$-variables in the tempered sequence space, using Hausdorff measure of noncompactness through Meir-Keeler condensing operator. At the end, we have provided a suitable example to verify obtained result.

**Keywords** Measure of noncompactness (MNC) · Hausdorff MNC · Condensing operators · Tempered space · Fixed point

## 1 Introduction

The initiality of MNC was done by Kuratowsi [1] in 1930. After that researchers resolved various type of integral or differential equations, using MNC and Meir-Keeler fixed point theory. In literature different type of MNC is defined in metric and topological space, for example Hausdorff MNC, Kuratowski MNC, Istratescu MNC (see [2]). The Hausdorff MNC was first introduced by Goldstein et al. [3] in 1957 and further studied was done by Goldenstein and Markus [4].

Recently, Banas and Krajewska [5] have introduced a new sequence space called tempered sequence space. These tempered sequence spaces are constructed from the known classical spaces with the help of a tempering sequence. For example, if we take the classical space $l_p$ and the tempering sequence $\beta_n$, then the new sequence space $\ell_p^\beta$ is understood as the space of all sequences $(x_n)$ such that the sequence $(\beta_n x_n)$ is in $l_p$. It is worthwhile to mention that the initial sequence $(x_n)$ may or may not be in $l_p$, but after tempering the new sequence $(\beta_n x_n)$ is in $l_p$.

Rahul · N. K. Mahato (✉)
Department of Mathematics, Indian Institute of Information Technology, Design, and Manufacturing, Jabalpur, India
e-mail: nihar@iiitdmj.ac.in

Rahul
e-mail: 1825602@iiitdmj.ac.in



B. Rushi Kumar et al. (eds.), *Mathematics and Computing*, Springer Proceedings in Mathematics & Statistics 415, https://doi.org/10.1007/978-981-19-9307-7_34

After that Das and Hazarika [6] proved MNC in the tempered sequences and has application on infinite systems of fractional differential equations. Recently, Reza et al. [7] proved solutions of infinite systems of integral equations ($\mathbb{IE}$) in $n$-variables in the tempered sequence spaces $c_0^{\beta}$ and $\ell_1^{\beta}$. So, motivated by Reza et al. [7], in this work, we have studied the solution of infinite system of Volterra $\mathbb{IE}$ in $n$-variables in the space of tempered sequence $\ell_p^{\beta}$, for $1 < p < \infty$ by using MNC and Meir Keeler condensing operator.

## 2 Preliminaries and Definitions

In this paper, we have used these notations, definitions, and preliminaries facts.

Let $(\mathbb{E}, \|\ .\ \|)$ be real Banach space and $\bar{\mathbb{B}}(\nu_0, \rho)$ be the closed ball with centered $\nu_0$ and radius $\rho$. Let $\bar{\mathbb{X}}$ and Conv$\mathbb{X}$ are the closure and convex closure of $\mathbb{X}$, for any nonempty subset $\mathbb{X}$ of $\mathbb{E}$. Also, let $\mathbb{M}_{\mathbb{E}}$ is the set of all nonempty and bounded subsets of $\mathbb{E}$ and $\mathbb{N}_{\mathbb{E}}$ is subsets of $\mathbb{M}_{\mathbb{E}}$ having all relatively compact sets. The definition of MNC was introduced by Banas and Lecko [10].

**Definition 1** A function $\Pi : \mathbb{M}_{\mathbb{E}} \to [0, \infty)$ is called a MNC if it satisfies:

($\Pi_1$) The family $\ker \Pi = \{\mathbb{X} \in \mathbb{M}_{\mathbb{E}} : \Pi(\mathbb{X}) = 0\} \neq \phi$ and $\ker \Pi \subset \mathbb{N}_{\mathbb{E}}$.
($\Pi_2$) $\mathbb{X} \subset \mathbb{Y} \implies \Pi(\mathbb{X}) \leq \Pi(\mathbb{Y})$.
($\Pi_3$) $\Pi(\bar{\mathbb{X}}) = \Pi(\mathbb{X})$.
($\Pi_4$) $\Pi(Conv\mathbb{X}) = \Pi(\mathbb{X})$.
($\Pi_5$) $\Pi(\lambda\mathbb{X} + (1-\lambda)\mathbb{Y}) \leq \lambda\Pi(\mathbb{X}) + (1-\lambda)\Pi(\mathbb{Y})$ for $\lambda \in [0, 1]$.
($\Pi_6$) If $\mathbb{X}_n \in \mathbb{M}_{\mathbb{E}}$, $\mathbb{X}_n = \bar{\mathbb{X}}_n$, $\mathbb{X}_{n+1} \subset \mathbb{X}_n$ for $n = 1, 2, 3, \ldots$ and $\lim_{n\to\infty} \Pi(\mathbb{X}_n) = 0$, then $\bigcap_{n=1}^{\infty} \mathbb{X}_n \neq \emptyset$.

**Definition 2** ([8]) A mapping $\mathbb{S}$ on $\mathbb{X}$ is called a Meir-Keeler contraction if for given $\epsilon > 0$, we can find $\delta > 0$ in such way that

$$\epsilon \leq d(u, v) < \epsilon + \delta \implies d(\mathbb{S}u, \mathbb{S}v) < \epsilon\ \forall u, v \in \mathbb{X}.$$

**Definition 3** ([9]) An operator $\mathbb{S} : \mathbb{X} \to \mathbb{X}$ is a Meir-Keeler condensing (MKC) operator if for given $\epsilon > 0$, search $\delta > 0$ s. t.

$$\epsilon \leq \Pi(\mathbb{X}) < \epsilon + \delta \implies \Pi(\mathbb{S}(\mathbb{X})) < \epsilon,$$

for every nonempty, bounded subset $\mathbb{X}$ of $\mathbb{E}$.

**Theorem 1** ([9]) *A continuous MKC operator $\mathbb{S} : \mathbb{C} \to \mathbb{C}$ has at least one $\mathbb{FP}$ for any nonempty, bounded, closed, and convex (NBCC) subset of $\mathbb{E}$.*

## 3 Hausdorff MNC in Tempered Space

The norm on $\ell_p^\beta$ for $1 < p < \infty$ is define as

$$\| h \|_{\ell_p^\beta}= \left(\sum_{i=1}^{\infty} \beta_i |h_i|^p\right)^{\frac{1}{p}}.$$

The Hausdorff MNC $\chi$ is defined as

$$\chi_{\ell_p^\beta}(\mathbb{D}) = \lim_{n\to\infty}\left[\sup_{h(\sigma_1,\sigma_2,\ldots,\sigma_n)\in\mathbb{D}}\left(\sum_{k\geq n}\beta_k|h_k(\sigma_1,\sigma_2,\ldots,\sigma_n)|^p\right)^{\frac{1}{p}}\right],$$

where $h(\sigma_1,\sigma_2,\ldots,\sigma_n) = (h_i(\sigma_1,\sigma_2,\ldots,\sigma_n))_{i=1}^{\infty} \in \ell_p^\beta$ for each $(\sigma_1,\sigma_2,\ldots,\sigma_n) \in \mathbb{R}_+^n$ and $\mathbb{D} \in \mathbb{M}_{\ell_p^\beta}$.

Consider the following infinite system of $\mathbb{IE}$

$$\begin{aligned} &h_n(\sigma_1,\ldots,\sigma_n)\\ &= f_n\Bigg(\sigma_1,\ldots,\sigma_n, \int_0^{a_1(\sigma_1)},\ldots,\int_0^{a_n(\sigma_n)} g_n(\sigma_1,\ldots,\sigma_n,\tau_1,\ldots,\tau_n,h(\tau_1,\ldots,\tau_n))\,d\tau_1,\ldots,d\tau_n,\\ &\qquad h(\sigma_1,\ldots,\sigma_n)\Bigg), \end{aligned} \tag{1}$$

where $h(\sigma_1,\sigma_2,\ldots,\sigma_n) = (h_i(\sigma_1,\sigma_2,\ldots,\sigma_n))_{i=1}^{\infty}$, $(\sigma_1,\sigma_2,\ldots,\sigma_n) \in \mathbb{R}_+^n$, $n \in \mathbb{N}$.

## 4 Solvability of the Infinite System of Volterra $\mathbb{IE}$ in the $\ell_p^\beta$ Space

We begin with the following assumptions to establish the solvability of (1) under the following assumptions:

(a) $a_1, a_2, \ldots, a_n : \mathbb{R}_+ \to \mathbb{R}_+$ are continuous functions.
(b) $f_n : \mathbb{R}_+^n \times \mathbb{R} \times \ell_p^\beta \to \mathbb{R}$ $(n \in \mathbb{N})$ are continuous functions along with

$$\sum_{n\geq 1}\beta_n\left|f_n\left(\sigma_1,\sigma_2,\ldots,\sigma_n,0,h^0(\sigma_1,\sigma_2,\ldots,\sigma_n)\right)\right|^p \to 0,$$

for any $\sigma_1,\sigma_2,\ldots,\sigma_n \in \mathbb{R}_+$ and $h^0(\sigma_1,\sigma_2,\ldots,\sigma_n) = \left(h_n^0(\sigma_1,\sigma_2,\ldots,\sigma_n)\right)_{n=1}^{\infty}$ $\in \ell_p^\beta$, where $h_n^0(\sigma_1,\sigma_2,\ldots,\sigma_n)=0$ $\forall n \in \mathbb{N}$. Also, $\exists\, \alpha_n, \gamma_n : \mathbb{R}_+^n \to \mathbb{R}_+$ $(n \in \mathbb{N})$ are continuous functions such that

$$\Bigg\{\Bigg|f_n\Big(\sigma_1,\ldots,\sigma_n,p(\sigma_1,\ldots,\sigma_n),h(\sigma_1,\ldots,\sigma_n)\Big)$$

$$-f_n\Big(\sigma_1,\ldots,\sigma_n,q(\sigma_1,\ldots,\sigma_n),\bar{h}(\sigma_1,\ldots,\sigma_n)\Big)\Bigg|^p\Bigg\}^{\frac{1}{p}}$$

$$\leq\Big(\alpha_n(\sigma_1,\ldots,\sigma_n)\Big|h_n(\sigma_1,\ldots,\sigma_n)-\bar{h}_n(\sigma_1,\ldots,\sigma_n)\Big|^p\Big)^{\frac{1}{p}}$$

$$+\Big(\gamma_n(\sigma_1,\ldots,\sigma_n)\Big|p(\sigma_1,\ldots,\sigma_n)-q(\sigma_1,\ldots,\sigma_n)\Big|^p\Big)^{\frac{1}{p}},$$

where $p,\ q:\mathbb{R}_+^n\to\mathbb{R}$, $h(\sigma_1,\ldots,\sigma_n)=(h_i(\sigma_1,\ldots,\sigma_n))_{i=1}^{\infty}$, $\bar{h}(\sigma_1,\ldots,\sigma_n)=\big(\bar{h}_i(\sigma_1,\ldots,\sigma_n)\big)_{i=1}^{\infty}\in\ell_p^{\beta}$.

(c) $g_n:\mathbb{R}_+^n\times\ell_p^{\beta}\to\mathbb{R}$ $(n\in\mathbb{N})$ are continuous functions and $Q_k$ is defined as

$$Q_k=\sup\left\{\sum_{n\geq k}\left(\beta_n\left|\gamma_n(\sigma_1,\ldots,\sigma_n)\int_0^{a_1(\sigma_1)},\ldots,\int_0^{a_n(\sigma_n)}g_n\left(\sigma_1,\ldots,\sigma_n,\tau_1,\ldots,\tau_n,h(\tau_1,\ldots,\tau_n)\right)d\tau_1,\ldots,d\tau_n\right|^p\right)^{\frac{1}{p}}\right\}.$$

Also, as $(\sigma_1,\ldots,\sigma_n)\to\infty$,

$$\Bigg[\sum_n\Bigg\{\beta_n\Bigg|\gamma_n(\sigma_1,\ldots,\sigma_n)\int_0^{a_1(\sigma_1)},\ldots,\int_0^{a_1(\sigma_n)}\Big[g_n\Big(\sigma_1,\ldots,\sigma_n,\tau_1,\ldots,\tau_n,h(\tau_1,\ldots,\tau_n)\Big)$$

$$-g_n\Big(\sigma_1,\ldots,\sigma_n,\tau_1,\ldots,\tau_n,\bar{h}(\tau_1,\ldots,\tau_n)\Big)\Big]d\tau_1,\ldots,d\tau_n\Bigg|^p\Bigg\}\Bigg]^{\frac{1}{p}}=0.$$

(d) Let an operator $\mathbb{S}$ on $\mathbb{R}_+^n\times\ell_p^{\beta}$ to $\ell_p^{\beta}$ as $(\sigma_1,\ldots,\sigma_n,h(\sigma_1,\ldots,\sigma_n))\to(\mathbb{S}h)(\sigma_1,\ldots,\sigma_n)$, where

$$(\mathbb{S}h)(\sigma_1,\ldots,\sigma_n)=(\beta_1f_1(\sigma_1,\ldots,\sigma_n,v_1(h),h(\sigma_1,\ldots,\sigma_n)),\beta_2f_2(\sigma_1,\ldots,\sigma_n,v_2(h),h(\sigma_1,\ldots,\sigma_n)),\ldots,),$$

where $$v_n(h)=\int_0^{a_1(\sigma_1)},\ldots,\int_0^{a_n(\sigma_n)}g_n\left(\sigma_1,\ldots,\sigma_n,\tau_1,\ldots,\tau_n,h(\tau_1,\ldots,\tau_n)\right)$$ $d\tau_1,\ldots,d\tau_n$.

(e) As $k\to\infty$, $Q_k\to 0$, $\sup_{k\in\mathbb{N}}\{Q_k\}=Q$ and $\sup\{\alpha_n(\sigma_1,\ldots,\sigma_n):\sigma_1,\ldots,\sigma_n\in\mathbb{R}_+,\ n\in\mathbb{N}\}=\alpha$, s. t. $0<2^p\alpha<1$ and for any $\sigma_1,\ldots,\sigma_n\in\mathbb{R}_+$, $\gamma=\sup_n\left\{\sum_n\beta_n\gamma_n(\sigma_1,\ldots,\sigma_n)\right\}<\infty$.

**Theorem 2** *The infinite system (1) with the assumptions (a)–(e) have at least one solution* $h(\sigma_1, \ldots, \sigma_n) = (h_i(\sigma_1, \ldots, \sigma_n))_{i=1}^{\infty} \in \ell_p^{\beta}$ *for all* $\sigma_1, \ldots, \sigma_n \in \mathbb{R}_+$.

***Proof*** By using (1) and applying assumptions (a)-(e), we have for every $\sigma_1, \ldots, \sigma_n \in \mathbb{R}_+$,

$$\begin{aligned}
&\|h(\sigma_1, \ldots, \sigma_n)\|_{\ell_p^{\beta}}^{p} \\
&= \sum_{n\geq 1} \beta_n \left| f_n\left(\sigma_1, \ldots, \sigma_n, \int_0^{a_1(\sigma_1)}, \ldots, \int_0^{a_n(\sigma_n)} g_n(\sigma_1, \ldots, \sigma_n, \tau_1, \ldots, \tau_n, h(\tau_1, \ldots, \tau_n))d\tau_1, \ldots, d\tau_n, h(\sigma_1, \ldots, \sigma_n)\right)\right|^{p} \\
&\leq 2^p \sum_{n\geq 1} \beta_n \left| f_n\left(\sigma_1, \ldots, \sigma_n \int_0^{a_1(\sigma_1)} \int_0^{a_n(\sigma_n)} g_n\left(\sigma_1, \ldots, \sigma_n, \tau_1, \ldots, \tau_n, h(\tau_1, \ldots, \tau_n)\right)d\tau_1, \ldots, d\tau_n, h(\sigma_1, \ldots, \sigma_n)\right)\right. \\
&\left. - f_n(\sigma_1, \ldots, \sigma_n, 0, h^0(\sigma_1, \ldots, \sigma_n))\right|^{p} + 2^p \sum_{n\geq 1} \beta_n \left| f_n(\sigma_1, \ldots, \sigma_n, 0, h^0(\sigma_1, \ldots, \sigma_n))\right|^{p} \\
&\leq 2^p \sum_{n\geq 1} \left\{ \alpha_n(\sigma_1, \ldots, \sigma_n)\beta_n \left| h_n(\sigma_1, \ldots, \sigma_n)\right|^{p} \right. \\
&\left. + 2^p \sum_{n\geq 1} \gamma_n(\sigma_1, \ldots, \sigma_n)\beta_n \left| \int_0^{a_1(\sigma_1)}, \ldots, \int_0^{a_n(\sigma_n)} g_n(\sigma_1, \ldots, \sigma_n, \tau_1, \ldots, \tau_n, h(\tau_1, \ldots, \tau_n))d\tau_1, \ldots, d\tau_n \right|^{p} \right\} \\
&\leq 2^p \alpha \sum_{n\geq 1} \beta_n \left| h_n(\sigma_1, \ldots, \sigma_n)\right|^{p} + 2^p Q \\
&= 2^p \alpha \parallel h(\sigma_1, \ldots, \sigma_n) \parallel \ell_p^{\beta} + 2^p Q
\end{aligned}$$

i.e., $(1 - 2^p\alpha) \parallel h(\sigma_1, \ldots, \sigma_n) \parallel \ell_p^{\beta} \leq 2^p Q$
$\Rightarrow \parallel h(\sigma_1, \ldots, \sigma_n) \parallel \ell_p^{\beta} \leq \frac{2^p Q}{1-2^p\alpha} = \rho^p$(say).

Therefore, we get $\parallel h(\sigma_1, \ldots, \sigma_n) \parallel_{\ell_p^{\beta}} \leq \rho$. Let $\bar{\mathbb{B}} = \bar{\mathbb{B}}\left(h^0(\sigma_1, \ldots, \sigma_n), \rho\right)$ be the closed ball having center $h^0(\sigma_1, \ldots, \sigma_n)$ and radius $\rho$, so $\bar{\mathbb{B}}$ is NBCC subset of $\ell_p^{\beta}$.

Suppose an operator $\mathbb{S} = (\mathbb{S}_i)$ on $BC\left(\mathbb{R}_+^n, \bar{\mathbb{B}}\right)$ defined as, for every $\sigma_1, \ldots, \sigma_n \in \mathbb{R}_+$,

$$(\mathbb{S}h)(\sigma_1, \ldots, \sigma_n) = \{(\beta_i \mathbb{S}_i h)(\sigma_1, \ldots, \sigma_n)\}_{i=1}^{\infty} = \{\beta_i f_i(\sigma_1, \ldots, \sigma_n, v_i(h), h(\sigma_1, \ldots, \sigma_n))\}_{i=1}^{\infty},$$

where $h(\sigma_1, \ldots, \sigma_n) = (h_i(\sigma_1, \ldots, \sigma_n)) \in \bar{\mathbb{B}}$ and $\forall i \in \mathbb{N}$.

Since for every $(\sigma_1, \ldots, \sigma_n) \in \mathbb{R}_+^n$, so by assumption (d) that

$$\left(\sum_{i\geq 1} \beta_i |(\mathbb{S}_i h)(\sigma_1, \ldots, \sigma_n)|^p\right)^{\frac{1}{p}} = \left(\sum_{i\geq 1} \beta_i |f_i(\sigma_1, \ldots, \sigma_n, v_i(h), h(\sigma_1, \ldots, \sigma_n))|^p\right)^{\frac{1}{p}} < \infty.$$

Hence, $(\mathbb{S}h)(\sigma_1, \ldots, \sigma_n) \in \ell_p^{\beta}$.

Therefore, $\parallel (\mathbb{S}h)(\sigma_1, \ldots, \sigma_n) - h^0(\sigma_1, \ldots, \sigma_n) \parallel_{\ell_p^{\beta}} \leq \rho$, so $\mathbb{S}$ is self mapping on $\bar{\mathbb{B}}$.

Next, we have to prove that $\mathbb{S}$ is continuous mapping on $\bar{\mathbb{B}}$.

Let $\epsilon > 0$ and any $h_x(\sigma_1, \ldots, \sigma_n) = \left(h_{x_i}(\sigma_1, \ldots, \sigma_n)\right)_{i=1}^{\infty}$, $h(\sigma_1, \ldots, \sigma_n) = (h_i(\sigma_1, \ldots, \sigma_n))_{i=1}^{\infty} \in \ell_p^{\beta}$ s. t. $\| h_x - h \|_{\ell_p^{\beta}} < \frac{\epsilon}{2^{1/p}\alpha^{1/p}}$.

We claim that $\| (\mathbb{S}h_x)(\sigma_1, \ldots, \sigma_n) - (\mathbb{S}h)(\sigma_1, \ldots, \sigma_n) \|_{\ell_p^{\beta}} \to 0$. Then we will prove that $\beta_n | (\mathbb{S}_n h_x)(\sigma_1, \ldots, \sigma_n) - (\mathbb{S}_n h)(\sigma_1, \ldots, \sigma_n)|_{\ell_p^{\beta}} \to 0$. For $(\sigma_1, \ldots, \sigma_n) \in \mathbb{R}_+^n$, we have

$$
\begin{aligned}
&\beta_n \left|(\mathbb{S}_n h_x)(\sigma_1, \ldots, \sigma_n) - (\mathbb{S}_n h)(\sigma_1, \ldots, \sigma_n)\right|^p \\
&= \beta_n \left| f_n(\sigma_1, \ldots, \sigma_n, v_n(h_x), h_x(\sigma_1, \ldots, \sigma_n)) - f_n(\sigma_1, \ldots, \sigma_n, v_n(h), h(\sigma_1, \ldots, \sigma_n))\right|^p \\
&\leq \beta_n \alpha \left|h_{x_n}(\sigma_1, \ldots, \sigma_n) - h(\sigma_1, \ldots, \sigma_n)\right|^p + \gamma_n \beta_n(\sigma_1, \ldots, \sigma_n) \left|v_n(h_x) - v_n(h)\right|^p \\
&= \alpha \beta_n \left|h_{x_n}(\sigma_1, \ldots, \sigma_n) - h(\sigma_1, \ldots, \sigma_n)\right|^p \\
&+ \gamma_n(\sigma_1, \ldots, \sigma_n)\beta_n \left| \int_0^{a_1(\sigma_1)} \ldots, \int_0^{a_n(\sigma_n)} \left[ g_n(\sigma_1, \ldots, \sigma_n, \tau_1, \ldots, \tau_n, h_x(\tau_1, \ldots, \tau_n)) \right.\right. \\
&\left.\left. - g_n(\sigma_1, \ldots, \sigma_n, \tau_1, \ldots, \tau_n, h(\tau_1, \ldots, \tau_n)) \right] d\tau_1, \ldots, d\tau_n \right|^p
\end{aligned}
$$

By assumption (c), we choose $T > 0$ as $\max(\sigma_1, \ldots, \sigma_n) > T$,

$$
\begin{aligned}
&\left( \sum_n \left\{ \gamma_n(\sigma_1, \ldots, \sigma_n)\beta_n \left| \int_0^{a_1(\sigma_1)} \cdots \int_0^{a_n(\sigma_n)} \left[ g_n(\sigma_1, \ldots, \sigma_n, \tau_1, \ldots, \tau_n, h_x(\tau_1, \ldots, \tau_n)) \right.\right.\right.\right. \\
&\left.\left.\left.\left. - g_n(\sigma_1, \ldots, \sigma_n, \tau_1, \ldots, \tau_n, h(\tau_1, \ldots, \tau_n)) \right] d\tau_1, \ldots, d\tau_n \right|^p \right\} \right)^{\frac{1}{p}} < \frac{\epsilon^p}{2}.
\end{aligned}
$$

Hence,

$$
\begin{aligned}
&\left( \sum_n \beta_n \left|(\mathbb{S}_n h_x)(\sigma_1, \ldots, \sigma_n) - (\mathbb{S}_n h)(\sigma_1, \ldots, \sigma_n)\right|^p \right)^{\frac{1}{p}} \\
&\leq \alpha \sum_n \beta_n \left|h_{x_n}(\sigma_1, \ldots, \sigma_n) - h(\sigma_1, \ldots, \sigma_n)\right|^p + \frac{\epsilon^p}{2} \\
&< \alpha \frac{\epsilon^p}{2\alpha} + \frac{\epsilon^p}{2}
\end{aligned}
$$

i.e., $\| (\mathbb{S}h_x)(\sigma_1, \ldots, \sigma_n) - (\mathbb{S}h)(\sigma_1, \ldots, \sigma_n) \|_{\ell_p^{\beta}} < \epsilon$. For $\sigma_1, \ldots, \sigma_n \in [0, T]$, let

$$A_1 = \sup\{a_1(\sigma_1) : \sigma_1 \in [0, T]\},$$

$$\vdots$$

$$A_n = \sup\{a_n(\sigma_n) : \sigma_n \in [0, T]\} \; and$$

$$g = \sup\Big\{ |g_n(\sigma_1, \ldots, \sigma_n, \tau_1, \ldots, \tau_n, h_x(\tau_1, \ldots, \tau_n)) - g_n(\sigma_1, \ldots, \sigma_n, \tau_1, \ldots, \tau_n, h(\tau_1, \ldots, \tau_n))|, \sigma_i \in [0, T], \tau_i \in [0, A_i], i = 1, 2, \ldots, n\Big\}.$$

Then,

$$\sum_n \beta_n |(\mathbb{S}_n h_x)(\sigma_1, \ldots, \sigma_n) - (\mathbb{S}_n h)(\sigma_1, \ldots, \sigma_n)| < \frac{\epsilon^p}{2} + g^p A_1{}^p, \ldots, A_n{}^p \sum_n \beta_n \gamma_n(\sigma_1, \ldots, \sigma_n)$$
$$< \frac{\epsilon^p}{2} + \gamma g^p A_1{}^p, \ldots, A_n{}^p.$$

Since $g_n$ is continuous on $[0, T] \times, \ldots, \times [0, T] \times [0, A_1] \times, \ldots, \times [0, A_n] \times \bar{\mathbb{B}}$, we have $g \to 0$ as $\epsilon \to 0$, therefore we have $\sum_n \beta_n |(\mathbb{S}_n h_x)(\sigma_1, \ldots, \sigma_n) - (\mathbb{S}_n h)(\sigma_1, \ldots, \sigma_n)|^p \to 0$ as $\| h_x(\sigma_1, \ldots, \sigma_n) - h(\sigma_1, \ldots, \sigma_n) \| \ell_p^\beta \to 0$. Thus, $\mathbb{S}$ is continuous on $\bar{\mathbb{B}} \subset \ell_p^\beta$.

Now, we have to show that $\mathbb{S}$ is a MKC operator.

Given $\epsilon > 0$, search $\delta > 0$ s. t. $\epsilon \le \chi(\bar{\mathbb{E}}) < \epsilon + \delta \implies \chi(\mathbb{S}(\bar{\mathbb{E}})) < \epsilon$.

We have

$$\chi(\mathbb{S}(\bar{\mathbb{E}}))$$
$$= \lim_{n\to\infty}\left[\sup_{h(\sigma_1,\ldots,\sigma_n)\in\bar{\mathbb{E}}} \left\{\sum_{k\ge n} \beta_k |f_k(\sigma_1, \ldots, \sigma_n, v_k(h), h(\sigma_1, \ldots, \sigma_n))|^p\right\}^{1/p}\right]$$
$$\le \lim_{n\to\infty}\left[\sup_{h(\sigma_1,\ldots,\sigma_n)\in\bar{\mathbb{E}}} \left\{2^p\left(\sum_{k\ge n} \beta_k\Big(\alpha_k(\sigma_1, \ldots, \sigma_n)|h_k(\sigma_1, \ldots, \sigma_n)|^p\right.\right.\right.$$
$$\left.\left.\left. + \gamma_k(\sigma_1, \ldots, \sigma_n)\left|\int_0^{a_1(\sigma_1)}, \ldots, \int_0^{a_n(\sigma_n)} g_k(\sigma_1, \ldots, \sigma_n, \tau_1, \ldots, \tau_n, h(\tau_1, \ldots, \tau_n))\, d\tau_1, \ldots, d\tau_n\right|^p\Big)\right)\right\}^{1/p}\right]$$
$$\le \lim_{n\to\infty}\left[\sup_{h(\sigma_1,\ldots,\sigma_n)\in\bar{\mathbb{E}}} \left\{2^p\left(\alpha\sum_{k\ge n} \beta_k |h_k(\sigma_1, \ldots, \sigma_n)|^p\right) + 2^p Q_n\right\}^{1/p}\right].$$

Observe that

$$\chi(\mathbb{S}(\bar{\mathbb{E}})) \le 2\alpha^{1/p}\chi(\bar{\mathbb{E}}) < \epsilon \Rightarrow \chi(\bar{\mathbb{E}}) < \frac{\epsilon}{2\alpha^{1/p}},$$

which gives

$$\sup_{(\sigma_1,\ldots,\sigma_n)\in\mathbb{R}_+^n}\left\{\chi(\mathbb{S}(\bar{\mathbb{E}}))\right\} \le 2\alpha^{1/p} \sup_{(\sigma_1,\ldots,\sigma_n)\in\mathbb{R}_+^n}\left\{\chi(\bar{\mathbb{E}})\right\} < \epsilon \Rightarrow \sup_{(\sigma_1,\ldots,\sigma_n)\in\mathbb{R}_+^n}\left\{\chi(\bar{\mathbb{E}})\right\} < \frac{\epsilon}{2\alpha^{1/p}},$$

i.e., $\chi_{BC(\mathbb{R}_+^n,\bar{\mathbb{B}})}(\mathbb{S}(\bar{\mathbb{E}})) \le 2\alpha^{1/p}\chi_{BC(\mathbb{R}_+^n,\bar{\mathbb{B}})}(\bar{\mathbb{E}}) < \epsilon \Rightarrow \chi_{BC(\mathbb{R}_+^n,\bar{\mathbb{B}})}(\bar{\mathbb{E}}) < \frac{\epsilon}{2\alpha^{1/p}}$.

If $\delta = \frac{\epsilon(1-2\alpha^{1/p})}{2\alpha^{1/p}}$, then we have

$$\epsilon \le \chi_{BC(\mathbb{R}_+^n,\bar{\mathbb{B}})}(\bar{\mathbb{E}}) < \epsilon + \delta.$$

Hence, $\mathbb{S}$ is a MKC operator on the set $\bar{\mathbb{B}}$ and fulfills all requirement of Theorem 1, therefore $\mathbb{S}$ have a $\mathbb{FP}$ in $\bar{\mathbb{B}}$. Hence, Eq. (1) have a solution in $\bar{\mathbb{B}}$.

**Example 1** Consider the following $\mathbb{IE}$ for any $\sigma, \sigma_1, \sigma_2, \sigma_3 \in \mathbb{R}_+$

$$h_n(\sigma_1, \sigma_2, \sigma_3) = \frac{\sin(\sigma_1^2 + \sigma_2^3 + \sigma_3^4)}{3} h_n(\sigma_1, \sigma_2, \sigma_3) + \frac{1}{e^{\sigma_1\sigma_2\sigma_3}} \int_0^{\sigma_1}\int_0^{\sigma_2}\int_0^{\sigma_3} \frac{\cos\left(\sum_{i=1}^{\infty} h_i(\tau_1, \tau_2, \tau_3)\right)}{5n + \sin\left(h_n(\tau_1, \tau_2, \tau_3)\right)} d\tau_1 d\tau_2 d\tau_3,$$

where $n \in \mathbb{N}$ here $a_1(\sigma) = a_2(\sigma) = a_3(\sigma) = \sigma$, and

$$f_n(\sigma_1, \sigma_2, \sigma_3, v_n(h(\sigma_1, \sigma_2, \sigma_3)), h(\sigma_1, \sigma_2, \sigma_3)) = \frac{\sin(\sigma_1^2 + \sigma_2^3 + \sigma_3^4)}{3} h_n(\sigma_1, \sigma_2, \sigma_3) + \frac{1}{e^{\sigma_1\sigma_2\sigma_3}} v_n(h(\sigma_1, \sigma_2, \sigma_3)),$$

$$v_n(h(\sigma_1, \sigma_2, \sigma_3)) = \int_0^{\sigma_1}\int_0^{\sigma_2}\int_0^{\sigma_3} g_n\left(\sigma_1, \sigma_2, \sigma_3, \tau_1, \tau_2, \tau_3, h(\tau_1, \tau_2, \tau_3)\right) d\tau_1 d\tau_2 d\tau_3, \ and$$

$$g_n\left(\sigma_1, \sigma_2, \sigma_3, \tau_1, \tau_2, \tau_3, h(\tau_1, \tau_2, \tau_3)\right) = \frac{\cos\left(\sum_{i=1}^{\infty} h_i(\tau_1, \tau_2, \tau_3)\right)}{5n + \sin\left(h_n(\tau_1, \tau_2, \tau_3)\right)}.$$

If $h(\sigma_1, \sigma_2, \sigma_3) \in \ell_p$, then

$$\begin{aligned} &\sum_{n=1}^{\infty} |f_n(\sigma_1, \sigma_2, \sigma_3, v_n(h(\sigma_1, \sigma_2, \sigma_3)), h(\sigma_1, \sigma_2, \sigma_3))|^p \\ &\leq \left(\frac{2}{3}\right)^p \sum_{n=1}^{\infty} \left|\sin(\sigma_1^2 + \sigma_2^3 + \sigma_3^4) h_n(\sigma_1, \sigma_2, \sigma_3)\right|^p + 2^p \left(\frac{\sigma_1\sigma_2\sigma_3}{e^{\sigma_1\sigma_2\sigma_3}}\right)^p \sum_{n=1}^{\infty} \frac{1}{n}. \\ &\leq \left(\frac{2}{3}\right)^p \| h(\sigma_1, \sigma_2, \sigma_3) \|_{\ell_p}^p + \left(\frac{2}{e}\right)^p \sum_{n=1}^{\infty} \frac{1}{n}. \end{aligned}$$

Therefore $(f_n(\sigma_1, \sigma_2, \sigma_3, v_n(h(\sigma_1, \sigma_2, \sigma_3)), h(\sigma_1, \sigma_2, \sigma_3))) \notin \ell_p$.

If $h(\sigma_1, \sigma_2, \sigma_3) \in \ell_p^{\beta}$, where $\beta_n = \frac{1}{n}$, then we have

$$\sum_{n=1}^{\infty}\beta_n\,|f_n(\sigma_1,\sigma_2,\sigma_3,v_n(h(\sigma_1,\sigma_2,\sigma_3)),h(\sigma_1,\sigma_2,\sigma_3))|^p$$
$$\le\left(\frac{2}{3}\right)^p\sum_{n=1}^{\infty}\beta_n\left|\sin(\sigma_1^2+\sigma_2^3+\sigma_3^4)h_n(\sigma_1,\sigma_2,\sigma_3)\right|^p+2^p\left(\frac{\sigma_1\sigma_2\sigma_3}{e^{\sigma_1\sigma_2\sigma_3}}\right)^p\sum_{n=1}^{\infty}\beta_n\left(\frac{1}{n}\right)$$
$$\le\left(\frac{2}{3}\right)^p\ \|\ h(\sigma_1,\sigma_2,\sigma_3)\ \|_{\ell_p^\beta}^p+\left(\frac{2}{e}\right)^p\sum_{n=1}^{\infty}\beta_n\left(\frac{1}{n}\right).$$

Therefore $(f_n(\sigma_1,\sigma_2,\sigma_3,v_n(h(\sigma_1,\sigma_2,\sigma_3)),h(\sigma_1,\sigma_2,\sigma_3)))\in\ell_p^\beta$.

Now, if $h_x(\sigma_1,\sigma_2,\sigma_3)=\left(h_{x_i}(\sigma_1,\sigma_2,\sigma_3)\right)\in\ell_p^\beta$, then

$$\beta_n\Big|f_n(\sigma_1,\sigma_2,\sigma_3,v_n(h_x(\sigma_1,\sigma_2,\sigma_3)),h_x(\sigma_1,\sigma_2,\sigma_3))$$
$$-f_n(\sigma_1,\sigma_2,\sigma_3,v_n(h_y(\sigma_1,\sigma_2,\sigma_3)),h_y(\sigma_1,\sigma_2,\sigma_3))\Big|^p$$
$$\le\left(\frac{2}{3n}\right)^p\left|h_{x_n}(\sigma_1,\sigma_2,\sigma_3)-h_{y_n}(\sigma_1,\sigma_2,\sigma_3)\right|^p$$
$$+\frac{2^p}{n^pe^{p\sigma_1\sigma_2\sigma_3}}\left|v_n(h_x(\sigma_1,\sigma_2,\sigma_3))-v_n(h_y(\sigma_1,\sigma_2,\sigma_3))\right|^p.$$

Here $\alpha_n(\sigma_1,\sigma_2,\sigma_3)=\left(\frac{2}{3n}\right)^p,\ \gamma_n(\sigma_1,\sigma_2,\sigma_3)=\frac{2^p}{n^pe^{p\sigma_1\sigma_2\sigma_3}}$. Also, $\alpha=\left(\frac{2}{3}\right)^p$.

We get $0<2^p\alpha<1$ and $\sum_{n\ge1}\beta_n\left|f_n\left(\sigma_1,\sigma_2,\sigma_3,\tau_1,\tau_2,\tau_3,0,h^0(\sigma_1,\sigma_2,\sigma_3)\right)\right|$ goes to zero for every $\sigma_1,\sigma_2,\sigma_3\in\mathbb{R}_+$. Again, we get

$$\sum_{n\ge k}\beta_n\,|\gamma_n(\sigma_1,\sigma_2,\sigma_3)v_n(h_x(\sigma_1,\sigma_2,\sigma_3))|^p$$
$$\le\left(\frac{2\sigma_1\sigma_2\sigma_3}{e^{p\sigma_1\sigma_2\sigma_3}}\right)^p\sum_{n\ge k}\frac{1}{n^{p+1}}$$
$$\le\left(\frac{2}{e}\right)^p\sum_{n\ge k}\frac{1}{n^{p+1}}.$$

*Also,* $$Q_k\le\sup_n\left\{\left(\frac{2}{e}\right)^p\sum_{n\ge k}\frac{1}{n^{p+1}}:\sigma_1,\sigma_2,\sigma_3,\tau_1,\tau_2,\tau_3\in\mathbb{R}_+\right\}.$$

As $k\to\infty$ we get $\sum\limits_{n\ge k}\frac{1}{n^{p+1}}\to0$. Thus, $Q_k\to0$ as $k\to\infty$ and $Q=\left(\frac{2}{e}\right)^pB_p$.

Now, we have

$$\sum_n \beta_n \Bigg| \gamma_n(\sigma_1, \sigma_2, \sigma_3) \int_0^{\sigma_1}\int_0^{\sigma_2}\int_0^{\sigma_3} [g_n(\sigma_1, \sigma_2, \sigma_3, \tau_1, \tau_2, \tau_3, h_x(\tau_1, \tau_2, \tau_3)) - g_n(\sigma_1, \sigma_2, \sigma_3, \tau_1, \tau_2, \tau_3, h_y(\tau_1, \tau_2, \tau_3))]d\tau_1 d\tau_2 d\tau_3 \Bigg|^p$$

$$\leq \left(\frac{2\sigma_1\sigma_2\sigma_3}{e^{p\sigma_1\sigma_2\sigma_3}}\right)^p \sum_n \frac{1}{n^{p+1}}$$

$$\leq \left(\frac{2\sigma_1\sigma_2\sigma_3}{e^{p\sigma_1\sigma_2\sigma_3}}\right)^p B_p.$$

As $\sigma_1, \sigma_2, \sigma_3 \to \infty$, we have

$$\lim_{\sigma_1,\sigma_2,\sigma_3\to\infty} \sum_n \beta_n \Bigg| \gamma_n(\sigma_1, \sigma_2, \sigma_3) \int_0^{\sigma_1}\int_0^{\sigma_2}\int_0^{\sigma_3} [g_n(\sigma_1, \sigma_2, \sigma_3, \tau_1, \tau_2, \tau_3, h_x(\tau_1, \tau_2, \tau_3)) - g_n(\sigma_1, \sigma_2, \sigma_3, \tau_1, \tau_2, \tau_3, h_y(\tau_1, \tau_2, \tau_3))]d\tau_1 d\tau_2 d\tau_3 \Bigg|^p = 0.$$

If for any $\sigma_1, \sigma_2, \sigma_3 \in \mathbb{R}_+$, we have $\gamma = \sup_n \left\{ \sum_n \beta_n \gamma_n(\sigma_1, \sigma_2, \sigma_3) \right\} \leq \left(\frac{2}{e}\right)^p B_p < \infty$.

Hence, $f_n$ and $g_n$ are continuous functions and satisfied all the assumptions of Theorem 2. Hence, Eq. (1) has a solution in $\ell_p^\beta$.

## 5 Conclusions

The present study focuses on a new sequence space called $\ell_p^\beta$ tempered space. First, we established the existence results for infinite system of Volterra $\mathbb{IE}$ of $n$-variables in $\ell_p^\beta$ space, using MNC and MKC operator. At the end, an example is constructed to support the newly achieved result.

## References


1. Kuratowski, K.: Sur les espaces complets. Fund. Math. **15**, 301–309 (1930)
2. Banas, J., Goebel, K.: Measure of noncompactness in Banach spaces. Lect. Notes Math. **60** (1980)
3. Goldenštein, L.S., Gohberg, I.T., Markus, A.S.: Investigations of some properties of bounded linear operators with their $q$-norms. Učen. Zap. Kish. Univ. **29**, 29–36 (1957)
4. Goldenštein, L.S., Markus, A.S.: On a measure of noncompactness of bounded sets and linear operators. Stud. Alg. Appl. Math. Kish. 45–54 (1965)

5. Banas, J., Krajewska, M.: Existence of solutions for infinite systems of differential equations in space of tempered sequences. Electron. J. Differ. Eqs. **60**, 1–28 (2017)
6. Das, A., Hazarika, B.: Measure of noncompactness in a new space of tempered sequences and its application on infinite systems of fractional differential equations with applications to differential and integral equations. Springer, New Delhi (2014)
7. Ghasemi, M., Khanehgir, M., Allahyari, R.: On solutions of infinite systems of integral equations in $n$- variables in the space of tempered sequences $c_0^\beta$ and $\ell_1^\beta$. J. Math. Anal. **9**(6), 1–16 (2018)
8. Meir, A., Keeler, E.: A theorem on contraction mappings. J. Math. Anal. Appl. **28**, 326–329 (1969)
9. Aghajani, A., Mursaleen, M., Haghighi, A.S.: Fixed point theorems for Meir-Keeler condensing operators via measure of noncompactness. Acta. Math. Sci. **35**(3), 552–566 (2015)
10. Banaś, J., Lecko, M.: Solvability of infinite systems of differential equations in Banach sequence spaces. J. Comput. Appl. Math. **137**, 363–375 (2001)

# On Generalizations of Integral Inequalities and Its Applications

**S. G. Latpate and S. V. Babar**

**Abstract** The present research paper obtains nonlinear generality of integral inequalities established by Pachpatte in [1]. These nonlinear integral inequalities are serviceable to study solutions of certain differential and integral equations. The discrete analogues of the main results are also given. Few applications are given to convey the significance of our results.

**Keywords** Differential and integral equations · Discrete analogues · Explicit bound · Integral inequality

## 1 Introduction

Integral inequalities perform a crucial role in the development of mathematical sciences. Most of integral inequalities are useful to study qualitative properties of solutions of differential and integral equations. To acquaint with the Gronwall–Bellman inequality [2, 3], the study of qualitative properties of the solutions of certain differential equations has been significant in the study of mathematical science. Many other results on its generalizations may be seen in [1, 4–12].

In this paper, we obtain nonlinear generalizations of integral inequalities established by Pachpatte in [1], which can be practicable to study the qualitative properties of solutions of specific differential and integral equations. Mere applications are also given to convey the significance of our results. Here $\mathbb{R}$ denotes the set of real numbers and $\mathbb{R}_+ := [0, \infty)$ is a subset of $\mathbb{R}$. The following lemma is a main tool in our paper.

**Lemma 1** *If* $x \geq 0$, $y \geq 0$ *and* $\frac{1}{p} + \frac{1}{q} = 1$ *with* $p > 1$, *then*

S. G. Latpate
Department of Mathematics, Nowrosjee Wadia College, Pune, Maharashtra 411001, India
e-mail: sglatpate@gmail.com

S. V. Babar (✉)
Department of Mathematics, Dr.Ghali College, Gadhinglaj, Kolhapur, Maharashtra 416502, India
e-mail: santosh.babar425@gmail.com

 
B. Rushi Kumar et al. (eds.), *Mathematics and Computing*, Springer Proceedings in Mathematics & Statistics 415, https://doi.org/10.1007/978-981-19-9307-7_35

$$x^{\frac{1}{p}} y^{\frac{1}{q}} \leq \frac{x}{p} + \frac{y}{q}$$

*with equality holds if and only if* $x = y$.

## 2 Main Results

Here, we state and prove nonlinear integral inequalities, which can be utilized in the analysis of properties of solutions of some differential and integral equations.

**Theorem 1** *Let* $u, x, y, g$, *and* $h$ *be nonnegative real-valued continuous functions defined on* $\mathbb{R}_+$ *and* $l \geq m \geq 1$, *where* $l$ *and* $m$ *are real constants. If*

$$u^l(p) \leq x(p) + y(p) \int_0^p \left[g(s)u^l(s) + h(s)u^m(s)\right] ds \tag{1}$$

*for* $p \in \mathbb{R}_+$, *then*

$$u(p) \leq \left(x(p) + y(p) \int_0^p \left[g(s)x(s) + h(s)\left(\frac{l-m}{l} + \frac{mx(s)}{l}\right)\right] \times \exp\left(\int_s^p y(\sigma)\left(g(\sigma) + \frac{mh(\sigma)}{l}\right) d\sigma\right) ds\right)^{\frac{1}{l}} \tag{2}$$

*for* $p \in \mathbb{R}_+$.

***Proof*** We define $z$ as follows:

$$z(p) = \int_0^p \left[g(s)u^l(s) + h(s)u^m(s)\right] ds, \quad p \in \mathbb{R}_+. \tag{3}$$

Subsequently $z(0) = 0$ and (1) becomes

$$u^l(p) \leq x(p) + y(p) z(p). \tag{4}$$

Applying Lemmas (1)–(4), we get

$$u^m(p) \leq \left(\frac{l - m + mx(p)}{l}\right) + \frac{my(p)z(p)}{l}. \tag{5}$$

Differentiating (3) and using (4) and (5), we get

$$z'(p) = g(p)u^l(p) + h(p)u^m(p)$$

and

$$z'(p) \le y(p)\left(g(p) + \frac{mh(p)}{l}\right)z(p) + g(p)x(p) + h(p)\left(\frac{l - m + mx(p)}{l}\right). \tag{6}$$

Inequality (6) gives

$$z'(p) - y(p)\left(g(p) + \frac{mh(p)}{l}\right)z(p) \le g(p)x(p) + h(p)\left(\frac{l - m + mx(p)}{l}\right).$$

Equivalently,

$$\left[\frac{z(p)}{\exp\left(\int_0^p y(s)\left(g(s) + \frac{mh(s)}{l}\right)ds\right)}\right]' \le g(p)x(p) + h(p)\left(\frac{l - m + mx(p)}{l}\right) \times \exp\left(-\int_0^p y(s)\left(g(s) + \frac{mh(s)}{l}\right)ds\right),$$

which yields

$$\frac{z(p)}{\exp\left(\int_0^p y(s)\left(g(s) + \frac{mh(s)}{l}\right)ds\right)} \le z(0) + \int_0^p \left[g(s)x(s) + h(s)\left(\frac{l - m + mx(s)}{l}\right)\right] \times \exp\left(-\int_0^s y(\sigma)\left(g(\sigma) + \frac{mh(\sigma)}{l}\right)d\sigma\right)ds.$$

Since $z(0) = 0$, we obtain

$$z(p) \le \int_0^p \left[g(s)x(s) + h(s)\left(\frac{l - m + mx(s)}{l}\right)\right] \times \exp\left(\int_s^p y(\sigma)\left(g(\sigma) + \frac{mh(\sigma)}{l}\right)d\sigma\right)ds. \tag{7}$$

Now inequality (2) is easily obtained from (4) and (7).

**Theorem 2** *Let $u$, $y$, $g$, and $h$ be nonnegative real-valued and continuous functions on $\mathbb{R}_+$ and $l \ge m \ge 1$, where $l$ and $m$ are real constants. Let $c$ be a positive real-valued continuous and nondecreasing function on $R_+$. If*

$$u^l(p) \le c^l(p) + y(p)\int_0^p \left[g(s)u^l(s) + h(s)u^m(s)\right]ds \tag{8}$$

*for $p \in \mathbb{R}_+$, then*

$$u(p) \leq c(p)\left(1 + y(p)\int_0^p \left[g(s) + \frac{h(s)c^m(s)}{c^l(s)}\right]\right. \tag{9}$$

$$\left.\times \exp\left(\int_s^p y(\sigma)\left(g(\sigma) + \frac{mh(\sigma)c^m(\sigma)}{l}\right)d\sigma\right)ds\right)^{\frac{1}{l}} \tag{10}$$

*for* $p \in \mathbb{R}_+$.

***Proof*** From (8), we have

$$\left(\frac{u(p)}{c(p)}\right)^l \leq 1 + y(p)\int_0^p \left[g(s)\left(\frac{u(s)}{c(s)}\right)^l + \frac{h(s)c^m(s)}{c^l(s)}\left(\frac{u(s)}{c(s)}\right)^m\right]ds. \tag{11}$$

Applying Theorems (1)–(11), we get

$$\frac{u(p)}{c(p)} \leq \left(1 + y(p)\int_0^p \left[g(s) + \frac{h(s)c^m(s)}{c^l(s)}\left(\frac{l-m}{l} + \frac{m}{l}\right)\right]\right.$$
$$\left.\times \exp\left(\int_s^p y(\sigma)\left(g(\sigma) + \frac{mh(\sigma)c^m(\sigma)}{lc^l(\sigma)}\right)d\sigma\right)ds\right)^{\frac{1}{l}}.$$

Hence

$$u(p) \leq c(p)\left(1 + y(p)\int_0^p \left[g(s) + \frac{h(s)}{c^{l-m}(s)}\right] \times \exp\left(\int_s^p y(\sigma)\left(g(\sigma) + \frac{mh(\sigma)}{lc^{l-m}(\sigma)}\right)d\sigma\right)ds\right)^{\frac{1}{l}}.$$

Thus, the proof is complete.

**Theorem 3** *Let* $u, x, y, g,$ *and* $h$ *be nonnegative real-valued continuous functions defined on* $\mathbb{R}_+$ *and* $l \geq m \geq 1$, *where* $l$ *and* $m$ *are real constants. Let* $k(\cdot,\cdot)$ *and its partial derivatives* $\frac{\partial}{\partial p}k(p,s)$ *be nonnegative real-valued continuous functions defined for* $0 \leq s \leq p < \infty$. *If*

$$u^l(p) \leq x(p) + y(p)\int_0^p k(p,s)\left[g(s)u^l(s) + h(s)u^m(s)\right]ds \tag{12}$$

*for* $p \in \mathbb{R}_+$, *then*

$$u(p) \leq \left\{x(p) + y(p)\int_0^p B(\sigma)exp\left(\int_\sigma^p A(\tau)d\tau\right)d\sigma\right\}^{\frac{1}{l}} \tag{13}$$

*for* $p \in \mathbb{R}_+$, *where*

$$A(p) := k(p,p)y(p)\left(g(p) + \frac{mh(p)}{l}\right) + \int_0^p \frac{\partial}{\partial p}k(p,s)y(s)\left(g(s) + \frac{mh(s)}{l}\right)ds \tag{14}$$

*and*

$$B(p) := k(p,p)\left(g(p)x(p) + h(p)\left(\frac{l-m+mx(p)}{l}\right)\right)$$
$$+\int_0^p \frac{\partial}{\partial p}k(p,s)\left(g(s)x(s) + h(s)\left(\frac{l-m+mx(s)}{l}\right)\right)ds. \tag{15}$$

***Proof*** We denote $z(p)$ by

$$z(p) = \int_0^p k(p,s)\left[g(s)u^l(s) + h(s)u^m(s)\right]ds. \tag{16}$$

So $z(0) = 0$ and (12) becomes

$$u^l(p) \le x(p) + y(p)z(p). \tag{17}$$

Making use of Lemmas (1)–(17), we obtain

$$u^m(p) \le \left(\frac{l-m+mx(p)}{l}\right) + \frac{my(p)z(p)}{l}. \tag{18}$$

Differentiating (16) and using (14), (15), (17), and (18), we obtain

$$z'(p) \le A(p)z(p) + B(p).$$

Equivalently,

$$\left(\frac{z(p)}{\exp\left(\int_0^p A(\tau)d\tau\right)}\right)' \le B(p)\exp\left(-\int_0^p A(\tau)d\tau\right).$$

This gives

$$\frac{z(p)}{\exp\left(\int_0^p A(\tau)d\tau\right)} \le z(0) + \int_0^p B(\sigma)exp\left(-\int_0^\sigma A(\tau)d\tau\right)d\sigma. \tag{19}$$

Now, Inequality (19) implies the estimate

$$z(p) \le \int_0^p B(\sigma)exp\left(\int_\sigma^p A(\tau)d\tau\right)d\sigma. \tag{20}$$

Substituting (20) in (17), we get (13) and hence proof is complete.

## 3 Discrete Analogues

Now, we state and prove discrete analogues of inequalities from Sect. 2. Let $\mathbb{N}_0 := \{p_0, p_0 + 1, p_0 + 2, \ldots\}$, where $p_0 \in \mathbb{N}_0$. For a function $u : \mathbb{N}_0 \to \mathbb{R}_+$, we define an operator $\Delta$ by $\Delta u(p) = u(p+1) - u(p)$, $p \in \mathbb{N}_0$ and for a function $k : \mathbb{N}_0^2 \to R_+$, we define an operator $\Delta_1$ by $\Delta_1 k(p,s) = k(p+1,s) - k(p,s)$ for $p, s \in \mathbb{N}_0$ with $p_0 \le s \le p$. For an empty set $\phi$, we let $\sum_{s\in\phi} u(s) = 0$ and $\prod_{s\in\phi} u(s) = 1$.

**Theorem 4** *Let $u, x, y, g$, and $h$ be nonnegative real-valued continuous functions defined on $\mathbb{N}_0$ and $l \ge m \ge 1$ be a real constants. If*

$$u^l(p) \le x(p) + y(p) \sum_{q=p_0}^{p-1} \left[g(q)u^l(q) + h(s)u^m(q)\right] \tag{21}$$

*for $p \in \mathbb{N}_0$, then*

$$u(p) \le \left( x(p) + y(p) \sum_{q=p_0}^{p-1} \left[ g(q)u(q) + h(q) \left( \frac{l - m + mx(q)}{l} \right) \right] \times \prod_{\sigma=q+1}^{p-1} \left[ 1 + y(\sigma) \left( g(\sigma) + \frac{mh(\sigma)}{l} \right) \right] \right)^{\frac{1}{l}} \tag{22}$$

*for $p \in \mathbb{N}_0$.*

***Proof*** Denoting $z(p)$ by

$$z(p) = \sum_{q=p_0}^{p-1} \left[g(q)u^l(q) + h(q)u^m(q)\right] \quad \text{for } p \in \mathbb{N}_0, \tag{23}$$

we obtain $z(p_0) = 0$ and

$$u^l(p) \le x(p) + y(p)z(p). \tag{24}$$

Applying Lemmas (1)–(24), we obtain

$$u^m(p) \le \left( \frac{l - m + mx(p)}{l} \right) + \frac{my(p)z(p)}{l}. \tag{25}$$

Using (24) and (25), we can write from (23) as follows:

$$\begin{aligned}
&z(p+1)-z(p) \\
&\leq g(p)\left[x(p)+y(p)z(p)\right]+h(p)\left[\frac{l-m+mx(p)+my(p)z(p)}{l}\right] \\
&=g(p)x(p)+g(p)y(p)z(p)+h(p)\left(\frac{l-m+mx(p)}{l}\right)+\frac{mh(p)y(p)z(p)}{l} \\
&=y(p)\left(g(p)+\frac{mh(p)}{l}\right)z(p)+g(p)x(p)+h(p)\left(\frac{l-m+mx(p)}{l}\right).
\end{aligned}$$

This gives

$$z(p+1)-\left(1+y(p)\left(g(p)+\frac{mh(p)}{l}\right)\right)z(p)\leq g(p)x(p)+h(p)\left(\frac{l-m+mx(p)}{l}\right). \tag{26}$$

Multiplying both sides of (26) by $\prod_{\sigma=p_0}^{p-1}\left[1+y(\sigma)\left(g(\sigma)+\frac{mh(\sigma)}{l}\right)\right]^{-1}$, we obtain

$$\begin{aligned}
&z(p+1)\prod_{\sigma=p_0}^{p-1}\left[1+y(\sigma)\left(g(\sigma)+\frac{mh(\sigma)}{l}\right)\right]^{-1}-z(p)\prod_{\sigma=p_0}^{p-2}\left[1+y(\sigma)\left(g(\sigma)+\frac{mh(\sigma)}{l}\right)\right]^{-1} \\
&\leq\left[g(p)x(p)+h(p)\left(\frac{l-m+mx(p)}{l}\right)\right]\times\prod_{\sigma=p_0}^{p-1}\left[1+y(\sigma)\left(g(\sigma)+\frac{mh(\sigma)}{l}\right)\right]^{-1}.
\end{aligned}$$

Taking $p=s$

$$\begin{aligned}
&z(s+1)\prod_{\sigma=p_0}^{s-1}\left[1+y(\sigma)\left(g(\sigma)+\frac{mh(\sigma)}{l}\right)\right]^{-1}-z(s)\prod_{\sigma=p_0}^{s-2}\left[1+y(\sigma)\left(g(\sigma)+\frac{mh(\sigma)}{l}\right)\right]^{-1} \\
&\leq\left[g(s)x(s)+h(s)\left(\frac{l-m+mx(s)}{l}\right)\right]\times\prod_{\sigma=p_0}^{s-1}\left[1+y(\sigma)\left(g(\sigma)+\frac{mh(\sigma)}{l}\right)\right]^{-1}.
\end{aligned}$$

By summing over $s$ from $p_0$ to $p-1$, we get

$$\begin{aligned}
&z(p_0+1)\prod_{\sigma=p_0}^{p_0-1}\left[1+y(\sigma)\left(g(\sigma)+\frac{mh(\sigma)}{l}\right)\right]^{-1}-z(p_0)\prod_{\sigma=p_0}^{p_0-2}\left[1+y(\sigma)\left(g(\sigma)+\frac{mh(\sigma)}{l}\right)\right]^{-1} \\
&\leq\left[g(p_0)x(p_0)+h(p_0)\left(\frac{l-m+mx(p_0)}{l}\right)\right]\times\prod_{\sigma=p_0}^{p_0-1}\left[1+y(\sigma)\left(g(\sigma)+\frac{mh(\sigma)}{l}\right)\right]^{-1}
\end{aligned}$$

$$z(p_0+2)\prod_{\sigma=p_0}^{p_0}\left[1+y(\sigma)\left(g(\sigma)+\frac{mh(\sigma)}{l}\right)\right]^{-1}-z(p_0+1)\prod_{\sigma=p_0}^{p_0-1}\left[1+y(\sigma)\left(g(\sigma)+\frac{mh(\sigma)}{l}\right)\right]^{-1}$$
$$\leq\left[g(p_0+1)x(p_0+1)+h(p_0+1)\left(\frac{l-m+mx(p_0+1)}{l}\right)\right]$$
$$\times\prod_{\sigma=p_0}^{p_0}\left[1+y(\sigma)\left(g(\sigma)+\frac{mh(\sigma)}{l}\right).\right]^{-1}$$

$$z(p_0+3)\prod_{\sigma=p_0}^{p_0+1}\left[1+y(\sigma)\left(g(\sigma)+\frac{mh(\sigma)}{l}\right)\right]^{-1}-z(p_0+2)\prod_{\sigma=p_0}^{p_0}\left[1+y(\sigma)\left(g(\sigma)+\frac{mh(\sigma)}{l}\right)\right]^{-1}$$
$$\leq\left[g(p_0+2)x(p_0+2)+h(p_0+2)\left(\frac{l-m+mx(p_0+2)}{l}\right)\right]$$
$$\times\prod_{\sigma=p_0}^{p_0+1}\left[1+y(\sigma)\left(g(\sigma)+\frac{mh(\sigma)}{l}\right)\right]^{-1}$$

and so on.

$$z(p)\prod_{\sigma=p_0}^{p-2}\left[1+y(\sigma)\left(g(\sigma)+\frac{mh(\sigma)}{l}\right)\right]^{-1}-z(p-1)\prod_{\sigma=p_0}^{p-3}\left[1+y(\sigma)\left(g(\sigma)+\frac{mh(\sigma)}{l}\right)\right]^{-1}$$
$$\leq\left[g(p-1)x(p-1)+h(p-1)\left(\frac{l-m+mx(p-1)}{l}\right)\right]$$
$$\times\prod_{\sigma=p_0}^{p-2}\left[1+y(\sigma)\left(g(\sigma)+\frac{mh(\sigma)}{l}\right)\right]^{-1}.$$

Adding all these inequalities, we get

$$z(p)\prod_{\sigma=p_0}^{p-2}\left[1+y(\sigma)\left(g(\sigma)+\frac{mh(\sigma)}{l}\right)\right]^{-1}-z(p_0)$$
$$\leq\sum_{s=p_0}^{p-1}\left[g(s)u(s)+h(s)\left(\frac{l-m+mx(s)}{l}\right)\right]\times\prod_{\sigma=p_0}^{s-1}\left[1+y(\sigma)\left(g(\sigma)+\frac{mh(\sigma)}{l}\right)\right]^{-1}.$$

Putting $z(p_0)=0$, we get

$$z(p)\prod_{\sigma=p_0}^{p-2}\left[1+y(\sigma)\left(g(\sigma)+\frac{mh(\sigma)}{l}\right)\right]^{-1}$$
$$\leq\sum_{s=p_0}^{p-1}\left[g(s)u(s)+h(s)\left(\frac{l-m+mx(s)}{l}\right)\right]\times\prod_{\sigma=p_0}^{s-1}\left[1+y(\sigma)\left(g(\sigma)+\frac{mh(\sigma)}{l}\right)\right]^{-1}.$$

Multiplying both sides by $\prod_{\sigma=p_0}^{p-2}\left[1+y(\sigma)\left(g(\sigma)+\frac{mh(\sigma)}{l}\right)\right]$ on both sides, we obtain

$$z(p) \le \sum_{s=p_0}^{p-1}\left[g(s)u(s)+h(s)\left(\frac{l-m+mx(s)}{l}\right)\right]\times\prod_{\sigma=s}^{p-2}\left[1+y(\sigma)\left(g(\sigma)+\frac{mh(\sigma)}{l}\right)\right].$$

Replace $\sigma$ by $\sigma-1$, we have

$$z(p) \le \sum_{s=p_0}^{p-1}\left[g(s)u(s)+h(s)\left(\frac{l-m+mx(s)}{l}\right)\right]\times\prod_{\sigma=s+1}^{p-1}\left[1+y(\sigma)\left(g(\sigma)+\frac{mh(\sigma)}{l}\right)\right]. \tag{27}$$

Using (27) in (24), we get the required inequality (22).

**Theorem 5** *Let $u, b, w, h$ be nonnegative real-valued continuous functions defined on $\mathbb{N}_0$ and $c$ be a positive real-valued nondecreasing function defined on $\mathbb{N}_0$. Let $l$ and $m$ be real constants such that $l \ge m \ge 1$. If*

$$u^l(p) \le c^l(p)+b(p)\sum_{s=p_0}^{p-1}\left[w(s)u^l(s)+h(s)u^m(s)\right] \tag{28}$$

*for $p \in \mathbb{N}_0$, then*

$$u(p) \le c(p)\left(1+b(p)\sum_{s=p_0}^{p-1}\left[\frac{w(s)u(s)}{c(s)}+\frac{h(s)c^m(s)}{c^l(s)}\right]\right.$$
$$\left.\times\prod_{\sigma=s+1}^{p-1}\left[1+b(\sigma)\left(w(\sigma)+\frac{mh(\sigma)c^m(\sigma)}{lc^l(\sigma)}\right)\right]\right)^{\frac{1}{l}}. \tag{29}$$

***Proof*** From (28), we have

$$\left(\frac{u(p)}{c(p)}\right)^l \le 1+b(p)\sum_{s=p_0}^{p-1}\left[w(s)\left(\frac{u(s)}{c(s)}\right)^l+\frac{h(s)c^m(s)}{c^l(s)}\left(\frac{u(s)}{c(s)}\right)^m\right]. \tag{30}$$

An application of Theorems (4)–(30) yields

$$\frac{u(p)}{c(p)} \le \left(1+b(p)\sum_{s=p_0}^{p-1}\left[\frac{w(s)u(s)}{c(s)}+\frac{h(s)c^m(s)}{c^l(s)}\left(\frac{l-m+m}{l}\right)\right]\right.$$
$$\left.\times\prod_{\sigma=s+1}^{p-1}\left[1+b(\sigma)\left(w(\sigma)+\frac{mh(\sigma)c^m(\sigma)}{lc^l(\sigma)}\right)\right]\right)^{\frac{1}{l}}.$$

This gives

$$u(p) \leq c(p) \left( 1 + b(p) \sum_{s=p_0}^{p-1} \left[ \frac{w(s)u(s)}{c(s)} + \frac{h(s)c^m(s)}{c^l(s)} \right] \right.$$
$$\left. \times \prod_{\sigma=s+1}^{p-1} \left[ 1 + b(\sigma) \left( w(\sigma) + \frac{mh(\sigma)c^m(\sigma)}{lc^l(\sigma)} \right) \right] \right)^{\frac{1}{l}}.$$

Hence, the proof.

**Theorem 6** *Let $u, a, b, g,$ and $h$ be nonnegative real-valued functions on $\mathbb{N}_0$ and $k(p, s)$ and $\Delta_1 k(p, s)$ be nonnegative real-valued continuous functions for $p, s \in \mathbb{N}_0$ with $p_0 \leq s \leq p$. Let $l$ and $m$ be real constants such that $l \geq m \geq 1$. If*

$$u^l(p) \leq a(p) + b(p) \sum_{s=p_0}^{p-1} k(p, s) \left[ g(s)u^l(s) + h(s)u^m(s) \right] \tag{31}$$

*for $p \in \mathbb{N}_0$, then*

$$u(p) \leq \left( a(p) + b(p) \sum_{\sigma=p_0}^{p-1} \bar{B}(\sigma) \prod_{\tau=\sigma+1}^{p-1} \left[ 1 + \bar{A}(\tau) \right] \right)^{\frac{1}{l}} \tag{32}$$

*for $p \in \mathbb{N}_0$, where*

$$\bar{A}(p) := k(p+1, p)b(p) \left( g(p) + \frac{mh(p)}{l} \right) + \sum_{s=p_0}^{p-1} \Delta_1 k(p, s)b(s) \left( g(s) + \frac{mh(s)}{l} \right) \tag{33}$$

*and*

$$\bar{B}(p) := k(p+1, p) \left[ g(p)a(p) + h(p) \left( \frac{l - m + ma(p)}{l} \right) \right]$$
$$+ \sum_{s=p_0}^{p-1} \Delta_1 k(p, s) \left[ g(s)u(s) + h(s) \left( \frac{l - m + ma(s)}{l} \right) \right]. \tag{34}$$

***Proof*** Define a function $z(p)$ by

$$z(p) = \sum_{s=p_0}^{p-1} k(p, s) \left[ g(s)u^l(s) + h(s)u^m(s) \right], \quad p \in \mathbb{N}_0. \tag{35}$$

Then $z(p_0) = 0$ and (31) can be written as

$$u^l(p) \leq a(p) + b(p)z(p). \tag{36}$$

Applying Lemma (1)–(36), we obtain

$$u^m(p) \leq \frac{l-m}{l} + \frac{ma(p)}{l} + \frac{mb(p)z(p)}{l}. \tag{37}$$

Using (33), (34), (36), and (37), we can write from (35) as follows:

$$z(p+1) - z(p) \leq \bar{A}(p)z(p) + \bar{B}(p).$$

This gives

$$z(p+1) - \left[1 + \bar{A}(p)\right] z(p) \leq \bar{B}(p). \tag{38}$$

Multiplying both sides of (38) by $\prod_{\sigma=p_0}^{p-1} \left[1 + \bar{A}(\sigma)\right]^{-1}$, we obtain

$$z(p+1) \prod_{\sigma=p_0}^{p-1} \left[1 + \bar{A}(\sigma)\right]^{-1} - z(p) \prod_{\sigma=p_0}^{p-2} \left[1 + \bar{A}(\sigma)\right]^{-1} \leq \bar{B}(p) \prod_{\sigma=p_0}^{p-1} \left[1 + \bar{A}(\sigma)\right]^{-1}.$$

Taking $p = s$ we get

$$z(s+1) \prod_{\sigma=p_0}^{s-1} \left[1 + \bar{A}(\sigma)\right]^{-1} - z(s) \prod_{\sigma=p_0}^{s-2} \left[1 + \bar{A}(\sigma)\right]^{-1} \leq \bar{B}(s) \prod_{\sigma=p_0}^{s-1} \left[1 + \bar{A}(\sigma)\right]^{-1}.$$

Putting $s = p_0, p_0 + 1, p_0 + 2, \ldots, p - 1$ and taking summation for all inequalities, we get

$$z(p) \prod_{\sigma=p_0}^{p-2} \left[1 + \bar{A}(\sigma)\right]^{-1} - z(p_0) \leq \sum_{s=p_0}^{p-1} \bar{B}(s) \prod_{\sigma=p_0}^{s-1} \left[1 + \bar{A}(\sigma)\right]^{-1}.$$

Since $z(p_0) = 0$, we have

$$z(p) \prod_{\sigma=p_0}^{p-2} \left[1 + \bar{A}(\sigma)\right]^{-1} \leq \sum_{s=p_0}^{p-1} \bar{B}(s) \prod_{\sigma=p_0}^{s-1} \left[1 + \bar{A}(\sigma)\right]^{-1}.$$

Multiplying both sides by $\prod_{\sigma=p_0}^{p-2} \left[1 + \bar{A}(\sigma)\right]$, we get

$$z(p) \leq \sum_{s=p_0}^{p-1} \bar{B}(s) \prod_{\sigma=s}^{p-2} \left[1 + \bar{A}(\sigma)\right].$$

That is,

$$z(p) \leq \sum_{\sigma=p_0}^{p-1} \bar{B}(\sigma) \prod_{\tau=\sigma+1}^{p-1} \left[1 + \bar{A}(\tau)\right]. \tag{39}$$

Now, using (39) in (36), we get (32) and the proof is complete.

## 4 Applications

**Example 1** Consider the nonlinear differential inequality

$$u^6(p) \leq p^2 + p^3 \int_0^p \left[\frac{1}{(1+s)} u^5(s) + \frac{1}{(1+s)^3} u^4(s)\right] ds, \quad p \in \mathbb{R}_+, \tag{40}$$

where $u$ is a nonnegative real-valued continuous function on $\mathbb{R}_+$.

Suppose that solution of (40) exists on $\mathbb{R}_+$. Then making use of Theorem 1 yields

$$u(p) \leq \Bigg(p^2 + p^3 \int_0^p \left[\frac{1}{6(1+s)} + \frac{5s^2}{6(1+s)} + \frac{1}{3(1+s)^3} + \frac{2s^2}{3(1+s)^3}\right]$$
$$\times \exp\left(\int_s^p \sigma^3 \left(\frac{5}{6(1+\sigma)} + \frac{2}{3(1+\sigma)^3}\right) d\sigma\right) ds\Bigg\}^{\frac{1}{6}} \tag{41}$$

for $p \in \mathbb{R}_+$. The right-hand side of (41) gives an exact bound on the solution of (40).

**Example 2** Consider the nonlinear integral inequality

$$y^8(p) \leq g(p) + \int_0^p k(p,s) D(s, y(s)) ds, \quad p \in \mathbb{R}_+, \tag{42}$$

where $y$ and $g$ are positive real-valued nondecreasing continuous functions defined on $\mathbb{R}_+$ such that $|g(p)| \leq p^8$ and $D, D$ are nonnegative real-valued continuous functions defined on $\mathbb{R}_+ \times \mathbb{R}_+$ such that $|D(s, y(s))| \leq s^4|y^6(s)| + s^6|y^4(s)|$ and $|K(p,s)| \leq 1$.

Suppose that solution of (42) exists on $\mathbb{R}_+$. Then (42) becomes

$$|y^8(p)| \leq p^8 + \int_0^p \left[s^4|y^6(s)| + s^6|y^4(s)|\right] ds.$$

Now, application of Theorem 2 yields

$$|y(p)| \le p\left(1+2\int_0^p s^2 \times \exp\left(\int_s^p \left(\frac{3\sigma^2}{4}+\frac{\sigma^2}{2}\right) d\sigma\right) ds\right)^{\frac{1}{8}}$$
$$\le p\left(1+2\int_0^p s^2 \times \exp\left(\frac{5(p^3-s^3)}{12}\right) ds\right)^{\frac{1}{8}}$$
$$\le p\left(1+\frac{8}{5}\left(\exp\frac{5(p^3-s^3)}{12}-1\right)\right)^{\frac{1}{8}}$$

for $p \in \mathbb{R}_+$. This gives an explicit bound for the solution of (42).

## References


1. Pachpatte, B.G.: On some new Inequalities related to a certain inequality arising in the theory of differential equations. J. Math. Anal. Appl. **251**, 736–751 (2000)
2. Bellman, R.: The stability of solutions of linear differential equations. Duke Math. J. **10**, 643–647 (1943)
3. Gronwall, H.T.: Note on the derivatives with respect to a parameter of the solutions of a system of differential equations. Ann. Math. **2**, 292–296 (1918–1919)
4. Abdeldaim, A., El-Deeb, A.A., Ahmed, R.G.: On retarded nonlinear integral inequalities of Gronwall and applications. J. Math. Inequalties **13**(4), 1023–1038 (2019)
5. El-Owaidy, H., Abdeldaim, A., El-Deeb, A.A.: On some new retarded nonlinear integral inequalities and their applications. Math. Sci. Lett. **3**, 157–164 (2014)
6. Agarwal, R.P., Kim, Y.H., Sen, S.K.: New retarded integral inequalities with applications. J. Inequalities Appl. 789–784 (2008)
7. Dhakne, M.B., Kendre, S.D.: On nonlinear Volterra integrodifferential equation in Banach spaces. Math. Inequalities Appl. **9**(4), 725–734 (2006)
8. El-Owaidy, H., Ragab, A., Abdeldaim, A.: On some new integral inequalities of Gronwall-Bellman type. Appl. Math. Comput. **106**, 289–303 (1999)
9. Kendre, S.D., Latpate, S.G.: On some nonlinear integral inequalities for Volterra integral equations. Anan. Alexandru Loan Cuza Univ.-Math. (2014)
10. Lakshmikantham, V., Leela, S.: Differential and Integral Inequalities, vol. I. Academic Press, New York (1969)
11. Pachpatte, B.G.: Inequalities for Differential and Integral Equations. Academic Press, New York and London (1998)
12. Kim, Y.-H.: Gronwall-Bellman and Pachpatte type integral inequalities with applications. Nonlinear Anal. **71**, e2641–e2656 (2009)

# Mathematical Modelling and Fluid Dynamics

# Solving Multi-objective Chance Constraint Quadratic Fractional Programming Problem

**Berhanu Belay and Adane Abebaw**

**Abstract** In this manuscript, the method of multi-objective chance constraint quadratic fractional programming problem is presented. A multi-objective chance constraint quadratic fractional programming problem is formulated by assuming some parameters as continuous random variables following logistic distribution. In the proposed mathematical model, only the right-hand side parameters are assumed to be random variables following logistic distribution. The chance constraints are handled by the concept of cumulative distribution function. After changing the proposed stochastic model into an equivalent deterministic model, the lexicography approach is used to get the Pareto optimal solution of the proposed model. The resulting single-objective quadratic fractional programming problem is solved by Dinkelbach algorithm together with LINGO 14.0 software. Finally, an example is provided to illustrate the proposed method.

**Keywords** Multi-objective optimization · Stochastic programming · Fractional programming · Quadratic programming · Lexicography method

## 1 Introduction

Many decision-making problems have multiple and conflicting objectives in real-life problems which is termed as multi-objective programming (MOP) problem. Some examples of multi-objective programming problem are maximizing profit and minimizing cost, maximizing production and minimizing risk, maximizing quality and minimizing cost in purchasing a car, etc. In an MOP problem, if the functions to be optimized are the ratio of affine functions, then the problem is termed as multi-objective linear fractional programming (MOLFP) problem. But if either of the objective function is not linear, then the MOFP problem is called multi-objective non-linear fractional programming problem. Quadratic fractional programming (QFP)

---

B. Belay (✉) · A. Abebaw
Department of Mathematics, Debre Tabor University, College of Natural and Computational Sciences, Debre Tabor, Ethiopia
e-mail: berhanubelay2@gmail.com



B. Rushi Kumar et al. (eds.), *Mathematics and Computing*, Springer Proceedings in Mathematics & Statistics 415, https://doi.org/10.1007/978-981-19-9307-7_36

problem is a nonlinear fractional programming problem where the objective function is quadratic function having a set of linear equality or inequality constraints. MOP problems may be uncertain due to randomness. In this case, the MOP problem is called multi-objective chance constrained programming (MOCCP) problem.

## 2 Literature Review

In MOCCP problem, it is challenging to get best compromise solution without finding its deterministic equivalent. To overcome this problem, [8] solved MOCCP problem using a fuzzy programming method where the parameters follow continuous distribution. Reference [7] suggested genetic algorithm for stochastic fuzzy problems, [3] derived the deterministic of the chance constraint, [12] explained fuzzy multiple objective programming and its solution, [11] suggested an approach for probabilistic programming by considering the data as a uniform random variable. Reference [5] solved MOCCP problem when the parameters follow generalized distribution. Reference [2] proposed a method for fuzzy probabilistic model with multiple objectives that involve log normal random variables. Reference [10] solved MOCCP problem by assuming the uncertain parameters as Weibull random variable. Reference [14] obtained solution of MOP problems that have uncertain random variables. Reference [6] proposed parametric approach for both nonlinear and linear fractional problem. Reference [9] proposed parametric approach for quadratic fractional programming problem. Reference [13] presented the application of fractional programming problem in economical and non-economical areas. Reference [1] proposed the methodology for fractional programming with uncertain parameters, [4] proposed a method for fuzzy fractional programming.

## 3 Mathematical Model of MOCC Quadratic Fractional

In any mathematical model if the objective function is the product of two affine functions, then the programming problem is called linear factorized quadratic programming problem. In an MOP problem if more than one quadratic fractional objectives are optimized at the same time subject to given constraints, then it is called multi-objective quadratic fractional programming problem. We consider a mathematical model where the fractional objective functions are quadratic, multiple, in commensurable and conflicting with each other. A multi-objective chance constraint quadratic fractional programming (MOCCQFP) problem occurs when the quadratic fractional objective functions are to be maximized or minimized subject to probabilistic constraints. Consider the following multi-objective chance constraint nonlinear fractional programming problem:

$$\max : Z_k = \frac{N_k(X)}{D_k(X)} \tag{1}$$

subject to

$$P(AX \leq b) \geq \eta \tag{2}$$

$$0 < \eta < 1 \tag{3}$$

$$X \geq 0, \tag{4}$$

where $N_k(X)$ and $D_k(X)$ are nonlinear functions. The probabilistic programming problem given in (1)–(4) is said to be MOCCQFP problem, if either $N_k(X)$ or $D_k(X)$ are quadratic functions. The objective function of QFP problem can be formulated in different models. Among these functions, linear factorized quadratic function is mostly known and mathematically expressed as $\frac{N_k(x_j)}{D_k(x_j)} = \frac{(c_{1k}^t X+\alpha_k)(c_{2k}^t X+\beta_k)}{(d_{1k}^t X+\delta_k)(d_{2k}^t X+\omega_k)}$. Therefore the MOCCQFP problem can be expressed as

$$\max : Z_k = \frac{(c_{1k}^t X + \alpha_k)(c_{2k}^t X + \beta_k)}{(d_{1k}^t X + \delta_k)(d_{2k}^t X + \omega_k)} \tag{5}$$

subject to

$$P(AX \leq b) \geq \eta \tag{6}$$

$$0 < \eta < 1 \tag{7}$$

$$X \geq 0, \tag{8}$$

where $c_{1k}, c_{2k}, d_{1k}, d_{2k} \in R^n, \alpha_k, \alpha_k, \alpha_k, \alpha_k \in R, A \in R^{mxn}, \eta, b \in R^m$ and the factors $c_{1k}^t X + \alpha_k, c_{2k}^t X + \beta_k, d_{1k}^t X + \delta_k, d_{2k}^t X + \omega_k \neq 0.$

## 4 Deterministic Model

Let the random variable $b$ follow logistic distribution with two parameters $\mu$ and $\gamma$ which are location and scale parameters, respectively. The deterministic equivalent of the chance constrained is obtained by using probability distribution function (PDF). The PDF of $b$ is expressed as

$$f(b) = \frac{e^{-\left(\frac{b-\mu}{\gamma}\right)}}{\gamma\left(1+e^{-\left(\frac{b-\mu}{\gamma}\right)}\right)^2} \tag{9}$$

$$-\infty < b < \infty, -\infty < \mu < \infty, \gamma > 0. \tag{10}$$

Now, to find the deterministic equivalent of the chance constraint, consider the following expression:

$$P\left(\sum_{j=1}^{n} a_{ij}x_j \le b_i\right) \ge \eta_i \tag{11}$$

$$P\left(b_i \ge \sum_{j=1}^{n} a_{ij}x_j\right) \ge \eta_i. \tag{12}$$

Let

$$y_i = \sum_{j=1}^{n} a_{ij}x_j.$$

Then the probabilistic constraint (12) is written as

$$1 - P(b_i \le y_i) \ge \eta_i \tag{13}$$

$$1 - \int_{-\infty}^{y_i} \frac{e^{-\left(\frac{b-\mu_i}{\gamma_i}\right)}}{\gamma_i\left(1 + e^{-\left(\frac{b-\mu_i}{\gamma_i}\right)}\right)^2} db_i \ge \eta_i, i = 1, 2, \ldots, m \tag{14}$$

using integration by substitution and substituting the limit of integrations, we have

$$1 - \frac{1}{1 + e^{-\left(\frac{y_i-\mu_i}{\gamma_i}\right)}} \ge \eta_i \tag{15}$$

which is simplified as

$$e^{-\left(\frac{y_i-\mu_i}{\gamma_i}\right)} \ge \frac{\eta_i}{1-\eta_i}. \tag{16}$$

Solving for $y_i$, we have

$$y_i \le -ln\left(\frac{\eta_i}{1-\eta_i}\right)\gamma_i + \mu_i \tag{17}$$

$$\Rightarrow \sum_{j=1}^{n} a_{ij}x_j \le -ln\left(\frac{\eta_i}{1-\eta_i}\right)\gamma_i + \mu_i. \tag{18}$$

Substituting (18) in (6), the deterministic equivalent of problems (5)–(8) is expressed as follows:

$$\max : Z_k = \frac{(c_{1k}^t x_j + \alpha_k)(c_{2k}^t x_j + \beta_k)}{(d_{1k}^t x_j + \delta_k)(d_{2k}^t x_j + \omega_k)} \tag{19}$$

subject to

$$\sum_{j=1}^{n} a_{ij}x_j \leq -ln\left(\frac{\eta_i}{1-\eta_i}\right)\gamma_i + \mu_i \tag{20}$$

$$0 < \eta_i < 1 \tag{21}$$

$$x_j \geq 0. \tag{22}$$

## 5 Solution Procedure

Since the MOP problem given in (5)–(8) involves uncertain parameter, several objectives, and fractional objectives, it is challenging to solve directly. The efficient solution is obtained by converting MOCCQFP problem into deterministic equivalent MOQFP problem. Then Lexicography approach is applied to get the efficient solution of deterministic MOQFP problem. Lexicography is used as it is simple to use and preferences are imposed by ordering the objective functions according to their importance rather than assigning weights. Hence we use Dinkelbach algorithm for finding solution of single-objective QFP problem. The algorithm is directly applied to solve QFP problems. Parametric approach is the most well-known method for fractional programming problem (not necessarily linear). It is developed by Dinkelbach [6]. The Dinkelbach algorithm for QFP problem is explained as follows. Consider the single-objective QFP problem

$$\max : Z = \frac{N(X)}{D(X)} \tag{23}$$

subject to

$$AX \leq b \tag{24}$$

$$X \geq 0 \tag{25}$$

suppose that

$$F(\lambda) = \max\{N(X) - \lambda D(X)\}, \lambda \in R. \tag{26}$$

According to Dinkelbach's vector $X$ is an optimal solution of the programming problem given in (23)–(25) if

$$F(\lambda^*) = \max_{x \in S}\{N(X) - \lambda^* D(X)\} = 0, \tag{27}$$

where $S$ denotes the feasible set and $\lambda^* = \frac{N(X^*)}{D(X^*)}$.

The algorithm of Dinkelbach is described as follows:

**step 0**. Take $x^0 \in S$ as a starting solution, compute $\lambda^1 = \frac{N(x^0)}{D(x^0)}$, set $k = 1$.
**step 1**. Determine $x^k = \max\{N(X) - \lambda^k D(X)\}$, $X \in S$.
**step 2**. If $F(\lambda^k) = 0$, then $X^* = X^k$ is an optimal solution, then stop the algorithm. In this case, the optimal solution is solved by using lingo software. The pseudocode for Lingo is expressed as follows:
Start:// Model:// max / min = $objectivefunction$;// evaluate the constraints;// set the non negativity criteria;// end;//
**step 3**. Let $\lambda^{k+1} = \frac{N(X^k)}{D(X^k)}$, Set $k = k + 1$, go to step 1.

## 6 Numerical Example

Consider the following MOPQFP problem:

$$\max Z_1 = \frac{(2x_1 + x_2 + 1)(2x_1 + x_2 + 2)}{2x_1 + 2x_2 + 2} \tag{28}$$

$$\max Z_2 = \frac{(-8x_1 - 4x_2 - 4)(6x_1 + 3x_2 + 6)}{5x_1 + 5x_2 + 5} \tag{29}$$

subject to

$$P(x_1 + 2x_2 \leq b_1) \geq 0.85 \tag{30}$$

$$P(3x_1 + x_2 \leq b_2) \geq 0.95 \tag{31}$$

$$x_1, x_2 \geq 0, \tag{32}$$

where $b_1$ and $b_2$ are random variables that follow logistic distribution with known parameters $\mu(b_1) = 8$, $\gamma(b_1) = 2$, $\mu(b_2) = 14$, $\gamma(b_2) = 3$.

Now, using Eqs. (19)–(22) the deterministic equivalent of the MOPQFP problem is expressed as follows:

$$\max Z_1 = \frac{(2x_1 + x_2 + 1)(2x_1 + x_2 + 2)}{2x_1 + 2x_2 + 2} \tag{33}$$

$$\max Z_2 = \frac{(-8x_1 - 4x_2 - 4)(6x_1 + 3x_2 + 6)}{5x_1 + 5x_2 + 5} \tag{34}$$

subject to

$$x_1 + 2x_2 \leq 4.531 \tag{35}$$

$$3x_1 + x_2 \leq 5.167 \tag{36}$$

$$x_1, x_2 \geq 0. \tag{37}$$

Apply lexicography method to obtain the efficient solution of the given programming problem. Now, optimize the first objective function $Z_1$ as follows:

$$\max Z_1 = \frac{(2x_1 + x_2 + 1)(2x_1 + x_2 + 2)}{2x_1 + 2x_2 + 2} \tag{38}$$

subject to

$$x_1 + 2x_2 \leq 4.531 \tag{39}$$

$$3x_1 + x_2 \leq 5.167 \tag{40}$$

$$x_1, x_2 \geq 0. \tag{41}$$

Now, let's apply the Dinkelbach algorithm for QFP problem as follows:

**step 0**: Take $X^0 = (x_1, x_2) = (0, 0)$ as a starting point which is the feasible solution and compute $\lambda^1 = \frac{N(X^*)}{D(X^*)}$, i.e., $\lambda^1 = 1$.

**step 1**: Determine $\max\{N(X) - \lambda^1 D(X)\}$, $X \in S$. This step is expressed as follows:

$$F(\lambda^1) = \max(2x_1 + x_2 + 1)(2x_1 + x_2 + 2) - \lambda^1(2x_1 + 2x_2 + 2) \tag{42}$$

subject to

$$x_1 + 2x_2 \geq 4.531 \tag{43}$$

$$3x_1 + x_2 \leq 5.167 \tag{44}$$

$$x_1, x_2 \leq 0. \tag{45}$$

**step 2**: Check $F(\lambda^1)$ is zero or not. Hence solving (42)–(45) using LINGO, we have $X^1 = (1.1606, 1.6852)$ and $F(\lambda^1) = 20.37884 \neq 0$.

**step 3**: Repeating steps 1 and 2 until $\max\{N(X) - \lambda^k D(X)\}$, $X \in S = 0$. Therefore optimize the following programming problem:

$$F(\lambda^2) = \max(2x_1 + x_2 + 1)(2x_1 + x_2 + 2) - \lambda^2(2x_1 + 2x_2 + 2), \lambda^2 = \frac{N(X^1)}{D(X^1)} = 3.909517 \tag{46}$$

subject to

$$x_1 + 2x_2 \geq 4.531 \quad (47)$$

$$3x_1 + x_2 \leq 5.167 \quad (48)$$

$$x_1, x_2 \leq 0. \quad (49)$$

Again solving (46)–(49) using LINGO, we have $X^2 = (1.72233, 0)$ and $F(\lambda^2) = 2.913712 \neq 0$.
Proceeding the same procedure we obtain $X^5 = (1.722333, 0)$ and $F(\lambda^5) = 0$.

Therefore, the optimal solution of the programming problem (38)–(41) is $(x_1, x_2) = (1.722333, 0)$ and the value of $Z_1$ is 4.444666. Similarly, optimize the second objective function subject to the original constraint and the maximization of the prior objective function max $Z_1(x)$ is considered as one of the constraints in addition to the original constraints.

$$\max Z_2 = \frac{(-8x_1 - 4x_2 - 4)(6x_1 + 3x_2 + 6)}{5x_1 + 5x_2 + 5} \quad (50)$$

subject to

$$x_1 + 2x_2 \leq 4.531 \quad (51)$$

$$3x_1 + x_2 \leq 5.167 \quad (52)$$

$$\frac{(2x_1 + x_2 + 1)(2x_1 + x_2 + 2)}{2x_1 + 2x_2 + 2} \geq 4.444666 \quad (53)$$

$$x_1, x_2 \geq 0. \quad (54)$$

Here $\frac{(2x_1+x_2+1)(2x_1+x_2+2)}{2x_1+2x_2+2} \geq 4.444666$ is included in the constraint. Apply the Dinkelbach algorithm to find the optimal solution of the QFP problem (50)–(54).

**step 0**: Take $X^0 = (x_1, x_2) = (1.722333, 0)$ as a starting solution which is the feasible solution and compute $\lambda^1 = \frac{N(X^0)}{D(X^0)}$, i.e., $\lambda^1 = -21.33439668$.

**step 1**: Determine $\max\{N(X) - \lambda^1 D(X)\}, X \in S$. This step is expressed as follows:

$$F(\lambda^1) = (-8x_1 - 4x_2 - 4)(6x_1 + 3x_2 + 6) - \lambda^1(5x_1 + 5x_2 + 5) \quad (55)$$

subject to

$$x_1 + 2x_2 \leq 4.531 \quad (56)$$

$$3x_1 + x_2 \leq 5.167 \quad (57)$$

$$\frac{(2x_1 + x_2 + 1)(2x_1 + x_2 + 2)}{2x_1 + 2x_2 + 2} \geq 4.444666 \tag{58}$$

$$x_1, x_2 \geq 0. \tag{59}$$

**step 2**: Check $F(\lambda^1)$ is zero or not. Hence solving (55)–(59) using LINGO, we have $X^1 = (1.722333, 0)$ and $F(\lambda^1) = -0.1633400E - 05 \neq 0$.

**step 3**: Repeating steps 1 and 2 until $\max\{N(X) - \lambda^k D(X)\}, X \in S = 0$. Therefore optimize the following programming problem:

$$F(\lambda^2) = (-8x_1 - 4x_2 - 4)(6x_1 + 3x_2 + 6) - \lambda^2(5x_1 + 5x_2 + 5), \lambda^2 = \frac{N(X^1)}{D(X^1)} = -21.33439668 \tag{60}$$

subject to

$$x_1 + 2x_2 \leq 4.531 \tag{61}$$

$$3x_1 + x_2 \leq 5.167 \tag{62}$$

$$\frac{(2x_1 + x_2 + 1)(2x_1 + x_2 + 2)}{2x_1 + 2x_2 + 2} \geq 4.444666 \tag{63}$$

$$x_1, x_2 \geq 0. \tag{64}$$

Again solving (46)–(49) using LINGO, we have $X^2 = (1.72233, 0)$ and $F(\lambda^2) = -0.1633400E - 05 \neq 0$.
Proceeding the same procedure we obtain the same value $X^k = (1.722333, 0)$ and $F(\lambda^k) = -0.1633400E - 05, k = 2, 3 \ldots$

Therefore, the optimal solution of the programming problem (50)–(54) is $(x_1, x_2) = (1.722333, 0)$ and the value of $Z_2$ is $-21.3343968$.

This gives one of the efficient solutions for the original MOPQFP problem which is $x_1 = 1.722333, 0$, with max $Z_1 = 4.444666.$, max $Z_2 = -21.3343968$.

In any multi-objective QFP problem, there exist number of good efficient solutions. These efficient solutions are equally acceptable. Choosing the efficient solution depends on the situation that decision-makers preferred. The preference of decision-maker depends on different conditions like budget, row material, resource, time limit, etc. Therefore, having more efficient solution to multi-objective QFP problem is necessary for decision-makers to select the best solution among the given alternatives which satisfies their need and capacity.

Hence, one can find more than one efficient solution for the above programming problem by applying the above procedure. In this case, first choose the second objective function and optimize it subject to the given constraints. Finally, optimizing the first objective function $Z_1$ subject to the original constraint and the maximization of the prior objective function max $Z_2(x)$ is considered as one of the constraints in addition to the original constraints.

## 7 Conclusion

In the manuscript, MOCCQFP problem is solved by finding the deterministic equivalent using cumulative distribution function. An MOCCQFP problem is formulated by taking the parameters in right-hand side as random variables following logistic distribution. In this model, randomness has been used in the right-hand side parameters and not in the objective functions, nor as the coefficient present in the constraints. The resulting MOQFP problem is directly solved by lexicography method. This is important for decision-makers to make a good decision by considering all the possible directions. The theoretical implication of the proposed method is that it is used to get alternative non-dominated solutions for which the decision-maker can easily select the best compromise solutions. Besides this theoretical implication, the method has managerial implication for real-life problems. The single-objective QFP problem is solved by Dinkelbach algorithm which is easy to find the optimal solution of fractional optimization problem. Once the Dinkelbach algorithm is applied, then Lingo is applied to find the optimal solution of the crisp problem. The method can be extended to multi-objective fuzzy probabilistic quadratic fractional programming problems.

## References

1. Acharya, S., Belay, B., Mishra, R.: Multi-objective probabilistic fractional programming problem involving two parameters Cauchy distribution. Math. Model. Anal. **24**(3), 385–403 (2019). https://doi.org/10.3846/mma.2019.024
2. Acharya, S., Ranarahu, N., Dash, J., Acharya, M.M.: Solving multi-objective fuzzy probabilistic programming problem. J. Intell. Fuzzy Syst. **26**(2), 935–948 (2014). https://doi.org/10.3233/IFS-130784
3. Biswal, M.P., Biswal, N., Li, D.: Probabilistic linear programming problems with exponential random variables: a technical note. Eur. J. Oper. Res. **111**(3), 589–597 (1998). https://doi.org/10.1016/S0377-2217(97)90319-2
4. Chakraborty, M., Gupta, S.: Fuzzy mathematical programming for multi objective linear fractional programming problem. Fuzzy Sets Syst. **125**(3), 335–342 (2002). https://doi.org/10.1016/S0165-0114(01)00060-4
5. Charles, V., Ansari, S., Khalid, M.: Multi-objective stochastic linear programming with general form of distributions. Int. J. Oper. Res. Optim. **2**(2), 261–278 (2011). https://optimization-online.org/?p=10963
6. Dinkelbach, W.: On nonlinear fractional programming. Manage. Sci. **13**(7), 492–498 (1967). https://doi.org/10.1287/mnsc.13.7.492
7. Dutta, S., Acharya, S., Mishra, R.: Genetic algorithm approach for solving multiobjective fuzzy stochastic programming problem. Int. J. Math. Oper. Res. **11**(1), 1–28 (2017). https://doi.org/10.1504/IJMOR.2017.085377
8. Hulsurkar, S., Biswal, M.P., Sinha, S.B.: Fuzzy programming approach to multiobjective stochastic linear programming problems. Fuzzy Sets Syst. **88**(2), 173–181 (1997). https://doi.org/10.1016/S0165-0114(96)00056-5
9. Ibaraki, T., Ishii, H., Iwase, J., Hasegawa, T., Mine, H.: Algorithms for quadratic fractional programming problems. J. Oper. Res. Soc. Jpn. **19**(2), 174–191 (1976). https://doi.org/10.15807/jorsj.19.174

10. Javaid, S., Ansari, S., Anwar, Z.: Multi-objective stochastic linear programming problem when bi's follow Weibull distribution. Opsearch **50**(2), 250–259 (2013). https://doi.org/10.1007/s12597-012-0101-6
11. Mohammed, W.: Chance constrained fuzzy goal programming with right-hand side uniform random variable coefficients. Fuzzy Sets Syst. **109**(1), 107–110 (2000). https://doi.org/10.1016/S0165-0114(98)00151-1
12. Ren, C.F., Li, R.H., Zhang, L.D., Guo, P.: Multiobjective stochastic fractional goal programming model for water resources optimal allocation among industries. J. Water Resour. Plan. Manag. **142**(10), 04016036 (2016)
13. Schaible, S.: Fractional programming. In: Handbook of Global Optimization, pp. 495–608. Springer (1995). https://doi.org/10.1007/978-1-4615-2025-2_10
14. Zhou, J., Yang, F., Wang, K.: Multi-objective optimization in uncertain random environments. Fuzzy Optim. Decis. Making **13**(4), 397–413 (2014). https://doi.org/10.1007/s10700-014-9183-3

# Higher Order Variational Symmetric Duality Over Cone Constraints

**Sony Khatri and Ashish Kumar Prasad**

**Abstract** The prime objective of our discussions moves around the higher order variational symmetric dual pairs for which constraints are defined over cones and to explore relevant duality relations for the constructed duals. Making use of higher order $\eta$-invexity, we derive appropriate duality results and validate the obtained results with the help of numerical examples. Further, we discuss the static case of the considered dual problems.

**Keywords** Higher order $\eta$-invexity · Symmetric duality · Cone constraints · Variational problem

## 1 Introduction

Variational principles provide a broad spectrum of mathematical theory for solving modeling problems studied in dynamical systems that have developed along with the study of optimization, stability, and control theory. Initially, it started with the expedition of variations around a point within the bounds specified by constraints. Several investigations were extensively made which required notions of generalized differentiability. The development of nonsmooth analysis opened various new dimensions rooted in the basic variational principles requiring duality results to be established for such problems.

Dantzig [3] was the first to introduce the concept of symmetric duality by extending the work of Dorn [4] to such problems. Bazaraa and Goode [1] examined these results taking convex and concave functions. Mangasarian [12] studied the duals

S. Khatri · A. K. Prasad (✉)
Vellore Institute of Technology, Vellore, TN 632014, India
e-mail: ashishprasa@gmail.com

S. Khatri
e-mail: sony.khatri2019@vitstudent.ac.in

A. K. Prasad
Presidency University, Bangalore 560064, India



B. Rushi Kumar et al. (eds.), *Mathematics and Computing*, Springer Proceedings in Mathematics & Statistics 415, https://doi.org/10.1007/978-981-19-9307-7_37

formulated by superimposing second-order as well as higher order terms with viewpoint of obtaining closer bounds. Mond and Weir [13] introduced the concept of another type of dual for nonlinear programming problems where both primal and dual were akin to each other. Chen [2] established duality results by incorporating support functions to deal higher order multiobjective programming problems. Khurana [11] examined the results needed to study dual relations for Mond–Weir-type multiobjective symmetric programming problems over arbitrary cones. Kaseem [8, 9] investigated the multiobjective nonlinear programming problems containing first-order symmetric dual program constraining the functions to be convex and concave and proposed all the three duality results over cone constraints utilizing $F$-convexity. Jayswal et al. [7] constructed a dual pair consisting of second-order multiobjective symmetric variational control problems. Prasad et al. [14] focused on second-order fractional variational problems and derived several duality theorems. Yang [17] focused on duality results for the first-order symmetric dual program with the help of invexity and derived suitable criteria for dual constructions.

Gupta and Gulati [5] studied higher order symmetric dual problems where constraints are defined over cones. Recently, Jayswal et al. [6] magnified the work to fractional symmetric dual model, whereas Suneja and Louhan [15] elongated the same using higher order cone invexity. Furthermore, Verma et al. [16] introduced a novel approach to study higher order multiobjective symmetric dual problems using cone invexity. Sharma and Kaur [10] focused on higher order symmetric fractional multiobjective problems over cones and explored various theorems of dual formulations with the help of higher order $(\phi, \rho)$ cone convex function.

In the next few sections, we concentrate on higher order variational symmetric problems constructed over cone constraints and introduce well-suited duality results using higher order $\eta$-invexity. The planning of the article is as per the following scheme. In Sect. 2, we compile the higher order $\eta$ invex function and some definitions based on which our investigations are carried out. A numerical example is also constructed in order to authenticate the definition used in this article. In Sect. 3, we get into a higher order variational symmetric fractional problems where constraints are depicted over cones and extract compulsory duality results and conclusions in the last two sections.

## 2 Preliminaries

$\mathbb{R}^n$ denotes ordered $n$-tuples of reals and $\mathbb{R}^n_+$ denotes the set of all elements of $\mathbb{R}^n$ whose each component is nonnegative. We use subscript "T" to denote transpose of a matrix.

**Definition 2.1** A subset $C \subset \mathbb{R}^n$ characterized by

$$0 \leq \lambda \in \mathbb{R},\ \pi \in C \Rightarrow \lambda\pi \in C$$

is known as a cone.

**Definition 2.2** For any cone $C$, its polar cone $C^*$ is mathematically represented as

$$C^* = \{\tilde{\pi} : \pi^T \tilde{\pi} \leqq 0, \ \forall \, \pi \in C\}.$$

**Definition 2.3** The functional $\int_{\tau_1}^{\tau_2} f(t, \varrho, \dot{\varrho})\, dt$ is known to be higher order $\eta$-invex at $\vartheta \in \mathbb{R}^n$ w.r.t. $h : I \times \mathbb{R}^n \times \mathbb{R}^n \mapsto \mathbb{R}$ provided there exists a suitable function $\eta : I \times \mathbb{R}^n \times \mathbb{R}^n \mapsto \mathbb{R}^n$ satisfying

$$\int_{\tau_1}^{\tau_2} f(t, \varrho, \dot{\varrho})\, dt - \int_{\tau_1}^{\tau_2} f(t, \vartheta, \dot{\vartheta})\, dt - \int_{\tau_1}^{\tau_2} h(t, \vartheta, \dot{\vartheta}, p) dt + \int_{\tau_1}^{\tau_2} p^t \nabla_p h(t, \vartheta, \dot{\vartheta}, p) dt$$

$$\geqq \int_{\tau_1}^{\tau_2} [\eta(t, \varrho, \vartheta)^T \{\nabla_{\varrho} f(t, \vartheta, \dot{\vartheta}) + D\nabla_{\dot{\varrho}} f(t, \vartheta, \dot{\vartheta}) + \nabla_p h(t, \vartheta, \dot{\vartheta}, p)\}] dt.$$

Now, we display a suitable example in order to show the existence of higher order $\eta$-invex functions that is not second-order $\eta$-invex.

**Example 2.1** Let $\aleph \subset \mathbb{R}_+^2$ and $\pi = (\pi_1, \pi_2) \in \aleph$, $\omega = (\omega_1, \omega_2) \in \aleph$ and $a = (a_1, a_2) \in \mathbb{R}^2$. Define $f : I \times \aleph \times \aleph \mapsto \mathbb{R}$ by $f(t, \pi, \dot{\pi}) = \cos \pi_1 + \cos \pi_2;$, $\eta : I \times \aleph \times \aleph \mapsto \mathbb{R}^2$ by $\eta(t, \pi, \omega) = (\pi_1 \omega_1 - 1, \pi_2 \omega_2 + 1)$ and $h : I \times \aleph \times \aleph \mapsto \mathbb{R}$ by $h(t, \omega, \dot{\omega}, p) = -p_1^2 e^{\omega_1} - p_2^2 e^{\omega_2}$. We take $\omega = (0, 0)$; $a = (1, 1)$ and I = [0, 1].

$$\int_0^1 f(t, \pi, \dot{\pi})\, dt - \int_0^1 f(t, \omega, \dot{\omega})\, dt - \int_0^1 h(t, \omega, \dot{\omega}, p) dt + \int_0^1 p^t \nabla_p h(t, \omega, \dot{\omega}, p) dt$$

$$- \int_0^1 [\eta(t, \pi, \omega)^T \{\nabla_{\pi} f(t, \omega, \dot{\omega}) - D\nabla_{\dot{\pi}} f(t, \omega, \dot{\omega}) - \nabla_p h(t, \omega, \dot{\omega}, p)\}] dt$$

$$= \int_0^1 (\cos \pi_1 + \cos \pi_2 - \cos \omega_1 - \cos \omega_2) dt - \int_0^1 (-p_1^2 e^{\omega_1} - p_2^2 e^{\omega_2})$$

$$+ (p_1, p_2) \begin{bmatrix} -2p_1 e^{\omega_1} \\ -2p_2 e^{\omega_2} \end{bmatrix} dt$$

$$= \int_0^1 (\cos \pi_1 + \cos \pi_2 - \cos \omega_1 - \cos \omega_2) dt - \int_0^1 (-3p_1^2 e^{\omega_1} - 3p_2^2 e^{\omega_2}) dt$$

$$= \int_0^1 (\cos \pi_1 + \cos \pi_2) dt + 4 \geq 0,$$

which agrees that $f$ is indeed higher order invex. The following discussion makes it clear that $f$ is not second-order invex.

$$\int_0^1 \left( f(t,\pi,\dot{\pi}) - f(t,\omega,\dot{\omega}) + \frac{1}{2}p^T \nabla_{\pi\pi} f(t,\omega,\dot{\omega})p - \eta(t,\pi,\omega)^T (\nabla_\pi f(t,\omega,\dot{\omega}) \right.$$

$$\left. + \nabla_{\pi\pi} f(t,\omega,\dot{\omega})p) \right) dt$$

$$= \int_0^1 (\cos\pi_1 + \cos\pi_2 - \cos\omega_1 - \cos\omega_2) + \frac{1}{2}(p_1, p_2) \begin{bmatrix} -\cos\omega_1 & 0 \\ 0 & -\cos\omega_2 \end{bmatrix} \begin{bmatrix} p_1 \\ p_2 \end{bmatrix}$$

$$- (\pi_1\omega_1 - 1, \pi_2\omega_2 + 1) \begin{bmatrix} -\sin\omega_1 \\ -\sin\omega_2 \end{bmatrix} + \begin{bmatrix} -\cos\omega_1 & 0 \\ 0 & -\cos\omega_2 \end{bmatrix} \begin{bmatrix} p_1 \\ p_2 \end{bmatrix}$$

$$= \int_0^1 (\cos\pi_1 + \cos\pi_2 - \cos\omega_1 - \cos\omega_2) + \frac{1}{2}[-{p_1}^2 \cos\omega_1 - {p_2}^2 \cos\omega_2]$$

$$-(\pi_1\omega_1 - 1, \pi_2\omega_2 + 1) \begin{bmatrix} \sin\omega_1 - p_1\cos\omega_1 \\ \sin\omega_2 - p_2\cos\omega_2 \end{bmatrix}$$

$$= \int_0^1 (\cos\pi_1 - \cos\pi_2 - 3)dt$$

$$\leq 0, \ \forall \pi \in \aleph.$$

In the coming sections, let us consider $C_1$ and $C_2$ stand for closed convex cones in $\mathbb{R}^n$ and $\mathbb{R}^m$, respectively, having nonempty interiors. Let $\aleph_1 \subset \mathbb{R}^n$ and $\aleph_2 \subset \mathbb{R}^m$ be open sets such that $C_1 \times C_2 \subset \aleph_1 \times \aleph_2$. Also, $\eta_1 : I \times \aleph_1 \times \aleph_1 \mapsto \mathbb{R}^n$ and $\eta_2 : I \times \aleph_2 \times \aleph_2 \mapsto \mathbb{R}^m$.

## 3 Higher Order Variational Symmetric Dual Formulations

In the present paper, we investigate the following order symmetric variational dual problem where constraints are defined over cones.

$$\text{(VSP)} \quad \text{Min} \int_{\tau_1}^{\tau_2} (f(t,\delta,\dot{\delta},\varrho,\dot{\varrho}) + h(t,\delta,\dot{\delta},\varrho,\dot{\varrho},p) - p^T \nabla_p h(t,\delta,\dot{\delta},\varrho,\dot{\varrho},p))dt$$

s.t

$$\delta(\tau_1) = 0 = \delta(\tau_2), \quad \dot{\delta}(\tau_1) = 0 = \dot{\delta}(\tau_2),$$

$$\varrho(\tau_1) = 0 = \varrho(\tau_2), \quad \dot{\varrho}(\tau_1) = 0 = \dot{\varrho}(\tau_2),$$

$$\nabla_\varrho f(t,\delta,\dot{\delta},\varrho,\dot{\varrho}) - D\nabla_{\dot{\varrho}} f(t,\delta,\dot{\delta},\varrho,\dot{\varrho}) + \nabla_p h(t,\delta,\dot{\delta},\varrho,\dot{\varrho},p) \in C_2^*, \quad (1)$$

$$\varrho^T[\nabla_\varrho f(t,\delta,\dot{\delta},\varrho,\dot{\varrho}) - D\nabla_{\dot{\varrho}} f(t,\delta,\dot{\delta},\varrho,\dot{\varrho}) + \nabla_p h(t,\delta,\dot{\delta},\varrho,\dot{\varrho},p) \geq 0, \quad (2)$$

$$\delta(t) \in C_1,\ t \in I.$$

(VSD) $\text{Max} \int_{\tau_1}^{\tau_2} (f(t,\vartheta,\dot{\vartheta},\sigma,\dot{\sigma}) + g(t,\vartheta,\dot{\vartheta},\sigma,\dot{\sigma},q) - q^T\nabla_q g(t,\vartheta,\dot{\vartheta},\sigma,\dot{\sigma},q))\,dt$

s.t

$$\vartheta(\tau_1) = 0 = \vartheta(\tau_2),\quad \dot{\vartheta}(\tau_1) = 0 = \dot{\vartheta}(\tau_2),$$

$$\sigma(\tau_1) = 0 = \sigma(\tau_2),\quad \dot{\sigma}(\tau_1) = 0 = \dot{\sigma}(\tau_2),$$

$$-[\nabla_\delta f(t,\vartheta,\dot{\vartheta},\sigma,\dot{\sigma}) - D\nabla_{\dot{\delta}} f(t,\vartheta,\dot{\vartheta},\sigma,\dot{\sigma}) + \nabla_q g(t,\vartheta,\dot{\vartheta},\sigma,\dot{\sigma},q)] \in C_1^*, \quad (3)$$

$$\vartheta^T[\nabla_\delta f(t,\vartheta,\dot{\vartheta},\sigma,\dot{\sigma}) - D\nabla_{\dot{\delta}} f(t,\vartheta,\dot{\vartheta},\sigma,\dot{\sigma}) + \nabla_q g(t,\vartheta,\dot{\vartheta},\sigma,\dot{\sigma},q)] \leq 0, \quad (4)$$

$$\sigma(t) \in C_2,\ t \in I,$$

where
(i) $f : I \times \aleph_1 \times \aleph_1 \times \aleph_2 \times \aleph_2 \to \mathbb{R}_+$,
(ii) $h, g : I \times \aleph_1 \times \aleph_1 \times \aleph_2 \times \aleph_2 \times \mathbb{R}^m \mapsto \mathbb{R}$ are differentiable functions, and
(iii) p $\in \mathbb{R}^m$ and q $\in \mathbb{R}^n$.
In order that the problem is suitably defined, we enforce nonnegativity on numerators and positivity on denominators.

# 4 Duality Results

In the present section, we derive the relevant duality relations for the dual pair considered in this paper.

**Theorem 1** (Weak Duality) *Let* $(\delta, \varrho, p)$ *and* $(\vartheta, \sigma, q)$ *be solutions feasible to primal* (VSP) *and dual* (VSD)*, respectively. Further, hypothesize the following conditions:*

*(a)* $\int_{\tau_1}^{\tau_2}(f(t,.,.,\sigma(t),\dot{\sigma}(t))\,dt$ *is higher order invex at* $\vartheta(t)$ *w.r. t.* $\eta_1$ *and* $g(t,\vartheta,\dot{\vartheta},\sigma,\dot{\sigma},q)$,
*(b)* $-\int_{\tau_1}^{\tau_2} f(t,\delta(t),\dot{\delta}(t),.,.)\,dt$ *is higher order invex at* $\varrho(t)$ *w.r.t.* $\eta_2$ *and* $-h(t,\delta,\dot{\delta},\varrho,\dot{\varrho},p)$,
*(c)* $(\eta_1(t,\delta(t),\vartheta(t)) + \vartheta)^T \in C_1, \forall\, \delta(t) \in C_1,\ t \in I$,
*(d)* $(\eta_2(t,\sigma(t),\varrho(t)) + \varrho)^T \in C_2, \forall\, \sigma(t) \in C_2,\ t \in I$.

*Then* $\int_{\tau_1}^{\tau_2}(f(t,\delta,\dot{\delta},\varrho,\dot{\varrho}) + h(t,\delta,\dot{\delta},\varrho,\dot{\varrho},p) - p^T\nabla_p h(t,\delta,\dot{\delta},\varrho,\dot{\varrho},p))dt$
$\geqq \int_{\tau_1}^{\tau_2}(f(t,\vartheta,\dot{\vartheta},\sigma,\dot{\sigma}) + g(t,\vartheta,\dot{\vartheta},\sigma,\dot{\sigma},q) - q^T\nabla_q g(t,\vartheta,\dot{\vartheta},\sigma,\dot{\sigma},q))\,dt.$

***Proof*** Let $(\delta, \varrho, p)$ and $(\vartheta, \sigma, q)$ be solutions feasible to (VSP) and (VSD). Using constraint (3) in condition $(c)$, we get

$$-(\eta_1(t,\delta,\vartheta)+\vartheta)^T[\nabla_\delta f(t,\vartheta,\dot{\vartheta},\sigma,\dot{\sigma}) - D\nabla_{\dot{\delta}} f(t,\vartheta,\dot{\vartheta},\sigma,\dot{\sigma}) + \nabla_q g(t,\vartheta,\dot{\vartheta},\sigma,\dot{\sigma},q)] \leqq 0.$$

On account of relation (4), the above relation turns out to be

$$(\eta_1(t,\delta,\vartheta))^T[\nabla_\delta f(t,\vartheta,\dot{\vartheta},\sigma,\dot{\sigma}) - D\nabla_{\dot{\delta}} f(t,\vartheta,\dot{\vartheta},\sigma,\dot{\sigma}) + \nabla_q g(t,\vartheta,\dot{\vartheta},\sigma,\dot{\sigma},q] \geqq 0. \quad (5)$$

Using supposition $(d)$ and constraint (1), we arrive at

$$(\eta_2(t,\sigma,\varrho)+\varrho)^T[\nabla_\varrho f(t,\delta,\dot{\delta},\varrho,\dot{\varrho}) - D\nabla_{\dot{\varrho}} f(t,\delta,\dot{\delta},\varrho,\dot{\varrho}) + \nabla_p h(t,\delta,\dot{\delta},\varrho,\dot{\varrho},p)] \leqq 0,$$

which by the virtue of (2) becomes

$$(\eta_2(t,\sigma,\varrho))^T[\nabla_\varrho f(t,\delta,\dot{\delta},\varrho,\dot{\varrho}) - D\nabla_{\dot{\varrho}} f(t,\delta,\dot{\delta},\varrho,\dot{\varrho}) + \nabla_p h(t,\delta,\dot{\delta},\varrho,\dot{\varrho},p)] \leqq 0. \quad (6)$$

Since $\int_{\tau_1}^{\tau_2} f(t,.,.,\sigma,\dot{\sigma})dt$ be higher order $\eta_1 - invex$ at $\vartheta$ for fixed $\sigma(t)$ with respect to g, the above inequality yields

$$\int_{\tau_1}^{\tau_2} f(t,\delta,\dot{\delta},\sigma,\dot{\sigma})dt - \int_{\tau_1}^{\tau_2} f(t,\vartheta,\dot{\vartheta},\sigma,\dot{\sigma})dt - \int_{\tau_1}^{\tau_2} g(t,\vartheta,\dot{\vartheta},\sigma,\dot{\sigma},q)dt + \int_{\tau_1}^{\tau_2} q^T\nabla_q g(t,\vartheta,\dot{\vartheta},\sigma,\dot{\sigma},q)dt \geqq 0. \quad (7)$$

Similarly,

$$\int_{\tau_1}^{\tau_2} f(t,\delta,\dot{\delta},\varrho,\dot{\varrho})dt - \int_{\tau_1}^{\tau_2} f(t,\delta,\dot{\delta},\sigma,\dot{\sigma})dt + \int_{\tau_1}^{\tau_2} h(t,\delta,\dot{\delta},\varrho,\dot{\varrho},p)dt - \int_{\tau_1}^{\tau_2} p^T\nabla_p h(t,\delta,\dot{\delta},\varrho,\dot{\varrho},p)dt \geqq 0. \quad (8)$$

From the inequalities (7) and (8) mentioned above, we get
$\int_{\tau_1}^{\tau_2}(f(t,\delta,\dot{\delta},\varrho,\dot{\varrho}) + h(t,\delta,\dot{\delta},\varrho,\dot{\varrho},p) - p^T\nabla_p h(t,\delta,\dot{\delta},\varrho,\dot{\varrho},p))dt \geqq$
$\int_{\tau_1}^{\tau_2}(f(t,\vartheta,\dot{\vartheta},\sigma,\dot{\sigma}) + g(t,\vartheta,\dot{\vartheta},\sigma,\dot{\sigma},q) - q^T\nabla_q g(t,\vartheta,\dot{\vartheta},\sigma,\dot{\sigma},q))\,dt.$
This configures the proof of the statement.

**Theorem 2** (Strong Duality) *Assume* $(\bar{\delta}, \bar{\varrho}, \bar{p})$ *be a solution locally optimal to* (VSP). *Presume the following conditions:*

(*i*) $\nabla_{\delta} h(t, \bar{\delta}, \dot{\bar{\delta}}, \bar{\varrho}, \dot{\bar{\varrho}}, 0) = \nabla_{q} g(t, \bar{\delta}, \dot{\bar{\delta}}, \bar{\varrho}, \dot{\bar{\varrho}}, 0),$
$\nabla_{\delta'} h(t, \bar{\delta}, \dot{\bar{\delta}}, \bar{\varrho}, \dot{\bar{\varrho}}, 0) = \nabla_{q'} g(t, \bar{\delta}, \dot{\bar{\delta}}, \bar{\varrho}, \dot{\bar{\varrho}}, 0),$
$\nabla_{\delta''} h(t, \bar{\delta}, \dot{\bar{\delta}}, \bar{\varrho}, \dot{\bar{\varrho}}, 0) = \nabla_{q''} g(t, \bar{\delta}, \dot{\bar{\delta}}, \bar{\varrho}, \dot{\bar{\varrho}}, 0)\ldots$
$\nabla_{\delta^{(2n)}} h(t, \bar{\delta}, \dot{\bar{\delta}}, \bar{\varrho}, \dot{\bar{\varrho}}, 0) = \nabla_{q^{(2n)}} g(t, \bar{\delta}, \dot{\bar{\delta}}, \bar{\varrho}, \dot{\bar{\varrho}}, 0).$

(*ii*) *the specified matrix* $\nabla_{pp} h(t, \bar{\delta}, \dot{\bar{\delta}}, \bar{\varrho}, \dot{\bar{\varrho}})$ *is negative definite or positive definite,*

(*iii*) $(\nabla_{\varrho} f - D\nabla_{\varrho'} f + \nabla_p h) \neq 0,$

(*iv*) *for choosen* $\bar{p} \in \mathbb{R}^m$,

$$\bar{p}^T (\nabla_{\varrho} f - D\nabla_{\varrho'} f + \nabla_p h) = 0$$

*implies* $\bar{p} = 0$, *and*

(*v*)

$$D\Big[(\nabla_{\delta'} f + \nabla_{q'} g) + D^2(\nabla_{\delta''} f + \nabla_{q''} g) - D^3(\nabla_{\delta'''} f + \nabla_{q'''} g) + \cdots +$$

$$D^{2n}(\nabla_{\delta^{(2n)}} f + \nabla_{q^{(2n)}} g)\Big] = 0.$$

*Then* $(\bar{\delta}, \bar{\varrho}, \bar{q} = 0)$ *is a solution feasible to* (VSD) *and both objectives produces equal output. If, in addition, the conditions of Theorem 1 are fulfilled, for every solution feasible to* (VSP) *and* (VSD), *then* $(\bar{\delta}, \bar{\varrho}, \bar{p} = 0)$ *and* $(\bar{\delta}, \bar{\varrho}, \bar{q} = 0)$ *is a solution globally optimal to* (VSP) *and* (VSD), *respectively.*

***Proof*** Since $(\bar{\delta}, \bar{\varrho}, \bar{p})$ is an absolute maximal or minimal solution to (VSP), there exists $\alpha \in \mathbb{R},\ \beta \in \mathbb{R},\ \gamma \in C_2$ and $\xi \in \mathbb{R}$ satisfying Fritz John optimality criteria at the point $(\bar{\varrho}(t), \bar{\delta}(t), \bar{p}(t))$ given below:

$$\beta\Big[(\nabla_{\delta} f + \nabla_{\delta} h - \bar{p}^T \nabla_{p\delta} h) - D(\nabla_{\delta'} f + \nabla_{\delta'} h - \bar{p}^T \nabla_{p\delta'} h) + D^2(\nabla_{\delta''} f + \nabla_{\delta''} h - \bar{p}^T \nabla_{p\delta''} h)$$

$$- D^3(\nabla_{\delta'''} f + \nabla_{\delta'''} h - \bar{p}^T \nabla_{p\delta'''} h) + \cdots + D^{2n}(\nabla_{\delta^{(2n)}} f + \nabla_{\delta^{(2n)}} h - \bar{p}^T \nabla_{p\delta^{(2n)}} h)\Big]$$

$$+(\gamma - \xi\bar{\varrho})^T\Big[(\nabla_{\varrho\delta} f - D\nabla_{\varrho'\delta} f + \nabla_{p\delta} h) \ - D(\nabla_{\varrho\delta'} f - D\nabla_{\varrho'\delta'} f + \nabla_{p\delta'} h)$$

$$+ D^2(\nabla_{\varrho\delta''} f - D\nabla_{\varrho'\delta''} f + \nabla_{p\delta''} h) - D^3(\nabla_{\varrho\delta'''} f - D\nabla_{\varrho'\delta'''} f + \nabla_{p\delta'''} h) + \ldots$$

$$+ D^{2n}(\nabla_{\varrho\delta^{2n}} f - D\nabla_{\varrho'\delta^{2n}} f + \nabla_{p\delta^{2n}} h)\Big](\delta(t) - \bar{\delta}(t)) \geqq 0,\ t \in I, \forall\, \delta \in C_1, \quad (9)$$

$$\beta\Big[(\nabla_{\varrho} f + \nabla_{\varrho} h - \bar{p}^T \nabla_{p\varrho} h) - D(\nabla_{\varrho'} f + \nabla_{\varrho'} h - \bar{p}^T \nabla_{p\varrho'} h) + D^2(\nabla_{\varrho''} f + \nabla_{\varrho''} h - \bar{p}^T \nabla_{p\varrho''} h)$$

$$- D^3(\nabla_{\varrho'''} f + \nabla_{\varrho'''} h - \bar{p}^T \nabla_{p\varrho'''} h) + \cdots + D^{2n}(\nabla_{\varrho^{2n}} f + \nabla_{\varrho^{2n}} h - \bar{p}^T \nabla_{p\varrho^{2n}} h)\Big]$$

$$+(\gamma-\xi\bar{\varrho})^T\Big[(\nabla_{\varrho\varrho}f-D\nabla_{\varrho'\varrho}f+\nabla_{p\varrho}h)-D(\nabla_{\varrho\varrho'}f-D\nabla_{\varrho'\varrho}f+\nabla_{p\varrho'}h)$$

$$+D^2(\nabla_{\varrho\varrho''}f-D\nabla_{\varrho''\varrho}f+\nabla_{p\varrho''}h)-D^3(\nabla_{\varrho\varrho'''}f-D\nabla_{\varrho'''\varrho}f+\nabla_{p\varrho'''}h)+\ldots$$

$$+D^{2n}(\nabla_{\varrho\varrho^{2n}}f-D\nabla_{\varrho^{2n}\varrho}f+\nabla_{p\varrho^{2n}}h)-\xi(\nabla_{\varrho}f-D\nabla_{\varrho'}f+\nabla_{p}h)\Big]=0,$$

$$t\in I,\ \forall\varrho\in\mathbb{R}^n, \tag{10}$$

$$\alpha=0,\ t\in I, \tag{11}$$

$$(\gamma-\xi\bar{\varrho}-\beta\bar{p})^T\nabla_{pp}h=0,\ t\in I, \tag{12}$$

$$\gamma^T(\nabla_{\varrho}f-D\nabla_{\varrho'}f+\nabla_{p}h)=0,\ t\in I, \tag{13}$$

$$-\xi\bar{\varrho}^T(\nabla_{\varrho}f-D\nabla_{\varrho'}f+\nabla_{p}h)=0,\ t\in I, \tag{14}$$

$$(\alpha,\beta,\gamma,\xi)\neq 0,\alpha>0,\gamma\in C_2,\xi\geq 0. \tag{15}$$

From the hypothesis $(ii)$, Eq. (12) turns out to be

$$(\gamma-\xi\bar{\varrho}-\beta\bar{p})=0. \tag{16}$$

We ascertain that $\beta\neq 0$. In case, if $\beta=0$, relation (16) produces

$$\gamma=\xi\bar{\varrho} \tag{17}$$

and Eq. (10) returns

$$\xi(\nabla_{\varrho}f-D\nabla_{\varrho'}f+\nabla_{p}h)=0, \tag{18}$$

which by hypothesis $(iii)$ yields $\xi=0$ and, from Eq. (17), we get $\gamma=0$ and, hence, from Eq. (11), we have $\alpha=0$. Thus, we get $(\alpha,\beta,\gamma,\xi)\neq 0,\ t\in I$ contradicting Eq. (15). Hence, $\beta>0$. On subtracting Eq. (14) from Eq. (13), we have

$$(\gamma-\xi\bar{\varrho})^T(\nabla_{\varrho}f-D\nabla_{\varrho'}f+\nabla_{p}h)=0.$$

Using Eq. (16) along with $\beta\neq 0$, the relation above settles down to

$$\bar{p}^T(\nabla_{\varrho}f-D\nabla_{\varrho'}f+\nabla_{p}h)=0.$$

By the hypothesis $(iv)$, we have $\bar{p}=0$. Using this in Eq. (16), we obtain $\gamma=\xi\bar{\varrho}$, which further gives $\gamma\in C_2$. Using the fact $\gamma=\xi\bar{\varrho}$, in Eq. (9), we arrive at

$$\beta\Big[(\nabla_{\delta}f+\nabla_{\delta}h)-D(\nabla_{\delta'}f+\nabla_{\delta'}h)+D^2(\nabla_{\delta''}f+\nabla_{\delta''}h)-D^3(\nabla_{\delta'''}f+\nabla_{\delta'''}h)$$

$$+\cdots+D^{2n}(\nabla_{\delta^{(2n)}}f+\nabla_{\delta^{(2n)}}h)\Big]^T(\delta(t)-\bar{\delta(t)})\geq 0, \forall\delta\in C_1. \tag{19}$$

From assumption $(i)$ along with $\bar{p}=0$, the inequality (19) returns

$$\Big[(\nabla_{\delta}f+\nabla_{q}g)-D(\nabla_{\delta'}f+\nabla_{q'}g)+D^2(\nabla_{\delta''}f+\nabla_{q''}g)-D^3(\nabla_{\delta'''}f+\nabla_{q'''}g)$$

$$+\cdots+D^{2n}(\nabla_{\delta^{(2n)}}f+\nabla_{q^{(2n)}}g)\Big]^T(\delta(t)-\bar{\delta(t)})\geq 0. \tag{20}$$

Suppose $\delta(t)\in C_1$ so that $\delta(t)+\bar{\delta(t)}\in C_1$. Hence, Eq. (20) implies

$$\Big[(\nabla_{\delta}f+\nabla_{q}g)-D(\nabla_{\delta'}f+\nabla_{q'}g)+D^2(\nabla_{\delta''}f+\nabla_{q''}g)-D^3(\nabla_{\delta'''}f+\nabla_{q'''}g)$$

$$+\cdots+D^{2n}(\nabla_{\delta^{(2n)}}f+\nabla_{q^{(2n)}}g)\Big]^T\delta(t)\geq 0, \forall\in C_1.$$

Using a property of polar cone, we obtain

$$-\Big[(\nabla_{\delta}f+\nabla_{q}g)-D(\nabla_{\delta'}f+\nabla_{q'}g)+D^2(\nabla_{\delta''}f+\nabla_{q''}g)-D^3(\nabla_{\delta'''}f+\nabla_{q'''}g)$$

$$+\cdots+D^{2n}(\nabla_{\delta^{(2n)}}f+\nabla_{q^{(2n)}}g)\Big]\in {C_1}^*.$$

Let $\delta(t)=0$ and $\delta(t)=2\bar{\delta(t)}$ in Eq. (20), we have

$$\bar{\delta(t)}^T\Big[(\nabla_{\delta}f+\nabla_{q}g)-D(\nabla_{\delta'}f+\nabla_{q'}g)+D^2(\nabla_{\delta''}f+\nabla_{q''}g)-D^3(\nabla_{\delta'''}f+\nabla_{q'''}g)$$

$$+\cdots+D^{2n}(\nabla_{\delta^{(2n)}}f+\nabla_{q^{(2n)}}g)\Big]=0. \tag{21}$$

Now, using assumption $(v)$ in above equation, we get

$$\bar{\delta(t)}^T(\nabla_{\delta}f+\nabla_{q}g)=0. \tag{22}$$

Thus, it becomes clear that $(\bar{\delta}(t),\bar{\varrho}(t),\bar{p}(t)=0)$ is a solution feasible to (VSD) giving equal output. Also, under additional conditions stated in Theorem 1, $(\bar{\delta},\bar{\varrho},\bar{p}=0)$ and $(\bar{\delta},\bar{\varrho},\bar{q}=0)$ becomes a solution globally optimal to (VSP) and (VSD), respectively.

**Theorem 3** (Converse Duality) *Assume* $(\bar{\vartheta},\bar{\sigma},\bar{q})$ *denotes a solution locally optimal to* (VSP). *Presume the following conditions:*

$(i)$ $\nabla_{\varrho}h(t,\bar{\vartheta},\dot{\bar{\vartheta}},\bar{\sigma},\dot{\bar{\sigma}},0)=\nabla_{p}g(t,\bar{\vartheta},\dot{\bar{\vartheta}},\bar{\sigma},\dot{\bar{\sigma}},0)$,
$\nabla_{\varrho'}h(t,\bar{\vartheta},\dot{\bar{\vartheta}},\bar{\sigma},\dot{\bar{\sigma}},0)=\nabla_{p'}g(t,\bar{\vartheta},\dot{\bar{\vartheta}},\bar{\sigma},\dot{\bar{\sigma}},0)$,

$\nabla_{\varrho''} h(t, \bar{\vartheta}, \dot{\bar{\vartheta}}, \bar{\sigma}, \dot{\bar{\sigma}}, 0) = \nabla_{p''} g(t, \bar{\vartheta}, \dot{\bar{\vartheta}}, \bar{\sigma}, \dot{\bar{\sigma}}, 0), \ldots$
$\nabla_{\varrho^{(2n)}} h(t, \bar{\vartheta}, \dot{\bar{\vartheta}}, \bar{\sigma}, \dot{\bar{\sigma}}, 0) = \nabla_{p^{(2n)}} g(t, \bar{\vartheta}, \dot{\bar{\vartheta}}, \bar{\sigma}, \dot{\bar{\sigma}}, 0).$

(*ii*) *the specified matrix* $\nabla_{qq} g(t, \bar{\vartheta}, \dot{\bar{\vartheta}}, \bar{\sigma}, \dot{\bar{\sigma}})$ *is negative definite or positive definite,*

(*iii*) $(\nabla_\delta f - D\nabla_{\delta'} f + \nabla_q g) \neq 0,$

(*iv*) *for choosen* $\bar{q} \in \mathbb{R}^n$,

$$\bar{q}^T (\nabla_\delta f - D\nabla_{\delta'} f + \nabla_q g) = 0$$

*implies* $\bar{q} = 0$, *and*

(*v*)

$$D(\nabla_{\varrho'} f + \nabla_{p'} g) + D^2(\nabla_{\varrho''} f + \nabla_{p''} g) - D^3(\nabla_{\varrho'''} f + \nabla_{p'''} g)$$
$$+ \cdots + D^{2n}(\nabla_{\varrho^{(2n)}} f + \nabla_{p^{(2n)}} g) = 0.$$

*Then* $(\bar{\vartheta}, \bar{\sigma}, \bar{p} = 0)$ *is a solution feasible to* (VSD) *and both objectives yields equal output. Additionally, if postulates of Theorem 1 are satisfied for feasible solutions* (VSP) *and* (VSD), *then* $(\bar{\vartheta}, \bar{\sigma}, \bar{q} = 0)$ *and* $(\bar{\vartheta}, \bar{\sigma}, \bar{p} = 0)$ *represent the absolute optimal solution of* (VSP) *and* (VSD), *respectively.*

## 5 Static Symmetric Dual Program

Dropping down the time co-ordinate in problem considered in this paper, the problem transforms into the given form:

**(VSP*)** Min $(f(\delta, \varrho) + h(\delta, \varrho, p) - p^T \nabla_p h(\delta, \varrho, p))$

s.t

$$\nabla_\varrho f(\delta, \varrho) - D\nabla_{\dot{\varrho}} f(\delta, \varrho) + \nabla_p h(\delta, \varrho, p) \in C_2^*,$$

$$\varrho^T [\nabla_\varrho f(\delta, \varrho) - D\nabla_{\dot{\varrho}} f(\delta, \varrho) + \nabla_p h(\delta, \varrho, p)] \geqq 0,$$

$$\delta(t) \in C_1.$$

**(VSD*)** Max $(f(\vartheta, \sigma) + g(\vartheta, \sigma, q) - q^T \nabla_q g(\vartheta, \sigma, q))$

s.t

$$-[\nabla_\delta f(\vartheta, \sigma) - D\nabla_{\dot{\delta}} f(\vartheta, \sigma) + \nabla_q g(\vartheta, \sigma, q)] \in C_1^*,$$

$$\vartheta^T [\nabla_\delta f(\vartheta, \sigma) - D\nabla_{\dot{\delta}} f(\vartheta, \sigma) + \nabla_q g(\vartheta, \sigma, q)] \leqq 0,$$

$$\sigma(t) \in C_2.$$

The weak and strong duality results can be easily established. One can refer for details Jayswal et al. [6].

## 6 Conclusions

In this paper, higher order $\eta$-invexity is mechanized to establish the duality results for a dual pair of higher order symmetric variational programs where constraints are defined over more general settings having cones. Also, we derived the results needed for dual formulations with the help of higher order invexity. In the future, this work can be extended to multiobjective problems and also to a nondifferentiable problem by additionally adjoining support functions in the objective making it nondifferentiable.

## References


1. Bazaraa, M.S., Goode, J.J.: On symmetric duality in nonlinear programming. Oper. Res. **21**, 1–9 (1973)
2. Chen, X.: Higher-order symmetric duality in nondifferentiable multiobjective programming problems. J. Math. Anal. Appl. **290**, 423–435 (2004)
3. Dantzig, G.B., Eisenberg, E., Cottle, R.W.: Symmetric dual nonlinear programs. Pac. J. Math. **15**, 809–812 (1965)
4. Dorn, W.S.: A symmetric dual theorem for quadratic programs. J. Oper. Res. Soc. **2**, 93–97 (1960)
5. Gulati, T.R., Gupta, S.K.: Higher-order symmetric duality with cone constraints. Appl. Math. Lett. **22**, 776–781 (2009)
6. Jayswal, A., Ahmad, I., Prasad, A.K.: Higher order fractional symmetric duality cone constraints. J. Math. Model Algor. **14**, 91–101 (2014)
7. Jayswal, A., Jha, S., Prasad, A.K., Ahmad, I.: Second-order symmetric duality in variational control problems over cone constraints. Asia-Pacific J. Oper. Res. **35**, 19 (2018)
8. Kassem, M.A.E.H.: Symmetric and self duality in vector optimization problem. Appl. Math. Comput. **183**, 1121–1126 (2006)
9. Kassem, M.A.E.H.: Higher-order symmetric duality in vector optimization problem involving generalized cone-invex functions. Appl. Math. Comp. **209**, 405–409 (2009)
10. Kaur, A., Sharma, M.K.: Higher order symmetric duality for multiobjective fractional programming problems over cones. Yugoslav J. Oper. Res. https://doi.org/10.2298/YJOR200615012K
11. Khurana, S.: Symmetric duality in multiobjective programming involving generalized cone-invex functions. Eur. J. Oper. Res. **165**, 592–597 (2005)
12. Mangasarian, O.L.: Second order and higher order duality in nonlinear programming. J. Math. Anal. Appl. **51**, 607–620 (1975)
13. Mond, B., Weir, T.: Generalized concavity and duality. In: Schaible, S., Ziemba, W.T. (eds.) Generalized Concavity in Optimization and Economics, pp. 263–279. Academic press, New York (1981)
14. Prasad, A.K., Singh, A.P., Khatri, S.: Duality for a class order symmetric nondifferentiable fractional variational problems. Yugoslav J. Oper. Res. **30**, 121–136 (2020)
15. Suneja, S.K., Louhan, P.: Higher-order symmetric duality under cone-invexity and other related concepts. J. Comput. Appl. Math. **255**, 825–836 (2014)

16. Verma, K., Verma, J.P., Ahmad, I.: A new approach on multiobjective higher-order symmetric duality under cone-invexity. Bull. Malaysian Math. Sci. Soc. **44**, 479–495 (2020)
17. Yang, X.M.: On symmetric and self duality in vector optimization problem. J. Indus. Manag. Optim. **7**, 523–529 (2011)

# On Generalized Energy Inequality of the Damped Navier–Stokes Equations with Navier Slip Boundary Conditions

**Subha Pal and Duranta Chutia**

**Abstract** In this article, we deal with a damped Navier–Stokes equations in $\mathbb{R}^3$ with slip boundary conditions. Sufficient conditions for the existence of the solutions to the Navier–Stokes system are established in a bounded domain $\Omega \subset \mathbb{R}^3$. Further, we show that the solutions derived by Rothe's method are satisfying the local energy inequality.

**Keywords** Navier–Stokes equation · Damping · Rothe method · Navier slip boundary condition

## 1 Introduction

Let us consider the following system of Navier–Stokes (N-S) equations with sufficiently smooth boundary in a simply connected bounded domain $\Omega$ of $\mathbb{R}^3$

$$\partial_t u - \Delta u + u \cdot \nabla u + \vartheta |u|^{\beta-1} u + \frac{1}{\rho} \nabla p = f \quad \text{in } \Omega \times (0, T), \tag{1}$$

$$\operatorname{div} u = 0 \quad \text{in } Q_T, \tag{2}$$

$$2D(u)\nu \cdot \tau + \xi u \cdot \tau = 0 \quad \text{on } \partial\Omega \tag{3}$$

$$u = u_0 \quad \text{in } \Omega \times \{0\}, \tag{4}$$

where $D(u)$ is the stress tensor of the form $\frac{1}{2}\left[\nabla u + (\nabla u)^T\right]$ and $\xi(x)$ $(> 0)$ is defined on the boundary $\partial\Omega$ with continuous differentiability. Here the unknown function $u$ corresponds to the velocity of the flow and $p$ is used to denote the pressure. $\vartheta |u|^{\beta-1} u$ is the damping term and $\vartheta > 0$ and $\beta \geq 1$ are the scalars appeared in the expression. Here $u_0$ is the initial velocity, $f$ represents the external force and $\rho$ is the density of

S. Pal (✉)
Tezpur University, Tezpur 784028, Assam, India
e-mail: sp234sp@gmail.com

D. Chutia
Dibru College, Boiragimoth, Dibrugarh 786003, Assam, India



B. Rushi Kumar et al. (eds.), *Mathematics and Computing*, Springer Proceedings in Mathematics & Statistics 415, https://doi.org/10.1007/978-981-19-9307-7_38

the fluid. $\tau$ and $\nu$ are unit tangent vector and unit exterior normal to the boundary. The Eq. (3) reflects the Navier slip boundary condition. We discuss slip boundary conditions in detail in [9].

The Navier–Stokes equations describe the motion of the fluid flows ranging from lubrication of ball bearings to large-scale atmospheric motions and reflect the conservation of mass as well as momentum. The system (1), in the case of no slip condition, has been studied by many authors in [3, 8, 11, 12, 15, 16]. We discuss the N–S equations with damping terms in detail in [8].

The results of Caffarelli et al. [2] and Scheffer [7] was regarding the partial regularity for the Navier–Stokes equations and they also introduced *Suitable Weak Solution* (SWS) which becomes more important in the theory of N–S equations. The following identity is satisfied by the weak solutions obtained by Leray and Hopf in a weak sense:

$$\int_0^\infty (u, \partial_t \phi) - (\nabla u, \nabla \phi) - ((u \cdot \nabla)u, \phi) dt = -(u_0, \phi(0)),$$

for all smooth, periodic, and divergence-free function $\phi$, such that $\phi(t, x) = 0$. The following energy inequality is also satisfied by the velocity $u$ :

$$\frac{1}{2}\|u(t)\|^2 + \int_0^T \|\nabla u(s)\|^2 ds \leq \frac{1}{2}\|u_0\|^2 \ \forall t \in [0, T]$$

In this short paper, we explore the requisite ideas to tackle the time discretization in the context of the construction of solutions to the system (1)–(4) satisfying the local energy inequality.

We organize the article in the following way. The assumptions and preliminaries are discussed in Sect. 2. In Sect. 3, we discuss the approximate solutions. The existence of the solutions is proved in Sect. 4. We show that the solution obtained in Sect. 4 satisfies local energy inequality in Sect. 6.

## 2 Assumptions and Preliminaries

Let us assemble some fundamental results, definitions, and notations from Sohr [10] and Temam [14], that are used in the remaining sections of the article. Throughout this article, $\Omega$ denotes a connected bounded domain in $\mathbb{R}^3$ having sufficiently smooth boundary and $C_0^\infty(\Omega)$ stands for the set of all $C^\infty$ vector functions $\phi$ with compact support in $\Omega$. Let $L^p(\Omega)$ $(1 \leq p \leq \infty)$ be the usual Lebesgue space and $H^r(\Omega)$ be the usual Sobolev space. We use $(\cdot, \cdot)$ to denote the usual $L^2$-inner product and define $((\cdot, \cdot))$ by

$$((v, w)) = \sum_{i=1}^{n} (D_i v, D_i w). \tag{5}$$

The notation $\|\cdot\|$ is used to denote the norm corresponding to the inner-product defined in (5). Motivated by Kelliher [5], we consider the following function spaces:

$$V = \{w \in H^1(\Omega) : \nabla \cdot w = 0 \text{ in } \Omega, \ \ w \cdot \nu = 0 \text{ on } \partial\Omega\},$$
$$H = \{w \in L^2(\Omega) : \nabla \cdot w = 0 \text{ in } \Omega, \ \ w \cdot \nu = 0 \text{ on } \partial\Omega\},$$
$$\mathcal{W} = \{w \in V \cap H^2(\Omega) : 2D(w)\nu \cdot \tau + \xi w \cdot \tau = 0 \text{ on } \partial\Omega\}.$$

For each $u_1, u_2 \in V$ we define the operator $A$ as [5]

$$(Au_1, u_2) = 2((u_1, u_2)) + \int_{\partial\Omega} \xi(u_1 \cdot \tau)(u_2 \cdot \tau). \tag{6}$$

Let $b$ be the trilinear form defined by

$$b(u_1, u_2, u_3) = \int_\Omega (u_1 \cdot \nabla u_2) \cdot u_3, \qquad \forall u_1, u_2, u_3 \in V. \tag{7}$$

For $u_1, u_2 \in V$, define $B(u_1, u_2)$ by

$$(B(u_1, u_2), u_3) = b(u_1, u_2, u_3), \ \ \forall u_3 \in V.$$

We put $B(u_1) = B(u_1, u_1) \in V'$, $\forall u_1 \in V$. So,

$$(Bu_1, u_2) = \int_\Omega (u_1 \cdot \nabla u_1) \cdot u_2. \tag{8}$$

The following Lemmas will be used to establish our main result.

**Lemma 1** ([4, Lemma 2.1, p. 759]) *Let $u_1, u_2, u_3 \in H^1(\Omega)$. Then*

$$|b(u_1, u_2, u_3)| \leq C\|u_1\|^{\frac{1}{4}}_{L^2(\Omega)}\|u_1\|^{\frac{3}{4}}_{H^1(\Omega)}\|u_2\|_{H^1(\Omega)}\|u_3\|^{\frac{1}{4}}_{L^2(\Omega)}\|u_3\|^{\frac{3}{4}}_{H^1(\Omega)}. \tag{9}$$

**Lemma 2** ([14, III Lemma 3.1]) *Let $u_1 \in L^2(0, T; V)$ and $d \leq 4$, where $d =$* dimension of the space. *Let the function $Bu_1$ defined as follows:*

$$(Bu_1(t), u_2) = b(u_1(t), u_1(t), u_2), \quad \forall u_2 \in V. \tag{10}$$

*Then $Bu_1 \in L^1(0, T; V')$ and satisfies*

$$\|Bu_1\|_{V'} \leq K\|u_1\|^2_{H^1}, \quad \forall u_1 \in V. \tag{11}$$

Let us state the following result, proved in [9, Lemma 2.2] which gives the existence of an orthonormal basis for $H$.

**Lemma 3** ([9, Lemma 2.2, p. 1631]) *There exists a basis, $\{v_n\} \subset H^3(\Omega)$ for $V$, also acts as an orthonormal basis for $H$ satisfying (3).*

## 3 The Approximate Solutions

In this section, we study whether the approximate solutions by numerical methods of the damped Navier–Stokes equations (1)–(3) with Navier slip boundary conditions can give solutions when the limit of the mesh size going to be zero [1]. Our main aim is the time discretization and using finite difference we approximate the time derivative. Let $M$ be an integer, later that will go to infinity and we put $k = T/M$. We define a family of elements $\{u^m, 1 \leq m \leq M\}$ from $V$ in a recursive way. Suppose $u^0, u^1, \ldots, u^M$, where $u^m$, $1 \leq m \leq M$ is an approximation of the function $u$. Our objective is to find $u$ on the interval $mk < t < (m+1)k$. We define the elements $f^1, \ldots, f^M$ of $V'$ as

$$f^m = \frac{1}{k}\int_{(m-1)k}^{mk} f(t)dt, \quad m = 1, \ldots, M; \quad f^m \in V'. \tag{12}$$

We start with $u^0 = u_0$, the given initial data; when $u^0, \ldots, u^{m-1}$ are known and we define $u^m$ as an element of $V$ which satisfies

$$\frac{u^m - u^{m-1}}{k} + Au^m + Bu^m + \vartheta|u^m|^{\beta-1}u^m = f^m. \tag{13}$$

We use the following finite-difference approximation to approximate the time-derivative

$$\partial_t u \sim d_t u^m := \frac{u^m - u^{m-1}}{k}, \text{ on } (t_{m-1}, t_m).$$

For every $m$, we associate pressure $p^m$ [1] to $u^m$. We associate the function $(w_M, u_M, q_M)$ defined in $[0, T]$, as follows for $m = 1, \ldots, M$

$$\left.\begin{array}{ll} w_M(t) = u^{m-1} + \frac{t-t_{m-1}}{k}(u^m - u^{m-1}) & \text{for } t \in [t_{m-1}, t_m), \\ w_M(t) = u^M & \text{for } t = t_M, \\ u_M(t) = u^0 & \text{for } t = t_0, \\ u_M(t) = u^m & \text{for } t \in (t_{m-1}, t_m], \\ q_M(t) = p^m & \text{for } t \in (t_{m-1}, t_m], \end{array}\right\} \tag{14}$$

We also assume that $w_M(t) = u_M(t)$, for all $m = 0, \ldots, M$.

## 4 Weak Solutions

Now, we establish a global existence to the solutions for the Navier–Stokes system. First, we give the definition of weak solutions.

**Definition 1** Suppose $w \in L^{\beta+1}(0, T; L^{\beta+1}(\Omega)) \cap L^\infty(0, T; H) \cap L^2(0, T; V)$ for $T > 0$, and $w$ satisfies the weak formulation of (1), i.e

$$(w', v) + 2((w, v)) + b(w, w, v) + (\vartheta |w|^{\beta-1} w, v) + \int_{\partial\Omega} \xi (w \cdot \tau)(v \cdot \tau) dS = (f, v) \; \forall \, v \in V. \tag{15}$$

Then $w$ is a weak solution of the Navier–stokes system.

**Theorem 1** *Suppose that $f \in L^2(0, T; V')$ and $u_0 \in H$. Then for any positive $T$, there is a solution $w$ and pressure $q$ to the Navier–Stokes system, such that*

$$w \in L^\infty(0, T; H) \cap L^2(0, T; V) \cap L^{\beta+1}(0, T; L^{\beta+1}(\Omega)), \tag{16}$$

$$w' \in L^2(0, T; V'), \tag{17}$$

$$q \in L^{5/3}(0, T; L^{5/3}). \tag{18}$$

*Further, $w_M$ and $u_M$ both converges to $u$ and $q_M$ converges to $q$ whenever $k \to 0$.*

We prove the following lemmas to complete the proof of the main Theorem. The idea of the proof is based on Temam [14] and Berselli and Spirito [1].

**Lemma 4** *There exists at least one $u^m$ satisfying*

$$\frac{u^m - u^{m-1}}{k} + Au^m + Bu^m + \vartheta |u^m|^{\beta-1} u^m = f^m, \tag{19}$$

*for fixed $k$ and $m \geq 1$. Moreover*

$$\|u^m\|_{L^2}^2 - \|u^{m-1}\|_{L^2}^2 + \|u^m - u^{m-1}\|_{L^2}^2 + 3k\|u^m\|_{H^1}^2 + 2\vartheta \|u^m\|_{L^{\beta+1}}^{\beta+1} \leq k\|f^m\|_{V'}^2. \tag{20}$$

***Proof*** Now (13) can be written as

$$(u^m, v) + 2k((u^m, v)) + (\vartheta |u^m|^{\beta-1} u^m, v) + kb(u^m, u^m, v) + k \int_{\partial\Omega} \xi (u^m \cdot \tau)(v \cdot \tau) dS = (u^{m-1} + kf^m, v) \quad \forall v \in V. \tag{21}$$

We apply the Galerkin method. We choose a sequence of elements $z_1, z_2, \ldots, z_i, \cdot$ from $Z$, where $Z = \{u \in \mathcal{D}(\Omega), \nabla \cdot u = 0\}$. Since $Z$ is dense in $V$ and separable, so element $z_i$ are linearly independent and spans $V$. For each $r$, we find an element $\phi_r$

$$\phi_r(x, t) = \sum_{i=1}^{r} \alpha_{i,r}(t) z_i, \tag{22}$$

where $\alpha_{i,r}(t)$ are to be determined. Putting $\phi_r$ in (21), we obtain

$$(\phi_r, v) + 2k((\phi_r, v)) + kb(\phi_r, \phi_r, v) + (\vartheta |\phi_r|^{\beta-1} \phi_r, v) + k \int_{\partial\Omega} \xi (\phi_r \cdot \tau)(v \cdot \tau) dS = (u^{m-1} + kf^m, v), \; \forall v \in sp(z_1, \ldots, z_r). \tag{23}$$

We put $v = \phi_r$ in (23) and obtain

$$(\phi_r - u^{m-1}, \phi_r) + 2k\|\phi_r\|_{H^1}^2 + \vartheta\|\phi_r\|_{L^{\beta+1}}^{\beta+1} + k\int_{\partial\Omega}\xi(\phi_r\cdot\tau)^2 dS = k(f^m, \phi_r). \tag{24}$$

We note that $|a_1 - a_2|^2 + |a_1|^2 - |a_2|^2 = 2(a_1 - a_2, a_1)$, $\forall a, b \in H$. Using these in (24), we get

$$\begin{aligned}
&\|\phi_r\|_{L^2}^2 + \|\phi_r - u^{m-1}\|_{L^2}^2 + 4k\|\phi_r\|_{H^1}^2 + 2\vartheta\|\phi_r\|_{L^{\beta+1}}^{\beta+1} + 2k\int_{\partial\Omega}\xi(\phi_r\cdot\tau)^2 dS \\
&= \|u^{m-1}\|_{L^2}^2 + 2k(f^m, \phi_r) \\
&\leq \|u^{m-1}\|_{L^2}^2 + 2k\|f^m\|_{V'}\|\phi_r\|_{L^2} \leq \|u^{m-1}\|_{L^2}^2 + k\|f^m\|_{V'}^2 + k\|\phi_r\|_{H^1}^2.
\end{aligned} \tag{25}$$

Hence

$$\|\phi_r\|_{L^2}^2 + \|\phi_r - u^{m-1}\|_{L^2}^2 + 3k\|\phi_r\|_{H^1}^2 + 2\vartheta\|\phi_r\|_{L^{\beta+1}}^{\beta+1} \leq \|u^{m-1}\|_{L^2}^2 + k\|f^m\|_{V'}^2. \tag{26}$$

From (26), we get the boundedness of sequence $\phi_r$ in $V$ as $r \to \infty$. So, we can extract a subsequence $\phi_{r'}$ satisfying the following convergence weakly in V: $\phi_{r'} \to \phi$, as $r' \to \infty$. We pass the limit in (23) and hence prove $\phi = u^m$ satisfies (21). Taking $v = u^m$ in (21), we obtain the lower limit in (25). Thus

$$\begin{aligned}
\|u^m\|_{L^2}^2 - \|u^{m-1}\|_{L^2}^2 + \|u^m - u^{m-1}\|_{L^2}^2 + 3k\|u^m\|_{H^1}^2 + 2\vartheta\|u^m\|_{L^{\beta+1}}^{\beta+1} \\
\leq k\|f^m\|_{V'}^2.
\end{aligned} \tag{27}$$

This completes the proof.

**Lemma 5** *Suppose $u_0 \in H$. We have the following bounds for $C_1$ and $C_2$:*

$$\|u^m\|_{L^\infty(0,T;H)\cap L^2(0,T;V)\cap L^{\beta+1}(0,T;L^{\beta+1}(\Omega))} \leq C_1, \tag{28}$$

$$\|p^m\|_{L^{5/3}(0,T;L^{5/3})} \leq C_2. \tag{29}$$

***Proof*** Summing up (27) over $m = 1, \ldots, M$, we get

$$\begin{aligned}
&\|u^M\|_{L^2}^2 + \sum_{1\leq m\leq M}\|u^m - u^{m-1}\|_{L^2}^2 + 3k\sum_{1\leq m\leq M}\|u^m\|_{H^1}^2 + 2\vartheta\sum_{1\leq m\leq M}\|u^m\|_{L^{\beta+1}}^{\beta+1} \\
&\quad\leq \|u^0\|_{L^2}^2 + 2k\sum_{1\leq m\leq M}\|f^m\|_{V'}.
\end{aligned} \tag{30}$$

It follows from the last inequality that $u^m \in L^\infty(0, T; H) \cap L^2(0, T; V) \cap L^{\beta+1}(0, T; L^{\beta+1}(\Omega))$ and using the interpolation result we obtain $u^m \in L^{10/3}(0, T; L^{10/3})$. We combine pressure $p^m$ to $u^m$ by De Rham's theorem. Since $\nabla \cdot u^m = 0$, we get from (1)

$$-\Delta p^m = \nabla \cdot (u^m \cdot \nabla) u^m = \sum_{i,j=1}^{3} \frac{\partial}{\partial x_i} \frac{\partial}{\partial x_i} u_i^m u_i^m, \qquad m = 1, \ldots, M. \quad (31)$$

Using the fact $u_i^m u_j^m \in L^{5/3}(0, T; L^{5/3})$ and the unique solution $p^m$, we obtain from (31) that

$$p^m \in L^{5/3}(0, T; L^{5/3}). \quad (32)$$

This can be seen by inverting the Laplace operator. This completes the proof.

**Lemma 6** *Suppose* $f^m = \frac{1}{k} \int_{(m-1)k}^{mk} f(t)dt$, $m = 1, \ldots, M$, $f^m \in V'$. *We have*

$$k \sum_{m=1}^{M} \|f^m\|_{V'}^2 \leq \int_0^T \|f(t)\|_{V'}^2 dt. \quad (33)$$

***Proof*** Using Schwarz inequality, we get

$$\|f^m\|_{V'}^2 = \frac{1}{k^2} \left\| \int_{(m-1)k}^{mk} f(t)dt \right\|_{V'}^2 \leq \frac{1}{k} \int_{(m-1)k}^{mk} \|f(t)\|_{V'}^2 dt. \quad (34)$$

Then (33) can be obtained by taking summation over the inequalities for $1 \leq m \leq M$.

**Lemma 7** *The approximate solution* $u^m$ *has the following estimates:*

$$\|u^m\|_{L^2}^2 \leq d_1, \ m = 1, \ldots, M, \quad (35)$$

$$k \sum_{1 \leq m \leq M} \|u^m\|_{H^1}^2 \leq d_1, \quad (36)$$

$$\sum_{1 \leq m \leq M} \|u^m - u^{m-1}\|_{L^2}^2 \leq d_1, \quad (37)$$

$$\sum_{1 \leq m \leq M} \|u^m\|_{L^{\beta+1}}^{\beta+1} \leq d_1, \quad (38)$$

*where* $d_1$ *depends only on the data*

$$d_1 = \|u_0\|_{L^2}^2 + \int_0^T \|f(s)\|_{V'}^2 ds. \quad (39)$$

***Proof*** Now we summing the inequality (27) for $m = 1, \ldots, M$. We get

$$\|u^M\|_{L^2}^2 + \sum_{1\leq m\leq M} \|u^m - u^{m-1}\|_{L^2}^2 + 3k \sum_{1\leq m\leq M} \|u^m\|_{H^1}^2 + 2\vartheta \sum_{1\leq m\leq M} \|u^m\|_{L^{\beta+1}}^{\beta+1}$$
$$\leq \|u^0\|_{L^2}^2 + k \sum_{1\leq m\leq M} \|f^m\|_{V'}^2. \tag{40}$$

Again we take the summation over (27) for $m = 1, \dots, r$ and removing $\|u^m\|_{H^1}^2$ and $\|u^m\|_{L^{\beta+1}}^{\beta+1}$, we get for $r = 1, \dots, M$.

$$\|u^r\|_{L^2}^2 \leq \|u_0\|_{L^2}^2 + k \sum_{1\leq m\leq r} \|f^m\|_{V'}^2 \leq \|u_0\|_{L^2}^2 + k \sum_{1\leq m\leq M} \|f^m\|_{V'}^2, \tag{41}$$

We can obtain (35) from (40)–(41) and using Lemma (6). In the similar way, we can obtain (36), (37), and (38).

**Lemma 8** *We have* $k \sum_{m=1}^{M} \|\frac{u^m - u^{m-1}}{k}\|_{V'}^2 \leq C_3$. *Where* $C_3$ *is a positive constant and not dependent on* $k$.

***Proof*** We can write (13) in such a way that

$$\left\|\frac{u^m - u^{m-1}}{k}\right\|_{V'}^2 \leq c_1\|Au^m\|_{V'}^2 + \|Bu^m\|_{V'}^2 + \|f^m\|_{V'}^2$$
$$\leq c_2(\|f^m\|_{V'}^2 + \|u^m\|_V^2 + \|Bu^m\|_{V'}^2). \tag{42}$$

for some positive constant $c_1, c_2$. From (35) and (9), we get

$$\|Bu^m\|_V^2 \leq c_3\|u^m\|_{L^2}^2\|u^m\|_{H^1}^2 \leq c_4\|u^m\|_{H^1}^2, \tag{43}$$

for some positive constant $c_3$ and $c_4$. We finally get

$$k\sum_{m=1}^{M} \left\|\frac{u^m - u^{m-1}}{k}\right\|_{V'}^2 \leq c_5 k \sum_{m=1}^{M} (\|f^m\|_{V'}^2 + \|u^m\|_{H^1}^2), \tag{44}$$

and we complete the proof using (36) and Lemma 6.

**Lemma 9** *We have the following estimate:*

$$\|u_M - w_M\|_{L^2(0,T;H)}^2 = \frac{k}{3}\sum_{m=1}^{M} \|u^m - u^{m-1}\|^2. \tag{45}$$

***Proof*** From (14), we have

$$w_M(t) - u_M(t) = \frac{t - t_{m-1}}{k}(u^m - u^{m-1}) + u^{m-1} - u^m,\ \forall t \in (t_{m-1}, t_m]. \tag{46}$$

Then

$$\begin{aligned}\int_0^T \|w_M - u_M\|^2 dt &= \sum_{m=1}^{M} \int_{t_{m-1}}^{t_m} \|w_M(t) - u_M(t)\|^2 dt \\ &= \sum_{m=1}^{M} \|u^m - u^{m-1}\| \int_{t_{m-1}}^{t_m} \left(\frac{t - t_{m-1}}{k} - 1\right)^2 dt \\ &= \frac{k}{3} \sum_{m=1}^{M} \|u^m - u^{m-1}\|^2. \end{aligned} \tag{47}$$

**Lemma 10** *The function $u_M$ and $w_M$ bounded in $L^\infty(0, T; H) \cap L^2(0, T; V) \cap L^{\beta+1}(0, T; L^{\beta+1}(\Omega))$ and $w'_M$ is bounded in $L^2(0, T; V')$. Further*

$$u_M - w_M \to 0 \text{ in } L^2(0, T; H) \text{ as } M \to \infty. \tag{48}$$

***Proof*** We have$\|u^m\|_{L^2}^2 \leq d_1$, and $k \sum_{m=1}^{M} \|u^m\|_{H^1}^2 \leq d_1$ and $\sum_{m=1}^{M} \|u^m\|_{L^{\beta+1}}^{\beta+1} \leq d_1$ from Lemma (7). Using the above relations, we can conclude that the function $u_M$ and $w_M$ bounded in $L^\infty(0, T; H) \cap L^2(0, T; V) \cap L^{\beta+1}(0, T; L^{\beta+1}(\Omega))$. Using Lemma (8), we ensure the boundedness of $w'_M$ in $L^2(0, T; V')$. Now we can obtain (48) from (37) and the Lemma (9).

**Lemma 11** *For $t \in [(m-1)k, mk]$, and $m = 1, \ldots, M$, we define $f_M(t) = f^m$, then $f_M \to f$ in $L^2(0, T; V')$ as $M \to \infty$*

***Proof*** The transformation $f_M \to f$ is a linear averaging mapping in $L^2(0, T; V')$. By Lemma 6, this mapping is continuous. From there, we can say that $\|f_M\|_{V'} \leq \|f\|_V$. Now

$$\|f_M - f\|_{V'} \leq 2\|f\|_V < K. \tag{49}$$

So, $f_M \to f$ in $L^2(0, T; V)$.

## 5 Proof of Theorem 1

***Proof*** From Lemma 10, we extract a subsequence $u_{M'}$ which satisfying $u_{M'} \to u$ weakly in $L^2(0, T; V)$, $u_{M'} \to u$ weak-star in $L^\infty(0, T; H)$, and $u_{M'} \to u$ weakly in $L^{\beta+1}(0, T; \Omega)$. We want to prove that $u$ is a solution. For passing the limit in (13), We need a strong convergence of $u_M$. The function $w_M$ will help for this. We have $w_{M'} \to u_*$ weakly in $L^2(0, T; V)$, $w_{M'} \to u_*$ weak-star in $L^\infty(0, T; H)$ and, $\frac{dw_{M'}}{dt} \to u'_*$ weakly in $L^2(0, T; V')$. Because of (48), $u = u_*$. From [14, III Theorem 2.1], it follows that $w_{M'} \to u$ in $L^2(0, T; H)$ strongly, thus by (48), $u_{M'} \to u$ in

$L^2(0, T; H)$ strongly. Using Lemma 3.2 [[14], Lemma III.3.2] and Lemma 2, we can say $Bu_{M'} \to Bu$ in $L^2(0, T; V)$ weakly. The Eq. (13) can be written as, for any $v \in V$

$$\left(\frac{dw_M}{dt}, v\right) + ((u_M, v)) + b(u_M, u_M, v) + (\vartheta |u_M|^{\beta-1} u_M, v) + \int_{\partial\Omega} \xi(u_M(t) \cdot \tau)(v \cdot \tau) dS = (f_M, v), \tag{50}$$

where for each $t \in [(m-1)k, mk)$ the function $f_M$ is defined by $f_M(t) = f^m$, for $m = 1, \ldots, M$. For $\phi \in C_0^\infty(0, T)$, we get

$$\int_0^T \phi(t) \left\{ \left(\frac{dw_M}{dt}, v\right) + ((u_M, v)) + b(u_M, u_M, v) + (\vartheta |u_M|^{\beta-1} u_M, v) + \int_{\partial\Omega} \xi(u_M(t) \cdot \tau)(v \cdot \tau) dS - (f_M, v) \right\} = 0 \tag{51}$$

Passing the limit in (50), we obtain

$$\int_0^T \phi(t) \{ (\frac{dw}{dt}, v) + ((u, v)) + b(u, u, v) + (\vartheta |u|^{\beta-1} u, v) + \int_{\partial\Omega} \xi(u(t) \cdot \tau)(v \cdot \tau) dS - (f, v) \} = 0. \tag{52}$$

Since $w_{M'}(0) = u_0$, we get $u(0) = u_0$. From (32), we can say $q_M$ is uniformly bounded in $L^{5/3}(0, T; L^{5/3})$. We get a subsequence $q_M$ which converges weakly in $L^{5/3}(0, T; L^{5/3})$. That is, we prove that $u$ is a weak solution of the damped Navier–Stokes equations with combined pressure $q$. This completes the proof.

## 6 Energy Inequality

We prove the following energy inequality for the solutions to the damped Navier–Stokes system. The idea of the proof is based on [1].

**Theorem 2** *For all positive $\phi \in C^\infty(\Omega \times [0, T])$ and $\phi(x, 0) = 0$, the solutions $(w, q)$ obtained by Theorem 1 satisfies the local energy inequality*

$$2\int_0^T \int_\Omega |\nabla w|^2 \phi dx dt \leq \int_0^T \int_\Omega \Big[ |w|^2(\Delta\phi + \partial_t \phi) + (|w|^2 + 2q) w \cdot \nabla\phi + 2(f \cdot w)\phi \Big] dx dt + 2\int_0^T \phi' \int_\Omega |w|^{\beta+1} dx dt. \tag{53}$$

***Proof*** We will proof that $(w, q)$ satisfies the energy inequality. We multiply the equation

$$\frac{dw_M}{dt} - \Delta u_M + u_M \cdot \nabla u_M + \vartheta_1 |u_M|^{\beta-1} u_M + \nabla q = f_M \tag{54}$$

by $u_M\phi$, where $\phi$ be a non-negative smooth function with compact support in $\Omega \times (0, T)$. We estimate the first term as follows:

$$\int_0^T (\partial_t w_M, u_M\phi)dt = \int_0^T (\partial_t w_M, (w_M - w_M + u_M)\phi)dt \tag{55}$$

$$= \int_0^T (\partial_t w_M, w_M)\phi dt + \int_0^T (\partial_t w_M, (u_M - w_M)\phi)dt = I_1 + I_2. \tag{56}$$

Let us consider the first term $I_1$. First, we split the interval over $[0, T]$ with the sum of integrals over $[t_{m-1}, t_m]$. Applying integration by parts, we get

$$\int_0^T (\partial_t w_M, w_M\phi)dt = \sum_{m=1}^M \int_{t_{m-1}}^{t_m} (\partial_t w_M, w_M\phi)dt = \sum_{m=1}^M \int_{t_{m-1}}^{t_m} \left(\tfrac{1}{2}\partial_t |w_M|^2, \phi\right) dt$$
$$= \tfrac{1}{2} \sum_{m=1}^M \left[(|u^m|^2, \phi(x, t_m)) - (|u^{m-1}|^2, \phi(x, t_{m-1}))\right] - \sum_{m=1}^M \int_{t_{m-1}}^{t_m} \left(\tfrac{1}{2}|w_M|^2, \partial_t\phi\right) dt$$

where we used that $\partial_t w_M(t) = \frac{u^m - u^{m-1}}{k}$ for $t \in [t_{m-1}, t_m)$. Also, we get

$$\int_0^T (\partial_t w_M, w_M\phi)dt$$
$$= \frac{1}{2}(|u^M|^2, \phi(x, T)) - \frac{1}{2}(|u_0|^2, \phi(x, 0)) - \sum_{m=1}^M \int_{t_{m-1}}^{t_m} \left(\frac{1}{2}|w_M|^2, \partial_t\phi\right) dt$$
$$= -\int_0^T \left(\frac{1}{2}|w_M|^2, \partial_t\phi\right) dt. \tag{57}$$

Since $w_M \to w$ in $L^2(0, T; H)$ strongly, we get

$$\lim_{M\to\infty} \int_0^T \int_\Omega \partial_t w_M w_M \phi dx dt = -\frac{1}{2} \int_0^T \int_\Omega |w|^2 \partial_t \phi dx dt. \tag{58}$$

Now, we consider $I_2$. Due to $u_M$ is constant in $[t_{m-1}, t_m)$, we can write

$$\int_0^T (\partial_t w_M, (u_M - w_M)\phi) = \sum_{m=1}^M \int_{t_{m-1}}^{t_m} (\partial_t (w_M - u_M), (w_M - u_M)\phi)dt$$
$$= -\sum_{m=1}^M \int_{t_{m-1}}^{t_m} \left(\partial_t \left(\frac{|w_M - u_M|^2}{2}\right), \phi\right) dt = \sum_{m=1}^M \int_{t_{m-1}}^{t_m} \left(\frac{|w_M - u_M|^2}{2}, \partial_t\phi\right) dt.$$

Since $u_M(t_m) = w_M(t_m)$ $\forall m = 0, \dots, M$, all boundary terms are vanishes in the last line. Since $u_M - w_M$ converge to 0 strongly in $L^2(0, T; H)$, we obtain that $I_2 \to 0$ as $M \to \infty$. Now we have

$$-\int_0^T \int_\Omega \Delta u_M u_M \phi dxdt = \int_0^T \int_\Omega |\nabla u_M|^2 \phi dxdt + \frac{1}{2}\int_0^T \int_\Omega \nabla |u_M|^2 \nabla \phi dxdt$$
$$= \int_0^T \int_\Omega |\nabla u_M|^2 \phi dxdt - \frac{1}{2}\int_0^T \int_\Omega |u_M|^2 \Delta \phi dxdt.$$

Since $\phi \geq 0$ and norm is lower semi-continuous, we obtain

$$\lim_{M\to\infty} \int_0^T \int_\Omega |\nabla u_M|^2 \phi dxdt \geq \int_0^T \int_\Omega |\nabla w|^2 \phi dxdt, \tag{59}$$

by using the strong convergence in $L^2(0, T; H)$, we deduce that

$$\lim_{M\to\infty} \frac{1}{2}\int_0^T \int_\Omega |u_M|^2 \Delta \phi dxdt = \frac{1}{2}\int_0^T \int_\Omega |w|^2 \Delta \phi dxdt. \tag{60}$$

We applying integration by parts for our nonlinear term. We obtain

$$\int_0^T \int_\Omega (u_M \cdot \nabla) u_M u_M \phi dxdt = \tfrac{1}{2}\int_0^T \int_\Omega \nabla |u_M|^2 u_M \phi dxdt \tag{61}$$
$$= -\tfrac{1}{2}\int_0^T \int_\Omega |u_M|^2 u_M \nabla \phi dxdt. \tag{62}$$

As $u_M$ converges to $w$ strongly in $L^2(0, T; H)$, we obtain

$$\lim_{M\to\infty} \int_0^T \int_\Omega (u_M \cdot \nabla) u_M u_M \phi dxdt = -\frac{1}{2}\int_0^T \int_\Omega |w|^2 w \nabla \phi dxdt. \tag{63}$$

For the damping term after passing limit, we get

$$\lim_{M\to\infty} \int_0^T \int_\Omega u_M u_M \phi dxdt = -\int_0^T \phi' \int_\Omega |w|^{\beta+1} dxdt. \tag{64}$$

Finally, We integrated by parts our pressure term and passing the limit. We get

$$\lim_{M\to\infty} \int_0^T \int_\Omega \nabla q_M u_M \phi dxdt = \int_0^T \int_\Omega q w \nabla \phi dxdt. \tag{65}$$

Using Lemma 11, we get

$$\lim_{M\to\infty} \int_0^T \int_\Omega f_M u_M \phi dxdt = \int_0^T \int_\Omega f \cdot w \phi dxdt. \tag{66}$$

We finally prove that

$$2\int_0^T\int_\Omega |\nabla w|^2\phi dxdt \le \int_0^T\int_\Omega\Big[|w|^2(\Delta\phi+\partial_t\phi)+(|w|^2+2q)w\cdot\nabla\phi$$
$$+2(f\cdot w)\phi\Big]dxdt+2\int_0^T\phi'\int_\Omega |w|^{\beta+1}dxdt. \quad (67)$$

# 7 Conclusion

We have established the existence of weak solutions of the system (1)–(4) in $\mathbb{R}^3$ by semi-discretization. Further, we have shown that the weak solution derived in Theorem 1 satisfies the generalized energy inequality (53).

**Funding** S. Pal was partially funded by National Board for Higher Mathematics (grant no—02011/9/2019NBHM(R.P.)/R & D II/1324).

# References

1. Berselli, L.C., Spirito, S.: Weak solution to the Navier-Stokes equations constructed by semi-discretization are suitable. Commun. Contemp. Math. **666**, 85–97 (2016)
2. Caffarelli, L., Kohn, R., Nirenberg, L.: Partial regularity of suitable weak solutions of the Navier-Stokes equations. Comm. Pure Appl. Math **35**(6), 771–831 (1982)
3. Cai, X., Jiu, Q.: Weak and strong solutions for the incompressible Navier-Stokes equations with damping. J. Math. Anal. Appl. **343**, 799–809 (2008)
4. Kashiwabara, T.: On a strong solution of the non-stationary Navier-Stokes equations under slip or leak boundary conditions of friction type. J. Differ. Eqs. **254**(2), 756–778 (2013)
5. Kelliher, J.P.: Navier-Stokes equations with Navier boundary conditions for a bounded domain in the plane. SIAM J. Math. Anal. **38**(1), 210–232 (2006)
6. Navier, C.L.M.H.: Sur les lois du mouvement des fluides. Mem. Acad. R. Sci. Inst. Fr. **6**, 389–440 (1827)
7. Scheffer, V.: Hausdorff measure and the Navier-Stokes equations. Comm. Math. Phys. **55**(2), 97–112 (1977)
8. Pal, S., Haloi, R.: Existence and uniqueness of solutions to the damped Navier-Stokes equations with Navier boundary conditions for three dimensional incompressible fluid. J. Appl. Math. Comput. **66**, 307–325 (2021)
9. Pal, S., Haloi, R.: On solution to the Navier-Stokes equations with Navier-slip boundary condition for three dimensional incompressible fluid. Acta Math. Sci. **39**(6), 1628–1638 (2019)
10. Sohr, H.: The Navier-Stokes equations. An elementary functional analytic approach, Modern Birkhäuser Classics, Birkhäuser/Springer Basel AG, Basel (2001)
11. Song, X., Hou, Y.: Attractors for the three dimensional incompressible Navier-Stokes equations with damping. Discret. Contin. Dyn. Syst. **31**, 239–252 (2012)
12. Song, X., Hou, Y.: Uniform attractors for three dimensional incompressible Navier-Stokes equation with nonlinear damping. J. Math. Anal. Appl. **422**, 337–351 (2015)

13. Necas, J.: Direct Methods in the Theory of Elliptic Equations. Springer, Berlin (2012)
14. Temam, R.: Navier-Stokes Equations. North-Holland, Amsterdam (1979)
15. Zhang, Z., Wu, X., Lu, M.: On the uniqueness of strong solution to the incompressible Navier-Stokes equation with damping. J. Math. Anal. Appl. **377**, 414–419 (2011)
16. Zhou, Y.: Regularity and uniqueness for the 3D incompressible Navier-Stokes equations with damping. Appl. Math. Lett. **25**, 1822–1825 (2012)

# The Diathermic Oils Over a Thin Liquid Film with $MOS_2$ Nano Particles: A Model with Analysis of Shape Factor Effects

**S. Suneetha, K. Subbarayudu, and P. Bala Anki Reddy**

**Abstract** A mathematical model is envisioned to depict and search out the report for different shapes of $MOS_2$ nanoparticles in a Casson nanofluid over an unsteady exponentially stretching sheet. The solid nanoparticles of Molybdenum disulphide are employed in different geometries such as bricks, cylinders, platelets, and blades in a porous medium. Also, Diathermic oil finds a remarkable application in mechanical engineering and industrial fields. By considering a non-uniform heat source/sink, it is possible to improve the rate of transferring of heat in diathermic oils, primarily Kerosene oil (KO) and Engine oil (EO). MATLAB's bvp4c function is used to compute the dimensionless forms of regulating flow expressions numerically. The role of relevant parameters on the fluid flow and heat transfer are debated by graphs and tables. It is significant that the heat transfer rate is more for blade-shaped $MOS_2$ nanoparticles when compared to other shapes.

**Keywords and Phrases** Thin liquid film · Shape effect · $MOS_2$ · Homogeneous—heterogeneous reactions · Porous media

---

S. Suneetha (✉)
Department of Applied Mathematics, Yogi Vemana University, Kadapa, AP 516005, India
e-mail: suneethayvu@gmail.com

K. Subbarayudu
Department of Sciences and Humanities, Sri Venkateswara Institute of Science and Technology, Kadapa, AP 516003, India

P. Bala Anki Reddy
Department of Mathematics, SAS, VIT, Vellore, Tamilnadu, India

 
B. Rushi Kumar et al. (eds.), *Mathematics and Computing*, Springer Proceedings in Mathematics & Statistics 415, https://doi.org/10.1007/978-981-19-9307-7_39

# 1 Literature Assessment

Heat transmission problem over a thin liquid film flow on an extending surface which is not steady has got vast applications in diverse fields like engineering, medical, and industrial themes such as wire varnishing, fiber coating, biophysics, films of tear in a human's eye, and also the condition of flow in a human's lungs and lubrication problems, are all unpredictable. Wang [1] was the first to discuss flow over a thin film on an extended sheet. Later, Andersson et al. [2, 3] adapted it for numerous physical natures of power-law fluid. However, Magyari and Keller [4] studied the effects of a stretching sheet on the flow field and boundary layers with exponentially rising velocity and temperature. As a result, many problems on an exponentially stretching sheet have been done by many investigators [5–8]. Nanofluid has nano-sized particles made of non-metallic or metallic materials and these fluids became a crucial item for researchers in many fields such as nano-drug delivery, microelectronics, nuclear power plants, heat exchanger, etc. The study of the mixed bag of fluid flow and electromagnetism is acknowledged as magneto-hydrodynamics (MHD), which is mainly used in the medical field, geophysics, designing MHD pumps, etc. Some motivating works on these nanofluids can be found in [9–11]. Molybdenum disulfide ($MOS_2$) is an inorganic compound made of layer by layer of molybdenum and sulphur atoms. Moreover, it is used in many mechanical applications due to its lubrication ability, slight frictional property, and robustness. Thermophysical properties depend on the size, shape, and volume fraction of nanoparticles and base fluid. Here four dissimilar nanoparticle shapes specifically cylinder, platelet, blade, and brick of $MOS_2$ are used. Hamilton and Crosser [12] observed an adequate amount of enhancement in the thermal conductivities with different nanoparticle shapes. Some other attempts on $MOS_2$ nanofluids with different shape effects are those made by [13–17]. In some chemically reacting processes like ignition, biochemical structures, explosion of fireworks, digestion of food, catalysis, and so forth homogeneous and heterogeneous reactions arise. Merkin [18] did the initial study on these reactions. Many scholars fed light on such type of reactions and are cited in Refs. [19–21].

Motivated by the above-cited literature, an effort has been made on a liquid thin film with EO and KO as the base fluids. $MOS_2$ nanoparticles of various shapes like cylinder, platelet, blade, and brick are appended in two separate individual source fluids. The model is resolved in MATLAB using the RK4S approach-bvp4c codes. The final results are displayed with graphs and tabular forms. This review extends the earlier studies and is also useful for different nanoparticle shape effects studies.

# 2 Formulation

In this work, a 2D porous nano liquid film of Magneto hydrodynamic nanofluid past an unsteady exponentially stretching sheet is considered. Here, $MOS_2$ is treated as a based nanoparticle. The stretching sheet switches on from a thin slot which is traced

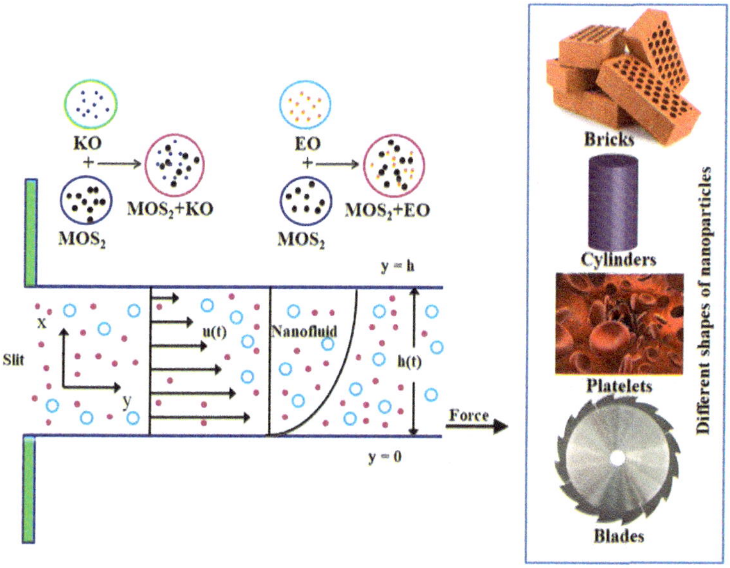


**Fig. 1** Flow framework

at the initial point of a two-dimensional coordinates system (x, y). The sheet stretches in the x-direction with velocity $U_w = \frac{cx}{1-\alpha t} e^{\frac{x}{l}}$ wherein c and $\alpha$ are not changeable and positive $c$ denotes expanding rate and $\alpha t < 1$. The plane stretches erect to the y-axis. A magnetic field of strength $B(t) = \frac{B_0}{(1-\alpha t)^{\frac{1}{2}}}$ is applied perpendicularly to the sheet on the outside as revealed in Fig. 1. Reynolds number is very small so the induced magnetic field is insignificant. The sheet temperature is designed as $T_w = T_0 + T_{ref}\left(\frac{cx^2}{2\upsilon_f}\right)(1-\alpha t)^{-\frac{3}{2}} e^{\frac{x}{2l}}$, where $\upsilon_f = \frac{\mu_f}{\rho_f}$ is the kinematic viscosity and $f$— the base fluid.

Chemical concentrations in the border flow for homogeneous and heterogeneous activities are assumed to be A** and B**.

$A** + 2B** \rightarrow 3B**,\ rate = k_c ab^2$, $A** \rightarrow B**,\ rate = k_s a$, where $k_c$ and $k_s$ are the rate constants, $a$ and $b$ are the chemical species concentrations, and Isothermal reactions are supposed for both processes.

The flow with the above attention is in the form

$$\frac{\partial u}{\partial x} + \frac{\partial v}{\partial y} = 0 \tag{1}$$

$$\rho_{nf}\left(\frac{\partial u}{\partial t} + u\frac{\partial u}{\partial x} + v\frac{\partial u}{\partial y}\right) = \mu_{nf}\left(1 + \frac{1}{\beta}\right)\frac{\partial^2 u}{\partial y^2} - u\left(\sigma_{nf} B^2(t) + \frac{\mu_{nf}}{k^*}\right), \tag{2}$$

$$(\rho C_p)_{nf}\left(\frac{\partial T}{\partial t} + u\frac{\partial T}{\partial x} + v\frac{\partial T}{\partial y}\right) = k_{nf}\frac{\partial^2 T}{\partial y^2} + \frac{k_f U_w}{x\upsilon_f}$$

$$\left(A * (T_w - T_0) f'(\eta) + B * (T - T_0)\right) \quad (3)$$

$$\left(\frac{\partial a}{\partial t} + u\frac{\partial a}{\partial x} + v\frac{\partial a}{\partial y}\right) = D_A\left(\frac{\partial^2 a}{\partial x^2} + \frac{\partial^2 a}{\partial y^2}\right) + \frac{D_T}{T_\infty}\left(\frac{\partial^2 T}{\partial x^2} + \frac{\partial^2 T}{\partial y^2}\right) - k_1 ab^2 \quad (4)$$

$$\left(\frac{\partial b}{\partial t} + u\frac{\partial b}{\partial x} + v\frac{\partial b}{\partial y}\right) = D_B\left(\frac{\partial^2 b}{\partial x^2} + \frac{\partial^2 b}{\partial y^2}\right) + \frac{D_T}{T_\infty}\left(\frac{\partial^2 T}{\partial x^2} + \frac{\partial^2 T}{\partial y^2}\right) + k_1 ab^2 \quad (5)$$

under the boundary conditions

$$u = U_w,\ v = V = 0,\ T = T_w,\ D_A\frac{\partial a}{\partial y} = -D_B\frac{\partial b}{\partial y} = ks \ \ at \ \ y = 0 \quad (6)$$

$$\frac{\partial u}{\partial y} = \frac{\partial T}{\partial y} = 0\,,\ a = a_0,\ b = 0 \ at \ y = H,\ \ v = \frac{\partial H}{\partial t} at \ y = H(t) \quad (7)$$

here, $H$(t) is the thickness of the fluid film.

Practically, the nanoparticle concentration of $MOS_2$ is little. With the help of Taylor's series, the nanofluid constants are

$$\frac{\mu_{nf}}{\mu_f} = \frac{1}{(1-2.5\chi)},\ \frac{\rho_{nf}}{\rho_f} = 1 - \chi + \chi\Upsilon,\ \frac{(\rho C_p)_{nf}}{(\rho C_p)_f} = 1 - \chi + \chi d,$$

$$\frac{k_{nf}}{k_f} = 1 + \frac{3(k-1)\chi}{(k+2)},\ \frac{\sigma_{nf}}{\sigma_f} = \frac{3(\sigma-1)\chi}{(\sigma+2)-(\sigma-1)\chi} + 1$$

$$\text{where} \quad ¡ = \frac{\rho_{sd}}{\rho_f},\ d = \frac{(\rho C_p)_{sd}}{(\rho C_p)_f},\ k = \frac{k_{sd}}{k_f},\ \sigma = \frac{(\sigma)_{sd}}{(\sigma)_f} \quad (8)$$

In the present investigation, the effective thermal conductive $k_{nf}$ can be estimated as [18]

$$\frac{k_{nf}}{k_f} = \frac{k_{sn} + (m-1)k_f + (m-1)(k_{sn} - k_f)\phi}{k_{sn} + (m-1)k_f - (k_{sn} - k_f)\phi}$$

Here, $k_f$ and $k_{sn}$ are the thermal conductivities of fluid and nanoparticles, respectively. The shape factor values are given in Table 1 (Table 2).

**Table 1** Nanoparticle's shape factor (m)

| Nanoparticles type | Shape factor (m) |
|---|---|
| Cylinders | 4.9 |
| Bricks | 3.7 |
| Blades | 8.6 |
| Platelets | 5.7 |

**Table 2** Shows the numerical values of the base fluids and nanoelements [21]

| Physical properties | $MOS_2$ | EO KO | |
|---|---|---|---|
| $Cp$ (J/kg K) | 397.21 | 2048 | 2090 |
| $\kappa$ (W/m K) | 904.4 | 0.1404 | 0.15 |
| $\sigma$(s/m) | $2.09 \times 10^{-4}$ | $55 \times 10^{-6}$ | $21 \times 10^{-6}$ |
| $\rho$ (kg/m$^3$) | 5060 | 863 | 783 |

Unveiling the dimensionless variables as

$$\eta = y\,e^{\frac{x}{2l}}\left(\frac{c}{\upsilon_f(1-\alpha t)}\right)^{\frac{1}{2}},\ \psi = \beta_1\left(\frac{c}{1-\alpha t}\right)^{\frac{1}{2}} f(\eta)e^{\frac{x}{2l}},\ \phi = \frac{a}{a_0},\ h = \frac{b}{a_0}$$

$$u = \beta_1\frac{c}{1-\alpha t}xf'(\eta)e^{\frac{x}{l}},\ v = -\beta_1\left(\frac{c\upsilon_f}{1-\alpha t}\right)^{\frac{1}{2}} f(\eta)e^{\frac{x}{2l}},\ T = T_0 + T_{ref}\left(\frac{cx^2}{2\upsilon_f}\right)(1-\alpha t)^{-\frac{3}{2}}\theta(\eta)e^{\frac{x}{2l}}, \tag{9}$$

Adopting Eqs. (8, 9) in Eqs. (1–5), we have

$$\phi_1 f'''\left(1+\frac{1}{\beta}\right) + \phi_2(\beta_1)^2\left[ff'' - (f')^2 - S\left(f' + \frac{\eta}{2}f''\right)\right] - \phi_3 Mf' - Df' = 0 \tag{10}$$

$$\phi_4\theta'' - \phi_5 \Pr(\beta_1)^2\left[2f'\theta - f\theta' + \frac{3}{2}S\theta + \frac{1}{2}S\eta\theta'\right] + A*f' + B*\theta = 0 \tag{11}$$

$$\frac{1}{Sc}\left(\phi'' + \frac{1}{N_{AT}}(\theta + \theta'')\right) + f\phi' - S\phi'\frac{\eta}{2} - K\phi h^2 = 0 \tag{12}$$

$$\frac{\delta}{Sc}\left(h'' + \frac{1}{N_{AT}}(\theta + \theta'')\right) + fh' - Sh'\frac{\eta}{2} + K\phi h^2 = 0 \tag{13}$$

$$\text{at } \eta = 0 \rightarrow f(0) = 0,\ f'(0) = 1,\ \theta(0) = 1,\ \phi'(0) = Ks\phi(0),\ \delta h'(0) = -Ks\phi(0) \tag{14}$$

$$\text{at } \eta = 1 \rightarrow f(1) = \frac{\beta_1 S}{2},\ f''(1) = 0,\ \theta'(1) = 0,\ \phi(1) = 1,\ h(1) = 0 \tag{15}$$

where

$$\phi_1 = \frac{(\mu)_{nf}}{(\mu)_f},\ \phi_2 = \frac{\rho_{nf}}{\rho_f},\ \phi_3 = \frac{\sigma_{nf}}{\sigma_f},\ \phi_4 = \frac{k_{nf}}{k_f},\ \phi_5 = \frac{(\rho C_p)_{nf}}{(\rho C_p)_f},$$

$$K = a_0^2 k_1 \frac{(1 - \alpha t)}{c},\ Ks = \frac{k_s}{D_A\, a}\left(\frac{c}{\upsilon_f(1 - \alpha t)}\right)^{-\frac{1}{2}},$$
$$S = \frac{\alpha}{c}, M = \frac{\sigma_f B_0^2}{c\rho_f},\ D = \frac{\upsilon}{\rho_f k^*}\frac{(1 - \alpha t)}{c},\ \Pr = \frac{(\mu C_p)_f}{k_f}, \delta = \frac{D_B}{D_A},$$
$$N_{AT} = \frac{D_A\, a_0}{\frac{D_T}{T_\infty}(T_w - T_\infty)}, \tag{16}$$

Thus, $\delta = 1$ for $D_A$ and $D_B$ is equal $\phi(\eta) + h(\eta) = 1$,
Equations (12) and (13) under this assumption decomposed to

$$\frac{1}{Sc}\left(\phi'' + \frac{1}{N_{AT}}(\theta + \theta'')\right) + f\phi' - S\phi'\frac{\eta}{2} - K\phi(1 - \phi)^2 = 0 \tag{17}$$

related boundary conditions: $\eta = 0 \to\ \phi'(0) = Ks\,\phi(0),\ \eta = 1 \to\ \phi(1) = 1$

The quantities of the engineering curiosity drag force on the surface and Nusselt number at local are given as

$$C_f = \frac{\tau_w}{\rho_f\, U_w^2},\ Nu = \frac{q_w x}{k_f(T_w - T_0)} \tag{18}$$

where $\tau_w$ (wall skin friction), and $q_w$ (wall heat flux) are given as

$$\tau_w = \mu_{nf}\left(1 + \frac{1}{\beta}\right)\left(\frac{\partial u}{\partial y}\right)_{y=0},\ q_w = -k_{nf}\left(\frac{\partial T}{\partial y}\right)_{y=0} \tag{19}$$

In view of Eqs. (9) and (19) in Eqs. (18), we acquire

$$C_f = (\mathrm{Re}_x)^{\frac{-1}{2}}\left(1 + \frac{1}{\beta}\right)\frac{1}{(1 - 2.5\chi)}f''(0),\ Nu_x = -(\mathrm{Re}_x)^{\frac{1}{2}}\frac{-1}{\beta_1}\frac{k_{nf}}{k_f}\phi'(0) \tag{20}$$

where $\mathrm{Re}_x = \frac{U_w x}{\upsilon_f}$ signifies the local Reynolds number.

## 3 Plan of Solution

By transforming partial differential expressions into ordinary differential equations by using suitable transformations, which results highly nonlinear Eqs. (10), (11) and (17) which cannot be solved analytically. Therefore, we employ the famous shooting technique with RKF method. The set of coupled nonlinear ODEs is renovated into the system of differential equations of the first order as follows:

$f = P_1,\ f' = P_2,\ f'' = P_3,\ f''' = P_3',\ \theta = P_4,\ \theta' = P_5,\ \theta'' = P_5',\ \phi = P_6,\ \phi'$

$$= P_7, \phi'' = P_7',$$

$$P_3' = \left(\frac{1}{\phi_1\left(1+\frac{1}{\beta}\right)}\right)\left(-\phi_2(\beta_1)^2 P_1 P_3 - P_2^2 - S(P_2 + 0.5\eta P_3)\right) + \phi_3 M P_2 + D P_2$$

$$P_5' = \frac{1}{\phi_4}\left[\phi_5 \Pr(\beta_1)^2\left(2P_2P_4 - P_1P_5 + \frac{3}{2}SP_4 + \frac{1}{2}S\eta P_5\right) + A * P_2 + B * P_4\right]$$

$$P_7' = -\frac{1}{N_{AT}}(P_4P_5') + Sc\left(-P_1P_7 + 0.5S\eta P_7 + KP_6(1-P_6)^2\right)$$

Associated boundary conditions are

$$P_1(0) = 0, P_2(0) = 1, P_4(0) = 1, P_7(0) = KsP_6(0), P_1(1) = \frac{\beta_1 S}{2}, P_3(1) = 0,$$
$$P_5(1) = 0, P_6(1) = 1$$

For solving boundary layer flow problems, this technique is very much useful. Assume two guesses $f''(0)$ and $-\theta'(0)$ to get an approximate solution. For all cases, the step size is 0.001, and the convergence criterion is $10^{-6}$.

## 4 Discussion

This portion designates momentous features of tangled flow parameters on velocity, temperature and nano particle concentration, the rate of heat transfer, surface drag force, and local convectional transmission coefficient. Computer codes have been propelled for these numerical results and are incited with graphs and tables assuming the default values for all the physical parameters $Sc = 0.5$, $K = 0.5$, $\beta_1 = 1.0$, $q = 0.5$,

$\phi_1 = 1.0$, $\phi_2 = 1.0$, $S = 0.5$, $\theta_w = 1.5$, $M = 0.5$, $\Pr = 0.72$, $A* = 0.05$, $B* = 0.05$, $N_{AT} = 1.0$, and $Ks = 10.0$are fixed. At this juncture, two cases: $MOS_2$ + KO and $MOS_2$ + EO are examined.

The upshot of the unsteadiness parameter $S$ on flow and thermal is displayed in Figs. 2 and 3. It is reported that the upgrading values of $S$ degrade the velocity and temperature profiles for two cases. It should also be noticed that $MOS_2$ + KO leads to better outcomes for the boundary layers of thermal and momentum than for the concentration. This is due to the influence of buoyancy on the fluid flow shrinks for escalating values of $S$. This results in a decrement in all three boundary layers.

Figure 4 fed light on the velocity variations for different porosity $D$ on thin nanoliquid film. It is observed that velocity hikes as $D$ hikes. The porosity should be increased slightly as the thickness of the film is very little, that is $D \to \infty$ corresponds to the case in which there does not exist any porous medium. The growing values of $D$ stand for the massive opening of the permeable gap, resulting in the retardation of the flow with elevated velocity.

**Fig. 2** $f'(\eta)$ for $S$

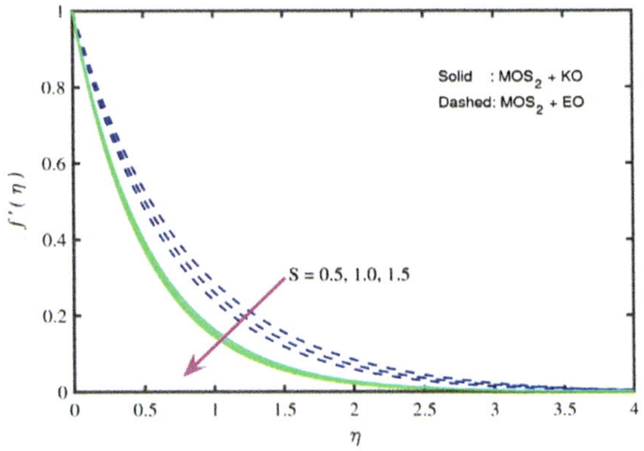


**Fig. 3** $\theta(\eta)$ for $S$

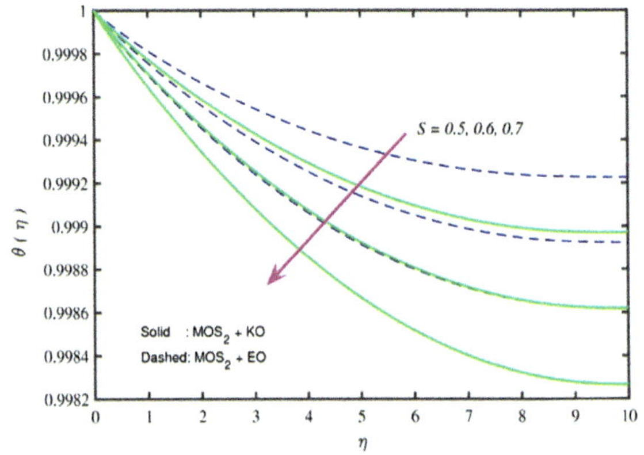


**Fig. 4** $f'(\eta)$ for $D$

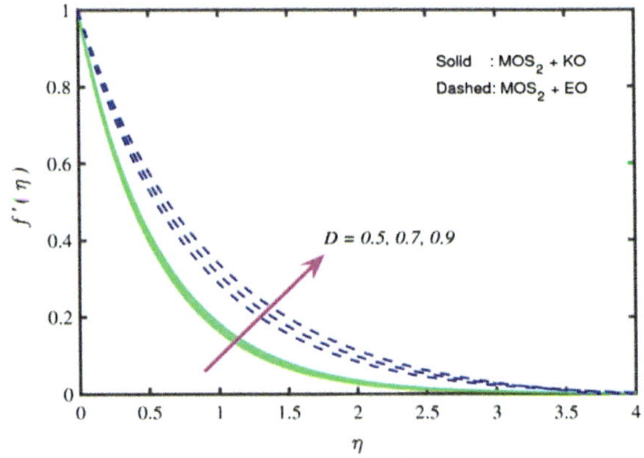

Concentration variations are displayed in Figs. 5 and 6 for changing the scale of the reactions $K$ and $Ks$. As $K$ and $Ks$ values mounted a decrement in concentration is inferred from the figures. When the strength of homogeneous reactions is increased, the consumption of chemical reactants improves as well, resulting in a large chemical reaction and, as a result, a smaller concentration distribution. For various base fluids, the thickness of the solutal boundary layer decreases. Because the heterogeneous reaction parameter Ks has an opposite relation with mass diffusivity, the concentration falls.

Figure 7 displays the outcome of *Sc* on concentration. Sc refers to simultaneous momentum and mass diffusion convection in a fluid flow. *Sc* is the proportion of the rates of viscous diffusion to molecular diffusion. Low concentration for high *Sc* is noted. This is due to the fact that strong diffusion species have a greater retarding impact on the concentration.

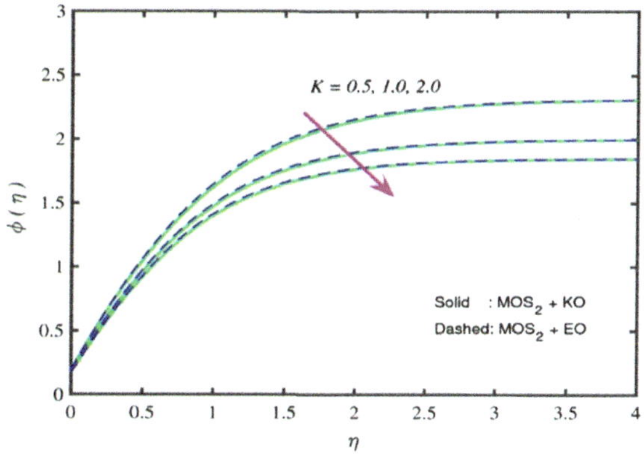


**Fig. 5** $\phi$ ($\eta$) for $K$

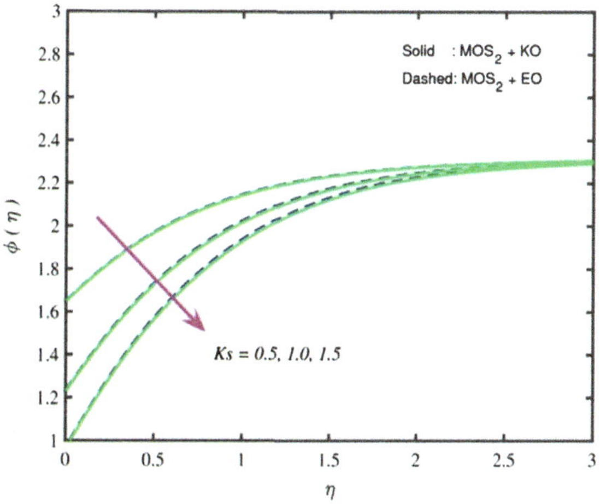


**Fig. 6** $\phi$ ($\eta$) for $Ks$

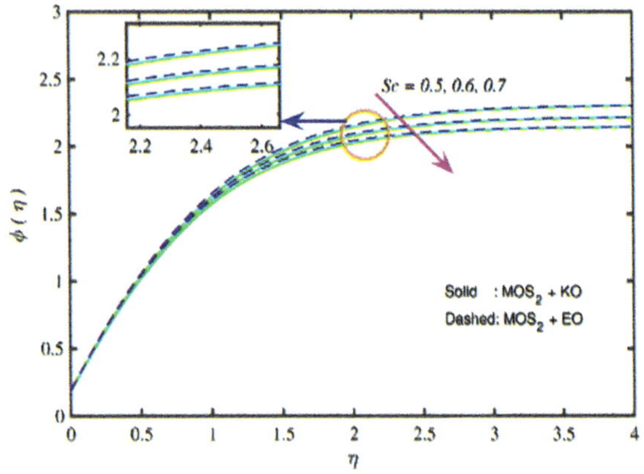


**Fig. 7** $\phi$ ($\eta$) for $Sc$

Figure 8 illustrates how the temperature distribution is affected by the space-dependent heat source/sink parameter A*. The boundary layer generates energy if A* > 0 (heat source), which boosts the liquid's thermal nature. The border layer absorbs energy for A* < 0 (absorption), consequences a drop in temperature. In Fig. 9, the effect of B* on temperature is shown. As energy is discharged into the fluid, a hike in fluid temperature is noticed when B* > 0, and a negativity is noted, i.e., a declivity in heat as the energy is taken up by the fluid when B* < 0.

Figure 10 emphasizes the variation in Surface drag force for different $D$. It brings out that escalating $D$ the drag force is added for $MOS_2$ + KO and $MOS_2$ + EO. The consequence of $S$ on drag force is shown in Fig. 11. It is uncovered that rising $S$ raises the drag force for both cases.

Figure 12 exemplifies the consequence of heat flux on $A*$. It is viewed from the figure that a rise in $A*$ has a plunge in the heat flux. It is detected from Fig. 13 that the heat flux reduces with $B*$. The increment for $MOS_2$ + EO is further with $MOS_2$ + KO.

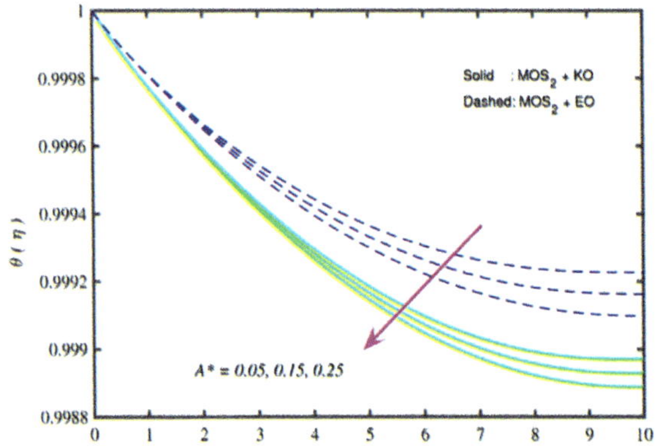


**Fig. 8** $\theta$ ($\eta$) for $A*$

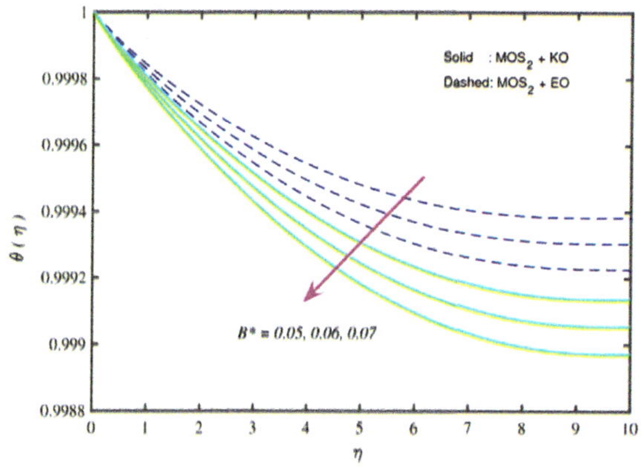


**Fig. 9** $\theta\ (\eta)$ for $B*$

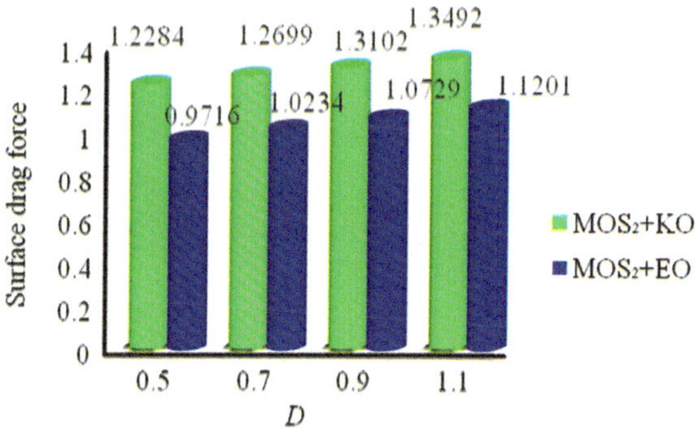


**Fig. 10** $f''\ (\eta)$ for $D$

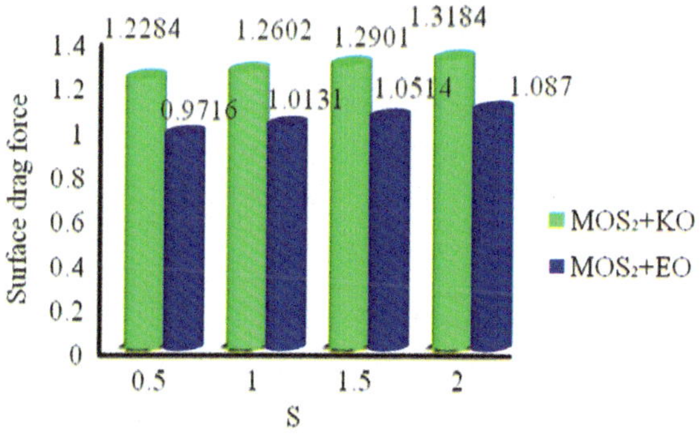


**Fig. 11** $f''\ (\eta)$ for $S$

For different shapes of $MOS_2$ nanoparticles, the rate of heat transfer and volume fraction in kerosene oil is debated in Fig. 14. Here, the heat transfer rate amplifies with $\phi$. The heat transfer rate of the nanofluid with particles of blade-shapes is bigger than brick, cylinder, and platelet shapes for the same volume fraction. The heat transfer rate with particles of blade shapes is 12.70, 23.93, and 34.07% greater

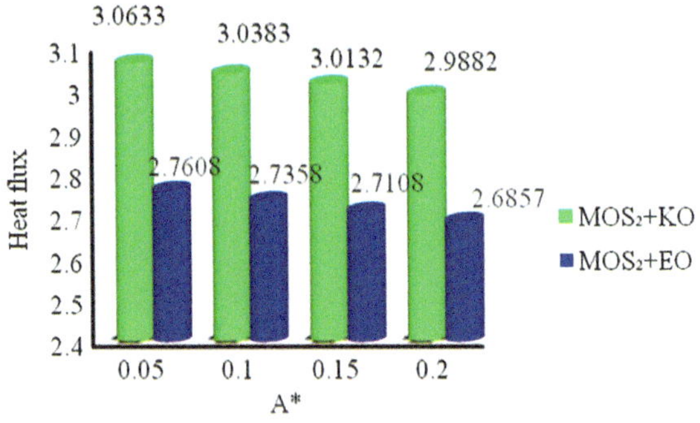


**Fig. 12** $-\theta'(\eta)$ for $A*$

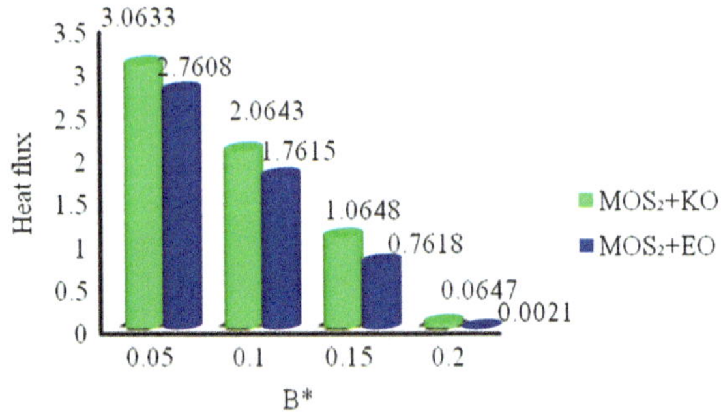


**Fig. 13** $-\theta'(\eta)$ for $B*$

than brick, cylinder, and platelet shapes in kerosene oil-based nanofluids. $MOS_2$ is more conductive when compared to other metals because of its special arrangement with diamagnetic nature. However, the heat transfer rate is low.

Figure 15 is displayed to report the increment percentage of unlike shapes of $MOS_2$ nanoparticles with Engine oil as a base fluid. The figure shows that the particles of blade shapes have the highest variation succeeded by brick, cylinder, and platelet shapes. Also, spotted that the thermal transfer rate of EO has been enhanced by 33.04% with blade-shaped particles. Also noted that 17.05%, 21.52%, and 24.28%

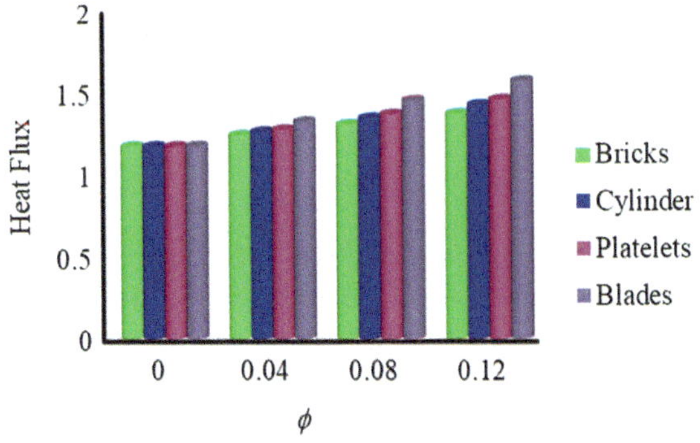


**Fig. 14** $-\theta'(\eta)$ for $\phi$ with different shapes on $MOS_2$ + KO

increment of the thermal transfer rate of EO with brick, cylinder, and platelet-shaped nanoparticles, respectively. Table 3 shows a deviancy of surface drag force for various denominations of $M$, $D$, $S$, and $\beta_1$. The drag force hastens with $M$, $D$, $S$ and decelerates along $\beta_1$. The increment of the rate of heat transfer with dissimilar shaped nanoparticles in EO and KO with divergent volume fractions are displayed in Tables 4 and 5. The blade-shaped particles show the utmost deviation caught on by brick, cylinder, and platelet-shaped particles in both base fluids.

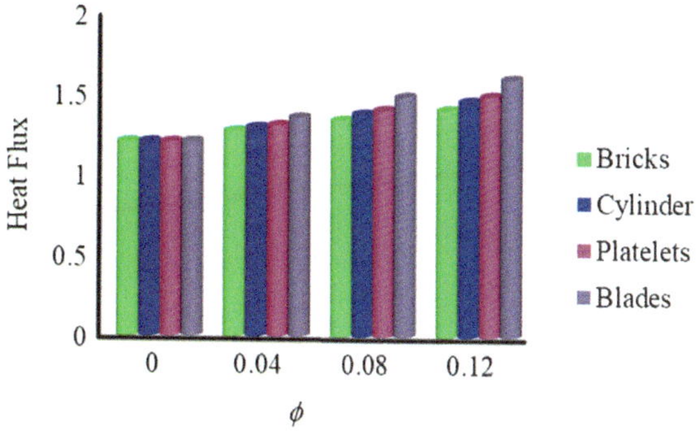


**Fig. 15** $-\theta'(\eta)$ for $\phi$ with different shapes on $MOS_2$ + EO

**Table 3** The values of surface drag force for various parameters

| | | | | $f''(0)$ | |
|---|---|---|---|---|---|
| $M$ | $D$ | $S$ | $\beta_1$ | $MOS_2$ + KO | $MOS_2$ + EO |
| 0.5 | 0.5 | 0.5 | 0.5 | 1.2284 | 0.9716 |
| 1.0 | 0.5 | 0.5 | 0.5 | 1.5398 | 1.1278 |
| 1.5 | 0.5 | 0.5 | 0.5 | 1.7988 | 1.2653 |
| 2.0 | 0.5 | 0.5 | 0.5 | 2.0253 | 1.3894 |
| 0.5 | 0.7 | 0.5 | 0.5 | 1.2699 | 1.0234 |
| 0.5 | 0.9 | 0.5 | 0.5 | 1.3102 | 1.0729 |
| 0.5 | 1.1 | 0.5 | 0.5 | 1.3492 | 1.1201 |
| 0.5 | 0.5 | 1.0 | 0.5 | 1.2602 | 1.0131 |
| 0.5 | 0.5 | 1.5 | 0.5 | 1.2901 | 1.0514 |
| 0.5 | 0.5 | 2.0 | 0.5 | 1.3814 | 1.0870 |
| 0.5 | 0.5 | 0.5 | 1.0 | 0.7967 | 0.6901 |
| 0.5 | 0.5 | 0.5 | 1.5 | 0.6746 | 0.6106 |
| 0.5 | 0.5 | 0.5 | 2.0 | 0.6150 | 0.5692 |

**Table 4** $\phi$ for $-\theta'(0)$ and percent enhancement with bricks, cylinders, platelets, and blades on $MOS_2 + EO$

| $\phi$ | $-\theta'(0)$ | | | | % | | | |
|---|---|---|---|---|---|---|---|---|
| | Bricks | Cylinders | Platelets | Blades | Bricks | Cylinders | Platelets | Blades |
| 0 | 1.2061 | 1.2061 | 1.2061 | 1.2061 | – | – | – | – |
| 0.04 | 1.2766 | 1.2978 | 1.3116 | 1.3589 | 5.84529 | 7.60302 | 8.7472 | 12.6689 |
| 0.08 | 1.3451 | 1.3841 | 1.4087 | 1.4900 | 11.5248 | 14.7583 | 16.7979 | 23.5387 |
| 0.12 | 1.4118 | 1.4657 | 1.4989 | 1.6046 | 17.055 | 21.5239 | 24.2766 | 33.0404 |

**Table 5** $\phi$ for $-\theta'(0)$ and percent enhancement with bricks, cylinders, platelets, and blades on $MOS_2 + KO$

| $\phi$ | $-\theta'(0)$ | | | | % | | | |
|---|---|---|---|---|---|---|---|---|
| | Bricks | Cylinders | Platelets | Blades | Bricks | Cylinders | Platelets | Blades |
| 0 | 1.1752 | 1.1752 | 1.1752 | 1.1752 | – | – | – | – |
| 0.04 | 1.2434 | 1.2642 | 1.2777 | 1.3244 | 5.80327 | 7.57318 | 8.72192 | 12.6957 |
| 0.08 | 1.3106 | 1.3493 | 1.3740 | 1.4564 | 11.5214 | 14.81450 | 16.91630 | 23.9278 |
| 0.12 | 1.3771 | 1.4315 | 1.4655 | 1.5756 | 17.1801 | 21.80910 | 24.70220 | 34.0708 |

## 5 Conclusions

The central concluded points are as follows:

- The $MOS_2$ nanoparticle is more active in EO than KO.
- The non-spherical elements (Platelet and Cylinder) within EO–KO-based fluids conclude better viscosity owing to their configuration and will intensely increase the heat transport capacity.
- The Surface drag force is high for high Porosity parameter.
- Heat transfer rates of EO having blade-shaped particles are 12.67, 23.54, and 33.04%; superior to the regular fluid with volume fraction $\phi = 0.04$, 0.08 and 0,12, respectively.
- The heat transfer rate of EO-based nanofluid suspended blade-shaped $MOS_2$ nanoparticles is 33.04, 17.05, 21.52, and 24.28%; bigger than brick, cylinder, and platelet-shaped nanoparticles.

**Graphical Trends**

(See Figs. 2 to 15. Table 2 to 5).

## Nomenclature

| | |
|---|---|
| $e_{ij}$ | (I, j)th element of the deformation rate |
| $B_0$ | Applied magnetic flux |
| $f(\eta)$ | Dimensionless velocity |
| $EO$ | Engine oil |
| $q_w$ | Heat flux from the surface |
| $q$ | Dimensionless parameter |
| $Ks$ | Heterogeneous reaction strength |
| $K$ | Homogeneous reaction strength |
| $U_w$ | Stretching velocity in X direction (m $s^{-1}$) |
| $V$ | Stretching velocity in Y direction (m $s^{-1}$) |
| $H(t)$ | Film size (m) |
| $T_w$ | Temperature of the fluid near the wall (K) |
| $T_0$ | Initial temperature of the fluid (K) |
| $KO$ | Kerosene oil |
| $C_{f_x}$ | Local skin friction in dimensionless form along x-direction |
| $Nu_x$ | Local Nusselt number |
| $\mathrm{Re}_x$ | Local Reynolds number |
| $M$ | Magnetic field parameter |
| Pr | Prandtl number |
| $p$ | Pressure (kg $m^{-1}s^{-2}$) |
| $D$ | Porosity parameter |
| $T_{ref}$ | Refered temperature of the fluid (K) |
| $T$ | Temperature (K) |
| $C_p$ | Specific heat at constant pressure (J $kg^{-1}K^{-1}$) |
| $T_s$ | Temperature of the fluid over surface (K) |
| $t$ | Time (s) |
| $c$ | Stretching rate ($s^{-1}$) |
| $k$ | Thermal conductivity ($m^2$ $s^{-1}$) |
| $S$ | Unsteadiness parameter |
| $Sc$ | Schmidt number |
| $u$ | Velocity component along x-axis (m $s^{-1}$) |
| $v$ | Velocity component along y-axis (m $s^{-1}$) |
| $x$ | $x-$ Coordinate (m) |
| $y$ | $y-$ Coordinate (m) |

## Greek Symbols

| | |
|---|---|
| $\alpha$ | Constant ($s^{-1}$) |
| $\pi_c$ | Critical value of the product of the deformation rate by itself |
| $\rho$ | Density (kg $m^{-3}$) |

| | |
|---|---|
| $\psi$ | Physical stream function ($m^2$ $s^{-1}$) |
| $\delta$ | Ratio of diffusion coefficient |
| $\beta_1$ | Dimensionless fluid thickness parameter |
| $\upsilon$ | Kinematic viscosity ($m^2$ $s^{-1}$) |
| $\phi_i (i = 1 - 5)$ | -Nanofluids constants |
| $\tau_w$ | Surface shear stress (kg $m^{-1} s^{-2}$) |
| $\eta$ | Similarity variable |
| $\sigma$ | Electrical conductivity |
| $\theta(\eta)$ | Dimensionless temperature |
| $\phi$ | Nano particles volume fraction of $MOS_2$ |

## *Subscripts*

| | |
|---|---|
| $f$ | Base fluid |
| $\infty$ | Fluid properties at ambient flow |
| $nf$ | Nanofluid |
| $s$ | Surface |

## References


1. Wang, C.Y.: Liquid film on an unsteady stretching surface. Q. Appl. Math. **48**, 601–610 (1990)
2. Andersson, H.I., Aarseth, J.B., Braud, N., Dandapat, B.S.: Flow of a power-law fluid film on unsteady stretching surface. J. Non-Newton. Fluid Mech. **62**, 1–8 (1996)
3. Andersson, H.I., Aarseth, J.B., Dandapat, B.S.: Heat transfer in a liquid film on an unsteady stretching surface. Int. J. Heat Mass Transf. **43**, 69–74 (2000)
4. Magyari, E., Keller, B.: Heat and mass transfer in the boundary layers on an exponentially stretching continuous surface. J. Phys. D: Appl. Phys. **32**, 577–585 (1999)
5. Tantry, I.A., Wani, S., Agrawal, B.: Study of MHD boundary layer flow of a casson fluid due to an exponentially stretching sheet with radiation effect. Int. J. Stat. Appl. Math. **6**, 138–144 (2021)
6. Nagaraja, B., Gireesha, B.J.: Exponential space-dependent heat generation impact on MHD convective flow of Casson fluid over a curved stretching sheet with chemical reaction. J. Therm. Anal. Calorim. **143**, 4071–4079 (2021). https://doi.org/10.1007/s10973-020-09360-0
7. Janareddy, S., Valsamy, P., Srinivasreddy, D.: Radiation and heat source/sink effects on MHD Casson fluid flow over a stretching sheet with slip conditions. J. Math. Comput. Sci. **11**, 6541–6556 (2021). https://doi.org/10.28919/jmcs/6385
8. Aloliga, G., Ibrahim Seini, Y., Musah, R.: Heat transfer in a magnetohydrodynamic boundary layer flow of a non-newtonian casson fluid over an exponentially stretching magnetized surface. J. Nanofluids **10**, 172–185 (2021). https://doi.org/10.1166/jon.2021.1777
9. Ahmad, F., Gul, T., Khan, I., Saeed, A., Selim, M.M., Kumam, P., Ali, I.: MHD thin film flow of the Oldroyd-B fluid together with bioconvection and activation energy. Case Stud. Therm. Eng. **27**, 101218 (2021). https://doi.org/10.1016/j.csite.2021.101218
10. Guled, C.N., Tawade, J.V., Priyanka, P.: The MHD flow of liquid film in the presence of dissipation and thermal radiation through an unsteady stretching sheet by HAM. Turkish J. Comput. Math. Educ. **12**, 949–959 (2021)

11. Gul, T., Rehman, M., Saeed, A., Khan, I., Khan, A., Nasir, S., Bariq, A.: Magnetohydrodynamic impact on carreau thin film couple stress nanofluid flow over an unsteady stretching sheet. Math. Probl. Eng. **2021**, 1–10 (2021). https://doi.org/10.1155/2021/8003805
12. Hamilton, R., Crosser, O.: Thermal conductivity of heterogeneous two-component systems. Ind. Eng. Chem. Fundam. **1**, 187–191 (1962)
13. Khan, I.: Shape effects of $MOS_2$ nanoparticles on MHD slip flow of molybdenum disulphide nanofluid in a porous medium. J. Mol. Liq. **233**, 442–451 (2017). https://doi.org/10.1016/j.molliq.2017.03.009
14. Gul, A., Khan, I., Makhanov, S.S.: Entropy generation in a mixed convection Poiseuille flow of molybdenum disulphide Jeffrey nanofluids. Results Phys. **9**, 947–954 (2018)
15. Hamid, M., Usman, M., Zubair, T., Haq, R.U., Wang, W.: Shape effects of $MOS_2$ nanoparticles on rotating flow of nanofluid along a stretching surface with variable thermal conductivity: a Galerkin approach. Int. J. Heat Mass Transf. **124**, 706–714 (2018)
16. Ali, F., Aamina, K.I., Sheikh, N.A., Gohar, M., Tlili, I.: Effects of different shaped nanoparticles on the performance of engine-oil and kerosene-oil: a generalized Brinkman-type fluid model with non-singular kernel. Sci. Rep. **8**,15285 (2018). https://doi.org/10.1038/s41598-018-33547-z
17. Dinarvand, S.: Mohammadreza Nademi Rostami: Three-dimensional squeezed flow of aqueous magnetite–graphene oxide hybrid nanofluid: a novel hybridity model with analysis of shape factor effects. Proc. IMechE Part E: J. Process. Mech. Eng. **234**(2), 193–205 (2020)
18. Merkin, J.H.: A model for isothermal homogeneous-heterogeneous reactions in boundary-layer flow. Math. Comput. Model. **24**, 125–136 (1996)
19. Bala Anki Reddy, P., Suneetha, S.: Effects of homogeneous-heterogeneous chemical reaction and slip velocity on mhd stagnation flow of a micropolar fluid over a permeable stretching/shrinking surface embedded in a porous medium. Front. Heat Mass Transf. (FHMT) **8**(24), (2017). https://doi.org/10.5098/hmt.8.24
20. Rana, S., Mehmood, R., Akbar, N.S.: Mixed convective oblique flow of a Casson fluid with partial slip, internal heating and homogeneous–heterogeneous reactions. J. Mol. Liq. **222**, 1010–1019 (2016)
21. Ali, F., Aamina, Khan, I., Sheikh, N.A., Gohar, M., Tlili, I.: Effects of different shaped nanoparticles on the performance of engine-oil and kerosene-oil: a generalized brinkman-type fluid model with non-singular kernel. Sci. Rep. **8**, 1–13 (2018)

# Wave Energy Dissipation by Multiple Permeable Barriers in Finite Depth Water

**Biman Sarkar and Soumen De**

**Abstract** A wave energy dissipation problem is solved for multiple permeable barriers in the water of finite depth. Applying Havelock's inversion formulae, this problem reduces to a set of first kind Fredholm integral equations involving potential differences across the barriers. The methodology utilized in this study is multi-term Galerkin's technique with a set of basis functions involving Chebychev's polynomials. A linear system has been solved for numerical estimations of the transmission and reflection coefficients. Dynamic wave force and wave energy dissipation have been computed both analytically and numerically. Also, at the end of the permeable barriers, square-root singularity of fluid velocity is tactfully handled. The numerical results for wave energy dissipation, dynamic wave force and reflection coefficients are depicted against wave numbers considering various values of parameters. Excellent ratification between previous results in the literature and present results is demonstrated.

**Keywords** Partially immersed permeable barriers · First kind fredholm integral equations · Galerkin's technique · Wave energy dissipation · Horizontal wave force

## 1 Introduction

Dissipation of wave energy over offshore platforms and harbours is a vital issue for researchers and ocean engineers. Different breakwater configurations have been built to defend coastal areas from the rough sea. Employing complex variable technique, an explicit solution of water wave scattering problems associated with an impermeable, thin barrier for normal incidence surface waves had been found for the first time in the

B. Sarkar (✉)
Department of Mathematics, Swami Vivekananda University,
Barrackpore, Kolkata 700121, India
e-mail: biman228sarkar@gmail.com

B. Sarkar · S. De
Department of Applied Mathematics, University of Calcutta, 92, A.P.C. Road,
Kolkata 700 009, India



B. Rushi Kumar et al. (eds.), *Mathematics and Computing*, Springer Proceedings in Mathematics & Statistics 415, https://doi.org/10.1007/978-981-19-9307-7_40

literature of Dean [1]. In the context of permeable breakwater, Sollitt and Cross [2] first investigated the problem of wave scattering by a thick rectangular porous bar. In the past few decades, many research works (Yu [3], Karmakar et al. [4], Karmakar and Soares [5], Lee and Chwang [6], Chanda and Bora [7] and the literature cited therein) had been carried out on permeable breakwaters. Recently, Sarkar et al. [8, 9] studied the non-uniform permeable barriers by considering different barrier configurations and revealed that this type of permeable barriers made a crucial impact on breakwater constructions.

In this paper, wave energy dissipation by multiple partially immersed thin vertical permeable barriers in the water of finite depth is explored by using inversion formulae of Havelock's and reducing the BVP into a set of Fredholm-type integral equations involving potential differences. Then Galerkin's approximation along with polynomials (Chebychev's) as basis has been used to obtain the solutions of these integral equations. Numerical results for wave energy dissipation, dimensionless wave force, reflection and transmission coefficients are depicted against wavenumber by adopting different parametric values. For the correctness of our numerical results, we validate our results with the existing results [6, 10, 11].

## 2 Mathematical Formulation

Considering the linearized water wave theory and irrotational fluid motion, the mathematical problem is to solve $\phi(x, y)$ which satisfies

$$\nabla^2\phi = 0, \;\; y \in [0, h], \text{(depth of the water)} \tag{2.1}$$

$$K\phi + \phi_y = 0 \text{ on } y = 0, \;\; x \in (-\infty, 0) \cup (0, \infty), \tag{2.2}$$

$$\phi_x = -\mathrm{i}k_0\mathcal{G}_j[\phi(\mp s_j + 0, y) - \phi(\mp s_j - 0, y)] \text{ on } y \in (0, d_j); \; (j = 1, 2), \tag{2.3}$$

where $\mathcal{G}_j$ represents porous effect parameter,

$$r^{1/2}\nabla\phi \text{ is bounded in the vicinity of the submerged sharp ends,} \tag{2.4}$$

$$\phi_y = 0 \text{ on } y = h \tag{2.5}$$

and

$$\phi(x, y) \sim \begin{cases} T\phi^{inc}(x, y) \text{ as } x \to -\infty, \\ \phi^{inc}(x, y) + R\phi^{inc}(-x, y) \text{ as } x \to \infty. \end{cases} \tag{2.6}$$

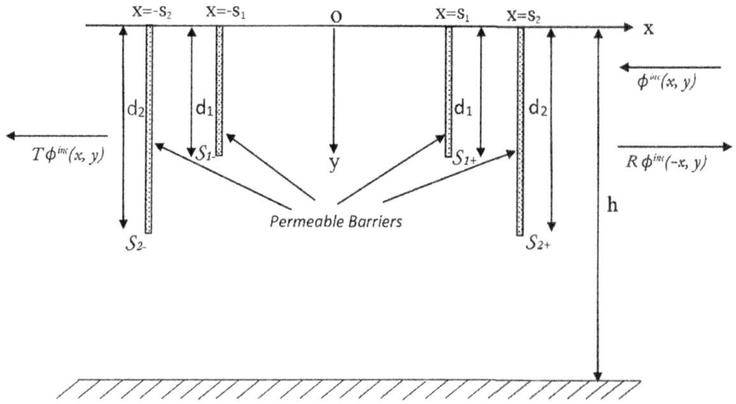


**Fig. 1** Geometry of the problem

Here $Re\{\phi^{inc}(x, y)e^{-\mathrm{i}\sigma t}\}$ represents velocity potential in the fluid region, $T$ and $R$ represent transmission and reflection coefficients and $\sigma$ denotes angular frequency of waves. Also, $\phi^{inc}(x, y) = \phi_0(y)e^{-\mathrm{i}k_0(x-s_2)}$, here $\phi_0(y) = \frac{\cosh k_0(h-y)}{\cosh k_0 h}$. The transcendental equation $K = k \tanh kh$ have a unique +ve real root $k_0$ and infinitely many purely imaginary roots $\mathrm{i}k_n$.

The barriers are arranged symmetrically with respect to the $y$-axis as shown in Fig. 1 and submerged parts occupy lines $\mathcal{S}_{j\mp} = \{x = \mp s_j; y \in l_j\}$ $(j = 1, 2)$ such that $\mathcal{S} = \mathcal{S}_{2-} \cup \mathcal{S}_{1-} \cup \mathcal{S}_{1+} \cup \mathcal{S}_{2+}$. So, the velocity potential can be split into symmetric and anti-symmetric parts follows:

$$\phi(x, y) = \phi^{sm}(x, y) + \phi^{am}(x, y) \tag{2.7}$$

where

$$\phi^{sm}(x, y) = \phi^{sm}(-x, y),\ \phi^{am}(x, y) = -\phi^{am}(-x, y). \tag{2.8}$$

Thus, we consider $x \geq 0$ region so that $\phi^{sm,am}(x, y)$ satisfies

$$\nabla^2 \phi^{sm,am} = 0, \quad x, y \in (0, \infty),$$

$$K\phi^{sm,am} + \phi_y^{sm,am} = 0 \text{ on } x \in (0, \infty),\ \ y = 0,$$

together with (2.3), (2.4), (2.5) and

$$\phi_x^{sm}(0, y) = 0, \quad \phi^{am}(0, y) = 0, \quad y \in (0, h). \tag{2.9}$$

## 3 Method of Solution

At infinity, $\phi^{sm,am}(x, y)$ satisfies

$$\phi^{sm,am}(x, y) \sim \frac{1}{2}\{\phi^{inc}(x, y) + R^{sm,am}\phi^{inc}(-x, y)\} \text{ as } x \to \infty \tag{3.1}$$

where $R^{sm,am}$ are connected with $T$ and $R$ by

$$T,\ R = \frac{R^{sm} \mp R^{am}}{2}. \tag{3.2}$$

$\phi^{sm}(x, y)$ expands as

$$\phi^{sm}(x, y) = \begin{cases} A_0^{sm}\cos k_0 x\, \phi_0(y) + \sum\limits_{n=1}^{\infty} A_n^{sm}\cosh k_n x\, \phi_n(y) & x \in (0, s_1),\ y \in (0, h), \\ (B_0^{sm}e^{ik_0 x} + C_0^{sm}e^{-ik_0 x})\phi_0(y) + \sum\limits_{n=1}^{\infty}(B_n^{sm}e^{k_n x} + C_n^{sm}e^{-k_n x})\phi_n(y) & x \in (s_1, s_2),\ y \in (0, h) \\ \phi^{inc}(x, y) + R^{sm}\phi^{inc}(-x, y) - \sum\limits_{n=1}^{\infty} D_n^{sm}e^{-k_n(x-s_2)}\phi_n(y) & x \in (s_2, \infty),\ y \in (0, h). \end{cases} \tag{3.3}$$

Here $\phi_n(y) = \cos k_n(h - y)$.

Let

$$p_j^{sm}(y) = \phi_x^{sm}(s_j, y), \quad j = 1, 2 \tag{3.4}$$

and

$$q_j^{sm}(y) = \phi^{sm}(\mp s_j + 0, y) - \phi^{sm}(\mp s_j - 0, y), \quad y \in (0, h). \tag{3.5}$$

Thus

$$p_j^{sm}(y) = -ik_0\mathcal{G}_j q_j^{sm}(y), \quad y \in l_j, \quad j = 1, 2 \tag{3.6}$$

Employing Havelock's inversion formulae on $q_j^{sm}(y)$ and using (3.6), we get

$$-k_0 A_0^{sm}\sin k_0 s_1 \phi_0(y) + \int_0^{d_1} q_1^{sm}(u)\mathcal{U}_{11}^{sm}(y, u)du + \int_0^{d_2} q_2^{sm}(u)\mathcal{U}_{12}^{sm}(y, u)du = -ik_0\mathcal{G}_1 q_1^{sm}(y) \tag{3.7}$$

$$-ik_0(1 - R^{sm}e^{2ik_0 s_2})\phi_0(y) + \int_0^{d_1} q_1^{sm}(u)\mathcal{U}_{21}^{sm}(y, u)du + \int_0^{d_2} q_2^{sm}(u)\mathcal{U}_{22}^{sm}(y, u)du = -ik_0\mathcal{G}_2 q_2^{sm}(y). \tag{3.8}$$

Here

$$\mathcal{U}_{11}^{sm}(y, u) = -\sum_{r=1}^{\infty} k_r\delta_r h^{-1}e^{-\alpha_r s_1}\sinh\alpha_r s_1 \phi_r(u)\phi_r(y) \tag{3.9a}$$

$$\mathcal{U}_{12}^{sm}(y,u)=\mathcal{U}_{21}^{sm}(y,u)=-\sum_{r=1}^{\infty}k_r\delta_r h^{-1}e^{-\alpha_r s_2}\sinh\alpha_r s_1\phi_r(u)\phi_r(y) \quad (3.9b)$$

$$\mathcal{U}_{22}^{sm}(y,u)=-\sum_{r=1}^{\infty}k_r\delta_r h^{-1}e^{-\alpha_r s_2}\sinh\alpha_r s_2\phi_r(u)\phi_r(y) \quad (3.9c)$$

and

$$\delta_r=\frac{4k_rh}{2k_rh+\sin 2k_rh}\quad (r=1,2,\ldots) \quad (3.10)$$

Setting

$$X_1^{sm}=0,\;\; X_2^{sm}=-1,\;\; Y_1^{sm}=-\mathrm{i}A_0^{sm}\sin k_0s_1,\;\; Y_2^{sm}=-R^{sm}e^{2\mathrm{i}k_0s_2} \quad (3.11)$$

Introducing the step function $\chi_j(y)$ defined as

$$\chi_j(y)=\begin{cases}0, & y\in\overline{l_j},\\ 1, & y\in l_j,\;\; j=1,2\end{cases} \quad (3.12)$$

where $\overline{l_j}$ = (0, h) - $l_j$.

Also, let

$$\mathbf{q}^{sm}(u)=\mathrm{i}k_0\mathrm{h}^2\mathbf{F}^{\mathrm{sm}}(\mathrm{u})(\mathbf{X}^{\mathrm{sm}}-\mathbf{Y}^{\mathrm{sm}}) \quad (3.13)$$

where $\mathbf{q}^{sm}(u)=\left(q_1^{sm}(u),q_2^{sm}(u)\right)^T$, $\mathbf{X}^{sm}=\left(X_1^{sm},X_2^{sm}\right)^T$, $\mathbf{Y}^{sm}=\left(Y_1^{sm},Y_2^{sm}\right)^T$, $\mathbf{F}^{sm}(u)=\left(F_{jl}^{sm}(u)\right)_{4\times 4}$.

Now, using (3.11) to combine the Eqs. (3.7) and (3.8) and then utilizing (3.12), the ranges of the combined equation converted into (0, h) and further using (3.13) to put it into a matrix form as follows:

$$\phi_0(y)\boldsymbol{\chi}(y)+h^2\int_0^h\boldsymbol{\chi}(y)\mathbf{U}^{sm}(y,u)\boldsymbol{\chi}(u)\mathbf{F}^{sm}(u)du=-\mathrm{i}k_0h^2\mathcal{G}\boldsymbol{\chi}(y)\mathbf{F}^{sm}(y),\;\; y\in(0,h) \quad (3.14)$$

where $\boldsymbol{\chi}(y)=\mathrm{diag}\left(\chi_j(y)\right)_{2\times 2}$, $\mathbf{U}^{sm}(y,u)=\left(\mathcal{U}_{jl}^{sm}(y,u)\right)_{2\times 2}$, $\mathcal{G}=\mathrm{diag}\left(\mathcal{G}_j\right)_{2\times 2}$.

Further employing Havelock's inversion formulae on $q_j^{sm}(y)$, we get

$$\mathrm{i}\csc k_0\tau\left(\mathrm{i}A_0^{sm}\sin k_0s_2+1-R^{sm}e^{2\mathrm{i}k_0s_2}\right)=\frac{\delta_0}{h}\int_0^{d_1}q_1^{sm}(y)\phi_0(y)dy \quad (3.15a)$$

$$\mathrm{i}\csc k_0\tau\left(-\mathrm{i}A_0^{sm}\sin k_0s_1-e^{\mathrm{i}\mu\tau}+R^{sm}e^{2\mathrm{i}k_0s_2}e^{-\mathrm{i}k_0\tau}\right)=\frac{\delta_0}{h}\int_0^{d_2}q_2^{sm}(y)\phi_0(y)dy \quad (3.15b)$$

where $\tau=s_2-s_1$ and $\delta_0=\frac{4k_0h\cosh^2k_0h}{2k_0h+\sinh 2k_0h}$.

Again using (3.11) to combine above equations and then using (3.13) to write the combined equation in matrix form as follows:

$$\left[\mathbf{Z}^{sm}\mathbf{X}^{sm} - \overline{\mathbf{Z}}^{sm}\mathbf{Y}^{sm}\right] = \delta_0 k_0 h \sin k_0\tau \; \mathbf{L}^{sm}(\mathbf{X}^{sm} - \mathbf{Y}^{sm}) \tag{3.16}$$

where

$$\mathbf{L}^{sm} = \int_0^h \chi(u)\mathbf{F}^{sm}(u)\phi_0(u)du, \tag{3.17}$$

$\mathbf{Z}^{sm} = \begin{pmatrix} \frac{\sin k_0 s_2}{\sin k_0 s_1} & -1 \\ -1 & e^{\mathrm{i}k_0\tau} \end{pmatrix}$ and $\overline{\mathbf{Z}}^{sm}$ is the conjugate transpose of $\mathbf{Z}^{sm}$.

By interchanging $\sin k_0 s_j$, $\sinh k_0 s_j$ $(j = 1, 2)$ by $\cos k_0 s_j$, $\cosh k_0 s_j$ $(j = 1, 2)$, respectively, in (3.3), we can get the corresponding expressions for $\phi^{am}(x, y)$. If we find $\mathbf{F}^{sm,am}(y)$ by solving (3.14) numerically, then $\mathbf{L}^{sm,am}$ are determined from (3.17) and hence numerical estimates for $R^{sm,am}$. Finally, $|T|$ and $|R|$ can be derived from (3.2).

## 4 Galerkin's Method

We consider $(N + 1)$-term approximation by Galerkin's method to solve (3.14) for $\mathbf{F}^{sm,am}(y)$ as

$$F_{jl}^{sm,am}(u) \simeq \sum_{n=0}^{N} a^{(n)_{sm,am}} \psi_j^{(n)}(u), \quad u \in (0, d_j). \tag{4.1}$$

We choose the suitable basis functions $\psi_j^{(n)}(u)$ as

$$\psi_j^{(n)}(u) = -\frac{d}{du}\left[e^{-Ku}\int_u^{d_j} \hat{\psi}_j^{(n)}(t)e^{Kt}dt\right], \; u \in (0, d_j) \quad (j = 1, 2) \tag{4.2}$$

with

$$\hat{\psi}_j^{(n)}(t) = \frac{2(-1)^n}{\pi(2n+1)hd_j}(d_j^2 - t^2)^{\frac{1}{2}} U_{2n}\left(\frac{t}{d_j}\right) \quad (j = 1, 2) \tag{4.3}$$

where $U_{2n}(x)$ is the 2n order Chebychev's polynomial.

Using (4.1)–(4.3) in Eq. (3.14) and introducing usual Kronecker delta function $\delta_{jl}$ $(j, l = 1, 2)$, we get

$$\delta_{jl}\chi_j(y)\phi_0(y) + h^2\chi_j(y)\Big[\sum_{n=0}^{N}\sum_{k=1}^{2} a_{kl}^{(n)sm,am}\int_{l_k}\mathcal{U}_{jk}^{sm,am}(y,t)\psi_k^{(n)}(t)dt\Big] = -\mathrm{i}k_0h^2\mathcal{G}_j\chi_j(y)$$
$$\sum_{n=0}^{N} a_{jl}^{(n)sm,am}\psi_j^{(n)}(y),\ y\in(0,h) \tag{4.4}$$

Multiplying (4.4) by $\psi_j^{(m)}(y)$ and integrating over $l_j$ $(j=1,2)$, respectively, and then substituting the values of $\mathcal{U}_{jl}^{sm,am}(y,u)$ from (3.9), we get the following system of equations:

$$\sum_{n=0}^{N}\sum_{l=1}^{2} a_{lk}^{(n)sm,am}\mathcal{V}_{mn}^{(jl)sm,am} = -\delta_{lk}\mathcal{W}_m^{(j)sm,am},\ (j,k=1,2), \tag{4.5}$$

where

$$\mathcal{V}_{mn}^{(jj)sm} = -\sum_{r=1}^{\infty}\frac{\delta_r k_r h\sinh k_r s_j\cos^2(k_r h)}{e^{k_r s_j}(k_r h)^2}\mathcal{J}_{2m+1}(k_r d_j)\mathcal{J}_{2n+1}(k_r d_j)$$
$$+\ \mathrm{i}k_0h^2\mathcal{G}_j\int_{l_j}\psi_j^{(m)}(y)\psi_j^{(n)}(y)dy, \tag{4.6a}$$

$$\mathcal{V}_{mn}^{(jl)sm} = -\sum_{r=1}^{\infty}\frac{\delta_r k_r h\sinh k_r s_1\cos^2(k_r h)}{e^{k_r s_2}(k_r h)^2}\mathcal{J}_{2m+1}(k_r d_j)\mathcal{J}_{2n+1}(k_r d_l),\ (j\neq l) \tag{4.6b}$$

$$\mathcal{V}_{mn}^{(jj)am} = -\sum_{r=1}^{\infty}\frac{\delta_r k_r h\cosh k_r s_j\cos^2(k_r h)}{e^{k_r s_j}(k_r h)^2}\mathcal{J}_{2m+1}(k_r d_j)\mathcal{J}_{2n+1}(k_r d_j)$$
$$+\ \mathrm{i}k_0h^2\mathcal{G}_j\int_{l_j}\psi_j^{(m)}(y)\psi_j^{(n)}(y)dy, \tag{4.6c}$$

$$\mathcal{V}_{mn}^{(jl)am} = -\sum_{r=1}^{\infty}\frac{\delta_r k_r h\cosh k_r s_1\cos^2(k_r h)}{e^{k_r s_2}(k_r h)^2}\mathcal{J}_{2m+1}(k_r d_j)\mathcal{J}_{2n+1}(k_r d_l),\ (j\neq l) \tag{4.6d}$$

$$\mathcal{W}_m^{(j)sm,am} = (-1)^m\frac{\mathcal{I}_{2m+1}(k_0 d_j)}{k_0 h}. \tag{4.6e}$$

Now substituting (4.1) into (3.17) and assuming $\mathbf{L}^{sm,am} = \{L_{jl}^{sm,am}\}_{2\times2}$, we determine $L_{jl}^{sm,am}$ as

$$L_{jl}^{sm,am} \simeq \sum_{n=0}^{N} a_{jl}^{(n)sm,am}\int_{l_j}\psi_j^{(n)}(t)\phi_0(t)dt = \sum_{n=0}^{N} a_{jl}^{(n)sm,am}\mathcal{W}_n^{(j)sm,am} \tag{4.7}$$

Equations (4.5) and (4.7) together imply the matrix $\mathbf{L}^{sm,am}$ as

$$\mathbf{L}^{sm,am} = \boldsymbol{\mathcal{W}}^{sm,am}(\boldsymbol{\mathcal{V}}^{sm,am})^{-1}(-\boldsymbol{\mathcal{W}}^{sm,am})^{T} \tag{4.8}$$

where $\mathbf{L}^{sm,am} = \left(L_{jl}^{sm,am}\right)_{2\times2}$, $\boldsymbol{\mathcal{W}}^{sm,am} = \mathrm{diag}\left(\boldsymbol{\mathcal{W}}^{(j)_{sm,am}}\right)_{2\times2}$, $\boldsymbol{\mathcal{V}}^{sm,am} = \left(\boldsymbol{\mathcal{V}}^{(jl)_{sm,am}}\right)_{2\times2}$, $\boldsymbol{\mathcal{W}}^{(j)_{sm,am}} = \{\mathcal{W}_0^{(j)_{sm,am}}, \mathcal{W}_1^{(j)_{sm,am}}, \ldots, \mathcal{W}_N^{(j)_{sm,am}}\}$, $\boldsymbol{\mathcal{V}}^{(jl)_{sm,am}} = \left(\mathcal{V}_{mn}^{(jl)_{sm,am}}\right)_{(N+1)\times(N+1)}$.

If $\mathbf{L}^{sm}$ can be obtained from (4.8), then $\mathbf{R}^{sm}$ can also be formulated from (3.16) and (3.11). In a similar manner, we also determine $\mathbf{R}^{am}$ too.

## 5 Wave Energy Dissipation and Wave Force

Employing Green's integral theorem, the energy relation for permeable walls can be obtained as follows:

$$|R|^2 + |T|^2 + \mathcal{J} = 1 \tag{5.1}$$

$$\mathcal{J}\,\text{(amount of wave energy dissipsation)} = \frac{\delta_0}{2}\sum_{j=1}^{2}\int_{l_j} \Re(\mathcal{G}_j)\{|q_j^{sm}(y)|^2 + |q_j^{am}(y)|^2\}dy \tag{5.2}$$

The horizontal wave force exerting upon the barriers as follows (cf. Li et al. [14])

$$F = 2i\rho\sigma\sum_{j=1}^{2}\int_{l_j} q_j^{(am)}(y)dy. \tag{5.3}$$

Thus, the non-dimensionless wave force is represented by

$$W_F = \frac{|F|}{F_0}, \tag{5.4}$$

$$F_0 = \frac{\rho g\sigma}{k_0}\tanh k_0 h. \tag{5.5}$$

## 6 Results and Interpretations

In this section, the numerical estimates are graphically represented. We consider $N = 2$ in Galerkin's approximation for the whole section.

Considering parametric values as $\frac{a_1}{h} = \frac{a_2}{h} = 0.2$, $\frac{b_1}{h} = 0.3$, $\frac{b_2}{h} = 0.301$, $\mathcal{G}_j = 0$ so that our barrier-configuration becomes almost Das et al's [10] configuration of

**Table 1** Numerical results of Das et al.'s [10] results for $R_1$ and $R_2$ and Present results for $R$ with $\frac{a_1}{h} = \frac{a_2}{h} = 0.2$, $\frac{b_1}{h} = 0.3$, $\frac{b_2}{h} = 0.301$, $\mathcal{G}_j = 0$

| $Kh$ | $R_1$ | $R_2$ | $R$ |
|---|---|---|---|
| 0.2 | 0.031799 | 0.032341 | 0.0315049 |
| 0.8 | 0.088445 | 0.089049 | 0.0871078 |
| 1.4 | 0.126188 | 0.128896 | 0.125453 |

**Table 2** Numerical results of Mandal and Dolai's [11] results for $R_1$ and $R_2$ and Present results for $R$ with $\frac{a_1}{h} = \frac{a_2}{h}$, $\frac{b_1}{h} = 0.001$, $\frac{b_2}{h} = 0.0011$, $\mathcal{G}_j = 0$

| $\frac{a}{h}$ | $R_1$ | $R_2$ | $R$ |
|---|---|---|---|
| 0.2 | 0.0176 | 0.0176 | 0.0172221 |
| 0.4 | 0.0712 | 0.0713 | 0.0716796 |
| 0.6 | 0.174 | 0.174 | 0.175369 |
| 0.8 | 0.3482 | 0.3521 | 0.361113 |

two barriers. Table 1 shows the correctness of our numerical results up to 2–3 decimal places.

To validate the present result with Mandal and Dolai [11], the parameters are taken as $\frac{a_1}{h} = \frac{a_2}{h}$, $\frac{b_1}{h} = 0.001$, $\frac{b_2}{h} = 0.0011$, $\mathcal{G}_j = 0$. An accuracy of 2–3 decimal places has been achieved in Table 2.

In Fig. 2, we recover figures of Lee and Chwang [6] for a single partially immersed barrier with $\mathcal{G}_j = 0.25, 1$. Other non-dimensional parameters are chosen as $\frac{a_1}{h} = \frac{a_2}{h} = 0.5$, $\frac{b_1}{h} = 0.001$, $\frac{b_2}{h} = 0.0011$. This establishes the exactness of the present result.

Figure 3 is plotted for single barrier, two-barrier and four-barrier configurations with $\mathcal{G}_j = 0.5$. From Fig. 3, we confirm that increase in number of barriers helps to reduce more wave energy.

The effect of different modulus of permeability parameter is shown in Fig. 4 with $\mathcal{G}_j = 0.5, 1, 1 + \mathrm{i}$. We assume the other parametric values as $\frac{a_1}{h} = 0.25$, $\frac{a_2}{h} = 0.45$, $\frac{b_1}{h} = 3.5$, $\frac{b_2}{h} = 4.5$. It is observed from Fig. 4 that non-dimensionless wave force exerted upon the barriers decreases as the modulus of $\mathcal{G}_j$ increases.

The influence of porosity upon the reflection coefficients is demonstrated in Fig. 5 with different values of $\mathcal{G}_j$. So, we consider parameters as $\frac{a_1}{h} = 0.15$, $\frac{a_2}{h} = 0.35$, $\frac{b_1}{h} = 3.5$, $\frac{b_2}{h} = 6.0$. This figure shows increase in magnitude of porosity implies the decrease in $|R|$.

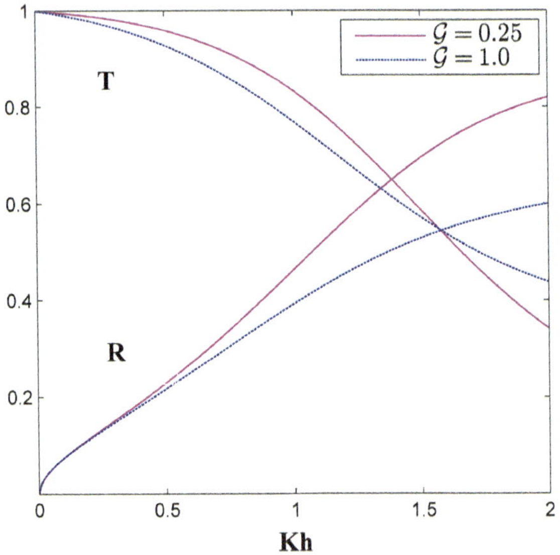


**Fig. 2** $|R|$ versus $Kh$ for $\frac{a_1}{h} = \frac{a_2}{h} = 0.5$, $\frac{b_1}{h} = 0.001$, $\frac{b_2}{h} = 0.0011$

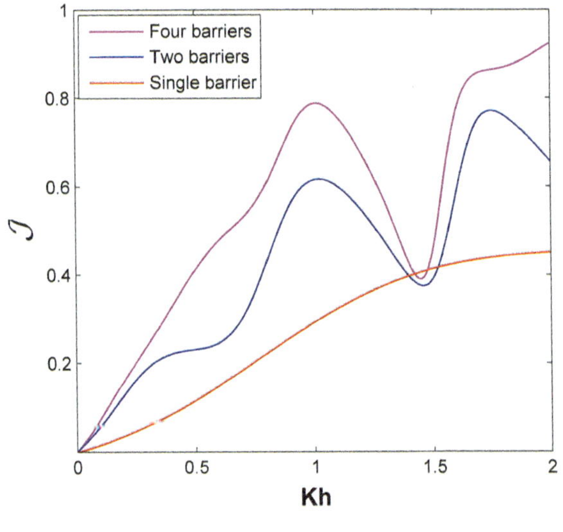


**Fig. 3** $\mathcal{J}$ versus $Kh$ for four, two and single barriers with $\mathcal{G}_j = 0.5$

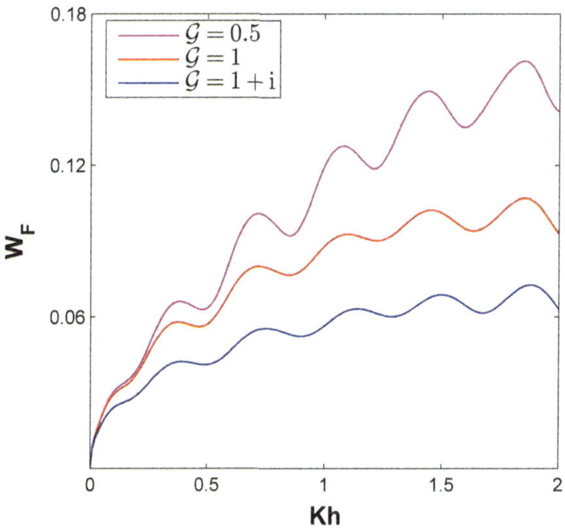


**Fig. 4** $W_F$ versus $Kh$ for $\frac{a_1}{h} = 0.25$, $\frac{a_2}{h} = 0.45$, $\frac{b_1}{h} = 3.5$, $\frac{b_2}{h} = 4.5$

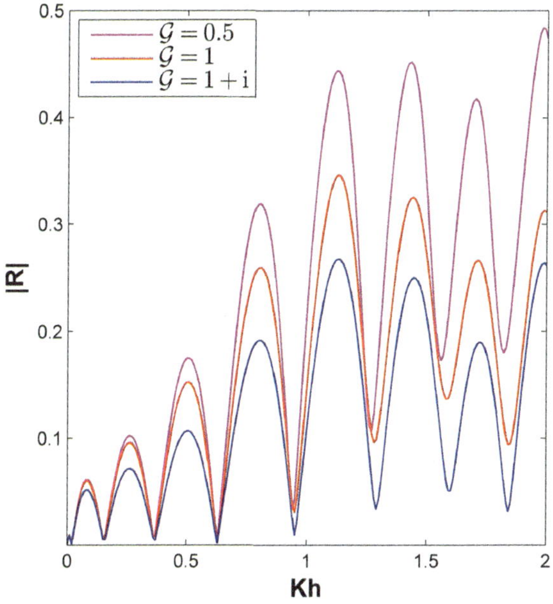


**Fig. 5** $|R|$ versus $Kh$ for $\frac{a_1}{h} = 0.15$, $\frac{a_2}{h} = 0.35$, $\frac{b_1}{h} = 3.5$, $\frac{b_2}{h} = 6.0$

# 7 Conclusions

In the present study, we consider the problem of wave energy dissipation by thin multiple partially immersed vertical porous barriers in the water of uniform finite depth. Some excellent conclusions are explored in our present work. The conclusions are recapitulated as follows:

1. Rise in the number of barriers is a major reason for the dissipation of more wave energy. This phenomenon signifies the crucial importance of the multiple partially immersed thin vertical porous barriers for constructing the breakwaters of various geometrical configurations.
2. It is also noticed that as the modulus of permeability parameter increases, dimensionless wave force reduces. Thus, the porosity of barriers helps to diminish wave load upon the barriers.
3. It is observed that $|R|$ decreases as the magnitude of permeability increases. This incident occurs in view of the wave energy dissipation by the holes of perforated barriers.

## References


1. Dean, W.R.: On the reflection of surface waves by submerged plane barriers. Proc. Camb. Phil. Soc. **41**, 231–238 (1945)
2. Sollitt, C.K., Cross, R.H.: Wave transmission through permeable breakwaters. Coast. Eng. Proc. **1**, 1827–1846 (1972)
3. Yu, X.: Diffraction of water waves by porous breakwaters. J. Waterw., Port, Coast., Ocean Eng. ASCE **121**(6), 275–282 (1995)
4. Karmakar, D., Bhattacharjee, J., Soares, C.: Scattering of gravity waves by multiple surface-piercing floating membrane. Appl. Ocean Res. **39**, 40–52 (2013)
5. Karmakar, D., Guedes Soares, C.: Wave transmission due to multiple bottom-standing porous barriers. Ocean Eng. **80**, 50–63 (2014)
6. Lee, M.M., Chwang, A.T.: Scattering and radiation of water waves by permeable barriers. Phys. Fluid **1**, 54–65 (2000)
7. Chanda, A., Bora, S.N.: Investigation of oblique flexural gravity wave scattering by two submerged thin vertical porous barriers with different porosities. J. Eng. Mech. **148**(2), 04021145 (2022)
8. Sarkar, B., De, S., Roy, R.: Oblique wave scattering by two thin non-uniform permeable vertical walls with unequal apertures in water of uniform finite depth. Waves Random Complex Media **31**(6), 2021–2039 (2021)
9. Sarkar, B., Paul, S., De, S.: Water wave propagation over multiple porous barriers with variable porosity in the presence of an ice cover. Meccanica **56**, 1771–1788 (2021)
10. Das, P., Dolai, D.P., Mandal, B.N.: Oblique water wave diffraction by two parallel thin barriers with gaps. J. Waterw., Port, Coast., Ocean Eng. ASCE **123**, 163–171 (1997)
11. Mandal, B.N., Dolai, D.P.: Oblique water wave diffraction by thin vertical barriers in water of uniform finite depth. Appl. Ocean Res. **16**, 195–203 (1994)
12. Mandal, B.N., Chakrabarti, A.: Water wave scattering by barriers. Southampton, 1st ed. WIT Press, U.K. (2000)
13. Evans, D.V., Porter, R.: Complementary methods for scattering by thin barriers. Int. Ser. Adv. Fluid Mech. **8**, 1–44 (1997)
14. Li, A.J., Liu, Y., Li, H.J.: Accurate solutions to water wave scattering by vertical thin porous barriers. Math. Probl. Eng. 1–11 (2015)

# Thermal Stress Analysis of Inhomogeneous Infinite Solid to 2D Elasticity of Thermoelastic Problems

**Abhijeet Adhe and Kirtiwant Ghadle**

**Abstract** This paper is developed to study an analytical solution of thermal stresses to the plane elasticity of thermoelastic problems for inhomogeneous materials with internal heat generation. Here, the original problems are reduced to set the governing equations by use of the method of direct integration. Further using the iteration techniques, the governing equations are reduced to integral equations. The numerical calculations have been performed with the aid of the iterative method, which gives the rapid convergence. The distribution of Young's modulus and shear modulus, and the dimensionless stresses, are shown graphically. An explicit solution is derived which will be more useful for analysis of stress field in an isotropic inhomogeneous solid.

**Keywords** 2D elasticity problems · Thermoelastic problems · Inhomogeneous solid · Direct integration method · Iterative technique · Analytical solution · Exact solution

## 1 Introduction

Thermoelasticity comprises the theory of heat conduction and the theory of stress and strain due to heat flow, when coupling of temperature and strain field takes place. Also, it contains the study of temperature distribution, stress, and strain developed in a material. A study of thermal stresses is essential in many applications. Thermal stresses in a material are one of the prime factors, which affect the life of a material. The determination of thermal stresses caused by an involvement in a medium is classical problem. The interest of researchers to study elasticity and thermoelastic

A. Adhe (✉)
Department of Basic Sciences and Humanities, Marathwada Institute of Technology, Aurangabad 431010, MS, India
e-mail: adhe.abhijeet@gmail.com

K. Ghadle
Department of Mathematics, Dr. Babasaheb Ambedkar Marathwada University, Aurangabad 431004, MS, India



B. Rushi Kumar et al. (eds.), *Mathematics and Computing*, Springer Proceedings in Mathematics & Statistics 415, https://doi.org/10.1007/978-981-19-9307-7_41

problems has grown very fast due to their wide applications to the real world. Among various inhomogeneous solids, FGM have fascinated academicians and researchers. Except for few particular cases, it is impossible to get the analytical solution. To get the better of this struggling, we need some clarification. FGM's have many applications in various fields of engineering sciences.

Gaikwad et al. [1] studied heat conduction problems in the case of nonhomogeneous hollow circular-type disk. Jafari et al. [3] discussed the stress analysis in an orthotropic infinite plate with a circular hole. Authors used a complex variable technique for the two-dimensional thermoelastic problem. Kalynyak [4] used a method of direct integration of equation of equilibrium and continuity in terms of stresses for inhomogeneous cylindrical bodies. A novel study by Kaminski [5] for hyperbolic heat conduction equations for nonhomogeneous materials.

Chien-ching Ma et al. [6] developed a fruitful analytical method for a full-field solution in an anisotropic multi-layered media. They have analyzed the steady-state temperature and heat conduction in each layer on the surfaces using Fourier transform and series expansion method to get the explicit solution in series form of the discussed problem. Manthena et al. [7] emphasized on the temperature distribution, displacement, and thermal stress of nonhomogeneous rectangular plate. Porter [9] discussed the procedure of the solution to integral equations with difference kernels applied on finite intervals. Rychachivskyy [10], Tokovyy et al. [16, 17] emphasized on solution of the 2D elasticity and thermoelasticity problems for inhomogeneous planes and semi-planes. Tanigawa et al. [12, 13] derived the basic equations for three-dimensional thermoelasticity problems with nonhomogeneous properties. Tokovyy et al. [14, 15, 19] extended the direct integration method for three-dimensional temperature and analysis of thermal stress in inhomogeneous solids. The same technique to study the construction of solution of the plane quasistatic thermoelasticity problems for cylindrically anisotropic hollow cylinders and disks satisfying inhomogeneous properties used by Tokovyy et al. [18].

Youssef et al. [23] developed a new model of three-dimensional generalized thermoelasticity by using the classic theory of Lord-Shulman. The double Fourier transform and Laplace technique had been applied to the governing equations subjected to rectangular traction-free surface, with the study of the temperature analysis, stresses, strain, and displacement in a three-dimensional half-space. Vigak [20, 22] and his followers in [21] developed a method to find solution of the elasticity and thermoelasticity problems using the method of direct integration.

Many engineering problems are concerned with the evaluation of the amount of heat transferred through surfaces and stresses due to coupling of temperature and strain field in a solid [8].

In this research article, we have extended our own work [2]. We considered an inhomogeneous half-plane and determined the explicit form analytical solution to the steady-state distribution of temperature, using the method of direct integration. The main focus of the present article is to determine the stress field for an infinite plane due to the generation of internal heat in inhomogeneous solids under steady-state temperature. Here, we have considered an isotropic inhomogeneous solid in an infinite plane.

The key points of this article are as below:
(a) The governing heat conduction problem for plane $R$ is formulated as a boundary value problem [8].
(b) The method of direct integration is applied to find the solution of the stated thermoelastic problem.
(c) Making use of iterative method, analytical solution to thermal stresses of an isotropic inhomogeneous solid is derived.
(d) The rapid convergence of iterations exists to solve the aforementioned problems.
(e) An explicit solution is derived which serves as better tool for analysis of stress field in an isotropic inhomogeneous solid.

## 2 Problem Formulation

Consider, an isotropic inhomogeneous solid in a plane $R = \{(x, y) \in (-\infty, \infty) \times (-\infty, \infty)\}$. Thermoelastic equilibrium of plane $R$ is ruled by the following equilibrium equations:

$$\begin{aligned} \frac{\partial \sigma_x}{\partial x} + \frac{\partial \sigma_{xy}}{\partial y} &= X \\ \frac{\partial \sigma_{xy}}{\partial x} + \frac{\partial \sigma_y}{\partial y} &= Y, \end{aligned} \tag{1}$$

strain-compatibility equations:

$$\frac{\partial^2 \tau_y}{\partial x^2} + \frac{\partial^2 \tau_x}{\partial y^2} = \frac{\partial^2 \tau_{xy}}{\partial x \partial y}, \tag{2}$$

stress–strain relations:

$$\begin{aligned} \tau_x &= \frac{\sigma_x}{E_1(x)} - \frac{V_1(x)\sigma_y}{E_1(x)} - d_1 + \alpha^*(x)T(x, y), \\ \tau_y &= -\frac{\sigma_y}{E_1(x)} - \frac{V_1(x)\sigma_x}{E_1(x)} - d_1 + \alpha^*(x)T(x, y), \\ \tau_{xy} &= \frac{\sigma_{xy}}{G(x)}. \end{aligned} \tag{3}$$

Here, $\sigma_x$, $\sigma_y$, and $\sigma_{xy}$ are the stress–tensor components; $\tau_x$, $\tau_y$, and $\tau_{xy}$ denotes the strain components; $X = X(x, y)$, and $Y = Y(x, y)$ are the stress-dimensional projections of forces in dimensionless cartesian coordinates $(x, y)$, respectively.

The steady temperature $T(x, y)$ can be found from following equation [2]:

$$\frac{\partial^2 T}{\partial x^2} + \frac{\partial^2 T}{\partial y^2} = -W(x, y), \tag{4}$$

under the condition

$$T(x, y) = T_0(y), \quad at \;\; x = 0; \tag{5}$$

where $W(x, y) = \frac{q(x,y)}{k}$ and $q(x, y)$ denoting the heat generated due to internal heat generated in $R$, $T_0(y)$ is known function, $k$ is coefficient of thermal conductivity,

$$E_1 = \frac{E}{1-v^2}, \quad for\ plain\ strain,$$
$$= E, \quad for\ plain\ stress,$$

$$V_1 = \frac{v}{1-v}, \quad for\ plain\ strain,$$
$$= v, \quad for\ plain\ stress,$$

$$\beta_1 = \beta(1+v), \quad for\ plain\ strain,$$
$$= \beta, \quad for\ plain\ stress,$$

$$d_1 = vd, \quad for\ plain\ strain,$$
$$= 0, \quad for\ plain\ stress,$$

where $E = E(x)$ and $v = v(x)$ are Young's modulus and Poisson's ratio, respectively; $d$ is the out-of-plane strain, which is treated as constant; $S = S(x)$ is shear modulus fulfilling the following equation:

$$S = \frac{E}{2(v+1)} = modulus\ of\ rigidity \tag{6}$$

By Hooke's law,

$$Ed = \sigma_z - v(\sigma_x + \sigma_y) + \beta ET, \tag{7}$$

where $\sigma_z$ is the plane stress and $T = T(x, y)$ is the distribution of temperature.

Boundary conditions imposed at boundary $x = 0$,

$$\sigma_x(0, y) = m(y), \quad \sigma_{xy}(0, y) = n(y). \tag{8}$$

Using equilibrium condition (1), the second condition (8) for shear stress for normal stress is reduced to

$$\frac{\partial \sigma_y}{\partial y} = Y(0, y) - \frac{\partial \sigma_{xy}(0, y)}{\partial x}, \tag{9}$$

at $x = 0$. With the aid of stress–strain relation (3) and Eq. (1), we can express Eq. (2) as

$$\Delta\left(\frac{\sigma}{E_1}+\beta_1 T\right)=\sigma_x\frac{d^2}{dx^2}\left(\frac{1}{2S}\right)+\frac{d^2 e_1}{dx^2}+X\frac{d}{dx}\left(\frac{1}{S}\right)+\frac{1}{2S}\left(\frac{\partial X}{\partial x}+\frac{\partial Y}{\partial y}\right) \tag{10}$$

where the total stress is $\sigma = \sigma_x + \sigma_y$ and;

$$\Delta=\frac{\partial^2}{\partial x^2}+\frac{\partial^2}{\partial y^2}$$

denotes the two-dimensional Laplace differential operator. Expression (10) represents the compatibility equation of stresses for a homogeneous solid. The stress components in plane $R$ are found by problems of heat conduction in terms of stresses, which is given by Eqs. (1) and (10).

## 3 Composition of Solution

To find the solution of the formulated problem, apply the Fourier Transform [11] w.r.t. $y$

$$\begin{aligned}\hat{F}(s)&=\int_{-\infty}^{\infty} f(y)exp(-isy)dy,\\ f(y)&=\frac{1}{\sqrt{2\pi}}\int_{-\infty}^{\infty}\hat{F}(s)exp(isy)ds,\end{aligned} \tag{11}$$

where $i^2 = -1$, $f$ is an arbitrary function and $s$ is a parameter. We choose $\sigma_x$ and $\sigma$ to be the governing functions.

We use Eq. (10) to determine the governing stresses.

$$\Delta\sigma_x=\frac{\partial^2\sigma}{\partial y^2}+\frac{\partial X}{\partial x}-\frac{\partial Y}{\partial y}. \tag{12}$$

Applying direct integral transform (11) to Eq. (12) yields

$$\left(\frac{d^2}{dx^2}-s^2\right)\hat{\sigma}_x=-s^2\hat{\sigma}-is\hat{Y}+\frac{d\hat{X}}{dx} \tag{13}$$

The particular solution to Eq. (13) in $R$ can be given as

$$\hat{\sigma}_x=\frac{|s|}{2}\int_{-\infty}^{\infty}\hat{\sigma}(\rho)exp(-|s||x-\rho|)d\rho+\frac{1}{2|s|}\int_{-\infty}^{\infty}\left(\frac{d\hat{\sigma}(\rho)}{d\rho}-is\hat{Y}(\rho)\right)exp(-|s||x-\rho|)d\rho, \tag{14}$$

where $|.|$ denotes the absolute value function.

Applying Fourier transform (11) to Eq. (10), we get

$$(\frac{d^2}{dx^2}-s^2)(\frac{\hat{\sigma}}{E_1}+\beta_1\hat{T})=\frac{\hat{\sigma}_x}{2}\frac{d^2}{dx^2}(\frac{1}{S})+\sqrt{2\pi}\delta(s)\frac{d^2d_1}{dx^2}+\hat{X}\frac{d}{dx}(\frac{1}{S})+\frac{1}{2S}(\frac{d\hat{X}}{dx}+is\hat{Y}),. \tag{15}$$

where $\delta(.)$ is the Dirac delta function.

The perticular solution to aformentioned Eq. (15) is

$$\begin{aligned}\hat{\sigma}=E_1[-\beta_1\hat{T}-\tfrac{1}{2|s|}\textstyle\int_{-\infty}^{\infty}(\hat{X}(\rho)\tfrac{d}{d\rho}(\tfrac{1}{S})+\tfrac{1}{2S(\rho)}(is\hat{Y}(\rho)+\tfrac{d\hat{X}(\rho)}{d\rho}))exp(-|s||x-\rho|)d\rho)\\ -\tfrac{1}{4|s|}\textstyle\int_{-\infty}^{\infty}\hat{\sigma}_x(\rho)\tfrac{d^2}{d\rho^2}(\tfrac{1}{S(\rho)})exp(-|s||x-\rho|)d\rho\\ -\tfrac{\pi\delta(s)}{|s|}\textstyle\int_{-\infty}^{\infty}\tfrac{d^2}{d\rho^2}d_1(\rho)exp(-|s||x-\rho|)d\rho].\end{aligned} \tag{16}$$

Now, using the value of $\hat{\sigma}_x$ of Eq. (14) into expression (16), we get the following integral equation:

$$\hat{\sigma}=E_1\left[-\beta_1\hat{T}+\theta-\gamma-\frac{1}{8}\int_{-\infty}^{\infty}\hat{\sigma}(\rho_1)P(x,\rho_1)d\rho_1\right], \tag{17}$$

where

$$\begin{aligned}\theta\quad&=\tfrac{1}{2|s|}\textstyle\int_{-\infty}^{\infty}\hat{X}(\rho)\tfrac{d}{d\rho}(\tfrac{1}{G(\rho)})+\tfrac{1}{2S(\rho)}(is\hat{Y}+\tfrac{d\hat{X}(\rho)}{d\rho}))exp(-|s||x-\rho|)d\rho\\ &-\tfrac{1}{8s^2}\textstyle\int_{-\infty}^{\infty}\int_{-\infty}^{\infty}(\tfrac{d\hat{X}(\rho_1)}{d\rho_1}-is\hat{Y}\rho_1))\tfrac{d^2}{d\rho^2}(\tfrac{1}{S(\rho)})exp(-|s|(|\rho_1-\rho|+|x+\rho_1|))d\rho_1 d\rho,\end{aligned}$$

$$\gamma=\frac{\pi\delta(s)}{|s|}\int_{-\infty}^{\infty}\frac{d^2}{d\rho^2}d_1(\xi)exp(-|s||x-\rho|)d\rho,$$

and

$$P(x,\rho_1)=\int_{-\infty}^{\infty}\frac{d^2}{d\rho^2}(\frac{1}{S(\rho)})exp(-|s|(|\rho-\rho_1|+|x-\rho_1|))d\rho.$$

To find the solution of Eq. (17), let us consider a limit

$$\hat{\sigma}=\lim_{n\to\infty}\hat{\sigma}_n, \tag{18}$$

where

$$\hat{\sigma}_n=E_1\left[-\beta_1\hat{T}+\theta-\gamma_1,n-\frac{1}{8}\int_{-\infty}^{\infty}\hat{\sigma}_{n-1}(\rho_1)P(x,\rho_1)d\rho_1\right] \tag{19}$$

$\hat{\sigma}_0\equiv 0$ and $n=1,2,\ldots$

Here,

$$\hat{\sigma}_1=E_1[-\gamma_1\hat{T}+\theta]$$

We can represent Eq. (19) for the $n$th iteration as

$$\hat{\sigma}_n = \hat{\sigma}_1 + \hat{s}_{n-1},$$

where

$$\hat{S}_{n-1} = \frac{-E_1}{8} \int_{-\infty}^{\infty} \hat{\sigma}_{n-1}(\rho_1) P(x, \rho_1) d\rho_1.$$

Let us consider

$$\beta_1 = 0.$$

The function $\hat{S}_{n-1}$ is based on $(n-1)$th approximation, and it represents the result of $n$th approximation after the successful implementation of the first iteration.

The stress $\sigma_x$ in a plane $R$ can be found by using expression (12). Then the stress $\sigma_y$ is given by

$$\sigma_y = \sigma - \sigma_x.$$

The shear stress can be determined from first Eq. (1).

$$\hat{\sigma}_{xy} = \frac{-i}{s} \left[ \frac{\partial \hat{\sigma}_x}{\partial x} - \hat{X} \right].$$

Making use of Eq. (14) for plane $R$, we get

$$\hat{\sigma}_{xy} = \frac{i}{s} \left[ \hat{X} - \frac{1}{2} \int_{-\infty}^{\infty} s^2 \hat{\sigma} \left( \rho - is\hat{Y} + \frac{d\hat{X}(\rho)}{d\rho} \right) exp(-|s||x-\rho|) sgn(x-\rho) d\rho \right]. \quad (20)$$

The expression (17) represents an analytic explicit solutions of the mentioned thermoelastic problem in $R$.

## 4 Numerical Examples

To find the thermal stress of $R$ stressed by external boundary conditions,

$$\begin{aligned} p &= \rho_0, |y| \leq y_0, \\ &= \zeta_0, y_1 \leq |y| \leq y_2 \\ &= 0, otherwise, \end{aligned} \quad (21)$$

where $\rho_0$, $\zeta_0 =$ constant, $q = 0$ and $0 < y_i$ is constant, where, $i = 0, 1, 2$. For this loading condition, we have considered temperature distribution $T = 0$ and body forces equals to zero.

Equation (21) fulfils the condition in Appendix 4 satisfying the relation

$$\zeta_0 = \frac{y_0 \rho_0}{y_2 - y_1}.$$

We observed that the boundary condition (21) and the essential integral condition mentioned in Appendix 2, 3 are fulfilled by the stresses.

### 4.1 Exact Solution

An exact solution can be found by considering the inhomogeneity case for

$$E = E_0 f(x),$$

where

$$f(x) = \begin{matrix} \frac{1}{1+ax}, & for\ x \le \frac{1}{a}, \\ \frac{1}{2}, & for\ x > \frac{1}{a}. \end{matrix}$$

$E_0 =$ constant, $a < 0 =$ constant. We have a relation

$$S = S_0 f(x)$$

where

$$S_0 = \frac{E_0}{2(1 + \upsilon)}$$

The distribution of Young's and shear modulus is depicted for $a = 1$ and $a = 5$ (see Fig. 1).

The $x$-distribution of the stresses $\sigma_x$ and $\sigma_y$, respectively, is shown (see Fig. 2a, b) for homogeneous and inhomogeneous properties in cross section $y = 0$. It is seen that the stress $\sigma_y$ is exponentially increasing along $y$-axis in homogeneous and inhomogeneous regions. We notice the effect of material on the stresses which is stronger for $\sigma_y$.

The y-distribution of the shear stress $\sigma_{xy}$ in the cross section of $x = 1$ is shown (see Fig. 3). we can see that y-distribution of $\sigma_{xy}$ is an exponential nature.

### 4.2 Case of Young's Modulus

Let us consider

$$E = E_\infty \left[1 - a\exp(-bx)\right] \tag{22}$$

of exponential form. Here, $E_\infty, a, b$ are positive constants and $\upsilon$ is also a constant, which gives, $E = E_\infty$ for $x$ tends to $\infty$. At $x = 0$, $E = E_0$ is constant, where

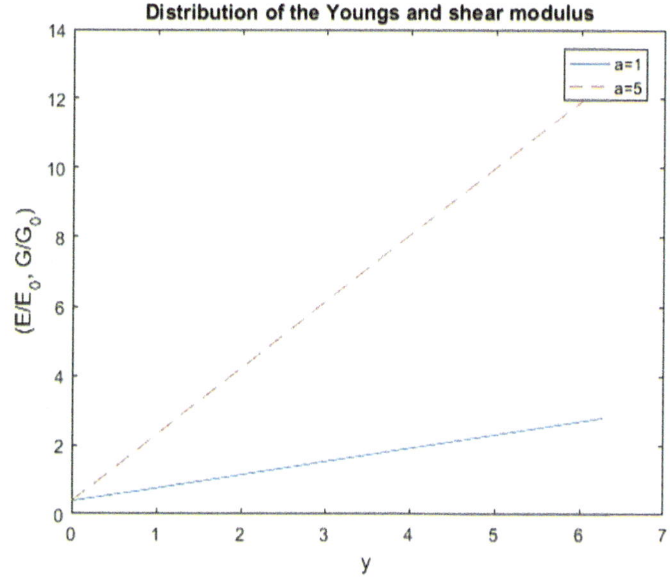


**Fig. 1** The Young's and shear modulus distribution

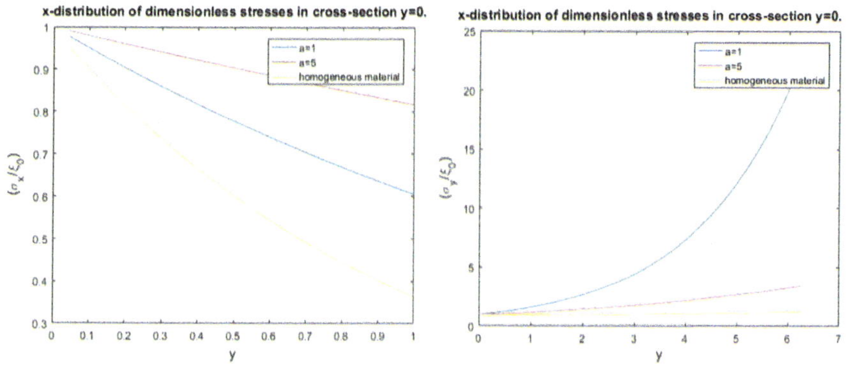


(a) x-distribution of $\sigma_x$.

(b) x-distribution of $\sigma_y$.

**Fig. 2** x-distribution of the stresses

$E_0 = E_\infty(1 - a)$. The shear modulus is appeared as

$$S = S_\infty \left[1 - a\exp(-bx)\right]$$

where

$$S_\infty = \frac{E_\infty}{2(1 + v)}.$$

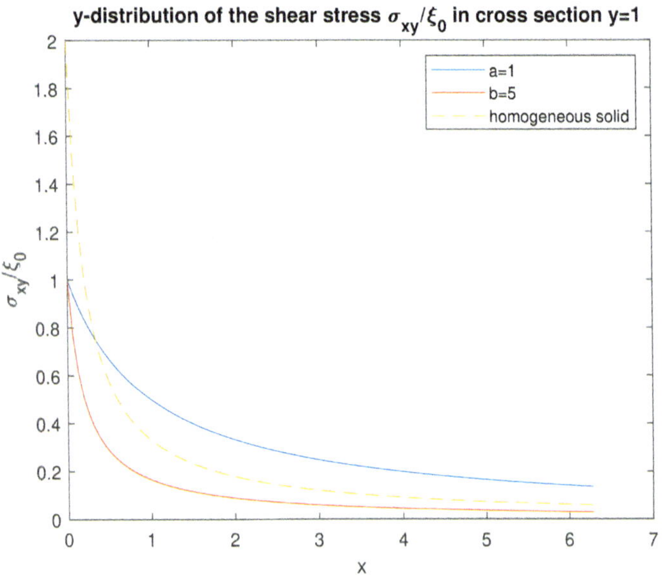


**Fig. 3** $y$-distribution of the stresses $\sigma_{xy}/\xi_0$

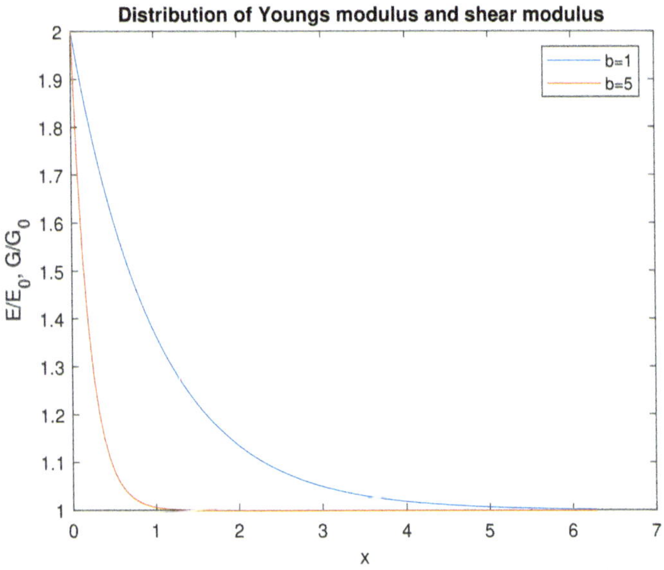


**Fig. 4** Dimensionless distribution for $b = 1, 5$

Elastic distribution of Young's and shear moduli for $a = 1$ and two different values of $b$ ($b = 1, 5$), which obey property (22) in the calculation of Eq. (19) (see Fig. 4). The result shows that the first iteration gives much accuracy in finding the total stress.

## 5 Conclusions

We construct an analytical solution to the 2D thermoelastic problems for inhomogeneous plane, making use of the direct integration method. The stresses have been found using the Fourier Integral Transform. The original problems are reduced in the form of integral equations. Further, an iterative technique is used to find their solutions.

The speedy convergence exists, which can be described by the successful trial of the first iteration.

So, the iterations set out an almost exact solution. The results show the vital outcome on stress distribution of solid of inhomogeneous material. It is essential for the analysis of stresses by taking material dependence properties into account. Due to its explicit nature, the formulated solution serves as better tool in solving thermoelastic problems, and also for verification of results by numerical or analytical meaning. The constructed solutions can be used for thermal stress field analysis on elastic solid having isotropic properties.

## 6 Appendix

The relations for the stresses and force in $R$

1. Stress–tensor relations:

$$\sigma_x = \int_{-\infty}^{\infty} \left( \frac{\partial \sigma_{xy}(\zeta, y)}{\partial y} + X(\zeta, y) \right) sgn(x - \zeta) d\zeta,$$

$$\sigma_y = \int_{-\infty}^{\infty} \left( \frac{\partial \sigma_{xy}(x, \rho)}{\partial x} + X(\zeta, y) \right) sgn(y - \rho) d\rho,$$

$$\sigma_{xy} = \int_{-\infty}^{\infty} \left( \frac{\partial \sigma_{y}(\zeta, y)}{\partial y} + Y(\zeta, y) \right) sgn(x - \zeta) d\zeta,$$

$$\sigma_{xy} = \int_{-\infty}^{\infty} \left( \frac{\partial \sigma_{x}(x, \rho)}{\partial x} + X(x, \rho) \right) sgn(y - \rho) d\rho.$$

2. The equilibrium conditions:

$$\int_{-\infty}^{\infty} \sigma_x dy = \iint\limits_R X(\zeta, y) sgn(x-\zeta)\, d\zeta .dy, \quad \int_{-\infty}^{\infty} \sigma_y dx = \iint\limits_R Y(x, \rho) sgn(y-\rho)\, d\rho .dx,$$

$$\int_{-\infty}^{\infty} \sigma_{xy} dy = \iint\limits_R Y(\zeta, y) sgn(x-\zeta)\, d\zeta .dy, \quad \int_{-\infty}^{\infty} \sigma_{xy} dx = \iint\limits_R X(x, \rho) sgn(y-\rho)\, d\rho .dx.$$

3. The resultant forces conditions:

$$\iint\limits_R \sigma_x \, dx .dy = \iint\limits_R xX \, dx .dy, \quad \iint\limits_R \sigma_y \, dx .dy = \iint\limits_R yY \, dx .dy,$$

$$\iint\limits_R \sigma_{xy} \, dx .dy = \iint\limits_R Xy \, dx .dy = \iint\limits_R xY \, dx .dy.$$

4. Equilibrium conditions for body forces:

$$\iint\limits_R X \, dx .dy = \iint\limits_R Y \, dx .dy = 0, \quad \iint\limits_R yX \, dx .dy = \iint\limits_R xY \, dx .dy.$$

## References


1. Gaikwad, K.R., Ghadle, K.P.: Nonhomogeneous heat conduction problem and it's thermal deflection due to internal heat generation in a thin hollow circular disk. J. Thermal Stresses **35**(6), 485–498 (2012). https://doi.org/10.1080/01495739.2012.671744
2. Ghadle, K.P., Adhe, A.B.: Steady-state temperature analysis to 2D elasticity and thermoelasticity problems for inhomogeneous solids in half-plane. J. Korean Soc. Ind. App. Math. **24**(1), 93–102 (2020). https://doi.org/10.12941/jksiam.2020.24.093
3. Jafari, M., Jafari, M.: Thermal stress analysis of otrhotropic plate containing a rectangular hole using complex variable method. Eur. J. Mech. A Solids **73**, 212–223 (2019). https://doi.org/10.1016/j.euromechsol.2018.08.001
4. Kalynyak, B.: Integration of equation of one-dimensional problems of elasticity and thermoelasticity for inhomogeneous cylindrical bodies. J. Math. Sci. **99**, 1662–1670 (2000). https://doi.org/10.1007/BF02674190
5. Kaminski, W.: Hyperbolic heat conduction equations for materials with a nonhomogeneous inner structure. J. Heat Trans. **112**, 555–560 (1990). https://doi.org/10.1115/1.2910422
6. Ma, C.C., Chang S.: Analytical exact solutions of heat conduction problems for anisotropic multi-layered media. Int. J. Heat Mass Trans. **47**, 1643–1655 (2004)

7. Manthena, V.R., Lamba, N.K., Kedar, G.D.: Thermoelastic analysis of a rectangular plate with nonhomogeneous material properties and internal heat source. J. Solid Mech. **10**(1), 200–215 (2018)
8. Ozisik, M.N.: Boundary Value Problems of Heat Conduction. Dover Publications INC, Mineola, New York (1968)
9. Porter, D.: The solution of integral equations with difference kernels. J. Integral Equ. Appl. **3**, 429–454 (1991)
10. Rychachivskyy, A.,Tokovyy, Y.: Correct analytical solutions to the thermoelasticity problem in a semi-plane. J. Thermal Stresses **31**, 1125–1145 (2008). https://doi.org/10.1080/01495730802250854
11. Sneddon, I.: Fourier Transform. McGraw-Hill Book Company, INC (1951)
12. Tanigawa, Y., Morishita, H., Ogaki, S.: Derivation of systems of fundamental equations for a three-dimensional thermoelastic field with nonhomogeneous material properties and it's applications to a semi-infinite body. J. Thermal Stresses **22**(7), 689–711 (1999). https://doi.org/10.1080/014957399280706
13. Tanigawa, Y.: Some basic thermoelastic problems for nonhomogeneous structural materials. Appl. Mech. Rev. **48**(6), 287–300 (1995). https://doi.org/10.1115/1.3005103
14. Tokovyy, Y.V., Ma, C.C.: An analytical solution to the three-dimensional problem on elastic equilibrium of an exponentially-inhomogeneous layer. J. Mech. **31**, 545–555 (2015). https://doi.org/10.1017/jmech.2015.17
15. Tokovyy, Y.V., Ma, C.C.: Three-dimensional temperature and thermal stress analysis of an inhomogeneous layer. J. Thermal Stresses **36**, 790–808 (2013). https://doi.org/10.1080/01495739.2013.787853
16. Tokovyy, Y.V., Ma, C.C.: Steady-state heat transfer and thermo-elastic analysis of inhomogeneous semi-infinite solids. Heat Conduction - Basic Research, pp. 249–268 (2011)
17. Tokovyy, Y.V., Ma, C.C.: Analytical solutions to the 2D elasticity and thermoelasticity problems for inhomogeneous planes and half-planes. Arch. Appl. Mech. **79**, 441–456 (2009). https://doi.org/10.1007/s00419-008-0242-5
18. Tokovyy, Y.V., Ma, C.C.: Analysis of 2D-Non-axisymmetric Elasticity and Thermoelasticity Problems for Radially Inhomogeneous Hollow Cyllinders, vol. 61, pp. 171–184. Springer, New York (2007). https://doi.org/10.1007/s10665-007-9154-6
19. Tokovyy, Y.V. , Dmytro Boiko, D., Gao, C.: Three-dimensional thermal-stress analysis of semi-infinite transversely isotropic composites. Trans. Nanjing Univ. Aeronaut. Astronaut. **38**(1), 18–28 (2021).https://doi.org/10.16356/j.1005-1120.2021.01.002
20. Vigak, V.: Correct Solutions of Plane Elastic Problems for a Semi-plane. Int. Appl. Mech. **40**, 283–289 (2004). https://doi.org/10.1023/B:INAM.0000031910.20827.19
21. Vigak, V., Rychachivskyy, A.: Bounded solutions of plane elasticity problems in semi-plane. J. Comp. Appl. Mech. **2**, 263–272 (2001)
22. Vigak, V.: Method for direct integration of the equations of an axisymmetric problem of thermoelasticity in stresses for unbounded regions. Int. Appl. Mech. **35**, 262–268 (1999)
23. Youssef, H., Al-Lehaibi, A.: The boundary value problem of a three dimensional generalized thermoelastic half-space subjected to moving rechangular heat source. Bound. Value Probl. **8**, 1–15 (2019). https://doi.org/10.1186/s13661-019-1119-y

# Study of Non-Newtonian Models for 1D Blood Flow Through a Stenosed Carotid Artery

**Mahesh Udupa and Sunanda Saha**

**Abstract** In this work, blood flow in human arteries is studied by considering different geometrical configurations and mechanical properties of the arteries. The arteries such as coronary, carotid, aorta, etc., are the main areas of focus that are carriers of oxygen and nutrients to or from the vital organs. Comparison of various non-Newtonian models is conducted for a better understanding of the effect of stenosis. In addition, variation in pressure waveform is observed numerically for the physically stenosed arteries. In the stenosed artery, the extent of stenosis, along with its length and position, have been treated as variables, amongst which, extent or degree of the stenosis is observed to be the predominant factor. Numerical schemes of Lax Friedrichs' scheme and HLL scheme have been used to obtain the solution. For clinical purposes, the present work can be used as an indicative index for pressure drop under stenotic conditions.

**Keywords** Non-Newtonian · Carotid artery · HLL scheme · Stenosis

## 1 Introduction

The cardiovascular system of humans is a highly complex system, performing tasks that form the basis for the life form to exist. Its main function being, the supply of oxygen and nutrients to all parts of the body, it also removes carbon dioxide and other toxin products from our system. So, any subtle changes, either in the channel or in the fluid, affects the other. This interdependence might eventually influence the evolution of severe cardiovascular pathology such as atherosclerosis. Atherosclerosis refers to the buildup of plaque, within the arteries causing an increase in wall thickness, and thus narrowing of the arteries, which is called Stenosis. Even a stenosis in its mild stage can develop arterial deformity, changing the regional blood rheology [1] and reaching its severity, it may even rupture the arterial walls under certain physiological conditions resulting in a stroke or heart attack and can prove even to be fatal [2, 3].

M. Udupa (✉) · S. Saha
Department of Mathematics, SAS, Vellore Institute of Technology, Vellore, India
e-mail: maheshudupa.c@vit.ac.in

 
B. Rushi Kumar et al. (eds.), *Mathematics and Computing*, Springer Proceedings in Mathematics & Statistics 415, https://doi.org/10.1007/978-981-19-9307-7_42

Thus understanding the stenotic flow becomes very crucial for clinical purposes. Deshpande et al. [4] have made numerical predictions on flow field due to the effect of stenosis in a cylindrical domain by making an assumption for the flow to be steady and laminar. Even, Smith [5] has made a similar assumption, for the flow to be steady and laminar, and has compared the obtained analytical solutions with the existing experimental data. But the blood flow is indeed unsteady for it is pulsatile in nature. Zendehbudi et al. [6] have compared numerical results of the physiological pulsatile flow, under the assumption that blood is Newtonian, with a simple pulsatile flow, having the same stroke volume, with no-slip boundary condition. On the contrary, Biswas et al. [7] have considered velocity slip condition at the stenotic wall, with the Newtonian assumption for the fluid, and have investigated the variations in flow parameters due to the effect of body acceleration, with its biological consequences. Zhong et al. [8] have performed numerical simulations of blood flow through stenosis by FFR (Fractional Flow Reserve) technique and have discussed their clinical implications with a variance in length, position, and degree of the stenosis.

In the above-mentioned works, blood is treated as Newtonian fluid, which is true for most part of the cardiovascular circuit. But the natural constrictions caused by the stenosis in the arteries narrow the path for fluid in the arteries, thus making its dimensions comparable to that of RBCs (Red Blood Corpuscles) that is majorly suspended in blood plasma, which consists of 90% of water. Under these conditions, opting for non-Newtonian models for blood flow is provoked for better accurate readings. Johnston et al. [9] have examined the contrast between classic Newtonian model and five non-Newtonian models of blood viscosity. Mandal et al. [10] have worked with the generalised power-law model with consideration of both shear-thinning and shear-thickening behaviours of blood through the stenosed geometry. Similarly, variations in flow parameters due to the asymmetric stenosis have been analysed by Sankar et al. [11] wherein they have modelled blood flow using Herschel–Bulkley model.

Stenosis usually occurs in various arteries throughout our body, only few of which that require medical attention, based on their proximity to the vital organs, such as heart or brain. So, the focus is on arteries such as coronary, carotid, aorta, etc., as they perform the tasks of providing oxygen and nutrients to or from the vital organs. In the present work, carotid artery has been considered as the domain of interest. Onaizah et al. [12] by experimental setup, have thoroughly analysed changes in the mechanical properties of the Carotid artery due to the formation of plaques and have also performed a theoretical study on Lumped Parameter or LP models, concluding with the reduction in flow due to stenosis. Zhang et al. [13] by adapting the FFR (Fractional Flow Reserve) technique of coronary artery, have performed CFD simulations by observing the ratio of distal pressure over proximal pressure in the stenosed carotid artery and showing a fine correlation with their invasive pressure-wired measurements.

The clinical implications have been the most important consequence of the study of blood flow in the above-mentioned arteries. In recent years, one-dimensional (1D) blood flow has been an efficient and valid model for averaged blood flow features. Upon a close examination made by Xiao et al. [14] of 1D and 3D models of blood flow, they have concluded 1D model provides a good approximation for the 3D model and, for the purpose of computational cost, 1D model for blood flow has been preferred. Hence, in the present work, 1D model for blood flow has been considered. Further, numerical settings have been opted with a couple of finite difference schemes, namely Lax Friedrichs [15] and Harten–Lax–Leer or HLL scheme [16]. Mathematical formulation of these schemes will be investigated in more detail in Sect. 5. Given to the best of author's knowledge, there has been no work done on 1D blood flow for a stenosed carotid artery (cosine shaped) considering the real physiological conditions with the numerical setting of the HLL scheme.

This article has been organised as follows. In Sect. 2, the mathematical formulation of 1D blood flow has been presented. In Sect. 3, non-Newtonian models along with the Newtonian model have been compared for different arterial domains. In Sect. 4, numerical results of Lax Friedrich's scheme have been compared with the results obtained by Mynard and Nithirasu [17], where they have considered locally conservative Galerkin or LCG method. In Sect. 5, the numerical model used to solve the problem has been presented. In Sects. 6 and 7, along with the geometry of the stenosed artery, variance in pressure parameter has been presented for different conditions of stenosis in the carotid artery and the results have been discussed.

## 2 Mathematical Modelling of Blood Flow

The one-dimensional system of hyperbolic partial differential equations, for blood flow [17], derived from the Navier–Stokes equations in terms of (A, u) is

$$\begin{aligned} \frac{\partial A}{\partial t} + \frac{\partial (uA)}{\partial x} &= 0, \\ \frac{\partial u}{\partial t} + \frac{\partial}{\partial x}\left(\frac{u^2}{2}\right) &= -\frac{1}{\rho}\frac{\partial p}{\partial x} + f, \end{aligned} \tag{1}$$

where $A(x, t)$: Cross-sectional area of the artery; $u(x, t)$: Velocity of the fluid; $p$: Pressure related to $A$ via a non-linear elastic wall law; $\rho$: density of the fluid; $f$: Source term.

Now, the system of equation can be written in the vector form as

$$\frac{\partial U}{\partial t} + \frac{\partial F(U)}{\partial x} = B(U), \tag{2}$$

where

$$U = \begin{bmatrix} A \\ u \end{bmatrix}, \qquad F(U) = \begin{bmatrix} uA \\ \dfrac{u^2}{2} + \dfrac{p(A)}{\rho} \end{bmatrix}, \qquad B(U) = \begin{bmatrix} 0 \\ f \end{bmatrix}.$$

The above system of equations has two equations and three variables; thus, use of an extended relation [18] between pressure, p, and Area, A, has been given below:

$$p(A) = p_{ext} + \beta\left(\sqrt{A} - \sqrt{A_0}\right), \tag{3}$$

where $p_{ext}$ is the external pressure; $A_0 = A(x, 0)$, is the initial area; $\beta$ accounts for physical and mechanical characteristics of vessel and is given as

$$\beta = \frac{\sqrt{\pi} h E}{A_0(1 - \sigma^2)}, \tag{4}$$

where h is the thickness of the arterial wall; $E$ is its Young's modulus; $\sigma$ is its Poisson's ratio, that has been considered to be 0.5 (wall material has been assumed to be incompressible). The source term on RHS of the Eq. (2) is given by

$$f = -\frac{8\pi\mu}{\rho}\frac{u}{A}. \tag{5}$$

In Eq. (5), $\mu$ represents the viscosity of the fluid, which refers to the internal friction between parallel layers of the fluid. The classification of fluids is done based on constancy and variability of this intrinsic property. The fluids in which the viscosity is independent of the stress applied, and hence remains constant, are called Newtonian fluids. Its stress–strain relation is given by

$$\tau = \mu\dot{\gamma}, \tag{6}$$

where $\tau$ is the stress on the fluid, $\dot{\gamma}$ is the shear rate, and $\mu$ is the constant of proportionality, called to be the Viscosity of the fluid. On contrary, in non-Newtonian fluid, the dependency of shear stress is neither linearly proportional nor can be said to be directly proportional to shear rate. The case, in which it is directly proportional, is called shear-thinning fluid and the case of inverse proportionality is shear-thickening fluid. There are very few fluids that fall in the latter category, and the fluid of interest, that is blood, amongst many other non-Newtonian fluids, falls in the former category. The mathematical crux of non-Newtonian fluid can be given as

$$\tau = \mu(\dot{\gamma})\dot{\gamma}, \tag{7}$$

where $\mu$ is a function of shear rate and is deterministic based on the non-Newtonian model considered. Here, in this article, the main emphasis has been given to four different types of non-Newtonian models [19], wherein the effective viscosity of each being different from the other has been presented in the table below, along with the standard Newtonian model.

Here, $\mu_0$ and $\mu_\infty$ represent the viscosity at low and high shear rates, respectively, $n_p$ represents the flow consistency, K is the flow behaviour index, $\lambda$ represents the relaxation time constant, $n$ represents the power index. $\tau_0$ is the yield stress, $\eta$ is the Casson rheological constant, $\Gamma_c$ represents the critical shear rate, that indicates the onset of shear thinning, and $n_c$ is the cross-rate constant. The values assumed for the above variables are listed in Table 2.

# 3 Comparison of Non-Newtonian Models

In this section, five different arteries have been considered, with varied settings in their lengths, cross-sectional area and elasticity of the vessel walls. Each of which has its own significance in capturing the viscous effect, and thus determining the true nature of pressure waveform across the arterial domain. The considered arteries with their properties [20] have been mentioned in Table 3. The pressure has been evaluated from Eq. (1) by using a 3-point scheme, namely Lax Friedrich scheme. This scheme is suitable for solving Hyperbolic PDEs due to its less computational cost. The step lengths in time and space have been chosen carefully without violating the Courant–Friedrichs–Lewy or CFL condition [21]. Unit velocity and cross-sectional area corresponding to $r_0$ given in Table 3 have been considered for the initial values. A pressure waveform has been considered at the inlet, which is generated by a Fourier series [22] with a time period of 0.9 s. Method of characteristics has been implemented to calculate the other variable at the inlet. Non-reflecting boundary conditions [23] have been implemented at the outlet.

For each arterial vessel mentioned in Table 3, each of the blood flow model mentioned in Table 1 has been considered, and the pressure waveform has been captured at the midsection of the artery and plotted against single time period. Since the uniform vessel length in humans are naturally short for the given speed that the blood flows with, the pressure measured at any given time throughout the arterial length is not of much interest, as it will almost be a flat line, representing a certain constant value. Thus, pressure vs time graphs have been plotted (Table 2).

**Table 1** .

| Model | Effective viscosity |
|---|---|
| Newtonian | $\mu = 0.0035$ Pa.s |
| Power law | $\mu = K(\dot{\gamma})^{n_p - 1}$ |
| Carreau model | $\mu = \mu_\infty + (\mu_0 - \mu_\infty)[1 + (\lambda\dot{\gamma})^2]^{(n-1)/2}$ |
| Casson model | $\mu = \frac{\tau_0}{\lvert\dot{\gamma}\rvert} + \frac{\sqrt{\tau_0 \eta}}{\lvert\sqrt{\dot{\gamma}}\rvert} + \eta$ |
| Cross law | $\mu = \mu_\infty + (\mu_0 - \mu_\infty)\left[1 + \left(\frac{\lvert\dot{\gamma}\rvert}{\Gamma_c}\right)^{n_c}\right]$ |

**Table 2** Parametric values for the non-Newtonian models

| Parameters | Values |
|---|---|
| $\mu_0$ | 0.056 Pa.s |
| $\mu_\infty$ | 0.0035 Pa.s |
| K | 0.017 $Pa^n$ |
| $n_p$ | 0.708 |
| $\lambda$ | 3.313 s |
| n | 0.3568 |
| $\tau_0$ | 0.005 Pa |
| $\eta$ | 0.035 Pa.s |
| $\Gamma_c$ | 2.63 $s^{-1}$ |
| $n_c$ | 1.45 |

**Table 3** .

| Vessel | L (cm) | $r_0$ (cm) | h (cm) | E $10^6$ (dyne/cm$^2$) |
|---|---|---|---|---|
| Ascending aorta | 4.0 | 1.470 | 0.163 | 4.0 |
| Left carotid | 13.9 | 0.6 | 0.063 | 7.0 |
| Left coronary | 3.0 | 0.259 | 0.08 | 8.0 |
| Left femoral | 44.3 | 0.314 | 0.05 | 8.0 |
| Left interosseous | 7.9 | 0.1 | 0.28 | 14.0 |

Given the amount and the rate at which the blood is pumped from the heart, wall of the arteries in the proximity of the heart, with evolutionary, are of naturally high thickness and very elastic in nature. These are classified into elastic arteries, contrary to the ones that are away from the heart, which are classified to be muscular arteries. Ascending aorta, given its huge diameter, allows blood to behave as a Newtonian fluid. Thus, it can be observed in Fig. 1 that even all the other general non-Newtonian models reduce to simple Newtonian models. But for the other four arterial domains, the Carreau model uniquely captures the viscous effect. Significantly, in the femoral vessel due to its long length, the viscous force becomes comparable with the inertial force, whereas, in the coronary vessel despite its cross-sectional area being less than that of femoral's, the inertial force dominates well over the viscous force, due its very short domain length. The interosseous vessel, one of the arteries that happens to be at the farther end and well away from the heart, is highly muscular in nature. This is due to high Young's modulus value of its walls and low radius. Thus, it can capture the viscous effect to a great extent. But usually, the arteries away from the heart or any vital organ, heal from stenosis over time with a bypass grafting that grows over the artery affected by stenosis, naturally and thus preventing any kind of hindrances to the blood flow. Hence, radiation treatments or bypass surgeries are typically known for coronary or carotid arteries. So, in the next section of this article, validation of the numerical scheme used to solve the equations has been carried out, following

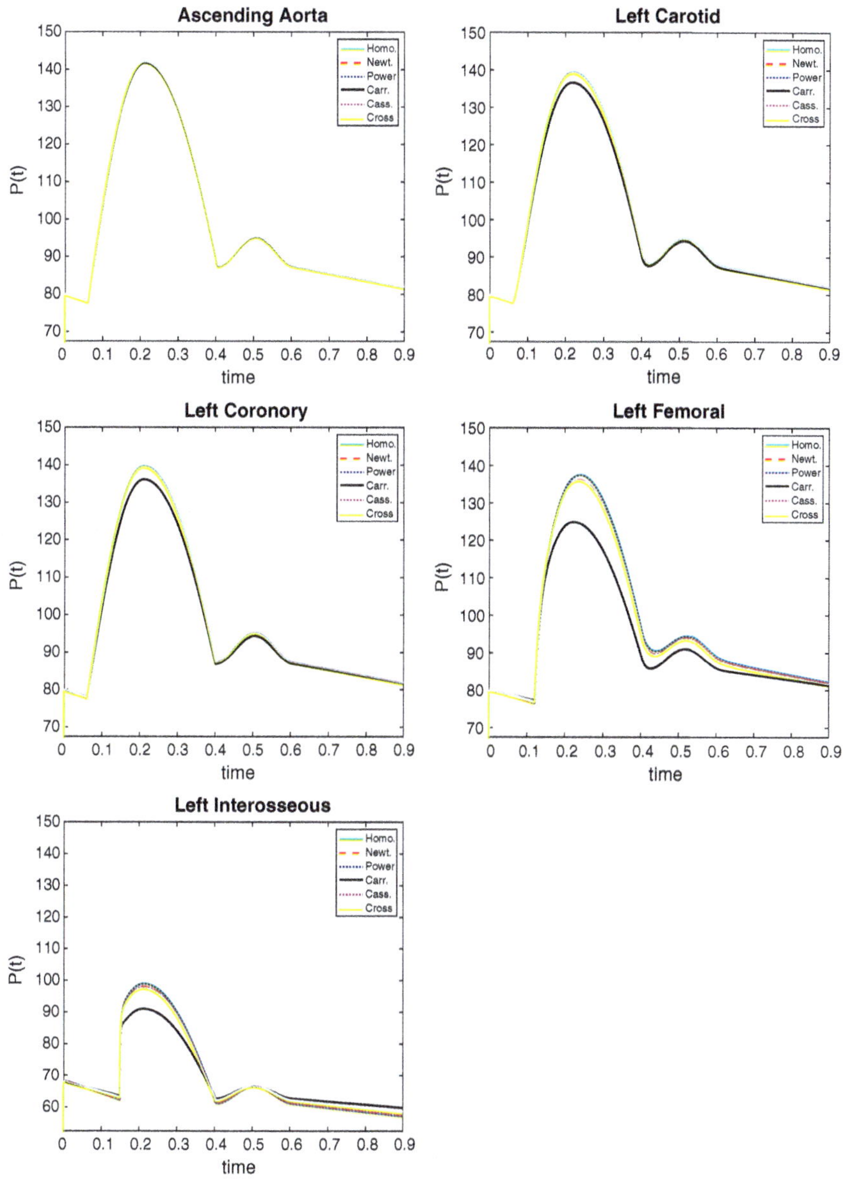


**Fig. 1** Variation of Pressure waveform captured at the mid-section of the respective arteries

which, carotid artery has been considered to observe the effects of stenosis on the pressure waveform and thus the blood flow.

## 4 Pulse Propagation in a Single Uniform Vessel

In this section, the work of Mynard and Nithirasu [17] has been validated, by considering a Gaussian pulse as the input pressure wave through a domain of length 20 cm for 0.1 s. An initial area of 1 $cm^2$ has been considered. A pulse width of 0.03 s with amplitude of $10^3$ dyne/$cm^2$ has been considered for the inlet pressure waveform. Assumption of Young's modulus as $10^6$ dyne/$cm^2$ and the thickness of the walls as 0.096 cm has been made. Lax Friedrich's Scheme has been considered for the numerical framework.

The red line in Fig. 2 represents an inviscid case, where the viscosity is zero, the graph in black represents the Newtonian case where the viscosity is non-zero and its value has been taken from Table 3, which captures the viscosity to a certain extent. The same work has been extended in this article, wherein the real physiological conditions of the vessels have been accounted with its measured physical variables. The work has been done with stenosed artery, that has been caused by a buildup of plaque (atherosclerosis) inside the artery wall, thus narrowing the region for blood flow. Carotid artery has been specifically chosen for this, which is referred as 'Carotid stenosis'. It is the narrowing of the carotid arteries, the two major arteries that carry oxygen-rich blood from the heart to the brain. Thus, narrowing of the artery results in the reduction of blood flow to the brain and hence studying Stenosis has been very pivotal since they block the bridge connecting the vital organs. Working with a stenosed geometry, due to its non-uniformity in the dimensions of the domain, demands for a more stable numerical setting, thus the following scheme has been considered for improved results.

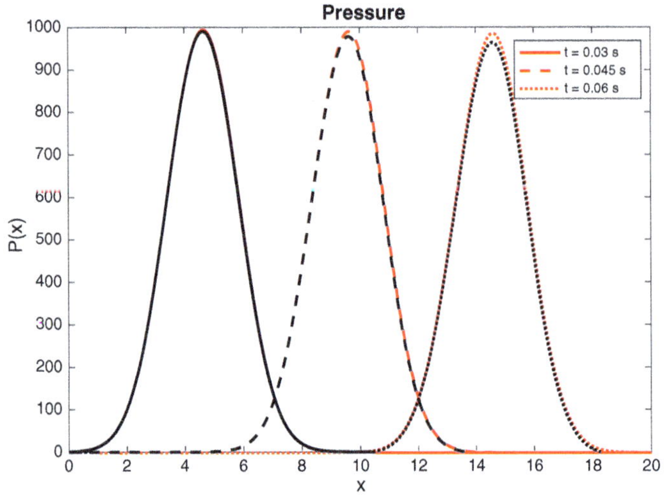


**Fig. 2** Validation of Numerical scheme by variations in pressure waveform measured in Pa

# 5 HLL Scheme

HLL scheme was introduced by Harten, Lax, van Leer [16] in the 1980s, is an upwind Riemann Solver.

$$U_i^{n+1} = U_i^n - \Lambda(F_{i+1/2}^n - F_{i-1/2}^n),$$

$$F_{i+1/2} = \begin{cases} F_i & 0 < S_L \\ F_{i+1/2}^{HLL} & S_L < 0 < S_R \\ F_{i+1} & 0 > S_R \end{cases}$$

where $\Lambda = {}^{\Delta t}/_{\Delta x}$; $S_L$ & $S_R$ are approximations of the smallest and the largest wave velocities at the interface $x_{i+1/2}$.

In the case when $S_L < 0 < S_R$, the flux function can be derived to be

$$F_{i+1/2}^{HLL} = \frac{S_R^+ F_i - S_L^- F_{i+1} + S_L^- S_R^+ (U_{i+1} - U_i)}{S_R^+ - S_L^-},$$

where $S_R^+ = max(S_R, 0)$; $S_L^- = min(S_L, 0)$. Due to its upwind nature, HLL Scheme has been very much suitable for hyperbolic problems even with discontinuity in its initial data.

Upon considering to work with this scheme, certain improvements have been done with regard to method of the characteristic. In hyperbolic PDEs, for the boundary values, linear extrapolation has been implemented. This involves extrapolating the boundary values at the current $n$th time level from the previous $(n-1)$th time level, based on the speed of the wave (information) travelling at the boundary at $n$th time level. This when multiplied with '$\Delta t$' gives a dimension of distance, whose value corresponds to the location of the wave on the spacial domain at $(n-1)$th time level. But, while numerically determining the solution, this may not be straightforward, as the linearly extrapolated position or distance might not be one of the discrete grid values. So, to resolve this, a simple enough technique has been developed, wherein, if the value falls in between, $i$th and $(i+1)$th node, then its distance from the $i$th node, becomes the weightage for the value at $(i+1)$th node and vice versa, resulting in a precisely weighted average of the two values. So, in the next section, with the introduction and description of the stenosis model, this scheme has been implemented.

# 6 Stenosis Model

The geometry of a stenosed uniform vessel of radius $r$ at zero transmural pressure and the arterial wall under zero stress has been expressed as (Fig. 3)

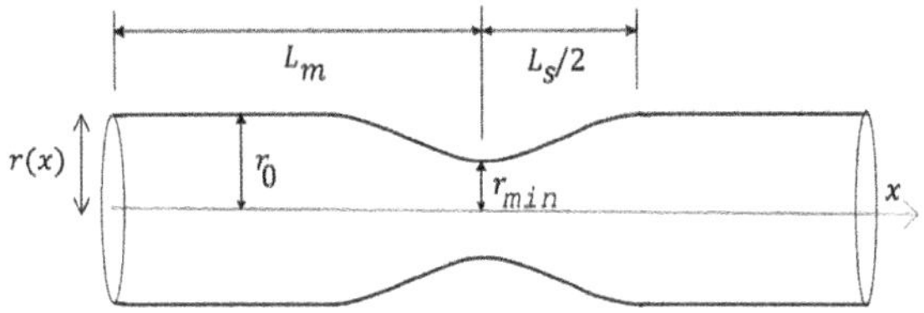


**Fig. 3** One-dimensional model of stenosis

$$r(x) = r_0 - \frac{r_{min}}{2}\left[1 - cos\left(2\pi\frac{x - L_m - L_s/2}{L_s}\right)\right], \tag{8}$$

$$\text{for} \qquad (L_m - L_s/2) < x < (L_m + L_s/2).$$

where $r_{min}$ denotes the mid-region of the stenosis where the arteries are narrowed the most, thus representing the region with the least radius. It is expressed as $r_{min} = r_0$ ($S/100$), with $S$ representing the severity of stenosis in percentage. The constant radius in the unstenosed region is denoted by $r_0$. $L_m$ represents the centre of the stenosis along the vessel, with $L_s$ being the length of the stenosis. In this study, the presence of thrombus and the variation of the wall thickness has been ignored. Furthermore, the variation of Young's modulus has been estimated to follow a similar pattern as that of change in radius, due to the fact that change in radius is due to the accumulation of the hard plaques, thus hardening the arteries near the stenosis.

$$E(x) = E_0 + \frac{E_m}{2}\left[1 - cos\left(2\pi\frac{x - L_m - L_s/2}{L_s}\right)\right], \tag{9}$$

$$\text{for} \qquad (L_m - L_s/2) < x < (L_m + L_s/2).$$

where $E_m$ denotes the maximum increment in Young's modulus and correspondingly the hardness in the arterial walls, hence affecting the value of $\beta$. It is expressed as $E_m = (\alpha - 1)E$, where $\alpha$ is a constant determining the stiffness variation and $\alpha > 1$ for hard plaques and chosen to be a real number very close to unit.

Now with all the variables defined, the three factors that mainly decide the effect of stenosis are the position of the stenosis ($L_m$), length of the stenosis ($L_s$) and mostly the severity of stenosis ($S$). An extensive work has been carried out on these three variables considering both Newtonian and non-Newtonian effects on the blood flow through a stenosed carotid artery and given the period of the cycle, i.e. T = 0.9 s, the beginning and the peak of the systole have been recorded to be 0.1 s and 0.25 s, respectively.

In Figs. 4, 5 and 6, while the length of the stenosis ($L_S$) has been kept constant (10% of the total length of the artery), the position of the stenosis ($L_m$) has been varied and thus its effects on the pressure has been observed, both in the beginning and during the peak of systole. It can be clearly seen that after a certain time, the position of the stenosis doesn't alter the variations in pressure waveform, as for the elapsed time in reaching the peak of systole, the flow would have developed. But,

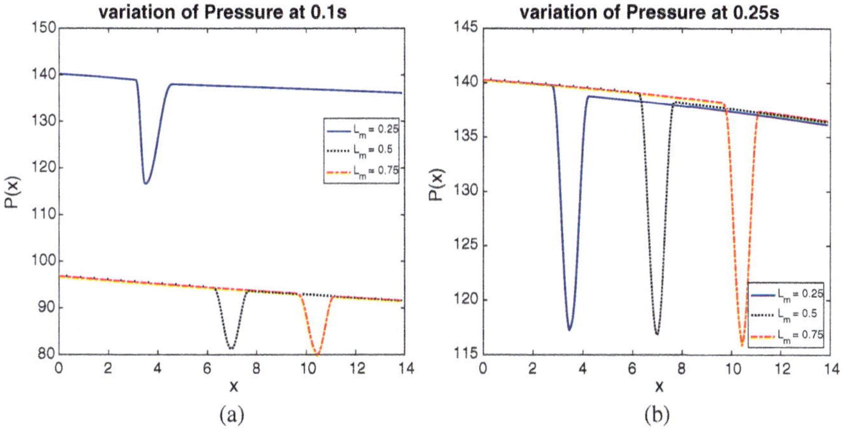


**Fig. 4** For 10% stenosis for varied positions of centre of stenosis ($L_m$)

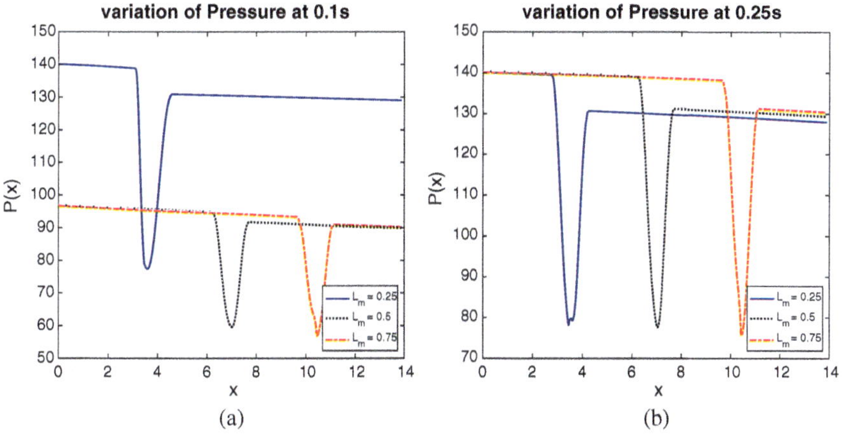


**Fig. 5** For 25% stenosis for varied positions of centre of stenosis ($L_m$)

for the initial time, the stenosis in the beginning of the domain, i.e. at $0.25L_m$, has an evident effect on the increase of pressure, along with extent of decrease in pressure in the stenosed region. With these findings, and the clinical readings, a prediction can be made as to where the stenosis is located in the given artery. The same can be observed in Figs. 7, 8 and 9, where the parameter change has been made in the initial setting with $L_s$, the length of the stenosis has been considered to be 20% of the total length of the domain and the rest has been carried out the same way. Similar results have been observed here as before, although to some small extent there has been an additional dip in the pressure waveform around the stenosed region, which has been specifically worked with and shown in Fig. 14. Here, the position of stenosis, $L_m$ is centred and maintained constant and in each case, keeping the percentage of

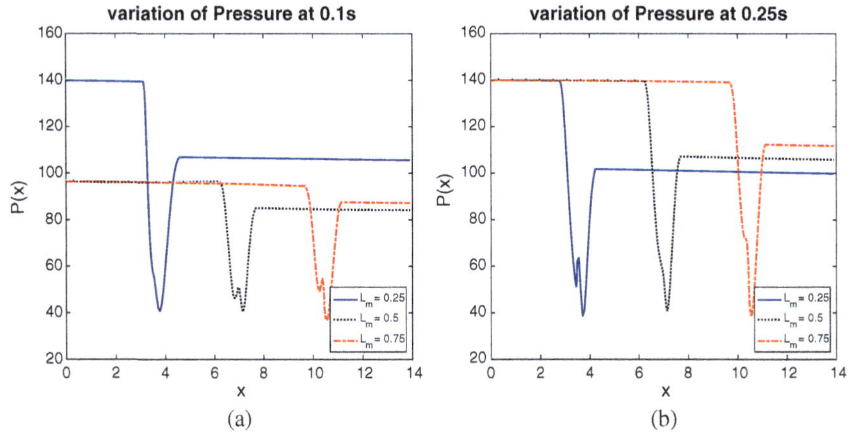


**Fig. 6** For 50% stenosis for varied positions of centre of stenosis ($L_m$)

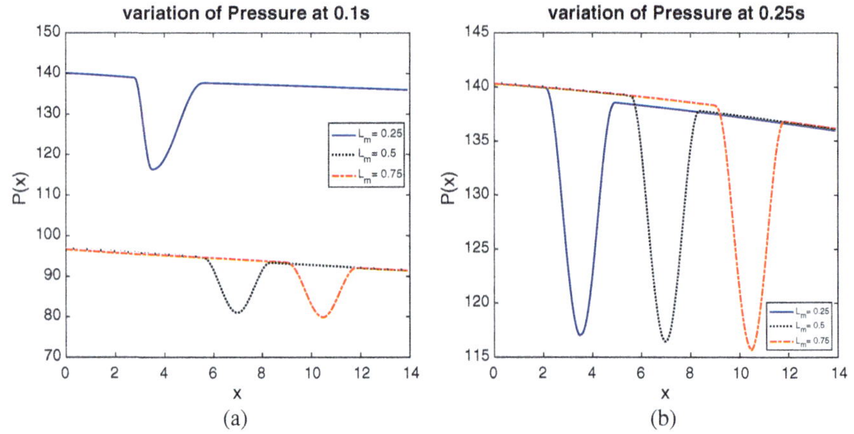


**Fig. 7** For 10% stenosis for varied positions of centre of stenosis ($L_m$)

stenosis, $S$ as constant, the length of the stenosis, $L_s$ has been treated as the variable, by considering, 10%, 30% and 50% of the total length of the domain. All the readings here have been taken at the peak of the systole.

It can be observed that with the increase in the percentage of the stenosis, the accountability of variations in the length of the stenosis, becomes less significant. This is because in each of the three sub-cases, with the variations in $L_s$, the ratio of distal pressure over the proximal pressure is almost the same. Whereas, over the three cases, wherein the percentage of stenosis, $S$ has been varied, the same ratio seems to change and decrease. So, the percentage of the stenosis, $S$ taking the prominence, its variations and thus its implication can be seen through Figs. 10, 11, 12 and 13. Position of the stenosis, $L_m$ is maintained constant and the reading has been taken at

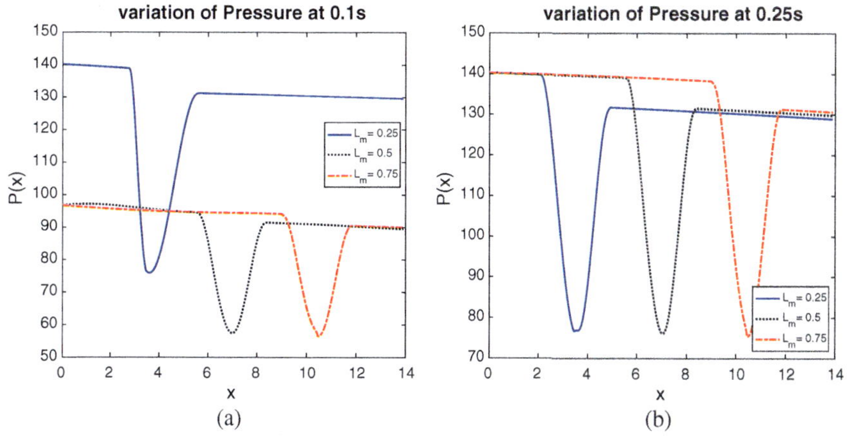


**Fig. 8** For 25% stenosis for varied positions of centre of stenosis ($L_m$)

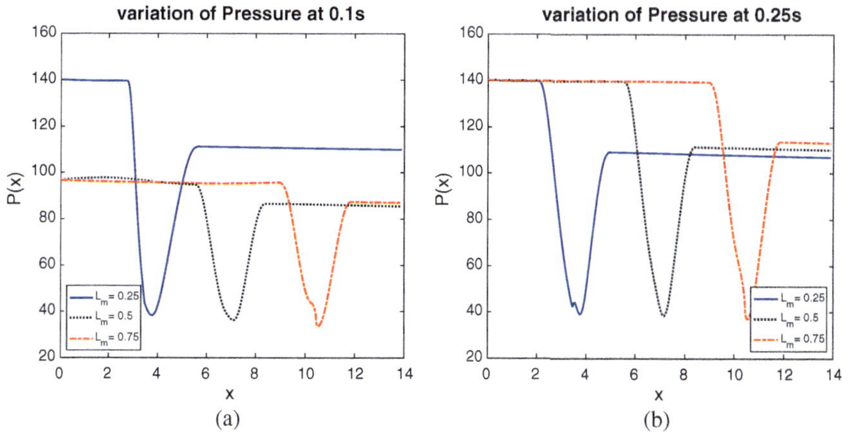


**Fig. 9** For 50% stenosis for varied positions of centre of stenosis ($L_m$)

the peak of the systole, and a comparison has been drawn between Newtonian and Carreau models for blood flow. It can be observed that the Carreau model compared to the Newtonian model, offers a smoother transition of pressure waveform between the pre-stenosis region and post-stenosis region, which presents, the distal and proximal pressure waveform.

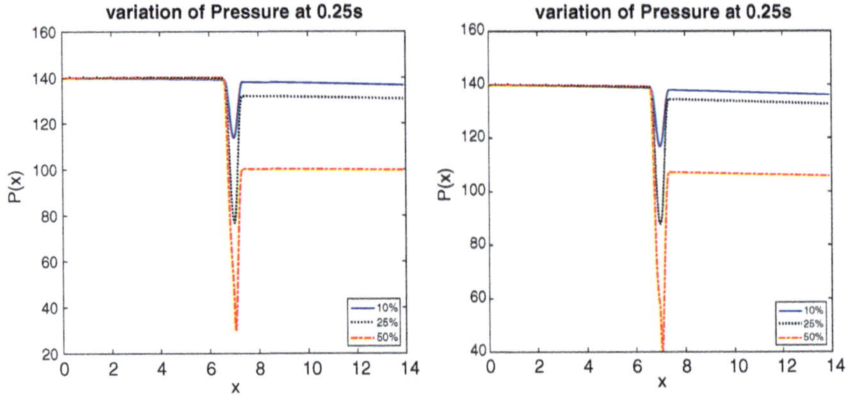


(a) Newtonian Model (b) Carreau Model

**Fig. 10** Length of Stenosis ($L_s$) is 5% of total length for varied percentage of stenosis ($S$)

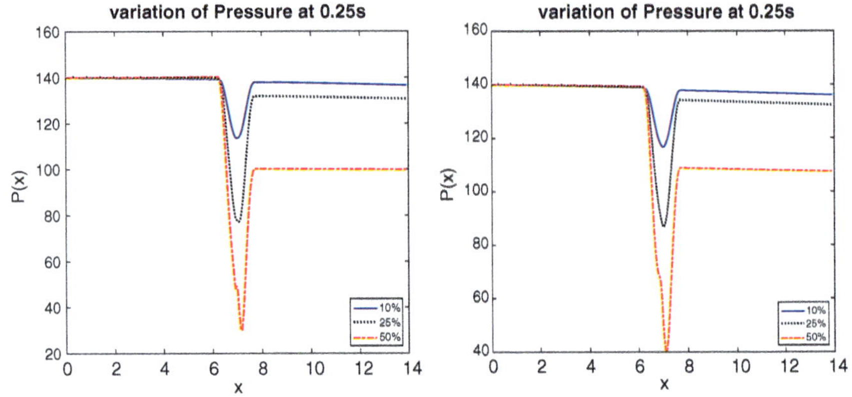


(a) Newtonian Model (b) Carreau Model

**Fig. 11** Length of Stenosis ($L_s$) is 10% of total length for varied percentage of stenosis ($S$)

## 7 Conclusion

In this paper, the importance of non-Newtonian models for studying blood flow has been investigated, depending on the dimensions of the physical arteries they have been transmitted through. It has been observed that length, diameter and the stiffness factor of the artery play a combined role in determining the appropriate non-Newtonian model for the flow. Based on the inference drawn from this work, a suitable non-Newtonian model that of Carreau, has been chosen for understanding the effect of stenosis in the carotid artery. Upon further analysis with carotid artery, which has been a major region of concern for stenosis amongst the arteries, three assigned variables to stenosis and their impact on pressure waveform have been compared.

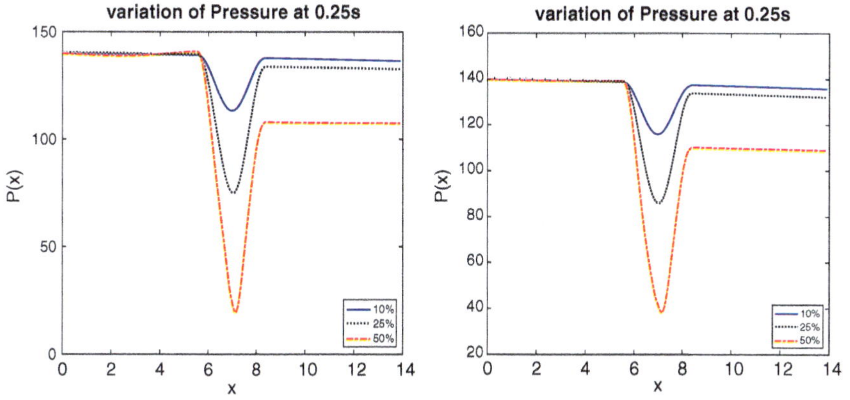


(a) Newtonian Model (b) Carreau Model

**Fig. 12** Length of Stenosis ($L_s$) is 20% of total length for varied percentage of stenosis ($S$)

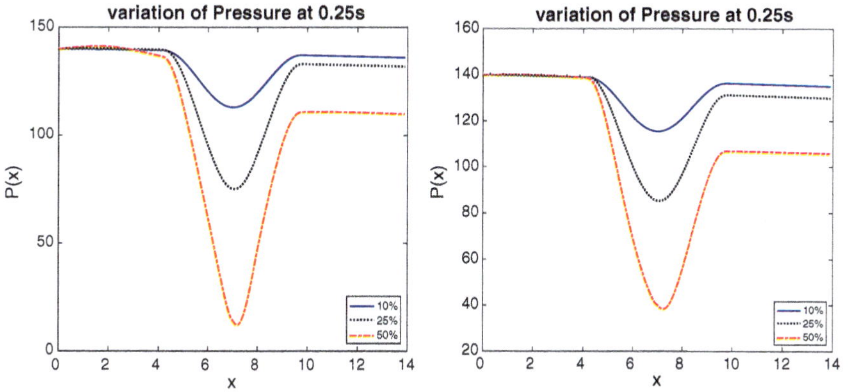


(a) Newtonian Model (b) Carreau Model

**Fig. 13** Length of Stenosis ($L_s$) is 40% of total length for varied percentage of stenosis ($S$)

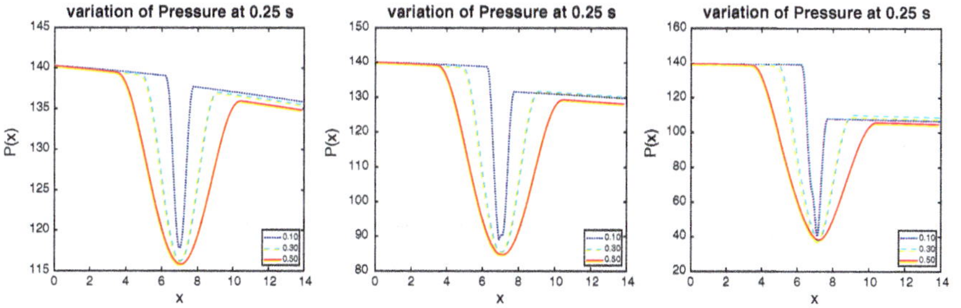


(a) 10% Stenosis (b) 25% Stenosis (c) 50% Stenosis

**Fig. 14** Length of the Stenosis ($L_s$) is varied in each case

The importance of extent of stenosis ($S$), being predominant over the position ($L_m$) and the length of the stenosis ($L_m$) has been observed by numerically computing the blood flow through the stenosed geometry using the HLL Scheme. It has been observed that the Carreau model provides a smoother solution compared to that provided by the Newtonian model. A much smoother solution can be expected with the implementation of finite element method, and also an investigation of stenosis in the coronary arteries is also equally vital, which are considered to be part of the future works. As for clinical application, the present work forms an indicative index for pressure drop under stenosis in the carotid artery.

**Acknowledgements** We thank the Department of Science and Technology—Science and Engineering Research Board (DST-SERB), India, for financial support under grant no. ECR/2017/002049.

## References

1. Liepsch, D.: An introduction to biofluid mechanics-basic models and applications. J. Biomech. **35**, 415–435 (2002)
2. Fuster V., Stein B., Ambrose J. A., Badimon L., Badimon J. J., Chesebro J. H.: Atherosclerotic Plaque Rupture and Thrombosis, Circulation, Supplement II, 82, No. 3, pp. II-47-II-59 (1990)
3. Burke, A.P., Farb, A., Malcom, G.T., Liang, Y.H., Smialek, J.E., Virmani, R.: Plaque Rupture and Sudden Death Related to Exertion in Men with Coronary Artery Disease. J. Am. Med. Assoc. **281**(10), 921–926 (1999)
4. Deshpande, M.D., Giddens, D.P., Mabon, F.R.: Steady laminar flow through modelled vascular stenoses. J. Biomech. **9**, 165–174 (1976)
5. Smith, F.T.: The separation flow through a severely constricted symmetric tube. J. Fluid Mech. **90**, 725–754 (1979)
6. Zendehbudi G. R., Moayer M. S.: Comparison of physiological and simple pulsatile flows through stenosed arteries. Journal of Biomechanics Volume 32, Issue 9 , Pages 959-965 https://doi.org/10.1016/S0021-9290(99)00053-6 (1999)
7. Biswas, D., Chakraborty, U.S.: Pulsatile Flow of Blood in a Constricted Artery with Body Acceleration **4**(2), 329–342 (2009)
8. Zhang J., Zhong L., Luo T., et al.: Numerical Simulation and Clinical Implications of Stenosis in Coronary Blood Flow Hindawi Publishing Corporation, BioMed Research International. Volume 2014, Article ID 514729, https://doi.org/10.1155/2014/514729 (2014)
9. Johnston, B.M., Johnson, P.R., Corney, S., Kilpatrick, D.: Non-Newtonian blood flow in human right coronary arteries: steady state simulations. J. of Biomechanics **37**, 709–720 (2004)
10. Mandal P., Chakravarty S., Mandal A., Norsarahaida A.: Effect of body acceleration on unsteady pulsatile flow of non-Newtonian fluid through a stenosed artery. Applied Mathematics and Computation. 189. 766-779. 10.1016/j.amc.2006.11.139. (2007)
11. Sankar S., Lee U.: Mathematical modeling of pulsatile flow of non-Newtonian fluid in stenosed arteries. Communications in Nonlinear Science and Numerical Simulation. 14. 2971-2981. 10.1016/j.cnsns.2008.10.015. (2009)
12. Onaizah O., et al.: A model of blood supply to the brain via the carotid arteries: Effects of obstructive vs. sclerotic changes, Medical Engineering and Physics https://doi.org/10.1016/j.medengphy.2017.08.009 (2017)
13. Zhang D., Xu P., Qiao H., et al.: Carotid DSA based CFD simulation in assessing the patient with asymptomatic carotid stenosis: a preliminary study. BioMed Eng OnLine 17, 31. https://doi.org/10.1186/s12938-018-0465-9 (2018)

14. Xiao N., Alastruey J., Figueroa C. A.: A Systematic Comparison between 1D and 3D Hemodynamics in Compliant Arterial Models. Int. J. Numer. Meth, Biomed. Engg.; 30: 204-231, Wiley Online Library. https://doi.org/10.1002/cnm.2598(2013)
15. Leveque R. J.: Finite Volume Methods for Hyperbolic Problems Cambridge University Press (2002)
16. Harten, A., Lax, P.D., van Leer, B.: On upstream differencing and Godunov-type schemes for hyperbolic conservation laws. SIAM Rev. **25**, 35–61 (1983). https://doi.org/10.1137/1025002
17. Mynard, J.P., Nithiarasu, P.: A 1D arterial blood flow model incorporating ventricular pressure, aortic valve and regional coronary flow using the locally conservative Galerkin (LCG) method. Commun. Nmer. Meth. Engng **24**, 267–417 (2008). https://doi.org/10.1002/cnm.1117
18. Smith N. P., Pullan A. J., Hunter P. J.: An approximation based model of coronary blood flow and myocardial mechanics. SIAM J. Appl. Math., (2002)
19. Rabby, M.G., Shupti, S.P., Molla Md. M.: Pulsatile Non-Newtonian Laminar Blood Flows through Arterial Double Stenoses Volume 2014, Article ID 757902, 13 pages. https://doi.org/10.1155/2014/757902 (2014)
20. Low, K., Loon, R.v., Sazonov, I., Bevan, R.L.T., Nithiarasu, P.: An improved baseline model for a human arterial network to study the impact of aneurysms on pressure-flow waveforms. Int. J. Numer. Meth. Biomed. Engng. **28**, 1224–1246 (2012)https://doi.org/10.1002/cnm.2533(2012)
21. Courant, R., Friedrichs, K., Lewy, H.: On the partial difference equations of mathematical physics. IBM J. Res. Devel. **11**(2), 215–234 (1967). https://doi.org/10.1147/rd.112.0215
22. Wiwatanapataphee, B., Poltem, D., Wu, Y.H., Lenbury, Y.: Simulation of pulsatile flow of Blood in Stenosed Coronary artery bypass with Graft. Math. Biosci. Eng. **3**(2), 371–383 (2006)
23. Thompson, K.W.: Time dependent boundary conditions for hyperbolic systems. J. Comp. Phys. **68**, 1–24 (1987)

# Two-Layered Flow of Ionized Gases Within a Channel of Parallel Permeable Plates Under an Applied Magnetic Field with the Hall Effect

**M. Nagavalli, T. LingaRaju, and Peri K. Kameswaran**

**Abstract** When a strong magnetic field and Hall currents are present, the flow of ionized gases between two parallel permeable plates in a horizontal channel is explored. Electrical conductivity and incompressibility are also expected of the two fluids. Analytical solution to the governing differential equations using the specified boundary and interface restrictions yields exact answers for velocity distributions—primary/secondary distributions in two locations. Their numerical findings for several sets of governing parameter values are derived to visually illustrate them and are examined.

**Keywords** MHD flows · Immiscible flow · Plasma · Hall effect · Porous plates (Non-Conducting and Conducting)

## 1 Introduction

Several experimental and theoretical studies on two-phase /or two-layered flow of classical hydrodynamic problems have been undertaken in the past by many investigators, including Zuber [1], Packham and Shail [2], Golding and Mah [3], Oshinowo and Charles [4], Jones and Zuber [5], Shipley [6], and many more. However, the need for current technology has increased interest in magnetohydrodynamic two-phase/two-layered flow research due to its vast applications in the industry linked to different energy conversion systems. As an example, fusion reactors and MHD

M. Nagavalli (✉)
Department of Engineering Mathematics, Andhra University College of Engineering for Women, Visakhapatnam 530017, India
e-mail: nagavallicf@andhrauniversity.edu.in; nagavalli.malisetty@gmail.com

T. LingaRaju
Department of Engineering Mathematics, AUCE (A), Andhra University, Visakhapatnam 530003, India
e-mail: prof.tlraju@andhrauniversity.edu.in

P. K. Kameswaran
Department of Mathematics, School of Advanced Sciences, Vellore Institute of Technology, Vellore 632014, India



B. Rushi Kumar et al. (eds.), *Mathematics and Computing*, Springer Proceedings in Mathematics & Statistics 415, https://doi.org/10.1007/978-981-19-9307-7_43

power producers are being explored on a conceptual level demanding an accurate and reliable knowledge of the thermohydraulic mechanics of two-phase/two-layered flow in the presence of an applied magnetic field. MHD two-phase flow phenomena have rapidly become an important subject of study for both academic and scientific groups, MHD power generation model, and so on [see 7–20].

All of the above-mentioned tests do not take Hall currents into account. Hall currents must be taken into consideration that the magnetic field is extremely strong, as is widely known in the literature. Ohm's law must be modified to account for these effects. There are numerous studies that have looked into the effects of Hall currents under the control of a powerful magnetic field on specific single-fluid flow systems [see 21–28]. According to Sato's [29] research, The effects of Hall currents on the viscous flow of an ionized gas between two parallel walls under the influence of an applied transverse magnetic field were investigated by LRaju and Rao [30]. LRaju's recent work [31] demonstrates that with Hall currents, MHD heat transfer of two-ionized fluids flowing between two parallel plates is possible. Accordingly, this work makes an attempt to solve the problem of two-layered ionized gas flow along a horizontal channel contained by two parallel porous plates and subjected to an applied transverse strong magnetic field by taking into account Hall currents in the light of what has been said thus far. An application in Hall accelerators, pumps and flow meters, MHD power generators, plasma jets and space craft, as well as in nuclear reactor cooling, is anticipated from the aforementioned theoretical investigation.

## 2 Formulation of the Problem

Using Hall currents, we model a "two-fluid, magnetohydrodynamic (MHD) two-dimensional steady flow of an ionize gas impelled by a common constant gradient of pressure $-{}^{\partial p}/_{\partial x}$ in a horizontal channel surrounded by two parallel porous plates" that extend along both $x$ and $z$ directions (Fig. 1). A constant "suction $v_0$ is applied normal to both plates, thus if $(u_i,\ v_i,\ w_i)$, $(\ i = 1, 2\ )$, are the velocity components in the two fluids, then the equation of continuity $\nabla.\overline{q_i} = 0$ provides $v_i = -v_0(v_0 > 0)$", where $\overline{q}_i = (u_i,\ v_i,\ w_i)$. A constant magnetic field $\mathbf{B_0}$ is applied in the "$y$-direction, that is, transverse to the flow field". Parallel to channel plates, the $x$-axis is measured in the direction of hydrodynamic pressure gradient, but not flow. "The regions I and II refer to the upper and lower fluids" in the areas $0 \leq y \leq h_1$ and $-h2 \leq y \leq 0$. "Two immiscible electrically conducting incompressible fluids occupy" regions I and II, each with a different density $\rho_1$, $\rho_2$, viscosity $\mu_1$,$\mu_2$, and electrical conductivity $\sigma_{01}$, $\sigma_{02}$. The assumption is that the channel width is much larger than the channel height. Aside from pressure, "all physical quantities will be functions of y" alone. A flat, stress-free, and undisturbed contact is assumed between the two immiscible fluids. The channel's edges are immovable. A low magnetic Reynolds number is also expected. To delineate the major equations for the two-layered flow, it is taken that the "velocity $\overline{V_i} = (u_i,\ v_0,\ w_i)$, magnetic

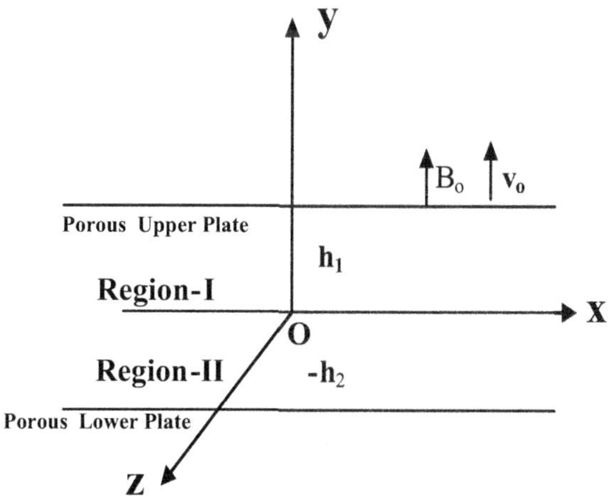


**Fig. 1** Flow diagram

field $\overline{B} = (0,\ B_0,\ 0)$, current density $\overline{J_i} = (J_{ix}, 0, J_{iz})$, and the electric field as $\overline{E_i} = (E_{ix}, 0, E_{iz}), (i = 1,\ 2)$" in basic equations.

## 3 Mathematical Study of the Problem

The motion and current equations for MHD two-layered flow of neutral fully-ionized gas legitimate under the suppositions above are simplified (Sato [29], LRaju and Rao [30] and LRaju [31]) as follows:

**Region-I**

$$-[1 - s(1 - \frac{\sigma_{11}}{\sigma_{01}})]\frac{\partial p}{\partial x} + \rho_1 \nu_1 \frac{d^2 u_1}{dy^2} - \rho_1 v_0 \frac{\partial u_1}{\partial y} + B_0[-\sigma_{11}(E_z + u_1 B_0) + \sigma_{21}(E_x - w_1 B_0)] = 0 \tag{1}$$

$$\left(s\frac{\sigma_{21}}{\sigma_{01}}\right)\frac{\partial p}{\partial x} + \rho_1 \nu_1 \frac{d^2 w_1}{dy^2} - \rho_1 v_0 \frac{\partial w_1}{\partial y} + B_0[\sigma_{11}(E_x - w_1 B_0) + \sigma_{21}(E_z + u_1 B_0)] = 0 \tag{2}$$

$$J_x = \sigma_{11} E_x - B_0 \sigma_{11} w_1 + \sigma_{21} E_z + B_0 \sigma_{21} u_1 + \frac{s\sigma_{21}}{\sigma_{01} B_0}\left(\frac{\partial p}{\partial x}\right) \tag{3}$$

$$J_z = \sigma_{11}\left(\frac{E_z}{B_0} + u_1\right) - \sigma_{21}\left(\frac{E_x}{B_0} - w_1\right) - \frac{s}{B_0}\left(1 - \frac{\sigma_{11}}{\sigma_{01}}\right)\left(\frac{\partial p}{\partial x}\right) \tag{4}$$

**Region-II**

$$-\left[1-s\left(1-\frac{\sigma_{12}}{\sigma_{02}}\right)\right]\frac{\partial p}{\partial x}+\rho_2 \nu_2\frac{d^2u_2}{dy^2}-\rho_2 \mathrm{v}_0\frac{\partial u_2}{\partial y} + B_0\left[-\sigma_{12}(E_z+u_2B_0)+\sigma_{22}(E_x-w_2B_0)\right]=0 \quad (5)$$

$$\left(s\frac{\sigma_{22}}{\sigma_{02}}\right)\frac{\partial p}{\partial x}+\rho_2 \nu_2\frac{d^2w_2}{dy^2}-\rho_2 \mathrm{v}_0\frac{\partial w_2}{\partial y} + B_0\left[\sigma_{12}(E_x-w_2B_0)+\sigma_{22}(E_z+u_2B_0)\right]=0 \quad (6)$$

$$J_x=\sigma_{12}E_x-B_0\sigma_{12}w_2+\sigma_{22}E_z+B_0\sigma_{22}u_2+\frac{s\sigma_{22}}{\sigma_{02}B_0}\left(\frac{\partial p}{\partial x}\right) \quad (7)$$

$$J_z=\sigma_{12}E_z+\sigma_{12}B_0u_2-\sigma_{22}E_x+\sigma_{22}B_0w_2-\frac{s}{B_0}\left(1-\frac{\sigma_{12}}{\sigma_{02}}\right)\left(\frac{\partial p}{\partial x}\right) \quad (8)$$

So, the boundary and interface conditions on $u_1$, $w_1$ and $u_2$, $w_2$ become

$$u_1(h_1)=0,\ w_1(h_1)=0,\ u_2(-h_2)=0,\ and\ w_2(-h_2)=0. \quad (9)$$

$$u_1(0)=\ u_2(0)\ and\ w_1(0)=w_2(0). \quad (10)$$

$$\mu_1\frac{du_1}{dy}=\mu_2\frac{du_2}{dy}\ and\ \mu_1\frac{dw_1}{dy}=\mu_2\frac{dw_2}{dy}at\ y=0. \quad (11)$$

Further to make Eqs. (1) to (8) and (9) to (11) dimensionless, the following non-dimensional variables are used:

$$u_1^\bullet=\frac{u_1}{u_p},\ u_2^\bullet=\frac{u_2}{u_p},\ y_i^\bullet=\left(\frac{y_i}{h_i}\right),\ w_1^\bullet=\frac{w_1}{u_p},\ w_2^\bullet=\frac{w_2}{u_p},\ u_p=\left(-\frac{\partial p}{\partial x}\right)\frac{h_1^2}{\rho_1\nu_1},$$

$$k_1=1-s\left(1-\frac{\sigma_{11}}{\sigma_{01}}\right),\ k_2=\frac{-s\sigma_{21}}{\sigma_{01}},\ m_{ix}=\frac{E_{ix}}{B_0u_p},\ m_{iz}=\frac{E_{iz}}{B_0u_p},$$

$$I_{ix}=\frac{J_{ix}}{\sigma_{0i}B_0u_p},\ h=\frac{h_2}{h_1},\ M^2=B_0^2h_1^2\left(\frac{\sigma_{01}}{\rho_1\upsilon_1}\right),\ \lambda=\frac{h_1\rho_1\upsilon_0}{\mu_1},$$

$$\alpha=\frac{\mu_1}{\mu_2},\ \sigma_0=\frac{\sigma_{01}}{\sigma_{02}},\ \sigma_1=\left(\frac{\sigma_{12}}{\sigma_{11}}\right),$$

$$\sigma_2=\left(\frac{\sigma_{22}}{\sigma_{21}}\right),\ \frac{1}{1+m^2}=\frac{\sigma_{11}}{\sigma_{01}},\ \frac{m}{1+m^2} =\frac{\sigma_{21}}{\sigma_{01}},\ m={}^{w_e}\!/_{(1/\tau+1/\tau_e)},\ (i=1,2). \quad (12)$$

The "mean collision time between electron and neutral particles and electron and ion" are given by the values of $\tau$ and $\tau_e$, respectively. When $\tau_e$ approaches infinity, the formula for Hall parameter m that is valid for partly ionized gas coincides with the expression for completely ionize gas. These "equations are transformed into non-dimensional forms by using the" transformations (12) and omitting asterisks for the sake of convenience.

**Region–I**

$$k_1 + \frac{d^2u_1}{dy^2} - \lambda\frac{du_1}{dy} - \left(\frac{M^2}{1+m^2}\right)(m_{1z}+u_1) + \left(\frac{mM^2}{1+m^2}\right)(m_{1x}-w_1) = 0 \quad (13)$$

$$k_2 + \frac{d^2w_1}{dy^2} - \lambda\frac{dw_1}{dy} + \left(\frac{M^2}{1+m^2}\right)(m_{1x}-w_1) + \left(\frac{mM^2}{1+m^2}\right)(m_{1z}+u_1) = 0 \quad (14)$$

$$I_{1x} = \frac{[m_{1x} - w_1 + (m_{1z}+u_1)m - sm/M^2]}{1+m^2} \quad (15)$$

$$I_{1z} = \frac{[m_{1z} + u_1 + (m_{1x}-w_1)m]}{1+m^2} + \frac{s}{M^2}\left(1 - \frac{m}{1+m^2}\right) \quad (16)$$

**Region-II**

$$\beta_1\alpha h^2 + \frac{d^2u_2}{dy^2} - \rho\alpha h\lambda\frac{du_2}{dy} - \left(\frac{1}{1+m^2}\right)\alpha\sigma_1 h^2M^2(m_{2z}+u_2) + \left(\frac{m}{1+m^2}\right)\alpha\sigma_2 h^2M^2(m_{2x}-w_2) = 0 \quad (17)$$

$$\beta_2\alpha h^2 + \frac{d^2w_2}{dy^2} - \rho\alpha h\lambda\frac{dw_2}{dy} + \left(\frac{1}{1+m^2}\right)\alpha\sigma_1 h^2M^2(m_{2x}-w_2) + \left(\frac{m}{1+m^2}\right)\alpha\sigma_2 h^2M^2(m_{2z}+u_2) = 0 \quad (18)$$

$$I_{2x} = \left(\frac{\sigma_0\sigma_1}{1+m^2}\right)(m_{2x}-w_2) + \left(\frac{m\sigma_0\sigma_2}{1+m^2}\right)(m_{2z}+u_2) - \frac{s\sigma_0^2\sigma_2}{M^2}\left(\frac{m}{1+m^2}\right) \quad (19)$$

$$I_{2z} = \left(\frac{\sigma_0\sigma_1}{1+m^2}\right)(m_{2z}+u_2) - \left(\frac{m\sigma_0\sigma_2}{1+m^2}\right)(m_{2x}-w_2) + \frac{s\sigma_0}{M^2}\left(1 - \frac{\sigma_0\sigma_1}{1+m^2}\right) \quad (20)$$

where

$$k_1 = 1 - \frac{m^2s}{1+m^2},\ k_2 = \frac{-ms}{1+m^2},\ \beta_1 = 1 - \left(1 - \frac{\sigma_0\sigma_1}{1+m^2}\right)s,\ \beta_2 = \frac{-\sigma_0\sigma_2 ms}{1+m^2}. \quad (21)$$

Conditions become

$$u_1(1) = 0,\ \ w_1(1) = 0,\ \ u_2(-1) = 0\, and\ w_2(-1) = 0. \tag{22}$$

$$u_1(0) = u_2(0),\ \ w_1(0) = w_2(0) \tag{23}$$

$$At\ \mathrm{y} = 0{:}\frac{du_1}{dy} = \frac{1}{\alpha h}\frac{du_2}{\mathrm{dy}},\ \frac{dw_1}{dy} = \frac{1}{\alpha h}\frac{dw_2}{dy} \tag{24}$$

## 4 Solution to the Problem

The "closed form solutions of the resulting governing differential Eqs. (13), (14), (17) and (18) with the help of (15), (16) and (19), (20) subject to the boundary and interface conditions (22), (24) for the primary and secondary velocities $u_1$, $u_2$ and $w_1$, $w_2$, respectively, also their corresponding mean velocities", viz., $u_{m_1}$, $u_{m_2}$ and $w_{m_1}$, $w_{m_2}$, respectively, in the two regions are obtained. The "solution for the investigated issue is attained in two cases as the plates are made up of non-conducting porous material and the other one conducting type".

### *4.1 Non-conducting Porous Plates*

When the z-direction side plates are maintained far apart and are formed of "the non-conducting porous material, the generated electric current does not exit the channel but circulates inside the fluid". In this way, a new non-dimensional condition for the current is established.

by $\int_0^1 I_{1z}dy = 0$ and $\int_0^1 I_{2z}dy = 0$. "Insulation at large x is also assumed, other relations are obtained as" $\int_0^1 I_{1x}dy = 0$ and $\int_0^1 I_{2x}dy = 0$, (see Sato [29]). Using the above two requirements, we can derive "solutions for $u_1$, $u_2$ and $w_1$, $w_2$, $I_1$ and $I_2$ also their corresponding mean velocity distributions $u_{m_1}$, $u_{m_2}$ and $w_{m_1}$, $w_{m_2}$ in the two regions". The "primary and secondary distributions", as well as currents, are represented by the combined form:

**Region-I**

$$
\begin{aligned}
q_1(y) &= u_1(y) + iw(y) = \frac{A_1(1+m^2)}{M^2(mi-1)} b_5 \\
&+ \frac{A_2(1+m^2)}{M^2(\sigma_1 - mi\sigma_2)} [b_3 e^{f_1 y} + \left(b_3 e^{f_1 - f_2}\right) e^{f_2 y}] \\
&+ iN_2 b_3 (e^{f_1 y} - e^{f y}) + iN_1[-b_4 e^{f_1 y} + \left(e^{-f_2} + b_4 e^{f_1 - f_2}\right) e^{f_2 y} + 1] \quad (25)
\end{aligned}
$$

**Region-II**

$$
\begin{aligned}
q_2(y) &= u_2(y) + iw_2(y) = \frac{A_2(1+m^2)}{M^2(\sigma_1 - mi\sigma_2)} [(b_6 - b_3 b_8) e^{f_3 y} + b_9 e^{f_4 y} + 1] \\
&+ \frac{A_1(1+m^2)}{M^2(mi-1)} [(-b_7 + b_4 b_8) e^{f_3 y} + b_{10} e^{f_4 y}] \\
&+ iN_2[(b_6 - b_3 b_8) e^{f_3 y} + b_{11} e^{f_4 y} + 1] + iN_1[(-b_7 + b_4 b_8) e^{f_3 y} + b_{12} e^{f_4 y}] \quad (26)
\end{aligned}
$$

## 4.2 *Case of Conducting Porous Plates*

The "induced electric current flows out of the channel when the two plates are formed of conducting porous materials and are short-circuited by an external conductor". There is no Electric Potential among the side plates in this situation. If "the electric field is assumed as zero in both the *x*- and *z*-directions, we obtain $m_x = 0$, $m_z = 0$". These two conditions determine the constants in the solution. The following are the "solutions for $u_1$, $u_2$ and $w_1$, $w_2$ in the two regions, as well as, $u_{1m}, u_{2m}$ and $w_{1m}, w_{2m}$, $I_1$ and $I_2$":

**Region-I**

$$
\begin{aligned}
q_1(y) &= u_1(y) + iw_1(y) = a_1 e^{c_7 y} + a_2 e^{c_8 y} + \frac{c_6}{c_5}, \text{ where} \\
u_1(y) &= \frac{q_1 + \overline{q_1}}{2}, \; w_1(y) = \frac{q_1 - \overline{q_1}}{2i} \quad (27)
\end{aligned}
$$

$$
\begin{aligned}
I_1 = I_{1x} + iI_{1z} &= \left(\frac{1}{1+m^2}\right)(iu_1 - w_1) + \left(\frac{m}{1+m^2}\right)(u_1 + iw_1) \\
&- \frac{s}{M^2}\left[\frac{m}{1+m^2} - i\left(1 - \frac{m}{1+m^2}\right)\right] \quad (28)
\end{aligned}
$$

The mean velocity is given by $q_{1m} = u_{1m} + iw_{1m} = \int_0^1 q_1 dy = a_1 a_3 + a_2 a_4 + \frac{c_6}{c_5}$

$q_{1m} = u_{1m} + iw_{1m} = \int_0^1 q_1 dy = a_1 a_3 + a_2 a_4 + \frac{c_6}{c_5}$

where

$$u_{1m} = \frac{q_{1m} + \overline{q_{1m}}}{2},\ w_{1m} = \frac{q_{1m} - \overline{q_{1m}}}{2i} \tag{29}$$

**Region-II**

$$q_2(y) = u_2(y) + i w_2(y) = a_5 e^{c_{12} y} + a_6 e^{c_{13} y} + \frac{c_{11}}{c_{10}} \tag{30}$$

$$I_2 = I_{2x} + iI_{2z} = \left(\frac{\sigma_0\sigma_1}{1+m^2}\right)(iu_2 - w_2) + \left(\frac{m\sigma_0\sigma_2}{1+m^2}\right)(u_2 + iw_2) - \frac{s\sigma_0^2\sigma_2}{M^2}\left(\frac{m}{1+m^2}\right) + \frac{s\sigma_0^2 i}{M^2} - \frac{s\sigma_0^2\sigma_1 i}{(1+m^2)M^2} \tag{31}$$

$$q_{2m} = a_5 a_7 + a_6 a_8 + \frac{c_{11}}{c_{10}},\ \text{where } u_{2m} = \frac{q_{2m} + \overline{q_{2m}}}{2},\ w_{2m} = \frac{q_{2m} - \overline{q_{2m}}}{2i}. \tag{32}$$

where the symbols $A_1$, $A_2$,…, $c_1$, $c_2$.., $d_1$, $d_2$.. are being utilized for simplicity, and their expressions are "omitted here as they are too lengthy".

## 5 Results and Discussion

The problem of two-fluid layering and transverse magnetic field effects on an ionize gas flow through porous plates in a horizontal channel is studied analytically in this paper under two cases "when the plates are made up of non-conducting and conducting porous materials". Differential "equations are solved to obtain closed-form solutions for both primary and secondary" velocities distributions; "for various sets of values of the governing parameters" are determined to represent their profiles, as depicted in Figs. 2 through 13. We also discussed "the effect of flow parameters, such as the Hartmann number M, Hall parameter m, porous parameter λ on the flow fields". Taken $\sigma_0 = 1, \sigma_1 = 1.2$, and $\sigma_2 = 1.5$ $\rho = 1$ in all the numerical estimations, the effect of other important parameters on the flow was analyzed. The solutions are found to be independent of s = ratio of "electron pressure to the total pressure in case of non-conducting porous plates and are dependent on 's' when the plates are conducting ($i = 1,\ 2$). The results coincide with those of LRaju [31] "when $\lambda = 0$ (non-porous plates) and the plates are non-conducting".

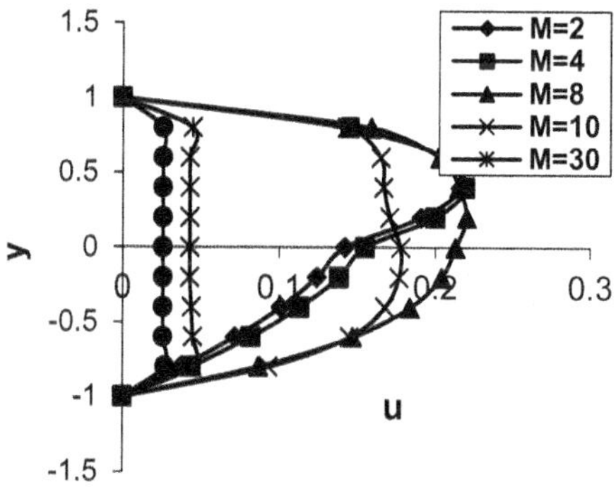


**Fig. 2** Primary velocity profile for various 'M' and $\alpha = 0.333$, $h = 0.75$, $m = 2$, $\rho = 1$, $\sigma 0 = 1$, $\sigma 1 = 1.2$, $\sigma 2 = 1.5$, $\lambda = 2$ (Non-Conducting porous plates)

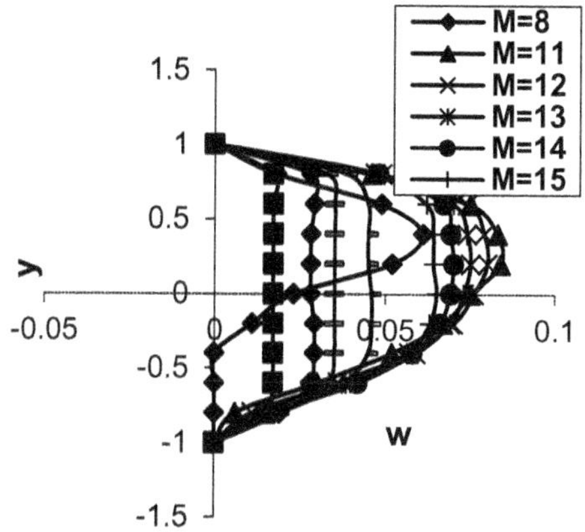


**Fig. 3** Scondary velocity profile for various'M' and $\alpha = 0.333$, $h = 0.75$, $m = 2$, $\rho = 1$, $\sigma_0 = 1$, $\sigma_1 = 1.2$, $\sigma_2 = 1.5$, $\lambda = 2$ (Non-Conducting porous plates)

### *5.1 Case of Non-conducting Porous Plates*

Figures 2, 3, 4, 5, 6 and 7 show the distribution of velocity profiles. Changing the Hartmann number M has a noticeable outcome on the velocity distribution in both regions, even when all other parameters are held constant, as seen in Figs. 2 and 3. The primary velocity distributions improve with an increase in Hartmann M as shown in the Figure. It can be expressed physically that with an increase in the Hartmann number increases the magnetic field's strength in both zones (as shown in Fig. 2), while $M > 8$ reduces them (as shown). Figure 3 shows that secondary velocity distributions increase in both areas when the Hartmann number M augments and decrease in both areas when $M > 11$. Increasing M causes a shift in the channel's most primary and secondary distributions toward the region-I. These findings show that the magnetic field has a stronger influence on the velocity profile.

Figures 4 and 5 display the impact of changing hall factor m on the distribution of primary and secondary components. While in the first region, it seems to be decreasing with an increase in 'm'. Figure 4 shows that it is increasing in the second zone with a rise in 'm' up to a certain point, say 3, and then decreasing. The secondary

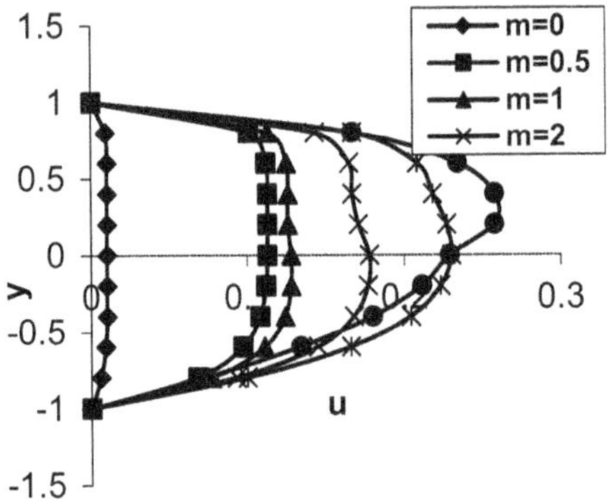


**Fig. 4** Primary velocity profile for various 'm' and M = 10, h = 0.75, $\rho = 1$, $\sigma_0 = 1$, $\sigma_1 = 1.2$, $\sigma_2 = 1.5$, $\alpha = 0.333$, $\lambda = 2$. (Non-Conducting porous plates)

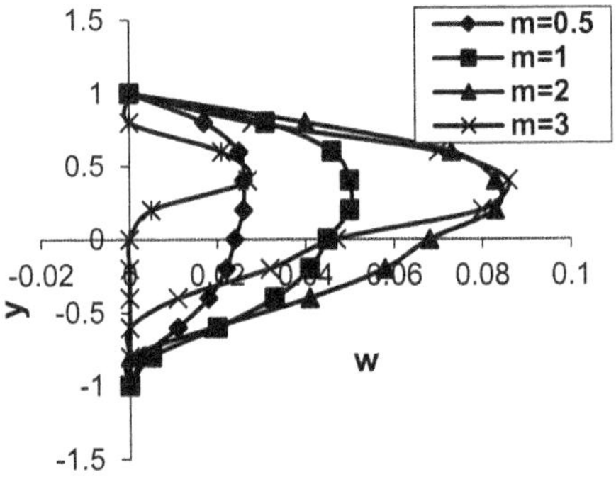


**Fig. 5** Secondary velocity profile for various 'm' and M = 10, h = 0.75, $\rho = 1$, $\sigma_0 = 1$, $\sigma_1 = 1.2$, $\sigma_2 = 1.5$, $\lambda = 2$, $\alpha = 0.333$ (Non-Conducting porous plates)

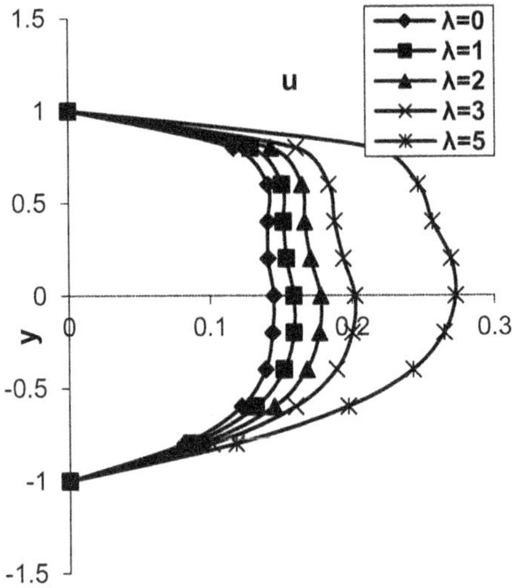


**Fig. 6** Primary velocity profile for various $\lambda$ and M = 10, m = 2, $\rho = 1$, $\sigma_0 = 1$, $\sigma_1 = 1.2$, $\sigma_2 = 1.5$, $\alpha = 0.333$, h = 0.75 (Non-Conducting porous plates)

**Fig. 7** Secondary velocity profile for various λ and M = 10, m = 2, ρ = 1, σ0 = 1, σ1 = 1.2, σ1 = 1.5,α = 0.333, h = 0.75 (Non-Conducting porous plates)

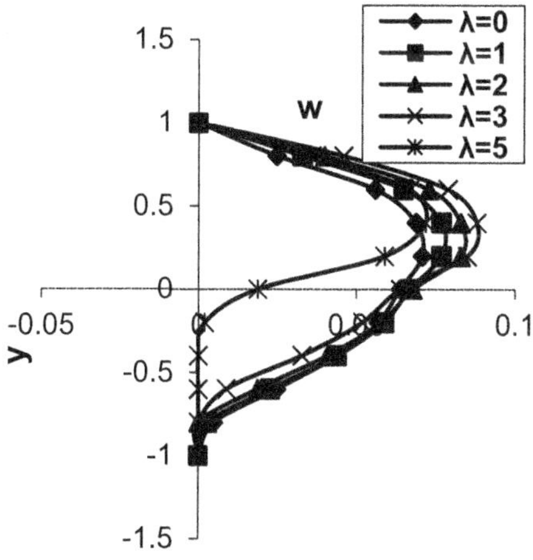


velocity distribution in Fig. 5 grows up to a value of 3 and then decreases, whereas it increases when m raises up to 2, and after that decreases in the second zone. This could be owing to the small retarding force caused by the interaction of the applied magnetic field and the Hall current in electrically conducting fluids acting in the y-direction.

Figures 6 and 7 show how the porosity parameter λ affects the "primary and secondary velocity distributions" within the zones. Increases in primary velocities in two areas up to porosity parameter λ = 5 are followed by decreases, as seen in Fig. 6. It can be explained that when the porosity parameter increases, the fluid has more space to move, and as a result, the velocity rises. With respect to the first area, it is clear from Fig. 7 that increasing secondary velocity distribution diminishes the same when λ > 3 in the first region, whereas in the second region it decreases as λ increases.

## 5.2 *Conducting Porous Plates*

Figures 8 to 13 illustrate profiles for distributions (primary and secondary velocity distributions) in circumstance, that is, when s = 0.

**(i) When the ionization parameter s = 0**.

Changing the Hartmann number M has a noticeable impact on the velocity distribution in both sites, as seen in Figs. 8 and 9. Figure 8 shows that as M grows, the major velocity distributions in the two zones become more skewed. Figure 9 shows that when M grows, the secondary velocity distributions also expand in size.

Variations in the hall parameter 'm' affect velocity distributions for both sites shown in Figs. 10 and 11. Figures 10 and 11 demonstrate that how increases in

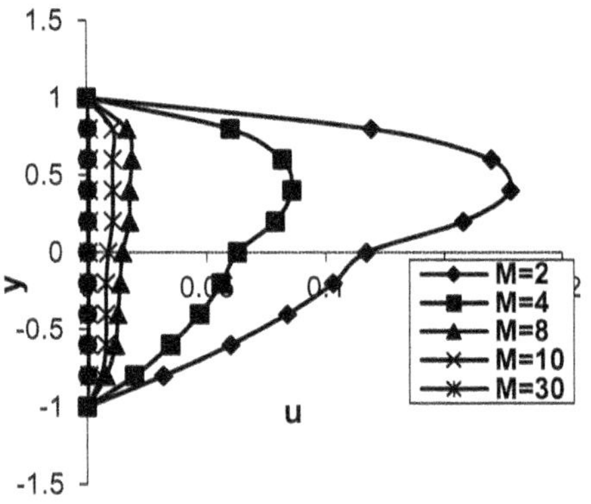


**Fig. 8** Primary velocity profile for various 'M' and h = 0.75, m = 2, $\alpha = 0.333$, $\rho = 1$, $\sigma_0 = 1$, $\sigma_1 = 1.2$, $\sigma_2 = 1.5$, $\lambda = 2$, s = 0 (Conducting porous plates)

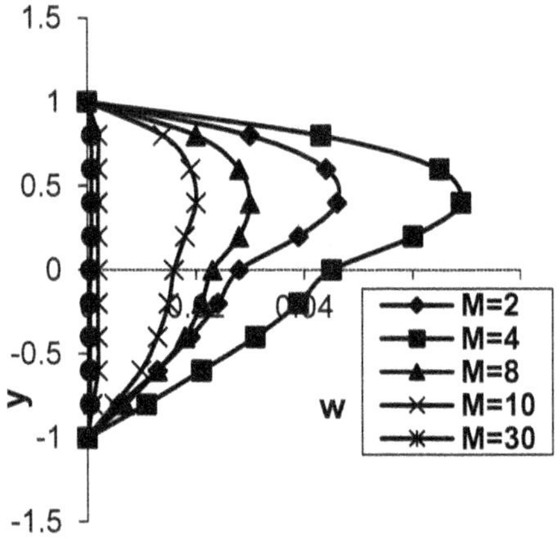


**Fig. 9** Secondary velocity profiles for various 'M' and h = 0.75, m = 2, $\alpha = 0.333$, $\rho = 1$, $\sigma_0 = 1$, $\sigma_1 = 1.2$, $\sigma_2 = 1.5$, $\lambda = 2$, s = 0 (Conducting porous plates)

'm' improve the main and secondary velocity distributions in the two locations. Two different locations are shown in Figs. 12 and 13 to demonstrate the impact of adjusting the porosity parameter on velocity patterns. Two sites demonstrate an increase in primary velocity distribution with an increase in porosity parameter. The increased porosity parameter in Fig. 13 causes the secondary velocity to increase in the top plate zone, but decreases everywhere else.

**For (ii) when the ionization parameter s = 1/2.**

It is observed that, when M grows, the main velocity distributions in both areas also climb. A fall in the secondary distribution is seen for M > 8, while a growth in the secondary velocity profile is shown for M > 8 in the first area. Variations in the hall parameter 'm' have an impact on both the main and secondary velocity profiles.

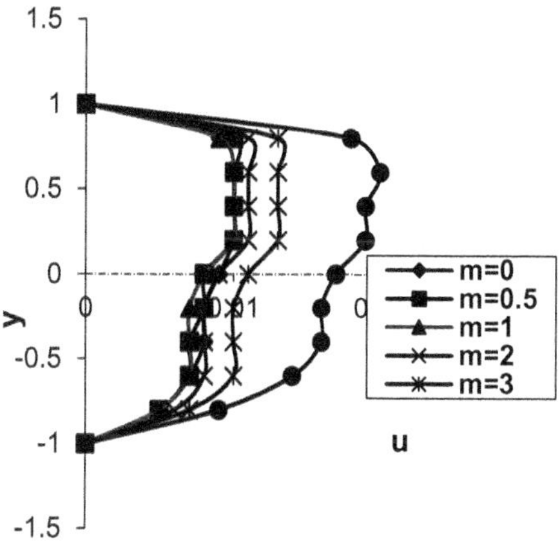


**Fig. 10** Primary velocity profile for various 'm' and M = 10, h = 0.75, ρ = 1, $\sigma_0$ = 1, $\sigma_1$ = 1.2, $\sigma_2$ = 1.5, α = 0.333, λ = 2, s = 0 (Conducting porous plates)

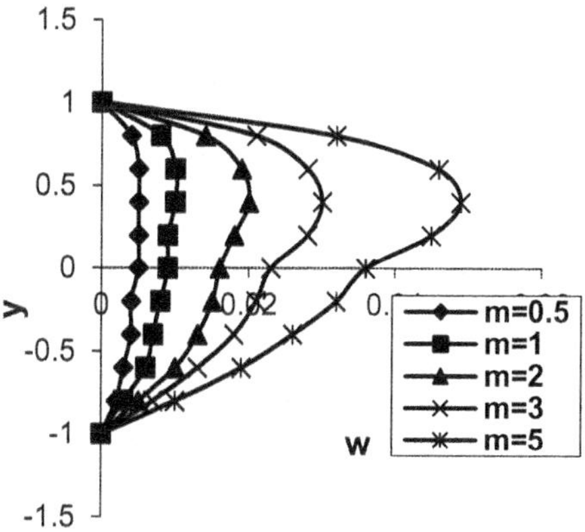


**Fig. 11** Secondary velocity profile for various 'm' and M = 10, h = 0.75, ρ = 1, $\sigma_0$ = 1, $\sigma_1$ = 1.2, $\sigma_2$ = 1.5, α = 0.333, λ = 2, s = 0 (Conducting porous plates)

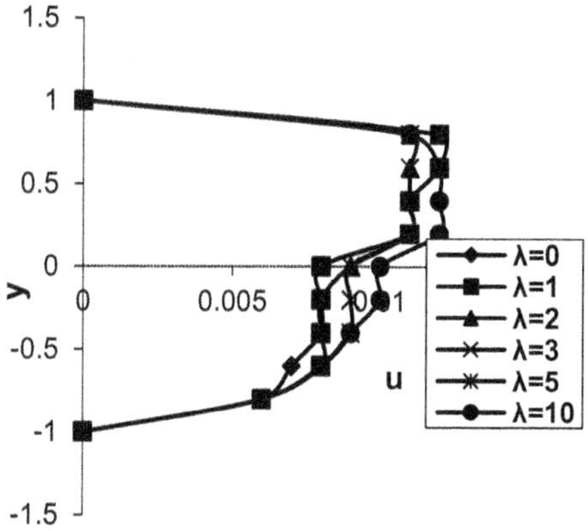


**Fig. 12** Primary velocity profile for various λ and M = 10, m = 2, ρ = 1,$\sigma_0$ = 1, $\sigma_1$ = 1.2, $\sigma_2$ = 1.5, α = 0.333,h = 0.75, s = 0 (Conducting porous plates)

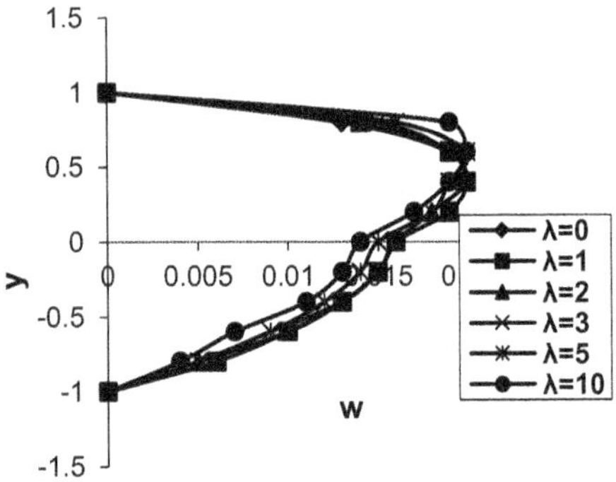


**Fig. 13** Secondary velocity profile for various λ and M = 10, m = 2, ρ = 1, $\sigma_0 = 1$, $\sigma_1 = 1.2$, $\sigma_2 = 1.5$, α = 0.333, h = 0.75, s = 0 (Conducting porous plates)

## 6 Conclusions

Hall Effect is used to investigate theoretically the MHD two-layered flow of ionized gases within the horizontal channel restricted by two parallel permeable plates. Non-conducting and conducting porous plates are assumed while solving the problem. It is shown that the flow parameters "Hartman number, Hall parameter, and porous parameter and the ratio of viscosities" have an impact on velocity fields in two liquid areas. The following are some of the most important outcomes:

- While plates are non-conducting, the velocity distributions climb to a particular value of Hartmann number before they begin to decline.
- Velocity profile increases in region-I as a result of increasing the Hall parameter value; in the second region, the profile increases and then decreases as this parameter grows in value.
- With growth in porosity parameter, velocity (primary and secondary) distributions rise initially but eventually fall.

## References


1. Zuber, N.: On the variable density single-fluid model for two-phase flow. J. Heat Trans. Trans. ASME82, 265–272 (1960)
2. Pakham, B.A., Shail, R.: Stratified laminar flow of two immiscible fluids. In: Proceedings of the Cambridge Philosophical Society, vol. 69, pp. 443–448 (1971)
3. Golding, J.A., Mah, C.C.: The Chemistry of Ethylene Oxide. Can. J. Chem. Eng. **52**(1), 36–42 (1974)
4. Oshinowo, T., Charles, M.E.: Vertical Two-Phase Flow. Part I. Flow Pattern Correlations. Can. J. Chem. Eng. **52**, 25–35 (1974)
5. Jones, O.C., Zuber, N.: First employed quantitative means for flow regime determination. Int. J. Mult. Flow **2**, 273 (1975)
6. Shipley, D.G.: Two-phase flow in large diameter pipes. Chem. Eng. Sci. **39**, 163–165 (1984)

7. LingaRaju, T.: Electro-Magnetohysrodynamic two fluid flow of ionized-gases with Hall and rotation effects. Int. J. Appl. Mech. Eng. **26**(4), 128–144 (2021). https://doi.org/10.2478/ijame-2021-0054
8. Zivojin Stamenkovic, M., Nikodijevic Dragis, D., Blagojevic Bratislav, D., Savic Slobodan, R.: MHD flow and heat transfer of two immiscible fluids between moving plates. Trans. Canadian Soc. Mech. Eng. **3–4**, 351–372 (2010)
9. Nikodijevic, D., Milenkovic, D., Stamenkovic, Z.: MHDCouette two-fluid flow and heat transfer in presence of uniform inclined magnetic field. Heat Mass Trans. **47**(12), 1525–1535 (2011)
10. StamenkovicZivojin, M., NikodijevicDragisa, Kocic Milos, M., NikodijevicJelena D.: MHD flow and heat transfer of two immiscible fluids with induced magnetic field effects. Thermal Sci. **16**(2), 323–336 (2012)
11. LingaRaju, T., NagaValli, M.: MHD two-layered unsteady fluid flow and heat transfer through a horizontal channel between parallel plates in a rotating system. Int. J. Appl. Mech. Eng. **19**(1), 97–121 (2014)
12. Mateen, A.: Transient magnetohydrodynamic flow of two immiscible fluids through a horizontal channel. Int. J. Eng. Res. **3**(1), 13–17 (2014)
13. Sharma, P.R., Kalpna, S.: Unsteady MHD two-fluid flow and heat transfer through a horizontal channel. Int. J. E. Sci. Invent. Res. Dev. **1**(3), 65–72 (2014)
14. Sivakami, L., Govindarajan, A.: Unsteady MHD flow of two immiscible fluids under chemical reaction in a horizontal channel. In: AIP Conference Proceedings, 020157 (2112). https://doi.org/10.1063/1.5112342 (2019)
15. Dobran, F.: On the consistency conditions of averaging operators in 2-phase flow models and on the formulation of magnetohydrodynamic 2-phase flow. Int. J. Eng. Sci. **19**(10), 353–1368 (1981)
16. Lohrasbi, J., Sahai, V.: Magnetohydrodynamic heat transfer in two-phase flow between parallel plates. Appl. Sci. Res. **45**, 53–66 (1989)
17. Malashetty, M.S., Leela, V.: Magnetohydrodynamivc heat transfer in two phase flow. Int. J. Eng. Sci. **30**, 371–377 (1992)
18. Mittal, M.L., Masapati, G.H., Rao, B.N.: Entrance flow in a MHD channel with Hall and ion-slip currents. AIAA J. **14**, 1768–1770 (1976)
19. Jana, R.N., Datta, N., Mazumder, B.S.: MagnetohydrodynamicCouette flow and heat transfer in a rotating system. J. Phys. Soc. Jpn. **42**, 1034 (1977)
20. Shail, R.: On laminar tow-phase flow in magnetohydrodynamics. Int. J. Eng. Sci. **11**, 1103 (1973)
21. Krishna, D.V., PrasadaRao, D.R.V.: Hall effect on free and forced convective in a rotating channel. Acta Mech. **43**, 49–59 (1982)
22. Raptis, A., Ram, P.C.: Role of rotation and hall currents on free convention and mass transfer flow through a porous medium. Int. Commun. Heat Mass Trans. **11**(4), 385–397 (1984)
23. Sharma, R.C., Rani, N.: Hall effects of thermosolutal instability of a plasma. Indian J. Pure Appl. Math. **19**(2), 202–207 (1988)
24. Ghosh, S.K.: Unsteady hydromagnetic flow in a rotating channel with oscillating pressure gradient. J. Phys. Soc. Jpn. **62**, 3893 (1993)
25. Aboeldahab, E.M., Elbarbary, E.M.E.: Hall current effects onmagnetohydrodynamic free-convection flow past a semi-infinitevertical plate with mass transfer. Int. J. Eng. Sci. **39**, 1641–1652 (2001)
26. Beg, O.A., Zueco, J., Takhar, H.S.: Unsteady magnetohydrodynamic Hartmann-Couette flow and heat transfer in a Darcian channel with Hall current, ionslip, viscous and Joule heating effects: network numerical solutions. Commun. Nonlinear Sci. Numer. Simul. **14**, 1082–1097 (2009)
27. Hazem Ali, A.: Effect of Hall current on the velocity and temperature distribution of Couette flow with variable properties and uniform suction and injection. Comput. Appl. Maths. **28**(2), 195–212 (2009)

28. Nikodijevic, M., Stamenkovic, Z., Petrovic, J.: Unsteady Fluid Flow and Heat Transfer Through a Porous Medium in a Horizontal Channel with an Inclined Magnetic Field. Transactions of FAMENA, vol. 44(4) (2020)
29. Sato, H.: The Hall effect in the viscous flow of ionized gas between parallel plates under transverse magnetic field. J. Phys. Soc. Japan **16**(7), 1427–1433 (1961)
30. LingaRaju, T., RamanaRao, V.V.: Hall effects on temperature distribution in a rotating ionized hydromagnetic flow between parallel walls. Int. J. Eng. Sci. **31**(7), 1073–1091 (1993)
31. LingaRaju, T.: MHD heat transfer two-ionized fluids flow between two parallel plates with Hall currents. Result. Eng. **4**, 100043 (2019). Elsevier BV. http://doi.org/https://doi.org/10.1016/j.rineng.2019.100043

# Influence of Heat Transfer, Chemical Reaction and Variable Fluid Properties on Oscillatory MHD Couette Flow Through a Partially-Porous Channel

**Sreedhara Rao Gunakala, Victor M. Job, and Jennilee Veronique**

**Abstract** In this work, we investigate the oscillatory magnetohydrodynamic Couette flow of a fluid that is incompressible and viscous with variable physical properties along a partially-porous channel. The impacts of heat transfer and first-order exothermic chemical reaction within the fluid are incorporated. We describe the flow through the porous region using the Darcy-Brinkman-Forchheimer model, whereas uniform wall suction/injection is considered. A numerical solution to the partial differential equations that model the transfer of heat and fluid flow is obtained using Galerkin's finite element technique. The impact of time t, Frank-Kamenetskii parameter λ, viscosity variation parameter b, suction/injection parameter S, and thermal conductivity variation parameter m on the flow velocity, wall shear stress, fluid temperature, and Nusselt number are investigated.

**Keywords** Chemical reaction · Finite element method · MHD Couette flow · Partially-porous channel · Suction · and Injection

## Nomenclature

| Roman symbols | | Greek symbols | |
|---|---|---|---|
| $A$ | Pre-exponential factor | $\kappa$ | Thermal conductivity |
| $B_0$ | Magnetic field | $\kappa_0$ | Thermal conductivity at temperature $T_0$ |
| $c$ | Specific heat capacity | $\mu$ | Viscosity |
| $c_F$ | Forchheimer coefficient | $\phi_0$ | Initial mass fraction |

(continued)

S. R. Gunakala · J. Veronique
Department of Mathematics and Statistics, The University of the West Indies, St. Augustine, Trinidad and Tobago
e-mail: Sreedhara.Rao@sta.uwi.edu

V. M. Job (✉)
Department of Mathematics, The University of the West Indies, Mona, Jamaica
e-mail: Victor.Job@uwimona.edu.jm



B. Rushi Kumar et al. (eds.), *Mathematics and Computing*, Springer Proceedings in Mathematics & Statistics 415, https://doi.org/10.1007/978-981-19-9307-7_44

(continued)

| Roman symbols | | Greek symbols | |
|---|---|---|---|
| $A$ | Pre-exponential factor | $\kappa$ | Thermal conductivity |
| $E_a$ | Activation energy | $\rho$ | Density |
| $h$ | Width of the channel | $\sigma$ | Electrical conductivity |
| $k_r$ | Reaction rate constant | $\tau_L$ | Shear Stress |
| $K$ | Permeability | $\omega$ | Oscillation Frequency |
| $p$ | Pressure | | |
| $R$ | Ideal gas constant | **Non-dimensional Parameters** | |
| $Q$ | Enthalpy of reaction | $b$ | Viscosity variation parameter |
| $T$ | Temperature | $Da$ | Darcy number |
| $T_0$ | Plate temperature | $Ha$ | Hartmann number |
| $t$ | Time | $k$ | Amplitude of pressure gradient |
| $x$ | Horizontal coordinate | $m$ | Thermal conductivity variation Parameter |
| $u$ | Velocity | $Pr$ | Prandtl number |
| $U_0$ | Plate velocity | $Re$ | Reynolds number |
| $v_0$ | Suction/injection velocity | $S$ | Suction/injection parameter |
| $z$ | Vertical coordinate | $\varepsilon$ | Activation energy parameter |
| $\mu_0$ | Viscosity at temperature $T_0$ | $\lambda$ | Frank-Kamenetskii parameter |

# 1 Introduction

Magnetohydrodynamics (MHD) focuses on the dynamics of magnetic fields in electrically conducting fluids such as in liquid metals and plasma [1]. In particular, the study of Couette flows with magnetic field effects is applicable to many areas of engineering and industry such as polymer technology, petroleum engineering, and the development of MHD power generators [2, 3]. Job and Gunakala [4] examined the time-dependent free convective magnetohydrodynamic Couette flow between two plates under the effects of thermal radiation and viscous and Joule dissipations. It was found that the Prandtl number and radiation parameter cause reductions in flow velocity and temperature at small time, and increases in velocity and temperature for large time. Also, when the Eckert number, Grashof number, and magnetic parameter increase, the temperature and velocity increase. Mosayebidorcheh et al. [5] examined heat transfer and time-dependent MHD Couette dusty fluid flow whose viscosity and electrical conductivity are temperature-dependent. The authors found that increasing the viscosity parameter results in an increased temperature and flow velocity, and increasing the Reynolds number causes the temperature to increase. Moreover, the Nusselt number on the lower plate and skin friction coefficient decrease when the magnetic field strength increases.

MHD Couette flows under the effects of suction and injection is an area of vast study among researchers. Attia [3] considered the unsteady MHD Couette flow that includes uniform suction/injection and heat transfer. The results indicated that the temperature of the fluid increases with increased magnetic field strength for small time, whereas the temperature decreases when the magnetic field strength increases for large time. Jha et al. [6] explored the time-dependent free convection MHD Couette flow between two permeable plates including thermal radiation effects. It was found that the fluid's temperature and velocity increase when time and the thermal radiation parameter increase. Uwanta and Hamza [7] examined the impacts of injection and suction on the unsteady hydromagnetic chemically-reactive convective flow between porous vertical plates with the impacts of variable viscosity and thermal diffusion. Their study showed that reaction consumption, thermal and solutal buoyancy, suction and injection, and thermal diffusion have a strong influence on the transport phenomena. The unsteady natural convective hydromagnetic Couette flow through a permeable-walled channel with thermal radiation and Joule and viscous dissipation effects was investigated by Job and Gunakala [8]. The results showed that the fluid temperature and velocity are significantly influenced by variations in the Grashof, Prandtl, radiation, Eckert numbers, and magnetic and suction parameters. Gupta and Jain [9] conducted an analysis on the unsteady heat transfer and hydromagnetic Couette flow along a horizontal rotating channel with wall suction/injection; the authors used an analytical approach by applying the perturbation technique in obtaining its solution. The study revealed that the influence of the rotation parameter, thermal slip, magnetic field, injection/suction, permeability, Prandtl number, and heat generation/absorption has a considerable impact on the heat transfer and hydromagnetic flow.

The complex phenomenon of mass transport in chemically reacting systems is applicable to geothermal and oil reservoir engineering [10], and can involve the consumption and production of chemically-reactant species at different reaction rates. Makinde and Chinyoka [10] conducted a numerical study on the unsteady reactive MHD Couette third-grade fluid flow having asymmetric convective cooling and temperature-dependent viscosity. The results showed that the fluid temperature and velocity are strongly impacted by the rate of reaction, viscous heating parameter, fluid viscosity parameter, magnetic parameter, and non-Newtonian parameter. VeeraKrishna and Reddy [11] examined the unsteady hydromagnetic Couette flow of a chemically-reactive second-grade fluid in a rotating channel and porous medium. The temperature dependence of fluid thermal conductivity was considered in their study. It was found that the temperature within the channel increases when the reaction rate parameter, magnetic parameter, rotation parameter, and Eckert number increase. However, the temperature decreases with increasing thermal conductivity variation parameters. Kareem and Gbadeyan [12] considered hydromagnetic Couette flow through a horizontal channel with an exothermic two-step chemical reaction and viscous dissipation. The impact of the Frank-Kamenetskii (reaction rate) parameter, exothermic reaction parameters, activation energy parameter, and chemical kinetic parameter on entropy generation and thermal criticality were investigated. Das et al. [13] investigated the unsteady MHD oscillatory reactive flow of a viscous fluid within a porous rotating channel with convective heat transfer and chemical reaction effects. It was determined that the flow characteristics in the channel are substantially influenced by the magnetic field, rotation, suction/injection, and convective heating.

To the best of the authors' knowledge, there is no existing work on unsteady reactive magnetohydrodynamic Couette flow along a partially-porous channel containing uniform suction/injection and an oscillating pressure gradient. Therefore in the present work, we investigate the MHD Couette flow of an incompressible viscous fluid in a partially-porous channel under the influence of an oscillating pressure gradient with heat transfer and a first-order exothermic chemical reaction. We consider the thermal conductivity and viscosity of the fluid to be temperature-dependent, and the Darcy-Brinkman-Forchheimer model is utilized for the fluid flow through the porous region. The impacts of pertinent parameters on convective fluid flow are explored.

## 2 Description of the Problem

The flow of an incompressible and viscous Newtonian fluid through two horizontal, parallel, and infinitely-long plates is considered. The lower plate is located at $z = -h$ and is stationary with a constant temperature $T_0$, whereas the upper plate is located at $z = h$ and moves with constant horizontal velocity $U_0$ and temperature $T_0$. The region between the lower and upper plates is comprised of a porous region with a thickness $h_p$ and an overlying free-fluid (non-porous) region. The fluid flow between the two plates is also influenced by uniform injection from below, suction from above, and an oscillating pressure gradient $-\frac{\partial p}{\partial x}$ in the $x$-direction. A constant magnetic field with strength $B_0$ is normal to the plates, and a first-order exothermic chemical reaction occurs within the fluid (Fig. 1).

We assume a negligible magnetic Reynolds number, and the influence of viscous dissipation and Joule dissipation and thermal radiation are neglected. The heat flux and shear stress are considered to be continuous at the shared boundary of the porous and free-fluid regions. Furthermore, we assume that the chemically-reacting species within the fluid is dilute with a uniform volume fraction.

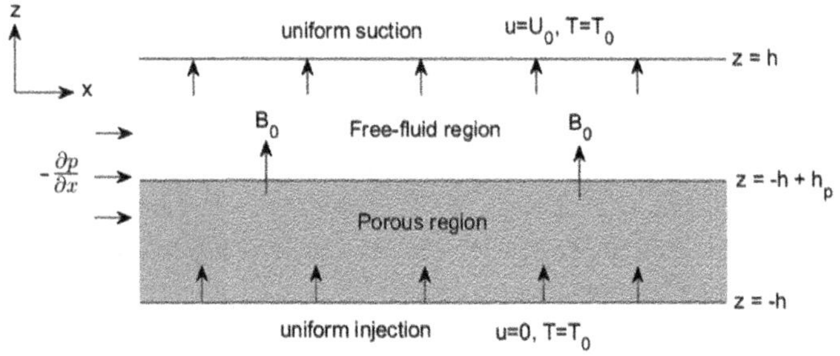


**Fig. 1** Geometrical diagram of the physical system

Following [10, 11], the fluid thermal conductivity and dynamic viscosity are described (respectively) by the equations

$$\mu(T) = \mu_0 e^{-b(T-T_0)} \tag{1}$$

$$\kappa(T) = \kappa_0 e^{m(T-T_0)} \tag{2}$$

where $\mu_0$, $\kappa_0$, $b$ and $m$ are (respectively) the viscosity at temperature $T_0$, thermal conductivity at temperature $T_0$, viscosity variation parameter, and thermal conductivity variation parameter. We express the reaction rate constant by the Arrhenius equation [14]

$$k_r(T) = Ae^{-E_a/RT} \tag{3}$$

where $E_a$ is the activation energy, $A$ is the pre-exponential factor and $R$ is the ideal gas constant.

Suppose that the flow velocity vector is $\vec{u}(z,t) = u(z,t)\vec{i} + v_0\vec{j}$. From the above assumptions and problem description, the heat transfer and convective fluid flow within the channel are described as follows [14–16]:

$$\rho\left(\frac{\partial u}{\partial t} + v_0\frac{\partial u}{\partial z}\right) = -\frac{\partial p}{\partial x} + \frac{\partial}{\partial z}\left(\mu(T)\frac{\partial u}{\partial z}\right) - \left(\sigma B_0^2 + \frac{1}{K}\right)u - \frac{\rho c_F}{\sqrt{K}}\sqrt{u^2 + v_0^2}u \tag{4}$$

$$\rho c\left(\frac{\partial T}{\partial t} + v_0\frac{\partial T}{\partial z}\right) = \frac{\partial}{\partial z}\left(\kappa(T)\frac{\partial T}{\partial z}\right) + Q\phi_0 k_r(T) \tag{5}$$

where $\rho$, $\sigma$, $c_F$, $K$, $c$, $Q$, and $\phi_0$ are the fluid density, electrical conductivity, Forchheimer coefficient, permeability, specific heat capacity, enthalpy of reaction, and initial mass fraction of the chemically-reacting species respectively.

The initial condition is

$$u = 0,\ \ T = T_0 \text{ at } t = 0 \tag{6}$$

and the boundary conditions are

$$u = 0,\ \ T = T_0 \text{ at } z = -h \tag{7}$$

$$u = U_0,\ \ T = T_0 \text{ at } z = h \tag{8}$$

Equations (4) and (5) are non-dimensionalized by using the dimensionless variables defined as:

$$\hat{x} = \frac{x}{h}, \hat{z} = \frac{z}{h}, \hat{h}_p = \frac{h_p}{h}, \hat{u} = \frac{u}{U_0}, \hat{p} = \frac{p}{\rho U_0^2}, \hat{t} = \frac{tU_0}{h},$$
$$\hat{T} = \frac{T - T_0}{T_0}, \hat{b} = bT_0, \hat{m} = mT_0 \tag{9}$$

On dropping all hats and taking the non-dimensional pressure gradient to be

$$-\frac{\partial p}{\partial x} = k(1 - \cos(\omega t)) \tag{10}$$

with amplitude $k$ of the pressure gradient and frequency $\omega$ of oscillation, we get the non-dimensional equations

$$\frac{\partial u}{\partial t} + S\frac{\partial u}{\partial z} = k(1 - \cos(\omega t)) + \frac{1}{Re}\frac{\partial}{\partial z}\left(e^{-bT}\frac{\partial u}{\partial z}\right)$$
$$- \frac{1}{Re}\left(Ha^2 + \frac{1}{Da}\right)u - \frac{c_F}{\sqrt{Da}}\sqrt{u^2 + S^2 u} \tag{11}$$

$$\frac{\partial T}{\partial t} + S\frac{\partial T}{\partial z} = \frac{1}{RePr}\frac{\partial}{\partial z}\left(e^{mT}\frac{\partial T}{\partial z}\right) + \frac{\lambda}{RePr}e^{\varepsilon T/(1+T)} \tag{12}$$

where $S = \frac{v_0}{U_0}$ is the suction parameter, $Ha = B_0 h\sqrt{\frac{\sigma}{\mu_0}}$ is the Hartmann number, $Re = \frac{\rho h U_0}{\mu_0}$ is the Reynolds number, $Da = \frac{K}{h^2}$ is the Darcy number, $Pr = \frac{\mu_0 c}{\kappa_0}$ is the Prandtl number, $\varepsilon = \frac{E_a}{RT_0}$ is the activation energy parameter and $\lambda = \frac{Q\phi_0 Ah^2}{\kappa_0 T_0}e^{-\varepsilon}$ is the Frank-Kamenetskii parameter. The non-dimensionalized boundary conditions and initial conditions are

$$u = T = 0 \text{ at } t = 0 \tag{13}$$

$$u = T = 0 \text{ at } z = -1 \tag{14}$$

$$u = 1, T = 0 \text{ at } z = 1 \tag{15}$$

The time-dependent shear stress and time-dependent Nusselt number on the stationary lower plate ($z = -1$) are given by

$$\tau_L(t) = \frac{\partial u}{\partial z}(-1, t) \tag{16}$$

$$Nu_L(t) = \frac{\partial T}{\partial z}(-1, t) \tag{17}$$

## 3 Numerical Solution Methodology

The coupled non-linear system of Eqs. (11)–(12) is numerically solved by Galerkin's finite element technique [17] with the prescribed boundary and initial conditions (13)–(15) to obtain the flow velocity, fluid temperature, shear stress on the lower plate, and Nusselt number on the lower plate. Spatial discretization was performed by finite element procedure with quadratic elements, and then the Crank–Nicholson scheme was used to perform the time discretization. After assembly of the elements, the resulting non-linear system of equations was iteratively solved using the computer software MATLAB with a $10^{-4}$ relative error tolerance. The numerical solution was obtained using 100 quadratic elements and 200 time steps.

## 4 Results and Discussion

The approximate solutions for the velocity, temperature, shear stresses, and Nusselt numbers were calculated for differing viscosity variation parameter $b$, suction/injection parameter $S$, thermal conductivity variation parameter $m$, Frank-Kamenetskii parameter $\lambda$, and time $t$. The values $S = 0, 0.5, 1, 2$; $b = 0, 2, 5, 10$; $m = 0, 5, 15, 30$; and $\lambda = 0, 1, 2, 5$ were used in analyzing the numerical results. The parameters $S$, $b$, $m$, $\lambda$, $Da$, $Re$, $Ha$, $h_p$, $c_F$, $k$, $\omega$, $Pr$, $\varepsilon$, and $t$ are taken to be 1, 2, 5, 2, 0.1, 10, 1, 1, 0.06, 1, $2\pi$ , 6.2, 0.1, and 4 (respectively) throughout the analysis unless otherwise stated.

Figures 2 and 3 display the velocity and temperature profiles with varying suction/injection parameter $S$. These figures show that the flow velocity and fluid temperature in the free-fluid and porous regions decrease with increased suction/injection parameters. We also observe that the $z$ values at which the maximum velocity and maximum temperature occur are increased with an increase in the suction/injection parameter; this occurs as a result of increased resistance to fluid flow through the channel as the advection of fluid from the plate below to the plate above is increased. These observations are consistent with the results obtained by Attia [3] on the influence of suction and injection through parallel permeable plates.

In Figs. 4 and 5, the impact of the viscosity variation parameter $b$ and thermal conductivity variation parameter is shown. We observe (Fig. 4) that raising the viscosity variation parameter causes the flow velocity to increase near the upper plate ($z > 0.65$), and decrease slightly in the lower part of the channel ($z < 0.65$). Increasing the viscosity variation parameter lowers the fluid viscosity, which enhances the flow velocity near the upper plate. Consequently, the drag on the fluid in the lower part of the channel is increased and leads to the observed reduction in flow velocity in this region. From Fig. 5, it is seen that raising the thermal conductivity variation parameter causes a reduction in the maximum fluid temperature. A similar result was obtained by VeeraKrishna and Reddy [11]. Furthermore, the value of $z$ at which the temperature is maximum decreases with increased thermal conductivity

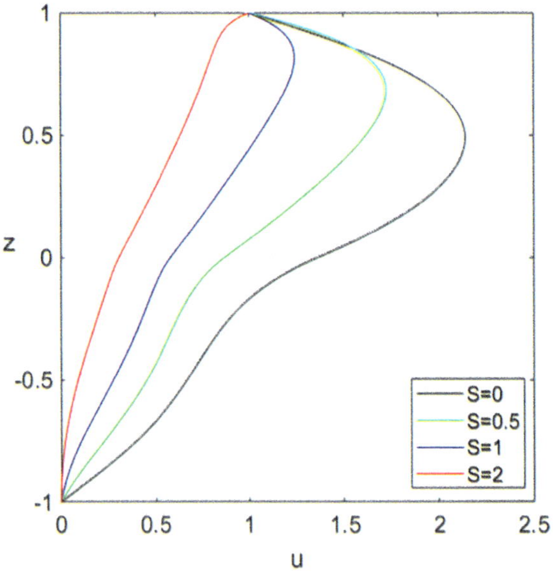


**Fig. 2** Velocity profiles for varying $S$

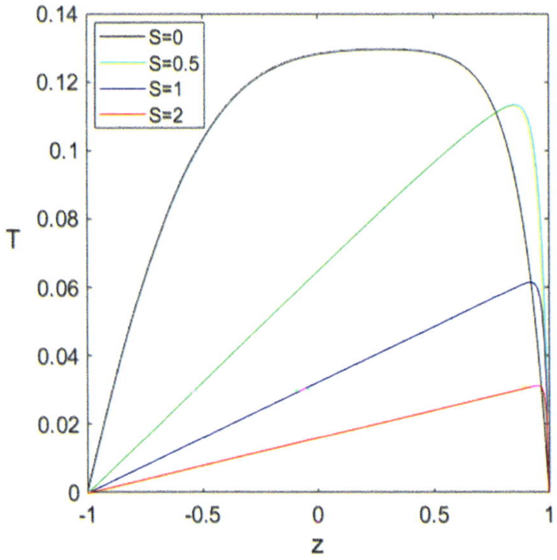


**Fig. 3** Temperature profiles for varying $S$

variation parameter. This is caused by an enhancement in the thermal conductivity of the fluid and an associated reduction in thermal advection through the upper and lower permeable plates.

Figures 6 and 7 display the fluid temperature and velocity for varying Frank-Kamenetskii parameter $\lambda$. Based on Fig. 7, we determined that the temperature of the fluid increases with increased Frank-Kamenetskii parameter as a result of enhanced

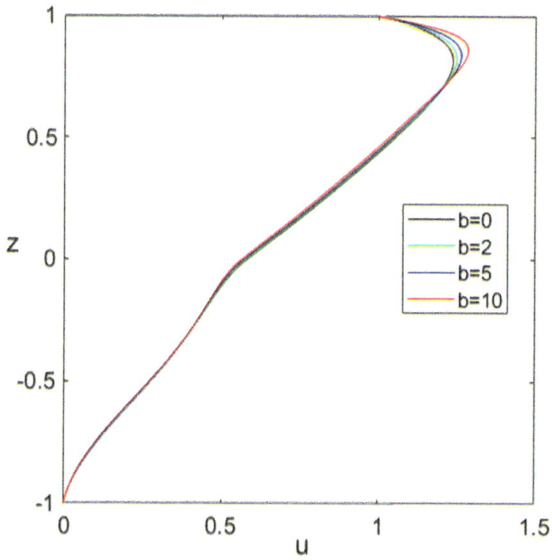


**Fig. 4** Velocity profiles for varying $b$

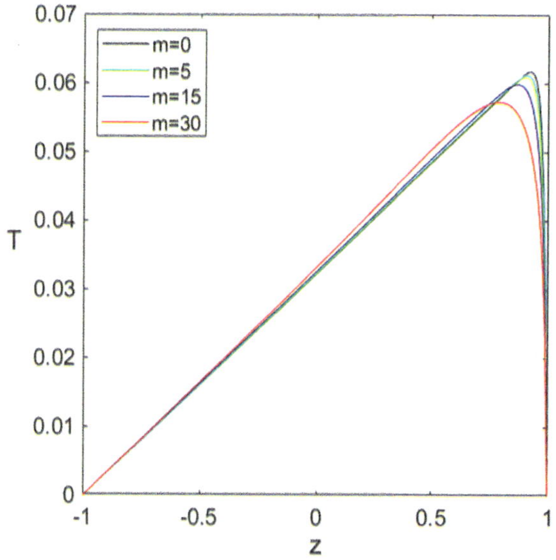


**Fig. 5** Temperature profiles for varying $m$

heat generation during the exothermic chemical reaction process. This finding is consistent with the works of Kareem and Gbadeyan [12] and Das et al. [13]. We also observe that the velocity of fluid near the upper plate (Fig. 6) increases with increased Frank-Kamenetskii parameter; this is caused by a reduction in fluid viscosity as the temperature within the channel increases.

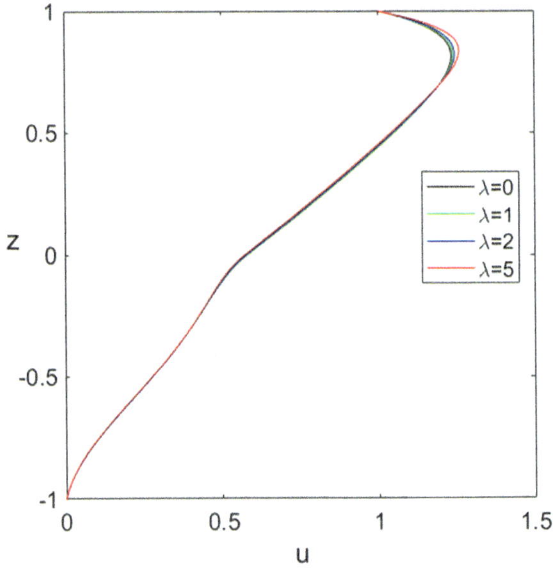


**Fig. 6** Velocity profiles for varying λ

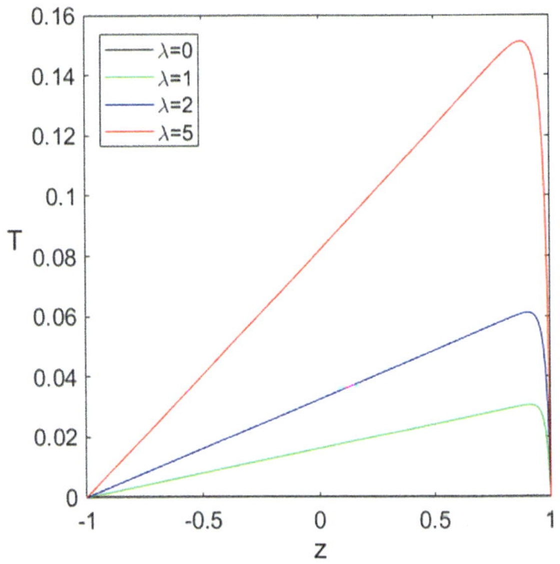


**Fig. 7** Temperature profiles for varying λ

Figures 8 to 10 show the influence of viscosity variation parameter $b$, suction/injection parameter $S$ and Frank-Kamenetskii parameter λ on the shear stress $\tau_L$ on the stationary lower plate over time $t$. From each of these figures, we see that the shear stress $\tau_L$ achieves a periodic-steady state as time increases. Furthermore, $\tau_L$ is decreased when the suction/injection parameter increases (Fig. 8), which is due to a reduction in flow velocity through the channel. When the viscosity variation parameter is increased (Fig. 9), the shear stress on the lower plate is reduced as a result of decreased flow resistance within the fluid near the lower plate. We also note that the shear stress $\tau_L$ decreases when the Frank-Kamenetskii parameter increases; this is caused by reduced fluid viscosity, as well as a corresponding reduction in fluid flow resistance near the stationary lower plate.

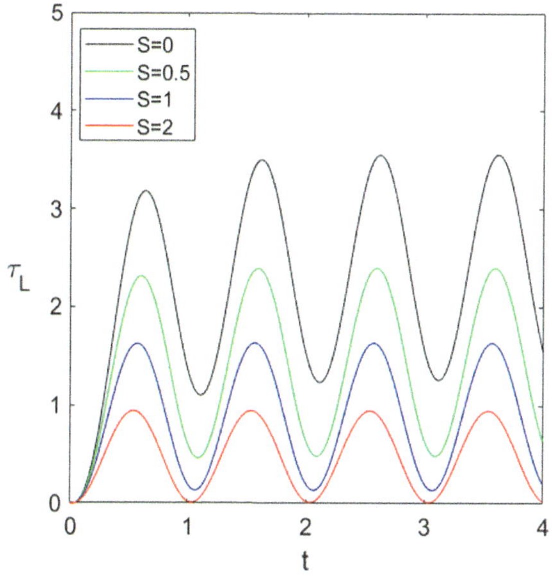


**Fig. 8** Shear stress for varying $S$

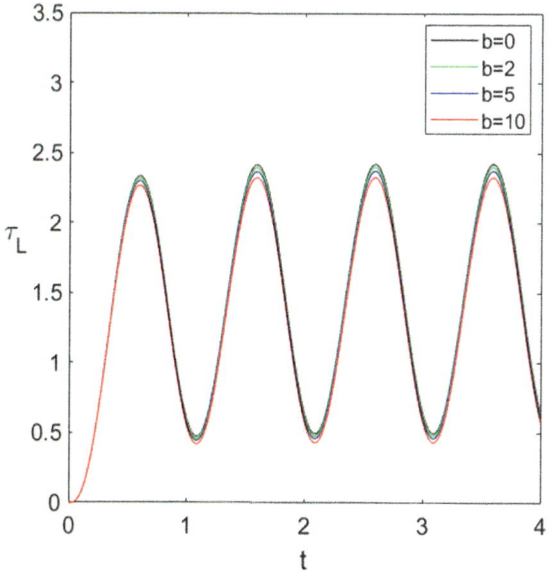


**Fig. 9** Shear stress for varying $b$

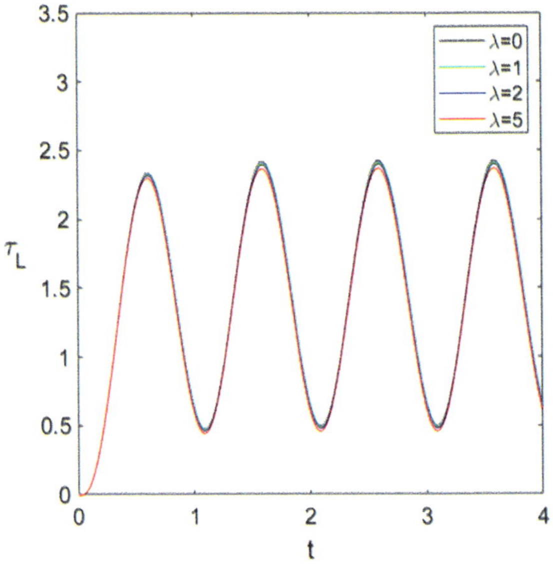


**Fig. 10** Shear stress for varying λ

The effects of the suction/injection parameter $S$, thermal conductivity variation parameter $m$ and Frank-Kamenetskii parameter $\lambda$ on the time variation of Nusselt number $Nu_L$ on the stationary lower plate are depicted in Figs. 11 to 13. We notice that for each of these parameters, the Nusselt number achieves a steady state over time. Figure 11 reveals a reduction in $Nu_L$ when the suction/injection parameter increases; this can be explained by a decrease in the loss of heat from the channel through the lower plate as fluid suction into the channel is increased. When the thermal conductivity variation parameter $m$ (Fig. 12) increases, the Nusselt number $Nu_L$ on the lower plate is enhanced due to increased heat conduction within the fluid near the stationary lower plate. It is also found (Fig. 13) that $Nu_L$ increases when the value of the Frank-Kamenetskii parameter is raised; this occurs as a result of enhanced heat generation by the chemically-reacting species within the fluid.

## 5 Conclusions

The unsteady heat transfer and hydromagnetic Couette flow under the influence of an oscillating pressure gradient, uniform suction/injection, and exothermic first-order chemical reaction were investigated in this study. The effects of the suction/injection, thermal conductivity variation, viscosity variation, and Frank-Kamenetskii parameters on the temperature, velocity, shear stress on the lower plate, and Nusselt number on the lower plate have been examined.

It was determined that the flow velocity through the channel can be enhanced with an increased Frank-Kamenetskii parameter and decreasing the suction/injection

**Fig. 11** Nusselt number for varying $S$

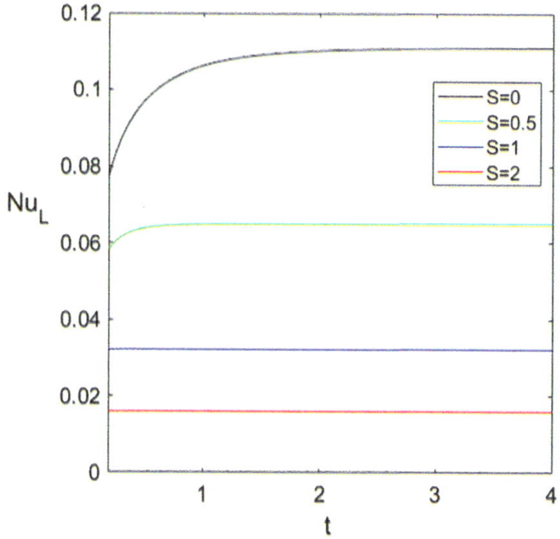


**Fig. 12** Nusselt number for varying $m$

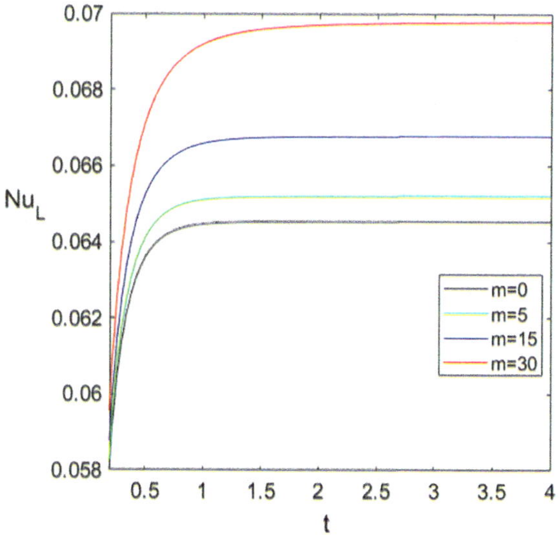


parameter. The flow velocity near the upper plate can be enhanced by increasing the viscosity variation parameter, whereas the flow velocity in the lower part of the channel can be increased by lowering the viscosity variation parameter value. The fluid temperature can be enhanced with increased Frank-Kamenetskii parameter, and by decreasing the suction/injection parameter and thermal conductivity variation parameter. The shear stress on the stationary lower plate can be increased by reducing the suction/injection parameter, viscosity variation parameter, and Frank-Kamenetskii parameter. Furthermore, the Nusselt number on the stationary lower

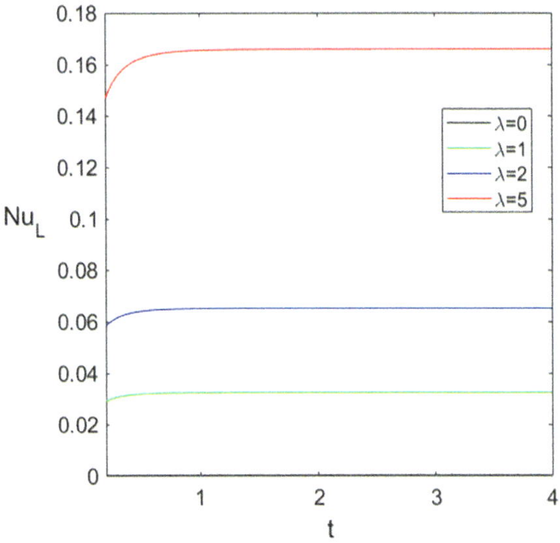


**Fig. 13** Nusselt number for varying λ

plate can be enhanced by increasing the thermal conductivity variation parameter and Frank-Kamenetskii parameter and by reducing the suction/injection parameter.

## References


1. Uwanta, I.J., Hamza, M.M.: Effect of suction/injection on unsteady hydromagnetic convective flow of reactive viscous fluid between vertical porous plates with thermal diffusion. Int. Sch. Res. Not. **2014** (2014)
2. Kala, B.: Numerical study of the effects of suction and pressure gradient on an unsteady MHD fluid flow between two parallel plates in a non-darcy porous medium. Asian Res. J. Math. **3**, 1–14 (2017)
3. Attia, H.A.: Unsteady MHD Couette flow with heat transfer in the presence of uniform suction and injection. Mech. Mech. Eng. **12**(2), 165–170 (2008)
4. Job, V.M., Gunakala, S.R.: Unsteady MHD free convection couette flow through a vertical channel in the presence of thermal radiation with viscous and joule dissipation effects using galerkin's finite element method. Int. J. Appl. Innov. Eng. Manag. **2**(9), 50–61 (2013)
5. Mosayebidorcheh, S., Makinde, O.D., Ganji, D.D., Abedian, M.: DTM-FDM hybrid approach to unsteady MHD Couette flow and heat transfer of dusty fluid with variable properties. Thermal Sci. Eng. Progress **2**, 57–63 (2017)
6. Jha, B.K., Isah, B.Y., Uwanta, I.J. Unsteady MHD free convective Couette flow between vertical porous plates with thermal radiation. J. King Saud Univ.—Sci. **27**(4), 338–348 (2015)
7. Uwanti, I.J., Hamza, M.M.: Unsteady hydromagnetic flow of a reactive viscous fluid in a vertical channel with thermal diffusion, diffusion-thermal and variable viscosity effects. Comput. Math. Model. **26**(3), 385–397 (2015)
8. Job, V.M., Gunakala, S.R.: Finite element analysis of unsteady radiative MHDmhd natural convection couette flow between permeable plates with viscous and joule dissipation. Int. J. Pure Appl. Math. **99**(2), 123–143 (2015)

9. Gupta, V.G., Jain, A.: An analysis of unsteady MHDmhd couette flow and heat transfer in a rotating horizontal channel with injection/suction. Int. J. Latest Technol. Eng. Manag. Appl. Sci. **6**, 28–45 (2016)
10. Makinde, O.D., Chinyoka, T.: Numerical study of unsteady hydromagnetic Generalized Couette flow of a reactive third-grade fluid with asymmetric convective cooling. Comput. Math. Appl. **61**, 1167–1179 (2011)
11. VeeraKrishna, M., Subba Reddy, G.: Unsteady MHD reactive flow of second grade fluid through porous medium in a rotating parallel plate channel. J. Anal. **27**, 103–120 (2019)
12. Kareem, R.A., Gbadeyan, J.A.: Entropy generation and thermal criticality of generalized Couette hydromagnetic flow of two-step exothermic chemical reaction in a channel. Int. J. Thermofluids **5–6**, 100037 (2020)
13. Das, S., Patra, R.R., Jana, R.N.: Hydromagnetic oscillatory reactive flow through a porous channel in a rotating frame subject to convective heat exchange under arrhenius kinetics. J. Eng. Phys. Thermophys. **94**, 702–713 (2021)
14. Frank-Kamenetskii, D.A., Albertovich, D.: Diffusion and Heat Exchange in Chemical Kinetics. Princeton University Press, NJ (2015)
15. Davidson, P.A.: An Introduction to Magnetohydrodynamics. Cambridge University Press, New York (2001)
16. Nield, D., Bejan, A.: Convection in Porous Media, 3rd edn. Springer Science+Business Media, Inc., New York (2006)
17. Reddy, J.N., Gartling, D.K.: The Finite Element Method in Heat Transfer and Fluid Dynamics. CRC Press, USA (2010)

# Effect of Heat Transfer on Peristaltic Transport of Prandtl Fluid in an Inclined Porous Channel

**Indira Ramarao, Priyanka N. Basavaraju, and Jagadeesha Seethappa**

**Abstract** A Prandtl fluid subjected to low Reynolds number and heat transfer is assumed to be flowing in an inclined porous channel is considered. Peristaltic waves are applied on the walls with the assumption of long-wave approximation. The equations governing the flow are highly non-linear and coupled. These are solved by the application of the regular perturbation method. The solutions for velocity, pressure, and temperature are obtained and numerically evaluated. The results are graphically depicted. The temperature in the inclined channel is higher than the horizontal one and the pressure gradient is less.

**Keywords** Peristalsis · Prandtl fluid · Heat transfer · Inclined channel · Perturbation

## 1 Introduction

Peristaltic pumping has a lot of industrial and biological applications. Transport of physiological fluid-like food bolus, colonic material in the intestine, transport of sperms, ovum, etc. is direct implications of peristaltic transport. Many studies have been conducted in this regard. Latham [1], Shapiro et al. [2] have conducted experimental studies and presented both theoretical and experimental results. Yin et al. [3], Gupta et al. [4] have considered Newtonian fluid flow under peristalsis. A power-law fluid was considered by Raju et al. [5]. A study on small blood vessels was considered by Misra et al. [6]. Peristaltic transport with the MHD effect was considered by Mekheimer et al. [7]. Kumari et al. [8] have studied heat transfer and flow of Jeffrey fluid in a verticle porous channel subject to peristalsis. Eldabe et al. [9] and Hayat et al. [10] have considered transport of power-law fluid under

I. Ramarao · J. Seethappa (✉)
Department of Mathematics, Nitte Meenakshi Institute of Technology, Bengaluru 560064, KA, India
e-mail: jagadeeshas31@gmail.com

P. N. Basavaraju
Department of Mathematics, ATME College of Engineering, Mysuru 570028, KA, India



B. Rushi Kumar et al. (eds.), *Mathematics and Computing*, Springer Proceedings in Mathematics & Statistics 415, https://doi.org/10.1007/978-981-19-9307-7_45

peristaltic motion with chemical reaction and heat transfer in an asymmetric channel. Shabaan et al. [11] have considered porous concentric annulus and studied the effect of MHD flow and heat transfer with peristalsis. Eldabe et al. [12] have considered Jeffrey fluid in a verticle porous tube and analysed the MHD effect on peristalsis. Selvi et al. [13] have conducted a study on the effect of heat transfer on Jeffrey fluid flow under peristalsis. Pandey et al. [14] have presented an analytical model of peristaltic transport of micropolar fluid in a porous medium. Tripathi [15] has studied the peristaltic flow through a finite porous channel. Ahmed et al. [16] have carried out a two-dimensional analysis of peristaltic transport of Jeffrey fluid in a curved channel under the influence of a magnetic field in the radial direction. Pandey et al. [14] and Ahmed et al. [16] have assumed the low Reynold's number and long wavelength approximations in their studies. Sreegowrav et al. [17] have considered peristaltic flow in an asymmetric channel with a couple stress fluid, and Rashmi et al. [18] have considered eccentric annulus. Nadeem et al. [19] have modeled the flow considering fixed and wave frame references. Indira et al. [20] have considered the effect of heat transfer on flow parameters considering flow of Prandtl fluid in a vertical annulus. Vajravelu et al. [21] have studied the viscous fluid flow in an annular region under a long-wavelength approximation.

The approach used by Selvi et al. [13] is adopted in the present study to understand the effect of heat transfer in a Prandtl fluid flowing in a porous channel and under peristaltic motion. The channel is considered to be inclined. The governing equations are solved using the regular perturbation technique.

## 2 Mathematical Formulation

A two-dimensional inclined porous channel is considered as shown in Fig. 1 whose walls are subjected to a sinusoidal wave motion is given by,

$$\overline{Y} = \eta(x,t) = \overline{a} + \overline{b}cos\left[\frac{2\pi}{\lambda}(x-ct)\right] \tag{1}$$

where $2\overline{a}$—width of channel, $\lambda$—wavelength and $\overline{b}$—amplitude.

The governing equations [see 19, 20] for a Prandtl fluid are given by,

$$\vec{T} = -P\vec{I} + \vec{S}, \tag{2}$$

$$\nabla.\vec{V} = 0 \tag{3}$$

$$\rho\frac{d\vec{V}}{dt} = \nabla.\vec{T} + \rho\vec{f}, \tag{4}$$

$$c\rho\frac{d\vec{T}}{dt} = k\nabla^2\vec{T} + \mu\phi + \frac{\mu}{k_0}v^2, \tag{5}$$

where $\vec{V}$—velocity, $\rho$—density, $\vec{T}$—Cauchy stress tensor, $S$—stress tensor.

Moving coordinates are introduced as:

$$p = P, u = U - \overline{c}, v = V, x = X - \overline{c}t \text{and} y = Y,$$

where $V, U$ are velocity in the fixed coordinates and $v, u$ are velocity components in moving coordinates.

Non-dimensional parameters used in the analysis are:

$$\begin{aligned}
&x^* = \frac{2\pi x}{\lambda},\ \phi = \frac{\overline{b}}{\overline{a}},\ \rho = \frac{\overline{a}}{\sqrt{k}},\ y^* = \frac{y}{\overline{a}},\ \delta = \frac{2\pi\overline{a}}{\lambda},\\
&\eta^* = \frac{\eta}{\overline{a}},\ p^* = \frac{2\pi\overline{a}^2 p}{\mu\overline{c}\lambda},\ v = \frac{\mu}{\rho_0},\ S^* = \frac{\overline{a}}{\mu\overline{c}}S,\ t^* = \frac{2\pi\overline{c}t}{\lambda},\\
&T = \theta(T_1 - T_0) + T_0,\ Pr = \frac{\mu c_p}{k_0},\ Gr = \frac{\alpha g(T_1 - T_0)\overline{a}^3}{v^2},\\
&u^* = \frac{\overline{u}}{c},\ Re = \frac{\overline{ac}}{v},\ G = \frac{Gr}{Re},\ v^* = \frac{v}{\overline{c}\delta},\ E_c = \frac{\overline{c}^2}{c_p(T_1 - T_0)},\\
&N = E_c Pr,\ = \frac{Re}{Fr},\ Fr = \frac{\overline{c}^2}{\overline{a}g}
\end{aligned} \tag{6}$$

where $Pr$—Prandtl number, $Gr$–Grashof number, $N$—perturbation parameter, $E_c$—Eckert number and $Fr$—Froude number.

Following [19], the flow is assumed to be under the effect of a low Reynolds number and long wavelength approximation $\delta \to 0$ is considered. Hence equation becomes,

$$\frac{\partial u}{\partial x} + \frac{\partial u}{\partial y} = 0, \tag{7}$$

$$-\frac{\partial p}{\partial x} + \frac{\partial}{\partial y}(S_{xy}) - \sigma^2(u+1) + G\theta + f sinv = 0, \tag{8}$$

$$S_{xy} = \alpha\frac{\partial u}{\partial y} + \beta\left(\frac{\partial u}{\partial y}\right)^3, \tag{9}$$

$$\frac{\partial^2\theta}{\partial y} + N\left(\frac{\partial u}{\partial y}\right)^2 + N\sigma^2(u+1)^2 = 0, \tag{10}$$

where $v$—inclination angle and $\sigma$—permeability.

Boundary conditions are given by,

$$at\ y = \eta(x),\ \theta = +1\,\text{and}\,u = -1,\ at\ y = 0,\ \frac{\partial \theta}{\partial y} = 0\text{and}\frac{\partial u}{\partial y} = 0 \tag{11}$$

where negative velocity is due to consideration of moving wave frame of reference $u = v - c$

## 2.1 Method of Solution

The equations are coupled and non-linear. To overcome this problem a regular perturbation is applied:

$$\begin{aligned} u &= [u_{00} + \beta u_{01} + \ldots] + N[u_{10} + \beta u_{11} + \ldots], \\ \theta &= [\theta_{00} + \beta\theta_{01} + \ldots] + N[\theta_{10} + \beta\theta_{11} + \ldots], \\ p &= [p_{00} + \beta p_{01} + \ldots] + N[p_{10} + \beta p_{11} + \ldots] \end{aligned} \tag{12}$$

Applying the above perturbation to governing Eqs. (7)–(10) we get the following zeroth and first-order equations and boundary conditions as given below.

### 2.1.1 Zeroth Order

$$\frac{\partial^2 u_{00}}{\partial y^2} = 0 \tag{13}$$

$$\frac{\partial^2 u_{00}}{\partial y^2} - \frac{\sigma^2}{\alpha}(u_{00} + 1) + \left[\frac{G + fsinv}{\alpha}\right] = \frac{1}{\alpha}\frac{\partial p_{00}}{\partial x} \tag{14}$$

Boundary conditions for zeroth order equations will be:

$$at\ y = \eta,\ \theta_{00} = 1\text{and}u_{00} = -1, \tag{15}$$

$$at\ y = 0,\ \frac{\partial \theta_{00}}{\partial y} = 0\text{and}\frac{\partial u_{00}}{\partial y} = 0 \tag{16}$$

### 2.1.2 First Order

$$\frac{\partial^2 \theta_{01}}{\partial y^2} = 0, \tag{17}$$

$$-\frac{\partial p_{01}}{\partial x} + \alpha\frac{\partial^2 u_{01}}{\partial y^2} + \left(\frac{\partial u_{00}}{\partial y}\right)^3 - \sigma^2 u_{01} + G\theta_{01} = 0 \tag{18}$$

$$\frac{\partial^2\theta_{10}}{\partial y^2} + \left(\frac{\partial u_{00}}{\partial y}\right)^2 = 0 \tag{19}$$

$$-\frac{\partial p_{10}}{\partial x} + \alpha\frac{\partial^2 u_{10}}{\partial y^2} - \sigma^2 u_{10} + G\theta_{10} = 0,$$

Boundary conditions for the first-order equations are as follows:

$$at\ \ y = \eta,\ \theta_{01} = 0, \theta_{10} = 0, u_{01} = 0, u_{10} = 0, \tag{21}$$

$$at\ \ y = \eta,\ \frac{\partial\theta_{01}}{\partial y} = 0,\ \frac{\partial\theta_{10}}{\partial y} = 0,\ \frac{\partial u_{01}}{\partial y} = 0,\ \frac{\partial u_{10}}{\partial y} = 0. \tag{22}$$

Solving the above equation we can obtain the solution as follows,

$$\theta_{00} = 1, \tag{23}$$

$$u_{00} = \frac{-\frac{\partial p}{\partial z} - G - f sinv}{\sigma^2}\left[1 - \frac{cosh\left(\frac{\sigma}{\sqrt{\alpha}}\right)y}{cosh\left(\frac{\sigma}{\sqrt{\alpha}}\right)\eta}\right] - 1, \tag{24}$$

$$\theta_{01} = 0, \tag{25}$$

$$\begin{aligned} u_{01} = &-\frac{1}{\sigma^2}\frac{\partial p_{01}}{\partial x}\left[1 - \frac{cosh\left(\frac{\sigma}{\sqrt{\alpha}}\right)y}{cosh\left(\frac{\sigma}{\sqrt{\alpha}}\right)H}\right] - \frac{9a_{11}^3}{32\sigma^2} \\ &\left[sinh\left(\frac{\sigma}{\sqrt{\alpha}}\right)y - tanh\left(\frac{\sigma}{\sqrt{\alpha}}\right)H cosh\left(\frac{\sigma}{\sqrt{\alpha}}\right)y\right] \\ &+\frac{a_{11}^3}{32\sigma^2}\left[\frac{sinh\left(\frac{3\sigma}{\sqrt{\alpha}}\right)\eta}{cosh\left(\frac{\sigma}{\sqrt{\alpha}}\right)\eta} - \frac{12\sigma\eta}{\sqrt{\alpha}}\right]cosh\left(\frac{\sigma}{\sqrt{\alpha}}\right)y, \end{aligned} \tag{26}$$

$$\theta_{10} = -a_{11}^2\left[\frac{\eta^2 - y^2}{4} - \frac{cosh\left(\frac{2\sigma}{\sqrt{\alpha}}\right)\eta - cosh\left(\frac{2\sigma}{\sqrt{\alpha}}\right)y}{\frac{8\sigma^2}{\alpha}}\right], \tag{27}$$

$$u_{10} = -\frac{1}{\sigma^2}\frac{\partial p_{10}}{\partial x}\left[1 - \frac{cosh\left(\frac{\sigma}{\sqrt{\alpha}}\right)y}{cosh\left(\frac{\sigma}{\sqrt{\alpha}}\right)\eta}\right] - \frac{a_{11}^2 G}{\left(\frac{24\sigma^4}{\alpha^2}\right)}\left[1 - \frac{cosh\left(\frac{2\sigma}{\sqrt{\alpha}}\right)\eta}{cosh\left(\frac{\sigma}{\sqrt{\alpha}}\right)\eta}\right]$$
$$cosh\left(\frac{\sigma}{\sqrt{\alpha}}\right)y + \frac{a_{11}^2 G}{\left(\frac{8\sigma^2}{\alpha^2}\right)} cosh\left(\frac{2\sigma}{\sqrt{\alpha}}\right) - \frac{a_{11}^2 G}{8\sigma^2} \tag{28}$$
$$\left[\frac{cosh\left(\frac{\sigma}{\sqrt{\alpha}}\right)y}{cosh\left(\frac{\sigma}{\sqrt{\alpha}}\right)\eta}\right] - \frac{a_{11}G}{16\sigma^2}(y^2 - 4\eta^2 + 2)$$

The volume flux can be obtained as,

$$Q = \int_0^n u(x, y)dy, \tag{29}$$

and mean flow is given by,

$$F = \int_0^n u\,dy, \tag{30}$$

where $F = (F_{00} + \beta F_{01} + \ldots) + N(F_{10} + \beta F_{11} + \ldots)$.

Solving for $F$ and rearranging to get the pressure gradient we obtain,

$$\frac{\partial p_{00}}{\partial x} = \frac{\sigma^2 q_{00}}{-\eta + \frac{\sigma}{\sqrt{\alpha}} tanh\left(\frac{\sigma}{\sqrt{\alpha}}\right)\eta} + G + f sinv, \tag{31}$$

$$\frac{\partial p_{01}}{\partial x} = \frac{\sigma^2 q_{01} - \sigma^2 a_{12}}{-\eta + \frac{\sigma}{\sqrt{\alpha}} tanh\left(\frac{\sigma}{\sqrt{\alpha}}\right)\eta} \tag{32}$$

$$\frac{\partial p_{10}}{\partial x} = \frac{\sigma^2 (q_{10} + a_{13})}{\eta \mid \frac{\sigma}{\sqrt{\alpha}} tanh\left(\frac{\sigma}{\sqrt{\alpha}}\right)\eta} \tag{33}$$

Hence the pressure gradient is given by,

$$\frac{\partial p}{\partial x} = \frac{\sigma^2 q + (a_{11}\beta + a_{13}N)\sigma^2}{-\eta + \frac{\sigma}{\sqrt{\alpha}} tanh\left(\frac{\sigma}{\sqrt{\alpha}}\right)\eta} + G + f sinv, \tag{34}$$

The pressure rise is calculated as,

$$\Delta p = \int_0^1 \frac{\partial p}{\partial x} dx. \tag{35}$$

The constants are listed in the appendix.

The above physical quantities are numerically evaluated and graphically depicted.

## 3 Results and Discussion

An inclined porous channel is considered to have peristaltic transport of a Prandtl fluid subjected to the temperature gradient. The pressure gradient, velocity, and pressure rise obtained from the regular perturbation are computed numerically and graphically depicted for various parameters arising out of the study. Figures 2, 3, 4, 5, 6, 7 and 8 show the velocity profile, 9–14 show the pressure gradient, 15–19 show the temperature profile, 20–24 depict the rise in pressure from the middle of the channel to the boundary.

The velocity profile is obtained using the perturbation method assuming symmetry and slip conditions at the boundaries. Velocity is influenced by heat transfer $(N\text{and}Gr)$, inclination angle $(\upsilon)$, amplitude $(\phi)$, permeability $(\sigma)$ and non-Newtonian parameter $\beta$ arising out of Prandtl fluid. The rate of flow $Q$ is also a factor influencing the flow. Figure 2 shows the variation of Axial velocity in $y$ direction with flow rate $Q$. As $Q$ increases velocity also increases which is an obvious effect. Velocity shows parabolic nature between the middle of the channel and upper boundary. The effect of $Q$ is very significant.

The effect of inclination is analysed in Fig. 3 by taking $\upsilon = 0$ (horizontal), $\upsilon = \frac{\pi}{4}$ (inclined) and $\upsilon = \frac{\pi}{2}$ (verticle). The velocity is maximum when $\upsilon = 0$ i.e., the

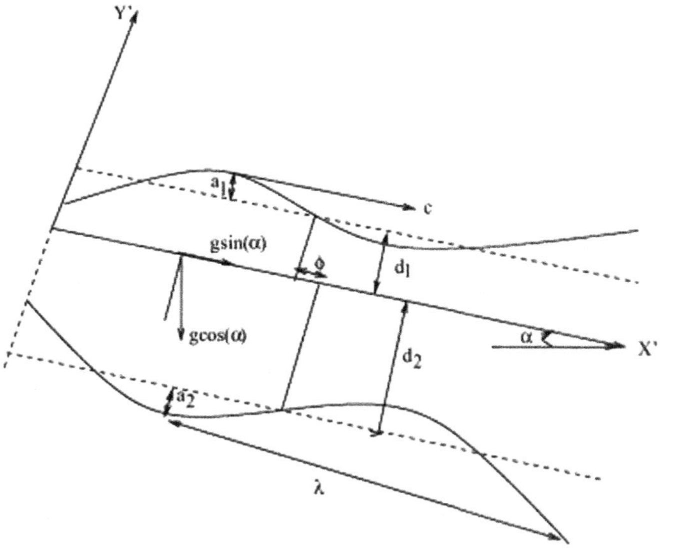


**Fig. 1** Physical configuration

Fig. 2 Axial velocity versus $y$

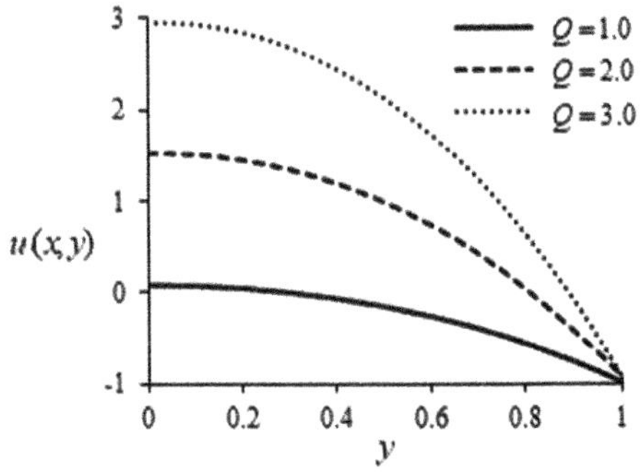


channel is horizontal and decreases with the inclination. The velocity in the verticle channel is the least. The effect is predominant in the middle of the channel than near the boundary of the channel. The effect of heat transfer is analysed by varying perturbation parameter N. The effect on velocity is not very significant but shows a decrease in velocity as $N$ increase. The effect of $G$ also analyses heat transfer effect by variation of Grashof number and the effect is very insignificant. The velocity shows a very slight increase as $G$ increases.

Fig. 3 Axial velocity versus $y$

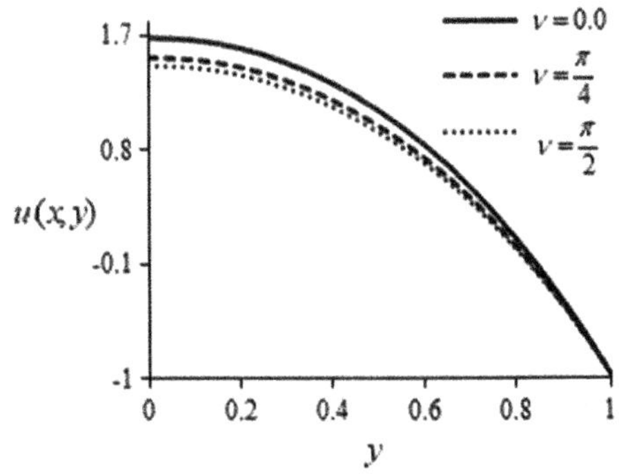


Fig. 4 Axial velocity versus $y$

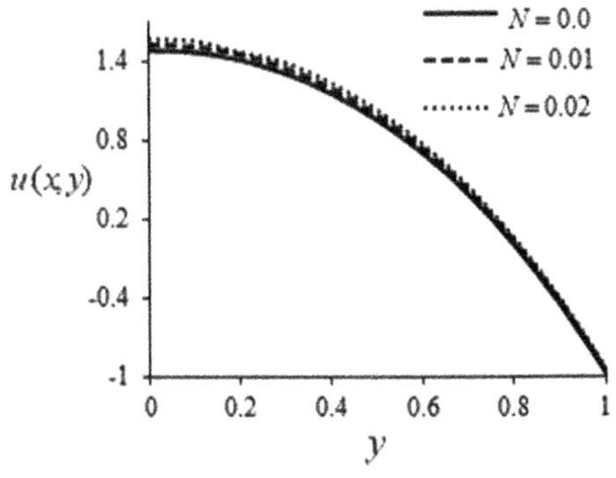

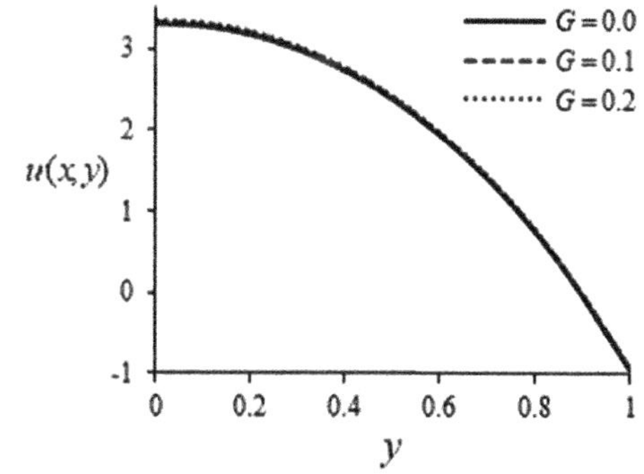


**Fig. 5** Axial velocity versus $y$

Figure 6 analyses the effect of amplitude of the sinusoid wave $\phi$ and again $\phi$ is not an influencing factor. The velocity shows different effects at the middle of the channel and at the wall. There is a slight decrease of velocity with increasing $\phi$ at the middle of the channel but an increase in $\phi$ results in an increase in velocity near the wall. The wall of the channel is under deformation due to the waveform where this effect is neutralized around the middle of the channel. Figure 7 shows the effect of the non-Newtonian parameter $\beta$ which also has some effects similar to $\phi$. As $\beta$ increases, the velocity decreases due to the effect of a non-linear parameter at the middle of the channel. The effect almost becomes negligent at $y = 0.8$ and reverses close to the wall. This is due to an increase in resistance to flow as $\beta$ increases and near the wall motion, the wall motion also affects.

Figure 8 shows velocity profile in the axial direction which is similar to the pressure gradient in geometry replicating sinusoidal waves. The effects of inclination and non-Newtonian parameter $\beta$ are analysed. As $\beta$ increases, there is a significant decrease in velocity and higher $\beta$ signifies more resistance to flow. $\upsilon = 0$ represents horizontal channel, $\upsilon = \frac{\pi}{2}$ represent verticle channel. Increase of $\upsilon$ results in reduced velocity. The effect is more evident for higher $\beta$ than at $\beta = 0.001$. This also indicates the resistance to flow enhances the effect of inclination on flow. Here we have taken the following as standard values: or$a \rightarrow \beta = 0.001, \upsilon = 0$, $b \rightarrow \beta = 0.001, \upsilon = \frac{\pi}{4}$, $c \rightarrow \beta = 0.01, \upsilon = \frac{\pi}{4}$ and $a \rightarrow \beta = 0.001, \upsilon = \frac{\pi}{2}$, $e \rightarrow \beta = 0.01, \upsilon = 0$, $f \rightarrow \beta = 0.01, \upsilon = \frac{\pi}{2}$.

The effect of different parameters $\sigma$, $Q$, $\upsilon$, $N$, and$\phi$ on pressure gradient is analysed in Figs. 9, 10, 11, 12, 13 and 14. The effect of an increase in the rate of flow $Q$, perturbation parameter $N$ and Grashof number $G$ is to increase pressure gradient. The effect of these parameters is more significant. This is due to the fact that for a given $Q$, velocity tries to be maintained by altering the pressure gradient. These effects are observed in Figs. 9, 11 and 12 respectively.

Figure 10 shows the effect of inclination on the pressure gradient. The pressure gradient was high for the horizontal than for the vertical channel. As $\upsilon$ increases $\frac{dp}{dx}$ also increase. The amplitude $\phi$ influence pressure gradient significantly as compared to velocity. The pressure gradient also varies as the sinusoidal wave propagates along an axis, a similar wave pattern is seen. As $\phi$ increases $\frac{dp}{dx}$ increases. This is evident in

Fig. 13. The effect of $\beta$ is seen in Fig. 14 which also influences in the same manner as $\phi$ but to a lesser magnitude.

The effect of different parameters on temperature $\theta$ is shown in Figs. 15, 16, 17, 18 and 19. $\theta$ slowly decreases between the middle of the channel and $y = 0.8$ and there is a decrease in $\theta$ near the boundary as the boundary is maintained at a constant temperature. The fluid regulates the temperature hence more heat is carried towards the middle. The effect of rate of flow is significant as $Q$ increases, $\theta$ also increase showcasing the effect of convection. The effect of inclination is small comparatively and more heat transfer is affected in verticle channel than horizontal. These effects are shown in Figs. 15 and 16 respectively.

Figure 17 shows the effect of the perturbation parameter $N$ and Grashof number $G$ is analysed in Fig. 18. Both cases as $N$ and $G$ increases, $\theta$ also increases. The effect of amplitude is analysed in Fig. 19 and the effect of $\phi$ is negligible and similar to that of velocity.

Figures 20, 21, 22, 23 and 24 showcase the rise of pressure along the channel versus different parameters. Figure 20 shows a drop of pressure rise $\Delta p$ with increasing mean rate of slow. The effect of $\beta$ and $v$ are very insignificant on pressure rise. The pressure rise versus $\beta$ is shown in Fig. 21. As $\beta$ increases, $\Delta p$ also increases due to an increase in resistance to flow. The effect of inclination is significant along increasing $\beta$. Increasing $\beta$ enhances the effect of inclination on pressure rise.

As permeability $\sigma$ increases, pressure rise also more which is seen in Fig. 22. The increase in permeability results in loss of fluid hence an increase in pressure. The effect of amplitude is seen in Fig. 23. As $\phi$ increases $\Delta p$ decreases. The same effect is seen with increasing perturbation parameter $N$. The effect of heat transfer on pressure rise is to decrease $\Delta p$ and $\Delta p$ also decreases from horizontal to verticle channel.

**Fig. 6** Axial velocity versus $y$

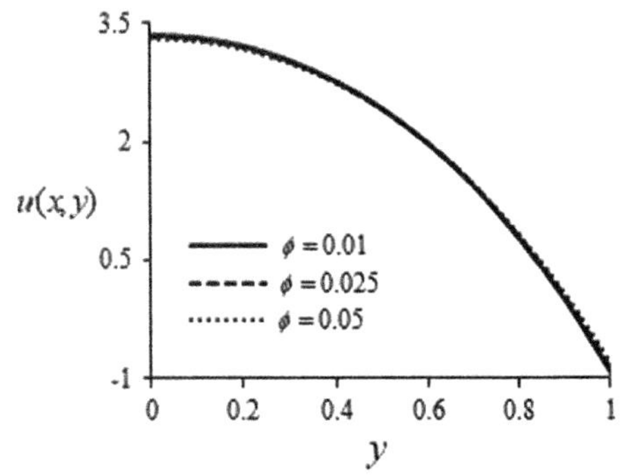

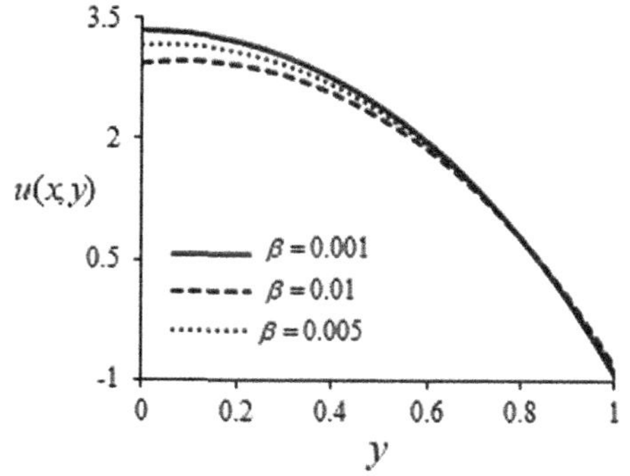


**Fig. 7** Axial velocity versus $y$

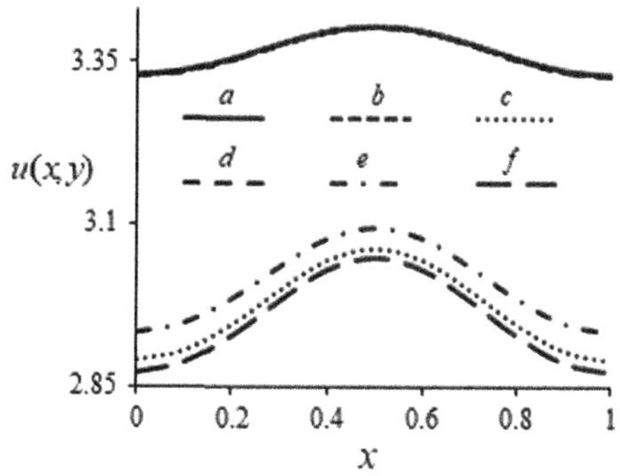


**Fig. 8** Axial velocity versus $x$

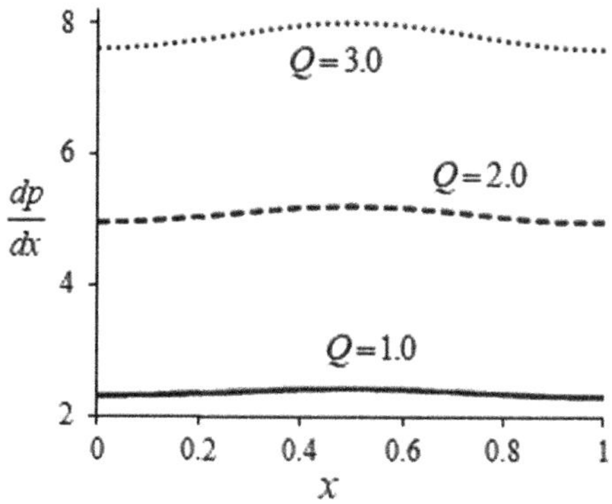


**Fig. 9** Pressure gradient in axial direction

## 3.1 Numerical Comparison

The table which is below showcases a comparison with a numerical solution. A numerical solution is obtained for a particular case where the channel is horizontal in absence of heat transfer. The wall motion is also absent as $\phi = 0$. The shooting

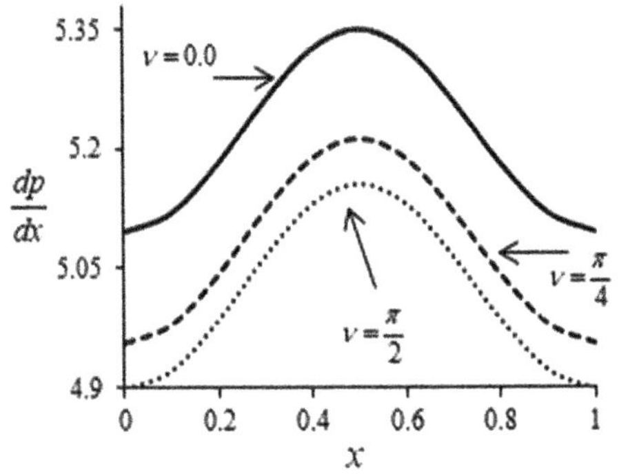


**Fig. 10** Pressure gradient in axial direction

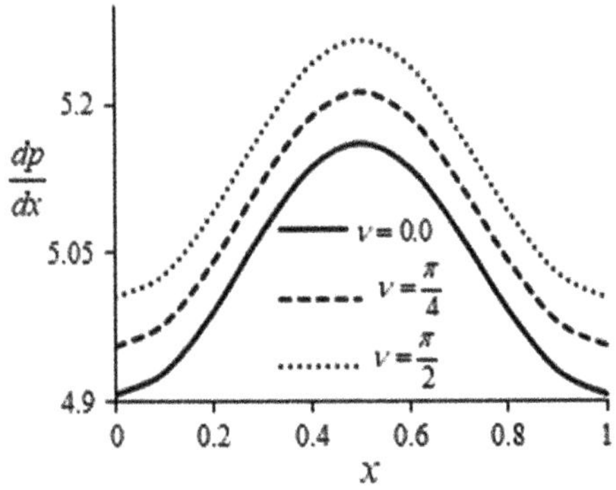


**Fig. 11** Pressure gradient in axial direction

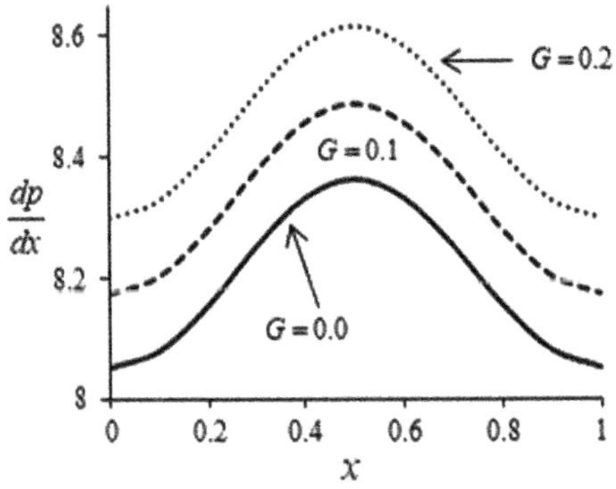


**Fig. 12** Pressure gradient in axial direction

method is adopted to solve the resulting equation: $\frac{\partial}{\partial y}\left[\alpha\frac{\partial u}{\partial y}+\beta\left(\frac{\partial u}{\partial y}\right)^3\right]=\frac{\partial p}{\partial x}$, A similar curve from the present study by taking $\phi = 0, G = 0, N = 0$ is also compared and compared for $\beta = 0.01, \alpha = 0.5, Q = 2.0$. The curves obtained are shown in Table 1. There is some error seen due to truncation otherwise results are in good agreement.

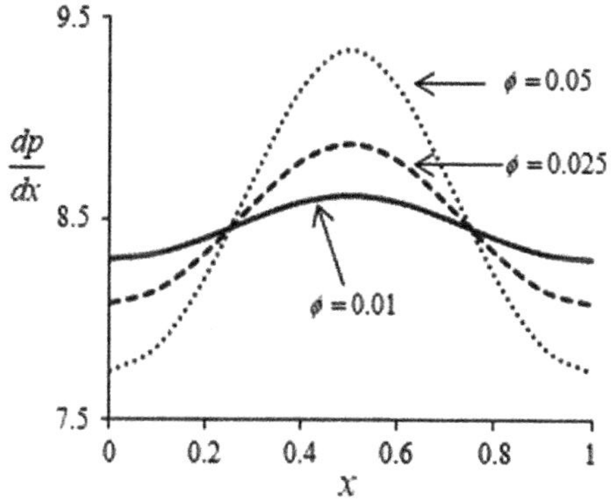


**Fig. 13** Pressure gradient in axial direction

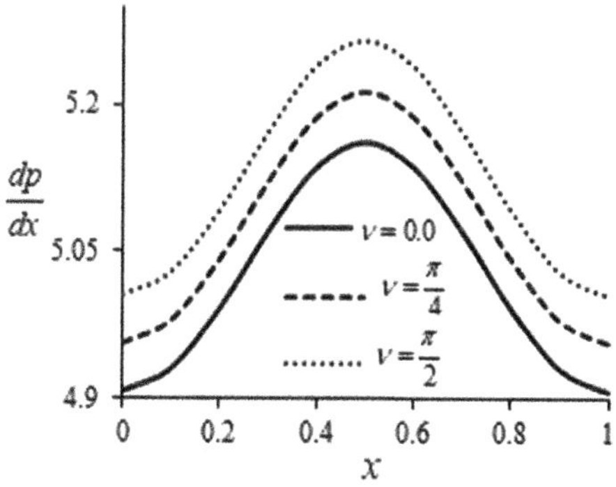


**Fig. 14** Pressure gradient in axial direction

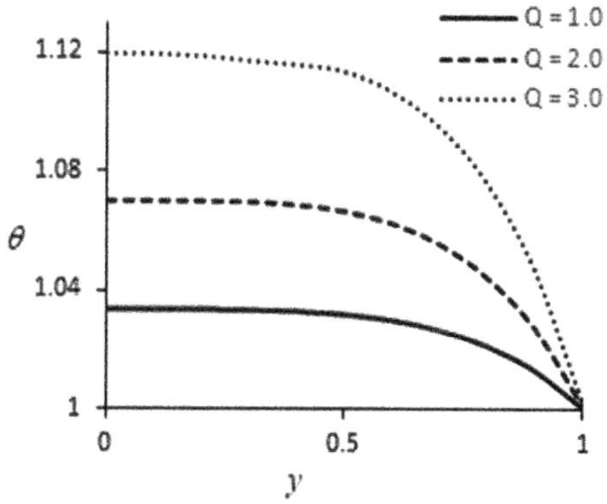


**Fig. 15** Temperature profile

## 4 Conclusions

The present study analyses the peristaltic flow and heat transfer in an inclined porous channel with Prandtl fluid. The non-linear equations are effectively reduced using the regular perturbation method. Analytical solutions are obtained and graphically

Fig. 16 Temperature profile

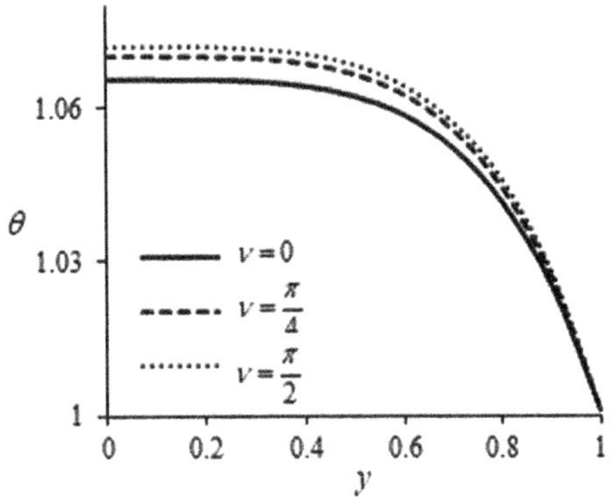


Fig. 17 Temperature profile

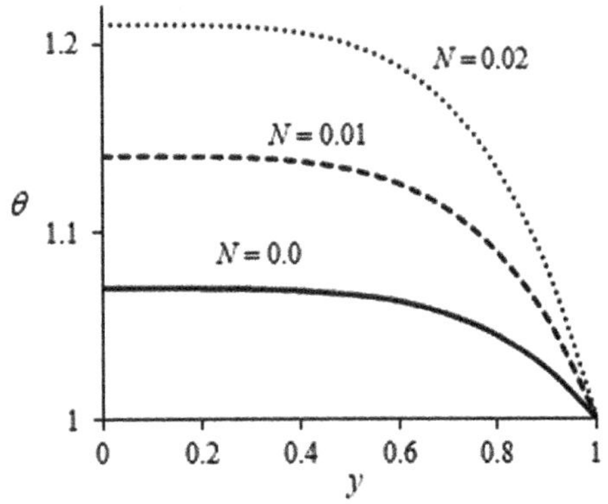


Fig. 18 Temperature profile

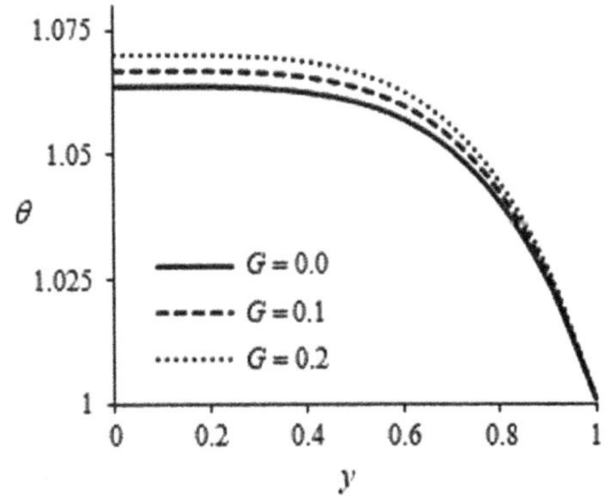


depicted. A numerical approximate solution by shooting method for a particular case of zero inclination is considered and the solution is compared. Velocity starts from −1 due to the moving frame of reference and shows parabolic nature in $y$—direction

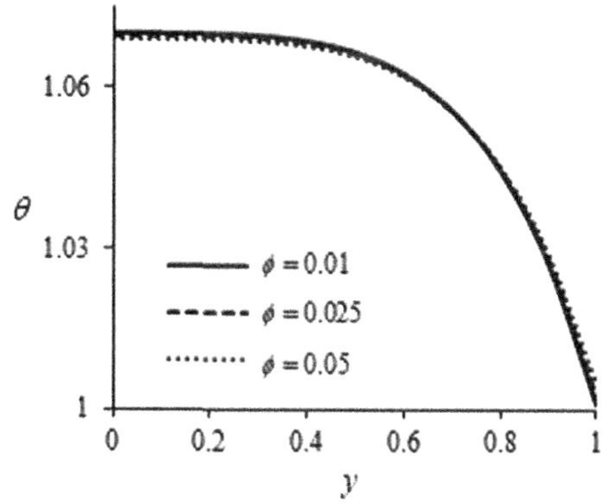


**Fig. 19** Temperature profile

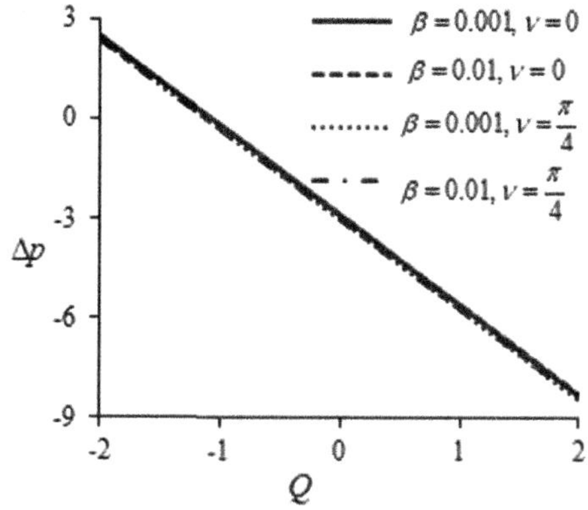


**Fig. 20** Rise in pressure versus rate of flow

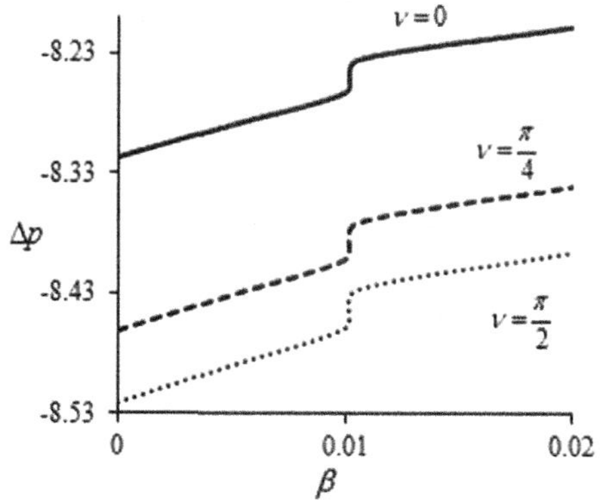


**Fig. 21** Rise in pressure versus non-newtonian parameter

and pressure pulse is replicated in $x$—direction. The effect of heat transfer is not so significant on velocity.

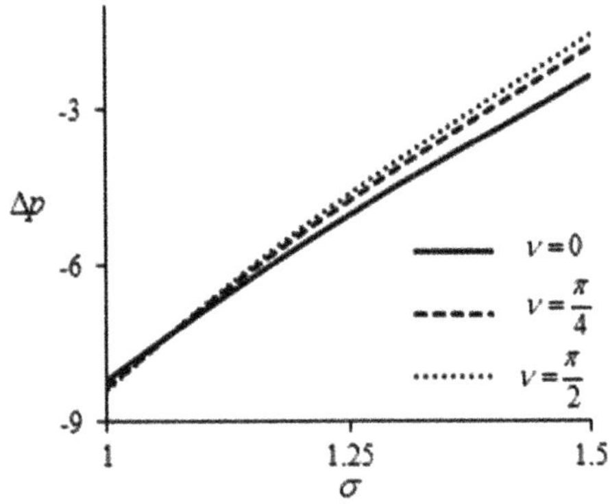


**Fig. 22** Rise in pressure versus permeability

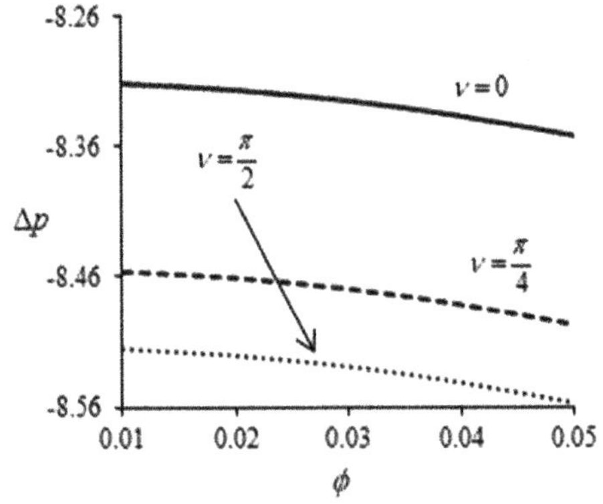


**Fig. 23** Rise in pressure versus inclination parameter

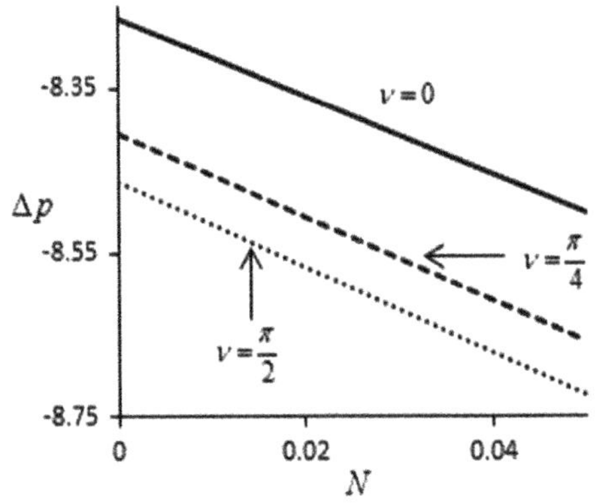


**Fig. 24** Rise in pressure versus perturbation parameter

The effect of a non-linear term is to increase resistance to flow and also to enhance the effects of inclination on velocity on velocity as well as heat transfer. The perturbation parameter indicates the effect of $\theta$ on pressure gradient and velocity. The effect of $\theta$ is more on pressure gradient than on velocity. As $\beta \to 0$, the fluid becomes

**Table 1** Comparison of Perturbation and Exact solution

| $Q$ | $\Delta p$ (Perturbation Solution) $\upsilon = 0$ | $\Delta p$ (Exact Solution) $\upsilon = 0$ | $\Delta p$ (Perturbation Solution) $\upsilon = \frac{\pi}{4}$ | $\Delta p$(Exact Solution) $\upsilon = \frac{\pi}{4}$ |
|---|---|---|---|---|
| −2 | 2.49029 | 2.49029 | 2.34887 | 2.34887 |
| −1 | −0.19971 | −0.19971 | −0.341136 | −0.341136 |
| 0 | −2.8897 | −2.8897 | −3.03114 | −3.03114 |
| 1 | −5.5797 | −5.5797 | −5.72115 | −5.72115 |
| 2 | −8.2697 | −8.2697 | −8.4112 | −8.4112 |

Newtonian. The physical quantities like velocity and pressure gradient show higher values for the horizontal channel than the inclined channel. The effect of wall motion is not so significant on the physical quantities.

## Appendix

$$a_{11} = \frac{1}{\sigma^2}\left(\frac{dp_{00}}{dx} - G - f sin\upsilon\right)\frac{\left(\frac{\sigma}{\sqrt{\alpha}}\right)}{cosh\left(\frac{\sigma}{\sqrt{\alpha}}\eta\right)},\ a_12 = -\frac{9a_1 1^3}{32\sigma^2}b_1 + \frac{a_{11}^3}{32\sigma^2}b_2,$$

$$a_{11} = \frac{1}{\sigma^2}\left(\frac{dp_{00}}{dx} - G - f sin\upsilon\right)\frac{\left(\frac{\sigma}{\sqrt{\alpha}}\right)}{cosh\left(\frac{\sigma}{\sqrt{\alpha}}\eta\right)},\ a_{12} = -\frac{9a_{11}^3}{32\sigma^2}b_1 + \frac{a_{11}^3}{32\sigma^2}b_2,$$

$$b_1 = \frac{\sigma}{\sqrt{\alpha}}cosh\left(\frac{\sigma}{\sqrt{\alpha}}\eta\right) - 1 - \frac{\sigma}{\sqrt{\alpha}}tanh\left(\frac{\sigma}{\sqrt{\alpha}}\eta\right)\frac{\sigma}{\sqrt{\alpha}}sinh\left(\frac{\sigma}{\sqrt{\alpha}}\eta\right),$$

$$b_2 = \left\{\frac{sinh\left(\frac{3\sigma}{\sqrt{\alpha}}\eta\right)}{cosh\left(\frac{3\sigma}{\sqrt{\alpha}}\eta\right)} - \frac{12\sigma\eta}{\sqrt{\alpha}}\right\}\frac{\sigma}{\sqrt{\alpha}}sinh\left(\frac{\sigma}{\sqrt{\alpha}}\eta\right),\ b_4 = \eta cosh\left(\frac{2\sigma}{\sqrt{\alpha}}\eta\right),$$

$$b_3 = \left\{1 - \frac{2cosh\left(\frac{2\sigma}{\sqrt{\alpha}}\eta\right)}{cosh\left(\frac{\sigma}{\sqrt{\alpha}}\eta\right)}\right\}\frac{\sigma}{\sqrt{\alpha}}sinh\left(\frac{\sigma}{\sqrt{\alpha}}\eta\right),\ b_5 = tanh\left(\frac{\sigma}{\sqrt{\alpha}}\eta\right),$$

$$b_6 = \frac{11\eta^3}{3} - 2\eta,\ a_{13} = \frac{a_{11}^2 G}{24\frac{\sigma^4}{\alpha^2}}b_3 - \frac{a_{11}^2 G}{8\frac{\sigma^4}{\alpha^2}}b_4 + \frac{a_{11}^2 G}{8\frac{\sigma}{\sqrt{\alpha}}}b_5 - \frac{a_{11}G}{16\sigma^2}b_6$$

## References

1. Latham, T.W.: Fluid Motion in a Peristaltic Pump—MS Thesis. Massachusetts Institute of Technology, Cambridge MA (1966)
2. Shapiro, A.H., Jaffrin, M.Y., Wienberg, S.L.: Peristaltic pumping with long wavelengths at low reynolds number. J. Fluid Mech. **37**, 799–825 (1969)
3. Yin, F., Fung, Y.C.: Peristaltic waves in circular cylindrical tubes. J. Appl. Mech. **36**(3), 579–587 (1969)
4. Gupta, B.B., Seshadri, V.: Peristaltic pumping in non-uniform tubes. J. Biomech. **9**(2), 105–109 (1976)
5. Raju, K.K., Devanathan, R.: Peristaltic motion of a non-newtonian fluid. Rheological Acta **11**(2), 170–178 (1972)
6. Misra, J.C., Pandey, S.K.: Persistaltic transport of blood in small vessels: study of mathematical model. Comput. Math. Appl. **43**(8–9), 1183–1193 (2002)
7. Mekheimer, K.S., Al-Arabi, T.H.: Nonlinear peristaltic transport of MHD flow through a porous medium. Int. J. Math. Sci. **26**, 1663–1682 (2003)
8. Kumari, S.V.H.N., Ravi Kumar, Y.V.K., Ramana Murthy, M.V., Sreenadh, S.: Peristaltic pumping of a conducting Jeffrey fluid in a vertical porous channel with heat transfer. Adv. Appl. Sci. Re. **6**, 439–453 (2011)
9. Eldabe, N.T., Zaghrout, A.S., Shawky, H.M.: Awad,: AS effects of chemical reaction with heat and mass transfer on peristaltic motion of power-law fluid in an asymmetric channel with wall's properties. Int. J. Res. Appl. Sci. Re. **15**, 280–292 (2013)
10. Hayat, T., Ali, N., Asghar, S.: Hall effects on peristaltic flow of a maxwell fluid in a porous medium. Phys. Lett. A. Re. **363**, 397–403 (2007)
11. Shaaban, A.A., Abou-Zeid, M.Y.: Effects of heat and mass transfer on MHD peristaltic flow of a non-newtonian fluid through porous medium between two coaxial cylinders. Math. Prob. Engg. **363**, 397–403 (2007)
12. Eldabe, N.T., Abou-Zeid, M.Y., Younis, Y.M.: Magnetohydrodynamic peristaltic flow of Jeffrey nanofluid with heat transfer through a porous medium in a vertical tube. Appl. Math. Inf. Sci. **11**(4), 1097–1103 (2017)
13. Selvi, C.K., Haseena, C., Srinivas, A.N.S., Sreenadh, S.: The effect of heat transfer on peristaltic flow of Jeffrey fluid in an inclined porous stratum. IOP Con. Ser.: Mater. Sci. Eng. **263** (2017)
14. Pandey, S.K., Chaube, M.K.: Peristaltic flow of a micropolar fluid through a porous medium in the presence of an external magnetic field. Commun. Nonlinear Sci. Numer. Simul. **16**(9), 3591–3601 (2011)
15. Tripathi, D.: Study of transient peristaltic heat flow through a finite porous channel. Math. Comput. Model. **57**, 1270–1283 (2012)
16. Abdulhadi, A.M., Ahmed, T.S.: Effect of radial magnetic field on peristaltic transport of Jeffrey fluid in curved channel with heat/mass transfer. J. Phys: Conf. Ser. **1003**(1), 012053 (2018)
17. Indira, R., Sreegowrav, KR., Dinesh, PA..: Effect of heat and transfer on peristaltic flow of couple stress fluid in oesophagus. Int. J. Pure Appl. Mech. **120**(6), 1321–1335 (2018)
18. Rashmi, K.R., Indira, K.R., Jagadeesha, S.: Peristaltic flow of couple-stress fluid in doubly connected region with reference to endoscope. Palest. J. Math. **10**, 1–5 (2021)
19. Nadeem, S., Sadat, H., Akbar, N.S.: Analysis of peristaltic flow for a prandtl fluid model in an endoscope. J. Power Technol. **94**(2), 1–11 (2012)
20. Indira, R., Priyanka, NB., Jagadeesha, S.: Peristaltic Flow and Heat Transfer Through a Prandtl Fluid in Vertical Annulus. Chapter 16, Recent Advances in Mechanical Engineering, pp. 173–186 (2023)
21. Vajravelu, K., Radhakrishmacharya, G., Radhakrishmurthy, V.: Peristaltic flow and heat transfer in a vertical porous annulus with long wave approximation. Int. J. of Non-Linear Mech. **42**(5), 754–759 (2007)

# A Multiscale Model of Stokes–Cahn–Hilliard Equations in a Porous Medium: Modeling, Analysis and Homogenization

**Nitu Lakhmara and Hari Shankar Mahato**

**Abstract** We consider a phase-field model for a mixture of two immiscible, incompressible porous media flow including surface tension effects. At micro-scale, the model comprises a strongly coupled system of Stokes–Cahn–Hilliard equations. An evolving diffuse interface having finite width independent of the scale parameter $\varepsilon$ is separating the fluids in the considered model. In order to investigate the well-posedness of system at micro-scale, we first derived some a-priori estimates. With the help of two-scale convergence and unfolding operator technique we rigorously derived the homogenized equations for the microscopic model. For our purpose, we have used extensions theorems and well-known theories available in the literature beforehand.

**Keywords** Phase-field model · Porous media flow · Stokes equations · Cahn–Hilliard equations · Existence of solution · Homogenization · Asymptotic expansion method · Two-scale convergence · Periodic unfolding

## 1 Introduction

We study a binary-fluid model where the considered fluids are incompressible and immiscible. The domain $U \subset \mathbb{R}^n$, $n = 2, 3$ is occupied by the binary-fluid mixture. On the time interval $S = (0, T)$, the model comprises a system of steady Stokes–Cahn–Hilliard equations

N. Lakhmara (✉) · H. S. Mahato
Indian Institute of Technology Kharagpur, West Bengal 721302, India
e-mail: nitulakhmara@gmail.com

H. S. Mahato
e-mail: hsmahato@maths.iitkgp.ac.in
URL: https://sites.google.com/view/nitulakhmara/home



B. Rushi Kumar et al. (eds.), *Mathematics and Computing*, Springer Proceedings in Mathematics & Statistics 415, https://doi.org/10.1007/978-981-19-9307-7_46

$$-\mu\Delta\mathbf{u} + \nabla p = \lambda w \nabla c \qquad \text{in } (0, T) \times U, \tag{1.1a}$$
$$\nabla.\mathbf{u} = 0 \qquad \text{in } (0, T) \times U, \tag{1.1b}$$
$$\partial_t c + \mathbf{u}.\nabla c = \Delta w \qquad \text{in } (0, T) \times U, \tag{1.1c}$$
$$w = -\Delta c + f(c) \qquad \text{in } (0, T) \times U, \tag{1.1d}$$

where $\mathbf{u}$ and $w$ are the unknown velocity and chemical potential, respectively. $\mu$ is the viscosity and $\lambda$ is the interfacial width parameter. Here $c$ represents microscopic concentration of one of the fluids with values lying in the interval $[-1, 1]$ in the considered domain and $(-1, 1)$ within the thin diffused interface of uniform width proportional to $\lambda$. The term $f(c) = F'(c)$, where $F$ is a homogeneous free energy functional that penalizes the deviation from the physical constraint $|c| \leq 1$. In our work, we consider $F$ to be a quadratic double-well free energy functional, i.e., $F(s) = \frac{1}{4}(s^2 - 1)^2$. One can choose $F$ as a logarithmic or a non-smooth (obstacle) free energy functional, cf. [3, 4]. The nonlinear term $c\nabla w$ in (1.1a) models the surface tension effects, and the advection effect is modeled by the term $\mathbf{u} \cdot \nabla c$ in (1.1c). The system (1.1a)-(1.1d) represent the steady Stokes equations for incompressible fluid and Cahn–Hilliard equations, respectively.

## 1.1 The Model

We consider $U$ as a bounded domain with a sufficiently smooth boundary $\partial U$ in $\mathbb{R}^n$, $n = 2, 3$, $S := (0, T)$ denotes the time interval for any $T > 0$, and the unit reference cell $Y := (0, 1)^n \subset \mathbb{R}^n$. $Y_p$ and $Y_s$ represent the pore and solid part of $Y$, respectively, which are mutually distinct, i.e., $Y_s \cap Y_p = \emptyset$, also $Y = Y_p \cup Y_s$. The solid boundary of $Y$ is denoted as $\Gamma_s = \partial Y_s$, see Fig. 1. The domain $U$ is assumed to be periodic and is covered by a finite union of the cells $Y$. In order to avoid technical difficulties, we postulate that: solid parts do not touch the boundary $\partial U$, solid parts do not touch each other and solid parts do not touch the boundary of $Y$. Let $\varepsilon > 0$

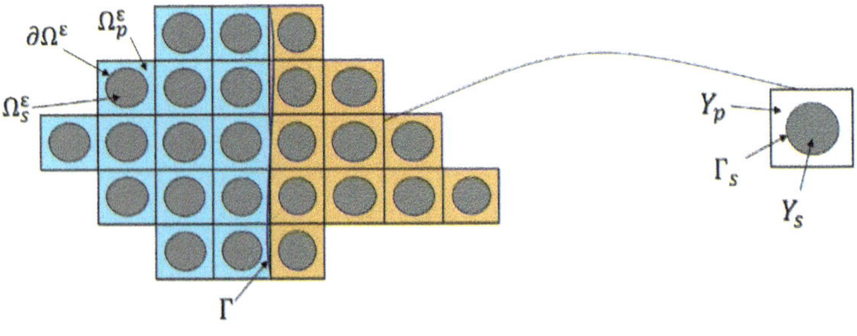


**Fig. 1** (left) Porous medium $U = U_p^\varepsilon \cup U_s^\varepsilon$ as a periodic covering of the reference cell $Y = Y_p \cup Y_s$ (right). The blue interface $\Gamma$ is the macroscopic interface between two fluids occupying the pore space $U_p^\varepsilon$

be the scale parameter. We define the pore space $U_p^\varepsilon := \bigcup_{\mathbf{k}\in\mathbb{Z}^n} Y_{p_k} \cap U$, the solid part as $U_s^\varepsilon := \bigcup_{\mathbf{k}\in\mathbb{Z}^n} Y_{s_k} \cap U = U \backslash U_p^\varepsilon$ and $\Gamma^\varepsilon := \bigcup_{\mathbf{k}\in\mathbb{Z}^n} \Gamma_{s_k}$, where $Y_{p_k} := \varepsilon Y_p + k$, $Y_{s_k} := \varepsilon Y_s + k$ and $\Gamma_{s_k} = \bar{Y}_{p_k} \cap \bar{Y}_{s_k}$.

Let $\chi(y)$ be the $Y$-periodic characteristic function of $Y_p$ defined by

$$\chi(y) = \begin{cases} 1 & y \in Y^p, \\ 0 & y \in Y - Y^p. \end{cases} \tag{1.2}$$

We assume that $U_p^\varepsilon$ is connected and has a smooth boundary. We consider the situation where the pore part $U_p^\varepsilon$ is occupied by the mixture of two immiscible fluids separated by an evolving macroscopic interface $\Gamma : [0, T] \to U$ represented by the blue part in Fig. 1, and includes the effects of surface tension on the motion of the interface. We model the flow of the fluid mixture on the pore-scale using a phase-field approach motivated by the Stokes–Cahn–Hilliard system (1.1) in [2]. The velocity of the fluid mixture is assumed to be $\mathbf{u}^\varepsilon = \mathbf{u}^\varepsilon(t, x), (t, x) \in S \times U_p^\varepsilon$ which satisfies the stationary Stokes equation. The order parameter $c^\varepsilon$ plays the role of microscopic concentration and the chemical potential $w^\varepsilon$ satisfies the Cahn–Hilliard equation. $p^\varepsilon$ is the fluid pressure. The term $\lambda c^\varepsilon \nabla w^\varepsilon$ models the surface tension forces which acts on the macroscopic interface between the fluids. Fluid density is taken to be 1. Then, the Stokes–Cahn–Hilliard system of equations is given by

$$-\mu\varepsilon^2 \Delta \mathbf{u}^\varepsilon + \nabla p^\varepsilon = -\lambda c^\varepsilon \nabla w^\varepsilon \qquad S \times U_p^\varepsilon, \tag{1.3a}$$

$$\nabla . \mathbf{u}^\varepsilon = 0 \qquad S \times U_p^\varepsilon, \tag{1.3b}$$

$$\mathbf{u}^\varepsilon = 0 \qquad S \times \partial U_p^\varepsilon, \tag{1.3c}$$

$$\partial_t c^\varepsilon + \varepsilon \mathbf{u}^\varepsilon . \nabla c^\varepsilon = \Delta w^\varepsilon \qquad S \times U_p^\varepsilon, \tag{1.3d}$$

$$w^\varepsilon = -\varepsilon^2 \Delta c^\varepsilon + f(c^\varepsilon) \qquad S \times U_p^\varepsilon, \tag{1.3e}$$

$$\partial_n c^\varepsilon = 0 \qquad S \times \partial U_p^\varepsilon, \tag{1.3f}$$

$$\partial_n w^\varepsilon = 0 \qquad S \times \partial U_p^\varepsilon, \tag{1.3g}$$

$$c^\varepsilon(0, x) = c_0(x) \qquad U_p^\varepsilon, \tag{1.3h}$$

where $\frac{\partial c^\varepsilon}{\partial \mathbf{n}} = \partial_n c^\varepsilon$ and $f(s) = s^3 - s = F'(s) = \frac{1}{4}(s^2 - 1)^2$ is the double-well free energy. The above scaling for the viscosity is such that the velocity $\mathbf{u}^\varepsilon$ has a nontrivial limit as $\varepsilon$ goes to zero. Also, $0 \le \alpha, \beta, \gamma \le 2$ where $\alpha, \beta, \gamma \in \mathbb{R}$. We denote (1.3a)–(1.3h) by $(\mathcal{P}^\varepsilon)$.

## 2 Preliminaries and Notation

Let $\theta \in [0, 1]$ and $1 \le r, s \le \infty$ be such that $\frac{1}{r} + \frac{1}{s} = 1$. Assume that $\Xi \in \{U, U_p^\varepsilon, U_s^\varepsilon\}$ and $l \in \mathbb{N}_0$, then as usual $L^r(\Xi)$ and $H^{l,r}(\Xi)$ denote the Lebesgue and Sobolev spaces with their usual norms and they are denoted by $||.||_r$ and $||.||_{l,r}$, cf. [5]. The extension and restriction operators are denoted by $E$ and $R$, respectively. The symbol $(.,.)_H$ represents the *inner product* on a *Hilbert space* $H$ and $||.||_H$ denotes the corresponding norm. For a Banach space $X$, $X^*$ denotes its dual and the duality pairing is denoted by $\langle . , .\rangle_{X^* \times X}$. By classical trace theorem on *Sobolev space* $H_0^{1,2}(\Xi)^* = H^{-1,2}(\Xi)$. The symbols $\hookrightarrow, \hookrightarrow\hookrightarrow$ and $\underset{\hookrightarrow}{d}$ denote the continuous, compact, and dense embeddings, respectively.

We define the function spaces:

$$\mathbf{H}^1(U) = H^1(U)^n, \quad \mathbf{H}_0^1(U) = H_0^1(U)^n,$$
$$\mathfrak{U}^\varepsilon := \mathbf{H}^1_{div}(U) = \{\eta : \eta \in \mathbf{H}_0^1(U), \nabla \cdot \eta = 0\},$$
$$\mathfrak{C}^\varepsilon = \{c^\varepsilon : c^\varepsilon \in L^\infty(S; H^1(U_p^\varepsilon)), \partial_t c^\varepsilon \in L^2(S; H^1(U_p^\varepsilon)^*)\},$$
$$\mathfrak{W}^\varepsilon = L^2(S; H^1(U_p^\varepsilon)) \text{ and } L_0^2(U) = \{\phi \in L^2(U) : \textstyle\int_U \phi\, dx = 0.\}.$$

We choose $\mathbf{u}^\varepsilon \in \mathfrak{U}^\varepsilon, c^\varepsilon \in \mathfrak{C}^\varepsilon, w^\varepsilon \in \mathfrak{W}^\varepsilon$ and $p^\varepsilon \in L^2(S \times U_p^\varepsilon)$. We will now state few results and lemmas which are used in this paper and proofs of these can be found in literature.

**Lemma 1** *Let $E$ be a Banach space and $E_0$ and $E_1$ be reflexive spaces with $E_0 \subset E \subset E_1$. Suppose further that $E_0 \hookrightarrow\hookrightarrow E \hookrightarrow E_1$. For $1 < p, q < \infty$ and $0 < T < 1$ define $X := \{u \in L^p(S; E_0) : \partial_t u \in L^q(S; E_1)\}$. Then $X \hookrightarrow\hookrightarrow L^p(S; E)$.*

**Lemma 2** (Restriction theorem) *There exists a linear restriction operator $R^\varepsilon$ : $L^2(S; H_0^1(U))^d \longrightarrow L^2(S; H_0^1(U_p^\varepsilon))^d$ such that $R^\varepsilon u(x) = u(x)|_{U_p^\varepsilon}$ for $u \in L^2(S; H_0^1(U))^d$ and $\nabla \cdot R^\varepsilon u = 0$ if $\nabla \cdot R^\varepsilon u = 0$ if $\nabla \cdot u = 0$. Furthermore, the restriction satisfies the following bound*

$$||R^\varepsilon u||_{L^2(S\times U_p^\varepsilon)} + \varepsilon||\nabla R^\varepsilon u||_{L^2(S\times U_p^\varepsilon)} \le C(||u||_{L^2(S\times U)} + \varepsilon||\nabla u||_{L^2(S\times U)}),$$

*where $C$ is independent of $\varepsilon$.*

Similarly, one can define the extension operator from $S \times U_p^\varepsilon$ to $S \times U$, cf. [1, 8].

**Definition 1** (*Two-scale convergence*) A sequence of functions $(u^\varepsilon)_{\varepsilon>0}$ in $L^p(S \times U)$ is said to be two-scale convergent to a limit $u \in L^p(S \times U \times Y)$ if

$$\lim_{\epsilon\to 0} \int_{S\times U} u^\varepsilon(t,x)\phi\left(t, x, \frac{x}{\varepsilon}\right) dx\, dt = \int_{S\times U\times Y} u(t,x,y)\phi(t,x,y)\, dx\, dt\, dy$$

for all $\phi \in L^q(S \times U; C_\#(Y))$.

**Lemma 3** *For $\varepsilon > 0$, let $(u^\varepsilon)_{\varepsilon>0}$ be a sequence of functions, then the following holds:*

*(i) for every bounded sequence $(u^\varepsilon)_{\varepsilon>0}$ in $L^p(S \times U)$ there exists a subsequence $(u^\varepsilon)_{\varepsilon>0}$ (still denoted by same symbol) and an $u \in L^p(S \times U \times Y)$ such that $u^\varepsilon \overset{2}{\rightharpoonup} u$.*

(ii) *let* $u^\varepsilon \to u$ *in* $L^p(S \times U)$*, then* $u^\varepsilon \overset{2}{\rightharpoonup} u$.
(iii) *let* $(u^\varepsilon)_{\varepsilon>0}$ *be a sequence in* $L^p(S; H^{1,p}(U))$ *such that* $u^\varepsilon \overset{w}{\rightharpoonup} u$ *in* $L^p(S; H^{1,p}(U))$*. Then* $u^\varepsilon \overset{2}{\rightharpoonup} u$ *and there exists a subsequence* $u^\varepsilon_{\varepsilon>0}$*, still denoted by same symbol, and an* $u_1 \in L^p(S \times U; H^{1,p}_\#(Y))$ *such that* $\nabla_x u^\varepsilon \overset{2}{\rightharpoonup} \nabla_x u + \nabla_y u_1$.
(iv) *let* $(u^\varepsilon)_{\varepsilon>0}$ *be a bounded sequence of functions in* $L^p(S \times U)$ *such that* $\varepsilon \nabla u^\varepsilon$ *is bounded in* $L^p(S \times U)^n$*. Then there exist a function* $u \in L^p(S \times U; H^{1,p}_\#(Y))$ *such that* $u^\varepsilon \overset{2}{\rightharpoonup} u$, $\varepsilon \nabla_x u^\varepsilon \overset{2}{\rightharpoonup} \nabla_y u$.

**Definition 2** (*Periodic Unfolding*) Assume that $1 \le r \le \infty$. Let $u^\varepsilon \in L^r(S \times U)$ such that for every $t$, $u^\varepsilon(t)$ is extended by zero outside of $U$. We define the unfolding operator $T^\varepsilon : L^r(S \times U) \to L^r(S \times U \times Y)$ as

$$T^\varepsilon u^\varepsilon(t, x, y) = u^\varepsilon\left(t, \varepsilon\left[\frac{x}{\varepsilon}\right] + \varepsilon y\right) \quad \text{for a.e. } (t, x, y) \in S \times U \times Y, \tag{2.1a}$$

$$= 0 \quad \text{otherwise.} \tag{2.1b}$$

For the following definitions and results, interested reader can refer to [7] and references therein.

**Definition 3** Assume that $1 \le r \le \infty$, $u^\varepsilon \in L^r(S \times U)$ and $T^\varepsilon$ is defined as in Definition 3. Then we say that:
(i) $u^\varepsilon$ is weakly two-scale convergent to a limit $u_0 \in L^r(S \times U \times Y)$ if $T^\varepsilon u^\varepsilon$ converges weakly to $u_0$ in $L^r(S \times U \times Y)$.
(ii) $u^\varepsilon$ is strongly two-scale convergent to a limit $u_0 \in L^r(S \times U \times Y)$ if $T^\varepsilon u^\varepsilon$ converges strongly to $u_0$ in $L^r(S \times U \times Y)$.

**Lemma 4** *Let* $(u^\varepsilon)_{\varepsilon>0}$ *be a bounded sequence in* $L^r(S \times U)$*. Then the following statements hold:*

(a) *if* $u^\varepsilon \overset{2}{\rightharpoonup} u$*, then* $T^\varepsilon u^\varepsilon \overset{w}{\rightharpoonup} u$*, i.e.,* $u^\varepsilon$ *is weakly two-scale convergent to a* $u$.
(b) *if* $u^\varepsilon \to u$ *, then* $T^\varepsilon u^\varepsilon \to u$*, i.e.,* $u^\varepsilon$ *is strongly two-scale convergent to* $u$.

**Lemma 5** *Let* $(u^\varepsilon)_{\varepsilon>0}$ *be strongly two-scale convergent to* $u_0$ *in* $L^r(S \times U \times \Gamma)$ *and* $(v^\varepsilon)_{\varepsilon>0}$ *be weakly two-scale convergent to* $v_0$ *in* $L^s(S \times U \times \Gamma)$*. If the exponents* $r, s, \nu \ge 1$ *satisfy* $\frac{1}{r} + \frac{1}{s} = \frac{1}{\nu}$*, then the product* $(u^\varepsilon v^\varepsilon)_{\varepsilon>0}$ *two-scale converges to the limit* $u_0 v_0$ *in* $L^\nu(S \times U \times Y)$*. In particular, for any* $\phi \in L^\mu(S \times U)$ *with* $\mu \in (1, \infty)$ *such that* $\frac{1}{\nu} + \frac{1}{\mu} = 1$ *we have*

$$\int_{S\times U} u^\varepsilon(t, x) v^\varepsilon(t, x) \phi(t, x)\, dx\, dt \overset{\varepsilon\to 0}{\longrightarrow} \int_{S\times U\times Y} u_0(t, x, y) v_0(t, x, y) \phi(t, x)\, dx\, dy\, dt.$$

Before we proceed with the weak formulation, we make the following assumptions for the sake of analysis of $(\mathcal{P}^\varepsilon)$.

**A1.** for all $x \in U$, $\mathbf{u}_0$, $c_0$ and $w_0 \geq 0$.
**A2.** $\mathbf{u}_0 \in L^\infty(U) \cap H^1(U)$, $c_0 \in L^\infty(U) \cap H^1(U)$ and $w^0 \in L^\infty(U) \cap H^1(U)$ such that $\sup_{\varepsilon>0} ||\mathbf{u}_0||_{L^\infty(U)\cap H^1(U)} < \infty$, $\sup_{\varepsilon>0} ||c_0||_{L^\infty(U)\cap H^1(U)} < \infty$, $\sup_{\varepsilon>0} ||w_0||_{L^\infty(U)\cap H^1(U)} < \infty$.
**A3.** $p^\varepsilon \in L^2(S; H^1(U_p^\varepsilon))$ such that $\sup_{\varepsilon>0} ||p^\varepsilon||_{L^2(S;H^1(U_p^\varepsilon))} < \infty$.

## 2.1 *Weak Formulation of* $(\mathcal{P}^\varepsilon)$

Let the assumptions A1–A4 be satisfied. A triple $(\mathbf{u}^\varepsilon, c^\varepsilon, w^\varepsilon) \in \mathfrak{U}^\varepsilon \times \mathfrak{C}^\varepsilon \times \mathfrak{W}^\varepsilon$ is said to be the weak solution of the model $(\mathcal{P}^\varepsilon)$ such that $(\mathbf{u}^\varepsilon, c^\varepsilon, w^\varepsilon)(0, x) = (\mathbf{u}_0, c_0, w_0)(x)$ for all $x \in U$, and

$$\mu\varepsilon^2 \int_{S\times U_p^\varepsilon} \nabla \mathbf{u}^\varepsilon : \nabla \eta \, dx\, dt = -\lambda \int_{S\times U_p^\varepsilon} c^\varepsilon \nabla w^\varepsilon \cdot \eta \, dx\, dt, \tag{2.2a}$$

$$\int_S \langle \partial_t c^\varepsilon, \phi \rangle \, dt - \varepsilon \int_{S\times U_p^\varepsilon} c^\varepsilon \mathbf{u}^\varepsilon \cdot \nabla \phi \, dx\, dt + \int_{S\times U_p^\varepsilon} \nabla w^\varepsilon \cdot \nabla \phi \, dx\, dt = 0, \tag{2.2b}$$

$$\int_{S\times U_p^\varepsilon} w^\varepsilon \psi \, dx\, dt = \varepsilon^2 \int_{S\times U_p^\varepsilon} \nabla c^\varepsilon \cdot \nabla \psi \, dx\, dt + \int_S \langle f(c^\varepsilon), \psi \rangle \, dx\, dt, \tag{2.2c}$$

for all $\eta \in L^2(S; \mathbf{H}^1_{div}(U_p^\varepsilon))$ and $\phi, \psi \in L^2(S; H^1(U_p^\varepsilon))$.

We are now going to state the two main theorems of this paper which are given below.

**Theorem 1** *Let the assumptions A1–A4 be satisfied, then there exists a unique positive weak solution* $(\mathbf{u}^\varepsilon, c^\varepsilon, w^\varepsilon) \in \mathfrak{U}^\varepsilon \times \mathfrak{C}^\varepsilon \times \mathfrak{W}^\varepsilon$ *of the problem* $(\mathcal{P}^\varepsilon)$ *which satisfies*

$$\begin{aligned} ||\mathbf{u}^\varepsilon||_{L^4(U_p^\varepsilon)} + \sqrt{\mu}\varepsilon||\nabla \mathbf{u}^\varepsilon||_{L^2(S\times U_p^\varepsilon)} + ||w^\varepsilon||_{L^2(S\times U_p^\varepsilon)} + \sqrt{\varepsilon\lambda}||\nabla w^\varepsilon||_{L^2(S\times U_p^\varepsilon)} \\ + ||c^\varepsilon||_{L^\infty(S;L^4(U_p^\varepsilon))} + \sqrt{\frac{\lambda}{2}}||\nabla c^\varepsilon||_{L^\infty(S);L^2(U_p^\varepsilon))} + ||\partial_t c^\varepsilon||_{L^2(S;H^1(U_p^\varepsilon)^*)} \\ \leq C < \infty \ \ \forall \varepsilon, \end{aligned} \tag{2.3}$$

*where the constant C is independent of* $\varepsilon$.

**Theorem 2** (Upscaled Problem $(\mathcal{P})$) *There exists* $(\mathbf{u}, c, w) \in \mathfrak{U} \times \mathfrak{C} \times \mathfrak{W}$ *which satisfies*

$$-\mu\Delta_y \mathbf{u} + \nabla_y p_1(x,y) + \nabla_x p(x) = -\lambda c\left(\nabla_x w(x) + \nabla_y w_1(x,y)\right), \quad S\times U\times Y_p, \tag{2.4a}$$

$$\nabla_y \cdot \mathbf{u}(x,y) = 0, \quad S\times U\times Y_p, \tag{2.4b}$$

$$\nabla_x \cdot \overline{\mathbf{u}}(x) = 0, \quad S\times U, \tag{2.4c}$$

$$\mathbf{u}(x,y) = 0, \quad S\times U\times \Gamma_s, \tag{2.4d}$$

$$\partial_t c(x,y) + \nabla_y \cdot c(x,y)\mathbf{u}(x,y) = \Delta_x w(x) + \nabla_x \cdot \nabla_y w_1(x,y), \quad S\times U\times Y_p, \tag{2.4e}$$

$$w(x,y) = -\Delta_y c(x,y) + f(c(x,y)), \quad S\times U\times Y_p, \tag{2.4f}$$

$$\nabla_y \cdot \{\nabla_x w(x) + \nabla_y w_1(x,y)\} = 0, \quad S\times U\times Y_p, \tag{2.4g}$$

$$\nabla_y \cdot \nabla_y w(x) = 0, \quad S\times U\times Y_p \tag{2.4h}$$

$$c(0,x) = c_0(x), \quad U. \tag{2.4i}$$

*where* $\bar{\kappa}(x) = \frac{1}{|Y_p|}\int_{\partial Y_p} \kappa(x,y)\,dy,\ x \in U$ *denotes the mean of the quantity* $\kappa$ *over the pore space* $Y_p$.

*The systems of equations* (2.4a)–(2.4i) *is the required homogenized (upscaled) model of* (1.3a)–(1.3h).

## 3 Anticipated Upscaled Model via Asymptotic Expansion Method

We consider the following expansions

$$\mathbf{u}^\varepsilon = \sum_{i=0}^{\infty} \varepsilon^i \mathbf{u}_i,\ c^\varepsilon = \sum_{i=0}^{\infty} \varepsilon^i c_i,\ w^\varepsilon = \sum_{i=0}^{\infty} \varepsilon^i w_i \text{ and } p^\varepsilon = \sum_{i=0}^{\infty} \varepsilon^i p_i, \tag{3.1}$$

where each term $\mathbf{u}_i$, $p_i$, $c_i$ and $w_i$ are $Y$-periodic functions in $y$-variable. We have $\nabla = \nabla_x + \frac{1}{\varepsilon}\nabla_y$. After the substitution of $\mathbf{u}^\varepsilon$, $c^\varepsilon$, $w^\varepsilon$, $p^\varepsilon$ in the problem $(\mathcal{P}^\varepsilon)$, we get from (1.3a)

$$\begin{aligned}&\varepsilon^{-1}(\nabla_y p_0)+\varepsilon^0(-\mu\Delta_y\mathbf{u_0}+\nabla_x p_0+\nabla_y p_1)\\&+\varepsilon[-\mu\{\Delta_y\mathbf{u_1}+(\nabla_x\cdot\nabla_y+\nabla_y\cdot\nabla_x)\mathbf{u_0}\}+\nabla_x p_1+\nabla_y p_2]\\&=\varepsilon^{-1}\{-\lambda(c_0\nabla_y w_0)\}+\varepsilon^0[-\lambda\{c_1\nabla_y w_0+c_0(\nabla_x w_0+\nabla_y w_1)\}]+\mathcal{O}(\varepsilon).\end{aligned}\tag{3.2}$$

We use (3.1) in (1.3b) then

$$\varepsilon^{-1}\nabla_y\cdot\mathbf{u_0}+\varepsilon^0(\nabla_x\cdot\mathbf{u_0}+\nabla_y\cdot\mathbf{u_1})+\varepsilon(\nabla_x\cdot\mathbf{u_1}+\nabla_y\cdot\mathbf{u_2})+\varepsilon^2(\ldots)=0.\tag{3.3}$$

From (1.3d), after plugging the expansions, we obtain

$$\begin{aligned}&\partial_t(c_0+\varepsilon c_1)+\varepsilon^0\{\nabla_y\cdot(c_0\mathbf{u_0})\}+\varepsilon\{\nabla_y\cdot(c_0\mathbf{u_1})+\nabla_x\cdot(c_0\mathbf{u_0})+\nabla_y\cdot(c_1\mathbf{u_0})\}\\&=\varepsilon^{-2}\Delta_y w_0+\varepsilon^{-1}\{\Delta_y w_1+(\nabla_x\cdot\nabla_y+\nabla_y\cdot\nabla_x)w_0\}\\&+\varepsilon^0\{\Delta_y w_2+(\nabla_x\cdot\nabla_y+\nabla_y\cdot\nabla_x)w_1+\Delta_x w_0\}+\mathcal{O}(\varepsilon).\end{aligned}\tag{3.4}$$

Next, we substitute the expansions for $w_\varepsilon$, $c_\varepsilon$ in (1.3e) and use the Taylor series expansion of $f$ around $c_0$ which leads to

$$w_0+\varepsilon w_1=-\Delta_y c_0+\varepsilon^1\{-\Delta_y c_1-(\nabla_x\cdot\nabla_y+\nabla_y\cdot\nabla_x)c_0\}+f(c_0)+\mathcal{O}(\varepsilon).\tag{3.5}$$

Now we substitute the expansions in the boundary conditions. From (1.3c), we obtain

$$\mathbf{u_0}+\varepsilon\mathbf{u_1}+\varepsilon^2\mathbf{u_2}+\cdots=0\quad\text{on }(0,T)\times\partial U_p^\varepsilon.\tag{3.6}$$

From (1.3f) and (1.3g), we get

$$\varepsilon^{-1}\nabla_y c_0\cdot\mathbf{n}+\varepsilon^0(\nabla_x c_0+\nabla_y c_1)\cdot\mathbf{n}+\varepsilon(\nabla_x c_1+\nabla_y c_2)\cdot\mathbf{n}+\cdots=0\tag{3.7}$$

and

$$\varepsilon^{-1}\nabla_y w_0\cdot\mathbf{n}+\varepsilon^0(\nabla_x w_0+\nabla_y w_1)\cdot\mathbf{n}+\varepsilon(\nabla_x w_1+\nabla_y w_2)\cdot\mathbf{n}+\cdots=0\tag{3.8}$$

respectively.

We compare the coefficient of $\varepsilon^0$ from (3.5) and integrate it over $Y_p$, then using (3.7) we get

$$w_0(t,x,y)=f(c_0(t,x,y))\quad\text{in }S\times U\times Y_p\tag{3.9}$$

We equate the coefficient of $\varepsilon^0$ from (3.4) and integrate it over $Y_p$, then using (3.8) we obtain

$$|Y_p|\{\partial_t c_0+\mathbf{u}_0\cdot\nabla_y c_0\}=\nabla_x\cdot\int_{Y_p}\{\nabla_y w_1+\nabla_x w_0\}\,dy.\tag{3.10}$$

The coefficients of $\varepsilon^{-2}$ and $\varepsilon^{-1}$ from (3.4) give The coefficient of $\varepsilon^{-1}$ from (3.4) gives

$$\Delta_y w_0 = 0 \qquad \text{and} \qquad \nabla_x \cdot \nabla_y w_0 + \nabla_y \cdot \{\nabla_x w_0 + \nabla_y w_1\} = 0 \tag{3.11}$$

From (3.8) and (3.11) we observe that

$$w_0 = w_0(t, x). \tag{3.12}$$

We equate the coefficients of $\varepsilon^{-1}$ from (3.2), then using (3.12) we get

$$\nabla_y p_0 = 0 \qquad \text{for } y \in Y_p. \tag{3.13}$$

The coefficient of $\varepsilon^0$ from (3.2) along with (3.12) gives

$$-\mu\Delta_y \mathbf{u_0} + \nabla_x p_0 + \nabla_y p_1 = -\lambda\, c_0\, (\nabla_x w_0 + \nabla_y w_1). \tag{3.14}$$

Again, using (3.3) and (3.6) one can deduce

$$\nabla_x \cdot \int_{Y_p} \mathbf{u_0}(x, y)\, dy = 0 \quad \text{in } S \times U. \tag{3.15}$$

Equating $\varepsilon$ coefficient from (3.5) we get using (3.7)

$$|Y_p| w_1 = -\nabla_x \cdot \int_{Y_p} \nabla_y c_0\, dy \tag{3.16}$$

# 4 Proof of Theorem 2.1

## *4.1 A Priori Estimates*

We put $\eta = \varepsilon \mathbf{u}^\varepsilon$, $\phi = \lambda w^\varepsilon$, $\psi = \lambda \partial_t c^\varepsilon$ in (2.2), and using $\nabla(c^\varepsilon w^\varepsilon) = c^\varepsilon \nabla w^\varepsilon + w^\varepsilon \nabla c^\varepsilon$ it yields

$$\sqrt{\mu}\varepsilon||\nabla \mathbf{u}^\varepsilon||_{L^2(S\times U_p^\varepsilon)} + \sqrt{\lambda}||\nabla w^\varepsilon||_{L^2(S\times U_p^\varepsilon)} + \sqrt{\frac{\lambda}{2}}\varepsilon||\nabla c^\varepsilon||_{L^\infty(S;L^2(U_p^\varepsilon))} \leq C \tag{4.1}$$

as $\varepsilon^{\frac{3}{2}} < \varepsilon$ for $\varepsilon \in (0, 1)$.

Next, Young's inequality gives

$$\int_{U_p^\varepsilon} F(c^\varepsilon(t))\,dx = \frac{1}{4}\int_{U_p^\varepsilon} ((c^\varepsilon)^2 - 1)^2\,dx \le C \quad \Rightarrow \int_{U_p^\varepsilon} |c^\varepsilon|^4\,dx \le C \quad \forall t$$

$$i.e., \quad \sup_{\varepsilon>0} ||c^\varepsilon||_{L^\infty(S;L^4(U_p^\varepsilon))} \le C. \quad (4.2)$$

We set $\psi = 1$ as a test function in (1.3e) and then using Poincare's inequality, we get

$$||w^\varepsilon - \int_{U_p^\varepsilon} w^\varepsilon\,dx||_{L^2(U_p^\varepsilon)} \le C||\nabla w^\varepsilon||_{L^2(U_p^\varepsilon)} \quad \Rightarrow ||w^\varepsilon||_{L^2(S\times U_p^\varepsilon)} \le C. \quad (4.3)$$

By Gagliardo–Nirenberg–Sobolev inequality for Lipschitz domain, $||u^\varepsilon||_{L^4(Y)} \le C||\nabla u^\varepsilon||_{L^2(Y)}$, where $C$ depend on $n$ and $Y$. By imbedding theorem, $||u^\varepsilon||_{L^2(Y)} \le C||u^\varepsilon||_{L^4(Y)} \le C$. By a straightforward scaling argument, we obtain

$$||\mathbf{u}^\varepsilon||_{L^4(U_p^\varepsilon)} \le C. \quad (4.4)$$

From (2.2b) we get,

$$||\partial_t c^\varepsilon||_{L^2(S;H^1(U_p^\varepsilon)^*)} \le C \quad \forall\varepsilon > 0 \quad (4.5)$$

From proposition III.1.1 in [10] and (2.2a), there exist a pressure $p^\varepsilon := \partial_t P^\varepsilon \in W^{-1,\infty}(S, L_0^2(U_p^\varepsilon))$ such that

$$\langle\nabla P^\varepsilon(t), \eta\rangle \le \mu\varepsilon^2 \int_S ||\nabla\mathbf{u}^\varepsilon||_{L^2(U_p^\varepsilon)}||\nabla\eta^\varepsilon||_{L^2(U_p^\varepsilon)}\,dt + \int_S ||c^\varepsilon||_{L^4(U_p^\varepsilon)}||\nabla w^\varepsilon||_{L^2(U_p^\varepsilon)}\,dt.$$

Thus by (4.1) and (4.2) it immediately follows that

$$\langle\nabla P^\varepsilon(t), \eta\rangle \le C||\eta||_{H_0^1(U_p^\varepsilon)^n} \Rightarrow \sup_{t\in[0,T]} ||\nabla P^\varepsilon(t)||_{H^{-1}(U_p^\varepsilon)^n} \le C \quad \forall\varepsilon > 0. \quad (4.6)$$

Now, with the help of a-priori estimates from (2.3), the existence of solution of $(\mathcal{P}^\varepsilon)$ can be shown using Galerkin's method, cf. [6] and references therein.

## 5 Proof of Theorem 2 (Homogenization of Problem $(\mathcal{P}^\varepsilon)$)

We start with the construction of an extension of solution from $U_p^\varepsilon$ to $U$ in the lemma below.

**Lemma 6** *There exists a positive constant $C$ depending on $c_0$, $\mathbf{u}_0$, $n$, $|Y|$, $\lambda$ and $\mu$ but independent of $\varepsilon$ and extensions $(\tilde{c^\varepsilon}, \tilde{w^\varepsilon}, \tilde{\mathbf{u}}^\varepsilon, \tilde{P^\varepsilon})$ of the solution $(c^\varepsilon, w^\varepsilon, \mathbf{u}^\varepsilon, P^\varepsilon)$ to $S \times U$ such that*

$$||\tilde{\mathbf{u}}^\varepsilon||_{L^\infty(S;L^2(U)^n)} + ||\tilde{c}^\varepsilon||_{L^\infty(S;L^4(U))} + ||\tilde{w}^\varepsilon||_{L^2(S;H^1(U))} + \sqrt{\mu}\varepsilon||\nabla\tilde{\mathbf{u}}^\varepsilon||_{L^2(S\times U)^{n\times n}}$$
$$+\sqrt{\frac{\lambda}{2}}\varepsilon||\nabla\tilde{c}^\varepsilon||_{L^\infty(S;L^2(U)^n)} + \sqrt{\lambda}||\nabla\tilde{w}^\varepsilon||_{L^2(S\times U)^n} + ||\partial_t\tilde{c}^\varepsilon||_{L^2(S;H^1(U)^*)}$$
$$+ \sup_{t\in[0,T]} ||\tilde{P}^\varepsilon(t)||_{L^2_0(U)} \le C. \tag{5.1}$$

**Lemma 7** *Let $(\mathbf{u}^\varepsilon, P^\varepsilon, c^\varepsilon, w^\varepsilon)_{\varepsilon>0}$ be the extension of the weak solution from Lemma 6 (denoted by the same symbol). Then there exists some functions $\mathbf{u} \in L^2(S \times U; H^1_\#(Y))^n$, $w \in L^2(S \times U)$, $P \in L^2(S \times U \times Y)$, $c$, $w_1 \in L^2(S \times U; H^1_\#(Y))$ and a subsequence of $(\mathbf{u}^\varepsilon, P^\varepsilon, c^\varepsilon, w^\varepsilon)_{\varepsilon>0}$, still denoted by the same symbol, such that the following convergences hold:*

*(i)* $(\mathbf{u}^\varepsilon)_{\varepsilon>0}$ *two-scale converges to* $\mathbf{u}$. *(ii)* $(c^\varepsilon)_{\varepsilon>0}$ *two-scale converges to* $c$.
*(iii)* $(w^\varepsilon)_{\varepsilon>0}$ *two-scale converges to* $w$. *(iv)* $(P^\varepsilon)_{\varepsilon>0}$ *two-scale converges to* $P$.
*(v)* $(\varepsilon\nabla_x c^\varepsilon)_{\varepsilon>0}$ *two-scale converges to* $\nabla_y c$. *(vi)* $(\varepsilon\nabla_x \mathbf{u}^\varepsilon)_{\varepsilon>0}$ *two-scale converges to* $\nabla_y \mathbf{u}$.
*(vii)* $(\nabla_x w^\varepsilon)_{\varepsilon>0}$ *two-scale converges to* $\nabla_x w + \nabla_y w_1$.

***Proof*** The convergences follow from the estimates (5.1), Lemmas 3 and 4.

In the next lemma we will discuss the convergence of nonlinear terms for $\varepsilon \to 0$.

**Lemma 8** *The following convergence results hold:*

*(i)* *$(c^\varepsilon)_{\varepsilon>0}$ is strongly convergent to $c$ in $L^2(S \times U)$. Thus, $\mathcal{T}^\varepsilon(c^\varepsilon)$ converges to $c$ strongly in $L^2(S \times U \times Y)$, i.e., $(c^\varepsilon)_{\varepsilon>0}$ is strongly two-scale convergent to $c$.*
*(ii)* *$\mathcal{T}^\varepsilon\mathbf{u}^\varepsilon$ is weakly convergent to $\mathbf{u}$ in $L^2(S \times U \times Y)^n$, i.e., $(\mathbf{u}^\varepsilon)_{\varepsilon>0}$ is weakly two-scale convergent to $\mathbf{u}$.*
*(iii)* *$\mathcal{T}^\varepsilon[\varepsilon\nabla_x c^\varepsilon]$ converges to $\nabla_y c$ weakly in $L^2(S \times U \times Y)^n$, i.e., $\varepsilon\nabla_x c^\varepsilon$ is weakly two-scale convergent to $\nabla_y c$.*
*(iv)* *The nonlinear terms $f(c^\varepsilon)$, $c^\varepsilon\nabla_x w^\varepsilon$ and $c^\varepsilon\mathbf{u}^\varepsilon$ two-scale converge to $f(c)$, $c(\nabla_x w + \nabla_y w_1)$ and $c\mathbf{u}$.*

***Proof*** We will prove step by step. From estimate (5.1) for $(c^\varepsilon)_{\varepsilon>0}$ and Theorem 2.1 in [9], there exists a subsequence of $(c^\varepsilon)_{\varepsilon>0}$, still denoted by same symbol, such that $(c^\varepsilon)_{\varepsilon>0}$ is strongly convergent to a limit $c$. The rest of (i) and the proofs of (ii) and (iii) follow from Lemma 4. Following the similar arguments as in [2] we can prove (iv).

***Proof*** *(Proof of Theorem 2)* (i) We choose a test function $\phi$ in (2.2b) defined as $\phi = \phi(t, x, \frac{x}{\varepsilon}) = \phi_0(t, x) + \varepsilon\phi_1(t, x, \frac{x}{\varepsilon})$, where the functions $\phi_0 \in C_0^\infty(S \times U)$ and $\phi_1 \in C_0^\infty(S \times U; C^\infty_\#(Y))$:

$$\int_S \langle\partial_t c^\varepsilon, \phi\rangle\, dt - \int_{S\times U_p^\varepsilon} c^\varepsilon\mathbf{u}^\varepsilon \cdot \varepsilon\nabla\phi\, dx\, dt + \int_{S\times U_p^\varepsilon} \nabla w^\varepsilon \cdot \nabla\phi\, dx\, dt = 0.$$

We extend the solution to $U$ and pass $\varepsilon \to 0$ in the two-scale sense and get

$$-\int_{S\times U} c(t,x,y)\partial_t\phi_0(t,x)\,dx\,dt - \int_{S\times U} c(t,x,y)\mathbf{u}(t,x)\cdot\nabla_y\phi_0(t,x)\,dx\,dt$$
$$+\int_{S\times U}\{\nabla_x w(t,x)+\nabla_y w_1(t,x,y)\}\cdot\Big(\nabla_x\phi_0(t,x)+\nabla_y\phi_1(t,x,y)\Big)\,dx\,dt = 0. \tag{5.2}$$

Setting $\phi_0 = 0$ and $\phi_1 = 0$ in (5.2) yield, respectively,

$$\nabla_y\cdot\{\nabla_x w(t,x)+\nabla_y w_1(t,x,y)\} = 0, \tag{5.3}$$
$$\partial_t c(t,x,y)+\nabla_y\cdot c(t,x,y)\mathbf{u}(t,x,y) = \Delta_x w(t,x)+\nabla_x\cdot\nabla_y w_1(t,x,y), \tag{5.4}$$

in $S\times U\times Y_p$. Similarly, choosing a function $\psi\in C_0^\infty(S\times U; C_\#^\infty(Y))$ in (2.2c) and passing the limit gives

$$w(t,x,y) = -\Delta_y c(t,x)+f(c(t,x,y)) \quad \text{in } S\times U\times Y_p. \tag{5.5}$$

(ii) We choose the test functions $\eta\in C_0^\infty(U; C_{\#)}^\infty(Y))^n$ and $\xi\in C_0^\infty(S)$ and proceed as in [2]. Then, using Lemmas 7 and 8, and passing to the two-scale limit

$$\lim_{\varepsilon\to 0}\int_{S\times U_p^\varepsilon} P^\varepsilon(t,x)\Big\{\nabla_x\cdot\eta(x,\frac{x}{\varepsilon})+\frac{1}{\varepsilon}\nabla_y\cdot\eta(x,\frac{x}{\varepsilon})\Big\}\partial_t\xi(t)\,dx\,dy\,dt$$
$$=\int_{S\times U\times Y_p} P(t,x,y)\nabla_y\cdot\eta(x,y)\partial_t\xi(t)\,dx\,dy\,dt$$
$$=0 \tag{5.6}$$

We get the y-variable independency of the two-scale limit of the pressure $P$ from (5.6). Further, we consider the function $\eta\in C_0^\infty(U; C_\#^\infty(Y))^n$ such that $\nabla_y\cdot\eta(x,y)=0$, so that

$$\mu\varepsilon^2\int_{S\times U_p^\varepsilon}\nabla\mathbf{u}^\varepsilon(t,x):\nabla\eta(x,y)\xi(t)\,dx\,dt+\int_{S\times U_p^\varepsilon}P^\varepsilon(t,x)\nabla\cdot\eta(x,y)\partial_t\xi(t)\,dx\,dt$$
$$=-\lambda\int_{S\times U_p^\varepsilon}c^\varepsilon(t,x)\nabla w^\varepsilon(t,x)\cdot\eta(x,y)\xi(t)\,dx\,dt. \tag{5.7}$$

We use the extensions of solution to $U$ (using the same notations), and pass to the two-scale limit.

$$
\begin{aligned}
-\lambda \int_{S\times U\times Y_p} & c(t,x,y)\{\nabla_x w(t,x) + \nabla_y w_1(t,x,y)\}\cdot \eta(x,y)\xi(t)\,dx\,dy\,dt \\
&= \mu \int_{S\times U\times Y_p} \nabla_y \mathbf{u}(t,x,y) : \nabla_y \eta(x,y)\xi(t)\,dx\,dy\,dt \\
&\quad + \int_{S\times U\times Y_p} P(t,x)\nabla_x \cdot \eta(x,y)\partial_t \xi(t)\,dx\,dy\,dt. \qquad (5.8)
\end{aligned}
$$

The existence of a pressure $P_1 \in L^\infty(S; L_0^2(U; L_\#^2(Y_p)))$ and two-scale convergence results are followed as in [2] for the final step of the upscaling of the model equations.

$$
\begin{aligned}
\int_{S\times U\times Y_p} & P(t,x)\nabla_x \cdot \eta(x,y)\partial_t\xi(t)\,dx\,dy\,dt + \int_{S\times U\times Y_p} P_1(t,x,y)\nabla_y \cdot \eta(x,y)\partial_t\xi(t)\,dx\,dy\,dt \\
&+\lambda \int_{S\times U\times Y_p} c(t,x,y)\{\nabla_x w(t,x) + \nabla_y w_1(t,x,y)\}\cdot\eta(x,y)\xi(t)\,dx\,dy\,dt \\
&+\mu \int_{S\times U\times Y_p} \nabla_y \mathbf{u}(t,x,y) : \nabla_y \eta(x,y)\xi(t)\,dx\,dy\,dt \\
&= 0. \qquad (5.9)
\end{aligned}
$$

for all $\eta \in C_0^\infty(U; C_\#^\infty(Y))^n$ and $\xi \in C_0^\infty(S)$.

From (5.9), we obtain

$$
-\mu\Delta_y \mathbf{u}(x,y) + \nabla_x p(x) + \nabla_y p_1(x,y) = -\lambda c(x,y)\{\nabla_x w(t,x) + \nabla_y w_1(t,x,y)\} \qquad (5.10)
$$

in $S \times U \times Y_p$.

## 6 Conclusion

A two fluids' mixture in strongly perforated domain is considered in which the fluids are separated by an interface of thickness of $\lambda$ in the pore part. From the modeling of such phenomena in the pore space, we got a strongly coupled system of Stokes–Cahn–Hilliard equations. The surface tension effects have been taken into account and the aforementioned interface is assumed to be independent of the scale parameter $\varepsilon$. Several a-priori estimates are derived and the well-posedness at the micro-scale is shown. Two-scale convergence, periodic unfolding, and the estimates after using extension theorems on them, yield the homogenized model.

## References


1. Allaire, G.: Homogenization and two scale convergence. SIAM J. Math. Anal. **23**(6), 1482–1518 (1992)
2. L'ubomír Baňas and Hari Shankar Mahato: Homogenization of evolutionary stokes-cahn-hilliard equations for two-phase porous media flow. Asympt. Anal. **105**(1–2), 77–95 (2017)
3. James F Blowey and Charles M Elliott. The cahn–hilliard gradient theory for phase separation with non-smooth free energy part i: Mathematical analysis. *European Journal of Applied Mathematics*, 2(3):233–280, 1991
4. Copetti, M.I.M., Elliott, C.M.: Numerical analysis of the cahn-hilliard equation with a logarithmic free energy. Numerische Mathematik **63**(1), 39–65 (1992)
5. Evans, L.C.: Partial Differential Equations. AMS Publication (1998)
6. Feng, Xiaobing, He, Yinnian, Liu, Chun: Analysis of finite element approximations of a phase field model for two-phase fluids. Math. Comput. **76**(258), 539–571 (2007)
7. Franců, Jan, Svanstedt, Nils EM.: Some remarks on two-scale convergence and periodic unfolding. Appl. Math. **57**(4), 359–375 (2012)
8. Hari Shankar Mahato and MICHAEL Böhm: Homogenization of a system of semilinear diffusion-reaction equations in an h 1, p setting. Electronic J. Diff. Equ. **2013**(210), 1–22 (2013)
9. Meirmanov, A., Zimin, R.: Compactness result for periodic structures and its application to the homogenization of a diffusion-convection equation. Electr. J. Diff. Equ. **2011**(115), 1–11 (2011)
10. Temam, R.: Navier-Stokes Equations: Theory and Numerical Analysis, vol. 343. American Mathematical Soc. (2001)

# Sensitivity and Directional Analysis of Two Mutually Competing Plant Population Under Allelopathy Using DDE

**Dipesh and Pankaj Kumar**

**Abstract** In this paper, we studied the mutual competition of plant development, with a focus on time-dependent change in concentrations. The effect of allelochemicals on plant populations is investigated with the help of a mathematical model using DDE. The effect of allelochemicals is studied by introducing the delay parameter in the term involving mutual competition. Stability was examined about the non-zero equilibrium point with the help of Routh-Hurwitz's theorem. The addition of delay distributed the system stability. The value $\tau = 0$ signifies the absence of delay at these points and keeps the system is stable. At $\tau < 6.9999$, the value of delay decreases from the threshold value, at this point the system shows asymptotic stability because it loses its stability. At $\tau \geq 6.9999$, the system shows hopf-bifurcation, when it crosses the threshold value. Directional and sensitivity analysis of the proposed model are performed. MATLAB is used to provide graphical help for theoretical results.

**Keywords** Allelopathy · Competing species · Delay · Sensitivity analysis · Hopf-bifurcation · Stability

## 1 Introduction

Ecologists create mathematical models at various levels of complexity to investigate ecosystems and plant populations dynamics. Models are constructed with inconsistencies in the values of variables, the modeling of the ecosystem, or the selection of mutually incompatible scenarios. Allelopathy is described as a plant's allelochemicals influence on another plant as a consequence of chemicals emitted into the environment. However, there has been a lot of uncertainty and variance in the definition and use of allelochemistry. The plan physiology analysis includes variations in plant population densities under mutual competing. Changes in the external environment of essential nutrients, as well as their interaction, are key variables that influence the size and density of plant populations [1]. Thornley created mathematical modeling

Dipesh · P. Kumar (✉)
Department of Mathematics, Lovely Professional University, Phagwara 144411, PB, India
e-mail: pankaj.kumar1@lpu.co.in



B. Rushi Kumar et al. (eds.), *Mathematics and Computing*, Springer Proceedings in Mathematics & Statistics 415, https://doi.org/10.1007/978-981-19-9307-7_47

of certain plant growth stages, and his models have been used to a variety of plant biology challenges [2]. Allelochemicals affect several physiochemical processes in plants, including photosynthetic rates, metabolism, root reduction, and so on [3]. P.Y. et al. demonstrated that modeling of the rise of different impacts of the first inter-competition on the significance of ecodiversity; competing between two phytoplankton species [4]. Tanveer et al. studied on Waste-land weeds in the fields that have an allelopathic effect on crops via their leaf leachates and rhizospheric soils [5]. Shovonlal Roy et al. observed that allelopathy is thought to minimize competitive exclusion and increase phytoplankton variety in aquatic environments where many species compete for a set of resources [6]. Abbas et al. worked on a fractional model for different phytoplankton species in which one species produced an allelochemical, which is stimulatory for another species [7]. Peng et al. tell us the impacts of secondary metabolites in plant invasion, and how we can save plant and remove the allelopathic effect in plants [8]. Wang et al. worked on the fractional order delayed model in paddy ecosystem and also analyzed the stability and hopf-bifurcation of the system [9]. To correctly interpret the model output, it is necessary to have a clear awareness of the sources of uncertainty that the methodology tackles. Saltelli et al. studied on the role of sensitivity analysis in ecological modeling and also tell us about the application of sensitivity analysis in ecological modeling [10]. Grzyb et al. worked on the environmental factors which affect the crop residue and talk about the application of crop residue [11]. Huang et.al examined a study of the global equilibrium of the non-linear DDE method concerning population development [12]. The existence of the zeros of the empirical given polynomial was studied in detail [13, 14]. Kalra and Kumar examined the effect of delay on plant maturing under the impact of hazardous mineral and also tell us about the Impact of delay in plant spreading [15, 16]. Russel and Mincheva create a parametric sensitivity analyzing the cyclic solution of DDE [17]. Rihan uses the adjoint equation and direct method for sensitivity analysis for the dynamic system with delay when the value of parameters change with time [18]. Even so, one of the most important aspects in nature known as allelopathy, where a single plant species can create a pollutant in the atmosphere affecting a plant species, received relatively little interest in its research.

## 2 Mathematical Model

### *2.1 Motivation of Work*

The standard two-species Lotka-Volterra competitive system is led by a set of nonlinear equations.

$$\frac{dP_1}{dt} = P_1(a_1 - a_2P_1 - \beta_1 P_2)$$

$$\frac{dP_2}{dt} = P_2(b_1 - b_2P_2 - \beta_2 P_1)$$

Let $P_1$ and $P_2$ be the competing plant populations. Assume that not only one plant species is in competition with each other and every plant species releases allelopathic substance to the other, whenever the other is present. The excretion of allelochemicals is not sudden, so some discrete time delay is needed for grown plants to develop. A time delay is induced in the excretion of allelochemicals by the 1st plant population. The model is given as:

$$\frac{dP_1}{dt} = a_1 P_1 - a_2 P_1^2 - \beta_1 P_1 P_2 + \gamma_1 P_1^2 P_2 \quad (1)$$

$$\frac{dP_2}{dt} = b_1 P_2 - b_2 P_2^2 - \beta_2 P_1(t-\tau) P_2 + \gamma_2 P_1(t-\tau) P_2^2 \quad (2)$$

where $P_1(0) > 0,\ P_2(0) > 0 \forall t \& P_1(t-\tau) = \text{constantfort} \epsilon [0, \tau]$

The parameters considered in the model are: $a_1, b_1$ are the rates of cell proliferation per hour, $a_2, b_2$ are the roots of intraspecific competition of 1st and 2nd plant population resp., $\beta_1, \beta_2$ are the roots of interspecific competition of 1st and 2nd plant population resp., $\gamma_1, \gamma_2$ are the allelochemical release of 1st and 2nd plant population niche respectively. The units of $a_2, b_2, \beta_1, \beta_2, and \gamma_1, \gamma_2$ are per hour per cell and unit of time is hours.

Equilibrium Point:

Various equilibrium points for the model (1)–(2) are $\mathrm{E}_{00}, \mathrm{E}_{a0}, \mathrm{E}_{0a}, \mathrm{E}^*$ existing with no restrictions on the variables that make the systems

$$\mathrm{E}_{00} : (0, 0)(\text{zeroequilibriumpoint, unstable})$$

$$\mathrm{E}_{a0} : \left(\frac{\mathrm{a}_1}{\mathrm{b}_1}, 0\right)(\text{axialequilibrumpoint, unstable})$$

$$\mathrm{E}_{0a} : \left(0, \frac{\mathrm{b}_1}{\mathrm{b}_2}\right)(\text{axialequilibrumpoint, unstable})$$

$$\mathrm{E}^* : \left(\mathrm{P}_1^*, \mathrm{P}_2^*\right)(\text{Non} - \text{zeroequilibriumpoint, Stable})$$

Further, we study and calculate the non-zero equilibrium point $\mathrm{E}^*\left(\mathrm{P}_1^*, \mathrm{P}_2^*\right)$

$$\frac{dP_1^*}{dt} = 0 \Rightarrow a_1 P_1^* - a_2 P_1^{*2} - \beta_1 P_1^* P_2^* + \gamma_1 P_1^{*2} P_2^* = 0$$

$$P_1^*(a_1 - a_2 P_1^* - \beta_1 P_2^* + \gamma_1 P_1^* P_2^*) = 0$$

$$P_1^* \neq 0 and P_1^* = \frac{a_1 - \beta_1 P_2^*}{a_2 - \gamma_1 P_2^*} \tag{3}$$

$P_1^*$ is positive when $a_2 - \gamma_1 P_2^* \neq 0 and a_1 > \beta_1 P_2^*$
And similarly, we can calculate $P_2^*$

$$P_2^* = \frac{b_1 - \beta_2 P_1^*}{b_2 - \gamma_2 P_1^*}$$

$P_2^*$ is positive when $b_2 - \gamma_1 P_1^* \neq 0 and b_1 > \beta_1 P_1^*$

$P_1^*$, $P_2^*$ being the real time concentration of plant population, minimum they can be zero but never be negative. This gives the idea of taking $P_1^*$, $P_2^*$ always positive.

Put the value of $P_2^*$ in Eq. (3), we get the quadratic equation in the form of $P_1^*$

$$(\gamma_1\beta_2 - a_2\gamma_2)P_1^{*2} + (a_2 b_2 - \gamma_1 b_1 + a_1\gamma_2 - \beta_1\beta_2)P_1^* + (a_1 b_2 - b_1\beta_1) = 0$$

$$P_1^* = \frac{-(a_2 b_2 - \gamma_1 b_1 + a_1\gamma_2 - \beta_1\beta_2) \pm \sqrt{(a_2 b_2 - \gamma_1 b_1 + a_1\gamma_2 - \beta_1\beta_2)^2 - 4(\gamma_1\beta_2 - a_2\gamma_2)(a_1 b_2 - b_1\beta_1)}}{2(\gamma_1\beta_2 - a_2\gamma_2)}$$

Stability & Hopf-Bifurcation of $E^*$ ($P_1^*$, $P_2^*$):

The dynamic behavior for equilibrium points $E^*$ ($P_1^*$, $P_2^*$) of the system given by Eqs. (1)–(2) is analyzed. The $E^*$ ($P_1^*$, $P_2^*$) equilibrium empirical characteristics equation is expressed as shown:

$$\frac{dP_1^*}{dt} = a_1 P_1^* - a_2 P_1^{*2} - \beta_1 P_1^* P_2^* + \gamma_1 P_1^{*2} P_2^* \tag{4}$$

$$\frac{dP_2^*}{dt} = b_1 P_2^* - b_2 P_2^{*2} - \beta_2 P_1^*(t-\tau) P_2^* + \gamma_2 P_1^*(t-\tau) P_2^{*2} \tag{5}$$

With the aid of a system of Eqs. (4)–(5), the characteristics equation is given by:

$$\lambda^2 + x\lambda + y + ze^{-\lambda\tau} = 0 \tag{6}$$

where $x = 2b_2 P_2^* - a_1 - b_1 + 2a_2 P_1^* + \beta_1 P_2^* - 2\gamma_1 P_1^* P_2^*$

Putting all the parametric values in the above, we get

$$x = 2 \times 0.08 \times 26.2837 - 1 - 2 + 2 \times 2 \times 17.6500 + 0.05 \times 26.2837$$
$$- 2 \times 0.0008 \times 17.6500 \times 26.2837 = 72.4$$

Which shows that $x = 72.4 > 0$

$$y = a_1 b_1 - 2a_1 b_1 P_2^* - 2a_2 b_2 P_1^* + 4a_2 b_2 P_1^* P_2^* - \beta_1 b_1 P_2^* - 2\beta_1 b_2 P_2^{*2} - 4\gamma_1 b_2 P_1^* P_2^{*2} + 2\beta_1 \gamma_1 P_2^* P_1^*$$

Putting all the parametric values in the above, we get

$$y = 2 \times 1 - 2 \times 2 \times 0.08 \times 26.2837 - 2 \times 0.07 \times 1 \times 17.6500 + 4 \times 0.07 \times 1 \times 26.2837 \times 17.6500 - 0.05 \times 26.2837 \times 1 - 2 \times 0.08 \times 0.05 \times 26.2837 \times 26.2837 - 4 \times 0.08 \times 0.0008 \times 17.6500 \times 26.2837 \times 26.2837 + 2 \times 0.0008 \times 0.05 \times 17.6500 \times 26.2837 = 111.14$$

Which shows that $y = 111.14 > 0$

$$z = P_1^* P_2^* (\beta_2 \gamma_1 P_1^* + \beta_1 \gamma_2 - \beta_1 \beta_2 - \gamma_1 \gamma_2 P_1^* P_2^*)$$

Putting all the parametric values in the above, we get

$$z = 17.6500 \times 26.2837(0.015 \times 0.0008 \times 17.6500 + 0.05 \times 0.003 - 0.05 \times 0.015 - 0.0008 \times 0.003 \times 17.6500 \times 26.2837 = -0.7$$

And when add the value of $.y + z = 110.44 > 0$
When we put $\tau = 0$ in Eq. (6), we get

$$\lambda^2 + x\lambda + y + z = 0 \quad (7)$$

With the help of Routh-Hurwitz criteria, root of Eq. (7) will be a negative real part if:

$(X_1) : x > 0$, $(X_2) : (y + z) > 0$, Which is true from the above calculated value.

Now we'll look at how the negative real elements of the roots shift towards the positive real elements when the value of $\tau$ varies.

Let $\lambda = i\omega$ be the root of equ of (6), then Eq. (6) becomes:

$$(i\omega)^2 + x(i\omega) + y + ze^{-(i\omega)\tau} = 0$$

$$\Rightarrow -\omega^2 + x(i\omega) + y + z(cos\omega\tau - isin\omega\tau) = 0$$

Separating real and imaginary parts we get:

$$\omega^2 - y = z\cos\omega\tau \quad (8)$$

$$x\omega = zsin\omega\tau \quad (9)$$

Squaring and adding (8) and (9), we get

$$\omega^4 + \left(x^2 - 2y\right)\omega^2 + \left(x^2 - z^2\right) = 0 \tag{10}$$

Equation (10) has two roots:

$$\omega_{1,2}^2 = \frac{\left(2y - x^2\right) \pm \sqrt{\left(x^2 - 2y\right)^2 - 4\left(x^2 - z^2\right)}}{2} \tag{11}$$

Putting the all-parametric value in Eq. (11), we get $.\omega_{1,2}^2 = \frac{-5019.5 \pm 5017.4}{2}$
None of the two roots $\omega_{1,2}^2$ is positive if:

$$(X_3) : \left(2y - x^2\right) < 0 \text{and} \left(x^2 - z^2\right) > 0 \text{or} \left(x^2 - 2y\right) < 4(x^2 - z^2)$$

$$(X_3) : \left(2 \times 111.14 - (72.4)^2\right) < 0 \text{ and } \left((72.4)^2 - (-0.7)^2\right) > 0 \text{ or } \left((72.4)^2 - 2 \times 111.14\right) < 4\left((72.4)^2 - (-0.7)^2\right)$$

$$(X_3) : -5019.5 < 0 \text{and} 5241.3 > 0 \text{or} 5019.48 < 4(5241.27)$$

$$(X_3) : -5019.5 < 0 \text{and} 5241.3 > 0 \text{or} 5019.48 < 20965.08$$

So, Eq. (11) doesn't have + ve root if condition $(X_3)$ holds.
We have the following the lemma [13]

**Lemma 1** *If* $(X_1) - (X_2)$ *hold, then all the roots of Eq.* (6) *have* $-ve$ *real part* $\forall \tau \geq 0$.

On the other hand, if

$$(X_4) : \left(x^2 - z^2\right) < 0or\left(2y - x^2\right) > 0and(x^2 - 2y)^2 = 4\left(x^2 - z^2\right)$$

Then, the positive root of Eq. (8) is $\omega_1^2$. On the other hand, if

$$(X_5) : \left(x^2 - z^2\right) > 0or\left(2y - x^2\right) > 0and(x^2 - 2y)^2 > 4\left(x^2 - z^2\right)$$

Then, Eq. (8) has two $+ve$ roots which are $\omega_{1,2}^2$.

In both $(X_4)and(X_5)$, Eq. (6) has a purely hypothetical root when $\tau$ takes different values. The threshold value $\tau_j^{\pm}$of$\tau$ can be evaluated from (8)–(9), given by

$$\tau_l^{\pm} = \frac{1}{\omega_{1,2}} \cos^{-1} \frac{(\omega_{1,2}^2 - b_1)}{z} + \frac{2l\pi}{\omega_{1,2}}, l = 0, 1, 2, \ldots \tag{12}$$

The above knowledge can be described in the lemma [13].

**Lemma 2** *(i) If* $(X_1) - (X_2)\text{and}(X_4)$ *hold and* $\tau = \tau_l^+$, *then Eq.* (6) *has a pair of imaginary roots* $\pm i\omega_1$.

(ii) If $(X_1) - (X_2)\text{and}(X_5)$ hold and $\tau = \tau_l^-(\tau = \tau_l^+\text{resp.})$, then Eq. (6) has a pair of imaginary roots $\pm i\omega_2(\pm i\omega_1)$ respe.

Our hypothesis is that the negative real component of some equation roots will move to the positive real component when $\tau > \tau_l^+$ & $\tau < \tau_l^+$. Let us have a look at this possibility:

$$\tau_l^{\pm} = \mu_l^{\pm}(\tau) + i\omega_l^{\pm}(\tau); l = 0, 1, 2, 3 \ldots.$$

The roots of Eq. (6) fulfil. $\mu_l^{\pm}(\tau_l^{\pm}) = 0, \omega_l^{\pm}(\tau_l^{\pm}) = \omega_{1,2}$
The preceding initial boundary criterion can be checked.

$$\frac{d}{d\tau}(Re\lambda_l^+(\tau_l^+)) > 0\text{and}\frac{d}{d\tau}(Re\lambda_l^-(\tau_l^-)) < 0$$

It deduces that $\tau_l^+$ are the bifurcating values. The distribution of the equation's (6) zeros is determined by the next hypothesis [Raun S.].

**Theorem 1** *Let* $\tau_l^+(l = 0, 1, 2, 3 \ldots)$ *be defined by Eq.* (12).

(1) If $(X_1), (X_2)$ hold, then all of root (6) have a negative real part $\forall\tau \geq 0$.

(2) If $(X_1), (X_2)\text{and}(X_4)$ hold and when $\tau\epsilon[0, \tau_0^+)$, then all of root (6) have a -ve element. When $\tau = \tau_0^+$, then (6) has a pair of hypothetical roots $\pm i\theta_1$. When $\tau > \tau_0^+$, (6) has at least one $+ve$ real part root.

(3) If $(X_1), (X_2)\text{and}(X_5)$ hold, then the positive integer n such that $0 < \tau_0^+ < \tau_0^- < \tau_1^+ < \tau_1^+ \ldots\ldots\ldots < \tau_{n-1}^- < \tau_n^+$ and there are n switches from stability to instability. Which show when $\tau\epsilon[0, \tau_0^+), (\tau_0^-, \tau_1^+) \ldots\ldots\ldots(\tau_{n-1}^+, \tau_n^+)$ all the roots of Eq. (6) have $-ve$ actual parts, & $\tau\epsilon[0, \tau_0^+), (\tau_0^-, \tau_1^+) \ldots\ldots\ldots(\tau_{n-1}^+, \tau_{n-1}^-)$ and $\tau > \tau_n^+$, Eq. (6) has a minimum single root with actual parts.

## 3 Sensitivity Analysis

The 'Direct Method' is used to estimate that how the different sources of uncertainty contribute in the model to overall uncertainty in the model. Assuming all the parameters $a_2, b_2, \beta_1, \beta_2, \gamma_1, \gamma_2$ in the formulated system (1)–(2) to be constant, further the solution's partial derivatives with regard to each parameter. For example, we assume $\beta_1$, then partial derivatives of the solution $(P_1, P_2)$ w.r.t. $\beta_1$ give the sensitivity equation:

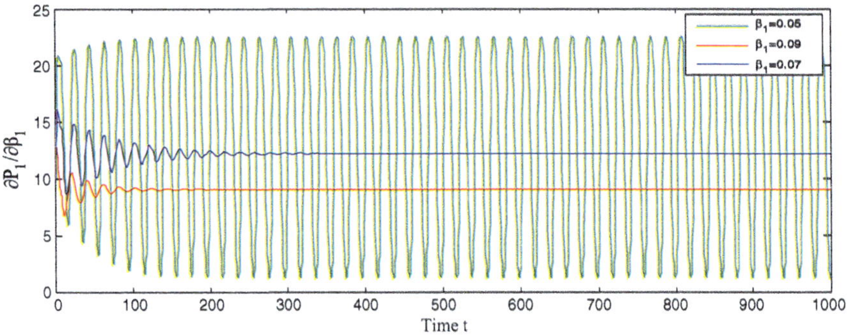


**Fig. 1** Time series graph between partial change in allelochemicals $P_1$ for different value of coefficient $\beta_1$

$$\frac{dS_1}{dt} = (a_1 - 2a_2)S_1 - P_1S_2(\beta_1 - \gamma_1 P_1) - P_2S_1(\beta_1 - 2\gamma_1) \tag{13}$$

$$\begin{aligned}\frac{dS_2}{dt} &= (b_1 - 2b_2 - \beta_2 P_1(t-\tau))S_2 - 2\gamma_2 P_1(t-\tau)S_2 \\ &\quad - [\beta_2 P_2 S_1(t-\tau) + P_2 P_1(t-\tau)](1 + P_2)\end{aligned} \tag{14}$$

$$whereS_1 = \frac{\partial P_1}{\partial \beta_1}, S_2 = \frac{\partial P_2}{\partial \beta_1},$$

Further, we studied (13)–(14) with the original system (1)–(2) to solve the variable $(P_{1,}, P_2)$ w.r.t $\beta_1$.

Sensitivity of Variable to Parameter $\beta_1$.

In Fig. 1., the sensitivity analyses of $P_1$ and $P_2$ variables w.r.t to $\beta_1$, putting all parameters, are constant. When we change the value of $\beta_1 = 0.05 to \beta_1 = 0.09$, the system becomes stable and remains stable.

Sensitivity of Variable to Parameter $\beta_2$

In Fig. 2., the sensitivity analysis of $P_1$ and $P_2$ variables w.r.t to $\beta_2$, putting all parameters are constant. When we change the value of $\beta_2 = 0.017 to \beta_2 = 0.021$, system become stable and remains stable.

## 4 Stability and Direction of Hopf-Bifurcating Solution

The result is a set of continued functions that bifurcate just at the threshold value of the positive stable state. We examine the stability and periods of bifurcation at the complex level, with the help of standard principals and different reduction given by Hassard, Kazarion and Wan 1981.

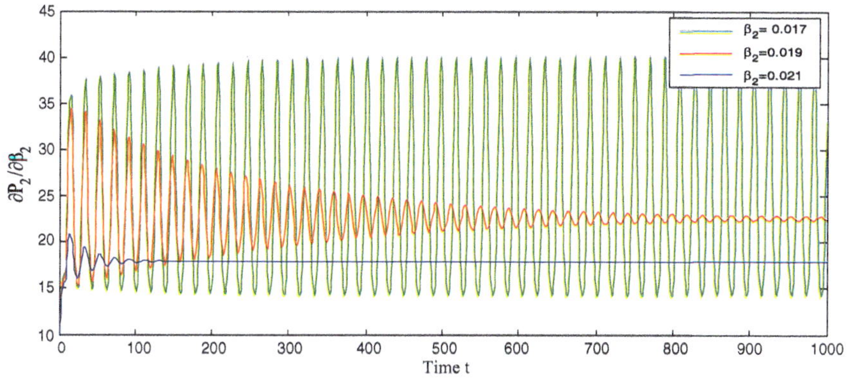


**Fig. 2** Time series graph between partial change in allelochemicals $P_2$ for different values of coefficient $\beta_2$

$Let u_1 = P_1 - P_1^*, u_2 = P_2 - P_2^*$, and normalizing the delay $\tau$ by time ascent, $t \rightarrow \frac{t}{\tau}$, altering into

$$\frac{du_1}{dt} = a_1u_1 + a_1P_1^* - a_1u_1^2 - a_2P_1^{*2} - 2a_2u_1P_1^* - \beta_1u_1P_2^* - \beta_1u_2P_1^* \\ - \beta_1u_1u_2 + \gamma_1P_2^*u_1^2 + \gamma_1P_1^{*2}u_2 + \gamma_1u_1u_2$$

$$\frac{du_2}{dt} = b_1u_2 + b_1P_2^* - b_2u_2^2 - b_2P_2^{*2} - 2b_2u_2P_2^* - \beta_2u_1(t-1)P_2^* \\ - \beta_2u_1(t-1)u_2 + \gamma_2P_2^*u_1(t-1) - \gamma_2u_1(t-1)u_2 \tag{15}$$

In this section, we can covenant $C = C\big((-1, 0), R_+^2\big)$. WLOG, denote the critical value $\tau_j by \tau_0$. Let $\tau = \tau_0 + \mu$, then$\mu = 0$ is the value of Hopf-bifurcation of the Eqs. (15–17). For the ease of sign, (15) then becomes

$$u'(t) = L_\mu(u_t) + F(\mu, u_t) \tag{16}$$

where $u(t) = (u_1(t), u_2(t),)^T \in R^2, u_t(\theta) \in C$ is defined by $u_t(\theta) = u_t(t+\theta)$, and

$L_\mu : C \rightarrow R, F : R \times C \rightarrow R$ are given, respectively, by

$$L_\mu\emptyset = (\tau_0 + \mu)\begin{bmatrix} a_1 - 2a_2P_1^* - \beta_1P_2^* & -\beta_1P_1^* + \gamma_1P_1^{*2} \\ 0 & b_1 \end{bmatrix}\begin{bmatrix} \phi_1(0) \\ \phi_2(0) \end{bmatrix} \\ + (\tau_0 + \mu)\begin{bmatrix} 0 & 0 \\ -\beta_2P_2^* + \gamma_2P_2^* & 0 \end{bmatrix}\begin{bmatrix} \phi_1(-1) \\ \phi_2(-1) \end{bmatrix}$$

And $.F(\mu, \varnothing) = (\tau_0 + \mu)\begin{bmatrix} F_1 \\ F_2 \end{bmatrix}$

where $F_1 = b_1\phi_1(0)\phi_2(0)$, $F_2 = (\gamma_2 - \beta_2)\phi_1(0)\phi_2(0)$

And $\varnothing(\theta) = (\varnothing_1(\theta), \varnothing_2(\theta))^T \in C((-1, 0), R)$.

According to the Riesz representation theorem, ∃ a function $\eta(\theta, \mu)$ of bounded variation for $\theta \in [-1, 0]$, s.t.

$L_\mu\varnothing = \int_{-1}^{0} d\eta(\theta, 0)\varnothing(\theta)$ for $\varnothing \in C$.

$$(\theta, \mu) = (\tau_0 + \mu)\begin{bmatrix} a_1 - 2a_2P_1^* - \beta_1P_2^* & -\beta_1P_1^* + \gamma_1P_1^{*2} \\ 0 & b_1 \end{bmatrix}\delta(\theta)$$

$$+ (\tau_0 + \mu)\begin{bmatrix} 0 & 0 \\ -\beta_2P_2^* + \gamma_2P_2^* & 0 \end{bmatrix}\delta(\theta + 1)$$

Here, $\delta$ is the Direct delta function, for $\phi \epsilon C([-1, 0], R_+^2)$, define as $\mathcal{A}(\mu)\varnothing = \begin{cases} \frac{d\varnothing(\theta)}{d\theta}, & \theta \in [-1, 0) \\ \int_{-1}^{0} d\eta(\theta, 0)\varnothing(\theta), & \theta = 0. \end{cases}$ and $R(\mu)\varnothing = \begin{cases} 0, & \theta \in [-1, 0) \\ F(\mu, \varnothing) & \theta = 0. \end{cases}$ & Eq. (16) is identical to

$$u'(t) = \mathcal{A}(\mu)\emptyset + R(\mu)u_t\text{For} \tag{17}$$

$\psi \in C^1([-1, 0], {R_+}^2)$, define

$$\mathcal{A}^*\psi(s) = \begin{cases} -\frac{d\psi(s)}{ds}, & s \in [-1, 0) \\ \int_{-1}^{0} d\eta^T(-t, 0)\psi(-t), & s = 0. \end{cases} \quad \text{\& bi} - \text{linear inner product}$$

$$\langle\psi(s), \emptyset(\theta)\rangle = \overline{\psi(0)}\emptyset(0) - \int_{-1}^{0}\int_{\xi=\theta}^{\theta} \overline{\psi}(\xi - \theta)d\eta(\theta)\phi(\xi)d\xi \tag{18}$$

$\mathcal{A}^*$ and $\mathcal{A} = \mathcal{A}(0)$ are computative operative & $i\omega_0$ are eigen values of $\mathcal{A}(0)$. As a result, they are coefficients of $\mathcal{A}^*$. Admit that $q(\theta) = q(0)e^{i\omega_0\theta}$ is an eigen vector of $\mathcal{A}(0)$ analogous to the eigen state $i\omega_0$. Then $\mathcal{A}(0) = i\omega_0 q(\theta)$. At $\theta = 0$, we obtain.

$\left[i\omega_0 I - \int_{-1}^{0} d\eta(\theta)e^{i\omega_0\theta}\right]q(0) = 0$, which option $q(0) = (1, \sigma_1, )^T$

where $\sigma_1 = \frac{a_1 - 2a_2P_1^* - \beta_1P_2^* + i\omega_0}{-\beta_1P_1^* + \gamma_1P_1^{*2}}$

Similarly, we can verify that $q^*(s) = D(1, \sigma_2)e^{i\omega_0\tau_0 s}$ is the eigen value of $\mathcal{A}^*$ corresponding to $-i\omega_0$,

where $\sigma_2 = \frac{a_1 - 2a_2P_1^* - \beta_1P_2^* - i\omega_0}{-\beta_1P_1^* + \gamma_1P_1^{*2}}$

In deeds $< q^*(s), q(\theta) >= 1$, we examined the value of D.

Using Eq. (7), $< q^*(s), q(\theta) >$

$$= \overline{D}(1, \overline{\sigma_2})(1, \sigma_1)^T - \int_{-1}^{0} \int_{\xi=\theta}^{\theta} \overline{D}(1, \overline{\sigma_2}) e^{-i\omega_0 \tau_0 (\xi - \theta)} d\eta(\theta)(1, \sigma_1)^T e^{i\omega_0 \tau_0} d\xi$$

$$= \overline{D}\left\{1 + \sigma_1 \overline{\sigma_2} - \int_{-1}^{0} (1, \overline{\sigma_2}) \theta e^{i\omega_0 \tau_0 \theta} (1, \sigma_1)^T \right\} \overline{D}\left\{1 + \sigma_1 \overline{\sigma_2} + \tau_0 \overline{\sigma_2} P_2^* (\gamma_2 - \beta_2) e^{i\omega_0 \tau_0}\right\}$$

Hence

$$\overline{D} = \frac{1}{\left(1 + \sigma_1 \overline{\sigma_2} + \rho_1 \overline{\rho_2} + \tau_0 \overline{\sigma_2} P_2^* (\gamma_2 - \beta_2) e^{i\omega_0 \tau_0}\right)}$$

Such that $< q^*(s), q(\theta) >= 1$, $< q^*(s), \overline{q(\theta)} >= 0$.

The method is proved by Hassard, B.D., Kazarinoff, N.D., Wan, Y.H and is used to compute the parameters, $C_0$ at $\mu = 0$. Let $u_t$ be the return of Eq. (11) with $\mu = 0$. Describe

$$z(t) = \langle q^*(s), u_t(\theta) \rangle, \ W(t, \theta) = u_t(\theta) - 2Re(z(t) q(\theta)) \tag{19}$$

On the Centre manifold $C_0$, $W(t, \theta) = W\left(z(t), \overline{z(t)}, \theta\right)$

where $W(z, \overline{z}, \theta) = W_{20}(\theta) \frac{z^2}{2} + W_{11}(\theta) z\overline{z} + W_{02}(\theta) \frac{\overline{z}^2}{2} + \ldots,$

$z$ and $\overline{z}$ are concordants for $C_0$ in the direction of $q^*$ and $\overline{q^*}$.And $W$ is +ve if $u_t$ is +ve and + ve results are taken. For solution $u_t \in C_0$ of Eq. (11), since $\mu = 0$,

$$z'(t) = i\omega_0 \tau_0 z + < \overline{q^*}(\theta), F(0, W(z, \overline{z}, \theta) + 2Re(z(t) q(\theta))) >$$

$$= i\omega_0 \tau_0 z + \overline{q^*}(0) F(0, W(z, \overline{z}, 0) + 2Re(z(t) q(\theta)))$$

$$\equiv i\omega_0 \tau_0 z + \overline{q^*}(0) F_0(z, \ \overline{z})$$

We rewrite this equation as

$$z'(t) = i\omega_0 \tau_0 z(t) + g(z, \overline{z}) \tag{20}$$

where $g(z, \overline{z}) = \overline{q^*}(0) F_0(z, \overline{z})$

$$g(z, \overline{z}) = g_{20}(\theta) \frac{z^2}{2} + g_{11}(\theta) z\overline{z} + g_{02}(\theta) \frac{\overline{z}^2}{2} + g_{21}(\theta) \frac{z^2 \overline{z}}{2} + \ldots \tag{21}$$

Noticing.

As $u_t(\theta) = (u_{1t}, u_{2t}) = W(t, \theta) + z q(\theta) + \overline{z q(\theta)}$ and $q(0) = (1, \sigma_1)^T e^{i\omega_0 \tau_0 \theta}$, we have.

$u_{1t}(0) = z + \overline{z} + W_{20}{}^{(1)}(0) \frac{z^2}{2} + W_{11}{}^{(1)}(0) z\overline{z} + W_{02}{}^{(1)}(0) \frac{\overline{z}^2}{2} + \ldots,$

$u_{2t}(0) = \sigma_1 z + \overline{\sigma_1 z} + W_{20}{}^{(2)}(0) \frac{z^2}{2} + W_{11}{}^{(2)}(0) z\overline{z} + W_{02}{}^{(2)}(0) \frac{\overline{z}^2}{2} + \ldots,$

$u_{1t}(-1) = ze^{-i\omega_0\tau_0} + \overline{z}e^{i\omega_0\tau_0} + W_{20}{}^{(1)}(-1)\frac{z^2}{2} + W_{11}{}^{(1)}(-1)z\overline{z} + W_{02}{}^{(1)}(-1)\frac{\overline{z}^2}{2} + \ldots,$

$u_{2t}(-1) = \sigma_1 e^{-i\omega_0\tau_0}z + \overline{\sigma_1}e^{i\omega_0\tau_0}\overline{z} + W_{20}{}^{(2)}(-1)\frac{z^2}{2} + W_{11}{}^{(2)}(-1)z\overline{z} + W_{02}{}^{(2)}(-1)\frac{\overline{z}^2}{2} + \ldots,$

Explanatory variables are compared using an equation.

$$g_{20} = \overline{D}(1,\sigma_1)f_{z^2},\ g_{02} = \overline{D}(1,\sigma_1)f_{\overline{z}^2}$$

$$g_{11} = \overline{D}(1,\sigma_1)f_{z\overline{z}},\ g_{21} = \overline{D}(1,\sigma_1)f_{z^2\overline{z}}$$

For calculating $g_{21}$, The calculation of must be prioritized of $W_{20}(\theta)$ and $W_{11}(\theta)$.

From Eqs. (17) and (19):

$$W' = u_t' - z'q - \overline{z}'q = \begin{cases} \mathcal{A}W - 2Re\left[\overline{q^*}(0)F_0 q(\theta)\right], & \theta \in [-1, 0) \\ \mathcal{A}W - 2Re\left[\overline{q^*}(0)F_0 q(0)\right] + F_0, & \theta = 0 \end{cases}$$

Let

$$W' = \mathcal{A}W + H(z,\overline{z},\theta) \tag{22}$$

where

$$H(z,\overline{z},\theta) = H_{20}(\theta)\frac{z^2}{2} + H_{11}(\theta)z\overline{z} + H_{02}(\theta)\frac{\overline{z}^2}{2} + H_{21}(\theta)\frac{z^2\overline{z}}{2} + \ldots \tag{23}$$

But at the other side, on $C_0$ close to the origin $.W' = W_z z' + W_z \overline{z}'$

We obtain by multiplying the above series by the variables.

$$[\mathcal{A} - 2i\omega_0 I]W_{20}(\theta) = -H_{20}(\theta),\ \mathcal{A}W_{11}(\theta) = -H_{11}(\theta) \tag{24}$$

By (16), $\theta \in [-1, 0)$,

$$H(z,\overline{z},\theta) = -\overline{q^*}(0)\overline{F_0}q(\theta) - \overline{q^*}(0)\overline{F_0}\overline{q}(\theta) = -gq(\theta) - \overline{g}\overline{q}(\theta)$$

Comparing the coefficient with (19) we get for $\theta \in [-1, 0)$, that.

$H_{20}(\theta) = -g_{20}q(\theta) - \overline{g_{02}}\,\overline{q}(\theta)$, $H_{11}(\theta) = -g_{11}q(\theta) - \overline{g_{11}}\,\overline{q}(\theta)$.

From (22), (24) and definition of $\mathcal{A}$ we obtain

$$W_{20}(\theta) = 2i\omega_0\tau_0 W_{20}(\theta) + g_{20}q(\theta) + \overline{g_{02}\overline{q}}(\theta)$$

Solving for $W_{20}(\theta)$:

$$W_{20}(\theta) = \frac{ig_{20}}{\omega_0\tau_0}q(0)e^{i\omega_0\tau_0\theta} + \frac{i\overline{g_{02}}}{3\omega_0\tau_0}\overline{q}(0)e^{-i\omega_0\tau_0\theta} + E_1 e^{2i\omega_0\tau_0\theta},$$

And similarly

$$W_{11}(\theta) = \frac{-ig_{11}}{\omega_0\tau_0} q(0)e^{i\omega_0\tau_0\theta} + \frac{i\overline{g_{11}}}{\omega_0\tau_0} \overline{q}(0)e^{-i\omega_0\tau_0\theta} + E_2$$

where $E_1$ and $E_2$ are the 3dim. vectors and examine by putting the $\theta = 0$ in $H$.

$H(z, \overline{z}, \theta) = -2Re\left[\overline{q^*}(0)F_0 q(0)\right] + F_0$, we have $H_{20}(\theta) = -g_{20}q(\theta) - \overline{g_{02}}\,\overline{q}(\theta) + F_{z^2}$,

$H_{11}(\theta) = -g_{11}q(\theta) - \overline{g_{11}}\,\overline{q}(\theta) + F_{z\overline{z}}$, Where $F_0 = F_{z^2}\frac{z^2}{2} + F_{z\overline{z}}z\overline{z} + F_{\overline{z}^2}\frac{\overline{z}^2}{2} + \ldots$ and adjust the defamation of $\mathcal{A}$,

$\int_{-1}^{0} d\eta(\theta)W_{20}(\theta) = 2i\omega_0\tau_0 W_{20}(0) + g_{20}q(0) + \overline{g_{02}}\,\overline{q}(0) - F_{z^2}$ and.

$\int_{-1}^{0} d\eta(\theta)W_{11}(\theta) = g_{11}q(0) - \overline{g_{11}}\,\overline{q}(0) - F_{z\overline{z}}$ Notice that.

$\left[i\omega_0\tau_0 I - \int_{-1}^{0} e^{i\omega_0\tau_0\theta} d\eta(\theta)\right]q(0) = 0$ and

$$\left[-i\omega_0\tau_0 I - \int_{-1}^{0} e^{-i\omega_0\tau_0\theta} d\eta(\theta)\right]\overline{q}(0) = 0 \Rightarrow$$

$\left[2i\omega_0\tau_0 I - \int_{-1}^{0} e^{2i\omega_0\tau_0\theta} d\eta(\theta)\right]E_1 = F_{z^2}$ and $-\left[\int_{-1}^{0} d\eta(\theta)\right]E_2 = F_{z\overline{z}}$

Hence

$$\begin{bmatrix} 2i\omega_0 + a_1 - 2a_2P_1^* - \beta_1P_2^* & -\beta_1P_1^* + \gamma_1P_1^{*2} \\ (-\beta_2P_2^* + \gamma_2P_2^{*2})e^{-2i\omega_0\tau_0\theta} & b_1 + 2i\omega_0 \end{bmatrix} E_1 = -2\begin{bmatrix} 0 \\ -P_2^*(\gamma_2 - \beta_2)e^{-i\omega_0\tau_0\theta} \end{bmatrix}$$

$$\begin{bmatrix} a_1 - 2a_2P_1^* - \beta_1P_2^* & -\beta_1P_1^* + \gamma_1P_1^{*2} \\ -\beta_2P_2^* + \gamma_2P_2^{*2} & b_1 \end{bmatrix} E_1 = -2\begin{bmatrix} 0 \\ -P_2^*(\gamma_2 - \beta_2)\sigma_1 e^{-i\omega_0\tau_0\theta} \end{bmatrix}$$

And $g_{21}$ can be shows by the parameters.

On the basis of the above calculations, every $g_{ii}$ can be calculated with the help of parameters. And these quantities can be calculated:

$$C_1(0) = \frac{i}{2\omega_0\tau_0}\left(g_{11}g_{20} - 2|g_{11}|^2 - \frac{|g_{02}|^2}{3}\right) + \frac{g_{21}}{2} and \mu_2 = -\frac{Re\{C_1(0)\}}{Re\{\lambda'(\tau_0)\}},$$

$$\beta_2 = Re\{C_1(0)\}$$

$$T_2 = -\frac{Im\{C_1(0)\} + \mu_2 Im\{\lambda'(\tau_0)\}}{\omega_0\tau_0} \tag{25}$$

**Theorem 2** *The direction of Hopf-bifurcation is calculated by the value of* $\mu_2$*: if* $\mu_2 > 0(\mu_2 < 0)$*, then the hopf-bifurcation is saturated and there is a regular bifurcation that endures for* $\tau > \tau_0(\tau < \tau_0)$*. With the help of* $\beta_2$*, we can calculate the stability of the bifurcating solutions: the bifurcation cyclic solutions are asymptotic if* $\beta_2 > 0(\beta_2 < 0)$*. The* $T_2$ *calculates the cycle of bifurcating cyclic solutions, as to whether the cyclic increase or decease* $T_2 > 0(T_2 < 0)$*.*

## 5 Numerical Example

The computation is carried out using MATLAB to coordinate the analytic result using a quantitative method. The system's behaviour is illustrated for the following value sets:

$$a_1 = 2,\ a_2 = 0.07,\ b_1 = 1,\ b_2 = 0.08,\ \beta_1 = 0.05,\ \beta_2 = 0.015,$$
$$\gamma_1 = 0.0008,\ \gamma_2 = 0.003,$$

ectional analysis of the Hopf-bifurcating has been accomplished.

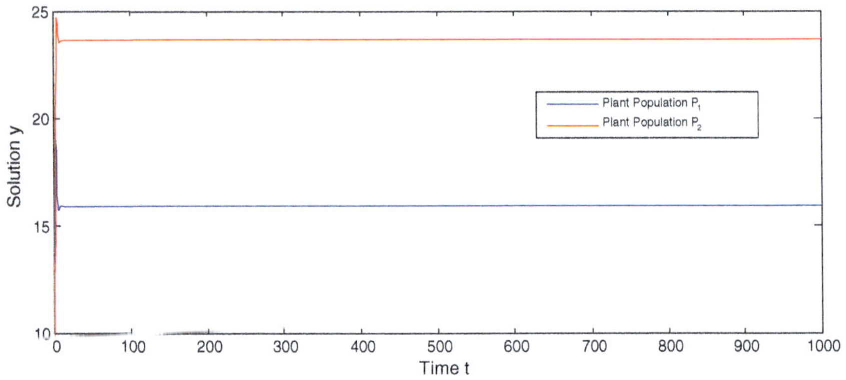


**Fig. 3** The Equilibrium point $E^*(P_1, P_2)$ is stable in the absence of delay i.e., $\tau = 0$

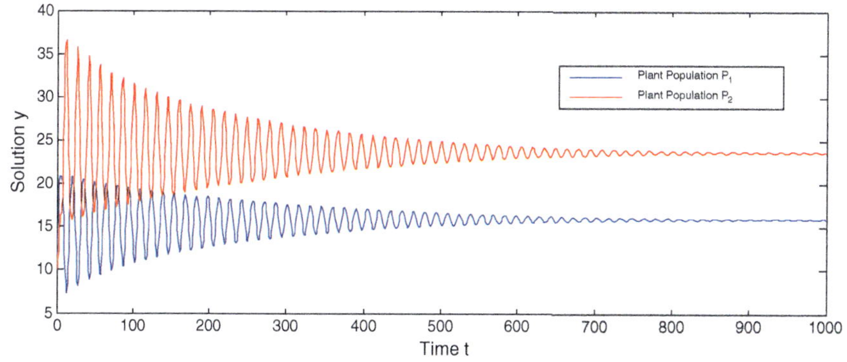


**Fig. 4** *At* $\tau < 6.9999$, *the equilibrium point* $E^*(P_1, P_2)$ *shows asymptotically stable*

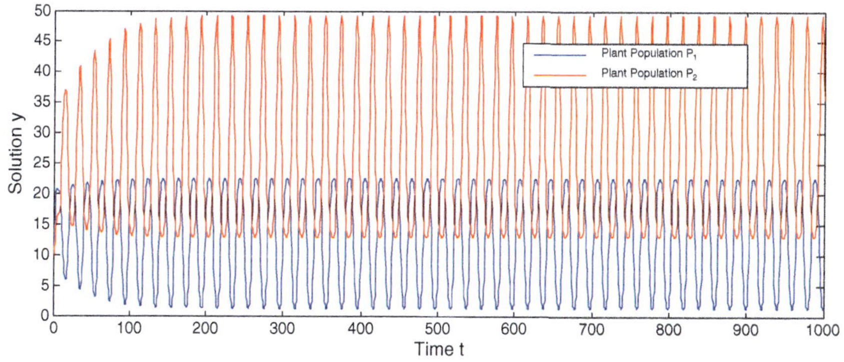


**Fig. 5** The equilibrium point $E^*(P_1, P_2)$ loses its asymptotical stability and Hopf-Bifurcation occurs at $\tau \geq 6.9999$

# 6 Conclusion

In this paper, a mathematical model is proposed to analyze the impact of allelochemicals on plant population development using delay differential equation. Stability and hopf-bifurcation determined about non-zero equilibriums point using Routh-Hurwitz's criteria. In the absence of delay none of the population affect adversely each other and grow at their normal rate, and the equilibrium point is stable as shown in Fig. 3. The system loses its stability when the value of delay decreases from the threshold value and goes for asymptotical stability, actually meaning if there is a delay involved in the allelochemicals realizes that still the system grows at the natural rate after few fluctuations in the beginning under the asymptotical stability as shown in Fig. 4. It loses its asymptoticality when the value of delay exceeds than the threshold value, where both the populations remain under the effect of allelochemicals forever. A repetition of limit cycles will always occur after a particular time period showing that the hopf-bifurcation is often shown in Fig. 5. In this article, the sensitivity of system variables w.r.t model parameters $\beta_1$and$\beta_2$ is done using the "Direct approach". It demonstrates how the sensitivity functions allows one to recognize particular parameters and enhance the view of the importance of the delay played by different parameters of the model as shown Figs. 1 and 2. When we vary the value of $\beta_1$and$\beta_2$ parameters in the system it shows hopf-bifurcation and asymptotical stability and then it shows stability of the model as shown in Figs. 1 and 2. As a result, the specific producer that explains the stability and dir

## References


1. Willis, R.J.: The History of Allelopathy. Springer Science & Business Media (2007)
2. Thornley, J.H.: Mathematical Models in Plant Physiology. Academic Press (Inc.) Ltd., London (1976)
3. Cheng, F., Cheng, Z.: Research progress on the use of plant allelopathy in agriculture and the physiological and ecological mechanisms of allelopathy. Front. Plant Sci. **6**, 1020 (2015)
4. Nwaoburu, A.O., Lgwe, P.Y., Asty, J.U., Ekaka-a, E.N.: Modeling the increasing differential effects of the first inter-competition coefficient on the biodiversity value; competition between two phytoplankton species. Int. J. Adv. Eng. Manag. Sci. **4**, 266189 (2018)
5. Hayyat, M.S., Safdar, M.E., Asif, M., Tanveer, A., Ali, L., Qamar, R., H Tarar, Z.: Allelopathic effect of waste-land weeds on germination and growth of winter crops. Planta Daninha **38** (2020)
6. Felpeto, A.B., Roy, S., Vasconcelos, V.M.: Allelopathy prevents competitive exclusion and promotes phytoplankton biodiversity. Oikos **127**(1), 85–98 (2018)
7. Abbas, S., Mahto, L., Favini, A., Hafayed, M.: Dynamical study of fractional model of allelopathic stimulatory phytoplankton species. Diff. Equ. Dynam. Syst. **24**(3), 267–280 (2016)
8. Chen, B.M., Liao, H.X., Chen, W.B., Wei, H.J., Peng, S.L.: Role of allelopathy in plant invasion and control of invasive plants. Allelopath. J. **41**, 155–166 (2017)
9. Zhou, X., Wu, Z., Wang, Z., Zhou, T.: Stability and Hopf bifurcation analysis in a fractional-order delayed paddy ecosystem. Adv. Diff. Equ. **2018**(1), 1–14 (2018)
10. Cariboni, J., Gatelli, D., Liska, R., Saltelli, A.: The role of sensitivity analysis in ecological modelling. Ecol. Model. **203**(1–2), 167–182 (2007)
11. Grzyb, A., Wolna-Maruwka, A., Niewiadomska, A.: Environmental factors affecting the mineralization of crop residues. Agronomy **10**(12), 1951 (2020)
12. Huang, G., Liu, A., Foryś, U.: Global stability analysis of some nonlinear delay differential equations in population dynamics. J. Nonlinear Sci. **26**(1), 27–41 (2016)
13. Ruan, S.: Absolute stability, conditional stability and bifurcation in Kolmogorov-type predator-prey systems with discrete delays. Q. Appl. Math. **59**(1), 159–173 (2001)
14. Ruan, S., Wei, J.: On the zeros of transcendental functions with applications to stability of delay differential equations with two delays. Dyn. Contin. Disc. Impuls. Syst. Ser. A **10**, 863–874 (2003)
15. Kalra, P., Kumar, P.: The study of time lag on plant growth under the effect of toxic metal: a mathematical model. Pertanika J. Sci. Technol. **26**(3) (2018)
16. Kalra, P., Kumar, P.: Role of delay in plant growth dynamics: a two compartment mathematical model. In: AIP Conference Proceedings, vol. 1860, pp. 020045. AIP Publishing LLC (2017)
17. Ingalls, B., Mincheva, M., Roussel, M.R.: Parametric sensitivity analysis of oscillatory delay systems with an application to gene regulation. Bull. Math. Biol. **79**(7), 1539–1563 (2017)
18. Rihan, F.A.: Sensitivity analysis for dynamic systems with time-lags. J. Comput. Appl. Math. **151**(2), 445–462 (2003)

# Pore Scale Analysis and Homogenization of a Diffusion-Reaction-Dissolution-Precipitation Model

**Nibedita Ghosh and Hari Shankar Mahato**

**Abstract** A pore scale model is explored where two types of mobile species having different non-constant diffusion coefficients react and then precipitate as crystals on the solid boundary. The reaction is reversible so dissolution also happens and it involves a discontinuous multivalued rate term. We start with establishing the existence of a unique positive global weak solution. After that, we derive the upscaled equations by applying periodic homogenization techniques relying on two-scale convergence and boundary unfolding operators.

**Keywords** Diffusion-reaction-dissolution-precipitation · Reactive transport · Porous media · Global weak solution · Periodic homogenization

## 1 Introduction

Transport through porous media is encountered in several engineering and biological applications such as oil production, soil erosion, groundwater pollution, polymer processing, filtration, tissue engineering and discussion of the dynamics of blood flow. A porous media is heterogeneous having porosity $\theta \ll 1$. It contains two parts: one is the pore space and another one is the solid parts. The heterogeneities inside the medium are smaller with respect to the size of the medium. Therefore to analyze what is happening within the domain we need to investigate the microscale description of the domain although it's not suited for numerical experiments due to the heterogeneity. Therefore we need to upscale the model to the macroscale from the microscale to study global behavior. Here we make an assumption that the solid parts are not connected and distributed in a periodic way in the given medium. However, in a natural porous medium, solid parts are connected whereas this periodicity

N. Ghosh (✉) · H. S. Mahato
Department of Mathematics, Indian Institute of Technology Kharagpur, Kharagpur 721302, WB, India
e-mail: nghosh.iitkgp@gmail.com

H. S. Mahato
e-mail: hsmahato@maths.iitkgp.ac.in



B. Rushi Kumar et al. (eds.), *Mathematics and Computing*, Springer Proceedings in Mathematics & Statistics 415, https://doi.org/10.1007/978-981-19-9307-7_48

assumption appears to be a good approximation to the main domain. The pore space contains the mobile species and the immobile species are present on the grain boundary. The process of transportation of the mobile species is modeled by Fick's law and governed by diffusion, dispersion or advection. We assume that there is a "constant activity" on the surface of the solids. We model the surface reaction phenomena with the help of nonlinear *Langmuir Kinetics* and the dissolution process is described by a monotone discontinuous multivalued rate term. We study the system for the case of constant different diffusion coefficients in [7]. The main challenge in the analysis here is to deal with the space and time-dependent different diffusion coefficients, multivalued dissolution rate term and nonlinear surface reaction rate term.

We consider a bounded porous medium $\Omega \subset \mathbb{R}^n (n \geq 2)$, which consists a pore space $\Omega^p$ and the union of solid parts $\Omega^s$ in a way that $\Omega := \Omega^p \cup \Omega^s$ where $\bar{\Omega}^s \cap \Omega^p = \phi$. The exterior boundary of the domain is denoted by $\partial\Omega$ and $\Gamma^*$ represents the union of solid boundaries. We choose the representative cell as $Y := (0, 1)^n \subset \mathbb{R}^n$ such that $Y = Y^s \cup Y^p$, where $Y^p$ is the pore part and $Y^s$ is the solid part with boundary $\Gamma$ so $\bar{Y}^s \cap \bar{Y}^p = \Gamma$. For each multi-index $l \in \mathbb{Z}^n$, let be the shifted sets are $Y_l := Y + l$, $Y_l^\mu := Y^\mu + l$ for $\mu \in \{p, s\}$ and $\Gamma_l := \Gamma + l$. Moreover, we make an assumption that $\Omega$ is $\varepsilon$-periodic where $\varepsilon$ is a positive small scaling parameter. That means the solid matrices in $\Omega$ are distributed periodically and the finite union of the representative cells $Y$ can be the cover of the domain $\Omega$. The geometry stated above satisfies the assumptions that: solid matrices never touch one another, solid matrices never touch the domain outer boundary $\partial\Omega$ and solid matrices never touch the boundary of $Y$. Since $\Omega$ is the finite union of translated version of $\varepsilon Y_l$ cells such that $\varepsilon Y_l \subset \Omega$ where $l \in \mathbb{Z}^n$, that is $\Omega \subset \underset{l\in\mathbb{Z}^n}{\cup} \varepsilon Y_l$, $\Omega^p \subset \underset{l\in\mathbb{Z}^n}{\cup} \varepsilon Y_l^p$, $\Omega^s \subset \underset{l\in\mathbb{Z}^n}{\cup} \varepsilon Y_l^s$ and $\Gamma^* \subset \underset{l\in\mathbb{Z}^n}{\cup} \varepsilon \Gamma_l$. We also define $\Omega_\varepsilon^p := \cup_{l\in\mathbb{Z}^n} \left\{\varepsilon Y_l^p : \varepsilon Y_l^p \subset \Omega\right\}$, $\Omega_\varepsilon^s := \cup_{l\in\mathbb{Z}^n} \left\{\varepsilon Y_l^s : \varepsilon Y_l^s \subset \Omega\right\}$, $\Gamma_\varepsilon^* := \cup_{l\in\mathbb{Z}^n} \{\varepsilon \Gamma_l : \varepsilon \Gamma_l \subset \Omega\}$, $\partial\Omega_\varepsilon^p := \partial\Omega \cup \Gamma_\varepsilon^*$, cf. Fig. 1. Let $S := [0, T)$ be the time interval for $T > 0$. We also denote the volume elements in $Y$ and $\Omega$ as $dy, dx$ and

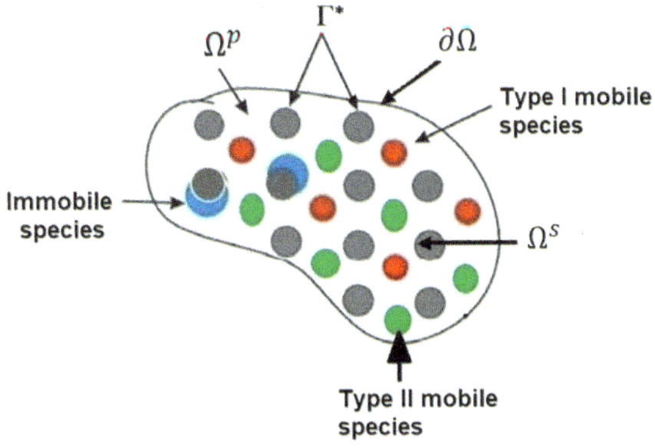


**Fig. 1** Crystal dissolution and precipitation on $\Gamma^*$ and mobile species in $\Omega^p$

the surface elements as $d\sigma_y$ and $d\sigma_x$ on $\Gamma$ and $\Gamma_\varepsilon^*$. The characteristic function of $\Omega_\varepsilon^p$ in $\Omega$ is defined by,

$$\chi_\varepsilon(x) = \chi(\frac{x}{\varepsilon}) \quad = \quad \begin{cases} 1 & \text{when } x \in \Omega_\varepsilon^p, \\ 0 & \text{otherwise.} \end{cases}$$

## 1.1 The Model

We consider two types of mobile species denoted by $\mathcal{I}_1$ and $\mathcal{I}_2$ are present in $\Omega_\varepsilon^p$ and the immobile species $\mathcal{I}_{12}$ is present on $\Gamma_\varepsilon^*$. Now let $\mathcal{I}_1, \mathcal{I}_2$ and $\mathcal{I}_{12}$ are connected via following reaction:

$$\mathcal{I}_1 + \mathcal{I}_2 \leftrightarrow \mathcal{I}_{12} \qquad \text{on} \quad \Gamma_\varepsilon^*. \tag{1}$$

In our situation, there is no reaction happening among the mobile species but at the outer boundary, we impose flux boundary conditions for the two mobile species. As by (1), the dissolution process supplied $\mathcal{I}_1$ and $\mathcal{I}_2$ to $\Gamma_\varepsilon^*$, therefore, the Neumann boundary condition for $\mathcal{I}_1$ and $\mathcal{I}_2$ on $\Gamma_\varepsilon^*$ will be the same as the rate of change of concentration of the mineral $\mathcal{I}_{12}$ on $\Gamma_\varepsilon^*$. According to the relation (1), one molecule of each $\mathcal{I}_1$ and $\mathcal{I}_2$ will provide one molecule of the crystal $\mathcal{I}_{12}$. We model the surface reaction phenomena with the help of *Langmuir kinetics*. On other hand, $\mathcal{I}_{12}$ will dissolve to produce $\mathcal{I}_1$ and $\mathcal{I}_2$. The dissolution process is modeled by the idea adopted from [9, 15]. Concerning dissolution at the surface of the solids, if the mineral is present then the dissolution rate is constant. For the situation when the mineral is absent, the dissolution rate can not be stronger than precipitation to mention the positivity of the surface concentration. This gives rise to a multivalued dissolution rate term $r_d(w_\varepsilon) \in k_d\psi(w_\varepsilon)$, such that

$$\psi(d) \quad = \quad \begin{cases} \{0\} & \text{when } d < 0, \\ [0, 1] & \text{when } d = 0, \\ \{1\} & \text{when } d > 0. \end{cases} \tag{2}$$

Let $u_\varepsilon$, $v_\varepsilon$ and $w_\varepsilon$ be the concentrations of the species $\mathcal{I}_1, \mathcal{I}_2$ and $\mathcal{I}_{12}$, respectively. Therefore, the mass-balance equations for $\mathcal{I}_1, \mathcal{I}_2$ and $\mathcal{I}_{12}$ are

$$\frac{\partial u_\varepsilon}{\partial t} + \nabla.(-\bar{D}_1^\varepsilon \nabla u_\varepsilon) = 0 \ \text{ in } S \times \Omega_\varepsilon^p, \tag{3a}$$

$$-\bar{D}_1^\varepsilon \nabla u_\varepsilon.\vec{n} = d_1(t, x) \ \text{ on } S \times \partial\Omega, \tag{3b}$$

$$-\bar{D}_1^\varepsilon \nabla u_\varepsilon.\vec{n} = \varepsilon\frac{\partial w_\varepsilon}{\partial t} \ \text{ on } S \times \Gamma_\varepsilon^*, \tag{3c}$$

$$u_\varepsilon(0, x) = u_0(x) \ \text{ in } \Omega, \tag{3d}$$

$$\frac{\partial v_\varepsilon}{\partial t} + \nabla.(-\bar{D}_2^\varepsilon \nabla v_\varepsilon) = 0 \text{ in } S \times \Omega_\varepsilon^p, \tag{4a}$$

$$-\bar{D}_2^\varepsilon \nabla v_\varepsilon.\vec{n} = d_2(t, x) \text{ on } S \times \partial\Omega, \tag{4b}$$

$$-\bar{D}_2^\varepsilon \nabla v_\varepsilon.\vec{n} = \varepsilon \frac{\partial w_\varepsilon}{\partial t} \text{ on } S \times \Gamma_\varepsilon^*, \tag{4c}$$

$$v_\varepsilon(0, x) = v_0(x) \text{ in } \Omega, \tag{4d}$$

$$\frac{\partial w_\varepsilon}{\partial t} = k_d(r_1(u_\varepsilon, v_\varepsilon) - z_\varepsilon) \text{ on } S \times \Gamma_\varepsilon^*, \tag{5a}$$

$$z_\varepsilon \in \psi(w_\varepsilon) \text{ on } S \times \Gamma_\varepsilon^*, \tag{5b}$$

$$w_\varepsilon(0, x) = w_0(x) \text{ in } \Omega, \tag{5c}$$

where $r_1 : \mathbb{R}^2 \to [0, \infty)$ is given by

$$r_1(u_\varepsilon, v_\varepsilon) = \begin{cases} k\dfrac{k_1 u_\varepsilon k_2 v_\varepsilon}{(1 + k_1 u_\varepsilon + k_2 v_\varepsilon)^2} & \text{if } (u_\varepsilon, v_\varepsilon) \in [0, \infty)^2, \\ 0 & \text{otherwise} \end{cases}$$

and $k = \frac{k_f}{k_d}$. $k_1$ and $k_2$ are the Langmuir constants for $\mathcal{I}_1$ and $\mathcal{I}_2$. The dissolution rate $r_d = k_d\psi(w_\varepsilon)$, where $k_d$ is the dissolution rate constant. Let $k_f$ denote the forward reaction rate constant. We represent the system (3a)–(5c) by $(\mathbb{S}_\varepsilon)$.

### 1.1.1 Function Space Setup

Following the usual definitions of Lebesgue spaces ($L^p$-spaces), Sobolev spaces ($H^{k,p}$-spaces) and Bochner spaces of time-space variables from [3, 14], we choose our solution space as $U_\varepsilon := \{u_\varepsilon \in L^2(S; H^{1,2}(\Omega_\varepsilon^p)) : \frac{\partial u_\varepsilon}{\partial t} \in L^2(S; H^{1,2}(\Omega_\varepsilon^p)^*)\} := H^{1,2}(S; H^{1,2}(\Omega_\varepsilon^p)^*) \cap L^2(S; H^{1,2}(\Omega_\varepsilon^p))$, $V_\varepsilon := \{v_\varepsilon \in L^2(S; H^{1,2}(\Omega_\varepsilon^p)) : \frac{\partial v_\varepsilon}{\partial t} \in L^2 (S; H^{1,2}(\Omega_\varepsilon^p)^*)\} := H^{1,2}(S; H^{1,2}(\Omega_\varepsilon^p)^*) \cap L^2(S; H^{1,2}(\Omega_\varepsilon^p))$, $W_\varepsilon := \{w_\varepsilon \in L^2(S; L^2(\Gamma_\varepsilon^*)) : \frac{\partial w_\varepsilon}{\partial t} \in L^2(S; L^2(\Gamma_\varepsilon^*))\} := H^{1,2}(S; H^{1,2}(\Gamma_\varepsilon^*))$, $Z_\varepsilon := \{z_\varepsilon \in L^\infty(S \times \Gamma_\varepsilon^*) : 0 \le z_\varepsilon \le 1\}$, $\mathcal{X}_p(\Xi) := (H^{1,q}(\Omega_\varepsilon^p)^*, H^{1,p}(\Omega_\varepsilon^p))_{1-\frac{1}{p},p}$. The spaces $L^2(S \times \Omega_\varepsilon^p)$, $L^2(S \times \Gamma_\varepsilon^*)$ and $H^{1,2}(\Omega_\varepsilon^p)$ equipped with the norms $\|\zeta\|^2_{(\Omega_\varepsilon^p)^T} := \int_0^T \int_{\Omega_\varepsilon^p} |\zeta|^2 dx dt$, $\|\zeta\|^2_{(\Gamma_\varepsilon^*)^T} := \varepsilon \int_0^T \int_{\Gamma_\varepsilon^*} |\zeta|^2 d\sigma_x dt$ and $\|\zeta\|_{H^{1,2}(\Omega_\varepsilon^p)} := \|\zeta\|_{\Omega_\varepsilon^p} + \|\nabla\zeta\|_{\Omega_\varepsilon^p}$, respectively.

**Weak Formulation**. A quadruple $(u_\varepsilon, v_\varepsilon, w_\varepsilon, z_\varepsilon) \in U_\varepsilon \times V_\varepsilon \times W_\varepsilon \times Z_\varepsilon$ is called weak solution of (3a)–(5c) if $(u_\varepsilon(0), v_\varepsilon(0), w_\varepsilon(0)) = (u_0, v_0, w_0) \in L^2(\Omega) \times L^2(\Omega) \times L^2(\Gamma^*)$ as well as

$$\langle\frac{\partial u_\varepsilon}{\partial t},\phi\rangle_{(\Omega_\varepsilon^p)^t}+\langle\bar{D}_1^\varepsilon\nabla u_\varepsilon,\nabla\phi\rangle_{(\Omega_\varepsilon^p)^t}=-\langle\frac{\partial w_\varepsilon}{\partial t},\phi\rangle_{(\Gamma_\varepsilon^*)^t}-\langle d_1,\phi\rangle_{(\partial\Omega)^t}, \tag{6a}$$

$$\langle\frac{\partial v_\varepsilon}{\partial t},\theta\rangle_{(\Omega_\varepsilon^p)^t}+\langle\bar{D}_2^\varepsilon\nabla v_\varepsilon,\nabla\theta\rangle_{(\Omega_\varepsilon^p)^t}=-\langle\frac{\partial w_\varepsilon}{\partial t},\theta\rangle_{(\Gamma_\varepsilon^*)^t}-\langle d_2,\theta\rangle_{(\partial\Omega)^t}, \tag{6b}$$

$$\langle\frac{\partial w_\varepsilon}{\partial t},\eta\rangle_{(\Gamma_\varepsilon^*)^t}=k_d\langle r_1(u_\varepsilon,v_\varepsilon)-z_\varepsilon,\eta\rangle_{(\Gamma_\varepsilon^*)^t},\ z_\varepsilon\in\psi(w_\varepsilon)\text{ almost everywhere on }(\Gamma_\varepsilon^*)^t, \tag{6c}$$

for all $(\phi,\theta,\eta)\in L^2(S;H^{1,2}(\Omega_\varepsilon^p))\times L^2(S;H^{1,2}(\Omega_\varepsilon^p))\times L^2(S;L^2(\Gamma_\varepsilon^*))$. We need to make few assumptions to do the analysis of the model:

(**A1.**) $u_0, v_0, w_0\geq 0$. (**A2.**) $r_1(u_\varepsilon,v_\varepsilon)=0$ for all $u_\varepsilon\leq 0$, $v_\varepsilon\leq 0$. (**A3.**) $d_1$, $d_2\in L^2(S\times\partial\Omega)$ and $d_1,d_2\leq 0$. (**A4.**) $u_0,v_0\in H^{1,2}(\Omega)$ and $w_0\in L^\infty(\Omega)$. (**A5.**) $r_1:\mathbb{R}^2\to[0,\infty)$ is Locally Lipschitz in $\mathbb{R}^2$ with Lipschitz constant $L_R>0$. (**A6.**) $\bar{D}_i^\varepsilon=diag(D_i(t,\frac{x}{\varepsilon}),D_i(t,\frac{x}{\varepsilon}),\ldots,D_i(t,\frac{x}{\varepsilon}))\in(L^\infty(S\times Y))^{n\times n}$, $i\in\{1,2\}$ such that $(D_i(t,y)\zeta,\zeta)\geq\alpha|\zeta|^2$ for every $\zeta\in\mathbb{R}^n$ and for $\alpha>0$ is a constant which does not depend on $\varepsilon$ and for every $(t,y)\in S\times Y$.

**Lemma 1** *For $v_{\varepsilon\delta},u_{\varepsilon\delta}\in H^{1,2}(\Omega_\varepsilon^p)$ there exist extension $\tilde{v}_{\varepsilon\delta},\tilde{u}_{\varepsilon\delta}$ to $\Omega$ such that*

*(i)* $\|\tilde{v}_{\varepsilon\delta}\|_{H^{1,2}(\Omega)}\leq C\|v_{\varepsilon\delta}\|_{H^{1,2}(\Omega_\varepsilon^p)}$, $\|\tilde{u}_{\varepsilon\delta}\|_{H^{1,2}(\Omega)}\leq C\|u_{\varepsilon\delta}\|_{H^{1,2}(\Omega_\varepsilon^p)}$.

***Proof*** The proof of the above lemma can be found in Lemma 5 of [8].

**Theorem 1** *Suppose that the assumptions* (**A1.**)–(**A6.**) *are satisfied and* $(u_\varepsilon,v_\varepsilon,w_\varepsilon)$ *satisfy the following a-priori bounds*

$$\|u_\varepsilon\|_{L^2(S\times\Omega_\varepsilon^p)}+\|\nabla u_\varepsilon\|_{L^2(S\times\Omega_\varepsilon^p)}+\left\|\frac{\partial u_\varepsilon}{\partial t}\right\|_{L^2(S;H^{1,2}(\Omega_\varepsilon^p)^*)}+\|v_\varepsilon\|_{L^2(S\times\Omega_\varepsilon^p)}+\|\nabla v_\varepsilon\|_{L^2(S\times\Omega_\varepsilon^p)}$$
$$+\left\|\frac{\partial v_\varepsilon}{\partial t}\right\|_{L^2(S;H^{1,2}(\Omega_\varepsilon^p)^*)}+\|w_\varepsilon\|_{L^2(S\times\Gamma_\varepsilon^*)}+\left\|\frac{\partial w_\varepsilon}{\partial t}\right\|_{L^2(S\times\Gamma_\varepsilon^*)}\leq C, \tag{7}$$

*where $C$ is a constant does not depend on $\varepsilon$ and $\delta$. Then there exists a unique positive global weak solution $(u_\varepsilon,v_\varepsilon,w_\varepsilon,z_\varepsilon)\in U_\varepsilon\times V_\varepsilon\times W_\varepsilon\times Z_\varepsilon$ of the system $(\mathbb{S}_\varepsilon)$:*

$$\frac{\partial u_\varepsilon}{\partial t}+\nabla.(-\bar{D}_1^\varepsilon\nabla u_\varepsilon)=0\ \ in\ S\times\Omega_\varepsilon^p, \tag{8a}$$

$$-\bar{D}_1^\varepsilon\nabla u_\varepsilon.\vec{n}=d_1(t,x)\ \ on\ S\times\partial\Omega, \tag{8b}$$

$$-\bar{D}_1^\varepsilon\nabla u_\varepsilon.\vec{n}=\varepsilon\frac{\partial w_\varepsilon}{\partial t}\ \ on\ S\times\Gamma_\varepsilon^*, \tag{8c}$$

$$u_\varepsilon(0,x)=u_0(x)\ \ in\ \Omega, \tag{8d}$$

$$\frac{\partial v_\varepsilon}{\partial t}+\nabla.(-\bar{D}_2^\varepsilon\nabla v_\varepsilon)=0\ \ in\ S\times\Omega_\varepsilon^p, \tag{9a}$$

$$-\bar{D}_2^\varepsilon\nabla v_\varepsilon.\vec{n}=d_2(t,x)\ \ on\ S\times\partial\Omega, \tag{9b}$$

$$-\bar{D}_2^\varepsilon\nabla v_\varepsilon.\vec{n}=\varepsilon\frac{\partial w_\varepsilon}{\partial t}\ \ on\ S\times\Gamma_\varepsilon^*, \tag{9c}$$

$$v_\varepsilon(0,x)=v_0(x)\ \ in\ \Omega, \tag{9d}$$

$$\frac{\partial w_\varepsilon}{\partial t} = k_d(r_1(u_\varepsilon, v_\varepsilon) - z_\varepsilon) \ \ on\ S \times \Gamma_\varepsilon^*, \tag{10a}$$

$$z_\varepsilon \in \psi(w_\varepsilon) \ \ on\ S \times \Gamma_\varepsilon^*, \tag{10b}$$

$$w_\varepsilon(0, x) = w_0(x) \ \ in\ \Omega. \tag{10c}$$

## 2 Proof of Theorem 1

**Lemma 2** (Positivity and $L^2$-estimates) *Under the assumptions* (**A1.**)–(**A6.**) *and for a.e.* $t \in S$ *the following estimates hold*

(*i*) $u_\varepsilon(t), v_\varepsilon(t), w_\varepsilon(t) \geq 0$ *a.e. in* $\Omega_\varepsilon^p$ *and on* $\Gamma_\varepsilon^*$, *respectively.*
(*ii*) (*a*) $\|u_\varepsilon(t)\|^2_{\Omega_\varepsilon^p} \leq M_u e^{\alpha T}$, $\|v_\varepsilon(t)\|^2_{\Omega_\varepsilon^p} \leq M_v e^{\alpha T}$, $\|w_\varepsilon(t)\|^2_{\Gamma_\varepsilon^*} \leq M_w e^{T}$ *a.e. in* $\Omega_\varepsilon^p$ *and on* $\Gamma_\varepsilon^*$, *respectively.*
(*b*) $\|\nabla u_\varepsilon\|^2_{(\Omega_\varepsilon^p)^t} \leq \bar{M}_u e^{\alpha T}$, $\|\nabla v_\varepsilon\|^2_{(\Omega_\varepsilon^p)^t} \leq \bar{M}_v e^{\alpha T}$ *a.e. in* $\Omega_\varepsilon^p$.
(*c*) $\left\|\frac{\partial u_\varepsilon}{\partial t}\right\|^2_{L^2(S;H^{1,2}(\Omega_\varepsilon^p)^*)} \leq C$, $\left\|\frac{\partial v_\varepsilon}{\partial t}\right\|^2_{L^2(S;H^{1,2}(\Omega_\varepsilon^p)^*)} \leq C$, $\left\|\frac{\partial w_\varepsilon}{\partial t}\right\|^2_{L^2(S\times\Gamma_\varepsilon^*)} \leq C$ *a.e. in* $\Omega_\varepsilon^p$ *and on* $\Gamma_\varepsilon^*$, *respectively.*

***Proof*** (*i*) We test (6a)–(6c) with $(\phi, \theta, \eta) = (-[u_\varepsilon]_-, -[v_\varepsilon]_-, -[w_\varepsilon]_-)$ and using (**A1.**) – (**A3.**) get

$$\|[u_\varepsilon(t)]_-\|^2_{\Omega_\varepsilon^p} + 2\alpha\|\nabla[u_\varepsilon]_-\|^2_{(\Omega_\varepsilon^p)^t} \leq 2\underbrace{k_d\langle z_\varepsilon, -[u_\varepsilon]_-\rangle_{(\Gamma_\varepsilon^*)^t}}_{\leq 0} - 2\underbrace{\langle d_1, -[u_\varepsilon]_-\rangle_{(\partial\Omega)^t}}_{\geq 0} \leq 0.$$

That means, $u_\varepsilon(t)$ is non-negative for almost everywhere $t \in [0, T)$. Likewise, we get $v_\varepsilon(t)$ is non-negative for almost everywhere $t \in [0, T)$. Now for the immobile species we see that

$$\|[w_\varepsilon(t)]_-\|^2_{\Gamma_\varepsilon^*} = \underbrace{\|[w_0]_-\|^2_{\Gamma_\varepsilon^*}}_{=0} + 2\underbrace{k_d\langle r_1(u_\varepsilon, v_\varepsilon), -[w_\varepsilon]_-\rangle_{(\Gamma_\varepsilon^*)^t}}_{\leq 0} - 2\underbrace{k_d\langle z_\varepsilon, -[w_\varepsilon]_-\rangle_{(\Gamma_\varepsilon^*)^t}}_{=0} \leq 0.$$

Since $u_\varepsilon$ and $v_\varepsilon$ are non-negative, we get $r_1(u_\varepsilon, v_\varepsilon) = k\frac{k_1k_2u_\varepsilon v_\varepsilon}{(1+k_1u_\varepsilon+k_2v_\varepsilon)^2} \geq 0$ and $z_\varepsilon = 0$ for $w_\varepsilon \leq 0$. Hence, $w_\varepsilon(t)$ is also non-negative for almost everywhere $t \in [0, T)$.
(*ii*) Let us consider the test function $u_\varepsilon$ in the weak form (6a) and calculate to deduce the estimate

$$\|u_\varepsilon(t)\|^2_{\Omega_\varepsilon^p} + 2\alpha\|\nabla u_\varepsilon\|^2_{(\Omega_\varepsilon^p)^t} \leq \|u_0\|^2_{\Omega_\varepsilon^p} + 2k_d|\langle r_1(u_\varepsilon, v_\varepsilon) - z_\varepsilon, u_\varepsilon\rangle_{(\Gamma_\varepsilon^*)^t}| + 2|\langle d_1, u_\varepsilon\rangle_{(\partial\Omega)^t}|$$

$$\Longrightarrow \|u_\varepsilon(t)\|^2_{\Omega_\varepsilon^p} + (2\alpha - 2C\gamma - 2C\gamma_1)\|\nabla u_\varepsilon\|^2_{(\Omega_\varepsilon^p)^t} \leq M_u + (2C\gamma + 2C\gamma_1)\int_0^t \|u_\varepsilon(s)\|^2_{\Omega_\varepsilon^p} ds, \tag{11}$$

where $M_u = \|u_0\|^2_{\Omega^p_\varepsilon} + \frac{k_d^2 T}{2\gamma}(1+\frac{k}{4})^2 \frac{|\Gamma||\Omega|}{|Y|} + \frac{1}{2\gamma_1}\|d_1\|^2_{(\partial\Omega)^t}$ as

$$r_1(u_\varepsilon, v_\varepsilon) = k\frac{k_1 k_2 u_\varepsilon v_\varepsilon}{(1+k_1 u_\varepsilon + k_2 v_\varepsilon)^2} \le \frac{k}{4}. \tag{12}$$

We use the trace theorem of Sect. 5.5 of [5] to estimate the boundary term and Lemma 1. Therefore for $\gamma = \frac{\alpha}{4C} = \gamma_1$ we have

$$\|u_\varepsilon(t)\|^2_{\Omega^p_\varepsilon} \le M_u + \alpha \int_0^t \|u_\varepsilon(s)\|^2_{\Omega^p_\varepsilon} ds.$$

Gronwall's inequality gives $\|u_\varepsilon(t)\|^2_{\Omega^p_\varepsilon} \le M_u e^{\alpha T}$. Similarly, $\|v_\varepsilon(t)\|^2_{\Omega^p_\varepsilon} \le M_v e^{\alpha T}$. Next, $\eta = w_\varepsilon$ in (6c) gives

$$\|w_\varepsilon(t)\|^2_{\Gamma^*_\varepsilon} \le M_w + \int_0^t \|w_\varepsilon(s)\|^2_{\Gamma^*_\varepsilon} ds \implies \|w_\varepsilon(t)\|^2_{\Gamma^*_\varepsilon} \le M_w e^T,$$

where $M_w = \|w_0\|^2_{\Gamma^*_\varepsilon} + k_d^2(1+\frac{k}{4})^2 T \frac{|\Omega||\Gamma|}{|Y|}$. With the choice of $\gamma = \frac{\alpha}{4C} = \gamma_1$, (11) yields

$$\|\nabla u_\varepsilon\|_{(\Omega^p_\varepsilon)^t} \le \bar{M}_u e^{\alpha T},$$

where $\bar{M}_u = \frac{M_u}{\alpha}$. Proceeding similarly for $v_\varepsilon$ we have, $\|\nabla v_\varepsilon\|_{(\Omega^p_\varepsilon)^t} \le \bar{M}_v e^{\alpha T}$, where $\bar{M}_v = \frac{M_v}{\alpha}$. We obtain from (6a)

$$\left\|\frac{\partial u_\varepsilon}{\partial t}\right\|_{H^{1,2}(\Omega^p_\varepsilon)^*} \le \left[\|D_1\|_{L^\infty(S\times Y)}\|\nabla u_\varepsilon\|_{\Omega^p_\varepsilon} + k_d\left(\frac{|\Gamma||\Omega|C}{|Y|}\right)^{\frac{1}{2}} + C\|d_1\|_{\partial\Omega}\right] \le C.$$

Squaring both sides and integrating w.r.t $t$ we get

$$\left\|\frac{\partial u_\varepsilon}{\partial t}\right\|^2_{L^2(S;H^{1,2}(\Omega^p_\varepsilon)^*)} \le C.$$

In the same way as above, we can also show that

$$\left\|\frac{\partial v_\varepsilon}{\partial t}\right\|^2_{L^2(S;H^{1,2}(\Omega^p_\varepsilon)^*)} \le C.$$

Now we use the test function $\eta = \frac{\partial w_\varepsilon}{\partial t}$ in (6c) and get

$$\left\|\frac{\partial w_\varepsilon}{\partial t}\right\|^2_{L^2(S\times\Gamma^*_\varepsilon)} \le k_d^2 T(1+\frac{k}{4})^2\frac{|\Gamma||\Omega|}{|Y|}.$$

## 2.1 Existence and Uniqueness

We tackle the multivalued dissolution rate term by introducing a regularization parameter $\delta > 0$ such that

$$\psi_\delta(w_\varepsilon) = \begin{cases} 0 & \text{if } w_\varepsilon < 0, \\ \frac{w_\varepsilon}{\delta} & \text{if } w_\varepsilon \in (0, \delta), \\ 1 & \text{if } w_\varepsilon > \delta. \end{cases}$$

Now the variational formulation of the regularized problem is

$$\langle \frac{\partial u_{\varepsilon\delta}}{\partial t}, \phi \rangle_{(\Omega_\varepsilon^p)^T} + \langle \bar{D}_1^\varepsilon \nabla u_{\varepsilon\delta}, \nabla \phi \rangle_{(\Omega_\varepsilon^p)^T} = -\langle \frac{\partial w_{\varepsilon\delta}}{\partial t}, \phi \rangle_{(\Gamma_\varepsilon^*)^T} - \langle d_1, \phi \rangle_{(\partial\Omega)^T}, \quad (13a)$$

$$\langle \frac{\partial v_{\varepsilon\delta}}{\partial t}, \theta \rangle_{(\Omega_\varepsilon^p)^T} + \langle \bar{D}_2^\varepsilon \nabla v_{\varepsilon\delta}, \nabla \theta \rangle_{(\Omega_\varepsilon^p)^T} = -\langle \frac{\partial w_{\varepsilon\delta}}{\partial t}, \theta \rangle_{(\Gamma_\varepsilon^*)^T} - \langle d_2, \theta \rangle_{(\partial\Omega)^T}, \quad (13b)$$

$$\langle \frac{\partial w_{\varepsilon\delta}}{\partial t}, \eta \rangle_{(\Gamma_\varepsilon^*)^T} = k_d \langle r_1(u_{\varepsilon\delta}, v_{\varepsilon\delta}) - \psi_\delta(w_{\varepsilon\delta}), \eta \rangle_{(\Gamma_\varepsilon^*)^T}, \quad (13c)$$

for all $(\phi, \theta, \eta) \in L^2(S; H^{1,2}(\Omega_\varepsilon^p)) \times L^2(S; H^{1,2}(\Omega_\varepsilon^p)) \times L^2(S \times \Gamma_\varepsilon^*)$.

We employ Rothe's method to show the existence of solution of the PDEs. For the ODE we consider (13c) along with the initial data $w_{\varepsilon\delta}(0, x) = w_0(x)$, then as $r_1(u_{\varepsilon\delta}, v_{\varepsilon\delta})$ is constant in $w_{\varepsilon\delta}$ and $\psi_\delta(w_{\varepsilon\delta})$ is Lipschitz with respect to $w_{\varepsilon\delta}$ therefore there exists a unique local solution by Picard-Lindelof theorem $w_{\varepsilon\delta} \in C^1(0, T_1(x))$ of the problem (13c) where $T_1(x) \leq T$. Partial integration of the strong form of (13c) and (12) gives

$$|w_{\varepsilon\delta}(t, x)| \leq \|w_0\|_{L^\infty(\Omega)} + k_d(1 + \frac{k}{4})T, \quad \text{for all } t \text{ and } x.$$

This implies the solution of the immobile species exists globally for every $t \in [0, T)$. We introduce two billinear forms on $\Omega_\varepsilon^p$ such that $b(u_{\varepsilon\delta}, \phi) = \langle \bar{D}_1^\varepsilon \nabla u_{\varepsilon\delta}, \nabla \phi \rangle_{\Omega_\varepsilon^p}$ and $b(v_{\varepsilon\delta}, \theta) = \langle \bar{D}_2^\varepsilon \nabla v_{\varepsilon\delta}, \nabla \theta \rangle_{\Omega_\varepsilon^p}$. Now for arbitrary $u_{\varepsilon\delta} \in U_\varepsilon$ we have to find $v_{\varepsilon\delta} : [0, T] \to H^{1,2}(\Omega_\varepsilon^p)$ such that

$$\langle \frac{\partial v_{\varepsilon\delta}}{\partial t}, \theta \rangle_{\Omega_\varepsilon^p} + b(v_{\varepsilon\delta}, \theta) = \langle f(v_{\varepsilon\delta}), \theta \rangle_{\Gamma_\varepsilon^*} - \langle d_2, \theta \rangle_{\partial\Omega}, \quad \forall \theta \in H^{1,2}(\Omega_\varepsilon^p) \text{ a.e. in } [0, T], \quad (14a)$$

$$v_{\varepsilon\delta}|_{t=0} = v_0, \quad (14b)$$

where $f(v_{\varepsilon\delta}) = k_d(\psi_\delta(w_{\varepsilon\delta}) - r_1(u_{\varepsilon\delta}, v_{\varepsilon\delta}))$.

## 2.2 Rothe's Method

We take a partition $\{0 = t_0 < t_1 < \cdots < t_{n-1} < t_n = T\}$ for the time interval $[0, T]$ with step size $h = (t_i - t_{i-1}) = \frac{T}{n}$. Time discretization to (14a) leads to

$$\langle \frac{v_i - v_{i-1}}{h}, \theta\rangle_{\Omega_\varepsilon^p} + b(v_i, \theta) = \langle f(v_{i-1}), \theta\rangle_{\Gamma_\varepsilon^*} - \langle d_{2i-1}, \theta\rangle_{\partial\Omega}, \ \forall i = 1, 2, \ldots, n, \tag{15}$$

where $f(v_{i-1}) = k_d(\psi_\delta(w_{\varepsilon\delta}) - r_1(u_{\varepsilon\delta}, v_{i-1}))$. Now we introduce two linear operators: one is $\mathcal{T}_h : H^{1,2}(\Omega_\varepsilon^p) \to L^2(\Omega_\varepsilon^p)$ such that $\langle \mathcal{T}_h v, \theta\rangle_{\Omega_\varepsilon^p} = \frac{1}{2}\langle v, \theta\rangle_{\Omega_\varepsilon^p} + b(v, \theta)$ and the other is $\langle l_{i-1}, \theta\rangle_{\Omega_\varepsilon^p} = \langle f(v_{i-1}), \theta\rangle_{\Gamma_\varepsilon^*} - \langle d_{2i-1}, \theta\rangle_{\partial\Omega} + \frac{1}{2}\langle v_{i-1}, \theta\rangle_{\Omega_\varepsilon^p}$. Then (15) can be rewritten as

$$\langle \mathcal{T}_h v, \theta\rangle_{\Omega_\varepsilon^p} = \langle l_{i-1}, \theta\rangle_{\Omega_\varepsilon^p}, \ \forall \theta \in H^{1,2}(\Omega_\varepsilon^p).$$

As $C_1\|v\|^2_{H^{1,2}(\Omega_\varepsilon^p)} \le \langle \mathcal{T}_h v, v\rangle_{\Omega_\varepsilon^p} \le C_2\|v\|^2_{H^{1,2}(\Omega_\varepsilon^p)}$ and $l_{i-1}$ is a bounded functional on $H^{1,2}(\Omega_\varepsilon^p)$, so there exists a unique $v_i \in H^{1,2}(\Omega_\varepsilon^p)$ by Lax-Milgram lemma satisfying (15). Next we define Rothe functions $v_n : [0, T] \to H^{1,2}(\Omega_\varepsilon^p)$ by

$$v_n(t) = v_i\left(\frac{t - t_{i-1}}{h}\right) - v_{i-1}\left(\frac{t - t_i}{h}\right)$$

and the step function $\overline{v}_n : [0, T] \to H^{1,2}(\Omega_\varepsilon^p)$ such that $\overline{v}_n(t) = v_i$, for all $t \in (t_{i-1}, t_i]$ and $\overline{v}_n(0) = v_0$. We need to find out some a-priori bounds to establish the convergence of Rothe's function to a solution of the continuous equation (14a).

**Lemma 3** *The difference* $(v_i - v_{i-1})$ *satisfy the inequality*

$$\left\|\frac{v_i - v_{i-1}}{h}\right\|^2_{\Omega_\varepsilon^p} + \frac{1}{2h^2}\|\nabla(v_i - v_{i-1})\|^2_{\Omega_\varepsilon^p} \le C,$$

*for all* $i = 1, 2, \ldots, n$.

***Proof*** For $i = 1$ and $\theta = \frac{v_1 - v_0}{h}$ from (15) we have

$$\left\|\frac{v_1 - v_0}{h}\right\|^2_{\Omega_\varepsilon^p} + \frac{\alpha}{h}\|\nabla(v_1 - v_0)\|^2_{\Omega_\varepsilon^p} \le \langle f(v_0), \frac{v_1 - v_0}{h}\rangle_{\Gamma_\varepsilon^*} - \langle d_{20}, \frac{v_1 - v_0}{h}\rangle_{\partial\Omega} - b(v_0, \frac{v_1 - v_0}{h}).$$

Application of the trace inequality and assumption (**A6.**) and (12) leads to

$$(1 - C\gamma - C\gamma_1)\left\|\frac{v_1 - v_0}{h}\right\|^2_{\Omega_\varepsilon^p} + (\frac{\alpha}{h} - \frac{C\gamma}{h^2} - \frac{C\gamma_1}{h^2} - \frac{\gamma_2}{h^2})\|\nabla(v_1 - v_0)\|^2_{\Omega_\varepsilon^p} \le C_3,$$

where $C_3 = \frac{k_d^2}{4\gamma}(1+\frac{k}{4})^2\frac{|\Gamma||\Omega|}{|Y|} + \frac{1}{4\gamma_1}\|d_{20}\|^2_{\partial\Omega} + \frac{1}{4\gamma_2}\|\nabla v_0\|^2_{\Omega_\varepsilon^p}\|D_2\|^2_{L^\infty(S\times Y)}$. Now there are three cases to consider: $(i)\alpha h < 1$, $(ii)\alpha h = 1$ and $(iii)\alpha h > 1$. For the first case we choose $\gamma = \frac{\alpha h}{4C} = \gamma_1$ and $\gamma_2 = \frac{\alpha h}{4}$ and for the remaining two cases taking $\gamma = \frac{1}{4C} = \gamma_1$ and $\gamma_2 = \frac{1}{4}$ we have our desired estimate. For $j \geq 2$ we subtract (15) for $i = j$ from for $i = j-1$ and test with $\theta = \frac{v_j - v_{j-1}}{h}$ to get

$$(1 - C\gamma - C\gamma_1 - \gamma_2)\left\|\frac{v_j - v_{j-1}}{h}\right\|^2_{\Omega_\varepsilon^p} + (\frac{\alpha}{h} - \frac{C\gamma}{h^2} - \frac{C\gamma_1}{h^2})\|\nabla(v_j - v_{j-1})\|^2_{\Omega_\varepsilon^p}$$
$$\leq C_4 + \frac{1}{4\gamma_2}\left\|\frac{v_{j-1} - v_{j-2}}{h}\right\|^2_{\Omega_\varepsilon^p},$$

where $C_4 = \frac{k^2k_d^2}{16\gamma}\frac{|\Gamma||\Omega|}{|Y|} + \frac{1}{4\gamma_1}\|d_{j-1} - d_{j-2}\|^2_{\partial\Omega}$. Now if we take $\sigma_j = \left\|\frac{v_j - v_{j-1}}{h}\right\|^2_{\Omega_\varepsilon^p} + \frac{1}{2h^2}\|\nabla(v_j - v_{j-1})\|^2_{\Omega_\varepsilon^p}$ and proceed like earlier then we have our required estimate as an implication of Gronwall's inequality.

**Lemma 4** *The following a-priori estimates hold for* $v_i$
*(a)* $\|v_i\|_{H^{1,2}(\Omega_\varepsilon^p)} \leq C$, $\left\|\frac{v_i - v_{i-1}}{h}\right\|_{\Omega_\varepsilon^p} \leq C$ *for all* $i = 1, 2, \ldots, n$.
*For Rothe's step functions this means*
*(b)* $\|\bar{v}_n(t)\|_{H^{1,2}(\Omega_\varepsilon^p)} \leq C$, $\left\|\frac{dv_n(t)}{dt}\right\|_{\Omega_\varepsilon^p} \leq C$ *for almost everywhere* $t \in [0, T]$.

***Proof*** We see $\|v_i\|_{\Omega_\varepsilon^p} \leq \|v_0\|_{\Omega_\varepsilon^p} + h\sum_{j=1}^i \left\|\frac{v_j - v_{j-1}}{h}\right\|_{\Omega_\varepsilon^p} \leq C$. We put $\theta = v_i$ in (15) and trace inequality in combination with (12) gives that

$$(\alpha - C\gamma - C\gamma_1)\|\nabla v_i\|^2_{\Omega_\varepsilon^p} \leq \frac{k_d^2}{4\gamma}(1+\frac{k}{4})^2\frac{|\Omega||\Gamma|}{|Y|} + \frac{\sigma_i}{2} + \frac{1}{4\gamma_1}\|d_{i-1}\|^2_{\partial\Omega} + (C\gamma + C\gamma_1 + \frac{1}{2})\|v_i\|^2_{\Omega_\varepsilon^p}.$$

So for $\gamma = \frac{\alpha}{4C} = \gamma_1$ we have, $\|\nabla v_i\|_{\Omega_\varepsilon^p} \leq C$. Thus we get the first estimate. Other estimates are a consequence of the Lemma 3.

**Theorem 2** *The sequence of Rothe's function converges to the unique solution of* (14a).

***Proof*** We get the estimates

$$\|v_n(t) - v_n(s)\|_{\Omega_\varepsilon^p} \leq \int_s^t \left\|\frac{dv_n(\sigma)}{d\sigma}\right\|_{\Omega_\varepsilon^p} d\sigma \overset{Lemma\ 4}{\leq} C|t-s|$$

and $\|v_n(t)\|_{H^{1,2}(\Omega_\varepsilon^p)} \leq C$.

Hence we can apply Arcela-Ascoli theorem and it follows that there exists $v \in H^{1,2}(\Omega_\varepsilon^p)$ such that upto a subsequence $v_n \to v$ in $C([0, T]; L^2(\Omega_\varepsilon^p))$. Now since $\|\bar{v}_n(t)\|_{H^{1,2}(\Omega_\varepsilon^p)} \leq C$ so upto a subsequence $\bar{v}_n(t) \rightharpoonup \bar{v}(t)$ in $H^{1,2}(\Omega_\varepsilon^p)$, for all $t \in [0, T]$. Now for $t \in (t_{i-1}, t_i]$

$$\|v_n(t)-\bar v_n(t)\|^2_{\Omega^p_\varepsilon}=\int_{\Omega^p_\varepsilon}|v_n(t)-\bar v_n(t)|^2dx=\int_{\Omega^p_\varepsilon}|v_{i-1}+\frac{v_i-v_{i-1}}{h}(t-t_{i-1})-v_i|^2dx$$

$$\leq (t-t_i)^2\int_{\Omega^p_\varepsilon}\left|\frac{v_i-v_{i-1}}{h}\right|^2dx\leq h^2C\to 0\ \text{ since } h\to 0$$

and so $\bar v(t)\equiv v(t)$ for all $t\in[0,T]$. The next target is to show $\frac{dv_n}{dt}\rightharpoonup\frac{dv}{dt}$ in $L^2([0,T];L^2(\Omega^p_\varepsilon))$.

As we know $\left\|\frac{dv_n}{dt}\right\|_{\Omega^p_\varepsilon}\leq C$ for almost everywhere $t$. Therefore $\frac{dv_n}{dt}$ is bounded in the space $L^2([0,T];L^2(\Omega^p_\varepsilon))$. So we get upto a subsequence $\frac{dv_n}{dt}\rightharpoonup\eta$ in $L^2([0,T];L^2(\Omega^p_\varepsilon))$. Claim: $\eta=\frac{dv}{dt}$ in the sense of distribution

$$\langle v,\frac{\partial\phi}{\partial t}\rangle_{\Omega^p_\varepsilon}=\lim_{n\to\infty}\langle v_n,\frac{\partial\phi}{\partial t}\rangle_{\Omega^p_\varepsilon}=-\lim_{n\to\infty}\langle\frac{\partial v_n}{\partial t},\phi\rangle_{\Omega^p_\varepsilon}=-\langle\eta,\phi\rangle_{\Omega^p_\varepsilon}\ \text{ for all smooth }\phi$$

$$\implies\eta=\frac{dv}{dt}\ \text{ in the sense of distribution.}$$

Again since

$$\|v_n(t)\|^2_{\Omega^p_\varepsilon}+\|\nabla v_n\|^2_{(\Omega^p_\varepsilon)^t}+\left\|\frac{dv_n}{dt}\right\|^2_{L^2(S;H^{1,2}(\Omega^p_\varepsilon)^*)}\leq C.$$

Hence $v_n\to v$ in $C([0,T];H^s(\Omega^p_\varepsilon)^*)\cap L^2((0,T);H^s(\Omega^p_\varepsilon))$ for $s\in(0,1)$ and in particular $v_n\to v$ strongly in $L^2(\Gamma^*_\varepsilon)^t$. Since $f$ is Lipschitz, $f(v_n)\to f(v)$ strongly in $L^2(\Gamma^*_\varepsilon)^t$ and pointwisely in $(\Gamma^*_\varepsilon)^t$. Now passing the limit as $n\to\infty$ in

$$\int_0^t\langle\frac{dv_n(t)}{dt},\theta\rangle_{\Omega^p_\varepsilon}dt+\int_0^t b(\bar v_n,\theta)dt$$
$$=\int_0^t\langle f(\bar v_{n-1}),\theta\rangle_{\Gamma^*_\varepsilon}dt-\int_0^t\langle d_{2i-1},\theta\rangle_{\partial\Omega}\,dt,\ \forall\theta\in H^{1,2}(\Omega^p_\varepsilon),$$

we get $v$ is a solution of

$$\int_0^t\langle\frac{dv(t)}{dt},\theta\rangle_{\Omega^p_\varepsilon}dt+\int_0^t b(v,\theta)dt$$
$$=\int_0^t\langle f(v),\theta\rangle_{\Gamma^*_\varepsilon}dt-\int_0^t\langle d_2,\theta\rangle_{\partial\Omega}\,dt,\ \text{ for all }\theta\in H^{1,2}(\Omega^p_\varepsilon).$$

Proceed in the same way we get the existence of $u_{\varepsilon\delta}$.

**Uniqueness**: Let $(u^1_{\varepsilon\delta},v^1_{\varepsilon\delta},w^1_{\varepsilon\delta})$ and $(u^2_{\varepsilon\delta},v^2_{\varepsilon\delta},w^2_{\varepsilon\delta})$ be two solutions of the system (13a)–(13c) and $W_{\varepsilon\delta}=w^1_{\varepsilon\delta}-w^2_{\varepsilon\delta}\geq 0$, $V_{\varepsilon\delta}=v^1_{\varepsilon\delta}-v^2_{\varepsilon\delta}\geq 0$ and $U_{\varepsilon\delta}=u^1_{\varepsilon\delta}-u^2_{\varepsilon\delta}\geq 0$. Now from (13c) we see that $W_{\varepsilon\delta}$ satisfy

$$\|W_{\varepsilon\delta}(t)\|^2_{\Gamma^*_\varepsilon} \le e^T k_d^2 L_R^2 \|U_{\varepsilon\delta} + V_{\varepsilon\delta}\|^2_{(\Gamma^*_\varepsilon)^t}. \tag{16}$$

Now we write (13a) for $U_{\varepsilon\delta}$ and (13b) for $V_{\varepsilon\delta}$ and adding up side by side leads to

$$\langle\frac{\partial}{\partial t}(U_{\varepsilon\delta} + V_{\varepsilon\delta}), \phi\rangle_{\Omega^p_\varepsilon} + \langle\bar{D}^\varepsilon_1 \nabla U_{\varepsilon\delta} + \bar{D}^\varepsilon_2 \nabla V_{\varepsilon\delta}, \nabla\phi\rangle_{\Omega^p_\varepsilon} \le 2\left|\langle\frac{\partial W_{\varepsilon\delta}}{\partial t}, \phi\rangle_{\Gamma^*_\varepsilon}\right|.$$

We denote $P(t) = U_{\varepsilon\delta}(t) + V_{\varepsilon\delta}(t)$ and test with $\phi = U_{\varepsilon\delta} + V_{\varepsilon\delta}$ and after simplification we have

$$\|P(t)\|^2_{\Omega^p_\varepsilon} \le (4k_d L_R C + \frac{C^2 k_d L_R}{\mu})\|P\|_{(\Omega^p_\varepsilon)^t}.$$

Then we get by Gronwall's inequality $u^1_{\varepsilon\delta}(t) = u^2_{\varepsilon\delta}(t)$ and $v^1_{\varepsilon\delta}(t) = v^2_{\varepsilon\delta}(t)$ and (16) gives $w^1_{\varepsilon\delta}(t) = w^2_{\varepsilon\delta}(t)$ for a.e. $t \in S$. So we get unique solution.

Now we send the regularization parameter $\delta \to 0$. By Corollary 4 and Lemma 9 of [13] we get, for $s \in (0, 1)$,

$$\begin{aligned}
&u_{\varepsilon\delta} \to u_\varepsilon \text{ strongly in } L^2(S; L^2(\Omega^p_\varepsilon)),\\
&u_{\varepsilon\delta} \to u_\varepsilon \text{ strongly in } L^2(S; H^{s,2}(\Omega^p_\varepsilon)) \cap C(S; H^{-s,2}(\Omega^p_\varepsilon)),\\
&v_{\varepsilon\delta} \to v_\varepsilon \text{ strongly in } L^2(S; L^2(\Omega^p_\varepsilon)) \text{ and}\\
&v_{\varepsilon\delta} \to v_\varepsilon \text{ strongly in } L^2(S; H^{s,2}(\Omega^p_\varepsilon)) \cap C(S; H^{-s,2}(\Omega^p_\varepsilon)).
\end{aligned}$$

After that, trace theorem (cf. Satz 8.7 of [16]) implies

$$u_{\varepsilon\delta} \to u_\varepsilon \text{ strongly in } L^2(\Gamma^*_\varepsilon)^t \text{ and } v_{\varepsilon\delta} \to v_\varepsilon \text{ strongly in } L^2(\Gamma^*_\varepsilon)^t.$$

As $r_1$ is Lipschitz therefore $r_1(u_{\varepsilon\delta}, v_{\varepsilon\delta}) \to r_1(u_\varepsilon, v_\varepsilon)$ in $L^2(\Gamma^{*t}_\varepsilon)$ and pointwise almost everywhere in $(\Gamma^*_\varepsilon)^t$. After that we follow the same arguments given in Theorem 2.21 of [15] to obtain the Eqs. (8a)–(10c).

## 3 Homogenization

**Lemma 5** *There exists a positive constant C does not depend on ε such that*

$$\sup_{\varepsilon>0}\big(\|u_\varepsilon\|_{L^2(S\times\Omega)} + \|\nabla u_\varepsilon\|_{L^2(S\times\Omega)} + \|\chi_\varepsilon \partial_t u_\varepsilon\|_{L^2(S;H^{1,2}(\Omega)^*)} + \|v_\varepsilon\|_{L^2(S\times\Omega)}$$
$$+\|\nabla v_\varepsilon\|_{L^2(S\times\Omega)} + \|\chi_\varepsilon \partial_t v_\varepsilon\|_{L^2(S;H^{1,2}(\Omega)^*)}\big) \le C < \infty. \tag{17}$$

***Proof*** This comes from the Lemma 1 and the estimate (7). The details can be seen in [11].

**Lemma 6** *The bounds* (17) *and* (7) *gives the following convergence results*

(i) $u_\varepsilon \rightharpoonup u$ *in* $L^2(S; H^{1,2}(\Omega))$, (ii) $\partial_t u_\varepsilon \rightharpoonup \partial_t u$ *in* $L^2(S; H^{1,2}(\Omega)^*)$,
(iii) $v_\varepsilon \rightharpoonup v$ *in* $L^2(S; H^{1,2}(\Omega))$, (iv) $\partial_t v_\varepsilon \rightharpoonup \partial_t v$ *in* $L^2(S; H^{1,2}(\Omega)^*)$,
(v) $u_\varepsilon \to u$ *in* $L^2(S \times \Gamma_\varepsilon^*)$, (vi) $v_\varepsilon \to v$ *in* $L^2(S \times \Gamma_\varepsilon^*)$,
(vii) *There exists* $u \in L^2(S; H^{1,2}(\Omega))$ *and* $u_1 \in L^2(S \times \Omega; H^{1,2}_{per}(Y)/\mathbb{R})$ *such that* $u_\varepsilon \overset{2}{\rightharpoonup} u$ *and* $\nabla u_\varepsilon \overset{2}{\rightharpoonup} \nabla_x u + \nabla_y u_1$,
(viii) *There exists* $v \in L^2(S; H^{1,2}(\Omega))$ *and* $v_1 \in L^2(S \times \Omega; H^{1,2}_{per}(Y)/\mathbb{R})$ *such that* $v_\varepsilon \overset{2}{\rightharpoonup} v$ *and* $\nabla v_\varepsilon \overset{2}{\rightharpoonup} \nabla_x v + \nabla_y v_1$,
(ix) $w_\varepsilon \overset{2}{\rightharpoonup} w$ *in* $L^2(S \times \Omega \times \Gamma)$, (x) $\partial_t w_\varepsilon \overset{2}{\rightharpoonup} \partial_t w$ *in* $L^2(S \times \Omega \times \Gamma)$,
(xi) $z_\varepsilon \overset{2}{\rightharpoonup} z$ *in* $L^2(S \times \Omega \times \Gamma)$.

***Proof*** The convergence $(i)$–$(iv)$ comes from the estimate (17) and rest follows from Proposition 1.14 of [1], Theorem 2.1 of [2] and Lemma 12 of [6].

**Lemma 7** (a) *The reaction rate term* $r_1(u_\varepsilon, v_\varepsilon) \to r_1(u, v)$ *in* $L^2(S \times \Gamma_\varepsilon^*)$. *From this we can deduce* $r_1(u_\varepsilon, v_\varepsilon) \overset{2}{\rightharpoonup} r_1(u, v)$ *in* $L^2(S \times \Omega \times \Gamma)$.
(b) $\mathcal{T}_\varepsilon^b(w_\varepsilon) \to w$ *in* $L^2(S \times \Omega \times \Gamma)$.

***Proof*** $(a)$Since $r_1$ is Lipschitz and by the help of Minkowski's inequality we get

$$\begin{aligned}\|r_1(u_\varepsilon, v_\varepsilon) - r_1(u, v)\|_{L^2(S\times\Gamma_\varepsilon^*)} &\le L_R\{\|u_\varepsilon - u\|_{L^2(S\times\Gamma_\varepsilon^*)} + \|v_\varepsilon - v\|_{L^2(S\times\Gamma_\varepsilon^*)}\} \\ &\to 0 \text{ as } \varepsilon \to 0, \text{ by } (v) \text{ and } (vi) \text{ of Lemma 6.}\end{aligned}$$

Then by Proposition 5.2 of [4] it follows that

$$\begin{aligned}&\mathcal{T}_\varepsilon^b(r_1(u_\varepsilon, v_\varepsilon)) \to r_1(u, v) \text{ in } L^2(S \times \Omega \times \Gamma), \\ &\Longrightarrow r_1(u_\varepsilon, v_\varepsilon) \overset{2}{\rightharpoonup} r_1(u, v) \text{ in } L^2(S \times \Omega \times \Gamma).\end{aligned}$$

$(b)$ This comes from [11] and Theorem 5.1 of [10].

**Theorem 3** *Under the assumptions* (**A1.**) – (**A6.**) *there exist* $(u, v, w, z) \in L^2(S; H^{1,2}(\Omega)) \times L^2(S; H^{1,2}(\Omega)) \times L^2(S; L^2(\Omega \times \Gamma)) \times L^\infty(S \times \Omega \times \Gamma)$ *in such a way that* $(u, v, w, z)$ *is the unique solution of the problem*

$$\frac{\partial u}{\partial t} + \nabla.(-A_1 \nabla u) + P_1(t, x) = 0 \;\; in\ S \times \Omega, \tag{18a}$$

$$-A_1 \nabla u.\vec{n} = \frac{d_1}{|Y^p|} \;\; on\ S \times \partial\Omega, \tag{18b}$$

$$u(0, x) = u_0(x) \;\; in\ \Omega, \tag{18c}$$

$$\frac{\partial v}{\partial t} + \nabla.(-B_1 \nabla v) + P_1(t,x) = 0 \ \ in\ S \times \Omega, \tag{19a}$$

$$-B_1 \nabla v.\vec{n} = \frac{d_2}{|Y^p|} \ \ on\ S \times \partial\Omega, \tag{19b}$$

$$v(0,x) = v_0(x) \ \ in\ \Omega, \tag{19c}$$

$$\frac{\partial w}{\partial t} = k_d(r_1(u,v) - z) \ \ in\ S \times \Omega \times \Gamma, \tag{20a}$$

$$z \in \psi(w) \ \ in\ S \times \Omega \times \Gamma, \tag{20b}$$

$$w(0,x) = w_0(x) \ \ on\ \Omega \times \Gamma, \tag{20c}$$

*where*

$$P_1(t,x) = \int_\Gamma \frac{1}{|Y^p|}\frac{\partial w}{\partial t}\, d\sigma_y.$$

*The elliptic and bounded homogenized matrix* $A_1(t,x) = (a_{ij})_{1\le i,j\le n}$ *and* $B_1(t,x) = (b_{ij})_{1\le i,j\le n}$ *are given by*

$$a_{ij}(t,x) = \frac{1}{|Y^p|}\int_{Y^p} D_1(t,y)\left(\delta_{ij} + \sum_{i,j=1}^{n}\frac{\partial k_j}{\partial y_i}\right) dy,$$

$$b_{ij}(t,x) = \frac{1}{|Y^p|}\int_{Y^p} D_2(t,y)\left(\delta_{ij} + \sum_{i,j=1}^{n}\frac{\partial k_j}{\partial y_i}\right) dy.$$

*Moreover,* $k_j \in L^\infty(\Omega; H^{1,2}_{per}(Y))$ *is the solutions of the cell problems*

$$\begin{cases} (div)_y(-D_i(t,y)(\nabla_y k_j + e_j)) = 0\ \forall y \in Y^p, \\ -D_i(t,y)(\nabla_y k_j + e_j).\vec{n} = 0 \ on\ \Gamma, \\ y \mapsto k_j(y) \ is\ Y - periodic, \end{cases} \tag{21}$$

*for* $j = 1, 2, \ldots, n$, $i = 1, 2$ *and for a.e.* $x \in \Omega$.

***Proof*** We utilize two-scale convergence to derive the macroscopic equations. First, we test the PDEs (8a) and (9a) with $\Psi_i(t,x,\frac{x}{\varepsilon}) = \psi_i(t,x) + \varepsilon\phi_i(t,x,\frac{x}{\varepsilon})$ such that $\psi_i \in C_0^\infty(S\times\Omega)$ and $\phi_i(t,x,\frac{x}{\varepsilon}) \in C_0^\infty(S\times\Omega; C^\infty_{per}(Y))$, for $i = 1, 2$. Now we pass the homogenization limit $\varepsilon$ to 0 for the each term separately and combining all terms we finally obtain

$$-\int_0^T\int_\Omega u(t,x)\frac{\partial\psi_1}{\partial t}dxdt + \frac{1}{|Y^p|}\int_0^T\int_\Omega\int_{Y^p} D_1(t,y)(\nabla u(t,x) + \nabla_y u_1(t,x,y))(\nabla_x\psi_1 + \nabla_y\phi_1)dxdydt$$
$$+\frac{1}{|Y^p|}\int_0^T\int_\Omega\int_\Gamma \frac{\partial w}{\partial t}\psi_1 dxd\sigma_y dt = 0. \tag{22}$$

We set $\psi_1 \equiv 0$ and choose $u_1(t,x,y) = \sum_{j=1}^{n} \frac{\partial u}{\partial x_j}(t,x)k_j(t,x,y) + q(x)$ to obtain the cell problem as (21) for $i = 1$. Now setting $\phi_1 \equiv 0$ we obtain the macroscopic equations for $\mathcal{I}_1$ as (18a)–(18c). Similarly, testing (9a) by $\Psi_2$ we have the homogenized equation for $\mathcal{I}_2$ as (19a)–(19c). Finally, we consider the test function $\psi(t, x, \frac{x}{\varepsilon}) \in C_0^\infty(S \times \Omega; C_{per}^\infty(Y))$ to test the ODE (10a) and using Lemmas 6 and 7 we get the strong form as

$$\frac{\partial w}{\partial t} = k_d(r_1(u,v) - z) \ \text{ on } S \times \Omega \times \Gamma.$$

Lastly, we will characterize the two-scale limit of the multivalued dissolution rate term. That can be seen in [12] and we obtain the homogenized equations for the ODE as (20a)–(20c). The proof of uniqueness follows the same line of arguments as the micro model.

**Remark.** We can get positivity and the same type of a-priori estimates for the macromodel by imposing similar kinds of assumptions to the macromodel. The existence of solution for the macromodel relies on Galerkin method as the source term $P_1(t,x) \in L^2(S \times \Omega)$ since $\|P_1(t,x)\|^2_{L^2(S\times\Omega)} \le \frac{k_d^2}{|Y^p|^2}(1+\frac{k}{4})^2|\Gamma|^2|\Omega|T = \text{constant}$.

## 4 Conclusion

We investigated a diffusion-reaction-dissolution-precipitation system. Having in mind the special choice of the multivalued discontinuous dissolution term, we insert a regularization parameter $\delta > 0$ to show the existence of a unique positive global weak solution. Further, we use homogenization techniques to obtain the upscaled model. As future plans, we wish to work with a general model with rate term comes from mass action kinetics.

## References


1. Allaire, G.: Homogenization and two-scale convergence. SIAM J. Math. Anal. **23**(6), 1482–1518 (1992). https://doi.org/10.1137/0523084
2. Allaire, G., Damlamian, A., Hornung, U.: Two-scale convergence on periodic structures and applications. In: Proceedings of the International Conference on Mathematical Modelling of Flow through Porous Media, pp. 15–25
3. Amann, H.: Linear and Quasilinear Parabolic Problems, 2nd ed., vol. I of Monographs in Mathematics. Birkhäuser Publication, Basel (1995)
4. Cioranescu, D., Donato, P., Zaki, R.: The periodic unfolding method in perforated domains. Portugaliae Mathematica **63**, 4 (2006)
5. Evans, L.C.: Partial differential equations. Am. Math. Soc. **19** (2010)
6. Fatima, T., Muntean, A.: Sulfate attack in sewer pipes: derivation of a concrete corrosion model via two-scale convergence. Nonlinear Anal.: Real World Appli. **15**, 326–344 (2014). https://doi.org/10.1016/j.nonrwa.2012.01.019

7. Ghosh, N., Mahato, H.S.: Diffusion–reaction–dissolution–precipitation model in a heterogeneous porous medium with nonidentical diffusion coefficients: analysis and homogenization. Asymptot. Anal. 1–35 (2022). DOI: https://doi.org/10.3233/ASY-221763
8. Hornung, U., Jäger, W.: Diffusion, convection, adsorption, and reaction of chemicals in porous media. J. Diff. Equ. **92**(2), 199–225 (1991). https://doi.org/10.1016/0022-0396(91)90047-D
9. Knabner, P., Van Duijn, C.J., Hengst, S.: An analysis of crystal dissolution fronts in flows through porous media. Part 1: compatible boundary conditions. Adv. Water Res. **18**, 3, 171–185(1995). DOI: https://doi.org/10.1016/0309-1708(95)00005-4
10. Kumar, K., Neuss-Radu, M., Pop, I.S.: Homogenization of a pore scale model for precipitation and dissolution in porous media. IMA J. Appl. Math. **81**(5), 877–897 (2016). https://doi.org/10.1093/imamat/hxw039
11. Mahato, H.S., Böhm, M.: Homogenization of a system of semilinear diffusion-reaction equations in an $H^{1,p}$ setting. Electron. J. Diff. Equ. **2013**(210), 1–22 (2013)
12. Mahato, H.S., Kräutle, S., Böhm, M., Knabner, P.: Upscaling of a system of semilinear parabolic partial differential equations coupled with a system of nonlinear ordinary differential equations originating in the context of crystal dissolution and precipitation inside a porous medium: existence theory and periodic homogenization. Adv. Math. Sci. Appl. **26**, 39–81 (2017)
13. Simon, J.: Compact sets in the space $L^p(0, T; B)$. Annali di Matematica pura ed Applicata **146**(1), 65–96 (1986). https://doi.org/10.1007/BF01762360
14. Triebel, H.: Interpolation Theory, Function Spaces and Differential Operators. Johann Ambrosius Barth Verlag (1995)
15. Van Duijn, C.J., Pop, I.S.: Crystal dissolution and precipitation in porous media: pore scale analysis. J. für die reine und angewandte Mathematik (Crelles Journal) **2004**(577), 171–211 (2004)
16. Wloka, J.: Partial Differential Equations. Cambridge University Press, New york(1987)

# Mathematical Modeling and Computing to Study the Influence of Quarantine Levels and Common Mitigation Strategies on the Spread of COVID-19 on a Higher Education Campus

**Raina Saha, Clarissa Benitez, Krista Cimbalista, Jolypich Pek, and Padmanabhan Seshaiyer**

**Abstract** In this work, we develop a mathematical model to study the COVID-19 dynamics on a higher education campus. The proposed model builds on successful compartmental models that describe the dynamics of the spread of disease between multiple student sub-populations within a closed environment. The model assumes no vaccinations and includes three different levels of quarantine adherence to represent student behavior with the common mitigation strategies of face mask usage and random testing. A detailed analysis of the model including boundedness and positivity of the solutions along with a derivation of the basic reproduction number for the model is presented. Additionally, we also create an interactive graphical user interface through a dashboard for public use.

**Keywords** COVID-19 · Basic reproduction number · Compartmental models · SEIR · Quarantine

## 1 Introduction

In 2019, the first case of Coronavirus disease 2019 (COVID-19) was detected in Wuhan, China [14]. In the following months, this infectious disease was found to be present with (symptomatic) or without symptoms (asymptomatic) respectively [3]. As the disease continued to spread, there have been a variety of mathematical models proposed to understand the dynamics of COVID-19 [9]. Most of these models build from foundational ideas involving compartments of sub-populations including Susceptible, Exposed, Infected and Recovered (SEIR) dynamics [2]. Many of these modified SEIR models have helped to provide insight into quantifying the spread,

R. Saha · C. Benitez · K. Cimbalista · J. Pek · P. Seshaiyer (✉)
George Mason University, Fairfax, VA 22030, USA
e-mail: pseshaiy@gmu.edu

R. Saha
e-mail: rsaha3@gmu.edu



B. Rushi Kumar et al. (eds.), *Mathematics and Computing*, Springer Proceedings in Mathematics & Statistics 415, https://doi.org/10.1007/978-981-19-9307-7_49

analyzing effective control strategies and developing measures to prevent the spread of disease [7, 9].

This paper's modified SEIR model looks specifically at the setting of higher level education campuses. Because of the diversity of social and economic background within the student body, the impact of student behavior on managing disease spread is considered in this model. As schools began to reopen, the question of mitigation strategy and enforcement had to be considered [5, 11].

Higher level education environments contain clear risks and rewards for cooperating in the form of grades, jobs, and projects. Because student risk perception of COVID-19 and perceived personal barriers vary amonge the student body, so do individual preventative measures. Thus, it is important to take into account differences in rule adherence to create a more accurate view of disease dynamics [1, 6, 12]. The purpose of this paper is therefore to develop a new model to accommodate various quarantining levels and to analyze the effect of common mitigation strategies and student behaviors on disease spread. The presented model is focused on populations within schools and is flexible to different school environments by changing the necessary parameter values. This model assumes a constant transmission rate and uniform mask use for all wearing masks. This model also does not assumes any comorbidities such as age or obesity which might affect the transmission rate or behavior.

This paper is organized as follows. In Sect. 2, we develop a new mathematical model which is a modified SEIR and include new compartments with justifications. In Sect. 3, we present the mathematical analysis for the model developed and present the derivation of the associated basic reproduction number using the next-generation matrix approach. Next, in Sect. 4, we perform numerical simulations of the proposed model to study a higher level education campus setup. Section 5 presents an overview of a graphical user interface (GUI) for this model. Finally, we conclude and present some recommendations based on the created model and associated simulations in Sect. 6.

## 2 Models and Methods

In this work, a modified SEIR compartmental model for understanding spread of COVID-19 along with the effects of levels of quarantining is introduced. The proposed model adds three infectious categories, asymptomatic semi-Quarantine, and Symptomatic semi-Quarantine and asymptomatic. The model is organized around the flow diagram in Fig. 1.

The total population $N = S + E + I + A + I^Q + Q + A^Q + R$ is assumed to be divided into the 8 mutually exclusive categories. We assume the only way for an infectious individual to mitigate spread is to go into strict Quarantine ($Q$). In our proposed model, the susceptible individuals ($S$) move into the exposed state ($E$) after being transmitted the virus by one of the four infectious categories Symptomatic ($I$), asymptomatic ($A$), symptomatic semi-Quarantine ($I^Q$), and asymptomatic semi-

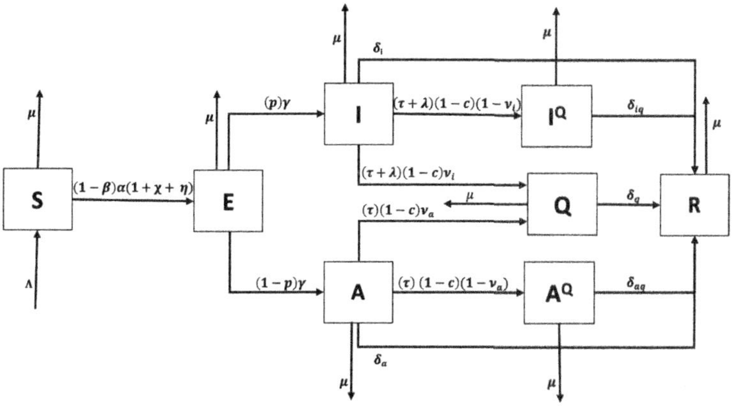


**Fig. 1** Flow Diagram of Quarantine COVID-19 model

Quarantine ($A^Q$). Transmission, $\alpha$, of the virus is assumed equal and constant for both asymptomatic and symptomatic infections. $\beta$, describes the rate Susceptible individuals wear masks. Susceptible individuals in masks are assumed to be completely protected against the spread of the virus. Once infected, the newly infected individuals move into the exposed category which represents the incubation time. A proportion $p$ and $(1-p)$ of exposed will move into symptomatic and asymptomatic, respectively. The symptomatic and asymptomatic categories represent undetected infectious groups. Both groups can be detected via random testing at the rate $\tau$. Symptomatic cases can also be detected at an additional rate of $\lambda$ representing the visual detection of symptoms. Of those who are detected to have the virus, a proportion $c$ will not self-isolate at all and will stay in their initial group. The rest, $(1-c)$, will go into one of two quarantine categories available to them. Asymptomatic and Symptomatic Semi-Quarantine represent quarantine behaviors where self-isolation is only followed for part of the required time. Individuals in this group might break quarantine because of a combination of their individual barriers for quarantining and their perception of the infectiousness [12]. Strict Quarantine is assumed to be the same for both groups and represents those who self-isolate for the entire recommended time lastly, given students can transfer into a campus during the semester we assume a recruitment of $\Lambda$. We also assume students may dropout or even die of natural death denoted by $\mu$. All five infected categories have a respective recovery rate. The governing equations for the model is

$$\frac{dS}{dt} = \Lambda - \alpha(1-\beta)S\left(\frac{I + A + \chi I^Q + \eta A^Q}{N}\right) - \mu S \tag{1}$$

$$\frac{dE}{dt} = \alpha(1-\beta)S\left(\frac{I + A + \chi I^Q + \eta A^Q}{N}\right) - \gamma E - \mu E \tag{2}$$

$$\frac{dI}{dt} = p\gamma E - (\tau + \lambda)I(1-c) - \delta_i I - \mu I \tag{3}$$

$$\frac{dA}{dt} = (1-p)\gamma E - \tau A(1-c) - \delta_a A - \mu A \tag{4}$$

$$\frac{dQ}{dt} = \tau(1-c)\nu_a A + (\tau + \lambda)(1-c)\nu_i I - \delta_q Q - \mu Q \tag{5}$$

$$\frac{dI^Q}{dt} = (\tau + \lambda)(1-c)(1-\nu_i)I - \delta_{iq} I^Q - \mu I^Q \tag{6}$$

$$\frac{dA^Q}{dt} = \tau(1-c)(1-\nu_a)A - \delta_{aq} A^Q - \mu A^Q \tag{7}$$

$$\frac{dR}{dt} = \delta_{iq} I^Q + \delta_{aq} I^Q + \delta_q Q + \delta_i I + \delta_a A - \mu R \tag{8}$$

The definitions of the parameters described in the system are given in Table 1.

**Table 1** Definition of parameters

| Parameter | Definition |
|---|---|
| $\alpha$ | Transmission rate |
| $\beta$ | Proportion of susceptible individuals wearing masks |
| N | Total population |
| $\chi$ | Proportion of the infectious period that symptomatic individuals do not self-isolate |
| $\eta$ | Proportion of the infectious period that asymptomatic individuals do not self-isolate |
| p | Proportion of Symptomatic cases |
| $\gamma$ | Duration of incubation period |
| $\tau$ | Positive rate from a random test |
| $\lambda$ | Visual detection rate of the virus |
| c | Proportion of individuals who do not self-isolate |
| $\nu_i$ | Symptomatic individuals going into Symptomatic Semi-Quarantine |
| $\nu_a$ | Asymptomatic individuals going into Asymptomatic Semi-Quarantine |
| $\delta_j$ | Recovery rate of Symptomatic (j = i), Asymptomatic (j = a), Strict Quarantine (j = q) |
| $\delta_{iq}$ | Recovery rate of Symptomatic Semi-Quarantine |
| $\delta_{aq}$ | Recovery rate of Asymptomatic Semi-Quarantine |
| $\Lambda$ | Recruitment rate |
| $\mu$ | Dropout rate |

## 3 Mathematical Analysis of the Model

In this section, we will show that the proposed system (1)–(8) is well-posed by proving it is non-negative and bounded for all values of $t$. We will also derive the basic reproduction number using the Next Generation Matrix approach.

### 3.1 Non-negativity of the Solution

**Theorem 1** *Let the initial conditions* $\left\{S_0, E_0, I_0, A_0, Q_0, I_0^Q, A_0^Q, R_0\right\} \geq 0$. *Then the solution to the system (1)–(8) is non-negative in* $[0, \infty)$.

***Proof*** Note that all the right-hand side terms of the system (1)–(8) are continuous and locally Lipschitzian on $\mathbb{R}$.

The solution $\{S(t), E(t), I(t), A(t), Q(t), I^Q(t), A^Q(t), R(t)\}$ with their initial conditions exist and are unique in the interval $[0, \infty)$ [8].
From Eq. (1) we have

$$\frac{dS}{dt} = \Lambda - S(\mu + g(t)) \tag{9}$$

where, $g(t) = \alpha(1-\beta)\left(\dfrac{I + A + \chi I^Q + \eta A^Q}{N}\right)$ Since $S_0 \geq 0$, by using an integrating factor we can integrate both sides of above relation with respect to $t$, we obtain

$$S(t_1) \geq e^{-\left(\mu t + \int_0^t g(s)ds\right)} \int_0^{t_1} \Lambda\, e^{\left(\mu t + \int_0^t g(s)ds\right)} dt$$

which proves $S(t) > 0$. Similarly, one can show $E(t), I(t), A(t), Q(t), I^Q(t)$, $A^Q(t)$, and $R(t)$ are also positive.

### 3.2 Boundedness of the Solution

Next, we prove that the system is bounded for all values of t.

**Theorem 2** *All solutions of the proposed system (1)–(8) are bounded inside the region* $\left\{\mathcal{X}(t) \in \mathbf{R}^8 : 0 \leq N(t) \leq \dfrac{\Lambda}{\mu}\right\}$.

***Proof*** Since $N(t) = S(t) + E(t) + I(t) + A(t) + Q(t) + I^Q(t) + A^Q(t) + R(t)$, we have

$$\frac{dN}{dt} = \Lambda - \mu N \tag{10}$$

Hence $\frac{dN}{dt} + \mu N = \Lambda$. This implies, $N(t) = \left(N(0) - \frac{\Lambda}{\mu}\right) e^{-\mu t} + \frac{\Lambda}{\mu}$, where $N(0) = S_0 + E_0 + I_0 + A_0 + Q_0 + I_0^Q + A_0^Q + R_0$. Letting $t \to 0$, we then get the solution $N(t) \subset \left[0, \frac{\Lambda}{\mu}\right]$.

### 3.3 Derivation of the Basic Reproduction Number

Next, we will derive the basic reproduction number using the Next Generation Matrix [2]. The basic reproduction number $\mathcal{R}_0$ is a powerful tool for predicting disease dynamics. An $\mathcal{R}_0 < 1$ indicates that an infection will die out and $\mathcal{R}_0 > 1$ indicates that the disease will continue to spread.

**Theorem 3** *The $\mathcal{R}_0$ is given by*

$$\mathcal{R}_0 = \mathcal{R}_0^1 + \mathcal{R}_0^2 + \mathcal{R}_0^3 + \mathcal{R}_0^4 \tag{11}$$

*where,*

$$\mathcal{R}_0^1 = \frac{\alpha(1-\beta)S}{N\,(\gamma+\mu)}\left(\frac{\chi\, p\, \gamma\, (\tau+\lambda)\,(1-c)\,(1-\nu_i)}{(\delta_{iq}+\mu)\,[(\tau+\lambda)(1-c)+\delta_i+\mu]}\right)$$

$$\mathcal{R}_0^2 = \frac{\alpha(1-\beta)S}{N\,(\gamma+\mu)}\left(\frac{\eta(1-p)\,\gamma\,\tau\,(1-c)\,(1-\nu_a)}{(\delta_{aq}+\mu)\,[\tau\,(1-c)+\delta_a+\mu]}\right)$$

$$\mathcal{R}_0^3 = \frac{\alpha(1-\beta)S}{N\,(\gamma+\mu)}\left(\frac{p\,\gamma}{[(\tau+\lambda)(1-c)+\delta_i+\mu]}\right)$$

$$\mathcal{R}_0^4 = \frac{\alpha(1-\beta)S}{N\,(\gamma+\mu)}\left(\frac{(1-p)\,\gamma}{[\tau\,(1-c)+\delta_a+\mu]}\right)$$

***Proof*** Given the infectious states: $E, I, A, I^Q, A^Q$, we first create the vector representing the new infections flowing only into the exposed given by

$$\mathcal{F} = \left\{\frac{S}{N}\alpha(1-\beta)(I + A + I^Q\chi + A^Q\eta), 0, 0, 0, 0\right\}$$

Along with the inflow, we also define the outflow from the five infectious groups:

$$\begin{aligned}\mathcal{V} = \{&E(\gamma+\mu), -p\gamma E + (\tau+\lambda)I(1-c) + \delta_i I + \mu I,\\ &-(1-p)\gamma E + \tau A(1-c) + \delta_a A + \mu A, -(\tau+\lambda)(1-c)(1-\nu_i)I\\ &+\delta_{iq}I^Q + \mu I^Q, -\tau(1-c)(1-\nu_a)A + \delta_{aq}A^Q + \mu A^Q\}\end{aligned}$$

Next, we find the Jacobian of vectors $\mathcal{F}$ and $\mathcal{V}$ which are given by: Next, we compute the next-generation matrix $FV^{-1}$ as

$$FV^{-1} = \begin{bmatrix} E_{1,1} & E_{1,2} & E_{1,3} & E_{1,4} & E_{1,5} \\ 0 & 0 & 0 & 0 & 0 \\ 0 & 0 & 0 & 0 & 0 \\ 0 & 0 & 0 & 0 & 0 \\ 0 & 0 & 0 & 0 & 0 \end{bmatrix} \quad (12)$$

The $\mathcal{R}_0$ corresponds to the dominant eigenvalue of the matrix $FV^{-1}$ in (12).

**Remark 1** Note that for (11), each expression for $\mathcal{R}_0^1$, $\mathcal{R}_0^2$,$\mathcal{R}_0^3$, and $\mathcal{R}_0^4$ are products of the transmission rate, the susceptible population and the inverse of the product of total population and the sum of the natural death rate and the incubation period. This product represents the movement from Susceptible to Exposed. The unique portion of each expression represents the duration of stay and the flow into each respective infectious group. Thus, each expression represents the ability of symptomatic semi-Quarantine, asymptomatic semi-quarantine, symptomatic and asymptomatic, respectively, to infect the susceptible category.

**Corollary 1** *Suppose the parameters $\chi$, $\eta$, $\beta$, $c$, $\mu$, $\lambda$ and $\tau$ are all equal to* 0, $p = 1$, $\delta_a = \delta_i = \delta_{aq} = \delta_{iq} = \delta$. *Then* $\mathcal{R}_0 = \dfrac{\alpha S}{\delta N}$ *which is the classic result [2]. By decreasing $\alpha$ and increasing $\delta$, one can control the $\mathcal{R}_0$.*

# 4 Computational Experiments

In this section, we will present a series of numerical simulations to show the effect of mitigation strategies and behaviors on the disease dynamics. For our simulations, we chose to study a *learning community* of freshmen students within the campus in one college within a higher education campus. We decided to choose this population as these selected students within the learning communities tend to live together in residence halls and engage with each other. Therefore, the opportunity for them to interact is more. We considered a total population of 388. Of these, 378 were assumed to be susceptible and 10 symptomatic. The other six compartments had an initial value of 0. The simulations were run using fourth-order Runge-Kutta methods in MATLAB. The parameter values were chosen as $\gamma^{-1} = 5.1$ days [3], $\delta_i^{-1} = 13$ days [4], $\delta_a^{-1} = 10$ days [4] $\delta_q^{-1} = 13$ days [4], $\delta_{iq}^{-1} = 13$ days [4], $\delta_{aq}^{-1} = 10$ days [4], $p = 0.6$ [13], $\tau = 0.05$ [11] and $\alpha = 0.07$ [11]. The remaining parameters $\beta$, $\nu_i$, $\nu_a$, $\lambda$, $\chi$, c, and $\eta$ have the chosen values 0.1, 0.3, 0.8, 0.5, 0.2, 0.85, and 0.6 because of lack of data respectively. Additionally, the parameters $\mu$ and $\Lambda$ are set to 0 for simplicity as the simulation was run over 90 days. Unless stated otherwise, these numbers will remain the same for all numerical computations.

## 4.1 Influence of the Parameters on $\mathcal{R}_0$

First, we will analyze the effect of changing parameters on the $\mathcal{R}_0$. Understanding the magnitude of effect of each parameter can help schools to decide what is important to communicate and enforce. Looking at Fig. 3, we can see that the transmission rate $\alpha$ and mask disuse rate $(1-\beta)$ causes the greatest changes in $\mathcal{R}_0$. Because of the significance of these two parameters, knowing the values of $\alpha$ and $(1-\beta)$ that cause an outbreak could be important for schools. Setting $\mathcal{R}_0 = 1$ and solving for $\alpha$ and $(1-\beta)$ in Eq. (13), we can determine the mask disuse-transmission rate $\alpha(1-\beta)$ threshold. $R_0^{n*}$ is $R_0^n$ from Eq. (11) divided by $\alpha(1-\beta)$. Using the given parameters, $\alpha(1-\beta)$ must be less then 0.1625 to mitigate spread.

$$\alpha(1-\beta) = \frac{1}{R_0^{1*} + R_0^{2*} + R_0^{3*} + R_0^{4*}} \tag{13}$$

The interactions between $\alpha$ and $(1-\beta)$ on $\mathcal{R}_0$ along with the threshold values are shown in Fig. 2. The portion above this threshold line represents disease propagation, below this line there is none. Because of its low placement of this line with respect to the peak value, we can determine that the $\mathcal{R}_0$ is sensitive to $\alpha$ and $(1-\beta)$. Additionally, Fig. 2 shows that mask disuse and the transmission rate are directly related and linear.

Next, we will analyze the effect of disease status on quarantine behaviors in Fig. 3. While not drastic, symptomatic quarantine behaviors tend to have a greater effect on the $\mathcal{R}_0$ the asymptomatic behaviors. Both $\nu_i$ and $\nu_a$ have negative linear growth

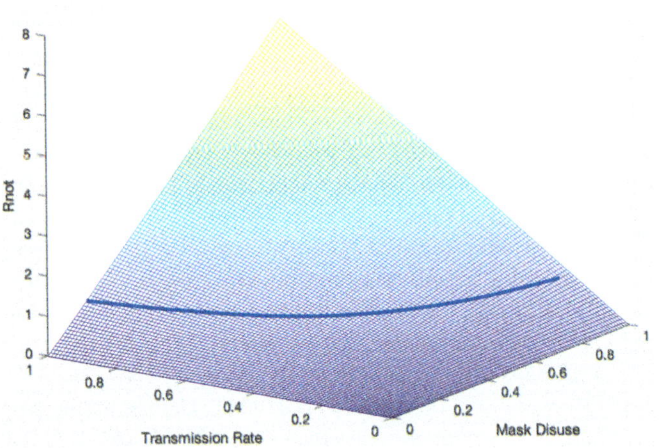


**Fig. 2** Influence of transmission rate ($\alpha$) and mask usage $(1-\beta)$ on $\mathcal{R}_0$. The line at $\mathcal{R}_0 = 1$ represents the pandemic thresh hold

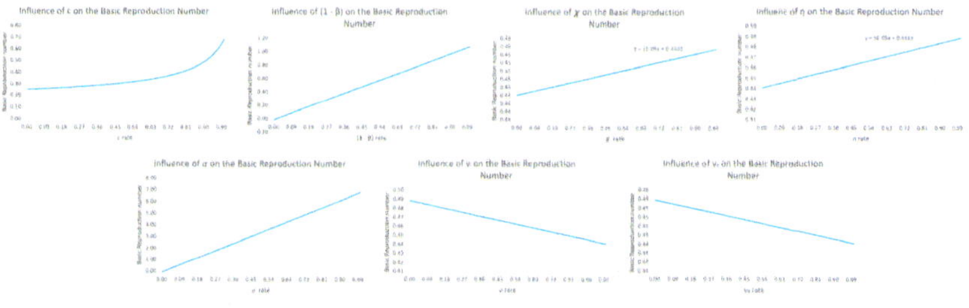

**Fig. 3** Influence of quarantine disobedience (c) on $\mathcal{R}_0$ (top most left), Influence of mask usage ($\beta$) on $\mathcal{R}_0$ (top second left), Influence of Asymptomatic semi-Quarantine ($\eta$) on $\mathcal{R}_0$ (top second left), Influence of Symptomatic semi-Quarantine ($\chi$) on $\mathcal{R}_0$ (top most left), Influence of transmission rate ($\alpha$) on $\mathcal{R}_0$ (bottom right), Influence of Strict Symptomatic Quarantine ($\nu_i$) on $\mathcal{R}_0$ (bottom middle), Influence of Strict Asymptomatic Quarantine ($\nu_i$) on $\mathcal{R}_0$ (bottom left)

with respect to the $\mathcal{R}_0$, however, $\nu_i$ has a greater effect. The difference in magnitude could be because of both the greater probability of symptomatic disease and the additional ability to visually detect the symptomatic cases. A similar difference was found for $\chi$ and $\eta$. Both graphs have a linear and positive growth with respect to the $\mathcal{R}_0$, however, $\chi$ has a greater effect. The greater effect of symptomatic quarantine behaviors overall could indicate a greater importance of detecting visible symptoms to prevent disease propagation. However because the difference in effect was not sizable, reducing barriers for self-isolation, such as assignments/grades, could be helpful in slowing spread.

Lastly looking at the third graph in the top panel of Fig. 3, we see a rapid positive growth in the $\mathcal{R}_0$ as you increase c. At greater c values, Asymptomatic and Symptomatic populations will grow larger. Because these categories are unrestricted in their spread, they can infect Susceptible individuals faster.

## 4.2 *Local and Global Sensitivity*

Next we will analyze the normalized local and global sensitivity. The normalized local sensitivity is the effect on $\mathcal{R}_0$ by changing a parameter by 1% in a fixed parameter space. Values closer to 1 will be more significant. We can find the normalized local sensitivity using $S^{G}_{\mathcal{R}_0} = \dfrac{\partial \mathcal{R}_0}{\partial G}\dfrac{G}{\mathcal{R}_0}$, where G is the interested parameter that is most sensitive in regard to the $\mathcal{R}_0$. The derivation of the sensitivity relations of $\alpha$ and $\beta$ are shown below. The remaining parameters in $\mathcal{R}_0$ can be found the same way

$$S^{\alpha}_{\mathcal{R}_0} = \frac{\partial \mathcal{R}_0}{\partial \alpha}\frac{\alpha}{\mathcal{R}_0} = 1 \qquad S^{\beta}_{\mathcal{R}_0} = \frac{\partial \mathcal{R}_0}{\partial \beta}\frac{\beta}{\mathcal{R}_0} = -\frac{\beta}{1-\beta}$$

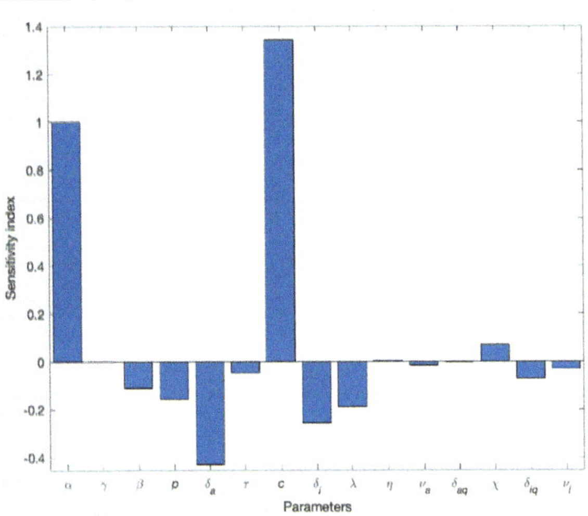


| Parameter | Sensitivity |
|---|---|
| α | 1.0000 |
| γ | 0.0000 |
| β | -0.1110 |
| p | -0.1570 |
| δa | -0.4304 |
| τ | -0.0470 |
| c | 1.3430 |
| δi | -0.2580 |
| λ | -0.1900 |
| η | 0.0040 |
| va | -0.0170 |
| δaq | -0.0040 |
| χ | 0.0710 |
| δiq | -0.0710 |
| vi | -0.0300 |

**Fig. 4** The normalized local sensitivity of the model with respect to the parameters

By using the given parameter values in $\mathcal{R}_0$ (excluding $\mu$), we can determine the normalized local sensitivity for the fixed parameter space in Fig. 4. The student behaviors parameters c, and $\alpha$ have a pronounced significance. The next most significant is $\beta$ by a great margin. Parameters $c$ and $\beta$ increase the $\mathcal{R}_0$ by 1.3% and 1% respectively. $\beta$ increases the $\mathcal{R}_0$ by $-0.11\%$. The quarantine behaviors $\chi$, $\eta$, $\nu_a$, and $\nu_i$ are small and do not vary to a significant degree. These results indicate the importance of quarantining at least to some degree and a minimal significance of disease type on quarantine. Next we will go over the global sensitivity of the system. The global sensitivity looks at the sensitivity over the entire parameter space. The values were calculated using the SOBOL or variance-based method in the Global Sensitivity Analysis toolbox (GSAT) [10].

Figure 5 shows the first order index, the effect of each parameter on the variance of $\mathcal{R}_0$, and the total order, which is the sum of the first order index and the interactions between parameters. Unlike the normalized local sensitivity, $\beta$, $\alpha$, and $c$ are the most significant behavioral parameters respectively. The quarantine behavior parameters retain a similar order as in the local sensitivity.

## 4.3 Dynamics of the State Variables

Finally, we illustrate the dynamics of the disease in Fig. 6 where we have used parameter values as described earlier. While the computational results were done for a learning community of 388 individuals, a similar approach can be done for typical higher education campuses in the United States with about 40, 000 students.

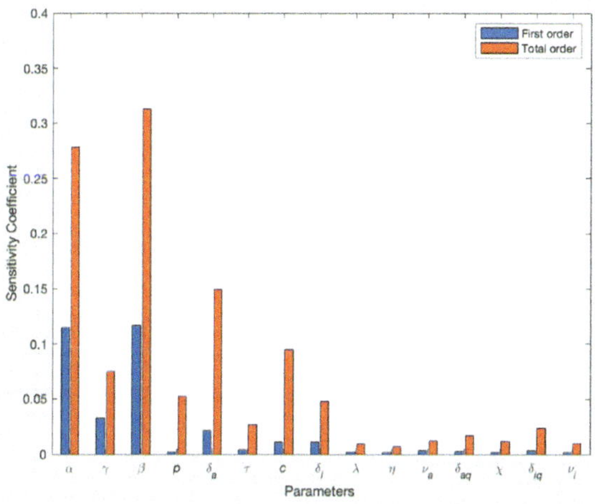


| Parameter | First order | Total order |
|---|---|---|
| α | 0.11459 | 0.27805 |
| γ | 0.03258 | 0.07476 |
| β | 0.11660 | 0.31281 |
| p | 0.00206 | 0.05250 |
| $\delta_a$ | 0.02134 | 0.14897 |
| τ | 0.00380 | 0.02677 |
| c | 0.01076 | 0.09496 |
| $\delta_i$ | 0.01090 | 0.04730 |
| λ | 0.00230 | 0.09496 |
| η | 0.00174 | 0.00700 |
| $\nu_a$ | 0.00309 | 0.01194 |
| $\delta_{aq}$ | 0.00268 | 0.01680 |
| χ | 0.00182 | 0.01130 |
| $\delta_{iq}$ | 0.00347 | 0.02351 |
| $\nu_i$ | 0.00154 | 0.00941 |

**Fig. 5** The first order and total global sensitivity of the model

The dynamics of a greater population with an initial susceptibility of 39, 000 and 1, 000 infected is shown in Fig. 7. Despite having different initial values, the overall dynamics of the state variables remained the same as in Fig. 6. For both graphs, the order of the peaks reflects the model in Fig. 1. Next, we analyzed the shape of the resulting disease dynamics when changing quarantine behaviors. Figure 8 illustrates the shape of the resulting disease dynamics when changing $c$. The categories exposed, Asymptomatic, Symptomatic, and Recovered have greater peaks for larger c values. This results in a greater infection of Susceptible. Thus, student adherence to quarantine procedure is a behavior schools should look at.

## 5 Graphical User Interphase (GUI)

In this section, we will briefly go over a dashboard that was created as a part of our quarantine model. This Dashboard was created in MATLAB using Matlab's UI design environment GUIDE. This graphical interface allows users to analyze disease dynamics visually in an interactive way. Additionally, users can determine the basic reproduction number $\mathcal{R}_0$ for their chosen parameters, as shown in Fig. 9. This straightforward dashboard allows users to interact and study the proposed model for future decision-making. The dashboard was organized for user ease. Behaviors and mitigation parameters, which include the parameters mask use and strict quarantine, are given both sliders and text boxes for data entry. This is because of the variability of behaviors across different school environments. Furthermore, this GUI allows users to directly compare the effect of different parameters. This is done by allowing users to save up to 3 sets of data. Users can then pick which saved data they want

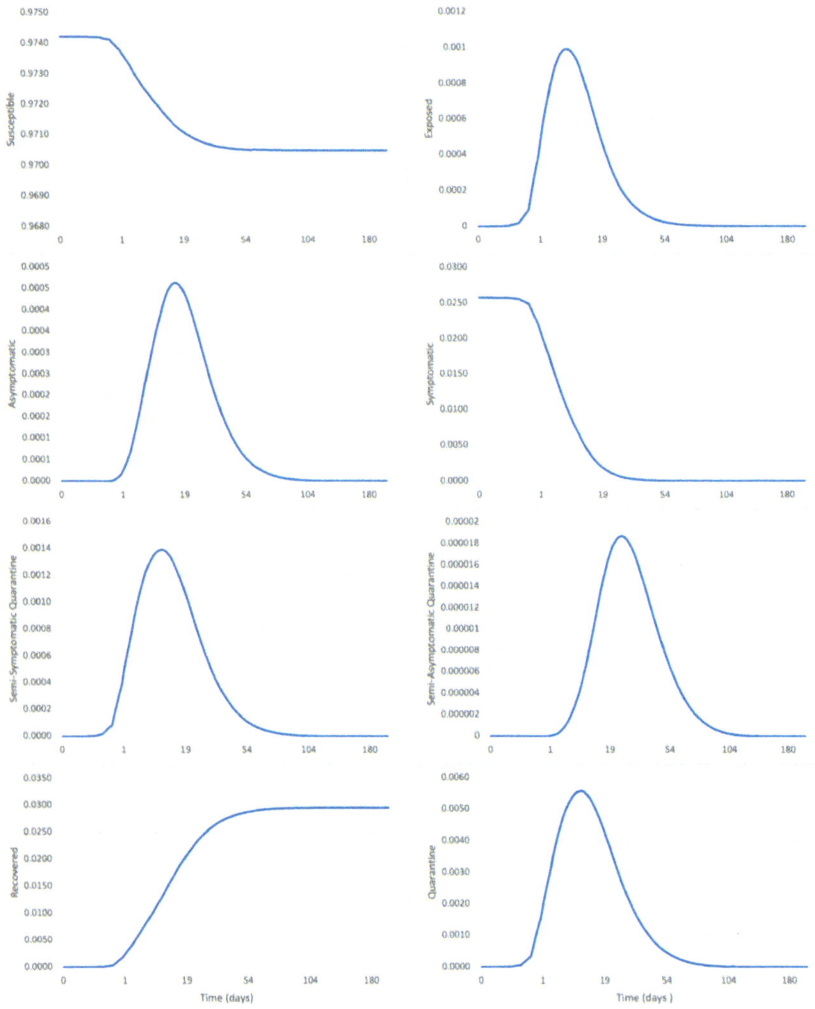


**Fig. 6** Quarantine model dynamics for N = 388. Each categories is shown individually

to graph on the same plot by clicking hold checkboxes under the save buttons. An example is shown in Fig. 10 where we compare the effect of variations of parameter values on Symptomatic and Asymptomatic. The dashboard will calculate the basic reproduction number $\mathcal{R}_0$ when the user clicks on the load, save, or graph buttons.

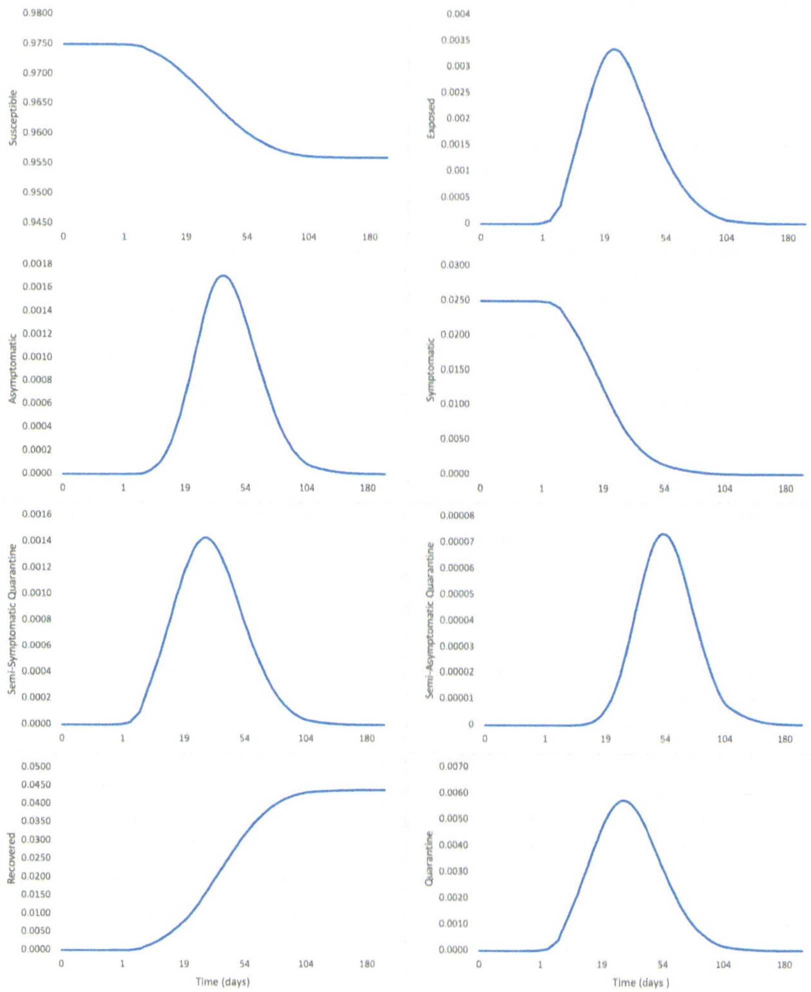


**Fig. 7** Quarantine model dynamics for $N = 40,000$. Each categories is shown individually

# 6 Discussion and Conclusions

In this paper, we considered an extended SEIR model that includes three levels of quarantine for understanding spread of COVID-19 within an upper-level education campus. This paper also discussed common mitigation strategies seen in this type of environment specifically applied to a learning community within a higher education campus. Next we showed that this model is well-posed by proving the positivity and boundedness of the system for all values of $t$. Then we calculated the basic reproduction number $\mathcal{R}_0$ using the Next Generation Matrix. Next, we performed

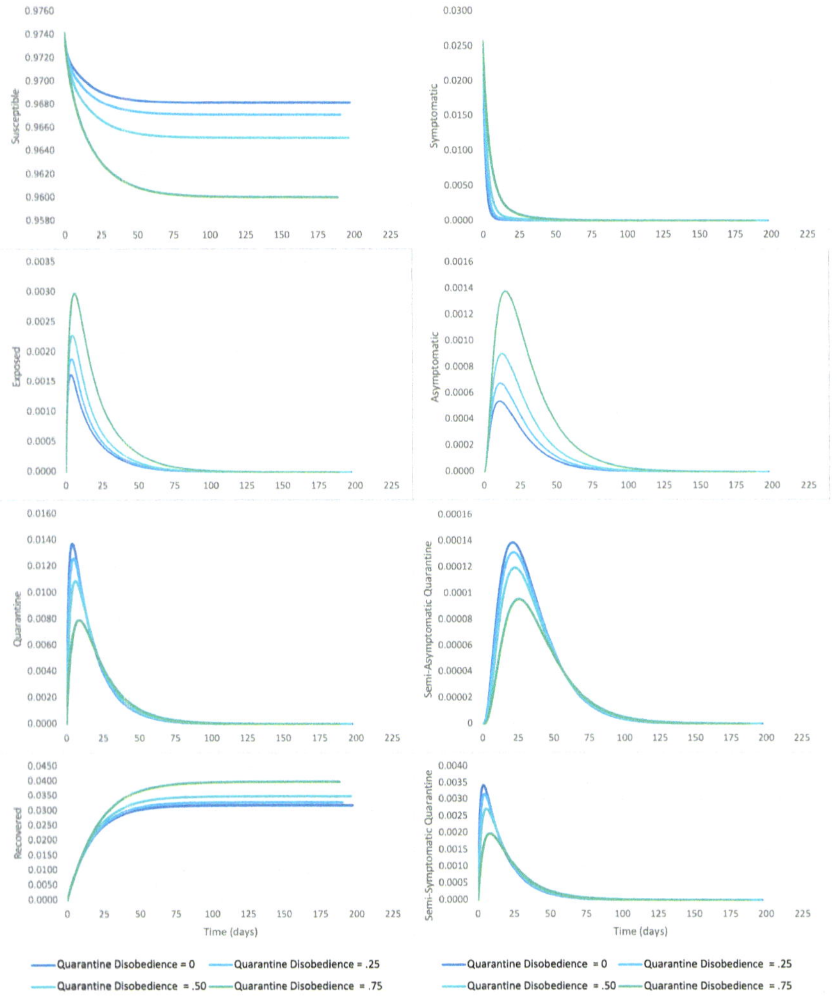


**Fig. 8** Quarantine model dynamics for each category when changing the proportion of the population who do not self-isolate (c) by 0.25

a series of numerical simulations and analysis to validate the model. Lastly, we developed a user-friendly dashboard that can be public use.

In this model, we demonstrated the importance of adherence to COVID-19 regulations. The model showed that managing the outbreak requires not only the use of mitigation strategies like masks and social distancing, but also the variance of student behaviors toward COVID-19 regulations. A combination of both mitigation strategies and student behaviors was found to be important in managing a pandemic.

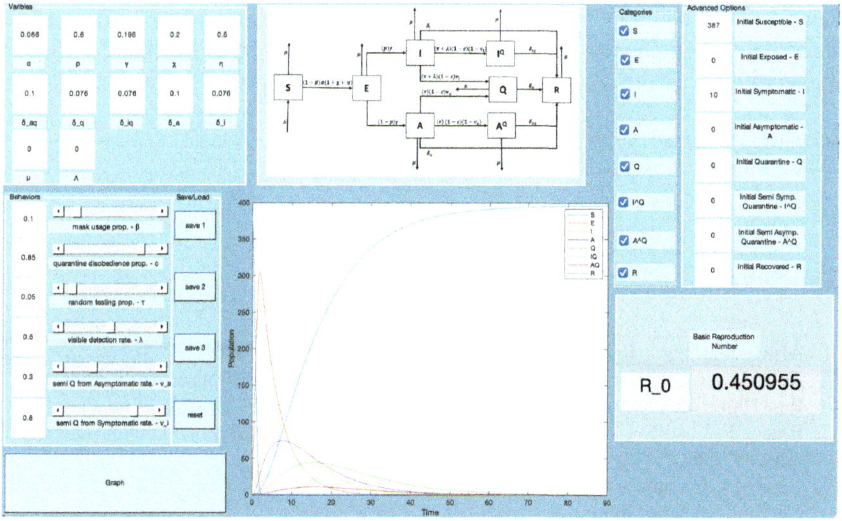


**Fig. 9** GUI model dynamics for the expanded SIR model. The calculated basic reproduction number is shown on the right

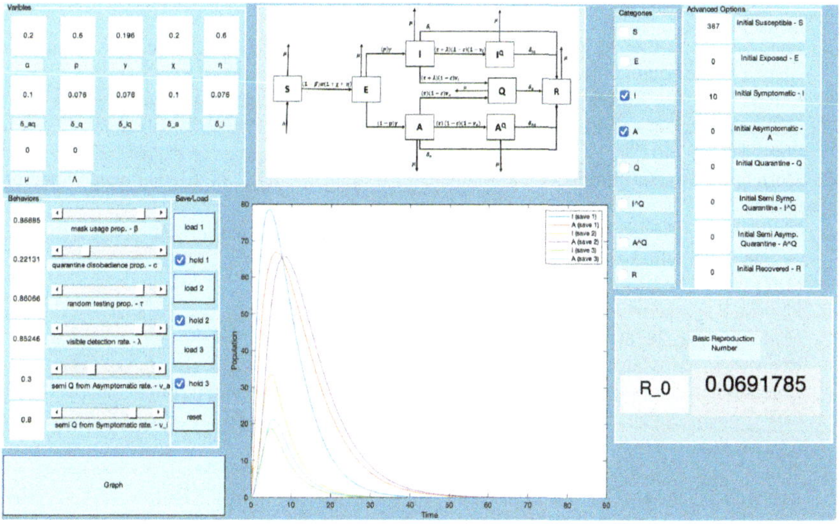


**Fig. 10** Comparison of three variations of the disease dynamics in regard to the Symptomatic and Asymptomatic populations

Future studies could include the effect of information transparency and incentives in increasing compliance. Additionally, this model might be useful in predicting the spread of other health concerns where adherence to regulations is pivotal.

**Acknowledgements** This work has been supported in part by the National Science Foundation DMS 2031027 and DMS 2031029.

## References


1. Alsulaiman, S.A., Rentner, T.L.: The use of the health belief model to assess US college students' perceptions of COVID-19 and adherence to preventive measures. J. Publ. Health Res. **10**(4) (2021)
2. Brauer, F., Castillo-Chavez, C., Castillo-Chavez, C.: Mathematical Models in Population Biology and Epidemiology, vol. 2, p. 508. Springer, New York (2012)
3. Covid-19 pandemic planning scenarios. https://www.cdc.gov/coronavirus/2019-ncov/hcp/planning-scenarios.html
4. Criteria for releasing COVID-19 patients from isolation. https://www.who.int/news-room/commentaries/detail/criteria-for-releasing-covid-19-patients-from-isolation
5. Guidance for institutions of Higher Education (IHES) (2022). https://www.cdc.gov/coronavirus/2019-ncov/community/colleges-universities/considerations.html. Accessed 03 Jan. 2022
6. Ju, C., Jiang, Y., Bao, F., Zou, B., Xu, C.: Online rumor diffusion model based on variation and silence phenomenon in the context of COVID-19. Front. Publ. Health **9** (2021)
7. López, L., Rodo, X., López, L., Rodo, X.: A modified SEIR model to predict the COVID-19 outbreak in Spain and Italy: simulating control scenarios and multi-scale epidemics. Results Phys. **21**, 103746 (2021)
8. Martcheva, M.: An Introduction to Mathematical Epidemiology, vol. 61. Springer, New York (2015)
9. Ohajunwa, C., Seshaiyer, P.: Mathematical Modeling, Analysis, and simulation of the COVID-19 pandemic with behavioral patterns and group mixing. Spora: A J. Biomath. **7**(1), 46–60 (2021)
10. Pianosi, F., Sarrazin, F., Wagener, T.: A Matlab toolbox for global sensitivity analysis. Environ. Modell. Softw. **70**, 80–85 (2015)
11. Rennert, L., McMahan, C., Kalbaugh, C.A., Yang, Y., Lumsden, B., Dean, D., Pekarek, L., Colenda, C.C.: Surveillance-based informative testing for detection and containment of SARS-CoV-2 outbreaks on a public university campus. Lancet Child Adolesc. Health **5**(6), 428–436 (2021)
12. Tam, C.C., Li, X., Li, X., Wang, Y., Lin, D.: Adherence to preventive behaviors among college students during COVID-19 pandemic in China: the role of health beliefs and COVID-19 stressors. Current Psychol 1–11 (2021)
13. Tupper, P., Colijn, C.: COVID-19 in schools: mitigating classroom clusters in the context of variable transmission. PLoS Comput. Biol. **17**(7) (2021)
14. Wang, C., Horby, P.W., Hayden, F.G., Gao, G.F.: A novel coronavirus outbreak of global health concern. The Lancet **395**(10223), 470–473 (2020)

# Numerical Analysis

# Numerical Solution of the Fredholm Integral Equations of the First Kind by Using Multi-projection Methods

**Subhashree Patel, Bijaya Laxmi Panigrahi, and Gnaneshwar Nelakanti**

**Abstract** The Fredholm integral equations (FIEs) of the first kind have been solved by Legendre spectral multi-projection methods by using Tikhonov regularized methods. The theoretical analysis utilizing this method under a priori parameter selection strategy has been explained and the best convergence rates obtained in $L^2$-norm. Next, in order to discover an appropriate regularization parameter, Arcangeli's discrepancy principle has been applied and the order of convergence has been deduced. Numerical example has been furnished which validates our theoretical findings.

**Keywords** Fredholm integral equation of the first kind · Ill-posed problems · Tikhonov regularization method · Legendre polynomials · Multi-Galerkin method · Arcangeli's Discrepancy

## 1 Introduction

We define the following FIEs of the first kind:

$$\int_{-1}^{1} \tau(u, v)x(v)\mathrm{d}v = f(u), \quad -1 \leq u \leq 1, \tag{1}$$

where $x$ is the unknown function in the Banach space $\mathbb{X} = L^2[-1, 1]$ to be estimated, and $f$ and $\tau(., .)$ are known functions. These types of Eq. (1) appear in several inverse problems in engineering and science such as geophysics (land-mining, oil

S. Patel (✉)
Department of Mathematics, Sambalpur University, Burla 768019, Odisha, India
e-mail: subhashreepatel22@gmail.com; subhashreepatel@suniv.ac.in

B. L. Panigrahi
Department of Mathematics, Gangadhar Meher University, Sambalpur 768004, Odisha, India

G. Nelakanti
Department of Mathematics, Indian Institute of Technology Kharagpur, Kharagpur 721302, India



B. Rushi Kumar et al. (eds.), *Mathematics and Computing*, Springer Proceedings in Mathematics & Statistics 415, https://doi.org/10.1007/978-981-19-9307-7_50

exploration, etc.), signal processing, medical imaging, electromagnetic field, and backward heat conduction problems (see [1, 7]).

Since the FIEs of the first kind (1) are ill-posed, conversion of ill-posed into a well-posed equation is highly desirable. In the literature, several regularization methods have been developed to handle the ill-posedness property and the Tikhonov regularization method is mostly used regularization approach. In the regularization methods, the choice of regularization parameters is the primary issue.

So, many works on the development of the regularization parameter have been carried out by the researchers and well documented in [4, 5, 13] and reference therein. In general, the regularized equations of the FIEs of the first kind (1) can not be solved explicitly. As a result, it is essential to develop numerical approximation methods for solving these regularized equations. Thus, the projection methods [10, 13], degenerate kernel method [6], multiscale methods [4], and wavelet methods [14] have been developed in the literature.

In [10, 11, 15], the projection-based methods for the FIEs of the first kind (1) utilizing Legendre polynomials to approximate the function space have been studied. However, the multi-projection methods have been employed for FIEs of the first kind in [12] and convergence analysis has been explained using infinity norm. In this article, we approximate Eq. (1) by its Tikhonov regularized equation using multi-projection method using similar approximation of function space as above. The basis of approximation subspace $\mathbb{X}_n$ is the Legendre polynomials with a maximum degree $n$. Legendre polynomials are iterative and can be constructed effortlessly, as well as having the orthogonal characteristic. Thus, the computational cost to calculate the matrix of the Tikhonov regularized equation of Eq. (1) is very less. The accuracy $\mathcal{O}(\delta^{\frac{2\nu}{2\nu+1}})$ for $\nu \in (0, 1]$ in $L^2$-norm has been established utilizing the above approximation of function space techniques under both a priori and a posteriori parameter strategies.

We denote $c$ as a generic constant throughout the paper.

## 2 Legendre Spectral Multi-projection Methods

Let the integral operator $\mathcal{A} : L^2[-1, 1] \to L^2[-1, 1]$ defined by

$$\mathcal{A}x(u) = \int_{-1}^{1} \tau(u, v)x(v)\,\mathrm{d}v, \quad -1 \leq u \leq 1,$$

where $\tau(.,.) \in \mathcal{C}([-1, 1] \times [-1, 1])$. Then the linear operator $\mathcal{A}$ on $L^2[-1, 1]$ is compact. The operator equation of Eq. (1) is rewritten as

$$\mathcal{A}x = f. \tag{2}$$

Eq. (2) is ill-posed. The solution of Eq. (2) occurs iff $f \in \mathcal{R}(\mathcal{A})$. By taking help of Moore–Penrose inverse $\mathcal{A}^{\dagger} : \mathcal{R}(\mathcal{A}) + \mathcal{R}(\mathcal{A})^{\perp} \to \mathbb{X}$ (see [9], P.-146) of the operator $\mathcal{A}$, the generalized solution of Eq. (2) is $\widehat{x} = \mathcal{A}^{\dagger} f$.

The adjoint $\mathcal{A}^*$ can be evaluated as $\mathcal{A}^* x(u) = \displaystyle\int_{-1}^{1} \tau(v, u)x(v)\mathrm{d}v.$
Denote

$$\mathcal{G}x(u) = \mathcal{A}^*\mathcal{A}x(u) = \int_{-1}^{1} \tau(u, v)\Big[\int_{-1}^{1} \tau(v, z)x(z)\mathrm{d}z\Big]\mathrm{d}v = \int_{-1}^{1} \widetilde{\tau}(u, z)x(z)\mathrm{d}z,$$

where $\widetilde{\tau}(u, z) = \int_{-1}^{1} \tau(u, v)\tau(v, z)\mathrm{d}v$. It can be straightforwardly shown that $\mathcal{G} : \mathbb{X} \to \mathbb{X}$ is self-adjoint. We quote the following lemma from [9].

**Lemma 1** ([9]) *Then* $(\mathcal{G} + \alpha\mathcal{I})$ *is invertible on* $L^2[-1, 1]$ *for every* $\alpha > 0$ *and*

$$\|(\mathcal{G} + \alpha\mathcal{I})^{-1}\|_{L^2} \le \frac{1}{\alpha}, \qquad \textit{and} \qquad \|(\mathcal{G} + \alpha\mathcal{I})^{-1}\mathcal{A}^*\|_{L^2} \le \frac{1}{2\sqrt{\alpha}},$$

*where* $\mathcal{G}$ *is positive self-adjoint.*

Let $x_\alpha, \alpha > 0$ be the regularized solution then the Tikhonov regularized equation of (2) is

$$(\mathcal{G} + \alpha\mathcal{I})x_\alpha = \mathcal{A}^* f. \tag{3}$$

Let $\widetilde{f}$ be the perturbed data such that $\|f - \widetilde{f}\|_{L^2} \le \delta$. Let $\widetilde{x}_\alpha$ be the regularized solution with respect to the perturbed data $\widetilde{f}$, then the Tikhonov regularized equation of (2) is

$$(\mathcal{G} + \alpha\mathcal{I})\widetilde{x}_\alpha = \mathcal{A}^* \widetilde{f}. \tag{4}$$

We will now talk about the approximation method using Legendre polynomial basis functions for Eq. (4). Let $\mathbb{X}_n$ represent the subspaces of $\mathbb{X}$ and the span of Legendre orthonormal polynomials, i.e., $\mathbb{X}_n = \text{span}\,\{\phi_0, \phi_1, \ldots, \phi_n\}$ and $\phi_i(s) = \sqrt{\frac{2i+1}{2}}L_i(s)$, where $L_i$'s represent the Legendre polynomials of maximum degree $i$ on $[-1, 1]$ for $i = 0, 1, \ldots, n$.

The orthogonal projection $\mathcal{P}_n : \mathbb{X} \to \mathbb{X}_n$ is then defined by $\mathcal{P}_n x = \sum_{j=0}^{n} \langle x, \phi_j\rangle \phi_j$, $x \in \mathbb{X},\ \phi_j \in \mathbb{X}_n$, where $\langle x, \phi_j\rangle = \int_{-1}^{1} x(t)\phi_j(t)\mathrm{d}t$.

**Lemma 2** ([2]) *Then the following results of the orthogonal projection hold:*

*(i)* $\|\mathcal{P}_n x\|_{L^2} \le p_1\|x\|_\infty$ *for any* $x \in \mathbb{X}$, *where* $p_1$ *is a constant independent of* $n$.
*(ii) For any* $x \in C^r[-1, 1]$, *there exists* $c > 0$ *independent of* $n$ *such that*

$$\|\mathcal{P}_n x - x\|_{L^2} \le c n^{-r} \|x^{(r)}\|_\infty.$$

Define $\mathcal{G}_n^M : \mathbb{X} \to \mathbb{X}$ by

$$\mathcal{G}_n^M x = \mathcal{P}_n \mathcal{G} x + \mathcal{G} \mathcal{P}_n x - \mathcal{P}_n \mathcal{G} \mathcal{P}_n x, \quad x \in \mathbb{X}. \tag{5}$$

Next, we approximate the operator $\mathcal{G}$ by $\mathcal{G}_n^M$. Using $\mathcal{G}_n^M$, Eq. (4) is approximated as to find $\widetilde{x}_{\alpha,n}^M \in \mathbb{X}$ such that

$$(\mathcal{G}_n^M + \alpha \mathcal{I}) \widetilde{x}_{\alpha,n}^M = \mathcal{A}^* \widetilde{f}. \tag{6}$$

This is the Legendre spectral multi-projection method for Eq. (4). To solve Eq. (6), we apply $\mathcal{P}_n$ and $(\mathcal{I} - \mathcal{P}_n)$ on both sides of Eq. (6). Then we get $\mathcal{P}_n \mathcal{G}_n^M \widetilde{x}_{\alpha,n}^M + \alpha \mathcal{P}_n \widetilde{x}_{\alpha,n}^M = \mathcal{P}_n \mathcal{A}^* \widetilde{f}$, i.e.,

$$\mathcal{P}_n \mathcal{G} \widetilde{x}_{\alpha,n}^M + \alpha \mathcal{P}_n \widetilde{x}_{\alpha,n}^M = \mathcal{P}_n \mathcal{A}^* \widetilde{f} \tag{7}$$

and $(\mathcal{I} - \mathcal{P}_n)(\mathcal{G}_n^M + \alpha \mathcal{I}) \widetilde{x}_{\alpha,n}^M = (\mathcal{I} - \mathcal{P}_n) \mathcal{A}^* \widetilde{f}$, i.e.,

$$\widetilde{x}_{\alpha,n}^M = \mathcal{P}_n \widetilde{x}_{\alpha,n}^M - \frac{1}{\alpha} (\mathcal{I} - \mathcal{P}_n) \mathcal{G} \mathcal{P}_n \widetilde{x}_{\alpha,n}^M + \frac{1}{\alpha} (\mathcal{I} - \mathcal{P}_n) \mathcal{A}^* \widetilde{f}, \tag{8}$$

respectively. Substituting Eqs. (8) in (7), we get

$$\mathcal{P}_n \widetilde{x}_{\alpha,n}^M = \frac{1}{\alpha} \Big[ \mathcal{P}_n - \frac{1}{\alpha} \mathcal{P}_n \mathcal{G} (\mathcal{I} - \mathcal{P}_n) \Big] \mathcal{A}^* \widetilde{f} - \frac{1}{\alpha} \Big[ \mathcal{P}_n - \frac{1}{\alpha} \mathcal{P}_n \mathcal{G} (\mathcal{I} - \mathcal{P}_n) \Big] \mathcal{G} \mathcal{P}_n \widetilde{x}_{\alpha,n}^M.$$

This indicates that, we look for $x_n^{M,1} = \mathcal{P}_n \widetilde{x}_{\alpha,n}^M$ from the following equation:

$$(\mathcal{S}_n^M \mathcal{G} + \alpha \mathcal{I}) x_n^{M,1} = \mathcal{S}_n^M \mathcal{A}^* \widetilde{f}, \tag{9}$$

where $\mathcal{S}_n^M = \mathcal{P}_n - \frac{1}{\alpha} \mathcal{P}_n \mathcal{G} (\mathcal{I} - \mathcal{P}_n)$. Now, by using Eq. (8), we can get $\widetilde{x}_{\alpha,n}^M = x_n^{M,1} + x_n^{M,2}$, where $x_n^{M,2} = \frac{1}{\alpha} (\mathcal{I} - \mathcal{P}_n)(\mathcal{A}^* \widetilde{f} - \mathcal{G} x_n^{M,1})$.

**Theorem 1** *Let* $\tau(.,.) \in C^{(r,r)}([-1,1] \times [-1,1])$, $r \ge 1$. *Then the following outcome is valid:*

$$\|\mathcal{G}_n^M - \mathcal{G}\|_{L^2} = \mathcal{O}(n^{-r}).$$

***Proof*** We have

$$\begin{aligned} \|(\mathcal{G}_n^M - \mathcal{G}) x\|_{L^2} &= \|(\mathcal{I} - \mathcal{P}_n) \mathcal{G} (\mathcal{I} - \mathcal{P}_n) x\|_{L^2} \\ &\le \|\mathcal{G} (\mathcal{I} - \mathcal{P}_n) x\|_{L^2} \|(\mathcal{I} - \mathcal{P}_n)\|_{L^2} \le \sqrt{2} (1 + p_1) \, \|\mathcal{G} (\mathcal{I} - \mathcal{P}_n) x\|_\infty. \end{aligned}$$

Now, using orthogonality of $(\mathcal{P}_n - \mathcal{I})$, Lemma 2 and Cauchy–Schwarz inequality, we get

$$\begin{aligned} \|\mathcal{G}(\mathcal{P}_n - \mathcal{I})x\|_\infty &= \sup_{u\in[-1,1]} \left| \int_{-1}^{1} \widetilde{\mathcal{T}}(u,z)(\mathcal{I} - \mathcal{P}_n)x(z)dz \right| \\ &\leq \sup_{u\in[-1,1]} \|(\mathcal{I} - \mathcal{P}_n)\widetilde{\mathcal{T}}(u,.)\|_{L^2}\|x\|_{L^2} \\ &\leq cn^{-r} \sup_{u\in[-1,1]} \|\widetilde{\mathcal{T}}_u^{(r,0)}\|_\infty \|x\|_{L^2} \leq B_1 cn^{-r}\|x\|_{L^2}, \end{aligned} \quad (10)$$

where $B_1 = \sup_{u\in[-1,1]} \|\widetilde{\mathcal{T}}_u^{(r,0)}\|_\infty$. Now, substituting estimate (10) in the above estimate of (10), we obtain the required result.

We show that for sufficiently large $n$, and for every $\alpha > 0$, $\mathcal{G}_n^M + \alpha\mathcal{I}$ is invertible in the next theorem.

**Theorem 2** *For $n$ large enough and for every $\alpha > 0$, the operator $\mathcal{G}_n^M + \alpha\mathcal{I} : L^2[-1,1] \to L^2[-1,1]$ is invertible and the following outcomes are true:*

$$\|(\mathcal{G}_n^M + \alpha\mathcal{I})^{-1}\|_{L^2} \leq \frac{2}{\alpha} \quad \text{and} \quad \|(\mathcal{G}_n^M + \alpha\mathcal{I})^{-1}\mathcal{A}^*\|_{L^2} \leq \frac{1}{\sqrt{\alpha}}.$$

***Proof*** From Lemma 1, we have $\mathcal{G} + \alpha\mathcal{I}$ is invertible. Then, we can write

$$\mathcal{G}_n^M + \alpha\mathcal{I} = \mathcal{G}_n^M - \mathcal{G} + \mathcal{G} + \alpha\mathcal{I} = (\mathcal{G} + \alpha\mathcal{I})[\mathcal{I} + (\mathcal{G} + \alpha\mathcal{I})^{-1}(\mathcal{G}_n^M - \mathcal{G})].$$

Since from Theorem 1, $\|\mathcal{G}_n^M - \mathcal{G}\|_{L^2}$ converges to 0 as $n \to \infty$ then, $n$ choosen as sufficiently large such that $\|\mathcal{G}_n^M - \mathcal{G}\|_{L^2} < \frac{\alpha}{2}$. Now, using Lemma 1 and above equation, we obtain

$$\|(\mathcal{G} + \alpha\mathcal{I})^{-1}(\mathcal{G}_n^M - \mathcal{G})\|_{L^2} \leq \|(\mathcal{G} + \alpha\mathcal{I})^{-1}\|_{L^2}\|\mathcal{G}_n^M - \mathcal{G}\|_{L^2} \leq \frac{1}{2} < 1. \quad (11)$$

Therefore, $\mathcal{I} + (\mathcal{G} + \alpha\mathcal{I})^{-1}(\mathcal{G}_n^M - \mathcal{G})$ is invertible. Thus, the invertibility of $\mathcal{G}_n^M + \alpha\mathcal{I}$ follows by using the invertibility of $\mathcal{I} + (\mathcal{G} + \alpha\mathcal{I})^{-1}(\mathcal{G}_n^M - \mathcal{G})$. Now, using Lemma 1 and estimate (11), we obtain

$$\begin{aligned} \|(\mathcal{G}_n^M + \alpha\mathcal{I})^{-1}\|_{L^2} &\leq \|[\mathcal{I} + (\mathcal{G} + \alpha\mathcal{I})^{-1}(\mathcal{G}_n^M - \mathcal{G})]^{-1}\|_{L^2}\|(\mathcal{G} + \alpha\mathcal{I})^{-1}\|_{L^2} \\ &\leq \frac{\|(\mathcal{G} + \alpha\mathcal{I})^{-1}\|_{L^2}}{1 - \|(\mathcal{G} + \alpha\mathcal{I})^{-1}\|_{L^2}\|\mathcal{G}_n^M - \mathcal{G}\|_{L^2}} \leq \frac{2}{\alpha}, \end{aligned}$$

which proves the first inequality. In the similar manner, by using estimate (11) and Lemma 1, we obtain $\|(\mathcal{G}_n^M + \alpha\mathcal{I})^{-1}\mathcal{A}^*\|_{L^2} \leq \frac{1}{\sqrt{\alpha}}$. This completes the proof.

## 3 Convergence Rates

Let $x^{M}_{\alpha,n}$ be the solution of equation

$$(\mathcal{G}^{M}_{n} + \alpha\mathcal{I})x^{M}_{\alpha,n} = \mathcal{A}^{*}f. \tag{12}$$

**Theorem 3** *Then, for* $\widehat{x} \in \mathcal{R}((\mathcal{A}^{*}\mathcal{A})^{\nu})$ *and* $\alpha = d_1\,\delta^{\frac{2}{2\nu+1}}$, *some constant* $d_1 > 0$ *and* $0 < \nu \leq 1$, *the following outcome is valid:*

$$\|\widehat{x} - \widetilde{x}^{M}_{\alpha,n}\|_{L^2} = \mathcal{O}(\delta^{\frac{2\nu}{2\nu+1}}).$$

***Proof*** Consider

$$\widehat{x} - \widetilde{x}^{M}_{\alpha,n} = (\widehat{x} - x_{\alpha}) + (x_{\alpha} - x^{M}_{\alpha,n}) + (x^{M}_{\alpha,n} - \widetilde{x}^{M}_{\alpha,n}). \tag{13}$$

Using Eqs. (3) and (12), we obtain

$$\begin{aligned} x_{\alpha} - x^{M}_{\alpha,n} &= (\mathcal{G} + \alpha\mathcal{I})^{-1}\mathcal{A}^{*}f - (\mathcal{G}^{M}_{n} + \alpha\mathcal{I})^{-1}\mathcal{A}^{*}f \\ &= (\mathcal{G}^{M}_{n} + \alpha\mathcal{I})^{-1}(\mathcal{G}^{M}_{n} - \mathcal{G})(\mathcal{G} + \alpha\mathcal{I})^{-1}\mathcal{A}^{*}f \\ &= (\mathcal{G}^{M}_{n} + \alpha\mathcal{I})^{-1}(\mathcal{G}^{M}_{n} - \mathcal{G})(x_{\alpha} - \widehat{x}) + (\mathcal{G}^{M}_{n} + \alpha\mathcal{I})^{-1}(\mathcal{G}^{M}_{n} - \mathcal{G})\widehat{x}. \end{aligned} \tag{14}$$

Using $\mathcal{G} = \mathcal{A}^{*}\mathcal{A}$, we get

$$\begin{aligned} (\mathcal{G}^{M}_{n} + \alpha\mathcal{I})^{-1}(\mathcal{G}^{M}_{n} - \mathcal{G})\widehat{x} &= (\mathcal{G}^{M}_{n} + \alpha\mathcal{I})^{-1}(\mathcal{P}_{n} - \mathcal{I})\mathcal{G}(\mathcal{P}_{n} - \mathcal{I})\widehat{x} \\ &= (\mathcal{G}^{M}_{n} + \alpha\mathcal{I})^{-1}(\mathcal{P}_{n} - \mathcal{I})\mathcal{A}^{*}\mathcal{A}(\mathcal{P}_{n} - \mathcal{I})\widehat{x}. \end{aligned} \tag{15}$$

From estimate (11) for sufficiently large $n$ and Theorem 2, we obtain

$$\|(\mathcal{G}^{M}_{n} + \alpha\mathcal{I})^{-1}\|_{L^2}\|\mathcal{G}^{M}_{n} - \mathcal{G}\|_{L^2} < 1. \tag{16}$$

Combining estimates (14), (15) and (16), and Theorem 2, we obtain

$$\begin{aligned} &\|x_{\alpha} - x^{M}_{\alpha,n}\|_{L^2} \\ &\leq \|(\mathcal{G}^{M}_{n} + \alpha\mathcal{I})^{-1}\|_{L^2}\|\mathcal{G}^{M}_{n} - \mathcal{G}\|_{L^2}\|x_{\alpha} - \widehat{x}\|_{L^2} + \|(\mathcal{G}^{M}_{n} + \alpha\mathcal{I})^{-1}(\mathcal{G}^{M}_{n} - \mathcal{G})\widehat{x}\|_{L^2} \\ &< \|x_{\alpha} - \widehat{x}\|_{L^2} + \|(\mathcal{G}^{M}_{n} + \alpha\mathcal{I})^{-1}\|_{L^2}\|(\mathcal{I} - \mathcal{P}_{n})\mathcal{A}^{*}\|_{L^2}\|\mathcal{A}(\mathcal{I} - \mathcal{P}_{n})\widehat{x}\|_{L^2} \\ &\leq \|x_{\alpha} - \widehat{x}\|_{L^2} + \frac{2}{\alpha}\|(\mathcal{I} - \mathcal{P}_{n})\mathcal{A}^{*}\|_{L^2}\|\mathcal{A}(\mathcal{I} - \mathcal{P}_{n})\widehat{x}\|_{L^2}. \end{aligned} \tag{17}$$

Using estimate (12) and Eq. (6) with Theorem 2 and $\|f - \widetilde{f}\|_{L^2} \leq \delta$, we obtain

$$\|x^M_{\alpha,n} - \widetilde{x}^M_{\alpha,n}\|_{L^2} = \|(\mathcal{G}^M_n + \alpha\mathcal{I})^{-1}\mathcal{A}^*(f - \widetilde{f})\|_{L^2} \leq \frac{\delta}{\sqrt{\alpha}}. \tag{18}$$

Now, combining estimates (17) and (18) with (13), we obtain

$$\|\widehat{x} - \widetilde{x}^M_{\alpha,n}\|_{L^2} \leq 2\|x_\alpha - \widehat{x}\|_{L^2} + \frac{2}{\alpha}\|(\mathcal{I} - \mathcal{P}_n)\mathcal{A}^*\|_{L^2}\|\mathcal{A}(\mathcal{I} - \mathcal{P}_n)\widehat{x}\|_{L^2} + \frac{\delta}{\sqrt{\alpha}}. \tag{19}$$

Next, applying Lemma 2 and Cauchy–Schwarz inequality, we obtain

$$\begin{aligned}\|(\mathcal{P}_n - \mathcal{I})\mathcal{A}^*x\|_{L^2} &\leq c\,n^{-r} \sup_{u\in[-1,1]} \left|\int_{-1}^{1} \frac{\partial^r}{\partial u^r}\tau(v,u)x(v)dv\right| \\ &\leq \sqrt{2}c\,n^{-r}\|\tau^{(0,r)}\|_\infty\|x\|_{L^2}.\end{aligned} \tag{20}$$

Similarly, the use of orthogonality of $(\mathcal{P}_n - \mathcal{I})$, Cauchy–Schwarz inequality and Lemma 2 yield

$$\begin{aligned}\|\mathcal{A}(\mathcal{P}_n - \mathcal{I})\widehat{x}\|_{L^2} &= \sqrt{2} \sup_{u\in[-1,1]} | < \tau(u,.), (\mathcal{I} - \mathcal{P}_n)\widehat{x}(.) > | \\ &\leq \sqrt{2} \sup_{u\in[-1,1]} \|(\mathcal{I} - \mathcal{P}_n)\tau(u,.)\|_{L^2}\|(\mathcal{I} - \mathcal{P}_n)\widehat{x}\|_{L^2} \\ &\leq c\,n^{-2r} \sup_{u\in[-1,1]} \|\tau^{(r,0)}(u,.)\|_\infty\|\widehat{x}\|_{r,\infty} \leq B_2c\,n^{-2r}\|\widehat{x}\|_{r,\infty},\end{aligned} \tag{21}$$

where $B_2 = \sup_{u\in[-1,1]} \|\tau^{(r,0)}(u,.)\|_\infty$. Hence, from estimates (20) and (21), we get

$$\|(\mathcal{I} - \mathcal{P}_n)\mathcal{A}^*\|_{L^2} = \mathcal{O}(n^{-r}) \quad \text{and} \quad \|\mathcal{A}(\mathcal{I} - \mathcal{P}_n)\widehat{x}\|_{L^2} = \mathcal{O}(n^{-2r}),$$

respectively. Thus, by selecting $n$ sufficiently large such that $n^{-r} < \delta$, we get

$$\|(\mathcal{P}_n - \mathcal{I})\mathcal{A}^*\|_{L^2} \leq c\,\delta \quad \text{and} \quad \|\mathcal{A}(\mathcal{P}_n - \mathcal{I})\widehat{x}\|_{L^2} \leq c\,\delta^2. \tag{22}$$

Now, combining estimate (19) with (22), we obtain

$$\|\widehat{x} - \widetilde{x}^M_{\alpha,n}\|_{L^2} \leq 2\|x_\alpha - \widehat{x}\|_{L^2} + \frac{2c\,\delta^3}{\alpha} + \frac{\delta}{\sqrt{\alpha}}.$$

We know from Theorem 4.15 of [9] that $\|x_\alpha - \widehat{x}\|_{L^2} = \mathcal{O}(\alpha^\nu)$ for $\widehat{x} \in \mathcal{R}((\mathcal{A}^*\mathcal{A})^\nu)$. Then, we obtain

$$\|\widehat{x} - \widetilde{x}^M_{\alpha,n}\|_{L^2} \leq c\left(\alpha^\nu + \frac{\delta^3}{\alpha} + \frac{\delta}{\sqrt{\alpha}}\right). \tag{23}$$

If $\alpha = d_1\, \delta^{\frac{2}{2\nu+1}}$ for some $d_1 > 0$ and $0 < \nu \leq 1$ in estimate (23), we obtain

$$\|\widehat{x} - \widetilde{x}^M_{\alpha,n}\|_{L^2} \leq c\, d_1\, \delta^{\frac{2\nu}{2\nu+1}} + \frac{c}{d_1}\, \delta^{3-\frac{2}{2\nu+1}} + \frac{c}{\sqrt{d_1}}\, \delta^{1-\frac{1}{2\nu+1}} \leq c\, \delta^{\frac{2\nu}{2\nu+1}}.$$

This concludes the theorem's proof.

**Remark 1** Under a priori parameter strategy, we obtain the optimal accuracy $\mathcal{O}\Big(\delta^{\frac{2\nu}{2\nu+1}}\Big)$ by picking $\alpha = d_1\, \delta^{\frac{2}{2\nu+1}}$ for some $d_1 > 0$ and $\nu \in (0, 1]$ in Theorem 3.

## 4 Arcangeli's Discrepancy Principle

To obtain the parameter $\alpha$ for the Tikhonov regularized Eq. (6), Arcangeli's discrepancy principle will be discussed and we will also evaluate the optimal accuracy in $L^2$ norm.

We choose $\alpha = \alpha(\delta, n)$ [3] which satisfy the equation

$$\frac{\delta^p}{\alpha^q} = \|\mathcal{G}^M_n \widetilde{x}^M_{\alpha,n} - \mathcal{A}^* \widetilde{f}\|_{L^2}, \quad p, q \in \mathbb{Q}^+, \tag{24}$$

for large $n$. Let $\{\gamma_n | n \in \mathbb{N}\}$ be a sequence of numbers such that $\gamma_n \to 0$ as $n \to \infty$, and satisfy

$$\|\mathcal{G}^M_n - \mathcal{G}\|_{L^2} \leq \|(\mathcal{I} - \mathcal{P}_n)\mathcal{G}(\mathcal{I} - \mathcal{P}_n)\|_{L^2} \leq d_0\, \gamma_n, \quad 0 < d_0 < 1. \tag{25}$$

**Theorem 4** *Then there exists a constant $\delta_0 > 0$ such that, for all $\alpha = \alpha(\delta, n)$, $n \geq N(\delta) \in \mathbb{N}$ and for each $\delta \in (0, \delta_0]$, the solution to Eq. (24) is unique.*

***Proof*** For $n \in \mathbb{N}$ and $\delta > 0$, we define $g$ on $\Big[\frac{\gamma_n}{d_0}, \infty\Big)$ as

$$g(\alpha) = \alpha^q \|\mathcal{G}^M_n \widetilde{x}^M_{\alpha,n} - \mathcal{A}^* \widetilde{f}\|_{L^2}. \tag{26}$$

So, it is enough to show that if $n \geq N$ and $N \in \mathbb{N}$ and $\delta \in (0, \delta_0]$ for some $\delta_0 > 0$, $\exists$ a positive number $\alpha = \alpha(\delta, n)$ such that $g(\alpha) = \delta^p$. Also, $d_4 < \|\mathcal{A}^* \widetilde{f}\|_{L^2} \leq d_5$, where $d_4$ and $d_5$ are constants. Now, using Eq. (6) and Theorem 2, we get

$$\begin{aligned} \|\mathcal{G}^M_n \widetilde{x}^M_{\alpha,n} - \mathcal{A}^* \widetilde{f}\|_{L^2} &= \| - \alpha \widetilde{x}^M_{\alpha,n}\|_{L^2} \\ &= \| - \alpha(\mathcal{G}^M_n + \alpha\mathcal{I})^{-1} \mathcal{A}^* \widetilde{f}\|_{L^2} \\ &\leq \alpha \|(\mathcal{G}^M_n + \alpha\mathcal{I})^{-1}\|_{L^2} \|\mathcal{A}^* \widetilde{f}\|_{L^2} \leq \alpha \frac{2}{\alpha} d_5 = d'. \end{aligned} \tag{27}$$

Next, consider $\|\mathcal{A}^*\widetilde{f}\|_{L^2} \leq \|(\mathcal{G}_n^M + \alpha\mathcal{I})(\mathcal{G}_n^M + \alpha\mathcal{I})^{-1}\mathcal{A}^*\widetilde{f}\|_{L^2}$ which gives

$$\|(\mathcal{G}_n^M + \alpha\mathcal{I})^{-1}\mathcal{A}^*\widetilde{f}\|_{L^2} \geq \frac{\|\mathcal{A}^*\widetilde{f}\|_{L^2}}{(\alpha + \|\mathcal{G}_n^M\|_{L^2})}. \tag{28}$$

Using estimate (28), we get

$$\begin{aligned}\|\mathcal{G}_n^M\widetilde{x}_{\alpha,n}^M - \mathcal{A}^*\widetilde{f}\|_{L^2} = \|\alpha\widetilde{x}_{\alpha,n}^M\|_{L^2} &= \|\alpha(\mathcal{G}_n^M + \alpha\mathcal{I})^{-1}\mathcal{A}^*\widetilde{f}\|_{L^2} \\ &\geq \alpha\frac{\|\mathcal{A}^*\widetilde{f}\|_{L^2}}{(\alpha + \|\mathcal{G}_n^M\|_{L^2})} \geq \frac{\alpha d_4}{\alpha + d_3}.\end{aligned} \tag{29}$$

Let $N_2 \geq N_1$ and $N_2$ be sufficiently large such that when $n \geq N_2$,

$$\frac{\gamma_n}{d_0} \leq \left(\frac{\delta^p}{d'}\right)^{\frac{1}{q}}. \tag{30}$$

Now, using estimates (27) and (30), we get

$$g\left(\frac{\gamma_n}{d_0}\right) = \left(\frac{\gamma_n}{d_0}\right)^q \|\mathcal{G}_n^M\widetilde{x}_{\alpha,n}^M - \mathcal{A}^*\widetilde{f}\|_{L^2} \leq \frac{\delta^p}{d'}d' = \delta^p. \tag{31}$$

Now denote $\widehat{\gamma} = \sup\{\gamma_n | n \in \mathbb{N}\}$, $d'' = \frac{\widehat{\gamma}d_4}{\widehat{\gamma}+d_0d_3}$ and $\alpha_0 = \max\left\{\frac{\widehat{\gamma}}{d_0}, \left(\frac{\delta^p}{d''}\right)^{\frac{1}{q}}\right\}$. Then $\alpha_0 \geq \frac{\widehat{\gamma}}{d_0}$ and $\alpha_0^q \geq \frac{\delta^p}{d''}$.

Now, using estimate (29), we obtain

$$\begin{aligned}g(\alpha_0) = \alpha_0^q\|\mathcal{G}_n^M\widetilde{x}_{\alpha,n}^M - \mathcal{A}^*\widetilde{f}\|_{L^2} &\geq \alpha_0^q\frac{\alpha_0 d_4}{\alpha_0 + d_3} \\ &\geq \frac{\delta^p}{d''}\left(\frac{\frac{\widehat{\gamma}}{d_0}d_4}{\frac{\widehat{\gamma}}{d_0} + d_3}\right) = \frac{\delta^p}{d''}\left(\frac{\widehat{\gamma}d_4}{\widehat{\gamma} + d_0d_3}\right) = \delta^p.\end{aligned} \tag{32}$$

Then, from estimates (31) and (32), we have $g\left(\frac{\gamma_n}{d_0}\right) \leq \delta^p \leq g(\alpha_0)$, on $\left[\frac{\gamma_n}{d_0}, \alpha_0\right]$. Since $g$ is continuous on $\left[\frac{\gamma_n}{d_0}, \alpha_0\right]$, by the use of Intermediate Value Theorem (IVT), $\exists\, \alpha \in \left(\frac{\gamma_n}{d_0}, \alpha_0\right)$ such that $g(\alpha) = \delta^p$. Thus, we obtain the desired result.

**Theorem 5** *Let the parameter $\alpha$ be selected using the discrepancy principle (24). Then there is a constant $d_1 > 0$ and $N'' \in \mathbb{N}$ such that $\alpha \leq d_1\delta^{\frac{p}{1+q}}$ and $\frac{\delta^p}{\alpha^q} \leq c\,\delta^\eta$, where $\eta = \min\left\{\frac{p}{1+q}, 1 + \frac{p}{2(1+q)}\right\}$.*

***Proof*** Multiplying $\alpha^q$ on both sides of Eq. (29) and then using Eq. (24), we obtain

$$\alpha^{q+1}\frac{d_4}{\alpha+d_3} \leq \alpha^q \|\mathcal{G}_n^M \widetilde{x}_{\alpha,n}^M - \mathcal{A}^*\widetilde{f}\|_{L^2} = \delta^p.$$

Since $\delta \in (0, \delta_0]$, we get $\alpha^{q+1}\frac{d_4}{\alpha\left(1+\frac{d_3}{\alpha}\right)} \leq \delta^p \leq \delta_0^p$. This implies

$$\alpha^q \leq \frac{1}{d_4}\left(1+\frac{d_3}{\alpha}\right)\delta_0^p \leq \frac{2}{d_4}\delta_0^p \quad \text{for } \alpha > d_3.$$

As a result, $\alpha$ is bounded by a constant that is independent of $\delta$ and $n$. Using Eqs. (6) and (24), we now obtain

$$\begin{aligned}
\|\mathcal{A}^*\widetilde{f}\|_{L^2} - \frac{\delta^p}{\alpha^q} &= \|\mathcal{A}^*\widetilde{f}\|_{L^2} - \|\mathcal{G}_n^M \widetilde{x}_{\alpha,n}^M - \mathcal{A}^*\widetilde{f}\|_{L^2} \\
&\leq \|\mathcal{G}_n^M\|_{L^2}\|\widetilde{x}_{\alpha,n}^M\|_{L^2} \\
&= \|\mathcal{G}_n^M\|_{L^2}\frac{\|\mathcal{G}_n^M \widetilde{x}_{\alpha,n}^M - \mathcal{A}^*\widetilde{f}\|_{L^2}}{\alpha} = \|\mathcal{G}_n^M\|_{L^2}\frac{1}{\alpha}\frac{\delta^p}{\alpha^q} = \|\mathcal{G}_n^M\|_{L^2}\frac{\delta^p}{\alpha^{q+1}}.
\end{aligned}$$

Hence, we get $\|\mathcal{A}^*\widetilde{f}\|_{L^2} \leq \frac{\delta^p}{\alpha^q} + \|\mathcal{G}_n^M\|_{L^2}\frac{\delta^p}{\alpha^{q+1}} = \frac{\delta^p}{\alpha^{q+1}}[\alpha + \|\mathcal{G}_n^M\|_{L^2}]$. This implies

$$\alpha^{q+1} \leq \frac{\delta^p}{\|\mathcal{A}^*\widetilde{f}\|_{L^2}}[\alpha + \|\mathcal{G}_n^M\|_{L^2}] \leq \frac{\alpha+d_3}{d_4}\delta^p.$$

Hence, there is a constant $d_1 > 0$ such that

$$\alpha \leq d_1\,\delta^{\frac{p}{1+q}}, \text{ where } d_1 = \left(\frac{\alpha+d_3}{d_4}\right)^{\frac{1}{1+q}}. \tag{33}$$

Next, using Eq. (6) and estimate (19), and since $\|\widehat{x}\|_{L^2}$ and $\|\hat{x} - x_\alpha\|_{L^2}$ are bounded, we obtain

$$\begin{aligned}
\frac{\delta^p}{\alpha^q} &= \|\mathcal{G}_n^M \widetilde{x}_{\alpha,n}^M - \mathcal{A}^*\widetilde{f}\|_{L^2} \leq \alpha[\|\widehat{x} - \widetilde{x}_{\alpha,n}^M\|_{L^2} + \|\widehat{x}\|_{L^2}] \\
&\leq \alpha\left[2\|x_\alpha - \widehat{x}\|_{L^2} + \frac{2}{\alpha}\|(\mathcal{I}-\mathcal{P}_n)\mathcal{A}^*\|_{L^2}\|\mathcal{A}(\mathcal{I}-\mathcal{P}_n)\widehat{x}\|_{L^2} + \frac{\delta}{\sqrt{\alpha}} + \|\widehat{x}\|_{L^2}\right] \\
&\leq c\,(\alpha + \|(\mathcal{I}-\mathcal{P}_n)\mathcal{A}^*\|_{L^2}\|\mathcal{A}(\mathcal{I}-\mathcal{P}_n)\widehat{x}\|_{L^2} + \delta\sqrt{\alpha}).
\end{aligned} \tag{34}$$

Using estimates (20) and (21), we see that $\|(\mathcal{I}-\mathcal{P}_n)\mathcal{A}^*\|_{L^2}\|\mathcal{A}(\mathcal{I}-\mathcal{P}_n)\widehat{x}\|_{L^2} = \mathcal{O}(n^{-3r}) \to 0$ as $n \to \infty$. Then, we select $N''$ sufficiently large such that when $n \geq N''$,

$$\|(\mathcal{I}-\mathcal{P}_n)\mathcal{A}^*\|_{L^2}\|\mathcal{A}(\mathcal{I}-\mathcal{P}_n)\widehat{x}\|_{L^2} \le c\,\delta^{\eta}, \text{ where } \eta = \left\{\frac{p}{1+q}, 1+\frac{p}{2(1+q)}\right\}. \tag{35}$$

Now, substituting estimates (35) and (33) in estimate (34), we obtain

$$\frac{\delta^p}{\alpha^q} \le c\,(\alpha+\delta^{\eta}+\delta\sqrt{\alpha}) \le c\left(\delta^{\frac{p}{1+q}}+\delta^{\eta}+\delta^{1+\frac{p}{2(1+q)}}\right) \le c\,\delta^{\eta},$$

where $\eta = \min\left\{1+\frac{p}{2(1+q)}, \frac{p}{1+q}\right\}$. Thus, we get the desired outcome.

**Theorem 6** *Assume $\alpha$ is chosen in accordance with the discrepancy principle (24). If $\widehat{x} \in \mathcal{R}((\mathcal{A}^*\mathcal{A})^{\nu})$, for some $0<\nu\le 1$, $p$ and $q$ satisfy $\frac{p}{2(1+q)}<1$, then there exists $\delta_0<1$ such that $\delta \in (0,\delta_0]$ and a constant $c>0$ and $N_3 \in \mathbb{N}$ such that*

$$\|\widehat{x}-\widetilde{x}^M_{\alpha,n}\|_{L^2} \le c\left(\delta^{\frac{p\nu}{1+q}}+\delta^{\mu-\frac{p}{1+q}}+\delta^{1-\frac{p}{2(1+q)}}\right),$$

*where $\mu = \min\left\{\frac{p(\nu+1)}{1+q}, 1+\frac{p}{2(1+q)}\right\}$. In particular, if $\frac{p}{1+q}=\frac{2}{2\nu+1}$, then*

$$\|\widehat{x}-\widetilde{x}^M_{\alpha,n}\|_{L^2} = \mathcal{O}(\delta^{\frac{2\nu}{2\nu+1}}).$$

***Proof*** We know from Theorem 4.15 of [9] that $\|x_\alpha-\widehat{x}\|_{L^2}=\mathcal{O}(\alpha^{\nu})$ for $\widehat{x}\in\mathcal{R}((\mathcal{A}^*\mathcal{A})^{\nu})$. Substituting $\|(\mathcal{P}_n-\mathcal{I})\mathcal{A}^*\|_{L^2}\|\mathcal{A}(\mathcal{P}_n-\mathcal{I})\widehat{x}\|_{L^2}=\mathcal{O}(n^{-3r})=\epsilon_n$ in estimate (19), we obtain

$$\|\widehat{x}-\widetilde{x}^M_{\alpha,n}\|_{L^2} \le 2\|x_\alpha-\widehat{x}\|_{L^2}+\frac{2\epsilon_n}{\alpha}+\frac{\delta}{\sqrt{\alpha}} \le c\left(\alpha^{\nu}+\frac{\epsilon_n}{\alpha}+\frac{\delta}{\sqrt{\alpha}}\right). \tag{36}$$

Since $\epsilon_n \to 0$ when $n\to\infty$, we pick $N_3$ to be sufficiently large such that when $n \ge N_3$,

$$\epsilon_n \le c\,\delta^{\mu}, \quad \text{where } \mu = \left\{\frac{p(\nu+1)}{1+q}, 1+\frac{p}{2(1+q)}\right\}. \tag{37}$$

Now, by combining estimates (33) and (37) with estimate (36), we get

$$\|\widehat{x}-\widetilde{x}^M_{\alpha,n}\|_{L^2} \le c\left(\delta^{\frac{p\nu}{1+q}}+\delta^{\mu-\frac{p}{1+q}}+\delta^{1-\frac{p}{2(1+q)}}\right). \tag{38}$$

If $\mu=\frac{p(\nu+1)}{1+q}$ or $\mu = 1+\frac{p}{2(1+q)}$, then from estimate (38), we obtain

$$\|\widehat{x}-\widetilde{x}^M_{\alpha,n}\|_{L^2} \le c\left(\delta^{\frac{p\nu}{1+q}}+\delta^{1-\frac{p}{2(1+q)}}\right). \tag{39}$$

If $\frac{p}{1+q} = \frac{2}{2\nu+1}$ in estimate (39), we obtain $\|\widehat{x} - \widetilde{x}^M_{\alpha,n}\|_{L^2} \leq c\left(\delta^{\frac{2\nu}{2\nu+1}} + \delta^{1-\frac{1}{2\nu+1}}\right) \leq c\,\delta^{\frac{2\nu}{2\nu+1}}$. Thus, we obtain the desired result.

## 5 Numerical Examples

An example has been included in this section that illustrates the theoretical study of Eq. (6) in both a priori and a posteriori parameter strategies under $L^2$-norm.

**Example 1** ([8]) Consider the FIE of the first kind:

$$\mathcal{A}x(u) = \int_0^1 \frac{(v+u)^2}{\sqrt{1+v^2}} x(v)dv = 0.266419u^2 + 0.390524u + 0.153738, \quad 0 \leq u \leq 1,$$

where the exact solution is given by $\widehat{x}(v) = v^2$.

Here, $\mathcal{A}$ is self-adjoint and $\widehat{x} = (\mathcal{A}^*\mathcal{A})^{1/2}v$. Thus, $\widehat{x} \in \mathcal{R}((\mathcal{A}^*\mathcal{A})^{1/2})$, i.e., $\nu = 1/2$.

**A priori parameter strategy**: Here, we choose $\alpha = d_1\delta^{\frac{2}{2\nu+1}} = d_1\delta$ for $\nu = \frac{1}{2}$ and $d_1 > 0$. For given $\delta$, we choose $\alpha = d_1\delta$, where $d_1 = 0.8$ and $0.55$ in Table 1.

Table 1 demonstrates that the estimated convergence rate is $\mathcal{O}(\delta^{\frac{1}{2}})$ which agrees to the theoretical conclusion in Theorem 3.

**A posteriori parameter choice strategy**: In Tables 2 and 3 we show the errors between $\widetilde{x}^M_{\alpha,n}$ and $\widehat{x}$ in $L^2$-norm under Arcangeli's discrepancy principle for various choices of $p$ and $q$, initial choice $\alpha_0 = 1$ and the tolerance $\epsilon$ =1.0e-04.

**Table 1** Numerical results for $\alpha = 0.8 * \delta$ and $\alpha = 0.55 * \delta$

| | $\delta = 0.12252$ and $\alpha = 0.8 * \delta = 0.098016$ | | $\delta = 0.117$ and $\alpha = 0.55 * \delta = 0.06435$ | |
|---|---|---|---|---|
| $n$ | $\|\widehat{x} - \widetilde{x}^M_{\alpha,n}\|_{L^2}$ | $\|\widehat{x} - \widetilde{x}^M_{\alpha,n}\|_{L^2}/\delta^{1/2}$ | $\|\widehat{x} - \widetilde{x}^M_{\alpha,n}\|_{L^2}$ | $\|\widehat{x} - \widetilde{x}^M_{\alpha,n}\|_{L^2}/\delta^{1/2}$ |
| 5 | 2.726069e-01 | 0.77881331 | 2.529041e-01 | 0.73937199 |
| 6 | 2.240525e-01 | 0.64009801 | 1.759486e-01 | 0.51439046 |
| 7 | 7.183671e-02 | 0.20523100 | 5.905306e-02 | 0.17264320 |

**Table 2** For $\delta$ = 7.101e-01, $p = 1, q = 1$, and $k = 5$ (convergence rate is $\delta^{1/4}$)

| $n$ | $\alpha$ | $\|\widehat{x} - \widetilde{x}^M_{\alpha,n}\|_{L^2}$ | $\|\widehat{x} - \widetilde{x}^M_{\alpha,n}\|_{L^2}/\delta^{1/4}$ | $\alpha/\delta^{\frac{p}{1+q}}$ |
|---|---|---|---|---|
| 3 | 7.677087e-01 | 6.666785e-01 | 0.726250 | 0.911038 |
| 4 | 7.718581e-01 | 6.534791e-01 | 0.711871 | 0.915962 |
| 5 | 8.168499e-01 | 5.225942e-01 | 0.569291 | 0.969354 |

**Table 3** For $\delta = 6.521e - 01$, $p = 2$, $q = 1$ and $k = 3$ (convergence rate is $\delta^{0.5}$)

| $n$ | $\alpha$ | $\|\widehat{x} - \widetilde{x}^M_{\alpha,n}\|_{L^2}$ | $\|\widehat{x} - \widetilde{x}^M_{\alpha,n}\|_{L^2}/\delta^{1/4}$ | $\alpha/\delta^{\frac{p}{1+q}}$ |
|---|---|---|---|---|
| 3 | 5.993167e-01 | 6.574515e-01 | 0.814154 | 0.919056 |
| 4 | 6.035490e-01 | 6.394046e-01 | 0.791805 | 0.925546 |
| 5 | 6.508798e-01 | 4.624276e-01 | 0.572646 | 0.998128 |

The numerical findings in Tables 2 and 3 demonstrate that, for $p = 2$ and $q = 1$, we get the optimal convergence rate $\delta^{\frac{1}{2}}$, which accords with our theoretical estimations in Theorem 6. The outcomes further demonstrate that $\alpha \leq d_1\, \delta^{\frac{p}{1+q}}$ for some $d_1 > 0$, which coincide with the conclusion of Theorem 5.

**Acknowledgements** The first author was supported by the INSPIRE Fellowship, Department of Science and Technology, Government of India, New Delhi.

## References

1. Adomian, G.: Solving Frontier Problems of Physics: The Decomposition Method. Kluwer, Boston (1994)
2. Canuto, C., Hussaini, M.Y., Quarteroni, A., Zang, T.A.: Spectral Methods: Fundamentals in Single Domains. Springer, Berlin (2006)
3. Chen, Z., Cheng, S., Nelakanti, G., Yang, H.: A fast multiscale Galerkin method for the first kind ill-posed integral equations via Tikhonov regularization. Int. J. Comput. Math. **87**(3), 565–582 (2010). https://doi.org/10.1080/00207160802155302
4. Chen, Z., Xu, Y., Yang, H.: Fast collocation methods for solving ill-posed integral equations of the first kind. Inverse Probl. **24**(6), 1–21 (2008). https://doi.org/10.1088/0266-5611/24/6/065007
5. Engl, H.W.: Discrepancy principles for Tikhonov regularization of ill-posed problems leading to optimal convergence rates. J. Optim. Theory Appl. **52**(2), 209–215 (1987). https://doi.org/10.1007/BF00941281
6. Groetsch, C.W.: Convergence analysis of a regularized degenerate kernel method for Fredholm integral equation of the first kind. Integral Equ. Oper. Theory **13**(1), 67–75 (1990). https://doi.org/10.1007/BF01195293
7. Kirsch, A.: Introduction to the Mathematical Theory of Inverse Problems. Springer, New York (1996)
8. Maleknejad, K., Mollapourasl, R., Nouri, K., Alizadeh, M.: Convergence of numerical solution of Fredholm integral equation of the first kind with degenerate kernel. Appl. Math. Comput. **181**(2), 1000–1007 (2006). https://doi.org/10.1016/j.amc.2006.01.074
9. Nair, M.T.: Linear Operator Equations: Approximation and Regularization. World Scientific, Singapore (2009)
10. Neggal, B., Boussetila, N., Rebbani, F.: Projected Tikhonov Regularization method for Fredholm integral equations of the first kind. J. Inequal. Appl. **195**, 1–21 (2016). https://doi.org/10.1186/s13660-016-1137-6
11. Patel, S., Panigrahi, B.L., Nelakanti, G.: Legendre spectral projection methods for Fredholm integral equations of first kind. J. Inverse Ill-Posed Probl. **30**(5), 677–691 (2022). https://doi.org/10.1515/jiip-2020-0104

12. Patel, S., Panigrahi, B.L., Nelakanti, G.: Legendre spectral multi-projection methods for Fredholm integral equations of the first kind. Adv. Oper. Theory **7**(51) (2022). https://doi.org/10.1007/s43036-022-00215-z
13. Rajan, M.P.: A modified convergence analysis for solving Fredholm integral equations of the first kind. Integral Equ. Oper. Theory **49**, 511–516 (2004). https://doi.org/10.1007/s00020-002-1213-9
14. Rostami, Y., Maleknejad, K.: Solving Fredholm integral equations of the first kind by using wavelet bases. Hacet. J. Math. Stat. **48**, 1–15 (2019). https://doi.org/10.15672/hujms.553433
15. Tahar, B., Nadjib, B., Faouzia, R.: A variant of projection-regularization method for ill-posed linear operator equations. Int. J. Comput. Methods **18**(4) (2021). https://doi.org/10.1142/S0219876221500080

# Local Convergence of a Family of Kurchatov Like Methods for Nonlinear Equations

**Abhimanyu Kumar and Soni Kumari**

**Abstract** The main purpose of this research paper is to establish the local convergence analysis of k-step Kurchatov methods for solving nonlinear equations. We have provided the sufficiently adequate convergence conditions for this purpose which gives us the the better convergence results. To elaborate the study done by us, we have also worked on a number of numerical examples.

**Keywords** Local convergence · Iterative methods · Kurchatov method · Nonlinear equations

## 1 Introduction

Consider the problem to approximate the nonlinear operator equation

$$H(r) = 0, \tag{1}$$

where $H : \Omega_0 \subseteq \text{A} \to \text{B}$. Here $\Omega_0$ specify the open convex domain of Banach space A; B is also a Banach space. It is not possible to find the solution of (1) in closed form always and therefore iterative methods have been generally used to approximate the solution. Many researchers [10, 23] have thus motivated towards this direction and extensively written many research articles and monographs for this purpose. They have also developed many iterative methods for this purpose. It is known that for different types of problems, different iterative methods have been being used. Convergence analysis of iterative methods are also important in order to ensure the applicability of these methods. Several types of convergence analysis have also been performed for this purpose. Semilocal and local convergence analysis are some of them. In semilocal convergence analysis [1, 5, 15, 16], we develop the domain from where the starting points can be chosen which ensures the convergence of the method. In local convergence analysis [2, 6, 8, 19, 21, 22, 24, 27], we find the radii

A. Kumar · S. Kumari (✉)
Department of Mathematics, Lalit Narayan Mithila University, Darbhanga 846004, BR, India
e-mail: sjha1414@gmail.com



B. Rushi Kumar et al. (eds.), *Mathematics and Computing*, Springer Proceedings in Mathematics & Statistics 415, https://doi.org/10.1007/978-981-19-9307-7_51

of convergence balls centered at the solution. Generally the radii of convergence balls are found small and one always try to enlarge it.

However, one of the most popular quadratic convergence method is known as Newton's method which is used for solving (1) and is given by

$$r_{n+1} = r_n - H'(r_n)^{-1}H(r_n), \tag{2}$$

Here $r_0 \in \Omega_0$ is the starting point.

**Remark 1** Here and throughout this paper, $H'(r_n)^{-1}$ denotes the inverse of the whole operator $(H'(r_n))^{-1}$ and $n$ denotes the integer starting from 0.

Many authors [4, 11, 12, 25] have developed the convergence analysis by weakening the convergence criteria of (2). However, it has also some limitations as it uses the Fréchet derivative at each iteration. Some researchers have used its alternate Secant method [9, 28] to avoid the Fréchet derivative at each iteration. This method is given by

$$r_{n+1} = r_n - [r_{n-1}, r_n; H]^{-1}H(r_n), \tag{3}$$

Here $r_{-1}, r_0 \in \Omega_0$ is the starting point. It converges superlinear and it's order of convergence is $\frac{1+\sqrt{5}}{2}$. Here $[:, :, H]$ is the divided difference and is defined by $[a, b; H](a-b) = H(a) - H(b)$. Another method which uses the divided differences is known as Kurchatov's method [26], given by

$$r_{n+1} = r_n - [2r_n - r_{n-1}, r_{n-1}; H]^{-1}H(r_n). \tag{4}$$

Here $r_{-1}, r_0 \in \Omega_0$ is the starting point. Many researchers have also studied (4) and established the local and semilocal convergence analysis using different convergence conditions. Many authors [2, 3, 7, 13, 17] have also constructed the multipoint version of (2) and established the convergence analysis. However the convergence analysis of some fixed version of (4) can be found in [18, 20, 26].

In this paper, we have established the multipoint version of (4) which generalizes the above method. The method is given by,

$$\begin{aligned} r_n^1 &= r_n^0 - [2r_n^0 - r_{n-1}^0, r_{n-1}^0; H]^{-1}H(r_n^0) \\ r_n^2 &= r_n^1 - [2r_n^0 - r_{n-1}^0, r_{n-1}^0; H]^{-1}H(r_n^1) \\ &\vdots \\ r_n^k &= r_n^{k-1} - [r_n^0 - r_{n-1}^0, r_{n-1}^0; H]^{-1}H(r_n^{k-1}) \end{aligned} \tag{5}$$

where $r_{n+1}^0 = r_n^k$. We have also developed the local convergence analysis of (5) in the later on section of this paper.

Finally the paper is constructed as follows: Introduction forms Sect. 1. In Sect. 2, some preliminaries and auxiliary results are presented. We have used some functions and estimated for the radii of convergence balls of (5) and then established the convergence theorem for this. Some numerical examples are also given in Sect. 3. Finally, conclusions and future scopes are included in Sect. 4.

## 2 Local Convergence Analysis

In this section, we present the local convergence of family of Kurchatov methods (5) for solving (1). For this purpose, we have constructed some auxiliary nonnegative parameters and constant that will be appear at the proof of main theorems. We start with the introduction of some functions and parameters. Let $\mathcal{L} > 0$, $\mathcal{L}_* > 0$ and $\alpha > 0$ be some nonzero and positive real numbers. We define a function $f_1(t, s)$ on the interval $\left(0, \frac{1}{4\mathcal{L}_*}\right)$, given by

$$f_1(t, s) = \frac{\mathcal{L}(t + 2s)}{1 - 2\mathcal{L}_*(t + s)}, \tag{6}$$

the function $g_1(t, s)$ is now defined by

$$g_1(t, s) = f_1(t, s) - 1.$$

For the instant, we have assumed that $t = s$ as for the convenience of the proof of our main theorem later. So that

$$\begin{aligned} g_1(t) &= f_1(t) - 1 \\ &= \frac{3\mathcal{L}t}{1 - 4\mathcal{L}_*t} - 1. \end{aligned}$$

We get that $g_1(0) = -1 < 0$ and $g_1(\frac{1}{4\mathcal{L}_*})^- \to \infty$. This shows that using intermediate value theorem that $g_1(t)$ has atleast one zero in $\left(0, \frac{1}{4\mathcal{L}_*}\right)$, we identify it as $x_1$ and we conclude that $0 < g_1 < 1$ in $(0, x_1)$. However, it may easily deduce that $x_1 = \frac{1}{3\mathcal{L}+4\mathcal{L}_*}$.

Now, we define another function $f_2(t, s)$ on the interval $(0, x_1)$, given by

$$f_2(t, s) = \frac{\mathcal{L}((2 + f_1(t, s))t + 2s)}{1 - 2\mathcal{L}_*(t + s)}, \tag{7}$$

the function $g_2(t, s)$ is now defined by

$$g_2(t, s) = f_2(t, s) - 1.$$

For the instant, we have assumed that $t = s$ as for the convenience of the proof of our main theorem later. So that

$$\begin{aligned} g_2(t) &= f_2(t) - 1 \\ &= \frac{\mathcal{L}((2 + f_1(t))t + 2t)}{1 - 4\mathcal{L}_* t} - 1. \end{aligned}$$

We get that $g_2(0) = -1$ and $g_2(\frac{1}{4\mathcal{L}_*})^- \to \infty$. This shows using intermediate value theorem that $g_2(t)$ has atleast one zero in $\left(0, \frac{1}{4\mathcal{L}_*}\right)$, we identify it as $x_2$. Now, using Mathematical induction on $'i'$, we define functions $f_i$ and $g_i$ on the interval $(0, x_{i-1})$, by

$$f_i(t, s) = \frac{\mathcal{L}((2 + f_1(t, s) f_2(t, s) ... f_{i-1}(t, s))t + 2s)}{1 - 2\mathcal{L}_*(t + s)}, \quad \text{and} \tag{8}$$

the function $g_i(t, s)$ is now defined by

$$g_i(t, s) = f_i(t, s) - 1.$$

For the instant, we have assumed that $t = s$ as for the convenience of the proof of our main theorem later. So that

$$\begin{aligned} g_i(t) &= f_i(t) - 1 \\ &= \frac{\mathcal{L}((2 + f_1(t) f_2(t) ... f_{i-1}(t))t + 2t)}{1 - 4\mathcal{L}_*(t)} - 1. \end{aligned}$$

We get that $g_i(0) = -1$ and $g_i(\frac{1}{4\mathcal{L}_*})^- \to \infty$. This shows using intermediate value theorem that $g_i(t)$ has atleast one zero in $\left(0, \frac{1}{4\mathcal{L}_*}\right)$, we identify it as $x_i$. Now, we take

$$x^* = \min\{x_i\} \text{ for } i = 1, 2, 3...$$

In this way, we assert that $0 < f_i(t) < 1$ on $(0, x^*)$.

Remaining of our work here, we denote the open and closed balls centered at $x$ and radius $y$ by $\mathcal{U}(x, y)$ and $\overline{\mathcal{U}}(x, y)$, respectively throughout.

We are now in position to present our main theorem of local convergence analysis here.

**Theorem 1** *Let $H : \Omega \subset A \to B$ be a Fréchet differentiable operator. Suppose there are some parameters $\mathcal{L} > 0$, $\mathcal{L}_* > 0$, $r^*$ such that $H(r^*) = 0$ and it satisfy the following conditions:*

$$
\begin{aligned}
&H'(r^*)^{-1} \in L(B, A),\\
&\|H'(r^*)^{-1}\left([a, b; H] - H'(r^*)\right)\| \leq \mathcal{L}_*(\|a - r^*\| + \|b - r^*\|), \qquad (9)\\
&\|H'(r^*)^{-1}([a, b; H] - [c, d; H])\| \leq \mathcal{L}(\|a - c\| + \|b - d\|). \qquad (10)\\
&\forall a, b, c, d \in \Omega_0 = \Omega \bigcap \mathcal{U}\left(r^*, \frac{1}{4\mathcal{L}_*}\right).
\end{aligned}
$$

*Then the iteration (5) executed by* $r_0^0,\ r_{-1}^0 \in \mathcal{U}(r^*, x^*) - r^*$ *is well defined, exist at* $\mathcal{U}(r^*, x^*)$ *and converges to* $r^*$. *Moreover, the following estimates hold true.*

$$
\|r_n^1 - r^*\| \leq f_1(\|r_n^0 - r^*\|, \|r_{n-1}^0 - r^*\|)\|r_n^0 - r^*\| \qquad (11)
$$

$$
\|r_n^i - r^*\| \leq f_i(\|r_n^0 - r^*\|, \|r_{n-1}^0 - r^*\|)\|r_n^{i-1} - r^*\| \ \text{ for } i \geq 2. \qquad (12)
$$

*Where the functions* $f_1,\ f_2,\ \ldots$ *are defined in (6), (7) and (8). Furthermore* $r^*$ *is the unique solution of (1) on* $\mathcal{U}(r^*, x)$ *where* $x < \frac{1}{\mathcal{L}_*}$ *and the uniqueness region can be established in* $\overline{\mathcal{U}}(r^*, x) \bigcap \Omega_0$.

***Proof*** We shall use the Mathematical induction on $'n'$ and $'i'$ to prove the above theorem. For this, we first abbreviate $[2r_n^0 - r_{n-1}^0, r_{n-1}^0; H]$ as $C_n$. Now, using (9) and triangle inequalities, we get

$$
\begin{aligned}
\|I - H'(r^*)^{-1}C_0\| &= \|H'(r^*)^{-1}(C_0 - H'(r^*))\|, \ \text{('I' denotes here the identity operator on A)}\\
&= \|H'(r^*)^{-1}([2r_0^0 - r_{-1}^0, r_{-1}^0; H] - H'(r^*))\|\\
&\leq \mathcal{L}_*(\|2r_0^0 - r_{-1}^0 - r^*\| + \|r_{-1}^0 - r^*\|)\\
&= \mathcal{L}_*(\|2r_0^0 - 2r^* - r_{-1}^0 + r^*\| + \|r_{-1}^0 - r^*\|)\\
&\leq \mathcal{L}_*(2\|r_0^0 - r^*\| + 2\|r_{-1}^0 - r^*\|) < 4\mathcal{L}_* r^* < 1.
\end{aligned}
$$

Using Banach Lemma on invertible operators [14], we get that

$$
\|C_0^{-1}H'(r^*)\| \leq \frac{1}{1 - 2\mathcal{L}_*(\|r_0^0 - r^*\| + \|r_{-1}^0 - r^*\|)}. \qquad (13)
$$

Using (5), we have

$$
\begin{aligned}
r_0^1 - r^* &= r_0^0 - r^* - [2r_0^0 - r_{-1}^0, r_{-1}^0; H]^{-1}H(r_0^0)\\
&= [2r_0^0 - r_{-1}^0, r_{-1}^0; H]^{-1}\left([2r_0^0 - r_{-1}^0, r_{-1}^0; H](r_0^0 - r^*) - H(r_0^0) - H(r^*)\right)\\
&= [2r_0^0 - r_{-1}^0, r_{-1}^0; H]^{-1}H'(r^*)H'(r^*)^{-1}\left([2r_0^0 - r_{-1}^0, r_{-1}^0; H](r_0^0 - r^*) - [r_0^0, r^*; H](r_0^0 - r^*)\right)\\
&= [2r_0^0 - r_{-1}^0, r_{-1}^0; H]^{-1}H'(r^*)H'(r^*)^{-1}\left([2r_0^0 - r_{-1}^0, r_{-1}^0; H] - [r_0^0, r^*; H]\right)(r_0^0 - r^*). \qquad (14)
\end{aligned}
$$

Taking norm on both sides of (14), we have

$$
\|r_0^1 - r^*\| \leq \|[2r_0^0 - r_{-1}^0, r_{-1}^0; H]^{-1}H'(r^*)\|\|H'(r^*)^{-1}\left([2r_0^0 - r_{-1}^0, r_{-1}^0; H] - [r_0^0, r^*; H]\right)\|\|r_0^0 - r^*\|. \qquad (15)
$$

From (10), (13) and (15), this provides

$$\begin{aligned}\|r_0^1 - r^*\| &\le \frac{\mathcal{L}(\|r_0^0 - r_{-1}^0\| + \|r_{-1}^0 - r^*\|)}{1 - 2\mathcal{L}_*(\|r_0^0 - r^*\| + \|r_{-1}^0 - r^*\|)}\|r_0^0 - r^*\| \\ &\le \frac{\mathcal{L}(\|r_0^0 - r^*\| + 2\|r_{-1}^0 - r^*\|)}{1 - 2\mathcal{L}_*(\|r_0^0 - r^*\| + \|r_{-1}^0 - r^*\|)}\|r_0^0 - r^*\| \\ &= f_1(\|r_0^0 - r^*\|, \|r_{-1}^0 - r^*\|)\|r_0^0 - r^*\|.\end{aligned}$$

This satisfies (11). Using the domain of our definition and hypothesis of the theorem we have that $f_1(\|r_0^0 - r^*\|, \|r_{-1}^0 - r^*\|) < 1$ and this gives $\|r_0^1 - r^*\| < \|r_0^0 - r^*\|$. This proves the theorem for $n = 0$ and $i = 1$. Now, we proceed the theorem for $n = 0$ and $i = 2$.

$$\begin{aligned}r_0^2 - r^* &= r_0^1 - r^* - [2r_0^0 - r_{-1}^0, r_{-1}^0; H]^{-1}H(r_0^1) \\ &= [2r_0^0 - r_{-1}^0, r_{-1}^0; H]^{-1}\left([2r_0^0 - r_{-1}^0, r_{-1}^0; H](r_0^1 - r^*) - H(r_0^1) - H(r^*)\right) \\ &= [2r_0^0 - r_{-1}^0, r_{-1}^0; H]^{-1}H'(r^*)H'(r^*)^{-1}\left([2r_0^0 - r_{-1}^0, r_{-1}^0; H](r_0^1 - r^*) - [r_0^1, r^*; H](r_0^1 - r^*)\right) \\ &= [2r_0^0 - r_{-1}^0, r_{-1}^0; H]^{-1}H'(r^*)H'(r^*)^{-1}\left([2r_0^0 - r_{-1}^0, r_{-1}^0; H] - [r_0^0, r^*; H]\right)(r_0^1 - r^*). \qquad (16)\end{aligned}$$

Taking norm on both sides of (16), we have

$$\|r_0^2 - r^*\| \le \|[2r_0^0 - r_{-1}^0, r_{-1}^0; H]^{-1}H'(r^*)\|\|H'(r^*)^{-1}\left([2r_0^0 - r_{-1}^0, r_{-1}^0; H] - [r_0^1, r^*; H]\right)\|\|r_0^1 - r^*\|. \qquad (17)$$

From (10), (13) and (17), this provides

$$\begin{aligned}\|r_0^2 - r^*\| &\le \frac{\mathcal{L}(\|2r_0^0 - r_{-1}^0 - r_0^1\| + \|r_{-1}^0 - r^*\|)}{1 - 2\mathcal{L}_*(\|r_0^0 - r^*\| + \|r_{-1}^0 - r^*\|)}\|r_0^1 - r^*\| \\ &= \frac{\mathcal{L}(\|2r_0^0 - 2r^* - r_{-1}^0 + r^* - r_0^1 + r^*\| + \|r_{-1}^0 - r^*\|)}{1 - 2\mathcal{L}_*(\|r_0^0 - r^*\| + \|r_{-1}^0 - r^*\|)}\|r_0^1 - r^*\| \\ &\le \frac{\mathcal{L}(2\|r_0^0 - r^*\| + \|r_0^1 - r^*\| + 2\|r_{-1}^0 - r^*\|)}{1 - 2\mathcal{L}_*(\|r_0^0 - r^*\| + \|r_{-1}^0 - r^*\|)}\|r_0^1 - r^*\| \\ &\le \frac{\mathcal{L}((2 + f_1(\|r_0^0 - r^*\|, \|r_{-1}^0 - r^*\|))\|r_0^0 - r^*\| + 2\|r_{-1}^0 - r^*\|)}{1 - 2\mathcal{L}_*(\|r_0^0 - r^*\| + \|r_{-1}^0 - r^*\|)}\|r_0^1 - r^*\| \\ &= f_2(\|r_0^0 - r^*\|, \|r_{-1}^0 - r^*\|)\|r_0^1 - r^*\|.\end{aligned}$$

This satisfies (12). Using the domain of our definition and hypothesis of the theorem we have that $f_2(\|r_0^0 - r^*\|, \|r_{-1}^0 - r^*\|) < 1$ and this gives $\|r_0^2 - r^*\| < \|r_0^1 - r^*\|$. This proves the theorem for $n = 0$ and $i = 2$. Now using induction hypothesis, suppose it is true for $n = 0$ and $i = l - 1$ which justifies $\|r_0^{l-1} - r^*\| \le f_{l-1}(\|r_0^0 - r^*\|, \|r_{-1}^0 - r^*\|)\|r_0^{l-2} - r^*\|$, $f_{l-1}(\|r_0^0 - r^*\|, \|r_{-1}^0 - r^*\|) < 1$ and $\|r_0^{l-1} - r^*\| < \|r_0^{l-2} - r^*\|$, to prove for $i = l$, we have

$$\begin{aligned}
r_0^l - r^* &= r_0^{l-1} - r^* - [2r_0^0 - r_{-1}^0, r_{-1}^0; H]^{-1} H(r_0^{l-1}) \\
&= [2r_0^0 - r_{-1}^0, r_{-1}^0; H]^{-1} \left([2r_0^0 - r_{-1}^0, r_{-1}^0; H](r_0^{l-1} - r^*) - H(r_0^{l-1}) - H(r^*)\right) \\
&= [2r_0^0 - r_{-1}^0, r_{-1}^0; H]^{-1} H'(r^*) H'(r^*)^{-1} \left([2r_0^0 - r_{-1}^0, r_{-1}^0; H](r_0^{l-1} - r^*) - [r_0^{l-1}, r^*; H](r_0^{l-1} - r^*)\right) \\
&= [2r_0^0 - r_{-1}^0, r_{-1}^0; H]^{-1} H'(r^*) H'(r^*)^{-1} \left([2r_0^0 - r_{-1}^0, r_{-1}^0; H] - [r_0^{l-1}, r^*; H]\right)(r_0^{l-1} - r^*). \qquad (18)
\end{aligned}$$

Taking norm on both sides of (18), we have

$$\|r_0^l - r^*\| \le \|[2r_0^0 - r_{-1}^0, r_{-1}^0; H]^{-1} H'(r^*)\| \|H'(r^*)^{-1} \left([2r_0^0 - r_{-1}^0, r_{-1}^0; H] - [r_0^{l-1}, r^*; H]\right)\| \|r_0^{l-1} - r^*\|. \qquad (19)$$

From (10), (13) and (19), this provides

$$\begin{aligned}
\|r_0^l - r^*\| &\le \frac{\mathcal{L}(\|2r_0^0 - r_{-1}^0 - r_0^{l-1}\| + \|r_{-1}^0 - r^*\|)}{1 - 2\mathcal{L}_*(\|r_0^0 - r^*\| + \|r_{-1}^0 - r^*\|)} \|r_0^{l-1} - r^*\| \\
&= \frac{\mathcal{L}(\|2r_0^0 - 2r^* - r_{-1}^0 + r^* - r_0^{l-1} + r^*\| + \|r_{-1}^0 - r^*\|)}{1 - 2\mathcal{L}_*(\|r_0^0 - r^*\| + \|r_{-1}^0 - r^*\|)} \|r_0^{l-1} - r^*\| \\
&\le \frac{\mathcal{L}(2\|r_0^0 - r^*\| + \|r_0^{l-1} - r^*\| + 2\|r_{-1}^0 - r^*\|)}{1 - 2\mathcal{L}_*(\|r_0^0 - r^*\| + \|r_{-1}^0 - r^*\|)} \|r_0^{l-1} - r^*\| \\
&\le \frac{\mathcal{L}((2\|r_0^0 - r^*\| + f_{l-1}(\|r_0^0 - r^*\|, \|r_{-1}^0 - r^*\|))\|r_0^{l-2} - r^*\| + 2\|r_{-1}^0 - r^*\|)}{1 - 2\mathcal{L}_*(\|r_0^0 - r^*\| + \|r_{-1}^0 - r^*\|)} \|r_0^{l-1} - r^*\| \\
&\le \frac{\mathcal{L}((2 + f_{l-1} \dots f_1)\|r_0^0 - r^*\| + 2\|r_{-1}^0 - r^*\|)}{1 - 2\mathcal{L}_*(\|r_0^0 - r^*\| + \|r_{-1}^0 - r^*\|)} \|r_0^{l-1} - r^*\| \\
&= f_l(\|r_0^0 - r^*\|, \|r_{-1}^0 - r^*\|)\|r_0^{l-1} - r^*\|.
\end{aligned}$$

This satisfies (12). Using the domain of our definition and hypothesis of the theorem we have that $f_l(\|r_0^0 - r^*\|, \|r_{-1}^0 - r^*\|) < 1$ and this gives $\|r_0^l - r^*\| < \|r_0^{l-1} - r^*\|$. This proves the induction hypothesis for all $i$ and $n = 0$. Thus, it is proved here that $\|r_0^k - r^*\| < \|r_0^{k-1} - r^*\| < \dots < \|r_0^0 - r^*\|$. Now in order to prove the induction for $n$, assume it is true for some $n = g$ which in turn provides $\|r_g^k - r^*\| < \|r_g^{k-1} - r^*\| < \dots < \|r_g^0 - r^*\|$ and $\|r_g^i - r^*\| \le f_i((\|r_g^0 - r^*\|, \|r_{g-1}^0 - r^*\|)\|r_g^{i-1} - r^*\|$.

$$\begin{aligned}
\|I - H'(r^*)^{-1} C_{g+1}\| &= \|H'(r^*)^{-1}(C_{g+1} - H'(r^*))\| \\
&= \|H'(r^*)^{-1}([2r_{g+1}^0 - r_g^0, r_g^0; H] - H'(r^*))\| \\
&\le \mathcal{L}_*(\|2r_{g+1}^0 - r_g^0 - r^*\| + \|r_g^0 - r^*\|) \\
&= \mathcal{L}_*(\|2r_{g+1}^0 - 2r^* - r_g^0 + r^*\| + \|r_g^0 - r^*\|) \\
&\le \mathcal{L}_*(2\|r_{g+1}^0 - r^*\| + 2\|r_g^0 - r^*\|) \\
&< \mathcal{L}_*(2\|r_0^0 - r^*\| + 2\|r_{-1}^0 - r^*\|) < 4\mathcal{L}_* r^* < 1.
\end{aligned}$$

It can be easily estimated using Banach lemma [14] that

$$\|C_{g+1}^{-1} H'(r^*)\| \le \frac{1}{1 - 2\mathcal{L}_*(\|r_{g+1}^0 - r^*\| + \|r_g^0 - r^*\|)}. \qquad (20)$$

Now,

$$\begin{aligned} r_{g+1}^{1}-r^{*} &= r_{g+1}^{0}-r^{*}-[2r_{g+1}^{0}-r_{g}^{0},r_{g}^{0};H]^{-1}H(r_{g+1}^{0}) \\ &= [2r_{g+1}^{0}-r_{g}^{0},r_{g}^{0};H]^{-1}\left([2r_{g+1}^{0}-r_{g}^{0},r_{g}^{0};H](r_{g+1}^{0}-r^{*})-H(r_{g+1}^{0})-H(r^{*})\right) \\ &= [2r_{g+1}^{0}-r_{g}^{0},r_{g}^{0};H]^{-1}H'(r^{*})H'(r^{*})^{-1}\left([2r_{g+1}^{0}-r_{g}^{0},r_{g}^{0};H](r_{g+1}^{0}-r^{*})-[r_{g+1}^{0},r^{*};H](r_{g+1}^{0}-r^{*})\right) \\ &= [2r_{g+1}^{0}-r_{g}^{0},r_{g}^{0};H]^{-1}H'(r^{*})H'(r^{*})^{-1}\left([2r_{g+1}^{0}-r_{g}^{0},r_{g}^{0};H]-[r_{g+1}^{0},r^{*};H]\right)(r_{g+1}^{0}-r^{*}). \end{aligned} \tag{21}$$

Taking norm on both sides of (21), we have

$$\|r_{g+1}^{1}-r^{*}\| \le \|[2r_{g+1}^{0}-r_{g}^{0},r_{g}^{0};H]^{-1}H'(r^{*})\|H'(r^{*})^{-1}\left([2r_{g+1}^{0}-r_{g}^{0},r_{g}^{0};H]-[r_{g+1}^{0},r^{*};H]\right)\|r_{g+1}^{0}-r^{*}\|. \tag{22}$$

From (10), (20) and (22), this provides

$$\begin{aligned} \|r_{g+1}^{1}-r^{*}\| &\le \frac{\mathcal{L}(\|r_{g+1}^{0}-r_{g}^{0}\|+\|r_{g}^{0}-r^{*}\|)}{1-2\mathcal{L}_{*}(\|r_{g+1}^{0}-r^{*}\|+\|r_{g}^{0}-r^{*}\|)}\|r_{g+1}^{0}-r^{*}\| \\ &\le \frac{\mathcal{L}(\|r_{g+1}^{0}-r^{*}\|+2\|r_{g}^{0}-r^{*}\|)}{1-2\mathcal{L}_{*}(\|r_{g+1}^{0}-r^{*}\|+\|r_{g}^{0}-r^{*}\|))}\|r_{g+1}^{0}-r^{*}\| \\ &= f_{1}(\|r_{g+1}^{0}-r^{*}\|,\|r_{g}^{0}-r^{*}\|)\|r_{g+1}^{0}-r^{*}\|. \end{aligned}$$

This satisfies (11). Using the domain of our definition and hypothesis of the theorem we have that $f_1(\|r_{g+1}^0-r^*\|,\|r_g^0-r^*\|)<1$ and this gives $\|r_{g+1}^1-r^*\|<\|r_{g+1}^0-r^*\|$. Using the similar lines and proceeding in the above manner, it can be easily verified that

$$\|r_{g+1}^{2}-r^{*}\| \le f_{2}(\|r_{g+1}^{0}-r^{*}\|,\|r_{g}^{0}-r^{*}\|)\|r_{g+1}^{1}-r^{*}\|$$

and

$$\|r_{g+1}^{i}-r^{*}\| \le f_{i}(\|r_{g+1}^{0}-r^{*}\|,\|r_{g}^{0}-r^{*}\|)\|r_{g+1}^{i-1}-r^{*}\| \;\; \forall\; i \ge 2.$$

Thus, using induction hypothesis this holds for all $n$. Now, we shall show the convergence of the iterates $\{r_n^l\}$ where l = 1, 2, ..., k, n = 0, 1, 2, .... For this,

$$\begin{aligned} \|r_{n+1}^{k}-r^{*}\| &\le f_{k}(\|r_{n+1}^{0}-r^{*}\|,\|r_{n}^{0}-r^{*}\|)\|r_{n+1}^{k-1}-r^{*}\| \\ &< f_{k}\|r_{n+1}^{k-1}-r^{*}\| < f_{k}f_{k-1}\|r_{n+1}^{k-2}-r^{*}\| \\ &< \ldots < f_{k}f_{k-1}\ldots f_{1}\|r_{n+1}^{0}-r^{*}\| \end{aligned}$$

Proceeding in this manner, we arrive at

$$\|r_{n+1}^{k}-r^{*}\| < (f_{k}f_{k-1}\ldots f_{1})^{n+1}\|r_{0}^{0}-r^{*}\|$$

As each of $f_i < 1$, so $r_{n+1}^k \to r^*$ for each $k$ as $n \to \infty$. Now to show the uniqueness part of the theorem. Suppose $q^*$ be the another solution of (1) so that $H(q^*) = 0$ and $H(r^*) = H(q^*)$ or $[r^*.q^*; H](r^* - q^*) = 0$. Now,

$$\begin{aligned}\|I - H'(r^*)^{-1}[r^*.q^*; H]\| &= \|H'(r^*)^{-1}(H'(r^*) - [r^*.q^*; H])\| \\ &\leq \mathcal{L}_* \|r^* - q^*\| < 1.\end{aligned}$$

Using Banach lemma on invertible operators [14] that $[r^*.q^*; H]$ exists, nonzero and therefore $r^* = q^*$.

## 3 Numerical Examples

**Example 1** Let $A = B = \mathcal{C}[0, 1]$ be the space on continuous functions defined on $[0, 1]$ and consider the integral equation

$$H(r)(s) = r(s) - \lambda \int_0^1 \frac{s}{s+t} r^2(t)dt, \tag{23}$$

where $r(s)$ is continuous function in $\mathcal{C}[0, 1]$ and $t, s \in [0, 1]$. Now,

$$H'(r)u(s) = u(s) - 2\lambda \int_0^1 r(t)u(t)dt$$

and

$$\begin{aligned}\|H'(r) - H'(s)\| &\leq \|2\lambda \int_0^1 \frac{s}{s+t}[r(t) - s(t)]\|dt \\ &\leq |\lambda| 2 \max_{s\in[0,1]} \left| \int_0^1 \frac{s}{s+t} \right| \|r(t) - s(t)\| \\ &\leq |\lambda| 2 log(2) \|r(t) - s(t)\|\end{aligned}$$

$$\begin{aligned}\|[a, b, H] - [c, d, H]\| &\leq \int_0^1 \|[H'(a(t) + \theta(b(t) - a(t))) - H'(c(t) + \theta(d(t) - c(t)))]\| d\theta \\ &\leq 2|\lambda| log(2) \int_0^1 \|(a(t) + \theta(b(t) - a(t))) - (c(t) + \theta(d(t) - c(t)))\| d\theta \\ &= 2|\lambda| log(2) \int_0^1 \|(1 - \theta)(a(t) - c(t)) + \theta(b(t) - d(t))\| d\theta \\ &\leq \lambda log(2)(\|a - c\| + \|b - d\|)\end{aligned}$$

**Table 1** List of constants appear in the examples

| | $r^*$ | $\mathcal{L}$ | $\mathcal{L}_*$ |
|---|---|---|---|
| Example-1($\lambda = 1$) | 0 | log(2) | log(2) |
| Example-2 | 0 | $\frac{e}{2}$ | $\frac{e-1}{2}$ |
| Example-3 | 0 | 3 | $\frac{3}{2}$ |

**Table 2** Radii of convergence balls

| | Example-1 | Example-2 | Example-3 |
|---|---|---|---|
| k = 1 | 0.206099291622195 | 0.133085149027196 | 0.066666666666667 |
| k = 2 | 0.166867446157297 | 0.102555747111161 | 0.050157358847771 |
| uniqueness ball | 1.442695040888963 | 1.163953413738653 | 0.666666666666667 |

**Example 2** Let $A = B = \mathbb{R}^3$, $\Omega_0 = \overline{\mathcal{U}(0, 1)}$, $r^* = (0, 0, 0)^T$. Define function $H$ on $\Omega$ for $r = (r_1, r_2, r_3)^t$ by

$$H(r) = \left(e^{r_1} - 1, \frac{e-1}{2}{r_2}^2 + r_2, r_3\right)^t$$

$$H'(r) = \begin{pmatrix} e^{r_1} & 0 & 0 \\ 0 & (e-1)r_2 & 0 \\ 0 & 0 & 1 \end{pmatrix}$$

**Example 3** Let $A = B = \mathcal{C}[0, 1]$ and consider the integral equation

$$H(r)(s) = r(s) - \int_0^1 sur^3(u)du,$$

where $r(s)$ is continuous function in $\mathcal{C}[0, 1]$.

In order to find the numerical parameters appearing in the theorem, we follow the same manner as shown in the first example and tabulated in Table 1. Now, using the tabulated values of the parameters in Table 1, we present the radii of convergence balls for the case $k = 1$ and $k = 2$ in Table 2.

## 4 Conclusions and Future Scope

In this paper, we present and intended to present a family of Kurchatov's like method. We have established the convergence theorem for computing the radii of convergence balls. In literature, we find the convergence theorems for this family of methods in some special cases. However, we have presented for multipoint method and the analysis found in the literature may be some special cases of our work. In this way, we have generalized the method and presented the local convergence analysis.

Due to the page limitation, we were not be able to cover the more features of this method. However, this is also a complete study for local convergence. More work is to be done on this method. In future, we shall estimate the order of convergence in each step and also found its computational order of convergence of this proposed work. We shall also try to establish its semilocal convergence analysis. Using domain of parameters, we will also try to enlarge its domain for the starting points.

**Conflict of Interest**
The authors have equally contributed and give their consent for publication. There is no any conflict of interest involved.

This paper does not contain any studies with animals performed by any of the authors.

**Acknowledgements** The authors are thankful to the reviewers for the contribution of their fruitful suggestions that uncovered the several weaknesses leading to the improvement of the paper.

## References

1. Alshomrani, A.S., Maroju, P., Behl, R.: Convergence of a stirling-like method for fixed points in banach spaces. J. Comput. Appl. Math. 354:183–190 (2019)
2. Amat, S., Argyros, I.K., Busquier, S., Hernández-Verón, M.A., Martínez, E.: On the local convergence study for an efficient k-step iterative method. J. Comput. Appl. Math. **343**, 753–761 (2018)
3. Amat, S., Bermúdez, C., Hernández-Verón, M.A., Martínez, E.: On an efficient k-step iterative method for nonlinear equations. J. Comput. Appl. Math. **302**, 258–271 (2016)
4. Argyros, I.K., Ezquerro, J.A., Hernández-Verón, M.Á., Kim, Y.I., Magreñán, Á.A.: Extending the choice of starting points for newton's method. Math. Methods Appl. Sci. **43**(14), 8042–8050 (2020)
5. Argyros, I.K., George, S.: Extended convergence of a two-step-secant-type method under a restricted convergence domain. Kragujevac J. Math. **45**(1), 155–164 (2021)
6. Argyros, I.K., Gupta, N., Jaiswal, J.P.: Extending the applicability of a two-step chord-type method for non-differentiable operators. Mathematics **7**(9), 804 (2019)
7. Argyros, I.K., Sharma, D., Parhi, S.K.: Generalizing the local convergence analysis of a class of $k$-step iterative algorithms with hölder continuous derivative in banach spaces. Applicationes Mathematicae 1–17
8. Argyros, I.K., Sharma, D., Parhi, S.K.: On the local convergence of weerakoon–fernando method with $\omega$ continuity condition in banach spaces. SeMA J. **77**(3), 291–304 (2020)
9. Argyros, I.K., Uko, L.U., Nathanson, E.: A new semi-local convergence analysis of the secant method. Int. J. Appl. Comput. Math. **3**(1), 225–232 (2017)

10. Dennis, J.E.: Toward a Unified Convergence Theory for Newton-like Methods (1970)
11. Ezquerro, J.A., Hernández-Verón, M.A., Magreñán, Á.A.: Starting points for newton's method under a center lipschitz condition for the second derivative. J. Comput. Appl. Math. **330**, 721–731 (2018)
12. Galántai, A.: The theory of newton's method. J. Computat. Appl. Math. **124**(1–2), 25–44 (2000)
13. Hernández-Verón, M.A., Molada, E.M., Teruel-Ferragud, C.: Semilocal convergence of a k-step iterative process and its application for solving a special kind of conservative problems. Numer. Algorithm. **76**(2), 309–331 (2017)
14. Kantorovich, L.V., Akilov, G.P.: Functional Analysis. Pergamon Press, Oxford (1982)
15. Kumar, A., Gupta, D.K., Martínez, E., Hueso, J.L.: Convergence and dynamics of improved chebyshev-secant-type methods for non differentiable operators. Numer. Algorithm. **86**(3), 1051–1070 (2021)
16. Kumar, A., Gupta, D.K., Martínez, E., Singh, S.: Semilocal convergence of a secant-type method under weak lipschitz conditions in banach spaces. J. Comput. Appl. Math. **330**, 732–741 (2018)
17. Kumar, A., Gupta, D.K., Molada, E.M., Singh, S.: Directional k-step newton methods in n variables and its semilocal convergence analysis. Mediter. J. Math. **15**(2), 15–34 (2018)
18. Kumar, H.: On semilocal convergence of three-step kurchatov method under weak condition. Arabian J. Math. **10**(1), 121–136 (2021)
19. Kumar, A., Gupta, D.K.: Local convergence of super halley's method under weaker conditions on fréchet derivative in banach spaces. J. Anal. **28**(1), 35–44 (2020)
20. Kumar, H., Parida, P.K.: Three step kurchatov method for nondifferentiable operators. Int. J. Appl. Comput. Math. **3**(4), 3683–3704 (2017)
21. Maroju, P., Magreñán, Á.A., Sarría, Í., Kumar, A.: Local convergence of fourth and fifth order parametric family of iterative methods in banach spaces. J. Math. Chem. **58**(3), 686–705 (2020)
22. Martínez, E., Singh, S., Hueso, J.L., Gupta, D.K.: Enlarging the convergence domain in local convergence studies for iterative methods in banach spaces. Appl. Math. Comput. **281**, 252–265 (2016)
23. Monsalve, M., Raydan, M.: Newton's method and secant methods: a longstanding relationship from vectors to matrices. Portugaliae Mathematica **68**(4), 431–475 (2011)
24. Ren, H., Argyros, I.K.: On the complexity of extending the convergence ball of wang's method for finding a zero of a derivative. J. Complex. **64**, 101526 (2021)
25. Rokne, J.: Newton's method under mild differentiability conditions with error analysis. Numerische Mathematik **18**(5), 401–412 (1971)
26. Schmidt, J.W., Schwetlick, H.: Ableitungsfreie verfahren mit höherer konvergenzgeschwindigkeit. Computing **3**(3), 215–226 (1968)
27. Sharma, D., Parhi, S.K.: On the local convergence of modified weerakoon's method in banach spaces. J. Anal. **28**(3), 867–877 (2020)
28. Wolfe, P.: The secant method for simultaneous nonlinear equations. Commun. ACM **2**(12), 12–13 (1959)

# An Effective Scheme for Solving a Class of Second-Order Two-Point Boundary Value Problems

**Saurabh Tomar, Soniya Dhama, and Kuppalapalle Vajravelu**

**Abstract** A piecewise Adomian decomposition approach is presented in this work to handle a class of nonlinear two-point boundary value problems effectively. The suggested technique enables quick convergence and helps to overcome the limitations of the traditional Adomian decomposition method in instances where it fails to produce a reasonably decent approximate solution or when a significant number of iterations are necessary to get a convergent series solution. Three numerical examples are given to demonstrate the method's applicability and efficacy.

**Keywords** Adomian decomposition method · Boundary value problems · Approximate solutions

## 1 Introduction

In this work, we consider the following nonlinear two-point boundary value problems (TPBVPs)

$$u''(x) - f(x, u(x), u'(x)) = 0, \qquad a \le x \le b, \tag{1}$$

subject to

$$\alpha_1 u(a) + \alpha_2 u'(a) = \alpha, \quad \beta_1 u(b) + \beta_2 u'(b) = \beta, \tag{2}$$

S. Tomar (✉)
Department of Mathematics and Statistics, Indian Institute of Technology Kanpur, Kanpur 208016, UP, India
e-mail: sauravtomar9793@gmail.com

S. Dhama
Department of Mathematical Sciences, Rajiv Gandhi Institute of Petroleum Technology, Jais Amethi 229304, UP, India

K. Vajravelu
Department of Mathematics, University of Central Florida, Orlando, FL 32816, USA
e-mail: kuppalapalle.vajravelu@ucf.edu



B. Rushi Kumar et al. (eds.), *Mathematics and Computing*, Springer Proceedings in Mathematics & Statistics 415, https://doi.org/10.1007/978-981-19-9307-7_52

where $\alpha_1, \alpha_2, \beta_1, \beta_2, \alpha, \beta$ satisfy $(\alpha_1\beta_2 - \beta_1\alpha_2 + \alpha_1\beta_1(b-a)) \neq 0$ and $f(x, u, u')$ is a continuous real valued function. These problems have a wide range of applications in applied science and engineering [13, 14, 20]. As a result of their relevance in real-world applications, these types of problems have garnered considerable attention from academics and scientists. Due to their nonlinearity, closed-form solutions to these problems are often hard to acquire. As a result, numerous analytical and numerical approximation approaches have been developed. Some well-known techniques such as shooting methods [24], finite difference techniques [9], finite element schemes [8], collocation techniques [25], Galerkin methods [19], variational iteration methods [15, 17, 21], homotopy perturbation method [26] and other methods for various boundary value problems are given in [1, 7, 10, 11, 16, 18, 22, 23, 27–30] have been introduced to solve these problems.

The Adomian decomposition method (ADM) [2–5] is a well-known and powerful approach for tackling problems of ordinary, partial, integro differential equations, and other types of problems using a direct recursive algorithm. As opposed to perturbation techniques, ADM solves beginning and BVPs without needing assumptions of linearization and the presence of a small parameter in physical problems. The nonlinearity of the issue is addressed by decomposing the nonlinear operator into a sequence of functions known as Adomian polynomials.

The primary purpose of this work is to develop a piecewise ADM (PADM) to get a convergent approximation solution for the TPBVPs (1). The PADM works well and improves the convergence rate for situations where the conventional ADM diverges, slows, or requires a large number of series terms of the approximation solution to achieve a convergent approximate series solution. The presented method's major goals are quick convergence and the use of a few terms to get a highly accurate approximation solution to problems. In the PADM, the interval $[a, b]$ is divided into equal-sized sub-intervals, and then the ADM is applied to the sub-intervals. The approximate analytical solution is then derived in terms of unknown constants in each sub-interval. The computation of unknown constants is then calculated by letting that $u(x)$ and $u'(x)$ are continuous on the boundary of each sub-interval, and the system of nonlinear equations is then obtained by applying these imposed continuity requirements. After that, the Newton-Raphson method is used to tackle the nonlinear system.

The draft of this article is presented as follows, Sect. 2 gives the description of standard ADM is illustrated. In Sect. 3, the proposed technique to get the approximate convergent series solution to (1) is given. In Sect. 4, numerical test examples are given to validate the present work. Finally, Sect. 5 is dedicated to the conclusion.

## 2 Review of Standard ADM

In this part, we go through the basics of ADM's problem-solving technique. Consider the form as

$$L[u] + R[u] + N[u] = g(x), \tag{3}$$

where $L$ represents an invertible linear operator and, in general, the highest order differential operator, the linear operator that is a remainder of the problem's linear operator is denoted by $R$, the nonlinear operator is denoted by $N$, and the system input is denoted by $g$.

Given that $L$ is invertible, we use $L^{-1}$ i.e. inverse linear operator to both sides of (3) to obtain

$$u = h - L^{-1}R[u] - L^{-1}N[u] + L^{-1}g, \tag{4}$$

where $h$ satisfies $L[h] \equiv 0$. In ADM, the solution $u$ is expressed by the decomposition series and the nonlinear operator $N[u]$ is decomposed in terms of the Adomian polynomials as

$$u = \sum_{n=0}^{\infty} u_n, \quad \text{and} \quad N[u] = \sum_{n=0}^{\infty} A_n, \tag{5}$$

where

$$A_n = \frac{1}{n!}\frac{d^n}{d\lambda^n}\Big[N\Big(\sum_{j=0}^{n}\lambda^j u_j\Big)\Big]_{\lambda=0}, \quad n \geq 0.$$

Now inserting (5) into (4) leads to

$$\sum_{n=0}^{\infty} u_n = p - L^{-1}R\Big[\sum_{n=0}^{\infty} u_n\Big] - L^{-1}A_n, \tag{6}$$

here $p = h + L^{-1}g$. From (6), we have the recursive formula as

$$u_0 = p, \qquad u_{n+1} = -L^{-1}R[u_n] - L^{-1}A_n, \qquad n \geq 0.$$

The $m$th term approximate solution is given by

$$U_m = \sum_{i=0}^{m-1} u_i. \tag{7}$$

For more detail see [6, 12].

## 3 Proposed Methodology

This section introduces a strategy for solving problem (1) efficiently. We do this by rewriting (1) in the following operator form

$$L[u] = N[u], \quad \text{where} \quad L(.) = u'' = \frac{d^2}{dx^2}(.) \quad \text{and} \quad N[u] = f(x, u, u'). \tag{8}$$

Note that here $L^{-1}(.) = \int_a^x \int_a^x (.)dxdx$. Next, we apply $L^{-1}$ on the both side of (8), after simplification we get

$$u = u(a) + (x - a)u'(a) + L^{-1}N[u]. \tag{9}$$

Now by combining (8) and (9) leads to

$$\begin{aligned} u_0 &= u(a) + (x - a)u'(a), \\ u_{n+1} &= L^{-1}A_n, \qquad n \geq 0. \end{aligned} \tag{10}$$

Here the values of $u(a)$ and $u'(a)$ are evaluated by imposing the corresponding boundary conditions, and then the $m$th term approximate solution is given by (7). Note that (10) is the standard ADM for (1). We note that the standard ADM requires a large number of iteration of the series solution to achieve reasonably good accuracy, slow or diverges in some cases as depicted by given numerical examples. To address these drawbacks, we provide a practical method for dealing with (1) by dividing the interval $[a, b]$ into $M$ evenly spaced sub-intervals as $h = (b - a)/M, x_i = a + ih, \ 0 \leq i \leq M$ with $a = x_0 < x_1 < \cdots < x_{M-1} < x_M = b$.

Taking $u(x_i) = k_i$ and $u'(x_i) = k_i'$ with $0 \leq i \leq M - 1$. Now, the following piece-wise ADM on $[x_i, x_{i+1}]$ is defined as follows according to (10).

For $[x_0, x_1]$, the approach is defined as

$$u_{0,0} = u(x_0) + (x - x_0)u'(x_0) = k_0 + (x - a)k_0', \tag{11}$$

$$u_{1,n+1} = L^{-1}A_n, \tag{12}$$

where $L^{-1}(.) = \int_{x_0}^x \int_{x_0}^x (.)dxdx$ and $m$th term approximate solution on $[x_0, x_1]$ is given by

$$U_{0,m} = \sum_{i=0}^{m-1} u_{0,i}.$$

For $[x_i, x_{i+1}]$, $1 \leq i \leq M - 1$, the scheme is constructed as

$$u_{i,0} = u(x_i) + (x - x_i)u'(x_i) = k_i + (x - x_i)k_i', \tag{13}$$

$$u_{i,n+1} = L^{-1}A_n, \tag{14}$$

where $L^{-1}(.) = \int_{x_i}^x \int_{x_i}^x (.)dxdx$ and $m$th term approximate solution on $[x_i, x_{i+1}]$ is given by

$$U_{i,m} = \sum_{i=0}^{m-1} u_{i,i}, \qquad 1 \leq i \leq M - 1.$$

Now by evaluating $2M$ unknown $k_i$ and $k_i'$ constants for $0 \le i \le M-1$, we can find the solutions for each sub-interval. All of these approximate solutions then matched together to generate a continuous solution on the interval $[a, b]$ by assuming the continuity of the solution and its derivative at the end points of the sub-intervals. As a result, at the grid points if $U_{i,m}(x)$ and $U_{i,m}'(x)$ have the same values, we can achieve a continuous solution. As a result of the approximate solution of (1) across the interval $[a, b]$, the following nonlinear system of $2M$ equations is solved.

$$\begin{cases} \alpha_1 U_{0,m}(a) + \alpha_2 U_{0,m}'(a) = \alpha, & \\ U_{i-1,m}(x_i) = U_{i,m}(x_i), & 1 \le i \le M-1, \\ U_{i-1,m}'(x_i) = U_{i,m}'(x_i), & 1 \le i \le M-1, \\ \beta_1 U_{M-1,m}(b) + \beta_2 U_{M-1,m}'(b) = \beta. & \end{cases} \tag{15}$$

Now by implementing the Newton-Raphson method on (15), the $2M$ unknowns coefficients $k_i$ and $k_i'$, $0 \le i \le M-1$ can be evaluated. Thus, the convergent series solution of (1) on the entire interval $[a, b]$ can be achieved.

## 4 Numerical Results

In this part, we look at three numerical test scenarios to show how successful the proposed technique is.

**Example 1** Consider the following problem [16]

$$u'' = \frac{1}{2}(1 + x + u)^3, \quad u'(0) - u(0) = -\frac{1}{2}, \quad u'(1) + u(1) = 1, \qquad 0 \le x \le 1. \tag{16}$$

The exact solution of (16) is given by $u(x) = \frac{2}{2-x} - x - 1$.

We solve Example 1 by using the standard ADM for $m = 4$ and the proposed PADM for $M = 10$ and $m = 4$. Table 1 demonstrates the absolute errors of these approaches. It is clear from Table 1 that the proposed PADM scheme is very effective and accurate.

**Example 2** Consider the following nonlinear BVP [15]

$$u'' = -(1 - \theta^2 u'^2), \quad u(0) = 0, \quad u(1) = 0, \qquad 0 \le x \le 1. \tag{17}$$

The true solution of (17) is given by $u(x) = \frac{1}{\theta^2} \ln\left(\frac{\cos\theta(x-\frac{1}{2})}{\cos\frac{\theta}{2}}\right)$.

We solve Example 2 with different values of $\theta$ by using the standard ADM for $m = 6$ and the proposed PADM for $M = 20$ and $m = 6$. The absolute errors are tabulated in

**Table 1** Absolute errors obtained by using the ADM and PADM approaches for $m = 4$ of Example 1

| $x$ | ADM | PADM ($M = 5$) | PADM ($M = 10$) |
|---|---|---|---|
| 0.0 | $7.24e-04$ | $3.01e-08$ | $1.58e-10$ |
| 0.2 | $8.94e-04$ | $3.72e-08$ | $1.95e-10$ |
| 0.4 | $1.13e-03$ | $4.69e-08$ | $2.45e-10$ |
| 0.6 | $1.47e-03$ | $6.05e-08$ | $3.16e-10$ |
| 0.8 | $1.96e-03$ | $7.91e-08$ | $4.11e-10$ |
| 1.0 | $2.23e-03$ | $9.81e-08$ | $4.90e-10$ |

**Table 2** Absolute errors obtained by using ADM and PADM approaches for $m = 6$ of Example 2

| x | $\theta = 0.5$ | | $\theta = 1.0$ | | $\theta = 2.0$ | |
|---|---|---|---|---|---|---|
| | ADM | PADM | ADM | PADM | ADM | PADM |
| 0 | 0 | 0 | 0 | 0 | 0 | 0 |
| 0.2 | $3.65e-05$ | $1.30e-12$ | $1.04e-03$ | $1.91e-10$ | $9.29e-02$ | $3.67e-07$ |
| 0.4 | $7.19e-05$ | $1.92e-12$ | $1.32e-02$ | $2.26e-10$ | $1.63e-01$ | $3.18e-07$ |
| 0.6 | $1.03e-05$ | $1.88e-12$ | $1.86e-02$ | $2.12e-10$ | $2.31e-01$ | $2.69e-07$ |
| 0.8 | $1.11e-05$ | $1.30e-12$ | $1.96e-02$ | $1.57e-10$ | $2.54e-01$ | $2.05e-07$ |
| 1.0 | 0 | 0 | 0 | 0 | 0 | 0 |

Table 2 of these approaches. It is clear from Table 2 that the ADM approximation for $\theta = 1, 2$ leading to unsatisfactory results while the convergent results of the proposed PADM can be observed.

**Example 3** Consider the following nonlinear boundary value problem [7]

$$u'' = \frac{3}{2}u^2, \quad u(0) = 4, \quad u(1) = 1, \qquad 0 \le x \le 1. \tag{18}$$

The true solution of (18) is given by $u(x) = \frac{4}{(1+x)^2}$.

We solve Example 3 by using the standard ADM for $m = 4$ and the proposed PADM for $M = 5, 10$ and $m = 4$. The absolute errors are tabulated in Table 3 of these approaches. It is clear from Table 3 that the ADM approximation for $\theta = 2$ leading to unsatisfactory results while the convergent results of the proposed PADM can be observed.

**Table 3** Absolute errors obtained by using ADM and PADM approaches for $m = 4$ of Example 3

| $x$ | ADM | PADM ($M = 10$) | PADM ($M = 20$) |
|---|---|---|---|
| 0.0 | 0 | 0 | 0 |
| 0.2 | $3.40e-02$ | $6.36e-09$ | $1.88e-11$ |
| 0.4 | $7.92e-02$ | $4.67e-09$ | $1.39e-11$ |
| 0.6 | $1.36e-01$ | $2.84e-09$ | $8.51e-11$ |
| 0.8 | $1.64e-01$ | $1.34e-09$ | $4.02e-11$ |
| 1.0 | 0 | 0 | 0 |

## 5 Conclusion

Standard ADM's approximate solutions of the second-order nonlinear TPBVPs may result in a sluggish convergence rate or a large number of iterative steps to reach a reasonable accuracy. We address these flaws by introducing the PADM method, a modified ADM that is both effective and efficient. The numerical findings show that PADM is a good analytical technique for solving second-order nonlinear TPBVPs, and it may be easily extended to other nonlinear problems.

## References


1. Abukhaled, M., Khuri, S.: A fast convergent semi-analytic method for an electrohydrodynamic flow in a circular cylindrical conduit. Int. J. Appl. Comput. Math. **7**(2), 1–15 (2021)
2. Adomian, G., Rach, R.: Inversion of nonlinear stochastic operators. J. Math. Anal. Appl. **91**(1), 39–46 (1983)
3. Adomian, G., Rach, R., Meyers, R.: Numerical algorithms and decomposition. Comput. Math. Appl. **22**(8), 57–61 (1991)
4. Adomian, G., Rach, R., Meyers, R.: Numerical integration, analytic continuation, and decomposition. Appl. Math. Comput. **88**(2–3), 95–116 (1997)
5. Adomian, G.: Solving Frontier Problems of Physics: the Decomposition Method, vol. 60. Springer Science & Business Media (2013)
6. Bigi, D., Riganti, R.: Solutions of nonlinear boundary value problems by the decomposition method. Appl. Math. Model. **10**(1), 49–52 (1986)
7. Cordero, A., Hueso, J.L., Martínez, E., Torregrosa, J.R.: Efficient high-order methods based on golden ratio for nonlinear systems. Appl. Math. Comput. **217**(9), 4548–4556 (2011)
8. Deacon, A.G., Osher, S.: A finite element method for a boundary value problem of mixed type. SIAM J. Numer. Anal. **16**(5), 756–778 (1979)
9. Doedel, E.J.: Finite difference collocation methods for nonlinear two point boundary value problems. SIAM J. Numer. Anal. **16**(2), 173–185 (1979)
10. Ghorbani, A., Gachpazan, M.: A spectral quasilinearization parametric method for nonlinear two-point boundary value problems. Bull. Malaysian Math. Sci. Soc. **42**(1), 1–13 (2019)
11. Ghorbani, A., Gachpazan, M., Saberi-Nadjafi, J.: A modified parametric iteration method for solving nonlinear second order BVPs. Comput. Appl. Math. **30**(3), 499–515 (2011)
12. Jang, B.: Two-point boundary value problems by the extended Adomian decomposition method. J. Comput. Appl. Math. **219**(1), 253–262 (2008)

13. Khan, U., Ahmed, N., Mohyud-Din, S.T.: Thermo-diffusion, diffusion-thermo and chemical reaction effects on MHD flow of viscous fluid in divergent and convergent channels. Chem. Eng. Sci. **141**, 17–27 (2016)
14. Khan, U., Ahmed, N., Mohyud-Din, S.T., Bin-Mohsin, B.: Nonlinear radiation effects on MHD flow of nanofluid over a nonlinearly stretching/shrinking wedge. Neural Comput. Appl. **28**(8), 2041–2050 (2017)
15. Khuri, S., Sayfy, A.: Generalizing the variational iteration method for BVPs: proper setting of the correction functional. Appl. Math. Lett. **68**, 68–75 (2017)
16. Lal, M., Moffatt, D.: Picard's successive approximation for non-linear two-point boundary-value problems. J. Comput. Appl. Math. **8**(4), 233–236 (1982)
17. Lu, J.: Variational iteration method for solving two-point boundary value problems. J. Comput. Appl. Math. **207**(1), 92–95 (2007)
18. Mary, M., Devi, M.C., Meena, A., Rajendran, L., Abukhaled, M.: Mathematical modeling of immobilized enzyme in porous planar, cylindrical, and spherical particle: a reliable semi-analytical approach. React. Kinet. Mech. Catal. **134**(2), 641–651 (2021)
19. Mohsen, A., El-Gamel, M.: On the Galerkin and collocation methods for two-point boundary value problems using sinc bases. Comput. Math. Appl. **56**(4), 930–941 (2008)
20. Mohyud-Din, S.T., Khan, S.I.: Nonlinear radiation effects on squeezing flow of a Casson fluid between parallel disks. Aerosp. Sci. Technol. **48**, 186–192 (2016)
21. Momani, S., Abuasad, S., Odibat, Z.: Variational iteration method for solving nonlinear boundary value problems. Appl. Math. Comput. **183**(2), 1351–1358 (2006)
22. Pandey, R.K., Tomar, S.: An efficient analytical iterative technique for solving nonlinear differential equations. Comput. Appl. Math. **40**(5), 1–16 (2021)
23. Pandey, R.K., Tomar, S.: An effective scheme for solving a class of nonlinear doubly singular boundary value problems through quasilinearization approach. J. Comput. Appl. Math. **392**, 113411 (2021)
24. Roberts, S.M., Shipman, J.S.: Two-Point Boundary Value Problems: Shooting Methods. North-Holland (1972)
25. Russell, R., Shampine, L.F.: A collocation method for boundary value problems. Numerische Mathematik **19**(1), 1–28 (1972)
26. Shivanian, E., Abbasbandy, S.: Predictor homotopy analysis method: two points second order boundary value problems. Nonlinear Anal.: Real World Appl. **15**, 89–99 (2014)
27. Sylvia, S.V., Salomi, R.J., Rajendran, L., Abukhaled, M.: Solving nonlinear reaction-diffusion problem in electrostatic interaction with reaction-generated ph change on the kinetics of immobilized enzyme systems using taylor series method. J. Math. Chem. **59**(5), 1332–1347 (2021)
28. Tomar, S., Pandey, R.K.: An efficient iterative method for solving Bratu-type equations. J. Comput. Appl. Math. **357**, 71–84 (2019)
29. Tomar, S.: A computationally efficient iterative scheme for solving fourth-order boundary value problems. Int. J. Appl. Comput. Math. **6**(4), 1–16 (2020)
30. Tomar, S., Singh, M., Vajravelu, K., Ramos, H.: Simplifying the variational iteration method: a new approach to obtain the Lagrange multiplier. Math. Comput. Simul. **204**, 640–644 (2023)

# An Analytic Solution for the Helmholtz-Duffing Oscillator by Modified Mickens' Extended Iteration Procedure

**M. M. Ayub Hossain and B. M. Ikramul Haque**

**Abstract** The Helmholtz-Duffing oscillator is a special type of problem in the field of nonlinear as well as engineering and science, due to its combined quadratic and cubic nonlinear terms. The analytic solution of the Helmholtz-Duffing oscillator has been obtained by modifying Mickens' Extended Iteration Procedure. The Fourier series has been used to find the solution. The second approximate frequencies show a good harmony with the exact result. Some researchers presented the solutions to the same oscillators by applying different methods. The obtained results have been compared with some previously published results. Also, the approximate solution obtained from the second iterated level gives extraordinary accuracy compared to the exact solution. Although the present modified Mickens' Extended Iteration Procedure has been applied to Helmholtz-Duffing Oscillator, it can be widely applicable to related problems in science and engineering.

**Keywords** Extended iteration procedure · Helmholtz-Duffing oscillator · Nonlinearity · Nonlinear oscillations · Fourier series

**AMS Subject Classification:** 34A34 · 34B99

## 1 Introduction

The Helmholtz-Duffing oscillator is an asymmetric nonlinear differential equation with two supplementary equations relevant in positive and negative orders. It has been widely applied in the mathematical formulation of engineering domains such as shallow arches, ship roll dynamics, electric circuits, panel absorber, symmetric gyroscope, the human eardrum, dynamics of a moving particle in a cubic potential, and one-dimensional structural systems [2, 3, 26, 27, 32]. The presence of quadratic and cubic nonlinear terms and asymmetric behavior makes it the most attractive for researchers. Perturbation Method [33, 34]; He's Homotopy Perturbation Method [5,

M. M. A. Hossain · B. M. I. Haque (✉)
Department of Mathematics, Khulna University of Engineering & Technology, Khulna 9203, Bangladesh
e-mail: ikramul@math.kuet.ac.bd



B. Rushi Kumar et al. (eds.), *Mathematics and Computing*, Springer Proceedings in Mathematics & Statistics 415, https://doi.org/10.1007/978-981-19-9307-7_53

6]; Harmonic Balance Method [7, 25, 28]; Iterative Method [12–23, 29–31]; Cubication Method [10]; He's Max–Min Method [4], Rational Energy Balance Method [8]; Energy Balance Method [1, 24], He's Energy Balance Method [9, 11] are the most effective methods to solve nonlinear equations, especially for highly nonlinear terms. A small number of researchers have done research on the Helmholtz-Duffing oscillator using different methods. For instance, Leung and Guo [26, 27] have used the homotopy perturbation method (HPM) and the iterative homotopy harmonic balance method (IHHBM), Askari et al. [3] have used He's energy balance method(HEBM) and He's frequency amplitude formulation (HFAF), Akbarzade et al. [32] have used the first-order of the Hamiltonian approach and coupled homotopy-variational formulation, Alal et al. [2] have used Modified Harmonic Balance Method (MHBM) to obtain the periodic solutions of the Helmholtz-Duffing oscillator.

In this paper, we have used modified Mickens' extended iteration method (MMEIM) to determine the approximate frequencies and periodic solutions of the Helmholtz-Duffing oscillator. An extended iteration technique has been attained by Mickens'. Later, the technique was developed by Lim, Hu Wu, and Haque. After applying the modified method, we have acquired fantastic results.

## 2 The Methodology

**1st Step:** Suppose a nonlinear differential equation of the form

$$F(\ddot{v}, v) = 0 \tag{1}$$

with the initial condition $v(0) = a$, $\dot{v}(0) = 0$

Equation (1) can be rewritten as

$$\ddot{v} + F_1(v) = 0 \tag{2}$$

**2nd Step:** Now the standard form of Eq. (2) is

$$\ddot{v} + \Omega^2 v = \Omega^2 v - F_1(v) = H(v, \Omega) \tag{3}$$

where the unknown symbol, $\Omega$ is the natural frequency.

**3rd Step:** The Iterative scheme of Eq. (3) is of the form

$$\ddot{v}_{k+1} + \Omega_k^2 v_{k+1} = H(v_k, \Omega_k); \quad k = 0, 1, 2, \cdots \tag{4}$$

$$v(t) = a\cos(\Omega t), \tag{5}$$

And

$$v_{k+1}(0) = a,\ \dot{v}_{k+1}(0) = 0, \tag{6}$$

where $v$ is the amplitude of the oscillator.

**4th Step:** The extended iteration scheme is of the form

$$\ddot{v}_{k+1} + \Omega_k^2 v_{k+1} = H(v_k, \ddot{v}_k) + H_v(v_0, \Omega_k)(v_k - v_0) \tag{7}$$

where $H_v = \frac{\partial H}{\partial v}$ and $v_{k+1}$ satisfies the conditions (6)

$v_1(t), v_2(t), v_3(t)$....and $\Omega_0, \Omega_1, \Omega_2, \cdots$ are the first, second, third, ...... approximate roots and corresponding frequencies of the oscillators respectively obtained by avoiding the secular terms in each step.

## 3 Solution Procedure

Consider the governing nonlinear equation as

$$\ddot{v} + v + (1-\beta)v^2 + \beta\, v^3 = 0. \text{ with } v(0) = a,\ \dot{v}(0) = 0 \tag{8}$$

where $\beta$ is an asymmetric parameter. When $\beta = 1$ then Eq. (8) is a cubic-Duffing oscillator and for $\beta = 0$, Eq. (8) is a Helmholtz oscillator with a single-well potential. Due to the characteristics of an asymmetric oscillator, it is dissimilar in positive and negative directions. That is why the Eq. (8) can be expediently considered in two parts

$$\ddot{v} + v + (1-\beta)v^2 sg(v) + \beta\, v^3 = 0, \quad \text{for } v \geq 0 \tag{9}$$

$$\ddot{v} + v - (1-\beta)v^2 sg(v) + \beta\, v^3 = 0, \quad \text{for } v \leq 0 \tag{10}$$

For oscillation of the above system an asymmetric limit zone can be assumed as $[-b,\ a]$, for positive $a$ and $b$. Both $v = a$ and $v = -b$ stand for the turning points in which $\dot{v} = 0$, $a$ and $b$ are unknown initial amplitudes to be obtained.

Introducing $\Omega^2 v$ in Eq. (9), we have

$$\ddot{v} + \Omega^2 v = \Omega^2 v - v - (1-\beta)v^2 - \beta\, v^3 \cong H(v,\ \Omega) \tag{11}$$

where

$$F(v,\ \Omega) = \Omega^2 v - v - (1-\beta)v^2 - \beta\, v^3 \tag{12}$$

And

$$H_v = \frac{\partial H}{\partial v} = \Omega^2 - 1 - 2(1-\beta)v - 3\beta\, v^2 \tag{13}$$

Applying the approximate technique (7), we get

$$\begin{aligned}\ddot{v}_{k+1} + \Omega_k^2 v_{k+1} &= (\Omega_k^2 v_0 - v_0 - (1-\beta)v_0^2 - \beta\, v_0^3) \\ &\quad + (\Omega_k^2 - 1 - 2(1-\beta)\, v_0 - 3\beta\, v_0^2)(v_k - v_0).\end{aligned} \tag{14}$$

For first iteration, we get

$$\ddot{v}_{a1} + \Omega_{a0}^2 v_{a1} = \Omega_{a0}^2 a\, \cos\theta - a\cos\theta - (1-\beta)a^2(\cos\theta)^2\ - \beta\ (a\, \cos\theta)^3 \tag{15}$$

Applying a suitable truncated Fourier series to make the right sides of Eq. (15) as a combination of linear harmonics, we get

$$\begin{aligned}\ddot{v}_{a1} + \Omega_{a0}^2 v_{a1} &= (\Omega_{a0}^2 a - a - 0.84882636\, a^2(1-\beta) - 0.75\beta\, a^3)\cos\theta \\ &\quad - (0.16976527\,(1-\beta)\, a^2 + 0.25\beta\, a^3)\cos 3\theta \\ &\quad + 0.02425218\, a^2\,(1-\beta)\cos 5\theta\end{aligned} \tag{16}$$

To avoid dominating terms, we obtain

$$\Omega_{a\,0} = \sqrt{1 + 0.84882636\, a\,(1-\beta) + 0.75\beta\, a^2} \tag{17}$$

Without applying the repeating procedure, it can be achieved in the negative direction for the trial function

$$v_b(t) = b\ \cos(\Omega_b t)\ \text{as}$$

$$\Omega_{b0} = \sqrt{1 + 0.84882636\, b\,(\beta - 1) + 0.75\beta\, b^2}$$

The first approximate frequency of the oscillator is

$$\Omega_0 = \frac{\Omega_{a0} + \Omega_{b0}}{2} \tag{18}$$

After simplification in Eq. (16) we have

$$\ddot{v}_{a1} + \Omega_a^2 v_{a1} = -(0.16976527\,(1-\beta)\, a^2 + 0.25\beta\, a^3)\cos 3\theta$$

$$+0.02425218\,a\,(1-\beta)\cos 5\theta \tag{19}$$

The particular solution of Eq. (9) is

$$\begin{aligned} v_{a1}^{p}(t) = {} & \frac{(0.16976527\,(1-\beta)\,a^2 + 0.25\beta\,a^3)}{8(1+0.84882636\,a\,(1-\beta)+0.75\beta\,a^2)^2}\cos 3\theta \\ & - \frac{0.02425218\,a^2\,(1-\beta)}{24(1+0.84882636\,a\,(1-\beta)+0.75\beta\,a^2)^2}\cos 5\theta \end{aligned} \tag{20}$$

The complete solution is

$$\begin{aligned} v_{a1}(t) = {} & c\,\cos\theta + \frac{(0.16976527\,(1-\beta)\,a^2 + 0.25\beta\,a^3)}{8(1+0.84882636\,a\,(1-\beta)+0.75\beta\,a^2)^2}\cos 3\theta \\ & - \frac{0.02425218\,a^2\,(1-\beta)}{24(1+0.84882636\,a\,(1-\beta)+0.75\beta\,a^2)^2}\cos 5\theta \end{aligned} \tag{21}$$

Using $v_{a1}(0) = a$ then

$$\begin{aligned} v_{a1}(t) = {} & \frac{(a+0.828616212\,(1-\beta)\,a^2 + 0.71875\beta\,a^3)}{(1+0.84882636\,a\,(1-\beta)+0.75\beta\,a^2)^2}\cos\theta \\ & + \frac{(0.02122066\,(1-\beta)\,a^2 + 0.03125\beta\,a^3)}{(1+0.84882636\,a\,(1-\beta)+0.75\beta\,a^2)^2}\cos 3\theta \\ & - \frac{0.00101051\,a^2\,(1-\beta)}{(1+0.84882636\,a\,(1-\beta)+0.75\beta\,a^2)^2}\cos 5\theta \end{aligned} \tag{22}$$

The first approximate solution of (9) is

$$v_{a1}(t) = \lambda_{a1}\cos\theta + \lambda_{a2}\cos 3\theta - \lambda_{a3}\cos 5\theta, \tag{23}$$

where

$$\lambda_{a1} = \frac{(a+0.828616212\,(1-\beta)\,a^2 + 0.71875\beta\,a^3)}{(1+0.84882636\,a\,(1-\beta)+0.75\beta\,a^2)^2}$$

$$\lambda_{a2} = \frac{(0.02122066\,(1-\beta)\,a^2 + 0.03125\beta\,a^3)}{(1+0.84882636\,a\,(1-\beta)+0.75\beta\,a^2)^2}$$

$$\lambda_{a3} = \frac{0.00101051\,a^2\,(1-\beta)}{(1+0.84882636\,a\,(1-\beta)+0.75\beta\,a^2)^2}$$

In case of a negative direction for the trial function $v_b(t) = b\,\cos(\Omega_b t)$, we have

$$v_{b1}(t) = \frac{(b+0.828616212\,(\beta-1)\,b^2 + 0.71875\beta\,b^3)}{(1+0.84882636\,a\,(\beta-1)\,+0.75\beta\,a^2)^2}\cos\theta$$

$$+\frac{(0.02122066\,(\beta-1)\,b^2+0.03125\beta\,b^3)}{(1+0.84882636\,a\,(\beta-1)\,+0.75\beta\,b^2)^2}\cos 3\theta$$
$$-\frac{0.00101051\,b^2\,(\beta-1)}{(1+0.84882636\,a\,(\beta-1)\,+0.75\beta\,b^2)^2}\cos 5\theta \tag{24}$$

The first approximate solution of (10) is

$$v_{b1}(t)=\lambda_{b1}\cos\theta+\lambda_{b2}\cos 3\theta-\lambda_{b3}\cos 5\theta, \tag{25}$$

where

$$\lambda_{b1}=\frac{(b+0.828616212\,(\beta-1)\,b^2+0.71875\beta\,b^3)}{(1+0.84882636\,b(\beta-1)+0.75\beta\,b^2)^2}$$
$$\lambda_{b2}=\frac{(0.02122066\,(\beta-1)\,b^2+0.03125\beta\,b^3)}{(1+0.84882636\,b\,(\beta-1)+0.75\beta\,b^2)^2}$$
$$\lambda_{b3}=\frac{0.00101051\,b^2\,(\beta-1)}{(1+0.84882636\,b\,(\beta-1)+0.75\,\beta\,b^2)^2}$$

For the second iteration of the oscillator (9), we get

$$\begin{aligned}\ddot{v}_{a2}+\Omega_{a1}^2 v_{a2}=\Omega_{a1}^2 v_{a1}&-v_{a1}-2(1-\beta)\,v_{a0}v_{a1}-3\beta\,v_{a0}^2 v_{a1}\\&+(1-\beta)\,v_{a0}^2+2\beta\,v_{a0}^3\end{aligned} \tag{26}$$

$$\begin{aligned}\ddot{v}_{a2}+\Omega_{a1}^2 v_{a2}=&\ \Omega_{a1}^2(\lambda_{a1}\cos\theta+\lambda_{a2}\cos 3\theta-\lambda_{a3}\cos\theta)\\&-(\lambda_{a1}\cos\theta+\lambda_{a2}\cos 3\theta-\lambda_{a3}\cos\theta)\\&-3\beta\,a^2(\cos\theta)^2(\lambda_{a1}\cos\theta+\lambda_{a2}\cos 3\theta-\lambda_{a3}\cos 5\theta)\\&+2\beta\,a^3\cos^3\theta\ +(1-\beta)a^2\cos^2\theta\\&-2(1-\beta)\,a\cos\theta(\lambda_{a1}\cos\theta+\lambda_{a2}\cos 3\theta-\lambda_{a3}\cos 5\theta)\end{aligned} \tag{27}$$

Applying a suitable truncated Fourier series to make the right sides of Eq. (25) as a combination of linear harmonics, we get

$$\begin{aligned}&\ddot{v}_{a2}+\Omega_{a1}^2 v_{a2}\\&=(\Omega_{a1}^2\lambda_{a1}-\lambda_{a1}+1.5\beta\,a^3+0.84882636\,a^2(1-\beta)-1.69765272\,a\,(1-\beta)\lambda_{a1}\\&-2.25\beta\,a^2\lambda_{a1}-0.33953054\,a\,(1-\beta)\lambda_{a2}-0.048504364\,a\,(1-\beta)\lambda_{a3}\\&-0.75\beta\,a^2\lambda_{a2})\cos\theta+(\Omega_{a1}^2\lambda_{a2}-\lambda_{a2}+0.5\beta\,a^3+0.16976527\,a^2(1-\beta)\\&-0.33953054\,a\,(1-\beta)\lambda_{a1}-0.75\beta a^2\lambda_{a1}-1.30961782a\,(1-\beta)\lambda_{a2}\\&+0.040420304\,a\,(1-\beta)\lambda_{a3}-1.5\beta a^2\lambda_{a2}+0.75\beta a^2\lambda_{a3})\cos 3\theta\\&+(-\Omega_{a1}^2\lambda_{a3}+\lambda_{a3}-0.02425218\,a^2(1-\beta)+0.04850436\,a\,(1-\beta)\lambda_{a1}\end{aligned}$$

$$- 0.40420304a\,(1-\beta)\lambda_{a2} + 1.28610056a\,(1-\beta)\lambda_{a3} - 0.75\beta a^2\lambda_{a2} + 1.5\beta a^2\lambda_{a3})\cos 5\theta \tag{28}$$

To avoid dominating terms, we obtain

$$\begin{aligned}\Omega_{a1}^2 &= 1 + 1.69765272\,a\,(1-\beta) + 2.25\beta\,a^2 \\ &+ (0.33953054\,a\,(1-\beta) + 0.75\beta\,a^2)\frac{\lambda_{a2}}{\lambda_{a1}} \\ &+ 0.048504364\,a\,(1-\beta)\frac{\lambda_{a3}}{\lambda_{a1}} \\ &- (1.5\beta\,a^3 + 0.84882636\,a^2(1-\beta))\frac{1}{\lambda_{a1}}\end{aligned} \tag{29}$$

Without applying the repeating procedure, it can be achieved in the negative direction

$$\begin{aligned}\Omega_{b1}^2 &= 1 + 1.69765272\,a\,(\beta-1) + 2.25\beta\,b^2 \\ &+ (0.33953054\,b\,(\beta-1) + 0.75\beta\,b^2)\frac{\lambda_{b2}}{\lambda_{b1}} \\ &+ 0.048504364\,b\,(\beta-1)\frac{\lambda_{b3}}{\lambda_{b1}} \\ &- (1.5\beta\,b^3 + 0.84882636\,b^2(\beta-1))\frac{1}{\lambda_{b1}}\end{aligned} \tag{30}$$

In the second iterated level, the approximate frequency of the oscillator is

$$\Omega_1 = \frac{\Omega_{a1} + \Omega_{b1}}{2} \tag{31}$$

After simplification in Eq. (28) we have

$$\ddot{v}_{a2} + \Omega_{a1}^2 v_{a2} = \lambda_{a4}\cos 3\theta + \lambda_{a5}\cos 5\theta \tag{32}$$

where

$$\begin{aligned}\lambda_{a4} &= \Omega_{a1}^2\lambda_{a2} - \lambda_{a2} + 0.5\beta\,a^3 + 0.16976527\,a^2(1-\beta) \\ &- 0.33953054\,a\,(1-\beta)\lambda_{a1} - 0.75\beta a^2\lambda_{a1} \\ &- 1.30961782a\,(1-\beta)\lambda_{a2} + 0.040420304\,a\,(1-\beta)\lambda_{a3} \\ &- 1.5\beta a^2\lambda_{a2} + 0.75\beta a^2\lambda_{a3}\end{aligned} \tag{33}$$

$$\lambda_{a5} = \Omega_{a1}^2 \lambda_{a3} - \lambda_{a3} - 0.02425218\, a^2(1-\beta) \\ + 0.04850436\, a\,(1-\beta)\lambda_{a1} - 0.40420304 a\,(1-\beta)\lambda_{a2} \\ + 1.28610056 a\,(1-\beta)\lambda_{a3} - 0.75\beta a^2 \lambda_{a2} + 1.5\beta a^2 \lambda_{a3} \quad (34)$$

The second approximate solution of (9) is

$$v_{a2}(t) = \frac{24\Omega_{a1}^2 a + 3\lambda_{a4} + \lambda_{a5}}{24\Omega_{a1}^2} \cos\theta - \frac{\lambda_{a4}}{8\Omega_{a1}^2} \cos 3\theta \\ - \frac{\lambda_{a5}}{24\Omega_{a1}^2} \cos 5\theta \quad (35)$$

For the negative direction, the second approximate solution of the oscillator (10) is

$$v_{b2}(t) = \frac{24\Omega_{b1}^2 a + 3\lambda_{b4} + \lambda_{b5}}{24\Omega_{b1}^2} \cos\theta - \frac{\lambda_{b4}}{8\Omega_{b1}^2} \cos 3\theta \\ - \frac{\lambda_{b5}}{24\Omega_{b1}^2} \cos 5\theta, \quad (36)$$

where

$$\lambda_{b4} = \Omega_{b1}^2 \lambda_{b2} - \lambda_{b2} + 0.5\beta\, b^3 + 0.16976527\, b^2(\beta - 1) \\ - 0.33953054\, b\,(\beta - 1)\lambda_{b1} - 0.75\beta\, b^2 \lambda_{b1} \\ - 1.30961782\, b\,(\beta - 1)\lambda_{b2} + 0.040420304\, b\,(\beta - 1)\lambda_{b3} \\ - 1.5\,\beta\, b^2 \lambda_{a2} + 0.75\beta b^2 \lambda_{b3} \quad (37)$$

$$\lambda_{b5} = \Omega_{b1}^2 \lambda_{b3} - \lambda_{b3} - 0.02425218\, b^2(\beta - 1) \\ + 0.04850436\, b\,(\beta - 1)\lambda_{b1} - 0.40420304 b\,(\beta - 1)\lambda_{b2} \\ + 1.28610056 b\,(\beta - 1)\lambda_{b3} - 0.75\beta a^2 \lambda_{b2} + 1.5\beta b^2 \lambda_{b3} \quad (38)$$

## 4 Results and Discussions

We have applied a modified Mickens' extended iteration method (MMEIM) to achieve the approximate solutions of the Helmholtz-Duffing oscillator. Here we have intended a sequence of the approximate frequencies and the corresponding analytical solutions of the oscillator. All the obtained frequencies are shown in the following Table 1. To compare with the approximate frequencies of the oscillator, we have also shown the existing values of frequencies obtained by Alal et al. [2], Askari et al. [3], Leung and Guo [26, 27]. Analyzing the results, we see that the percentage errors are

**Table 1** Comparison of the approximate frequencies with the existing and the exact values of frequencies for $\beta = 0.9$

| $a$ | $\Omega_0$ Er(%) | $\Omega_1$ Er(%) | ${\Omega_{[2]}}^{2nd\ MHBM}$ Er(%) | ${\Omega_{[3]}}^{EBM}$ Er(%) | ${\Omega_{[3]}}^{EBM\ (Galerkin-Petrov)}$ Er(%) | $\Omega_{[26,27]}^{IHHBM\ and\ HPM}$ Er(%) | ${\Omega_{[3]}}^{ex}$ |
|---|---|---|---|---|---|---|---|
| 0.01 | 1.00003365<br>2.8999e-007 | 1.00003365<br>2.8999e-005 | 1.00003365<br>2.8999e-005 | 1.00003335<br>1.6399e-006 | 1.00003113<br>2.2300e-004 | 1.00003335<br>1.6399e-006 | 1.00003336 |
| 0.1 | 1.00336040<br>1.0724e-003 | 1.00335899<br>9.3187e-004 | 1.00335895<br>9.27897e-004 | 1.00335001<br>3.5992e-005 | 1.00312738<br>0.022152 | 1.00335106<br>1.4068e-004 | 1.00334964 |
| 0.5 | 1.08090914<br>0.109295 | 1.08043296<br>0.153301 | 1.08044075 0.152581 | 1.08256404<br>0.043640 | 1.07725564<br>0.446928 | 1.08258341<br>0.045430 | 1.08209182 |
| 1.0 | 1.29380212<br>0.309771 | 1.28934491<br>0.653209 | 1.28954738<br>0.637608 | 1.30241057<br>0.353527 | 1.28444299<br>1.030913 | 1.30244785<br>0.356400 | 1.29782241 |
| 5.0 | 4.22758464<br>1.293254 | 4.14504306<br>0.684471 | 4.15285021<br>0.497411 | 4.25577625<br>1.968703 | 4.11954570<br>1.295389 | 4.25579387<br>1.969126 | 4.17361022 |
| 10 | 8.27631381<br>1.786028 | 8.09982073<br>0.384566 | 8.11709681<br>0.172097 | 8.30612634<br>2.152677 | 8.02862044<br>1.260221 | 8.30613546<br>2.152789 | 8.13109020 |
| 50 | 41.09132917<br>2.142131 | 40.189214<br>0.100287 | 40.27851570<br>0.121692 | 41.12172027<br>2.217675 | 39.71819248<br>1.246264 | 41.12172212<br>2.217680 | 40.22955916 |
| 100 | 82.16445296<br>2.181946 | 80.35897117<br>0.063400 | 80.53776157<br>0.158948 | 82.19486707<br>2.219770 | 79.40820415<br>1.245799 | 82.19486800<br>2.219771 | 80.40995076 |

*Note* where $\Omega_0$ and $\Omega_1$ represent the first and the second approximation frequencies of the adopted method, ${\Omega_{[2]}}^{2nd\ MHBM}$, ${\Omega_{[3]}}^{EBM}$, ${\Omega_{[3]}}^{EBM\ (Galerkin-Petrov)}$ and $\Omega_{[26,27]}^{HFAF\ and\ HPM}$ represent approximation frequencies obtained by Alal et al. [2], Askari et al. [3], and Leung and Guo [26, 27] respectively. ${\Omega_{[3]}}^{ex}$ is the exact frequency obtained by Askari et al. [3]. $Er(\%) = \left|\frac{\Omega_{ex}-\Omega_k}{\Omega_{ex}}\right| \times 100, \quad k = 0, 1, \cdots\cdots$ is the percentage Error lower for all small values of initial amplitude and almost the same for all methods. But the percentage errors are larger for large values of initial amplitude for all existing methods [2, 3, 26, 27] except our proposed method. In our proposed method, the percentage errors of the second approximate frequency are less for large values of initial amplitude and all calculated values of the second approximate frequency are proximate to the exact values. A comparison of errors is made between the proposed method for $a = 50, 100$ together with the existing methods, which are shown in Fig. 1. The compare between the second approximate solutions for the asymmetric parameter $\beta = 0.9$ and the initial amplitude $a = 10$ together with the exact solutions is presented in Fig. 2.

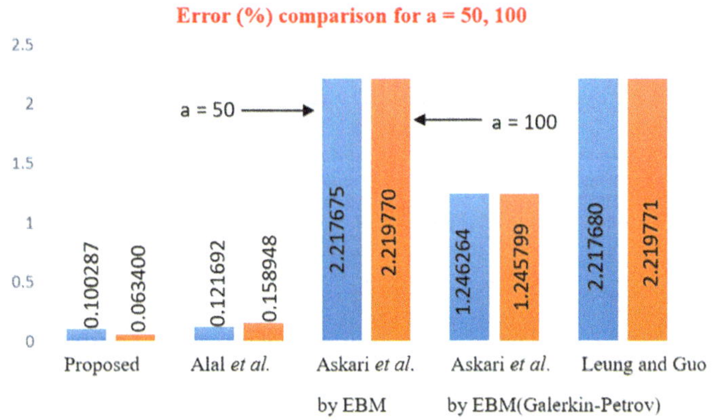


**Fig. 1** Errors comparison among the proposed method for $a = 50,\ 100$ together with the existing methods

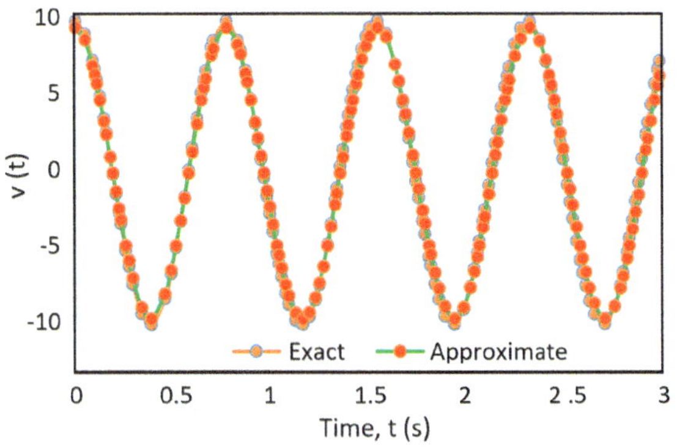


**Fig. 2** A Comparison between the second-order approximate solutions of the Eq. (8) for $\beta = 0.9$ and $a = 10$ together with the corresponding exact solutions

## 5 Conclusion

After consideration of He's energy balance method (HEBM), He's frequency amplitude formulation (HFAF), first-order of the Hamiltonian approach, coupled homotopy-variational formulation, and Modified harmonic balance method (MHBM), it can be clearly shown that the proposed method, modified Mickens' extended iteration method (MMEIM), gives the excellent results that are very close to the exact values of the approximate frequencies, especially the second approximate frequency, and the obtained results are better than all existing results. Also, using the proposed method, the periodic solutions of the oscillator are very simple, easy, and straightforward compared to other existing methods. It is observed that a good number of the researchers have concentrated to modify the method to achieve further improvement of the solutions, but in our research, we have given concentration to rearranging the principal oscillators with their own merit and taking appropriate harmonic terms. It has been accomplished that the applied two themes of approach are also a crucial issue for determining the better improvement of the analytical solutions in an iteration method.

## References


1. Alal, M.H., Chowdhury, M.S.H., Yeakub, M.A., Faris, A.I.: An analytical approximation technique for the duffing oscillator based on the energy balance method. Italian J. Pur. App. Math. **37**, 455–466 (2016)
2. Alal, M.H., Chowdhury, M.S.H.: Analytical approximate solutions for the helmholtz-duffing oscillator. ARPN J. Eng. Appl. Sci. **10**(23), 17363–17369 (2015)
3. Askari, H., Saadatnia, Z., Esmailzadeh, E., Younesian, D.: Approximate periodic solution for the Helmholtz-Duffing equation. Comput. Math. Appl. **62**, 3894–3901
4. Azami, R., Ganji, D.D., Babazadeh, H., Dvavodi, A.G., Ganji, S.S.: He's Max-min method for the relativistic oscillator and high order Duffing equation. Int. J. Mod. Phys. B **23**(32), 5915–5927 (2009)
5. Beléndez, A., Hernamdez, A., Beléndez, T., Fernandez, E., Alvarez, M. L., Neipp, C.: Application of He's homotopy perturbation method to Duffing-harmonic Oscillator. Int. J. Nonlinear Sci. and Numer. Simul., 8(1), 79–88 (2007).
6. Beléndez, A., Pascual, C., Ortuno, M., Beléndez, T., Gallego, S.: Application of a modified He's homotopy perturbation method to obtain higher order approximations to a nonlinear oscillator with discontinuities. Nonlinear Anal. Real World Appl. **10**(2), 601–610 (2009)
7. Chowdhury, M.S.H., Alal, M.H., Kartini, A., Ali, M.Y., Ismail, A.F.: High-order approximate solutions of strongly nonlinear cubic-quintic Duffing oscillator based on the harmonic balance method. Results Phys. **7**, 3962–3967 (2017)
8. Daeichin, M., Ahmadpoor, M.A., Askari, H., Yildirim, A.: Rational energy balance method to nonlinear oscillators with cubic term. Asian Eur. J. Math. **6**(02), 13500–13519 (2013)
9. Durmaz, S., Kaya, M.O.: High-order energy balance method to nonlinear oscillators. J. Appl. Math. (2012)
10. Elias-Zuniga, A., Oscar, M.-R., Rene, K.C.-D.: Approximate solution for the Duffing-harmonic oscillator by the Enhanced Cubication Method. Math. Probl. Eng. (2012)
11. Ganji, D.D., Gorji, M., Soleimani, S., Esmaeilpour, M.: Solution of nonlinear cubic-quintic Duffing oscillator using He's Energy Balanced Method. J. Zhejiang Univ. Sci. A **10**(9), 1263–1268 (2009)

12. Haque, B.M.I., Alam, M.S., Majedur, R.M.: Modified solutions of some oscillators by iteration procedure. J. Egyptian Math. Soci. **21**, 68–73 (2013)
13. Haque, B.M.I.: A new approach of Mickens' iteration method for solving some nonlinear jerk equations. Glob. J. Sci. Front. Res. Math. Decis. Sci. **13**(11), 87–98 (2013)
14. Haque, B.M.I., Alam, M.S., Majedur, R. M., Yeasmin I. A.: Iterative technique of periodic solutions to a class of non-linear conservative systems. Int. J. Concept. Comput. Inf. Technol. **2**(1), 92–97 (2014)
15. Haque, B.M.I.: A new approach of Mickens' extended iteration method for solving some nonlinear jerk equations. British J. Math. Comput. Sci. **4**(22), 3146–3162 (2014)
16. Haque, B.M.I., Bayezid, B.M., Ayub, H.M.M., Hossain, M.R., Rahman, M.M.: Mickens iteration like method for approximate solution of the inverse cubic nonlinear oscillator. British J. Math. Comput. Sci. **13**, 1–9 (2015)
17. Haque, B.M.I., Ayub, H.M.M., Bayezid, B.M., Hossain, M.R.: Analytical approximate solutions to the nonlinear singular oscillator: an iteration procedure. British J. Math. Comput. Sci. **14**, 1–7 (2016)
18. Haque, B.M.I., Asifuzzaman, M., Kamrul, H.M.: Improvement of analytical solution to the inverse truly nonlinear oscillator by extended iterative method. Commun. Comput. Inf. Sci. **655**, 412–421 (2017)
19. Haque, B.M.I., Selim, R.A.K.M., Mominur, R.M.: On the analytical approximation of the nonlinear cubic oscillator by an iteration method. J. Adv. Math. Comput. Sci. **33**, 1–9 (2019)
20. Haque, B.M.I., Ayub, H.M.M.: A modified solution of the nonlinear singular oscillator by extended iteration procedure. J. Adv. Math. Comput. Sci. **34**, 1–9 (2019)
21. Haque, B.M.I., Afrin, F.S.: On the analytical approximation of the quadratic nonlinear oscillator by modified extended iteration Method. Appl. Math. Nonlinear Sci. **6**, 1–10 (2020)
22. Haque, B.M.I, Zaidur, R.M., Iqbal, H.M.: Periodic solution of the nonlinear jerk oscillator containing velocity times acceleration-squared: an iteration approach. J. Mech. Continua Math. Sci. **15**(6), 419–433 (2020)
23. Haque, B.M.I., Iqbal, H.M.: An analytical approach for solving the nonlinear Jerk Oscillator containing velocity times acceleration-squared by an extended iteration method. J. Mech. Continua Math. Sci. **16**(2), 35–47 (2021)
24. Hosen, M.A., Chowdhury, M.S.H., Ali, M.Y., Ismail, A.F.: A new analytic approximation technique for highly nonlinear oscillations based on energy balanced method. Results Phys. **6**, 496–504 (2016)
25. Hosen, M.A., Chowdhury, M.S.H.: A new reliable analytic solution for strongly nonlinear oscillator with cubic and harmonic restoring force. Results Phys. **5**, 111–114 (2015)
26. Leung, A.Y.T., Guo, Z.J.: Homotopy perturbation for conservative Helmholt-Duffing oscillator. J. Sound Vib. **325**, 287–296 (2009)
27. Leung, A.Y.T., Guo, Z.J.: The iterative homotopy harmonic balance method for conservative Helmholt-Duffing oscillator. Appl. Math. Comput. **215**, 3163–3169 (2010)
28. Mickens, R.E.: Comments on the method of harmonic balance. J. Sound Vib. **94**, 456–460 (1984)
29. Mickens, R.E.: Iteration Procedure for determining approximate solutions to nonlinear oscillator equation. J. Sound Vib. **116**, 185–188 (1987)
30. Mickens, R.E.: A general procedure for calculating approximation to periodic solutions of truly nonlinear oscillators. J. Sound Vib. **287**, 1045–1051 (2005)
31. Mickens, R.E.: Truly Nonlinear Oscillations. World Scientific, Singapore (2010)
32. Akbarzade, M., Khan, Y., Kargar, A.: Determination of periodic solution for the Helmholtz-Duffing oscillators by Hamiltonian approach and coupled homotopy-variational formulation. Int. J. Phys. Sci. **7**, 560–565 (2012)
33. Nayfeh, A.H.: Perturbation Method. Wiley, New York (1973)
34. Nayfeh, A.H., Mook, D.T.: Nonlinear Oscillations. Wiley, New York (1979)

# Crank-Nicolson Finite Difference Scheme for Time-Space Fractional Diffusion Equation

**Kalyanrao C. Takale and Veena V. Sangvikar (Kshirsagar)**

**Abstract** This paper aims in developing the Crank-Nicolson type of finite difference scheme for space-time fractional order diffusion equation (TSFDE) with a non-linear term. The proof for scheme to be unconditionally stable and also convergent is been discussed. Further, an application in terms of numerical solution is solved and graph simulated using Mathematica.

**Keywords** Finite difference scheme · Caputo derivative · Space-time fractional diffusion equation · Stableness of scheme · Convergence · Mathematica software

## 1 Introduction

Recently, fractional order partial differential equations have been widely used by researchers to represent any physical phenomina and study its minute and diversed applications in science and technology, fluid mechanics, control systems, biology, viscoelasticity, physics, dynamical systems, etc. [4, 12, 14]. Major benefit that the fractional derivatives provide is that of being a best estimate for minute elaboration of memory as well as hereditary properties of different processes and involved materials [6, 9, 13]. But, it is very difficult to tackle partial differential equations of fractional order for exact solution. Researchers find variety of essential dynamical systems, exhibit fractional order behaviour which could change with space, time or both space-time and hence the analytical solution becomes difficult. This provoked many researchers to develop numerical methods.

We consider the space-time fractional heat-transfer/diffusion equation. The space-time fractional equation of diffusion is obtained using the standard equation of diffusion by replacing second order derivative in space variable by fractional derivative of order $\beta$, $1 < \beta < 2$ [3, 13] and the first order derivative in time variable by frac-

K. C. Takale
Department of Mathematics, NSC Bytco Science College, Nashik, MS, India

V. V. Sangvikar (Kshirsagar) (✉)
School of Mathematics and Statistics, MIT World Peace University, Pune, MS, India
e-mail: kshirsagar.v.p@gmail.com

 
B. Rushi Kumar et al. (eds.), *Mathematics and Computing*, Springer Proceedings in Mathematics & Statistics 415, https://doi.org/10.1007/978-981-19-9307-7_54

tional derivative of order $\alpha,\ 0 < \alpha < 1$. Thus, we develop the time-space fractional Crank-Nicolson finite difference scheme for diffusion equation with a non-linear source term.

The following space-time fractional equation of diffusion (TSFDE) with a non-linear term is considered.

$$\frac{\partial^{\alpha} U(x,t)}{\partial t^{\alpha}} = d\frac{\partial^{\beta} U(x,t)}{\partial x^{\beta}} + f(U,x,t),\ 0 < x < L,\ t > 0 \tag{1}$$

$$initial\ condition : U(x,0) = \phi(x),\ 0 \le x \le L \tag{2}$$

$$boundary\ conditions : U(0,t) = U_L,\ U(L,t) = U_R,\ 0 \le t \le T \tag{3}$$

where diffusion coefficient $d > 0,\ 0 < \alpha \le 1,\ 1 < \beta \le 2$.

Below are few definitions of fractional derivatives which would be useful for our subsequent development of scheme [7, 9–11, 13].

**Definition 1.1** The definition of Caputo time fractional derivative of order $\alpha$, $(0 < \alpha \le 1)$ is

$$\frac{\partial^{\alpha} U(x,t)}{\partial t^{\alpha}} = \begin{cases} \frac{1}{\Gamma(1-\alpha)} \int_0^t \frac{\partial U(x,t)}{\partial \xi} \frac{d\xi}{(t-\xi)^{\alpha}}, & 0 < \alpha < 1 \\ \frac{\partial U(x,t)}{\partial t}, & \alpha = 1 \end{cases}$$

**Definition 1.2** The definition of Grunwald-Letnikov space fractional derivative of order $\beta$, $(1 < \beta \le 2)$ is

$$\frac{\partial^{\beta} U(x,t)}{\partial x^{\beta}} = \frac{1}{\Gamma(-\beta)} \lim_{N\to\infty} \frac{1}{h^{\beta}} \sum_{j=0}^{N} \frac{\Gamma(j-\beta)}{\Gamma(j+1)} U(x-(j-1)h,t)$$

where $\Gamma(.)$ is the gamma function.

The paper is planned in the following way: In Sect. 2, the Crank-Nicolson finite difference scheme is advanced for one dimensional time-space fractional order equation of diffusion. The schemes stability is discussed in Sect. 3 and the convergence is proved in Sect. 4. In the last session we have the numerical solution of the time-space fractional equation of diffusion which is graphically represented using Mathematica software.

## 2 Finite Difference Scheme

We now develop fractional order Crank-Nicolson type finite difference scheme for time-space fractional equation of diffusion [8, 15–19]. We consider following time-space fractional diffusion equation having non-linear term along with initial and boundary conditions.

$$\frac{\partial^\alpha U(x,t)}{\partial t^\alpha} = d\frac{\partial^\beta U(x,t)}{\partial x^\beta} + f(U,x,t),\ 0 < x < L,\ t > 0 \tag{4}$$

$$initial\ condition : U(x,0) = \phi(x),\ 0 \le x \le L \tag{5}$$

$$boundary\ conditions : U(0,t) = U_L\ and\ U(L,t) = U_R,\ 0 \le t \le T \tag{6}$$

where $0 < \alpha \le 1; 1 < \beta \le 2$ and d is diffusivity constant. For the implicit numerical approximation scheme, we define $h = \frac{(x_R - x_L)}{N} = \frac{L}{N}$ and $\tau = \frac{T}{N}$ the space and time steps respectively, such that $t_k = k\tau$; k = 0, 1,...,N be the integration time $0 \le t_k \le T$ and $x_i = x_L + ih$ for i = 0,1, ..., N. Let $U(x_i, t_k),\ i = 1, 2, ...N,\ k = 1, 2, ...n$, be the exact solution of the fractional partial differential equation (4)–(6) at the node point $(x_i, t_k)$. Let $U_i^k$ be the numerical approximation to $U(x_i, t_k)$. We discretise the time fractional derivative of equation (4) by the following scheme:

$$\begin{aligned}
\frac{\partial^\alpha U(x_i,t_{k+1})}{\partial t^\alpha} &\approx \frac{1}{\Gamma(1-\alpha)}\int_0^{t_{k+1}} \frac{1}{(t_{k+1}-\xi)^\alpha}\frac{\partial U(x_i,\xi)}{\partial \xi}d\xi \\
&= \frac{1}{\Gamma(1-\alpha)}\sum_{j=0}^{k}\frac{U(x_i,t_{j+1}) - U(x_i,t_j)}{\tau}\int_{j\tau}^{(j+1)\tau}\frac{d\xi}{(t_{k+1}-\xi)^\alpha} \\
&= \frac{1}{\Gamma(1-\alpha)}\sum_{j=0}^{k}\frac{U(x_i,t_{j+1}) - U(x_i,t_j)}{\tau}\int_{(k-j)\tau}^{(k-j+1)\tau}\frac{d\eta}{\eta^\alpha} \\
&= \frac{1}{\Gamma(1-\alpha)}\sum_{j=0}^{k}\frac{U(x_i,t_{k+1-j}) - U(x_i,t_{k-j})}{\tau}\int_{j\tau}^{(j+1)\tau}\frac{d\eta}{\eta^\alpha} \\
&= \frac{\tau^{1-\alpha}}{\Gamma(2-\alpha)}\sum_{j=0}^{k}\frac{U(x_i,t_{k+1-j}) - U(x_i,t_{k-j})}{\tau}[(j+1)^{1-\alpha} - j^{1-\alpha}]
\end{aligned}$$

$$\begin{aligned}
\frac{\partial^\alpha U(x_i,t_{k+1})}{\partial t^\alpha} = &\frac{\tau^{-\alpha}}{\Gamma(2-\alpha)}[U(x_i,t_{k+1}) - U(x_i,t_k)] + \\
&\frac{\tau^{-\alpha}}{\Gamma(2-\alpha)}\sum_{j=1}^{k} b_j[U(x_i,t_{k+1-j}) - U(x_i,t_{k-j})]
\end{aligned}$$

where $b_j = (j+1)^{1-\alpha} - j^{1-\alpha},\ j = 1, 2, ..., k$.

For $\frac{\partial^\beta U(x,t)}{\partial x^\beta} =_0 D_x^\beta U(x,t)$, we use the shifted *Grünwald* finite difference formula at all time levels as follows

$$\frac{\partial^\beta U(x_i, t_{k+1})}{\partial x^\beta} = oD_x^\beta U(x_i, t_{k+1})$$
$$= \frac{1}{h^\beta} \sum_{j=0}^{i+1} g_{\beta,j} U[x_i - (j-1)h, t_{k+1}] + O(h^2)$$

Here the *Grünwald* normalized weights are defined by

$$g_{\beta,0} = 1, \ \ g_{\beta,j} = \frac{\Gamma(j-\beta)}{\Gamma(-\beta)\Gamma(j+1)}, \ \ j = 0, 1, ...$$

On substituting *Grünwald* estimates in the superdiffusion equation (4) to obtain the Crank-Nicolson type numerical approximation, the obtained finite difference equation is

$$\frac{\tau^{-\alpha}}{\Gamma(2-\alpha)}[U_i^{k+1} - U_i^k] + \frac{\tau^{-\alpha}}{\Gamma(2-\alpha)} \sum_{j=1}^{k} b_j [U_i^{k-j+1} - U_i^{k-j}] = \frac{d}{2} (\delta_{\beta,x} U_i^{k+1} + \delta_{\beta,x} U_i^k) + f_i^k \tag{7}$$

where $f_i^k = f(U_i^k, x_i, t_k)$ and the above operator which is a fractional partial differential, is defined as

$$\delta_{\beta,x} U_i^k = \frac{1}{h^\beta} \sum_{j=0}^{i+1} g_{\beta,j} U_{i-j+1}^k \tag{8}$$

Therefore, from (2.4) and (2.5) we get

$$\frac{\tau^{-\alpha}}{\Gamma(2-\alpha)}[U_i^{k+1} - U_i^k] + \frac{\tau^{-\alpha}}{\Gamma(2-\alpha)} \sum_{j=1}^{k} b_j [U_i^{k-j+1} - U_i^{k-j}] = \frac{d}{2h^\beta} \{\sum_{j=0}^{i+1} g_{\beta,j} U_{i-j+1}^{k+1} + \sum_{j=0}^{i+1} g_{\beta,j} U_{i-j+1}^k\} + f_i^k$$

$$U_i^{k+1} - U_i^k + \sum_{j=1}^{k} b_j [U_i^{k-j+1} - U_i^{k-j}] = \frac{d\tau^\alpha \Gamma(2-\alpha)}{2h^\beta} \{\sum_{j=0}^{i+1} g_{\beta,j} U_{i-j+1}^{k+1} + \sum_{j=0}^{i+1} g_{\beta,j} U_{i-j+1}^k\} + \tau^\alpha \Gamma(2-\alpha) f_i^k$$

$$U_i^{k+1} - U_i^k + \sum_{j=1}^{k} b_j [U_i^{k-j+1} - U_i^{k-j}] = r\{\sum_{j=0}^{i+1} g_{\beta,j} U_{i-j+1}^{k+1} + \sum_{j=0}^{i+1} g_{\beta,j} U_{i-j+1}^k\} + \tau^\alpha \Gamma(2-\alpha) f_i^k \tag{9}$$

$$where \ r = \frac{d\tau^\alpha \Gamma(2-\alpha)}{2h^\beta} \ \ for\ i = 0, 1, 2, ...N, k = 0, 1, 2, ...$$

After further simplification, we get

$$(1 - r g_{\beta,1}) U_i^{k+1} - r \sum_{j=0, j\neq 1}^{i+1} g_{\beta,j} U_{i-j+1}^{k+1} = (1 - b_1 + r g_{\beta,1}) U_i^k + \sum_{j=1}^{k-1} (b_j - b_{j+1}) U_i^{k-j}$$
$$+ r \sum_{j=0, j\neq 1}^{i+1} g_{\beta,j} U_{i-j+1}^k + b_k U_i^0 + \tau^\alpha \Gamma(2-\alpha) f_i^k$$

The approximation to initial condition is as $U_i^0 = \phi(x_i),\ i = 0, 1, 2, \ldots$ The approximations to boundary conditions are as $U_0^k = U_L,\ U_N^k = U_R,\ k = 0, 1, 2, \ldots$ Hence, the complete discretised scheme to IBVP (4)–(6) is

$$(1+\beta r)U_i^1 - r\sum_{j=0, j\neq 1}^{i+1} g_{\beta,j}U_{i-j+1}^1 = (1-\beta r)U_i^0 + r\sum_{j=0, j\neq 1}^{i+1} g_{\beta,j}U_{i-j+1}^0 + \tau^\alpha\Gamma(2-\alpha)f_i^0,\ for\ k=0 \tag{10}$$

$$\begin{aligned}(1+\beta r)U_i^{k+1} - r\sum_{j=0, j\neq 1}^{i+1} g_{\beta,j}U_{i-j+1}^{k+1} &= (1-r\beta-b_1)U_i^k + \sum_{j=1}^{k-1}(b_j-b_{j+1})U_i^{k-j}\\ &\quad + r\sum_{j=0, j\neq 1}^{i+1} g_{\beta,j}U_{i-j+1}^k + b_kU_i^0 + \tau^\alpha\Gamma(2-\alpha)f_i^k,\ for\ k\geq 1\end{aligned} \tag{11}$$

$$initial\ condition:\ U_i^0 = \phi(x_i),\ i = 0, 1, 2, \ldots \tag{12}$$

$$boundary\ conditions:\ U_0^k = U_L,\ U_N^k = U_R,\ k = 0, 1, 2, \ldots. \tag{13}$$

where $r = \frac{d\tau^\alpha\Gamma(2-\alpha)}{2h^\beta}$, $g_{\beta,0} = 1$, $g_{\beta,1} = (-\beta)$, $g_{\beta,j} = \frac{\Gamma(j-\beta)}{\Gamma(-\beta)\Gamma(j+1)}$, and $b_j = (j+1)^{1-\alpha} - j^{1-\alpha}$, $j = 0, 1, 2, .., k$. The finite-difference Eqs. (10) to (13) are expressed in the matrix form as:

$$AU^1 = BU^0 + \tau^\alpha\Gamma[2-\alpha]f_i^0 \tag{14}$$

$$AU^{k+1} = BU^k + \sum_{j=1}^{k-1}(b_j-b_{j+1})U^{k-j} + b_kU^0 + \tau^\alpha\Gamma[2-\alpha]f_i^k + D \tag{15}$$

where $U^k = (U_1^k, U_2^k, \ldots, U_{N-1}^k)^T$, $k = 0, 1, 2\ldots, N$ $A = (a_{ij})$ is a (N − 1) ordered square matrix of coefficients

$$A = \begin{pmatrix} (1+r\beta) & (-r) & & & \\ (-r)g_{\beta,2} & (1+r\beta) & (-r) & & \\ (-r)g_{\beta,3} & (-r)g_{\beta,2} & (1+r\beta) & (-r) & \\ \vdots & \vdots & \ddots & \ddots & \ddots \\ (-r)g_{\beta,m-1} & (-r)g_{\beta,m-2} & (-r)g_{\beta,m-3} & \cdots & \cdots & (1+r\beta) \end{pmatrix}$$

$B = (b_{ij})$ is a (N − 1) ordered square matrix of coefficients

$$B = \begin{pmatrix} (1-b_1-r\beta) & r & & & \\ rg_{\beta,2} & (1-b_1-r\beta) & r & & \\ rg_{\beta,3} & rg_{\beta,2} & (1-b_1-r\beta) & r & \\ \vdots & \vdots & \ddots & \ddots \ddots & \\ rg_{\beta,m-1} & rg_{\beta,m-2} & rg_{\beta,m-3} & \cdots\cdots & (1-b_1-r\beta) \end{pmatrix}$$

and D is a constant column matrix given by

$$D = \begin{pmatrix} rg_{\beta,2}(U_0^k + U_0^{k+1}) \\ rg_{\beta,3}(U_0^k + U_0^{k+1}) \\ rg_{\beta,4}(U_0^k + U_0^{k+1}) \\ \vdots \\ \vdots \\ rg_{\beta,N-1}(U_0^k + U_0^{k+1}) \\ rg_{\beta,N}(U_0^k + U_0^{k+1}) + r(U_N^k + U_N^{k+1}) \end{pmatrix}$$

where $r = \frac{d\tau^\alpha \Gamma(2-\alpha)}{2h^\beta}$, $g_{\beta,j} = \frac{\Gamma(j-\beta)}{\Gamma(-\beta)\Gamma(j+1)}$, $b_j = (j+1)^{1-\alpha} - j^{1-\alpha}$, $j = 0, 1, 2, ..., k$.

The system above is of algebraic equations which will be solved using some mathematical software tool preferably Mathematica.

Next, we work on stability of solution that would be obtained from the developed time-space fractional Crank-Nicolson finite difference scheme (10)–(13) for the time-space fractional diffusion equation (TSFDE) (4)–(6).

## 3 Stability

**Lemma 3.1** *For $i = 1, 2, ..., N$, $k = 1, 2, ..., N$, $0 < \alpha \leq 1$, $1 < \beta \leq 2$ the coefficients $b_j$ and $g_{\beta,j}$ for $j = 0, 1, 2, ....$ satisfy*
*(i) $b_j > b_{j+1}$, $j = 0, 1, 2,...$*
*(ii) $b_0 = 1$, $b_j > 0$, $j = 0, 1, 2,...$*
*(iii) $g_{\beta,0} = 1$, $g_{\beta,1} = -\beta$, $g_{\beta,j} \geq 0\,(j \neq 1)$, $\sum_{j=0}^{\infty} g_{\beta,j} = 0$*
*(iv) We have $\sum_{j=1}^{n} g_{\beta,j} < 0$, for any positive integer $n$.*

**Definition 3.1** For $E^0$, being some initial rounding error arbitrarily, if there exists $c$ a positive number, independent of $h$ and $\tau$ such that $\|E^k\| \leq c\|E^0\|$ or $\|E^k\| \leq c$, then the difference approximation is stable.

**Theorem 3.2** *Solution obtained from the Crank-Nicolson finite approximation scheme defined by* (10)–(13) *is unconditionally stable.*

***Proof*** We assume that $\bar{U}_i^k$ is a vector of exact solution of TSFDE (4)–(6). Denote, $E_i^k = \bar{U}_i^k - U_i^k$ for $i = 0, 1, ...N$; $k = 0, 1, ...N$. where $E^0 = 0$ and $E^k = (\varepsilon_1^k, \varepsilon_2^k, ..., \varepsilon_{N-1}^k)^T$. Furthermore, we assume that

$$|E_l^k| = \max_{1 \leq i \leq N-1} |\varepsilon_i^k| = \|E^k\|_\infty, \ for\ l = 1, 2, ...$$

Therefore, from Eq. (10), we get

$$\begin{aligned}
|E_l^1| &= |(1 + r\beta)\varepsilon_i^1 - r \sum_{j=0, j\neq 1}^{i+1} g_{\beta,j}\varepsilon_{i-j+1}^1| \\
&\leq |(1 - r\beta)||\varepsilon_i^0| + r \sum_{j=0, j\neq 1}^{i+1} g_{\beta,j}|\varepsilon_{i-j+1}^0| + \tau^\alpha \Gamma(2-\alpha)|f[U(x_i, t_0), x_i, t_0] - f[U_i^0, x_i, t_0]| \\
&\leq |\varepsilon_i^0| + \tau^\alpha \Gamma[2-\alpha]L|U(x_i, t_0) - U_i^0| \\
&\leq |\varepsilon_i^0| + \tau L|\varepsilon_i^0| \\
&\leq (1 + \tau L)|E_l^0| \\
\Rightarrow \|E^1\|_\infty &\leq (1 + \tau L)\|E^0\|_\infty \\
&\leq e^{\tau L}\|E^0\|_\infty
\end{aligned}$$

We assume that, $|E_l^k| = \|E^k\|_\infty \leq (1 + \tau L)^k \|E^0\|_\infty \leq e^{k\tau L}\|E^0\|_\infty$.
From Eq. (11) we get

$$\begin{aligned}
|E_l^{k+1}| &= |(1 + r\beta)\varepsilon_i^{k+1} - r \sum_{j=0, j\neq 1}^{i+1} g_{\beta,j}\varepsilon_{i-j+1}^{k+1}| \\
&\leq |(1 - b_1 - r\beta)\varepsilon_i^k + r \sum_{j=0, j\neq 1}^{i+1} g_{\beta,j}\varepsilon_{i-j+1}^k + \sum_{j=1}^{k-1}(b_j - b_{j+1})\varepsilon_i^{k-j} + b_k\varepsilon_i^0 + \\
&\qquad \tau^\alpha \Gamma[2-\alpha] f[U(x_i, t_k), x_i, t_0] - f[U_i^k, x_i, t_0]| \\
&\leq (1 - b_1)|\varepsilon_i^k| + (b_1 - b_k)|\varepsilon_l^k| + b_k|\varepsilon_i^k| + \tau L|\bar{U}_i^k - U_i^k| \\
&\leq (1 - b_1 + b_1 - b_k + b_k)|E_l^k| + \tau L|E_l^k| \\
&\leq (1 + \tau L)|E_l^k| \\
&\leq (1 + \tau L)^{k+1}\|E^0\|_\infty \\
\Rightarrow \|E^{k+1}\|_\infty &\leq e^{\tau L(k+1)}\|E^0\|_\infty
\end{aligned}$$

Hence, by mathematical induction this shows that the Crank-Nicolson finite approximation scheme defined by (10)–(13) is unconditionally stable.

Proceeding further to the next section, we discuss the convergence of the approximate scheme.

## 4 Convergence

**Theorem 4.1** *Let the problem* (4)–(6) *has smooth solution* $U(x,t)\varepsilon C^{1+\alpha,2+\beta}_{x,t}(\Omega)$. *Let* $U^k_i$ *be the numerical approximate computed from* (10)–(13). *Then there exists a positive constant* $C$ *independent of* $i, k, h$ *and* $\tau$ *such that* $|U(x_i,t_k)-U^k_i| \le cO(\tau^2+h^2)$ *for* $i = 1, 2...N-1$; $k = 1, 2, ...N$.

***Proof*** Define $e^k_i = U(x_i,t_k)-U^k_i$ for $i = 0, 1, ...N$; $k = 0, 1, ...N$. Where $E^0 = 0$ and $E^k = (e^k_1, e^k_2, ..., e^k_N)^T$. Furthermore, we assume that $|e^k_l| = \max\limits_{1\le i\le N-1} |e^k_i| = \|E^k\|_\infty$, *for* $l = 1, 2, ...$ and $T^k_l = \max\limits_{1\le i\le N-1} |T^k_i|$ then using $\sum\limits_{j=0}^{\infty} g_{\beta,j} = 0$ and $\sum\limits_{j=1}^{i+1} g_{\beta,j} < 0$, from Eq. (10), we get

$$
\begin{aligned}
|e^1_l| &= |(1+r\beta)e^1_i - r\sum_{j=0,j\ne 1}^{i+1} g_{\beta,j}e^1_{i-j+1}| \\
&\le |(1-r\beta)||e^0_i| + r\sum_{j=0,j\ne 1}^{i+1} g_{\beta,j}|e^0_{i-j+1}| + \\
&\qquad \tau^\alpha\Gamma(2-\alpha)|f[U(x_i,t_0),x_i,t_0] - f[U^0_i,x_i,t_0]| + |T^1_i| \\
&\le |e^0_i| + \tau^\alpha\Gamma[2-\alpha]L|U(x_i,t_0)-U^0_i| + |T^1_i| \\
&\le |e^0_i| + \tau L|e^0_i| + |T^1_i| \\
&\le (1+\tau L)|e^0_l| + |T^1_l| \\
\Rightarrow |e^1_l| &\le (1+\tau L)|e^0_l| + c_1O(\tau^2+h^2) \\
\Rightarrow \|E^1\|_\infty &\le (1+\tau L)\|E^0\|_\infty + cO(\tau^2+h^2)
\end{aligned}
$$

Assume that

$$\|E^k\|_\infty \le (1+\tau L)^k\|E^0\|_\infty + cO(\tau^2+h^2)$$

From Eq. (11), we get

$$
\begin{aligned}
|e^{k+1}_l| &= |(1+r\beta)e^{k+1}_i - r\sum_{j=0,j\ne 1}^{i+1} g_{\beta,j}e^{k+1}_{i-j+1}| \\
&\le (1-b_1-r\beta)|e^k_i| + r\sum_{j=0,j\ne 1}^{i+1} g_{\beta,j}|e^k_{i-j+1}| + \sum_{j=1}^{k-1}(b_j-b_{j+1})|e^{k-j}_i| + b_k|e^0_i| + \\
&\qquad \tau^\alpha\Gamma[2-\alpha]|f[U(x_i,t_k),x_i,t_k] - f[U^k_i,x_i,t_k]| + |T^{k+1}_i| \\
&\le (1-b_1)|\varepsilon^k_l| + (b_1-b_k)|\varepsilon^k_l| + b_k|\varepsilon^k_l| + \tau L|\bar{U}^k_i - U^k_i| + |T^{k+1}_l| \\
&\le (1-b_1+b_1-b_k+b_k)|E^k_l| + \tau L|E^k_l| + |T^{k+1}_l| \\
&\le (1+\tau L)|E^k_l| + c_2O(\tau^2+h^2) \\
\Rightarrow \|E^{k+1}\|_\infty &\le (1+\tau L)^{k+1}\|E^0\|_\infty + c_1O(\tau^2+h^2) + c_2O(\tau^2+h^2) \\
&\le (1+\tau L)^{k+1}\|E^0\|_\infty + cO(\tau^2+h^2)
\end{aligned}
$$

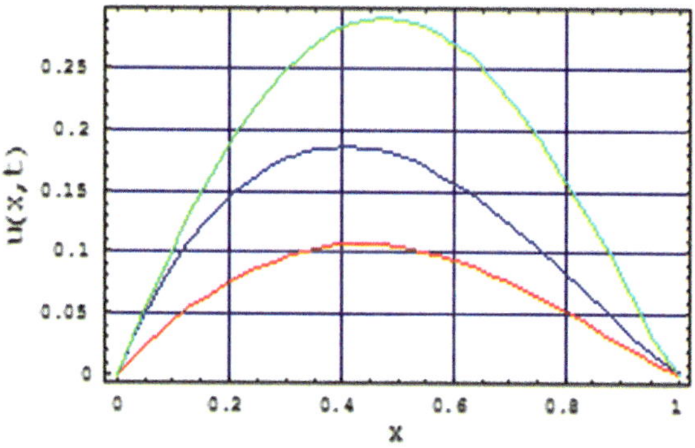


**Fig. 1** The diffusion profile with $t = 0.05,\ h = 0.1,\ \alpha = 0.7,\ \beta = 1.7(blue),\ \alpha = 0.8, \beta = 1.8(red)\ and\ \alpha = 0.9,\ \beta = 1.9(green)$

Hence, by induction we prove $\|E^k\|_\infty \le (1 + \tau L)^k \|E^0\|_\infty + cO(\tau^2 + h^2)$, for all $k = 1, 2, ...N$.

Therefore, we observe that for any $x$ and $t$, as $(h, \tau) \to (0, 0)$, $U_i^k$ converges to $U(x_i, t_k)$. Hence proof completed (Fig. 1).

## 5 Numerical Solutions

We now obtain the numerical solution of one dimensional time-space fractional diffusion equation by the discrete scheme developed in Eqs. (10)–(13). The following time-space fractional diffusion equation with initial and boundary conditions and a non-linear term is considered.

$$\frac{\partial^\alpha U(x,t)}{\partial t^\alpha} = \frac{\partial^\beta U(x,t)}{\partial x^\beta} + \sin U;\ 0 < x < 1,\ 0 < \alpha \le 1, 1 < \beta \le 2,\ t > 0$$

$$initial\ condition : U(x, 0) = \sin \pi x,\ 0 \le x \le 1$$

$$boundary\ conditions : U(0, t) = U_L = 0,\ U(1, t) = U_R = 0,\ t > 0$$

with the diffusion coefficient $d = 1$.

The numerical solution is obtained at $t = 0.05$ by considering the parameters $\tau = 0.005$ and $h = 0.1$, which are simulated using Mathematica Software for three different values of $\alpha$ and $\beta$ that is, $\alpha = 0.7,\ \beta = 1.7(blue)$, next $\alpha = 0.8,\ \beta = 1.8(red)$ and next $\alpha = 0.9,\ \beta = 1.9(green)$ followed by the solution graphically.

**Conclusions**
(i) We have successfully developed the Crank-Nicolson fractional order finite difference scheme for time-space fractional diffusion equation in a bounded domain.
(ii) We observe that the developed scheme is unconditionally stable.
(iii) Analysis shows clearly that the finite difference scheme is numerically stable and the results are compatible with our theoretical analysis. Therefore, these solution techniques can be applicable to other fractional partial differential equations.

## References


1. Baeumer, B., Meerschaert, M.M., Mortensen, J.: Space-time fractional derivative operators. Proc. Am. Math. Soc. **133**, 2273–2282 (2005)
2. Ben Adda, F.: Geometric interpritaion of the fractional derivative. J. Fract. Calc. **11**, 21–52 (1997)
3. Diethelm, K., Ford, N.J., Freed, A.D., Luchko, Yu.: Algorithms for the fractional calculus: a selection of numerical methods. Comput. Methods Appl. Mech. Engrg. **194**, 743–773 (2005)
4. Hilfer, R.: Applications of Fractional Calculus in Physics. World Scientific, Singapore (2000)
5. Jain, M.K., Iyengar, S.R.K., Jain, R.K.: Numerical Methods for Scientific and Engineering Computation, 3rd edn. New AGE International (P) Limited, Publishers, New Delhi (1992)
6. Strikwerda, J.C.: Finite Difference Schemes and Partial Differential Equations, 2nd edn. SIAM (2004)
7. Lavoie, J.L., Osler, T.J., Tremblay, R.: Fractional derivatives and special functions. SIAM Rev. **18**(2), 240–268
8. Mainardi, F., Luchko, Yu., Pagnini, G.: The fundamental solution of the space-time fractional diffusion equation. Fract. Calc. Appl. Anal. **4**, 153–192 (2001)
9. Miller, K., Ross, B.: An Introduction to the Fractional Calculus and Fractional Differential Equations. Eiley, New York (1993)
10. Munkhammar, J.D.: Riemann-Liouville Fractional Derivatives and the Taylor-Riemann Series, UUDM Project Report (2004)
11. Murio, D.A.: On stable numerical evaluation of caputo fractional derivatives. Comput. Math. Appl. **51**, 1539–1550 (2006)
12. Nishimoto, K. (ed.): Fractional Calculus and its Applications. Nihon University, Koriyama (1990)
13. Podlubny, I.: Fractional Differential Equations. Academic Press, San Diago (1999)
14. Pskhu, A.V.: Partial Differential Equations of Fractional Order. Nauka, Moscow (2005)
15. Yang, Q., Turner, I., Liu, F.: Analytical and numerical solutions for the time and space-symmetric fractional diffusion equation. ANZIAM J. **50**(CTA 2008), C800–C814 (2009)
16. Richtmeyer, R.D., Morton, K.W.: Difference Method in Initial Value Problems, 2nd edn. Interscience, New York (1967)
17. Sankara Rao, K.: Numerical Methods for Scientist and Engineers. Printice-Hall of India, New Delhi (2004)
18. Shen, S., Liu, F.: Error analysis of an explicit finite difference approximation for the space fractional diffusion equation with insulated ends. ANZIAM J. **46**(E), C871–C887 (2005)
19. Diego Murio, A.: Implicit finite difference approximation for time fractional diffusion equations. Comput. Math. Appl. **56**, 1138–1145 (2008)

# Gauss-Newton-Secant Method for the Solution of Non-linear Least-Square Problems Using $\omega$-Condition

**Naveen Chandra Bhagat, P. K. Parida, Chandresh Prasad, Sapan Kumar Nayak, Babita Mehta, and P. K. Sahoo**

**Abstract** The convergence of iterative process, based on the combination of Gauss-Newton and Secant's method, for the solution of nonlinear least-square problems in Banach space under $\omega$-condition for the first and second order divided difference and first order derivative is provided. To demonstrate the efficiency of proposed method, numerical experiments are carried out.

**Keywords** Least-square problems · Gauss-Newton-Secant method · $\omega$-condition · Jacobian · Divided difference

## 1 Introduction

Finding the numerical solution of non-linear least square problems is one of the important problems in computational mathematics. Non-linear least square problems are generally arise while solving nonlinear regression models, overdetermined system of nonlinear equations, solving engineering problems etc. We consider the least square problem of the type:

$$\min_{i^* \in D} \frac{1}{2} P(i^*)^T P(i^*) \tag{1}$$

where $P : D \subset \mathbb{R}^m \to \mathbb{R}^n$ is nonlinear in $i^*$ and is continuously differentiable function, $D$ is an open convex domain. Gauss-Newton's method [1, 2] is the most used technique for solving (1).

These authors contributed equally to this work.

N. C. Bhagat (✉) · P. K. Parida · C. Prasad · S. K. Nayak · B. Mehta · P. K. Sahoo
Department of Mathematics, Central University of Jharkhand, Ranchi 235205, JH, India
e-mail: navi.bgt@gmail.com

P. K. Parida
e-mail: pkparida@cuj.ac.in



B. Rushi Kumar et al. (eds.), *Mathematics and Computing*, Springer Proceedings in Mathematics & Statistics 415, https://doi.org/10.1007/978-981-19-9307-7_55

In this study let us consider the least square problem

$$\min_{i^* \in D} \frac{1}{2}(P(i^*) + Q(i^*))^T (P(i^*) + Q(i^*)), \tag{2}$$

where $P$ is continuously differentiable and $Q$ is continuous function. Also, $P + Q$ is nonlinear function which maps from $\mathbb{R}^m$ to $\mathbb{R}^n$, $n > m$. The domain $D$ is open convex set in $\mathbb{R}^m$ and differentiablity of $Q$ is not required. To solve (2), we propose Gauss-Newton-Secant method [3] which is combination of Gauss-Newton [4, 5] and Secant's method [6].

$$\left.\begin{array}{l} i_{j+1} = i_j - (\mathcal{N}^T \mathcal{N})^{-1} \mathcal{N}^T (P(i_j) + Q(i_j)) \\ \mathcal{N} = P'(i_j) + Q[i_j, i_{j-1}], \quad j = 0, 1, 2, \ldots \end{array}\right\}, \tag{3}$$

where, $P'$ is jacobian of $P$ and $Q[i_j, i_{j-1}]$ is first order divided difference of $Q$ with two arguments and $i_0, i_{-1}$ are given.

Relation (3) will reduce to Gauss-Newton-Kurchatov [7] method if we take $\mathcal{N} = P'(i_j) + Q[2i_j - i_{j-1}, i_{j-1}]$ and will become Gauss-Newton-potra method [8] when we take $\mathcal{N} = P'(i_j) + Q[i_j, i_{j-1}] + Q[i_{j-2}, i_j] - Q[i_{j-2}, i_{j-1}]$.

In this study, we use $\omega$-condition to provide a local convergence analysis of Gauss-Newton-Secant's method (3), where differentiablity of non-linear function is not required in the solution. Numerical experiments are also provided to verify conditions used in the convergence analysis of the method.

## 2 Conergence Analysis

In this section, by using sufficient conditions, we determine the local convergence analysis of our method.

Let $D$ is an open convex set in $\mathbb{R}^m$. Let us define $\delta D(i_0, a^*) = \{i^* : \|i^* - i_0\| < a^*\}$ be an open ball with center $i_0$ and radius $a^*(a^* > 0)$.

Let us assume the continuous function $Q$ and continuously differentiable function $P$ in the domain $D \subset \mathbb{R}^m$ and $P + Q : D \subset \mathbb{R}^m \to \mathbb{R}^n$ is the given function. Further we assume that $\omega_0 : \mathbb{R}_+ \to \mathbb{R}_+$ is a non negative continuous function for which the Fréchet derivative $P'$ satisfies the condition.

$$\|P'(m) - P'(n)\| \leq \omega_0(\|m - n\|). \tag{4}$$

Also, their exist a continuous non-negative function $g : [0, 1] \to \mathbb{R}_+$ such that $\omega_0(tz) \leq g(t)\omega_0(z)$ for $t \in [0, 1]$ and $z \in [0, \infty)$ and $T = \int_0^1 g(t)dt$.

Also, let $\omega_1 : \mathbb{R}_+ \to \mathbb{R}_+$, a continuous non-negative function for which the function $Q$ have first order divided difference, that satisfies the condition

$$\|[m, n; Q] - [p, q; Q]\| \leq \omega_1(\|m - p\|, \|n - q\|). \tag{5}$$

We now provide the following theorem under above assumptions, which gives the sufficient condition for the local convergence of our iterative method.

**Theorem 1** *Assume that $i^* \in D$ is a solution of the problem and let for some $i$, their exist $(N)^T(N)^{-1}$, where $(N) = P'(i^*) + Q[i, i^*]$, such that $\|i - i^*\| = \lambda > 0$ and $\|((N)^T(N))^{-1}\| \leq A$. Moreover, $\|(N)\| \leq \gamma$ and their exist $a^* \in \mathbb{R}_+$, such that $\delta D(i^*, a^*) \subset D$ and*

$$\alpha(a^*) + \tilde{\alpha}(a^*) < 1, \tag{6}$$

*where*

$$\alpha(a^*) = A[2\gamma + \omega_0(a^*) + \omega_1(0, a^*) + \omega_1(a^* + \lambda, 0)] \times [\omega_0(a^*) + \omega_1(0, a^*) + \omega_1(a^* + \lambda, 0)],$$

*and*

$$\tilde{\alpha}(a^*) = A[T\omega_0(a^*) + \omega_1(0, a^*)] \times [\gamma + \omega_0(a^*) + \omega_1(0, a^*) + \omega_1(a^* + \lambda, 0].$$

*Then, for each $i_0, i_{-1} \in D$ the iterative process is correctly defined and the sequence $\{i_j\}$, $j = 0, 1, 2, \ldots$ generated by this process belongs to $\delta D(i^*, a^*)$ and converges to the solution $i^*$. Further the following estimate holds for $j \geq 0$*

$$\|i_{j+1} - i^*\| \leq \frac{\tilde{\alpha}(a^*)}{1 - \alpha(a^*)} \|i_j - i^*\|. \tag{7}$$

***Proof*** The inequalities $\alpha(a^*) < 1$ and $\tilde{\alpha}(a^*) < 1$ holds, as $\alpha(a^*) + \tilde{\alpha}(a^*) < 1$ and hence $\frac{\tilde{\alpha}(a^*)}{1-\alpha(a^*)} < 1$. As per our assumption $i_0, i_{-1} \in D$. To prove the result we apply mathematical induction.

For $j = 0$, we will have,

$$\begin{aligned} \|\mathcal{N}_0 - \mathcal{N}_*\| &= \|P'(i_0) + Q(i_0, i_{-1}) - P'(i^*) - Q(\tilde{i}, i^*)\| \\ &\leq \|\mathcal{N}_0 - \mathcal{N}_*\| + \|Q(i_0, i_{-1}) - Q(i_0, i^*) + Q(i_0, i^*) - Q(\tilde{i}, i^*)\| \\ &\leq \omega_0(\|i_0 - i^*\|) + \omega_1(0, \|i_{-1} - i^*\|) + \omega_1(\|i_0 - i^*\| + \|i^* - \tilde{i}\|, 0). \end{aligned} \tag{8}$$

Therefore,

$$\begin{aligned}\|\mathcal{N}_0\| &= \|\mathcal{N}_* + \mathcal{N}_0 - \mathcal{N}_*\| \\ &\le \alpha + \omega(\|i_0 - i^*\|) + \omega_1(0, \|i_{-1} - i^*\|) + \omega_1(\|i_0 - i^*\| + \|i^* - \tilde{i}\|, 0).\end{aligned} \tag{9}$$

Also,

$$\begin{aligned}&\|I - (\mathcal{N}_*^T \mathcal{N}_*)^{-1} \mathcal{N}_0^T \mathcal{N}_0\| \\ &\le \|(\mathcal{N}_*^T \mathcal{N}_*)^{-1}\| \|\mathcal{N}_*^T (\mathcal{N}_* - \mathcal{N}_0) + (\mathcal{N}_*^T - \mathcal{N}_0^T)(\mathcal{N}_0 - \mathcal{N}_*) + (\mathcal{N}_*^T - \mathcal{N}_0^T)\mathcal{N}_*\| \\ &\le A[2\alpha + \omega_0(\|i_0 - i^*\|) + \omega_1(0, \|i_{-1} - i^*\|) + \omega_1(\|i_0 - i^*\| + \|i^* - \tilde{i}\|, 0)] \\ &\quad \times [\omega_0(\|i_0 - i^*\|) + \omega_1(0, \|i_{-1} - i^*\|) + \omega_1(\|i_0 - i^*\| + \|i^* - \tilde{i}\|, 0)] \\ &\le A[2\alpha + \omega_0(a^*) + \omega_1(0, a^*) + \omega_1(a^* + \lambda, 0)] \times [\omega_0(a^*) + \omega_1(0, a^*) + \omega_1(a^* + \lambda, 0)].\end{aligned} \tag{10}$$

Using (10), we can have

$$\begin{aligned}\|(\mathcal{N}_0^T \mathcal{N}_0)^{-1}\| &\le A\{1 - A[2\alpha + \omega(\|i_0 - i^*\|) + \omega_1(0, \|i_{-1} - i^*\|) \\ &\quad + \omega_1(\|i_0 - i^*\| + \|i^* - \tilde{i}, 0)] \times [\omega(\|i_0 - i^*\|) + \omega_1(0, \|i_{-1} - i^*\|) \\ &\quad + \omega_1(\|i_0 - i^*\| + \|i^* - \tilde{i}, 0)]\}^{-1} \\ &\le A\{1 - A[2\alpha + \omega_0(a^*) + \omega_1(0, a^*) + \omega_1(a^* + \lambda, 0)] \\ &\quad \times [\omega_0(a^*) + \omega_1(0, a^*) + \omega_1(a^* + \lambda, 0)]\}^{-1}.\end{aligned} \tag{11}$$

By using the equality $P(i^*) + Q(i^*) = 0$, we can show that $i_1 \in \delta D(i^*, a^*)$, as

$$\begin{aligned}\|i_1 - i^*\| &= \|i_0 - i^* + \{-(\mathcal{N}_0^T \mathcal{N}_0)^{-1} \mathcal{N}_0^T (P(i_0) + Q(i_0))\} + \mathcal{N}_*^T (P(i^*) + Q(i^*))\| \\ &\le \| - (\mathcal{N}_0^T \mathcal{N}_0)^{-1}\| \| - (\mathcal{N}_0^T \mathcal{N}_0)(i_0 - i^*) + \mathcal{N}_0^T (P(i_0) + Q(i_0)) \\ &\quad - \mathcal{N}_*^T (P(i^*) + Q(i^*))\| \\ &\le \| - (\mathcal{N}_0^T \mathcal{N}_0)^{-1}\| \| - \mathcal{N}_0^T\| \|\mathcal{N}_0 - \int_0^1 P'(i^* + t(i_0 - i^*))dt - Q(i_0, i^*)\| \|i_0 - i^*\|.\end{aligned} \tag{12}$$

Now,

$$\begin{aligned}&\|\mathcal{N}_0 - \int_0^1 P'(i^* + t(i_0 - i^*))dt - Q(i_0, i^*)\| \\ &\le \int_0^1 \|P'(i_0) - P'(i^* + t(i_0 - i^*))\|dt + \|Q(i_0, i_{-1}) - Q(i_0, i^*)\| \\ &\le T\omega_0(\|i_0 - i^*\|) + \omega_1(0, \|i_{-1} - i^*\|).\end{aligned} \tag{13}$$

Using (11), (9) and (13) in (12), we get

$$\begin{aligned}\|i_1 - i^*\| \le{}& A\{1 - A[2\alpha + \omega_0(\|i_0 - i^*\|) + \omega_1(0, \|i_{-1} - i^*\|)\\&+ \omega_1(\|i_0 - i^*\| + \|i^* - \tilde{i}\|, 0)] \times [\omega_0(\|i_0 - i^*\|) + \omega_1(0, \|i_{-1} - i^*\|)\\&+ \omega_1(\|i_0 - i^*\| + \|i^* - \tilde{i}\|, 0)]\}^{-1} \times [\alpha + \omega_0(\|i_0 - i^*\|)\\&+ \omega_1(0, \|i_{-1} - i^*\|) + \omega_1(\|i_0 - i^*\| + \|i^* - \tilde{i}\|, 0)] \times [T\omega_0(\|i_0 - i^*\|)\\&+ \omega_1(0, \|i_{-1} - i^*\|)]\|i_0 - i^*\|\\\le{}& A\{1 - A[2\alpha + \omega_0(a^*) + \omega_1(0, a^*) + \omega_1(a^* + \lambda, 0)]\\&\times [\omega_0(a^*) + \omega_1(0, a^*) + \omega_1(a^* + \lambda, 0)]\}^{-1} \times [\alpha + \omega_0(a^*) + \omega_1(0, a^*)\\&+ \omega_1(a^* + \lambda, 0)]\|i_0 - i^*\|\\\le{}& a^*.\end{aligned} \tag{14}$$

This shows that $i_1 \in \delta D(i^*, a^*)$. Further we will show that $i_{n+1} \in \delta D(i^*, a^*)$. For this, we find

$$\begin{aligned}\|\mathcal{N}_n - \mathcal{N}_*\| \le{}& \|P'(i_n) - P'(i^*)\| + \|Q(i_n - i_{n-1}) - Q(i_n - i^*)\| + \|Q(i_n - i^*) - Q(\tilde{i}, i^*)\|\\\le{}& \omega_0(\|i_n - i^*\|) + \omega_1(0, \|i_{n-1} - i^*\|) + \omega_1(\|i_n - i^*\| + \|i^* - \tilde{i}\|, 0)\\\le{}& \omega_0(a^*) + \omega_1(0, a^*) + \omega_1(a^* + \lambda, 0).\end{aligned} \tag{15}$$

Also,

$$\begin{aligned}&\|I - (\mathcal{N}_*^T\mathcal{N}_*)^{-1}(\mathcal{N}_n^T\mathcal{N}_n)\|\\&\le \|(\mathcal{N}_*^T\mathcal{N}_*)^{-1}\|[\|\mathcal{N}_*^T\|\|\mathcal{N}_* - \mathcal{N}_n\| + \|\mathcal{N}_*^T - \mathcal{N}_n^T\|\|\mathcal{N}_n - \mathcal{N}_*\| + \|\mathcal{N}_*^T - \mathcal{N}_n^T\|\|\mathcal{N}_*\|]\\&\le A[2\alpha + \|\mathcal{N}_n - \mathcal{N}_*\|]\|\mathcal{N}_n - \mathcal{N}_*\|.\end{aligned} \tag{16}$$

Therefore,

$$\begin{aligned}\|(\mathcal{N}_n^T\mathcal{N}_n)^{-1}\| \le{}& \|(\mathcal{N}_*^T\mathcal{N}_*)\|[1 - \|(\mathcal{N}_*^T\mathcal{N}_*)^{-1}\mathcal{N}_n^T\mathcal{N}_n - I\|]^{-1}\\\le{}& A\{1 - A[\omega_0(\|i_n - i^*\|) + \omega_1(0, \|i_{n-1} - i^*\|\\&+ \omega_1(\|i_n - i^*\| + \|i^* - \tilde{i}\|, 0)] \times [\omega_0(\|i_n - i^*\|)\\&+ \omega_1(0, \|i_{n-1} - i^*\| + \omega_1(\|i_n - i^*\| + \|i^* - \tilde{i}\|, 0)]\}^{-1}\\\le{}& A\{1 - A[2\alpha + \omega_0(a^*) + \omega_1(0, a^*) + \omega_1(a^* + \lambda, 0)]\\&\times [\omega_0(a^*) + \omega_1(0, a^*) + \omega_1(a^* + \lambda, 0)]\}^{-1}.\end{aligned} \tag{17}$$

Therefore, iteration $i_{n+1}$ is correctly defined and the following estimate is true:-

$$
\begin{aligned}
\|i_{n+1} - i^*\| &= \|i_{n+1} - i_n + i_n - i^*\| \\
&\le \| - (\mathcal{N}_n^T \mathcal{N}_n)^{-1}\| \| - (\mathcal{N}_n^T \mathcal{N}_n)(i_n - i^*) + \mathcal{N}_n^T (P(i_n) + Q(i_n)) \\
&\quad - \mathcal{N}_*^T (P(i^*) + Q(i^*))\| \\
&\le \| - (\mathcal{N}_n^T \mathcal{N}_n)^{-1}\| \| - \mathcal{N}_n^T \| \|\mathcal{N}_n - \int_0^1 P'(i^* + t(i_n - i^*))dt \\
&\quad - Q(i_n, i^*)\| \|i_n - i^*\|. \qquad (18)
\end{aligned}
$$

Now,

$$
\begin{aligned}
&\|\mathcal{N}_n - \int_0^1 P'(i^* + t(i_n - i^*))dt - Q(i_n, i^*)\| \\
&\le \int_0^1 \|P'(i_n) - P'(i^* + t(i_n - i^*))\|dt + \|Q(i_n, i_{n-1}) - Q(i_n, i^*)\| \\
&\le T\omega_0(\|i_n - i^*\|) + \omega_1(0, \|i_{n-1} - i^*\|). \qquad (19)
\end{aligned}
$$

Also,

$$
\|\mathcal{N}_n\| \le \alpha + \omega_0(\|i_n - i^*\|) + \omega_1(0, \|i_{n-1} - i^*\|) + \omega_1(\|i_n - i^*\| + \|i^* - \tilde{i}, 0)\|. \qquad (20)
$$

Using inequalities (17), (19) and (20) in inequality (18), we have

$$
\begin{aligned}
\|i_{n+1} - i^*\| &\le A\{1 - A[2\alpha + \omega_0(\|i_n - i^*\|) + \omega_1(0, \|i_{n-1} - i^*\|) \\
&\quad + \omega_1(\|i_n - i^*\| + \|i^* - \tilde{i}\|, 0)] \times [\omega_0(\|i_n - i^*\|) + \omega_1(0, \|i_{n-1} - i^*\|) \\
&\quad + \omega_1(\|i_n - i^*\| + \|i^* - \tilde{i}\|, 0)]\}^{-1} \times [\alpha + \omega_0(\|i_n - i^*\|) \\
&\quad + \omega_1(0, \|i_{n-1} - i^*\|)\omega_1(\|i_n - i^*\| + \|i^* - \tilde{i}\|, 0)] \times [T\omega_0(\|i_n - i^*\|) \\
&\quad + \omega_1(0, \|i_{n-1} - i^*\|)]\|i_n - i^*\| \\
&\le A\{1 - A[2\alpha + \omega_0(a^*) + \omega_1(0, a^*) + \omega_1(a^* + \lambda, 0)] \\
&\quad \times [\omega_0(a^*) + \omega_1(0, a^*) + \omega_1(a^* + \lambda, 0)]\}^{-1} \times [\alpha + \omega_0(a^*) + \omega_1(0, a^*) \\
&\quad + \omega_1(a^* + \lambda, 0)] \times [T\omega_0(a^*) + \omega_1(0, a^*)]\|i_n - i^*\| \\
&\le a^*. \qquad (21)
\end{aligned}
$$

Thus, $i_{n+1} \in \delta D(i^*, a^*)$. Hence, by method of mathematical induction, we proved the theorem.

## 3 Numerical Examples

**Example 1** Consider the system of equations

$$\left.\begin{array}{l}\frac{1}{50}x^2 - \frac{1}{50}y^2 + 0.001 + \frac{1}{45} \mid x - 2 \mid = 0 \\ \frac{1}{25}xy + \frac{1}{25}y - 0.11 + \frac{1}{45} \mid y - 3 \mid = 0\end{array}\right\}. \tag{22}$$

Here we take jacobian of differentiable part and taking max norm, we get $\|P'(X) - P'(U)\| \leq \frac{1}{25}\|X - U\|$, which satisfies the condition $\|P'(X) - P'(U)\| \leq \omega_0(\|X - U\|)$. Again, we apply first order divided difference in non differentiable part and taking max norm, we get $\|[s, t; Q] - [u, v; Q]\| \leq \frac{2}{45}$, which satisfy the condition $\|[s, t; Q] - [u, v; Q]\| \leq \omega_1(\|s - u\|, \|t - v\|)$. For the initial guesses $(5, 5)$ and $(1, 1)$, we have the approximate solution $(0.31776355, 1.42131035)$. Using these results and taking $a^* = 0.2$ we get, $\lambda = 0.19849982 > 0$, $\omega_0(a^*) = 0.00444444$, $\omega_1(0, a^*) = 0.04444444$, $\omega_1(a^* + \delta, 0) = 0.04444444$, $\gamma = 0.093873969$, $A = 4.268811885$. Also we get $\alpha(a^*) = 0.1119890846$, $\tilde{\alpha}(a^*) = 0.072811694$ and hence $\alpha(a^*) + \tilde{\alpha}(a^*) = 0.1848007788 < 1$. Thus all conditions of our theorem 1 is satisfied Eq. (22).

**Example 2** Let us consider a non-linear integral equation

$$t(u) = \mathbf{0.5} + \frac{1}{29}\int_{-1}^{1} \mathcal{K}(u, v)\left(\frac{1}{17}(t(v)\right)^2 + \mid t(v) - 3 \mid)dv, \quad u \in [-1, 1] \tag{23}$$

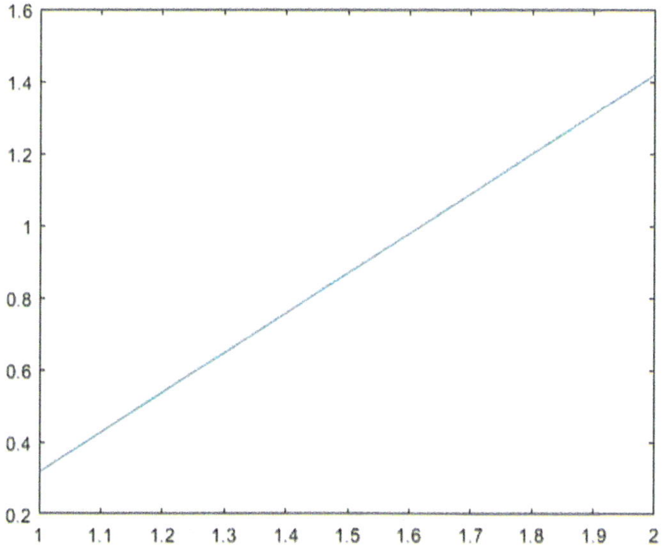


**Fig. 1** Approximate solution plot of (22)

where the unknown function $t$ is to be determined and the Green's function $\mathcal{K}$ is defined over $[-1, 1] \times [-1, 1]$. Now, solving Eq. (23) is equivalent to solving $\mathcal{F}(t) = 0$, where $\mathcal{F} : C[-1, 1] \to C[-1, 1]$, is a non-linear operator, given as

$$\mathcal{F}(t(u)) = t(u) - \mathbf{0.5} - \frac{1}{29}\int_{-1}^{1} \mathcal{K}(u, v)\left(\frac{1}{17}(t(v)\right)^2 + \mid t(v) - 3 \mid), \quad u \in [-1, 1]. \tag{24}$$

Now, to approximate the integral part of the above equation we use Gauss-Legendre quadrature formula with $m$-nodes and taking $t(v_i) = t_i$ and $g(v(i)) = g_i$ for $i = 1, 2, \dots m$, Eq. (24) can be transformed into system of non-linear equations of the form:

$$\mathcal{F}_i = t_i - \mathbf{0.05} - \sum_{j=1}^{m} a_{ij} H(v_j, t_j) = 0, \qquad i = 1, 2, \dots m, \tag{25}$$

where, $t_i = (t_1, t_2, \dots t_m)^T$, $\mathbf{0.5} = (0.5, 0.5, \dots 0.5)^T$, $\mathcal{F} : \mathbb{R}^m \to \mathbb{R}^n$,

$$a_{ij} = w_j \mathcal{K}(v_i, v_j) = \begin{cases} \frac{w_j[1-e^{3v_i+3}][4-e^{3v_j-3}]}{3e^{3v_j[4e^3-3^{-3}]}}, & j \le i \\ \frac{w_j([1-e^{3v_j+1}][4-e^{3v_i-3}]}{3v_j[4e^3-3^{-3}]}, & j > i \end{cases}$$

Now for $m = 4$, by using jacobian and taking max norm to the differential part we get, $\|P'(t) - P'(s)\| \le 0.0018637809\|t - s\|$, which satisfies the condi-

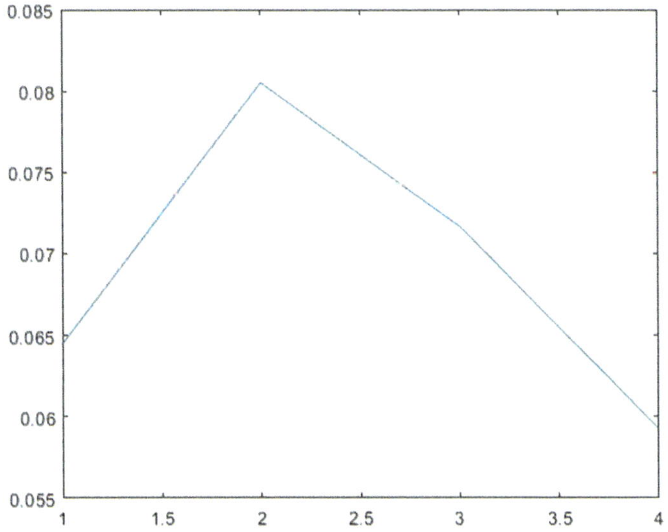


**Fig. 2** Approximate solution plot of (24)

tion $\|P'(t) - P'(s)\| \leq \omega_0(\|t - s\|)$, which leads to $\omega_0(tz) \leq t\omega_0(z)$. Thus $h(t) = t$ and hence $T = \int_0^1 t dt = 0.5$.

Also by using divided difference and taking max norm to the non-differentiable part we have, $\|[s, t; Q] - [p, q; Q]\| \leq 0.03168428$. Taking initial guesses as $(2.0, 2.0, 2.0, 2.0)$ and $(2.5, 2.5, 2.5, 2.5)$ we get the approximate solution $(0.064488, 0.080534, 0.071712, 0.059241)$. Using these results and taking $a^* = 0.1$ we get, $\lambda = 0.041219, \omega_0(a^*) = 0.000186378, \omega_1(0, a^*) = 0.03168428 = \omega_1(a^* + \lambda, 0)$, $\gamma = 1.012913742$, $A = 1.001908274$. Also we get $p(a^*) = 0.133043984$, $\tilde{p}(a^*) = 0.034272727$ and hence, $p(a^*) + \tilde{p}(a^*) = 0.167316711 < 1$. Thus all conditions of our theorem 1 is satisfied for thiss problem Eq. (24).

## 4 Conclusion

In this paper, we have studied Gauss-Newton-Secant method for solving non linear least-square problems with non-differentiable function. We have used $\omega$-condition for convergence analysis of the proposed method. We have done some numerical experiments to check the efficiency of the method, and found that our method is suitable for these kind of problems.

## References


1. Argyros, I.K.: Convergence and Application of Newton-Type Iterations. Springer, New York (2008)
2. Dennis, J.E., Schnabel, R.B.: Numerical Methods for Unconstrained Optimization and Nonlinear Equations. SIAM, Philadelphia (1996)
3. Argyros, I.K., Shakhno, S., Shunkin, Y.: Improved convergence analysis of Gauss-Newton-Secant method for solving nonlinear least squares problems. Mathematics **7**(1), 99 (2019). https://doi.org/10.3390/math7010099
4. Ortega, J.M., Rheinboldt, W.C.: Iterative Solution of Nonlinear Equations in Several Variables. Academic, New York (1970)
5. Argyros, I.K., Magreñán, Á.A.: A Contemporary Study of Iterative Methods. Elsevier (Academic Press), New York, NY, USA (2018)
6. Argyros, I.K.: The secant method and fixed points of nonlinear operators. Monatshefte für Mathematik **106**, 85–94 (1988)
7. Shakhno, S.M.: Gauss-Newton-Kurchatov method for the solution of nonlinear least-square problems. J. Math. Sci. **247**(1), 58–72 (2020)
8. Shakhno, S.M., Yarmola, H.P., Shunkin, Yu.V.: Convergence analysis of the Gauss-Newton-Potra method for nonlinear least squares problems. Mat. Stud. **50**, 211–221 (2018)

SPRINGER NATURE

# GPSR Compliance

*The European Union's (EU) General Product Safety Regulation (GPSR) is a set of rules that requires consumer products to be safe and our obligations to ensure this.*

*If you have any concerns about our products, you can contact us on ProductSafety@springernature.com*

In case Publisher is established outside the EU, the EU authorized representative is:

Springer Nature Customer Service Center GmbH
Europaplatz 3
69115 Heidelberg, Germany

The manufacturer's authorised representative in the EU is Springer Nature Customer Service Centre GmbH, Europaplatz 3, 69115 Heidelberg, Germany. If you have any concerns regarding our products, please contact ProductSafety@springernature.com

Printed and bound by CPI Group (UK) Ltd, Croydon, CR0 4YY
01/10/2026
02225860-0020